W0079851

DNA Repair Mechanisms and Their Biological Implications in Mammalian Cells

NATO ASI Series

Advanced Science Institutes Series

A series presenting the results of activities sponsored by the NATO Science Committee, which aims at the dissemination of advanced scientific and technological knowledge, with a view to strengthening links between scientific communities.

The series is published by an international board of publishers in conjunction with the NATO Scientific Affairs Division

A	**Life Sciences**	Plenum Publishing Corporation
B	**Physics**	New York and London
C	**Mathematical and Physical Sciences**	Kluwer Academic Publishers Dordrecht, Boston, and London
D	**Behavioral and Social Sciences**	
E	**Applied Sciences**	
F	**Computer and Systems Sciences**	Springer-Verlag
G	**Ecological Sciences**	Berlin, Heidelberg, New York, London,
H	**Cell Biology**	Paris, and Tokyo

Recent Volumes in this Series

Volume 177—Prostanoids and Drugs
edited by B. Samuelsson, F. Berti, G. C. Folco, and G. P. Velo

Volume 178—The Enzyme Catalysis Process: Energetics, Mechanism, and Dynamics
edited by Alan Cooper, Julien L. Houben, and Lisa C. Chien

Volume 179—Immunological Adjuvants and Vaccines
edited by Gregory Gregoriadis, Anthony C. Allison, and George Poste

Volume 180—European Neogene Mammal Chronology
edited by Everett H. Lindsay, Volker Fahlbusch, and Pierre Mein

Volume 181—Skin Pharmacology and Toxicology: Recent Advances
edited by Corrado L. Galli, Christopher N. Hensby, and Marina Marinovich

Volume 182—DNA Repair Mechanisms and Their Biological Implications in Mammalian Cells
edited by Muriel W. Lambert and Jacques Laval

Volume 183—Protein Structure and Engineering
edited by Oleg Jardetzky

Series A: Life Sciences

DNA Repair Mechanisms and Their Biological Implications in Mammalian Cells

Edited by

Muriel W. Lambert

University of Medicine and Dentistry of New Jersey
New Jersey Medical School
Newark, New Jersey

and

Jacques Laval

Institut Gustave Roussy
Villejuif, France

Plenum Press
New York and London
Published in cooperation with NATO Scientific Affairs Division

Proceedings of a NATO Advanced Research Workshop on DNA Repair Mechanisms and Their Biological Implications in Mammalian Cells, held October 2-7, 1988, in Fontevraud, France

Library of Congress Cataloging-in-Publication Data

NATO Advanced Research Workshop on DNA Repair Mechanisms and Their Biological Implications in Mammalian Cells (1988 : Fontevrault-l'Abbaye, France)
DNA repair mechanisms and their biological implications in mammalian cells / edited by Muriel W. Lambert and Jacques Laval.
p. cm. -- (NATO ASI series. Series A, Life sciences ; v. 182)
"Proceedings of a NATO Advanced Research Workshop on DNA Repair Mechanisms and Their Biological Implications in Mammalian Cells held October 2-7, 1988, in Fontevraud, France"--T.p. verso.
"Published in cooperation with NATO Scientific Affairs Division."
Includes bibliographical references.

1. DNA repair--Congresses. I. Lambert, Murial W. II. Laval, Jacques. III. North Atlantic Treaty Organization. Scientific Affairs Division. IV. Title. V. Series.
[DNLM: 1. Cells--physiology--congresses. 2. DNA Repair--congresses. QH 467 N279d 1988]
QH467.N38 1988
599'.0873282--dc20
DNLM/DLC
for Library of Congress 89-72122
CIP

ISBN 978-1-4684-1329-8 ISBN 978-1-4684-1327-4 (eBook)
DOI 10.1007/978-1-4684-1327-4

Softcover reprint of the hardcover 1st edition 1989
A Division of Plenum Publishing Corporation
233 Spring Street, New York, N.Y. 10013

Paul Howard-Flanders
(1919-1988)

Dr. Paul Howard-Flanders, Professor of Therapeutic Radiobiology at Yale University, died of cancer during September, 1988. Paul was born in Bristol, England in 1919. He received his BSc in mathematics and physics from London University in 1940, and obtained a Ph.D. in biophysics in 1956. Subsequently, he was involved in the development of X-ray equipment at Hammersmith Hospital. During this period, he became increasingly interested in the effects of radiation on cells and performed pioneering work on oxygen radicals and radiosensitiser chemicals.

At the age of 40, Paul was stimulated by the power of bacterial genetics to change the area of his research and search for radiation sensitive mutants in _E. coli_ K-12. His isolation and characterization of the _uvrA_, _uvrB_, and _uvrC_ mutants in 1962 was a milestone for the development of the field of DNA repair. During the 1960's, Paul subsequently discovered two main cellular pathways of DNA repair, excision repair of radiation-induced lesions such as pyrimidine dimers, and postreplicational repair of DNA damage. The delineations of these pathways are classic examples of his insight and willingness to look for new problems in biology. He was a man who was not limited by his own expertise, a physicist who learned genetics, and then when he needed it, biochemistry.

In the last ten years of his life, Paul did some of his most elegant work. He realised the importance of the spiral filaments that RecA protein forms around DNA prior to homologous pairing and strand exchange, and proposed a mechanism for genetic recombination in which the interacting DNA molecules are interwound with the RecA filament. This molecular model for recombination is gaining increased acceptance and its proposal was characteristic of Paul's scientific intuition and conviction.

Widely respected and admired by his colleagues and students, Paul Howard-Flanders had a powerful influence on the field of DNA repair and many of its most successful investigators. He is survived by his wife, June, and his two sons, Mark and Rob. He will be greatly missed.

PREFACE

This volume contains edited contributions from the speakers at the NATO Advanced Research Workshop on "DNA Repair Mechanisms and Their Biological Implications in Mammalian Cells" held October 1-6, 1988, at the Abbaye Royale de Fontevraud, Fontevraud France.

The meeting was dedicated to Paul Howard-Flanders (Yale University, New Haven, CT., 1919-1988), whose seminal contributions to the DNA repair field include the co-discovery of the excision repair pathway, the elucidation of post-replication repair in E. coli, the isolation of the lexA and recC mutants, and his extensive work on the enzymology of RecA.

A plethora of recent developments in DNA repair mechanisms and related processes in mammalian cells have advanced our understanding of this field in a number of different areas and have given new emphasis to the ways these systems both resemble DNA repair processes in other groups of organisms in some respects yet are strikingly different from them in others. Within the past decade there have been a number of international conferences on DNA damage and repair mechanisms but none has been focused on these processes in mammalian cells. Because of the rapid progress that has been made in this area over the past several years, and because of the complexity of the problem itself coupled with the increasing awareness of the relevance of DNA repair to human health and disease processes, the organizers of the Fontevraud meeting felt it extremely timely to bring together researchers in the field to discuss some of the important recent advances in DNA repair as it relates in particular to mammalian cells.

The meeting focused on specifically targeted topics in which intensive research is ongoing and exciting new findings are being generated. Detailed examinations of repair of alkylation damage in DNA have led to much new and often unexpected information on these repair processes. There is evidence that O-alkyl purines and pyrimidines play an important role in mutagenesis and carcinogenesis. New methodologies have been developed for comparing the mutagenic efficiency of different O-alkyl lesions at single sites. Enzymes involved in repair of some of these lesions, the O^6-alkyl-DNA alkyltransferases, are being extensively studied and cDNAs coding for them are being cloned. Studies have also been carried out on repair of other alkylation products and advances have been made in the purification and characterization of specific mammalian cDNA glycosylases. In addition, advances have been made in understanding the molecular basis of alkylating agent resistance in mammalian cells.

Inducible responses to DNA damage have been examined in mammalian cells. At least two DNA repair activities, a DNA methyltransferase specific for O^6-methylquanine and a DNA glycosylase specific for N^3-methylguanine, have been shown to increase after a single dose of chemical or physical agents. In addition a "SOS"-like function has been described in mammalian cells in which protection from alkylation-induced cytotoxicity is provided by overexpression of metallothionein.

Molecular studies of mutagenesis have been carried out utilizing a number of different systems. Through the use of plasmid shuttle vectors, mutation frequency has been found to be greater in cells derived from patients with the cancer-prone genodermatosis, xeroderma pigmentosum (XP) than in normal cells with unique mutagenic hotspots present in cells from several XP complementation groups. A number of studies have examined the mutagenic and site specificity of various adducts in the hprt gene in mammalian cells. A transcribed or non-transcribed strand specificity in mutation-induction has been found in this gene in lines of Chinese hamster cells, depending upon the UV sensitivity of the cells.

Investigations into the involvement of homologous recombination in DNA repair in mammalian cells have led to the identification of factor(s) in human cell nuclei which promote recombination between two homologous duplex DNAs. A protein has been found in human cells which specifically binds to Holliday junctions formed during strand exchange and a strand transferase activity has been characterized in human cells which promotes homologous pairing and strand exchange. During recombination, mismatches form. A mismatch binding protein has been identified from human cells which recognizes specific mispaired bases in DNA.

Recent investigations from a number of diverse laboratories have shown that chromatin structure and the nuclear matrix play critical roles in mammalian DNA repair process. Nucleosome structure modulates sites of damage and repair as well as activity of repair enzymes. Posttranslational modifications of histones, such as poly(ADP-ribosyl)ation, modulate chromatin structure. Important differences in the repair of transcriptionally active versus inactive genes have been discovered and examined in mammalian cells.

Much new interest has developed in the last several years in a number of cell lines derived from patients with rare genetic diseases. In several of these diseases, such as xeroderma pigmentosum, earlier studies suggested the existence of relatively straightforward molecular deficiencies which more recent investigations have proven to be much more complex. A number of new investigations, at both the cellular and molecular levels, in cells derived form patients with such other inherited diseases as ataxia-telangiectasia, Fanconi's anemia, Bloom's syndrome and trichothiodystrophy, as well as xeroderma pigmentosum and the more common neurodegenerative disease, amyotrophic lateral sclerosis, are reported here. Progress has been made in understanding the specific enzymatic defects associated with these diseases. In addition, a number of induced mutant cell lines have been developed in animal cells and studies of these mutants have confirmed and extended the complexity of these mammalian repair processes. These studies have important implications not only for understanding repair deficiencies in these specific cell lines but also for understanding a number of such more common disorders as aging, cancer, and neurodegenerative diseases. At this meeting there was an ongoing lively discussion of the possible interrelationships between the mechanisms responsible for deficiencies in these mutant cells and the etiopathogenesis of these rare and common disorders, at least some of which is reflected in this volume.

Recent progress has been made in the isolation, cloning and characterization of cultured animal cell (Chinese hamster ovary, CHO) and human excision-repair genes. Several of these genes have been found to correct repair deficiencies in selected CHO mutant and repair-deficient human cell lines. Studies are also being carried out on oncogene activation in tumors from repair-deficient patients, and amplification and rearrangement of specific oncogenes has been found in XP skin tumors.

The emerging concepts in complexity of DNA repair process in mammalian cells have also altered strategies to exploit DNA damage therapeutically. For example, DNA repair processes have been shown to hamper chemotherapy by alkylating agents. A number of chapters in this volume one devoted to new approaches to chemotherapy based on these new concepts.

Although a number of recent meetings devoted to DNA repair processes have been held, some with published proceedings, the Fontevraud assembly was unique in its focus on mammalian systems. We believe that the goal to bring together a number of different investigators working in this emerging field was achieved. The NATO Scientific Affairs Division is gratefully acknowledged for granting an award that made the origination of this workshop possible. Contributions were also made by Laboratories Servier, L'Oreal, Pharmacia, Rhone-Poulenc Sante, Sandoz Pharmaceuticals Corporation, Schering-Plough Corporation, and Smith, Klein, Beckman Corporation.

We would particularly like to thank the organizing committee of the workshop consisting of W. Clark Lambert (UMDNJ-New Jersey Medical School, Newark), Tomas Lindahl (Imperial Cancer Research Fund, Clare Hall Laboratories, South Mimms, U.K.), and Paul H. Lohman (University of Leiden, The Netherlands). We would also like to thank Gregory Tsongalis for typing the manuscripts.

Pleasant memories of the meeting itself, which was held in a eleventh-fifteenth century abbey in the picturesque Loire Valley, remain with all who took part. The most vivid recollections, however, are of the extensive productive discussions among people with diverse interests and backgrounds. The success of the meeting was a result of the enthusiastic interactions of all of the participants. We hope that some of their enthusiasm and excitement has been captured in the contents of this volume. We wish to thank all of the participants who so willingly shared their data, knowledge and ideas.

Muriel W. Lambert

Jacques Laval

CONTENTS

REPAIR OF ALKYLATION DAMAGE IN DNA

INDUCIBLE RESPONSES TO DNA DAMAGE

DNA REPAIR RELATED EVENTS

MUTAGENESIS

RECOMBINATION AND MISMATCH REPAIR

CHROMATIN PROTEINS AND CHROMATIN STRUCTURE IN DNA REPAIR

REPAIR DEFECTIVE CELL TYPES

ANTICANCER DRUGS IN DNA REPAIR

DNA REPAIR GENES AND ONCOGENES IN DNA REPAIR

STRUCTURE-FUNCTION RELATIONSHIPS IN ALKYLATION DAMAGE AND REPAIR

B. Singer

Donner Laboratory, Lawrence Berkeley Laboratory
University of California, Berkeley, CA 94720

INTRODUCTION

Since 1962, when 7-methylguanine was found as an *in vivo* product of dimethylnitrosamine administered to rats (Magee & Farber, 1962), the list of sites of alkylation by simple alkylating agents has increased so that now we know that all oxygens and nitrogens of nucleosides in polynucleotides, DNA and RNA can be modified (Singer, 1976, 1982) (Figure 1). However, each reagent produces characteristic patterns of alkylation, both *in vitro* and *in vivo* (Singer & Grunberger, 1983). The enzymatic repair or loss *in vivo* of each of these products has been the focus of research for many scientists for two reasons. Firstly, enzymologists have been interested in purifying such activities and elucidating their mechanism of action. Secondly, it was apparent that adducts can cause changed base incorporation or act as blocks to replication. Because repair is not completed before replication in all situations, remaining adducts can be either mutagenic or lethal. Most recently, in order to separate the potential biological effects of each derivative, defined oligonucleotide sequences containing a single adduct have been synthesized and used to study both replication and mutation. These defined oligomers have also been of great value in measuring repair, particularly of O^6-alkyl G and O^4-alkyl T. This paper will focus on the mutagenic potential of selected O-alkyl derivatives, their repair in eukaryotes, and the effects of their presence on replication and nucleic acid structure.

IN VIVO RESPONSE TO N-NITROSO CARCINOGENS

The two types of N-nitroso alkylating agents most studied are the nitrosoureas that do not require metabolic activation, and the nitrosamines that do. In the case of ethylation, proportionally more O-alkylation takes place than with methylation, although the absolute extent of reaction is lower (Singer, 1975, 1986; Pegg, 1984). Analyses of the proportion of products measured in DNA after administration of various N-nitroso alkylating agents is in Figure 2. The earliest time

DNA Repair Mechanisms and Their Biological Implications in Mammalian Cells
Edited by M.W. Lambert and J. Laval
Plenum Press, New York

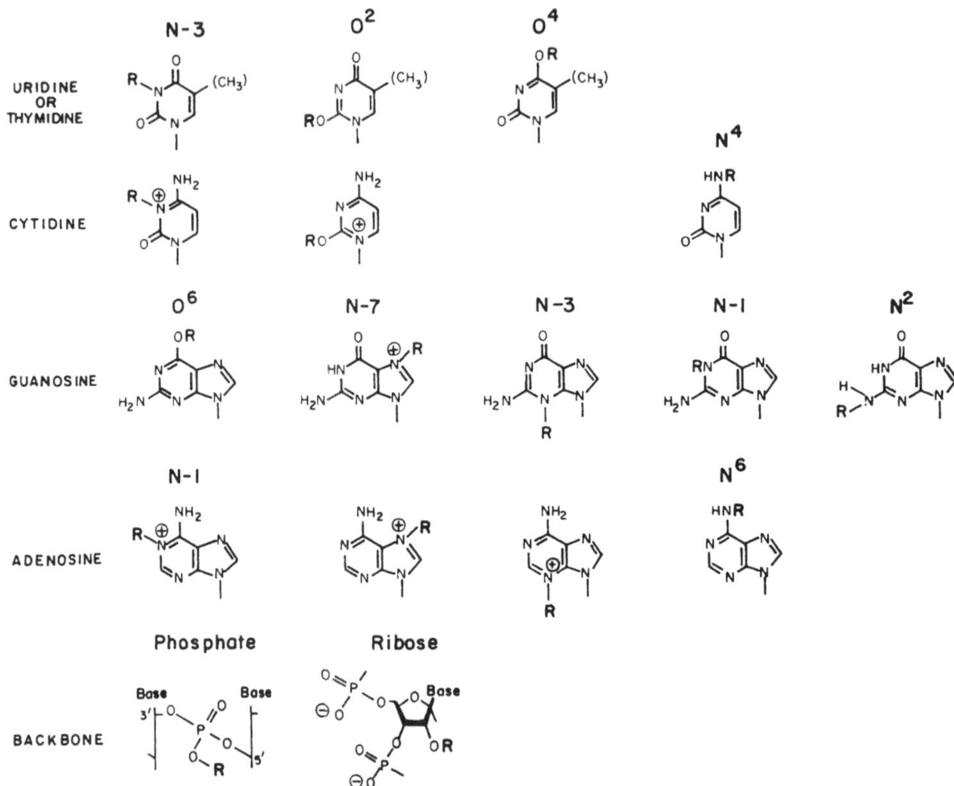

Figure 1. Sites of reaction of N-nitroso alkylating agents with nucleic acids or polynucleotides under physiological conditions. The N^4 of C, N-1 and N^2 of G and the N^6 of A have been found to react in polynucleotides. These latter four sites of reaction have not been investigated <u>in vivo</u>. The R indicates any alkyl group.

point taken is at 1-5 hours, so that repair is observed for some derivatives, particularly 3-methyl A and O^6-methyl G. About 80% of the ethyl groups are on the O^6 of G, O^2 of C and T, O^4 of T and the phosphodiesters (Singer, 1985). The complete structures of these derivatives are shown in Figure 1. Their individual roles in initiating carcinogenesis are difficult to assess. This is due not only to the multiplicity of derivatives and the size of the alkyl group which affects repair and replication rates, but also to the diversity of target organs for each mammal studied (Table 1), the age of the animal (Figure 3) (Druckrey, <u>et al.</u>, 1969) and the extensive range of the repair capability of a cell (Pegg, 1983) or tissue type (Lewis & Swenberg, 1980). An example of differential repair in two cell types from rat liver is shown in

INITIAL *IN VIVO* ALKYLATION

		DMN MNU SDMH	DEN ENU
		% of total Alkylation	
ADENINE:	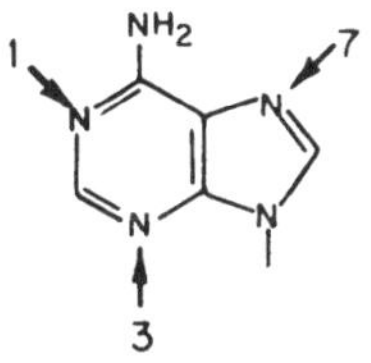 N-1	0.8	~0.1
	N-3	~4 (9)	4
	N-7	1.5	0.6
GUANINE:	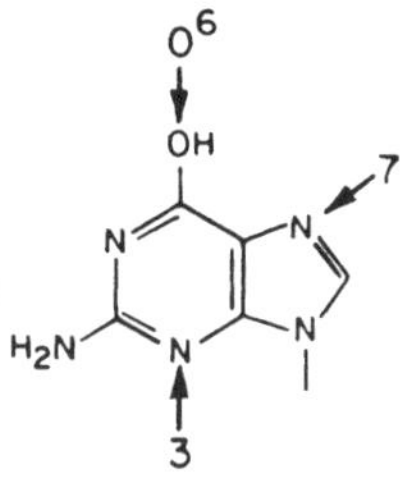 N-3	0.6	1.5
	O^6	3 – 6 (6)	8
	N-7	69	12
CYTOSINE:	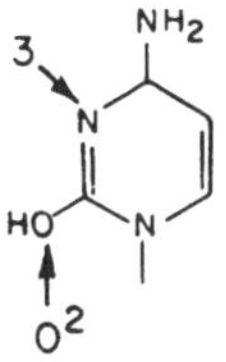 O^2	~0.1	2
	N-3	0.5	~0.3
THYMINE:	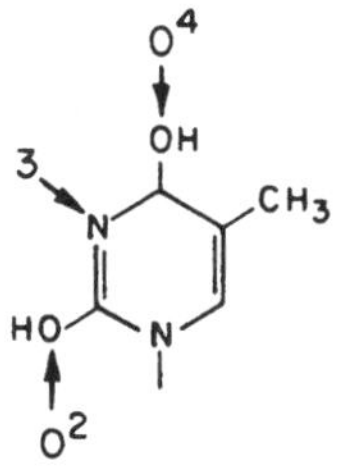 O^2	~0.1	7
	N-3	0.3	0.4
	O^4	~0.1	2.5
PHOSPHATE:	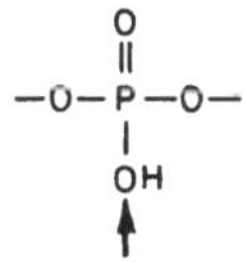 Triester	12	58
	N total	82 %	20 %
	O total	18 %	80 %

DMN Dimethylnitrosamine
MNU Methylnitrosourea
SDMH 1, 2-Dimethylhydrazine
DEN Diethylnitrosamine
ENU Ethylnitrosourea

Figure 2. The arrows on the structural formulas indicate sites of modification of nucleic acids. The proportion of alkylation in vivo at each site is shown for both methylating and ethylating agents. The numbers in parentheses are expected alkylation, if no repair occurs. The initial time point (up to 5 hours) varies with the carcinogen.

Table 1. Effect of Single Exposure to Ethylnitrosourea

Animals	Primary Site(s) of Tumors
Perinatal	
Rat	Brain, central nervous system
Mouse	Lung, liver, kidney
Syrian hamster	Peripheral nervous system
Rabbit	Nervous system, kidney
Gerbil	Melanocytes
Adult	
Rat	Brain, haemopoietic system, kidney
Mouse (M)	Lymph nodes, lung
Mouse (F)	Liver, kidney, lung
Syrian hamster	Forestomach
Rabbit	Kidney
Gerbil	Melanocytes

Table 2. Obviously, other factors are involved in carcinogenesis, including the carcinogen dosage, mode of exposure,cell cycle, state of differentiation, rate of proliferation, etc.

In a set of experiments (Figure 4), occurrence and repair of several ethyl derivatives was studied in [^{14}C]-ethylnitrosourea-exposed human fibroblasts (Bodell, et al., 1979) and in 10-day-old rat brain (Singer, et al., 1981), the target organ for this age and species after ethylnitrosourea administration (Table 1). Fibroblasts efficiently repaired or removed all derivatives measured except ethyl phosphotriesters, while rat brain was poor in repair of all derivatives except for 3-ethyl A, which is depurinated both by chemical hydrolysis as well as by a glycosylase. In particular, O^4-ethyl T (e^{4}T) did

Table 2. Alkylation of Hepatocyte and Nonparenchymal Cell (NPC) DNA Following Oral Exposure of Rats to 1, 2-[^{14}C]Dimethylhydrazine

		Alkylation of DNA per 10^6 Guanine	
Cell type	Hours after Exposure	m^{6}G	m^{7}G
Hepatocyte	2	89	1032
	24	9	753
NPC	2	65	783
	24	35	525

Single administration of 3 mg/kg. Data from Lewis and Swenberg, 1980.

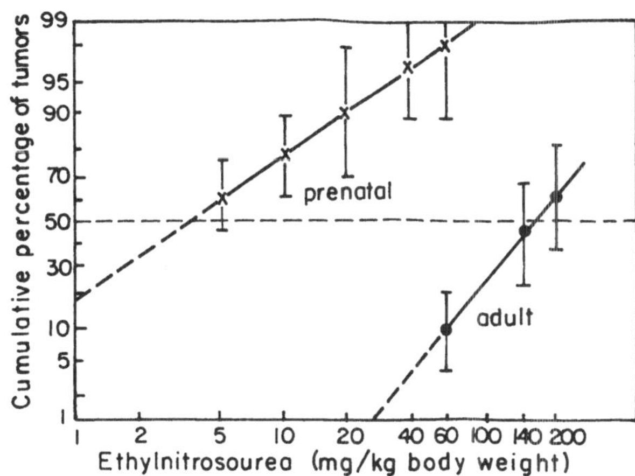

Figure 3. Dose-response relationships for the prenatal induction of malignant tumors of the nervous system compared with that in adult rats. Ethylnitrosourea was administered as a single dose (Druckrey, et al., 1969).

not appear to be significantly repaired in rat brain (Singer, et al., 1981). The use of monoclonal antibodies can be more sensitive than measuring changes in radioactivity of a derivative. With this method of detection, O^4-ethyl T persists over the 48 hour time period studied, while only about 2% of the O^6-ethyl G (e^6G) remained after 12 hours (Figure 5) (Huh & Rajewsky, 1986). Specific studies directed toward the biochemical effects of O^4-alkylation are discussed in the following sections.

In rat hepatocytes, after a single dose of dimethyl- or diethyl-nitrosamine (DEN), a half life of O^4-methyl T (m^4T) and O^4-ethyl T could be measured. Although O^4-ethyl T was lost very slowly ($t^1/_2$ 11 days), compared to O^4-methyl T ($t^1/_2$ 20 hours), this experiment clearly showed that slow repair did occur (Figure 6) (Richardson, et al., 1985). A significantly longer half life for ethylated adducts compared to methylated is found in all systems studied to date. On continuous administration of DEN, a regimen which leads to hepatocellular carcinoma, the hepatocytes were able to continuously repair O^6-ethyl G, but O^4-ethyl T accumulated and reached a steady state about 50-fold greater than the initial level (Figure 7) (Swenberg, et al., 1984). These experiments and related ones (e.g., Belinsky, et al., 1986) provide strong evidence that the persistence of O^4-alkyl T in target organs may be an initiating event.

IN VITRO STUDIES OF O-ALKYL PYRIMIDINE REPAIR

The initial work on the mechanism of repair used well

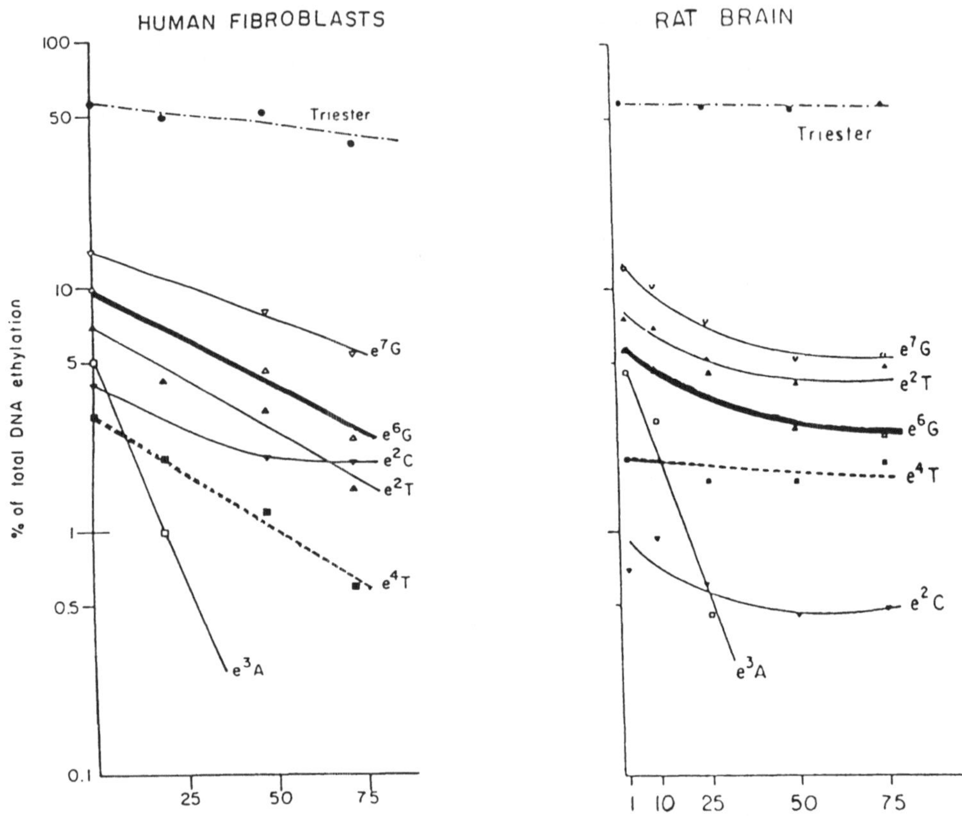

Figure 4. (Left) Formation and loss of seven ethyl derivatives from [^{14}C]-ethylnitrosourea-exposed human fibroblasts in cell culture. The initial time point was 1 hour (Bodell, et al., 1979). (Right) Formation and loss of the same derivatives in the brain of 10-day-old BDIX rats after administration of a single dose of the carcinogen (Singer, et al., 1981).

characterized enzymes from E. coli, 3-alkyladenine-DNA glycosylase and O^6-alkyl-guanine-DNA alkyltransferase (reviewed by Lindahl, 1982; Lindahl, et al., 1988). The former enzyme acts on other N-3 and N-7 alkylpurines (Singer & Brent, 1981; Margison & Pegg, 1981; Laval, et al., 1981). Later a second glycosylase from E. coli was found to act on O^2-methyl C and O^2-methyl T, albeit at a slower rate than on 3-methyl A (Ahmmed & Laval, 1984; McCarthy, et al., 1984; Hall & Karran, 1986). The O^6-alkylguanine-DNA alkyltransferase from prokaryotes also removed the alkyl group from the O^4 of T (McCarthy, et al., 1984) and, stereospecifically, from the alkyl phosphotriesters (McCarthy & Lindahl, 1985; Weinfeld, et al., 1985), again more slowly than from O^6-alkyl G.

In sharp contrast, the analogous enzymes isolated from

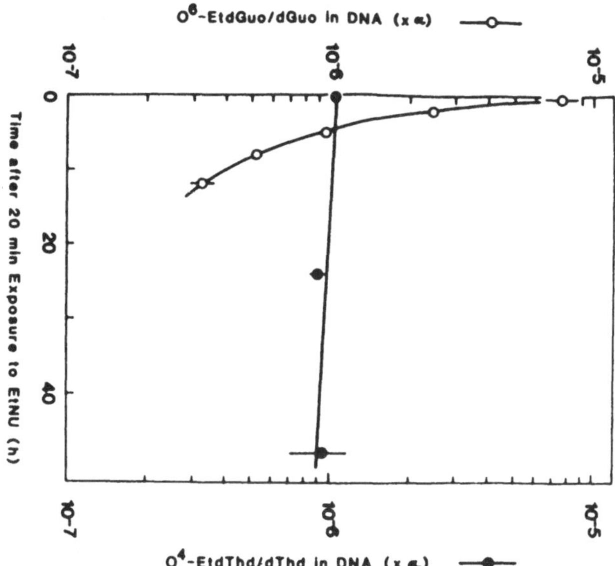

Figure 5. Elimination kinetics of O^6-ethylguanine (O) and O^4-ethylthymidine (●) from the DNA of the malignant neural BDIX rat cell line (BT3Ca) (Huh & Rajewsky, 1986).

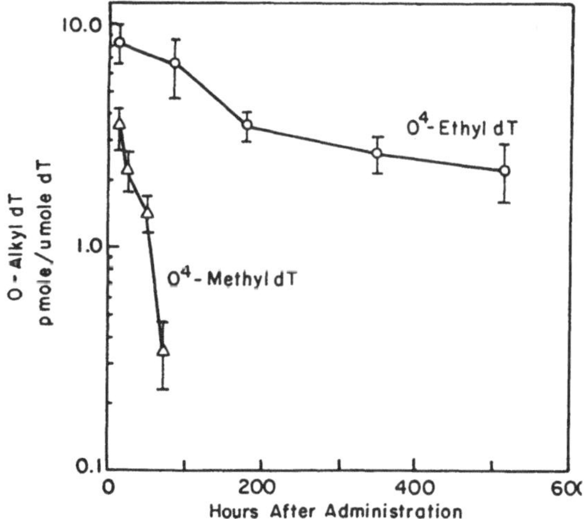

Figure 6. Loss of O^4-methyl dT and O^4-ethyl dT from rat liver after a single dose of diethylnitrosamine or 1,2-dimethylhydrazine administration to adult rats. The dose of the two carcinogens was adjusted to give approximately equal O^4-alkyl dT concentrations. Data from Richardson, <u>et al.</u>, 1985.

a variety of mammalian sources did not measurably use any of the alkyl pyrimidines as substrates, while active against 3-alkyl A or O^6-alkyl G (Brent, et al., 1988). The most sensitive analytical data for detection of repair of m^4T used a doublestranded dodecamer containing a single m^4T. This assay could have measured 0.01% of the enzyme activity found with the prokaryotic enzymes (Pegg & Dolan, 1989). It was thus concluded that the observed in vivo loss of this derivative can not be due to the characterized mammalian alkyl transferase. Similarly, the mammalian purine glycosylases did not act on O^2-alkyl T or O^2-alkyl C. It appears that, while there

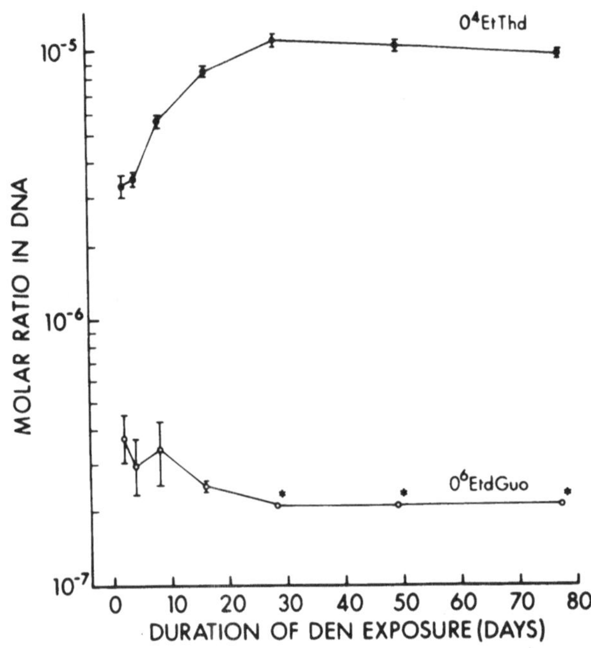

Figure 7. Relative concentrations of O^4-EtdThd and O^6-EtdGuo in DNA of hepatocytes of F-344 male rats exposed to DEN at 40 ppm in their drinking water for 2-77 days. Data are given as mean ± SEM except for the points marked by asterisks, for which the amounts in all samples were below the limit of detection (O^6-EtdGuo to deoxyguanosine molar ratio $<2 \times 10^{-7}$). Reprinted with permission from Swenberg, et al., 1984.

are many parallels in mechanism between prokaryotic and eukaryotic enzymes, other repair enzymes must occur in eukaryotes to account for the observed losses in vivo (Table 3). Suggestions have been made that a type of excision repair is a possible mechanism for repair of alkyl pyrimidines in eukaryotes.

Table 3. Range of Observed Half-lives of O-alkylpyrimidines and Representative Alkyl Triesters *in vivo*.

	$t^1/_2$ (hours, days, or weeks)	
Derivative	Whole Rat Liver or Hepatocytes	Cultured Mammalian Cells
m^2C	<4 h	
m^2T	12 h	
m^4T	~20 h; ~80 h	2-3 h; ~13 h
dThd-P(Me)-dThd	7-10 days	>20 h
e^2C	40-78 h	5, 50 h
e^2T	>2; 11-20 days	40 h
e^4T	>2-3; 11-19 days	1-3 days
dThd-P(Et)-dThd	5-15 weeks	>8 days

The approximate proportion of initial total alkylation of each discussed derivative in DNA alkylated *in vivo* is as follows: m^2C, m^2T, m^4T, ~0.1%; m^6G, 6.3%; e^2C, 3%; e^2T, 7%; e^4T, 2%; e^6G, 8%.

EFFECTS ON DNA REPLICATION OF UNREPAIRED ALKYL BASES

Both adducts that can be rapidly repaired, such as O^6-alkyl G, and those found to persist, such as O^4-alkyl T, may be present in a mammalian cell at a critical time of replication or transcription, or in a specific sequence that is repaired slowly or replicated rapidly.

In vivo, modification of a normal base occurs either when basepaired, e.g., T·A -> m^4T·A, or in a single strand, e.g., T -> m^4T. Mutation is expressed when the modified base can direct incorporation of a different base, e.g. , m^4T·A -> m^4T·G, leading to an A -> G transition. While it is obvious that the dNTP pools can also be alkylated, the K_m of polymerases accepting modified dNTPs is usually at least an order of magnitude larger than for the unmodified dNTPs, using polymerases with relatively high fidelity. Nevertheless, polymerases can be used to insert one or many specific alkyl dNTPs in polymeric sequences. These sequences can then be replicated either *in vitro*, e.g., poly[dA-dT, m^4dT], or *in vivo*, e.g., M13 or ϕX174 DNA containing an alkyl base at a predetermined site.

Our initial focus was *in vitro* replication of DNA and polynucleotides using O^2-methyl dTTP and O^2-methyl dTTP as substitutes for dTTP. Although not as efficient substrates as dTTP, both could be incorporated (Singer, et al., 1986a, 1986b) and, once incorporated, could direct occasional incorporation of G. This methodology provided a means of studying the mutagenic potential of a single type of derivative in a defined system.

Based on these data, the same technique was used to insert O^4-methyl-, ethyl-, and isopropyl-dTTP in poly[dA-dT] (Singer, et al., 1986b). Although the extent of substitution for T can be quantified, the location of the alkyl derivative is not known. However, structural changes in such polymers are reflected in the parameters associated with helix-coil transition. Poly[dA-dT] structure was not detectably altered by the presence of 2-7% of any O^4-alkyl T in place of T. On replication using DNA polymerase I (Pol I), all directed increased levels of dGTP incorporation (Singer, et al., 1983; 1986a, 1986b).

Advances in oligonucleotide synthesis, coupled with genetic engineering, have now made it possible to observe the effects of a single adduct on kinetics of basepair formation, mutation, or the function of a gene. Using these techniques with a series of O^4-alkyl dTTPs we showed that, regardless of the size of the alkyl group (methyl, ethyl, isopropyl), all could substitute for dTTP opposite A in primer extension (Preston, et al., 1986). Similarly, O^2-methyl dTTP was an acceptable substrate. The ϕX174 DNA primer containing O^4-alkyl T at the selected am3 site (Figure 8) (Preston, et al., 1987) could be replicated in E. coli spheroplasts and the progeny mutant phage, when sequenced, showed A -> G transitions at the site of substitution, supporting in vitro data for mutation (Singer, et al., 1983, 1986a, 1986b).

In parallel with these in vivo mutagenesis studies, we have used various template-primer complexes in vitro to determine the kinetic parameters of polymerases accepting O^2- and O^4-alkyl dNTps and when the modified nucleotide is in the primer, the effect on further replication. Although the K_m^{apps} were relatively higher for the alkyl derivatives, once inserted they all were readily extended (Singer, et al., 1989). This indicated that O-alkyl T·A pairs were not blocks to replication.

The several in vivo systems used for studying the mutagenic potential of O^6-methyl G (m^6G) and O^2- and O^4-methyl-, ethyl- and isopropyl T in vivo and in vitro are all based on putting an adduct in a phage or plasmid at a defined site, replicating in E. coli spheroplasts, selecting mutants, and sequencing these to prove that the mutant progeny has a changed base. In general, these site-specific experiments agree with in vitro mutation and replication studies, which yield kinetic data and information about the type and frequency of changed basepairing. An illustration of a primer-template used and the technique are in Figures 9 and 10. In addition, sequence dependence of mutation, modification and repair can be measured with this experimental design by varying the primer sequence and strain background.

Turning to O^6-methyl G mutation, Essigmann and collaborators built this derivative into M13 phage and, under conditions where repair was suppressed, found C -> T transitions (Loechler, et al., 1984). The same sequence containing O^6-methyl G has been incorporated in a 26-mer template. After annealing to a 17-mer primer, replication kinetics using DNA polymerase I can be studied for both $m^6G{\cdot}C$ and $m^6G{\cdot}T$ pairing. This polymerase has a relatively high preference for inserting T opposite m^6G, but both $m^6G{\cdot}T$ and $m^6G{\cdot}C$ formation are less favored than is formation of G·C pairs. As with $m^4T{\cdot}A$ termini, extension to the next normal basepair is rapid. In contrast to mismatches such as A·C, A·A, etc., or apurinic/apyrimidinic sites, which significantly distort the terminus, slowing extension by several orders of magnitude (Kunkel, et al., 1983; Petruska, et al., 1988), these alkylated basepairings present a near-normal terminus for extension.

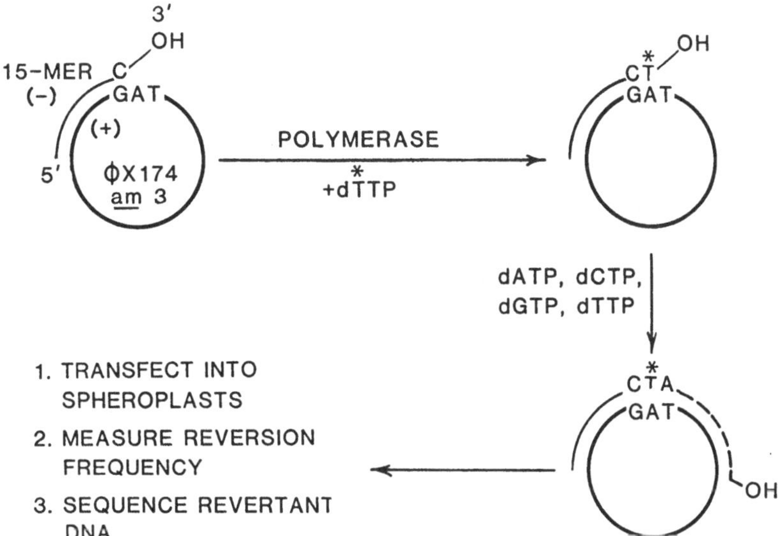

Figure 8. Protocol for site-specific insertion of dTTP analogues using oligonucleotide-primed single-stranded φX174 DNA containing an amber mutation. DNA polymerase I was used to elongate the primer (Preston, et al., 1986).

EFFECT OF ALKYL GROUP SIZE AND POSITION ON STRUCTURE

The destabilization of $m^6G{\cdot}C$ or $m^6G{\cdot}T$ in oligomers is observed by NMR and other physical methods (Gaffney, et al.,

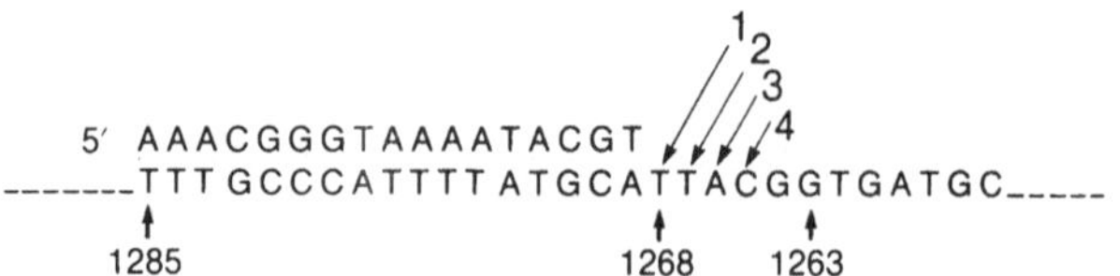

Figure 9. Diagram of the system used to measure kinetics of incorporation of modified dTTPS opposite A (site 3) in the template. The synthetic primer strand is 17 nucleotides long, labeled at the 5'-end with ^{32}P and annealed to a complementary section (bases 1285-1267) of circular M13 DNA template. Substrate dATP is added to allow primer extension from site 0 to site 2 by insertion of A opposite T in sites 1 and 2 (Singer, et al., 1989).

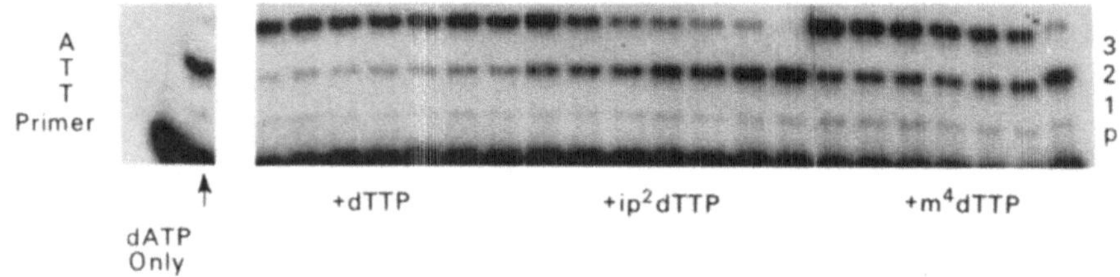

Figure 10. Example of the gel assay used to measure kinetics of insertion of O-alkyl dTTPs opposite A (site 3, Figure 9) in primed M13 DNA. The first two lanes on the left are the ^{32}P-end labeled primer alone, and after dATP addition with 90 second reaction using DNA polymerase I. This terminates synthesis at site 2. In the presence of both dATP and dTTP or an analogue, further extension can be demonstrated, as indicated by the additional bands at site 3. For each dTTP, a series of concentrations is used, decreasing from left to right. The seven dTTP lanes contain concentrations above the K_m so that extension is relatively concentration-independent, while those with ip^2dTTP or m^4dTTP increase over the triphosphate concentration range (1-200 µM). Details of the experimental method and calculation of kinetic constants from such data are given in Singer, et al., 1989.

1984; Patel, et al., 1986a, 1986b). Two hydrogen bonds are detected by NMR for $m^6G \cdot C$ and $m^6G \cdot T$, but the kinetic data strongly suggest that the ability to form hydrogen bonds alone does not determine the facility to basepair correctly and, that stacking effects may be of prime importance. Although NMR and thermal denaturation measurements furnish valuable information on structure, replication differs in its requirements for basepair formation, in that the polymerase can play an active role in positioning the incoming nucleotide.

In the case of O^4-methylthymine, only one hydrogen bond is formed on annealing (Kalnik, et al., 1988a, 1988b), yet this derivative can pair with A or G and then act as good primer termini (Singer, et al., 1989). For neither O^6-alkyl G nor O^2- or O^4-alkyl T does proofreading, a form of repair, appear to be a major factor in the decreased affinity of DNA polymerase I for alkylated dNTps.

Studies on repair of bulky adducts such as the N2-acetoxyamino-fluorene (AAF) adduct on the C-8 of G indicate that this adduct is lost extremely slowly in vivo (Kriek, 1972). Similarly O^6-alkylguanine-DNA methyltransferase is progressively less rapid in dealkylating DNA containing a series of O^6-alkylguanines with increasing chain size (Morimoto, et al., 1985). We expected that a polymerase would show decreased affinity for the alkylated dNTP as a function of size of alkyl group. This was the case when determining kinetic constants for the series of O^4-alkyl dTTPS (1-3 carbons) (Singer, et al., 1986a). Surprisingly, the largest O^2-alkyl derivative, O^2-isopropyl dTTP, was accepted almost as readily as O^4-methyl dTTP in the synthesis of DNA (Singer, et al., 1989), while O^4-isopropyl dTTP was very poorly utilized (Singer, et al., 1986a, 1986b).

These kinetic and mutation data are being used as the basis for molecular modeling studies of the DNA helix with a modified base incorporated and extended to continue the double strand. In collaboration with E. L. Loechler, the O^2-alkyl bonding with A in our sequence has been modeled. An unusual interaction is possible with the isopropyl group forming a hydrophobic pocket with the C2 of A. Neither methyl nor ethyl groups at the O^2 position are capable of this interaction. Thus interactions other than hydrogen bonding can stabilize unusual basepairs. This type of structural interaction may be relevant to recognition by both polymerases and repair enzymes.

SUMMARY

The two O-alkyl bases most often associated with initiation of carcinogenesis and mutation by N-nitroso alkylating agents are O^6-alkyl G and O^4-alkyl T. The former is rapidly repaired in most normal eukaryotic cells while the latter is more slowly repaired. In contrast to the specific repair enzymes in E. coli, no mechanism or specific enzyme has been identified for O-alkyl pyrimidine repair in eukaryotes. Kinetic and site-specific studies both agree that O^6-methyl G causes G -> A transitions, while O^4-alkyl T leads to T -> C transitions. However, the rate of insertion by DNA polymerase I is slow for either C or T opposite O^6-MeG while O^4-MeT substitutes quite readily for T. Hydrogen bonding alone cannot be the basis for these results. Using modeling studies it appears that the orientation of the modified base in a helix and stacking are important determinants for efficient replication and repair.

ACKNOWLEDGEMENT

A portion of this work was supported by Grant CA 12316 from the National Cancer Institute, National Institutes of Health. Additional support came from Grant CA 42736, administered by Lawrence Berkeley Laboratory under DOE Contract #DE-AC03-76SF00098.

REFERENCES

Ahmmed, Z. and Laval, J., 1984, Enzymatic repair of O-alkylated thymidine residues in DNA: Involvement of an O^4-methylthymine-DNA methyltransferase and an O^2-methylthymine DNA glycosylase, Biochem. Biophys. Res. Comm., 120:1-8.

Belinsky, S. A., White, C. M., Boucheron, J. A., Richardson, F. C., Swenberg, J. A., and Anderson, M., 1986, Accumulation and persistence of DNA adducts in respiratory tissue of rats following multiple administrations of the tobacco specific carcinogen 4-(methyl-N-nitrosamino)-1-(3-pyridyl)-1-Butanone, Cancer Res., 46:1280-1284.

Bodell, W. J., Singer, B., Thomas, G. H. and Cleaver, J. E. 1979, Evidence for removal at different rates of O-ethyl pyrimidines and ethyl-phosphotriesters in two human fibroblast cell lines, Nucleic Acids Res., 6:2819-2829.

Brent, T. P., Dolan, M. E., Fraenkel-Conrat, H., Hall, J., Karran, P., Laval, F., Laval, J., Margison, G. P., Montesano, R., Pegg, A. E., Potter, F. M., Singer, B., Swenberg, J. A. and Yarosh, D. B., 1988, Repair of O-alkylpyrimidines in mammalian cells: A present consensus, Proc. Natl. Acad. Sci. USA, 85:1759-1762.

Druckrey, H., Preussman, R. and Ivanovic, S., 1969, N-Nitroso compounds in organotropic and transplacental carcinogenesis, Ann. N. Y. Acad. Sci., 163:676-696.

Gaffney, B. L., Marky, L. A. and Jones, R. A., 1984, Synthesis and characterization of a set of four dodecadeoxyribonucleoside undecaphosphates containing O^6-methylguanine opposite adenine, cytosine, guanine and thymine, Biochemistry, 23:5686-5691.

Hall, J. and Karran, P., 1986, O-Methylated pyrimidines important lesions in cytotoxicity and mutagenicity in mammalian cells, in: M. B. Myrnes and H. Krokan, eds. , "Repair of DNA Lesions Introduced by N-nitroso Compounds," Norwegian University Press, Oslo, pp. 77- 88.

Huh, N. and Rajewsky, M. F., 1986, Enzymatic elimination of O^6-ethylguanine and stability of O^4-ethylthymine in the DNA of malignant neural cell lines exposed to N-ethyl-N-nitrosourea in culture, Carcinogenesis, 7:435-439.

Kalnik, M. W., Kouchakdjian, M., Li, B. F. L. , Swann, P. F. and Patel, D. J., 1988, Base mismatches and carcinogen modified bases in DNA: An NMR study of A·C and A·O^4meT pairing in dodecanucleotide duplexes, Biochemistry, 27: 100-108.

Kalnik, M.W., Kouchakdjian, M., Li, B. F. L., Swann, P. F. and Patel, D. J., 1988, Base mismatches and carcinogen-modified bases in DNA: An NMR study of G.T and G.O^4meT pairing in dodecanucleotide duplexes, Biochemistry, 27: 108-115.

Kriek, E., 1972, 1972, Persistent binding of a new reaction product of the carcinogen N-hydroxy-N-2-acetylaminofluorene with guanine in rat liver DNA in vivo, Cancer Res., 32:2042-2048.

Kunkel, T. A., Schaaper, R. M. and Loeb, L. A., 1983, Depurination-induced infidelity of deoxyribonucleic acid synthesis with purified deoxyribonucleic acid replication proteins in vitro, Biochemistry, 22:2378-2384.

Laval, J., Pierre, J. and Laval, F., 1981, Release of 7-methylguanine residues from alkylated DNA by extracts of Micrococcus luteus and Escherichia coli, Proc. Natl. Acad. Sci. USA, 78:852-855.

Lewis, J. G. and Swenberg, J. A., 1980, Differential repair of O^6-methylguanine in DNA of rat hepatocytes and nonparenchymal cells, Nature, 228:185-187.

Lindahl, T., 1982, DNA repair enzymes, Ann. Rev. Biochem., 51: 61-87.

Lindahl, T., Sedgwick, B., Sekiguchi, M. and Nakabeppu, Y., 1988, Regulation and expression of the adaptive response to alkylating agents, Ann. Rev. Biochem., 57:133-157.

Loechler, E. L., Green, C. L. and Essigmann, J. M., 1984, In vivo mutagenesis by O^6-methylguanine built into a unique site in a viral genome, Proc. Natl. Acad. Sci. USA, 81:6271-6275.

Magee, P. N. and Farber, E., 1962, Toxic liver injury and carcinogenesis. Methylation of rat-liver nucleic acids by dimethylnitrosamine in vivo, Biochem. J., 83: 114-124.

Margison, G. P. and Pegg, A. E., 1981, Enzymatic release of 7-methylguanine from methylated DNA by rodent liver extracts, Proc. Natl. Acad. Sci. USA, 78:861-865.

McCarthy, T. V., Karran, P. and Lindahl, T., 1984, Inducible repair of O-alkylated DNA pyrimidines in Escherichia coli, Embo J., 3:545- 550.

McCarthy, T. V. and Lindahl, T., 1985, Methyl phosphotriesters in alkylated DNA are repaired by the Ada regulatory protein of E. coli, Nucl. Acids Res., 13:2683-2698.

Morimoto, K., Dolan, M. E., Schicchitano, D. and Pegg, A.C., 1985, Repair of O^6-propylguanine and O^6-butylguanine in DNA by O^6-alkylguanine-DNA-alkyltransferase from rat liver and E. coli, Carcinogenesis, 6:1027-1031.

Patel, D. J., Shapiro, L., Kozlowski, S. A., Gaffney, B. L. and Jones, R. A., 1986a, Structural studies of the O^6meG·C interaction in the d(C-G-C-G-A-A-T-T-C-O^6meG-C-G) duplex, Biochemistry, 25:1027-1036.

Patel, D. J., Shapiro, L., Kozlowski, S. A., Gaffney, B. L. and Jones, R. A., 1986b, Structural studies of the O^6meG·T interaction in the d(C-C>T-G-A,A-T-T-C-O^6meG-C-G) duplex, Biochemistry, 25:1036-1042.

Pegg, A. E., 1983, Alkylation and subsequent repair of DNA after exposure to dimethylnitrosamine and related carcinogens, Rev. Biochem. Toxicol. 5:83-133.

Pegg, A. E., 1984, Methylation of the O^6 position of guanine in DNA is the most likely initiating event in carcinogenesis by methylating agents, Cancer Invest., 2(3):223-231.

Pegg, A. E. and Dolan, M. E., 1989, Investigation of sequence specificity in DNA alkylation and repair using oligodeoxynucleotide substrates, this volume.

Petruska, J., Goodman, M. F., Boosalis, M. S., Sowers, L. C., Cheong, C. and Tinoco, I., Jr., 1988, Comparison between DNA melting thermodynamics and DNA polymerase fidelity, Proc. Natl. Acad. Sci. USA, 85:6252-6256.

Preston, B. D., Singer, B. and Loeb, L. A., 1986, Mutagenic potential of O^4-methylthymine in vivo determined by an enzymatic approach to site-specific mutagenesis, Proc. Natl. Acad. Sci. USA, 83:8501-8505.

Preston, B. D., Singer, B. and Loeb, L. A., 1987, Comparison of the relative mutagenicities of O-alkylthymines site-specifically incorporated into ϕX174 DNA, J. Biol. Chem., 262:13821-13827.

Richardson, F. C., Dyroff, M. C., Boucheron, J. A. and Swenberg, J. A., 1985, Differential repair of O^4-alkylthymidine following exposure to methylating and ethylating hepatocarcinogens, Carcinogenesis, 6:625-629.

Singer, B., 1975, Chemical effects of nucleic acid alkylation and their relation to mutagenesis and carcinogenesis, Progr. Nucleic Acids Res. and Mol. Biol, 15:219-284, 330-332.

Singer, B., 1976, All oxygens in nucleic acids react with carcinogenic ethylating agents, Nature, 264:333-339.

Singer, B., 1982, Mutagenesis from a chemical perspective: Nucleic acid reactions, repair, translation and transcription, in: J. F. Lemontt and W. M. Generoso, eds., "Molecular and Cellular Mechanisms of Mutagenesis, " plenum Publishing, New York, pp. 1-42.

Singer, B., 1985, In vivo formation and persistence of modified nucleosides resulting from alkylating agents, Environ.Health Perspect., 62:41-48.

Singer, B., 1986, Perspectives in cancer research. O-Alkyl pyrimidines in mutagenesis and carcinogenesis: Occurrence and significance, Cancer Res., 46:4879-4885.

Singer, B. and Brent, T. P., 1981, Human lymphoblasts contain DNA glycosylase activity excising N-3 and N-7 methyl and ethyl purines but not O^6-alkylguanine or 1-alkyladenine, Proc. Natl. Acad. Sci. USA, 78:856-860.

Singer, B. and Grunberger, D., 1983, "Molecular Biology of Mutagens and Carcinogens, " Planum Publishing, New York.

Singer, B., Spengler, S. and Bodell, W. J., 1981, Tissue-dependent enzyme-mediated repair or removal of O-ethyl pyrimidines and ethyl purines in carcinogen-treated rats, Carcinogenesis, 2:1069-1073.

Singer, B., Sagi, J. and Kuśmierek, J. G., 1983, Escherichia coli polymerase I can use O^2-methyldeoxythymidine or O^4-methyldeoxythymidine in place of deoxythymidine in primed poly (dA-dT) poly (dA-dT) synthesis, Proc. Natl. Acad. Sci. USA, 80:4584-4588.

Singer, B., Chavez, F. and Spengler, S. J., 1986a, O^4-Methyl-, ethyl-, and isopropyl deoxythymidine triphosphates as analogues of deoxythymidine triphosphate: Kinetics of incorporation by Escherichia coli DNA polymerase I, Biochemistry, 26:1201-1205.

Singer, B., Spengler, S. J., Fraenkel-Conrat, H. and Kuśmierek, J. T., 1986b, O^4-Methyl-, ethyl-, or isopropyl substituents on thymidine in poly(dA-dT) all lead to transitions upon replication. Proc. Natl. Acad. Sci. USA, 83:28-32.

Singer, B., Chavez, F., Spengler, S., Kuśmierek, J. T., Mendelman, L. and Goodman, M. F., 1989, Comparison of polymerase insertion and extension kinetics of a series of O^2-alkyl deoxythymidine triphosphates with O^4-methyl deoxythymidine triphosphate, Biochemistry, 28:1478-1483.

Swenberg, J. A., Dyroff, M. C., Bedell, M. A., Popp, J. A. , Huh, W., Kirstein, U. and Rajewsky, M. F., 1984, O^4-Ethyldeoxythymidine, but not O^6-ethyldeoxyguanosine, accumulates in hepatocyte DNA of rats exposed continuously to diethylnitrosamine, Proc. Natl. Acad. Sci. USA, 81:1692-1695.

Weinfeld, M., Drake, A, F., Saunders, J. K. and Paterson, M. C., 1985, Stereospecific removal of methyl phosphotriesters from DNA by an Escherichia coli ada+ extract, Nucl. Acids Res., 13:7067-7077.

THE ROLES OF BETA- AND DELTA-ELIMINATIONS IN THE REPAIR OF AP SITES IN DNA

Véronique Bailly and Walter G. Verly

Biochimie, Faculté des Sciences
Université de Liège, Sart Tilman B6
B-4000 Liège 1, Belgium

In principle, an AP site could be released by two hydrolytic events, one nicking the phosphodiester bond 3', the other the phosphodiester bond 5' to the AP site. Since in each phosphodiester bond, there are two linkages, $C_{3'}$-O-P and $C_{5'}$-O-P, that can be hydrolyzed, four different hydrolases might participate in the excision of AP sites. In fact, we know a single class of AP endonucleases; these enzymes hydrolyze the $C_{3'}$-O-P bond 5' to AP sites and the resulting nicks are limited by 3'-OH and 5'-phosphate ends.

But the phosphodiester bonds, on each side of the AP site, can be broken in other ways. Beta-elimination removes a H^+ from the 2' position of the base-free deoxyribose: the electron redistribution leads to the elimination of the phosphate attached to $C_{3'}$ and the formation of a 2',3'-double bond. The enzymes that catalyze such a reaction are not hydrolases. They can thus not be called AP endonucleases, and we propose to name them AP lyases since they belong not to the third, but to the fourth category of the international enzyme classification. The beta-elimination can be followed by delta-elimination: removal of a H^+ from the 4' position of the base-free sugar leads to the elimination of the phosphate attached to $C_{5'}$ and the formation of a 4',5'-double bond. An unsaturated derivative of deoxyribose is released and a 3'-phosphate end is formed. The delta-elimination is a second beta-elimination; the enzymes that catalyze a beta-delta-elimination can also be called AP lyases.

Beta-elimination breaks the $C_{3'}$-O-P bond 3' to the AP site; delta-elimination breaks the $C_{5'}$-O-P bond 5' to the AP site. The two reactions excise the AP site and leave a gap limited by 3'-phosphate and 5'-phosphate ends identical to gaps formed by ionizing radiations. A 3'-phosphatase is needed to start the repair of such gaps.

THE REPAIR OF AP SITES IN BACTERIA

DNA containing AP sites was repaired _in vitro_ with _E._

DNA Repair Mechanisms and Their Biological Implications in Mammalian Cells
Edited by M.W. Lambert and J. Laval
Plenum Press, New York

coli endonuclease VI, DNA polymerase I, and T4 DNA ligase (Verly, *et al.*, 1974). To analyze the excision step, the AP sites were labelled with ^{3}H by reduction with tritiated $NaBH_4$, whereas the DNA strands were randomly labelled with ^{32}P(Gossard and Verly, 1978). The AP endonuclease hydrolyzed the $C_{3'}$-O-P bond 5' to the AP sites, and the nicking was followed by the degradation of the DNA strand in the 3'-5' direction by the exonucleolytic activity (exonuclease III) of endonuclease VI; this can be followed by the release of ^{32}P and the analysis of the acid-soluble fragments. But the labelled AP sites remained in the DNA to be quickly released on addition of *E. coli* DNA polymerase I; DEAE-Sephadex chromatography showed that the deoxyribose-5-phosphate was set free bound to one nucleotide. In other words, it was the second phosphodiester bond 3' to the 5'-terminal AP site that was hydrolyzed by the 5'-3' exonuclease activity of DNA polymerase I. The general picture of the AP site excision is thus the following which is consistent with the work of Lundquist and Olivera (1982) on nick translation: starting from the 3'-OH, the DNA polymerase fills the gap produced by exonuclease III, then displaces the strand terminated by an AP site and its 5'-3' exonuclease hydrolyzes the second phosphodiester bond; this is followed by the translation of a nick which is finally closed by DNA ligase.

More recently, we have investigated the AP site repair using *E. coli* endonuclease IV instead of endonuclease VI, and we used as substrate a DNA containing AP sites labelled with ^{32}P on their 5' side. To our surprise, the acid-soluble ^{32}P was rapidly above what was expected from a simple fragmentation of the DNA strands at AP sites. Using 5'-labelled and 3'-labelled oligonucleotides and gel electrophoresis, we could observe two successive reactions: the hydrolysis of the $C_{3'}$-O-P bond 5' to the AP site was followed by the rupture of the $C_{3'}$-O-P bond 3' to the AP site. In order to analyze the nature of this second reaction, we used DNA with doubly-labelled AP sites: with ^{3}H on the 1' and 2' positions of the base-free deoxyribose, and with ^{32}P 5' to this base-free sugar. The second reaction was associated with the release of volatile ^{3}H, likely from the 2' position of the base-free deoxyribose. When the reaction products were analyzed on DEAE-Sephadex, two peaks of ^{3}H and one peak of ^{32}P were observed. The first ^{3}H peak corresponded to the volatile ^{3}H; the second coincided with the ^{32}P peak. The doubly-labelled sugar-phosphate emerged after a deoxyribose-5-phosphate marker and the $^{3}H/^{32}P$ ratio was below that of the labelled AP sites in the substrate DNA. All these observations indicate that the second reaction could not be the hydrolysis of the $C_{3'}$-O-P bond 3' to the AP site; on the other hand, they all agree with the hypothesis that the 3' nicking was a beta-elimination reaction.

But *E. coli* endonuclease IV does not play a role in this second reaction: the half-life of a 5'-terminal base-free deoxyribose-5'-phosphate is the same whether *E. coli* endonuclease IV is there or not. In absence of beta-elimination catalysts, this half-life is 2 hours at $37^{0}C$; this contrasts with the 200 hour half-life of the $C_{3'}$-O-P bond 3' to an internal AP site (Lindahl and Anderson, 1972). Beta elimination catalysts, such as polyamines, considerably shorten the half-life of 5'-terminal AP sites.

The above results mean that, after nicking 5' to an AP site with an AP endonuclease, it is not necessary to use a 5'-3' exonuclease to excise the lesion; it simply falls off. We have indeed repaired AP site-containing DNA with E. coli endonuclease IV, the Klenow fragment of DNA polymerase I (which is devoid of 5'-3' exonuclease activity), and T4 DNA ligase. This does not mean that the excision step can not be accelerated. In an experiment where the AP site had been labelled by reduction in order to follow its excision (Gossard and Verly, 1978), after the 5' incision by the AP endonuclease, beta-elimination could no longer excise the AP site, but, when DNA polymerase I was added, the half-life of the 5'-terminal reduced AP site was a few minutes. This is much shorter than the 2 hour half-life corresponding to the spontaneous release of non-reduced 5'-terminal AP sites, but we do not know how much this half-life is shortened by the beta-elimination catalysts present in the bacterial cell. It is very likely that the two mechanisms (catalyzed beta-elimination and hydrolysis by the 5'-3' exonuclease activity of DNA polymerase I) participate in the AP site excision. We do not know yet which is the more important.

We have also examined the nicking activity near AP sites by E. coli endonuclease III (Bailly and Verly, 1987), M. luteus and T4 UV endonucleases, and E. coli formamidopyrimidine-DNA glycosylase. In all cases, it is the $C_{3'}$-O-P bond 3' to the AP site which is broken, and the mechanism is a beta-elimination reaction. Thus, none of these enzymes can be called an endonuclease. They are not AP endonucleases, but AP lyases.

With E. coli endonuclease III, the process stops after the 3' nicking, and we have shown that the 3'-terminal base-free unsaturated sugar-5'- phosphate can be released either by E. coli endonuclease VI or endonuclease IV. With T4 UV endonuclease, the 3' nicking by beta-elimination can be followed by a 5' nicking by delta-elimination. With E. coli formamidopyrimidine-DNA glycosylase, the 3' nicking by beta-elimination is immediately followed by a 5' nicking by delta-elimination. The intermediate product is undetectable, an unsaturated derivative of deoxyribose is released, leaving a gap limited by 3'-phosphate and 5'-phosphate ends. The repair thus needs a 3'-phosphatase, but both E. coli endonuclease VI and endonuclease IV are 3'-phosphatases. We have repaired DNA containing AP sites with the formamidopyrimidine-DNA glycosylase, a 3'-phosphatase (T4 polynucleotide kinase), Klenow polymerase and T4 DNA ligase.

We underscore that the sterile 3' ends produced by beta-elimination or beta-delta-elimination acting on AP sites, can both be activated by E. coli endonuclease VI (exonuclease III) or endonuclease IV to produce 3'-OH able to prime DNA polymerases, explaining what has been published on the so-called AP endonucleases, class I, which, as we said previously, are not endonucleases.

REPAIR OF NUCLEAR DNA IN MAMMALIAN CELLS

Mammalian cells contain an AP endonuclease which is mainly located in chromatin (Verly and Paquette, 1973;

Thibodeau and Verly, 1980). The enzyme hydrolyzes the $C_{3'}$-O-P bond 5' to AP sites (Verly, *et al.*, 1981). But we do not know of any enzyme capable of hydrolyzing a phosphodiester bond 3' to an AP site; in other words, we do not know of a mammalian equivalent of the 5'-3' exonuclease activity of *E. coli* DNA polymerase I. We could excise AP sites by the successive actions of rat-liver AP endonuclease and rat-liver DNase IV, a 5'-3' exonuclease acting from nicks on double-stranded DNA (Lindahl, *et al.*, 1969). The broken phosphodiester bond 3' to the 5'-terminal AP site, was not the second, as with the 5'-3' exonuclease of *E. coli* DNA polymerase I, but the first one. Moreover, the nicking was not the result of a hydrolysis, but of a beta-elimination reaction (Grondal-Zocchi and Verly, 1985). The excision of the 5'-terminal AP site in the presence of DNase IV was rather slow, and we presently wonder whether the enzyme did not simply wait for the 5'-terminal AP site to drop off spontaneously before enlarging uselessly the one-nucleotide gap. Indeed, we repaired AP site-containing DNA in the absence of any 5'-3' exonuclease activity; for that, we used only rat-liver AP endonuclease, rat-liver DNA polymerase beta, and DNA ligase.

We thus came to the hypothesis that, in the mammalian cell nucleus, the nicking 3' to an AP site was always a beta-elimination reaction (Bailly and Verly, 1988). Chromatin is very rich in beta-elimination catalysts, especially histones and polyamines. Two different pathways of AP site repair were considered depending on whether the beta-elimination reaction was the first or the second step of the repair.

We successfully repaired, *in vitro,* AP site-containing DNA using the four following steps: 5' nicking with rat-liver AP endonuclease; 3' nicking with histones or polyamines; filling the one-nucleotide gap with DNA polymerase beta; closing the last nick with DNA ligase (Bailly and Verly, 1988).

But the 3' nicking could occur first, and we were surprised to find that, after a 3' nicking by beta-elimination, the rat-liver AP endonuclease could not excise the AP site. This is radically different from what we observed with *E. coli* endonuclease VI and endonuclease IV. It does not mean that the lesion becomes unrepairable; indeed, histones and polyamines also catalyze a delta-elimination. Thus, those beta-delta-elimination catalysts are alone sufficient to excise the base-free sugar, leaving a gap limited by 3'-phosphate and 5'-phosphate ends. We were able to repair AP site-containing DNA *in vitro* using the following steps: 3' nicking by beta-elimination and 5' nicking by delta-elimination catalyzed by spermine: hydrolysis of the 3'-phosphate (chromatin contains a 3'-phosphatase; Habraken and Verly, 1983); filling the one-nucleotide gap with DNA polymerase beta; reestablishing the strand continuity with DNA ligase.

The two pathways that we have explored *in vitro* with mammalian enzymes are probably those that work *in vivo* to repair AP sites in nuclear DNA. In both of them, the nicking 3' to the AP site is by beta-elimination; in one of them (probably the minor route and the slower one), the nicking 5' to the AP site is by delta-elimination.

REFERENCES

Bailly, V. and Verly, W.G., 1987, E.coli endonuclease III is not an endonuclease but a beta-elimination catalyst, Biochem. J., 242:565-572.

Bailly, V. and Verly, W.G., 1988, Possible roles of beta-elimination and delta-elimination reactions in the repair of DNA containing AP (apurinic/apyrimidinic) sites in mammnalian cells, Biochem. J., 253:553-559.

Gossard, F. and Verly, W.G., 1978, Properties of the main endonuclease specific for apurinic sites of E. coli (endonuclease VI). Mechanism of apurinic site excision from DNA, Eur. J. Biochem., 82:321-332.

Grondal-Zocchi, G. and Verly, W.G., 1985, Deoxyribonuclease IV from rat-liver chromatin and the excision of AP sites from depurinated DNA, Biochem. J. 225:535-542.

Habraken, Y. and Verly, W.G., 1983, The DNA 3'-phosphatase and 5'-hydroxylkinase of rat-liver chromatin, FEBS Lett., 160:46-50.

Lindahl, T. and Anderson, A., 1972, Rate of chain breaks at apurinic sites in double-stranded DNA, Biochemistry, 11:3618-3623.

Lindahl, T., Gally, J.A. and Edelman, G.M., 1969, Deoxyribonuclease IV: a new exonuclease from mammalian tissues, Proc.Natl Acad.Sci.USA., 62: 597-603.

Lundquist, R.C. and Olivera, B.M., 1982, Transient generation of displaced single-stranded DNA during nick translation, Cell, 31:53-60.

Thibodeau, L. and Verly, W.G., 1980, Cellular localization of the AP (apurinic/apyrimidinic) endodeoxyribonucleases in rat liver, Eur. J. Biochem., 107:555-563.

Verly, W.G., Colson, P., Zocchi, G., Goffin, C., Liuzzi, M., Buchsenschmidt, G. and Muller, M., 1981, Localization of the phosphoester bond hydrolyzed by the major apurinic/apyrimidinic endodeoxyribonuclease from rat liver chromatin, Eur. J. Biochem., 118:195-201.

Verly, W.G., Gossard, F. and Crine, P., 1974, In vitro repair of apurinic sites in DNA, Proc.Natl Acad.Sci.USA., 71: 2273-2275.

Verly, W.G. and Paquette, Y., 1973, An endonuclease for depurinated DNA in rat liver, Can. J. Biochem., 51: 1003-1009.

IMIDAZOLE RING-OPENED PURINES: OCCURENCE AND REPAIR IN <u>ESCHERICHIA COLI</u> AND MAMMALIAN CELLS

Jacques Laval, Timothy O'Connor, Claudine d'Hérin-Lagravere, Patricia Auffret Van der Kemp, and Serge Boiteux

Groupe "Réparation des lésions radio- et chimioinduites" Institut Gustave Roussy, 94805 Villejuif Cédex, France

ABSTRACT

Upon alkylation of DNA by alkylating agents, the main reaction product is 7-methylguanine (me7G), which is believed to be a harmless lesion. However, guanine methylated at the N7 position, is susceptible to depurination yielding apurinic sites, and to cleavage of the imidazole ring yielding: 2,6-diamino-4-hydroxy-5N-methyl-formamidopyrimidine (Fapy). HPLC and NMR analysis show that, at the base level, Fapy is an equimolecular mixture of two rotameric forms.

DNA synthesis catalysed by <u>Escherichia coli</u> DNA polymerase I, using as template M13 single stranded DNA containing Fapy, shows that Fapy blocks DNA chain elongation and that the arrest of DNA synthesis occurs one base 3' to template Fapy residues. This implies that Fapy lesions are potentially lethal.

We have cloned and sequenced the gene coding for the Fapy-DNA glycosylase of <u>Escherichia coli</u>. The <u>fpg</u> gene codes for a protein of 269 amino acids with a calculated molecular weight of 30.2 Kd. The protein was purified to apparent homogeneity and exhibited two different enzymatic activities: a DNA glycosylase and an activity incising DNA at apurinic/apyrimidinic (AP) sites. Moreover, based on restriction maps, the <u>fpg</u> gene is not associated with known <u>E. coli</u> genes coding for activities which cleave DNA at AP-sites. Termini generated following cleavage by the Fapy-DNA glycosylase: 1) do not prime <u>in vitro E. coli</u> DNA polymerase I synthesis unless they are further treated with Exonuclease III or alkaline phosphatase, and 2) are not substrates for 5' phosphorylation using T4 polynucleotide kinase unless they are pre-treated by phosphatase. In addition, the action of Fapy-DNA glycosylase on DNA containing [^{3}H]AP-sites generates radioactivity which is ethanol soluble and non-adsorbable on DEAE cellulose.

DNA Repair Mechanisms and Their Biological Implications in Mammalian Cells
Edited by M.W. Lambert and J. Laval
Plenum Press, New York

The Fapy-DNA glycosylase activity has been demonstrated in several bacterial systems. We have partially purified an enzymatic activity from calf thymus which removes the Fapy base. This protein has a molecular weight between 26 and 32 kdalton and is active in reaction solutions without Mg^{2+}.

INTRODUCTION

Alkylation of the N7 of guanine residues is the major reaction product formed in vitro and in vivo following treatment with alkylating agents (Singer and Grunberger, 1983). This N7 alkylation labilizes the glycosylic bond and the imidazole ring of the guanine which can result in the formation of an apurinic (AP)-site or in the opening of the imidazole ring to yield a formamidopyrimidine (Fapy) residue (Haines, et al., 1962). Although the N7 methylguanine is the major methylation product, this lesion does not inhibit in vitro DNA synthesis and persists in bacterial DNA over generations and therefore is probably not harmful to the cell. In contrast, the formation of AP-sites and Fapy lesions are not as benign. The introduction of an AP-site by destabilization of the glycosylic bond of the N7 modified guanine can lead to mutation (Boiteux and Laval, 1982). The opening of the imidazole ring of N7 methylguanine yields a Fapy base, which blocks in vitro DNA synthesis (Boiteux and Laval, 1983, and O'Connor, et al., 1988) and may therefore cause cell death if not repaired.

Presumably as a response to the hazard of Fapy lesions, E. coli has developed an enzyme to excise this lesion from DNA (Chetsanga and Lindahl, 1979). The Fapy-DNA glycosylase, which effects the repair of this lesion, is not expressed at high levels in the cell. Therefore, we have cloned the fpg gene coding for the Fapy-DNA glycosylase by screening for a plasmid overproducing the enzyme (Boiteux, et al., 1987). Thus, the physical properties of the enzyme, as well as the structure and function of the gene can now be studied. In this paper we summarize the properties of Fapy residues and of the Fapy-DNA glycosylase. We also show that the Fapy-DNA glycosylase of E. coli possesses an associated activity which cleaves DNA at AP-sites and that this cleavage reaction may involve a β-elimination mechanism. We also report the partial purification of an enzyme from calf thymus which removes Fapy lesions from DNA.

RESULTS

Characterization of the Fapy base, the ring-opened form of 7-methylguanine

Fapy was prepared from 7-methylguanosine by alkali cleavage of the imidazole ring and further elimination of the ribosyl residue by formic acid treatment (Boiteux and Laval, 1983). When Fapy was analysed by HPLC using a reversed phase column, two peaks (FI and FII) of the same magnitude were resolved, eluting at 4 and 5 minutes, respectively (Figure 1A). When Fapy is lyophylized, dissolved in water, and immediately analysed, only peak FII was observed (Figure 1B). Rechromatography of each isolated component indicated that they are slowly interconverted to give a 1:1 mixture as in Figure 1A. The interconversion kinetics of the two isolated species were measured. The rate of convertion of FI => FII and FII => FI increased with increasing temperature. The half life of FI or FII was 120 minutes, 35 minutes, 22 minutes and 8 minutes at 20^0C, 25^0C, 30^0C, and 37^0C, respectively. These results suggest that the FI and FII species are isomeric forms of the same molecule rather than two distinct compounds (Boiteux, et al., 1984). NMR studies confirmed unambiguously that there are two rotational isomers of Fapy. Thermodynamic measurements strongly suggested that the equilibrium can be assigned to rotation around the N-methyl formamido bond (Boiteux, et al., 1984). The two species, FI and FII, separated by HPLC were identified as rotamers E and Z, respectively (Figure 2).

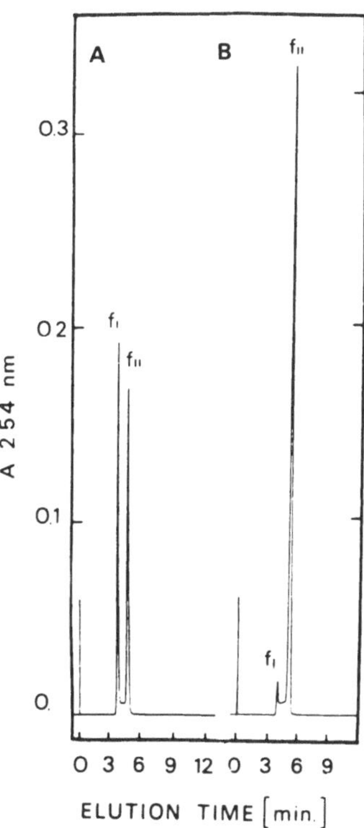

Figure 1. Reversed phase HPLC analysis of Fapy. Chromatography was performed with an HPLC system model 6000 Waters Associate (Milford, Ma.). Separation was obtained on a C_{18} Bondapak column (Waters), isocratically run at 1.5 ml/minute. The mobile phase was 50 mM $NH_4H_2PO_4$, pH 4.5, containing 5% methanol (v/v). A. elution profile of Fapy. B. elution profile of Fapy lyophilized and dissolved in water immediately before HPLC analysis. 7-methylguanine and 7-methylguanosine eluted at 11 and 8 minutes, respectively.

Chemical properties of Fapy lesions in polynucleotides

The chemical properties of Fapy lesions in polynucleotides were studied to determine the rate of formation and stability of these lesions in nucleic acids. Therefore, we analysed the products obtained after alkaline treatment of poly(dG-dC)·poly(dG-dC) methylated with tritiated dimethylsulfate. Formic acid hydrolysis of the alkylated polymer showed that 97% of the radioactivity eluted with 7-methylguanine, and 2-3% with 3-methylguanine. The methylated poly(dG-dC)·poly(dG-dC) was further incubated under alkaline conditions for increasing lengths of time. Formic acid hydrolysis of these polymers shows that the amount of 7-methylguanine decreases as the amount of Fapy increases, whereas the percentage of 3-methylguanine residues remains unchanged. Complete conversion of 7-methylguanine into Fapy was obtained after 40 hours of incubation in Na_2HPO_4-NaOH buffer, pH 11.4. We have also determined that Fapy residues are completely released from DNA after heating for two hours at 100^0C, at neutral pH. Therefore, Fapy is much more stable in DNA than

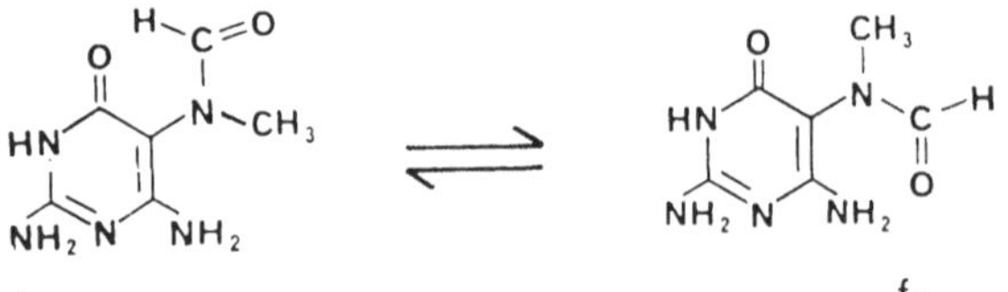

Figure 2. Structure of the two rotamers of Fapy.

me7G, which is completely released after 15 minutes under the same conditions.

The ring opened form of 7-methylguanine is a block to DNA synthesis

In vitro DNA synthesis with *E. coli* DNA polymerase I was measured using as template either M13 DNA treated with DMS thus containing 1-methyladenine, 3-methyladenine, 7-methylguanine a and few other minor unidentified bases or the same modified M13 DNA but further treated under alkaline conditions. As shown in Figure 3, the only modification due to alkaline treatment was the transformation of 7-methylguanine into Fapy. Using these modified DNAs as templates for nucleotide incorporation during *in vitro* DNA synthesis by *E. coli* DNA polymerase (Klenow fragment), we showed that 1) DNA synthesis was reduced compared to the unmethylated template and 2) additional alkaline treatment further reduced *in vitro* DNA synthesis compared to the synthesis on methylated templates. Thus, the presence of Fapy is a strong block to DNA elongation.

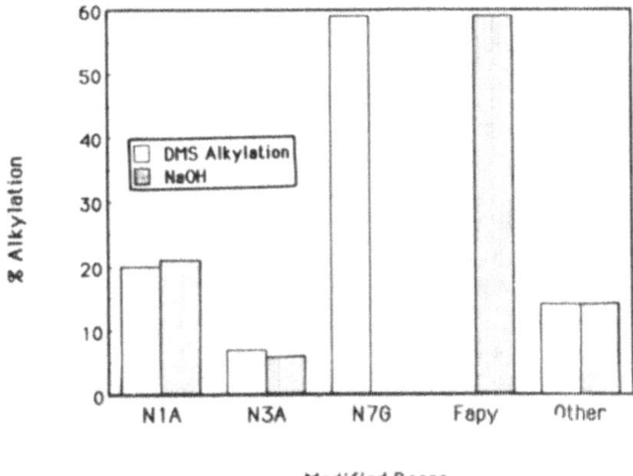

Figure 3. Relative amount of alkylated bases obtained after alkylation of M13 DNA either by dimethylsulfate or by dimethylsulfate followed by alkaline treatment. The modified DNAs were acid depurinated and the products separated using reverse phase HPLC chromatography as described in O'Connor, et al., 1988.

In order to obtain more information, we analysed the products of the reaction on sequencing gels. The results shown in Figure 4 are consistent with the fact that synthesis stops one base 3' to template adenine residues in agreement with Larson, et al., 1985. These results strongly suggest that N3-methyladenine residues if unrepaired are potentially lethal lesions. If the template is modified only by transforming the N7-methylguanine residues to Fapy residues, synthesis by DNA polymerase Klenow fragment stops one base 3' to guanine residues. Thus, since the N3-methyladenine is a lethal lesion (Boiteux, et al., 1984) which stops *in vitro* DNA synthesis, by analogy, Fapy residues are potentially lethal.

Molecular cloning, purification and genetic evidence that the Fapy-DNA glycosylase of Escherichia coli possesses an associated activity which cleaves DNA at AP-sites

E. coli possesses a DNA glycosylase which excises Fapy residues from DNA substrates *in vitro* (Chetsanga and Lindahl, 1979). In order to obtain significant amounts of pure enzyme, we cloned the Fapy-DNA glycosylase gene of *E. coli* into a multicopy plasmid. We screened a plasmid library of *E. coli* DNA for an increase in glycosylase activity in individual crude lysates (Boiteux, et al., 1987) and isolated a 15 Kb recombinant plasmid which allowed the overproduction of the Fapy-DNA glycosylase by a factor of 10. After subcloning, a 1.4 Kb fragment of *E. coli* DNA containing the *fpg* gene coding for the Fapy-DNA glycosylase was isolated. Finally, the *fpg* gene placed under the control of the lac promoter in the pFPG60 plasmid displayed a 50-100 fold increase in glycosyl-

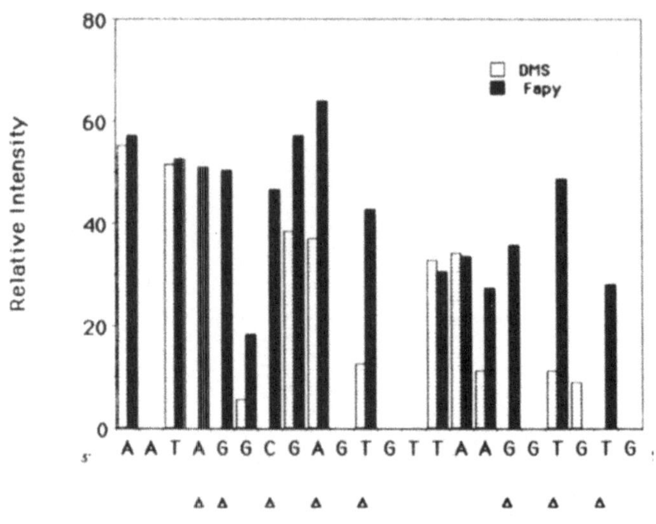

Figure 4. Scan of the sequencing gel of the products of *in vitro* DNA synthesis using Klenow fragment DNA polymerase I with M13mp18 methylated templates containing N7-methylguanine or Fapy residues. (▤) refers to M13mp18 DNA containing N7-methylguanine residues, (■) refers to M13mp18 DNA containing Fapy residues. The direction of DNA synthesis is indicated by an arrow and new stops found one base before Fapy lesions in the template are indicated by triangles (Δ).

Table I. Physical and catalytic parameters of *E. coli* formamidopyrimidine-DNA glycosylase

Parameter and method	Value
Molecular weight	
-gel filtration	30,000 ± 1,000
-sodium dodecylsulfate gel electrophoresis	30,000 ± 1,000
-calculated from DNA sequence	30,200
Stokes radius gel filtration	2.45 nm
Isoelectric point chromatofocalization	7.9 ± 0.1
Km	
-Fapy-poly(dG-dC)	5.4×10^{-9} M.L^{-1}
-Fapy-poly(dG-m^{5}dC)	5.6×10^{-9} M.L^{-1}
Activation Energy	9.2 Kcal·mol^{-1}
Inhibition by Fapy (50%)	3.75 mM

ase activity. From these cells, the Fapy-DNA glycosylase was purified to apparent physical homogeneity and some of the main properties of the enzyme are reported in Table 1. The nucleotide sequence of the fpg^{+} gene was also determined. The plasmids depicted in Figure 5 indicate the position of the gene, and other E. coli sequences which are in the region of the fpg gene. The fpg gene of E. coli consists of 807 bp which code for a protein of 269 amino acids with a calculated molecular weight of 30.2 kd. Since the T4 UV endonuclease and E. coli Endonuclease III are both glycosylases which possess activities which incise DNA at AP-sites and since there was amino acid homology between the T4 UV endonuclease and the Fapy-DNA glycosylase (Boiteux, et al., 1987), we decided to

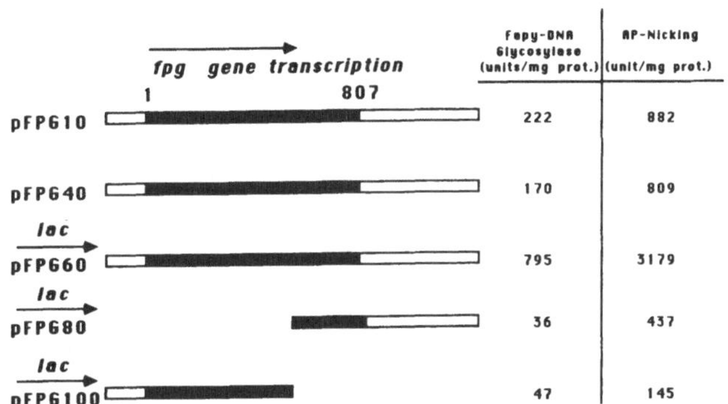

Figure 5. Fapy-DNA glycosylase activity and nicking activity at AP-sites associated with Escherichia coli harboring plasmids containing constructs of the fpg gene. The pFPG10 and pFPG40 plasmids are cloned into pBR322 and the pFPG60, PFPG80, and pFPG100 plasmids are cloned into pUC19. The pFPG10 plasmid is composed of approximately an 11 kb insert into the BamHI site of pBR322. The pFPG40 plasmid was formed by digesting the pFPG10 plasmid with Cla I and re-ligating. The pFPG60 plasmid was generated by inserting the EcoRI, SalI DNA fragment into EcoRI, SalI cleaved pUC19 DNA. The pFPG80 and pFPG100 plasmids were generated by cloning the SalI, BamHI and the EcoRI, BamHI fragments from pFPG40 into pUC19 cut DNA. The specific activities of the Fapy-DNA glycosylase and AP-nicking found in crude lysates of E. coli harboring each of the plasmids is indicated next to each of the plasmids.

investigate the possibility that the Fapy-DNA glycosylase possesses an associated activity which cleaves DNA at AP-sites. We determined Fapy-DNA glycosylase activity and the activity which cleaves DNA at AP-sites in crude lysates from E. coli harboring the plasmids shown in Figure 5. When the entire fpg gene is present (pFPG10, pFPG40, pFPG60), increased levels of AP-nicking activity are also observed. Moreover, the ratio of the Fapy-DNA glycosylase activity:AP-nicking activity is constant for HB101 harboring pFPG10, pFPG40 or pFPG60. Plasmids which do not contain the entire fpg gene do not show substantial increases in AP-nicking activity. Thus, this genetic evidence suggests that the Fapy-DNA glycosylase has an associated activity which cleaves DNA at AP-sites.

Since the genetic evidence suggests that the activity which incises DNA at AP-sites is associated with the Fapy-DNA glycosylase activity, we purified the Fapy-DNA glycosylase of E. coli using the protocol, described in Boiteux, et al., 1987, yielding the electrophoretically homogeneous protein and assayed each fraction from the various columns for Fapy-DNA glycosylase activity and activity incising DNA at AP-sites. During various steps of the purification, the Fapy-DNA glycosylase and AP-nicking activities co-eluted and in each column, the ratio of the two activities remained constant accross the peak. Therefore, since the biochemical evidence supports the genetic evidence, we conclude that the Fapy-DNA glycosylase has an activity which incises DNA at AP-sites.

Analysis of the 5' terminus using the phosphorylating activity of bacteriophage T4 polynucleotide kinase

DNA containing AP-sites was cleaved using the nicking activity at AP-sites of Fapy-DNA glycosylase and then reacted in the presence of T4 polynucleotide kinase or first reacted with calf intestine phosphatase and then T4 polynucleotide kinase as indicated in Table II.

The results in Table II are consistent with the presence of a 5' nucleotide following cleavage with the activity of the Fapy-DNA glycosylase which incises DNA at AP-sites. Since this is the same terminus left by β-elimination catalysts such as the T4 UV endonuclease, Ade-Z-Acr (Constant, et al., 1988), and Endonuclease III, we tried without success to cleave DNA containing reduced AP-sites with the nicking activity at AP-sites of Fapy-DNA glycosylase. The fact that the Fapy-DNA glycosylase does not cleave AP-sites which are reduced also is consistent with a mechanism of β-elimination for the nicking activity at AP-sites of this enzyme.

Table II. Polynucleotide kinase labelling of termini generated by various agents nicking DNA at AP-sites

	Femtomoles ^{32}P incorporated	
Incising Agent	Kinase	CIP/Kinase
Blank	6	12
Fapy-DNA glycosylase	4	27
Endonuclease III	5	27
Endonuclease IV	6	10
Exonuclease III	2	7
T4 UV Endonuclease	7	28
Ade-Z-Acr	7	35

Analysis of the 3' terminus using nick translation by DNA polymerase I

The 3' terminus left following cleavage by the activity of the Fapy-DNA glycosylase was analyzed using nick translation with E. coli DNA polymerase I on pBR322 DNA containing AP-sites which have been cleaved by the nicking activity of the Fapy-DNA glycosylase. The results shown in Figure 6 are not consistent with the cleavage of the DNA containing AP-sites by a simple mechanism of β-elimination. If only a mechanism of β-elimination was implicated, the calf intestine phosphatase would not be expected to activate the terminus for nick-translation. However, this result does not exclude the possibility that there may be another mechanism yielding the same product.

However, cleavage by the Fapy-DNA glycosylase of M13mp18 RF DNA containing radioactively labelled AP-sites generates labelled products which are not retained on DEAE cellulose columns. This last piece of data is not consistent with a simple mechanism of β-elimination.

Partial purification of the Fapy-DNA glycosylase of calf thymus

A variety of bacterial and mammalian systems, including Salmonella thyphimurium, Bacillus subtilis, Micrococcus luteus, rat liver and calf thymus, were found to have activities which are associated with the presence of an enzyme removing the Fapy lesion from DNA (data not shown).

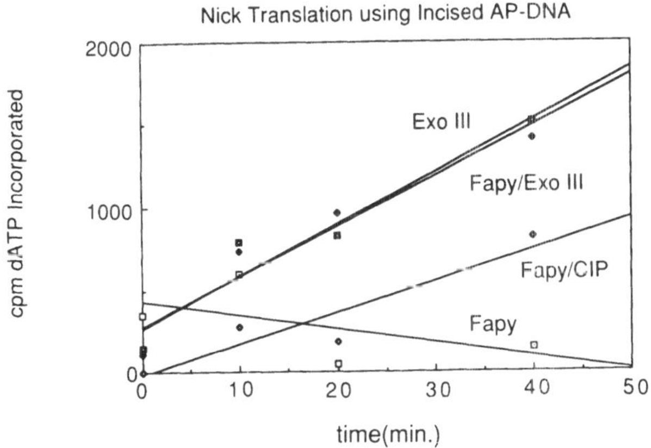

Figure 6. Nick translation of pBR322 DNA using DNA polymerase I following incision at AP-sites using various cleavage agents. Exo III (◆) corresponds to DNA incised by Exonuclease III of E. coli, Exonuclease III, Fapy/Exo III (■) corresponds to DNA first cleaved by the Fapy-DNA glycosylase followed by incision by Exonuclease III of E. coli. Fapy/CIP (◇) corresponds to Fapy-DNA glycosylase cleavage followed by reaction using calf intestine phosphatase. Fapy (□) corresponds to cleavage using the Fapy-DNA glycosylase.

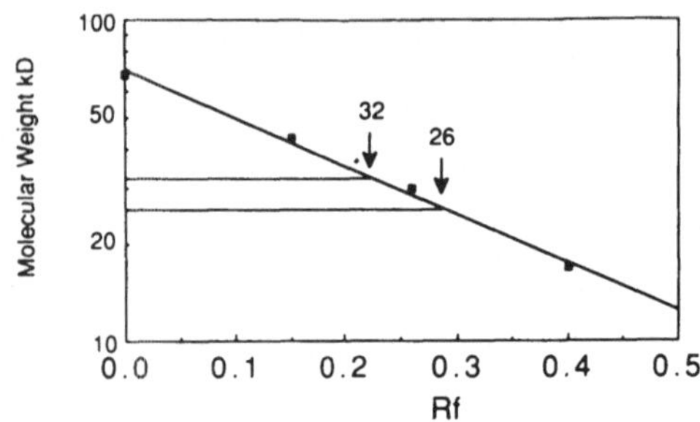

Figure 7. Gel filtration chromatography on an AcA54 column of Fapy-DNA glycosylase of calf thymus. The four points are the molecular weight calibration of the column. The arrows indicate the approximate molecular weight of the Fapy-DNA glycosylase as calculated from the activity of the fractions.

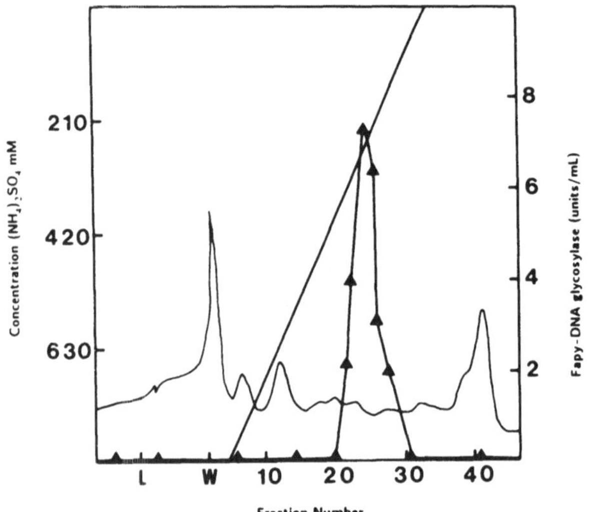

Figure 8. Fractions of the calf thymus Fapy-DNA glycosylase isolated from the Phenyl-Sepharose column. The triangles represent the activity of the Fapy-DNA glycosylase with the axis on the right of the figure. The trace corresponds to the 280 nm optical density. The straight line gives the gradient applied. The mark indicating L and W correspond to the load and washing fractions of the column, respectively.

We chose to purify the Fapy-DNA glycosylase activity of calf thymus. Table II shows the partial purification of this enzymatic activity. The protein was purified 600-fold and the gel filtration column shown in Figure 7 indicates that the molecular weight of the protein is between 26 and 32 kd. The elution profile of the Phenyl Sepharose column in Figure 8 shows that there is a peak of Fapy-DNA glycosylase activity which elutes at approximately 200 mM Am_2SO_4. We are currently in the process of further purifying this protein.

DISCUSSION

We have demonstrated that the Fapy-DNA glycosylase possesses an associated enzymatic activity which cleaves DNA at AP-sites. The gene coding for this protein is different from the known genes coding for enzymes acting at AP-sites: exonuclease III (xth) endonuclease IV (nfo) endonucleage III (nth) (Cunningham, et al., 1986; Cunningham and Weiss, 1985). The activity cleaving DNA at AP-sites was found to increase in E. coli harboring the fpg gene on recombinant plasmids. In addition, purification of the Fapy-DNA glycosylase also results in the co-purification of an activity cleaving DNA at AP-sites. Analysis of the reaction mechanism of this enzyme suggests that cleavage may occur by a mechanism of β-elimination, but this is not entirely clear. There is also an enzymatic activity which removes Fapy residues in calf thymus which we have partially purified 600-fold.

Coupled DNA-glycosylase/AP-nicking activities have been reported for bacteriophage T4 UV-endonuclease, E. coli endonuclease III, and M. luteus pyrimidine dimer DNA glycosylase (Bailly and Verly, 1987). Endonuclease III and the Fapy-DNA glycosylase have several similar properties which suggest that the functions of these two E. coli enzymes may be complementary (Cunningham and Weiss, 1985). In addition to the similarity of the physical properties, the Endonuclease III and the Fapy-DNA glycosylase recognizes substrates which include damaged rings (Sancar and Sancar, 1988; Boiteux, et al., 1988 this volume).

The fact that this enzyme is found also in eukaryotic systems suggests that this enzyme is ubiquitous. The purification of the calf thymus enzyme is the first step in cloning the mammalian gene. The overall role of this enzyme in biological systems and the relevance of the Fapy lesion, however, remain unanswered questions at this stage.

ACKNOWLEDGMENTS

We would like to thank Dr. Walter Verly for discussions and Janine Seité for preparing this manuscript. This work was supported by grants from CNRS URA 158, INSERM U 140 and ARC. T.R.O. was supported by grants from INSERM (U 140), l'Association de la Recherche sur le Cancer, and la Ligue Nationale contre le Cancer.

REFERENCES

Bailly, V., and Verly, W.G., 1987, Escherichia coli endonuclease III is not an endonuclease but a β-elimination catalyst, Biochem. J., 242:565.

Boiteux, S., Belleney, J., Roques, B.P., and Laval, J., 1984, Two rotameric forms of open ring 7-methylguanine are present in alkylated polynucleotides, Nucl. Acids. Res., 12:5429.

Boiteux, S., Bichara, M., Fuchs, R.P.P., and Laval, J., 1988, Substrate specificity of the formamidopyrimidine-DNA glycosylase of E. coli: Repair of the imidazole ring-opened form of N-hydroxy-2-amino-fluorene-guanine adducts in DNA, This volume.

Boiteux, S., Huisman, O., and Laval, J., 1984, 3-methyladenine residues in DNA induce the SOS function sfiA in Escherichia coli, EMBO J., 3:2569.

Boiteux, S., and Laval, J., 1982, Coding properties of poly(deoxycytidilic acid) templates containing uracil or apyrimidinic sites: In vitro modulation of mutagenesis by deoxyribonucleic acid repair enzymes, Biochemistry, 21:6746.

Boiteux, S., and Laval, J., 1985, Imidazole open ring 7-methylguanine:An inhibitor of DNA synthesis, Biochem.Biophys. Res. Commun., 110:552.

Boiteux, S., O'Connor, T.R., and Laval, J., 1987, Formamidopyrimidine-DNA glycosylase of Escherichia coli: Cloning and sequencing of the fpg structural gene and overproduction of the protein, EMBO J., 6:3377.

Chetsanga, C.J., and Lindahl, T., 1979, Release of 7-methylguanine residues whose imidazole ring have been opened from damaged DNA, by a DNA-glycosylase from E. coli, Nucl. Acids. Res., 6:3673.

Constant, J.F., O'Connor, T.R., Lhomme, J., and Laval, J., 1988,9-[(10-(aden-9-yl)-4,8-diazadecyl)amino]-6-chloro-2-methoxy-acridine incises DNA at apurinic sites, Nucl. Acids. Res., 16:2691.

Cunningham, R.P., Saporito, S.M., Spitzer, S.G., and Weiss, B., 1986, Endonuclease IV (nfo) mutant of Escherichia coli, J. Bact., 168:1120.

Cunningham, R.P., and Weiss, B., 1985, Endonuclease III (nth) mutants of Escherichia coli, Proc. Natl. Acad. Sci. USA, 82:474.

Haines, J.A., Reese, C.B., and Todd, L., 1962, Methylation of guanosine and related compounds with diazomethane, J. Chem. Soc., 5281.

Larson, K., Sahm, J., Shenkar, R., and Strauss, B., 1985, Methylation-induced blocks to in vitro DNA replication, Mutation Res., 150:77.

O'Connor, T.R., Boiteux, S., and Laval, J., 1988, Ring-opened 7-methylguanine residues in DNA are a block to in vitro DNA synthesis, Nucl. Acids. Res., 16:5879.

Sancar, A., and Sancar, G.B., 1988, DNA Repair Enzymes in: Ann. Rev. Biochem., 57:29.

Singer, B., and Grunberger, D., 1983, in: Molecular Biology of Mutagens and Carcinogens, Plenum Press NY, p. 45-96.

SUBSTRATE SPECIFICITY OF THE FORMAMIDOPYRIMIDINE-DNA GLYCOSYLASE OF E. COLI: REPAIR OF THE IMIDAZOLE RING-OPENED FORM OF N-HYDROXY-2-AMINO-FLUORENE-GUANINE ADDUCTS IN DNA

Serge Boiteux[1], Marc Bichara[2], Robert P.P. Fuchs[2], and Jacques Laval[1]

[1]Groupe Reparation des lésions radio- et chimioinduites URA158 CNRS, U140 INSERM, Institut Gustave Roussy, 94805 Villejuif Cédex, France

[2]Groupe Cancérogénesè et Mutagénesè Moléculaire et Structurale, IBMC du CNRS 67084 Strasbourg, France

SUMMARY

A polynucleotide containing G-C8-AF residues was obtained by treatment of poly(dG-dC) with the carcinogen N-hydroxy-2-aminofluorene. The resulting product [^{3}H]-AF-poly(dG-dC) was further incubated in 0.1 N NaOH for 24 hours at 37^{0}C, which resulted in the conversion of 60% of the G-C8-AF residues to their imidazole ring-opened derivative (iro-G-C8-AF). This modified polynucleotide was used as substrate for the Fapy-DNA glycosylase of E. coli. H.P.L.C. analysis of the products of the reaction shows that the pure Fapy-DNA glycosylase excised the ring-opened derivative (iro-G-C8-AF). In contrast, the primary lesion (G-C8-AF) was not removed. These results show that the Fapy-DNA glycosylase of E. coli excises imidazole ring-opened purines which are modified at the C8 position. These observations suggest that the Fapy-DNA glycosylase may have a broad substrate specificity which includes all imidazole ring-opened purines modified at the N7 or C8 position in DNA.

INTRODUCTION

The N7 and C8 positions of guanine in DNA are major targets for mutagens and carcinogens (Singer and Grunberger, 1983). Modified guanine residues are considerably less stable than the parent bases. This instability may result in the formation of ring-opened derivatives after cleavage of the imidazole ring across N7-C8 or C8-N9 bonds. Such decomposition products have been identified in the DNA of animals treated by chemical carcinogens, including: N-methylnitrosourea (Kadlubar, et al., 1984), aflatoxin B1 (Essigman, et al.,

DNA Repair Mechanisms and Their Biological Implications in Mammalian Cells
Edited by M.W. Lambert and J. Laval
Plenum Press, New York

1984), and 2-naphthylamine (Kadbular, et al., 1981).

The ring-opened form of N7-methylguanine (iro-me7-G or Fapy) is removed from the DNA by a specific DNA glycosylase in E. coli (Chetsanga and Lindahl, 1979; Boiteux, et al., 1984) and in mammalian cells (Margisson and Pegg, 1981; Laval, et al., 1988). The gene coding for the Fapy-DNA glycosylase of E. coli has been cloned and sequenced (Boiteux, et al., 1987). The Fapy DNA glycosylase of E. coli exhibits a substrate specificity which is not limited to iro-me7-G, since it excises the ring-opened form of guanine residues modified at the N7 position by aflatoxin B1 (Chetsanga and Frenette, 1983), or phosphoramide mustard (Chetsanga, et al., 1982). These observations suggest that the ring-opened form of alkylated purines might play a significant role in biological processes leading to mutagenesis and/or cell death by chemical carcinogens.

In this paper, we show that the Fapy-DNA glycosylase of E. coli removes the imidazole ring-opened form of guanine residues modified at the C8 position by the carcinogen N-hydroxy-2-aminofluorene.

MATERIAL AND METHODS

Preparation of poly(dG-dC) containing [^{3}H]-G-C8-AF and [^{3}H]-iro-G-C8-AF residues

Poly(dG-dC) (Boehringer Mannheim) was reacted with [^{3}H]-N-OH-AF (140 mCi/mmol) as previously described (Fuchs and Seeberg, 1984). The specific activity of the [^{3}H]-AF-poly-(dG-dC) was 1650 cpm/μg. The imidazole ring-opened derivative was obtained after incubation of [^{3}H]-AF- poly(dG-dC) with 0.1 N NaOH for 24 hours at 37^{0}C. Polynucleotides containing [^{3}H]-me7-G or [^{3}H]-iro-me7-G were prepared as described (Boiteux, et al., 1984).

Analysis of the reaction products

The authentic markers (G-C8-AF and iro-G-C8-AF) were prepared as described by Kriek and Westra (1980). The products were separated by H.P.L.C. using a C18 μBondapack column (Waters). The mobile phase was 20 mM $NH_4H_2PO_4$, pH 4.5, containing 60% methanol (V/V). The column was isocratically eluted at 1.0 ml/minute. The products were detected by UV absorption at 280 nm and by scintillation counting of fractions.

Purification of the Fapy-DNA glycosylase of E. coli

LB-broth medium (2.5 l) containing 50 μg/ml ampicillin was inoculated with 50 ml of an overnight culture of E. coli strain JM 105 carrying the pFPG220 (fpg^+) plasmid (O'Connor, et al., 1988). The bacteria were grown at 37^{0}C for 2 hours, then supplemented with 0.5 mM IPTG and further incubated for 17 hours under vigorous agitation. The cells (10 g wet weight) were harvested and the Fapy-DNA glycosylase was purified according to Boiteux, et al., 1987.

Fapy-DNA glycosylase assay

The standard incubation mixture (total volume 50 µl) contained 70 mM Hepes-KOH, pH 7.6, 100 mM KCl, 2 mM Na_2 EDTA, 10% glycerol, 3000 cpm [^{3}H]-iro-AF-poly(dG-dC) and limited amounts of Fapy-DNA glycosylase. The assay mixture was incubated for 10 minutes at 37°C. The radioactivity in the ethanol-soluble fraction was quantified by scintillation counting. Enzyme unit: 1 unit released 1 pmol of iro-me7-G in 5 minutes at 37°C.

RESULTS AND DISCUSSION

Purification of the Fapy-DNA glycosylase of E. coli

The structural gene coding for the Fapy DNA glycosylase of *E. coli* was cloned in a multicopy plasmid. The nucleotide sequence of the *fpg*$^+$ gene is composed of 807 base pairs and coded for a protein of 269 amino acids with a molecular weight of 30.2 kd (Boiteux, *et al.*, 1987). In order to overproduce the Fapy-DNA glycosylase, the *fpg*$^+$ gene was placed under the control of the *lac* promoter yielding the pFPG220 plasmid (O'Connor, *et al.*, 1988). Figure 1A shows that the lysates from cells containing the pFPG220 plasmid exhibited a reinforced 31 kd protein band on SDS-polyacrylamide gel. The addition of

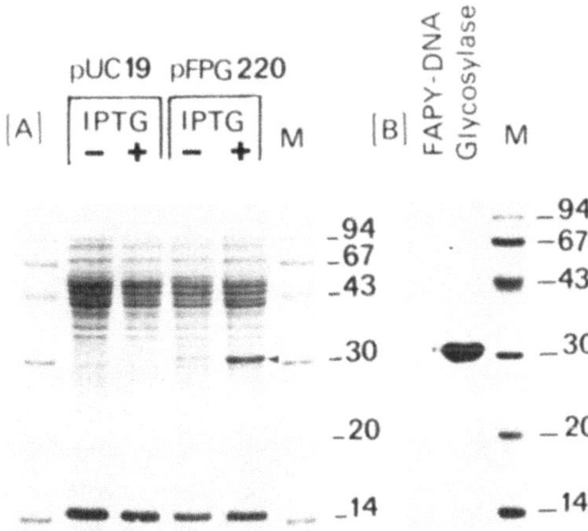

Figure 1. Overproduction and purification of the Fapy-DNA glycosylase of *E. coli*.
A. Thirty micrograms of the total soluble proteins from crude cell lysates of *E. coli* JM 105 harboring either pUC19 or pFPG220 plasmids supplemented (+) or not (-) with 0.5 mM IPTG were loaded on a 15% SDS-polyacrylamide gel.
B. Five micrograms of purified Fapy-DNA glycosylase were loaded on a 15% SDS-polyacrylamide gel. This fraction corresponded to F.P.L.C. Mono S HR5/5 fraction as described by Boiteux, *et al.*, 1987. M; molecular weight standards.

0.5 mM IPTG further stimulates the synthesis of the Fapy-DNA glycosylase which represents more than 10% of total soluble proteins in E. coli (Figure 1A). These bacteria were used to purify the Fapy-DNA glycosylase according to the protocol previously described by Boiteux, et al., (1987). Figure 1B shows the purified Fapy-DNA glycosylase on a denaturating polyacrylamide SDS gel. The lysis of 10 g of bacteria allow the preparation of 10 mg of apparently homogeneous protein (Figure 1B) with a specific activity of 80,000 u/mg.

Excision of the ring-opened form of N-OH-2-aminofluoreneguanine adduct (iro-G-C8-AF) from DNA by the Fapy-DNA glycosylase of E. coli

The preparation of a polynucleotide containing G-C8-AF and/or iro-G-C8-AF residues is summarized in Scheme 1. The poly(dG-dC) was reacted with [^{3}H]-N-OH-AF under conditions which lead exclusively to the formation of the G-C8-AF adduct (Tang and Lieberman, 1983). The [^{3}H]-AF-poly(dG-dC) was further incubated in 0.1 N NaOH for 24 hours at 37^{0}C. The resulting template was depurinated by formic acid hydrolysis and the bases separated by H.P.L.C. chromatography. Figure 2A shows that the radioactivity eluted as two peaks at RT=9 minutes and RT=21 minutes, respectively. The first peak contained 60% of total radioactivity and was assigned to the imidazole ring-opened form (iro-G-C8-AF). The other peak (RT=19 minutes) was assigned to the primary lesion (G-C8-AF). The modified polynucleotide containing both G-C8-AF and iro-G-C8-AF residues (Figure 2A) was used as substrate for the Fapy-DNA glycosylase of E. coli. The radioactive material released by the Fapy-DNA glycosylase in the ethanol-soluble fraction was recovered and the bases separated by H.P.L.C. chromatography. Figure 2B shows that the purified glycosylase specifically removes the

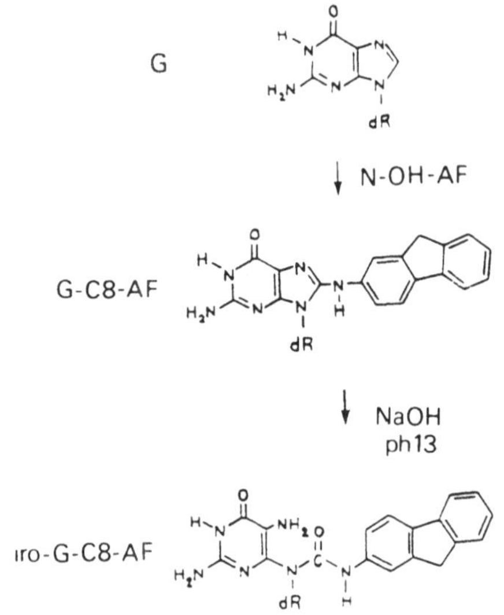

G

N-OH-AF

G-C8-AF

NaOH
ph13

iro-G-C8-AF

Scheme 1. The iro-G-C8-AF residues were generated in a two step process. The poly(dG-dC) was reacted with N-hydroxy-2-aminofluorene and further incubated under alkaline conditions.

ring-opened derivative (iro-G-C8-AF). The primary adduct (G-C8-AF) is not excised at a detectable rate (Figure 2B).

Substrate specificity of the Fapy-DNA glycosylase of *E. coli*

The experiments reported in this paper show that the imidazole ring-opened form of a guanine modified at the C8 position is repaired by the pure Fapy-DNA glycosylase of *E. coli*. Imidazole ring-opened purines excised by the Fapy-DNA glycosylase of *E. coli* are listed in Table 1. There are several features which are suggested by Table 1:

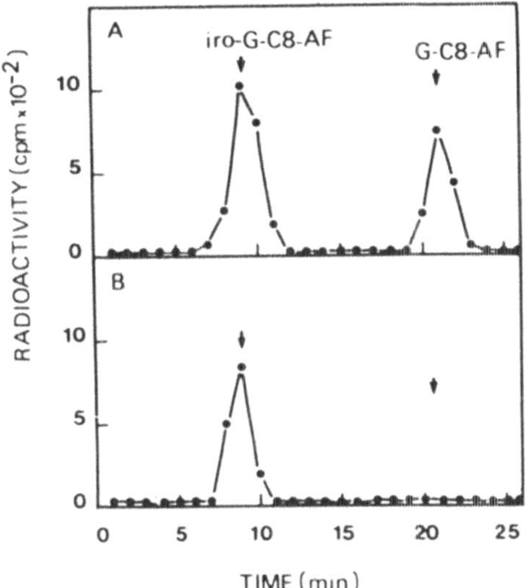

Figure 2. Excision of $[^3H]$-iro-G-C8-AF residues by the Fapy-DNA glycosylase of *E. coli*. Poly(dG-dC) was reacted with $[^3H]$-NOH-AF and further incubated in 0.1 N NaOH for 24 hours at 37^0C. The modified poly(dG-dC) was hydrolysed by formic acid or incubated in the presence of 6 units of Fapy-DNA glycosylase for 10 minutes at 37^0C. The bases were separated by H.P.L.C. chromatography. Panel A shows the elution profile of radioactive material released after formic acid hydrolysis of $[^3H]$-iro-AF-poly(dG-dC). Panel B shows the elution profile of the radioactive material released in the ethanol-soluble fraction by the Fapy-DNA glycosylase of *E. coli* using $[^3H]$-iro-AF-poly-(dG-dC) as substrate.

1. The Fapy-DNA glycosylase excises imidazole ring-opened purines modified either at N7 or C8 positions. In contrast, purines modified at N7 or C8 with the imidazole ring intact are not excised.

2. The Fapy-DNA glycosylase excises the ring-opened derivatives of purines modified either at N7 or C8 positions by small or bulky adducts.

3. The Fapy-DNA glycosylase excises purines whose im-

Table 1. Imidazole ring-opened purines excised by the Fapy-DNA glycosylase of *E. coli*.

Inducing agent	Modified purine	Position of modification	Position of ring-breakage	References
γ-radiation	adenine	-	C8-N9	Breimer (1984)
methylating agents	guanine	N7	C8-N9	Chetsanga and Lindahl, 1979 Boiteux, *et al.*,1984
phosphoramide mustard	guanine	N7	C8-N9	Chetsanga, *et al.*,1982
aflatoxin B1	guanine	N7	C8-N9	Chetsanga and Frenette, 1983
N-hydroxy-2-aminofluorene	guanine	C8	N7-C8	This study

idazole ring has been opened through the breakage of either C8-N9 or N7-C8 chemical bond.

These conclusions lead us to propose that the Fapy-DNA glycosylase might remove all purine derivatives whose imidazole ring has been opened. Thus, recognition of a structural defect in double-stranded DNA, as a result of imidazole ring cleavage, may provide a molecular mechanism by which a single enzyme could remove many lesions. Another example of this strategy is provided by the thymine glycol-DNA glycosylase encoded by the *nth+* gene of *E. coli* which excised the pyrimidine derivatives lacking the 5-6 endocyclic double bond (Breimer and Lindahl, 1983; Cunningham and Weiss, 1985). Finally, the ability of the Fapy-DNA glycosylase to remove many substrates suggests that this glycosylase might contribute to DNA repair in cells treated with chemical carcinogens.

ACKNOWLEDGEMENTS

We are very grateful to P. Auffret Van der Kemp and C. Lagravére for their excellent technical assistance. We thank T.R. O'Connor for helpful discussions. This work was supported by CNRS (URA 158), INSERM (U 140) and Association pour la Recherche sur le Cancer (ARC).

Abbreviations: G, guanine; G-C8-AF, N-(guanine-8-yl)-2-aminofluorene; iro-G-C8-AF, imidazole ring-opened-N-(guanine-8-yl)-2-aminofluorene; me7-G, N7-methylguanine; iro-me7-G, imidazole ring opened N7-methylguanine; N-OH-AF, N-hydroxy-2-aminofluorene; AF-poly(dG-dC), poly(dG-dC) reacted with N-OH- AF;iro-AF-poly(dG-dC), poly(dG-dC) reacted with N-OH-AF and further incubated in 0.1 N NaOH.

REFERENCES

Boiteux, S., Belleney, J., Roques, B.P., and Laval, J., 1984, Two rotameric forms of open ring 7-methylguanine are present in alkylated polynucleotides, Nucleic Acids Res., 12:5429.

Boiteux, S., O'Connor, T.R., and Laval, J., 1987, Formamidopyrimidine-DNA glycosylase of Escherichia coli: cloning and sequencing of the fpg structural gene and overproduction of the protein, EMBO J., 6:3177.

Breimer, L., and Lindahl, T., 1984, DNA glycosylase activities for thymine residues damaged by ring saturation, fragmentation, or ring concentration are functions of endonuclease III in E. coli, J. Biol. Chem., 259:5543.

Breimer, L.H., 1984, Enzymatic excision from γ-irradiated polynucleotides of adenine residues whose imidazole ring have been ruptured, Nucleic Acids Res., 12:6359.

Chetsanga, C.J., and Lindahl, T., 1979, Release of 7-methylguanine residues whose imidazole rings have been opened from damaged DNA by a DNA glycosylase from Escherichia coli, Nucleic Acids Res., 6:3673.

Chetsanga, C.J., Polidori, G., and Mainwaring, M., 1982, Analysis and excision of ring-opened phosphoramide mustard deoxyguanosine adducts in DNA, Cancer Res., 42: 2616.

Chetsanga, C.J. and Frenette, G.P., 1983, Excision of aflatoxin Bl-imidazole ring-opened guanine adducts from DNA by formamidopyrimidine DNA glycosylase, Carcinogenesis, 4:997.

Cunningham, R.P., and Weiss, B., 1985, Endonuclease III (nth) mutants of Escherichia coli, Proc. Natl. Acad. Sci. USA, 82:474.

Essigman, J.M., Green, C.L., Croy, R.C., Fowler, K.W., Buchi, G.H., and Wogan, G.M., 1983, Interaction of Aflatoxin Bl and alkylating agents with DNA: structural and functional studies, CSH Symp. on Quant. Biol., XLVII, p. 327.

Fuchs, R.P.P., and Seeberg, E., 1984, pBR322 plasmid DNA modified with 2-acetylaminofluorene derivatives; transforming activity and in vitro strand cleavage by the E. coli UvrABC endonucleases, EMBO J., 3:757.

Kadlubar, F.F., Anson, J.F., Dooley, K.L., and Beland, F. A., 1981, Formation of urothelial and hepatic DNA adducts from the carcinogen 2-naphtylamine, Carcinogenesis,2:467.

Kadlubar, F.F., Beranek, D.T., Weiss, C.C., Evans, E.F., Cox, R., and Irving, C.C., 1984, Characterization of the purine ring-opened 7-methylguanine and its persistance in rat bladder epithelial DNA after treatment with the carcinogen N-methyl-nitrosourea, Carcinogenesis, 5:587.

Kriek, E., and Westra, J.G., 1980, Structural identification of the pyrimidine derivatives formed from N(deoxyguanosin-8-yl)-2-amino-fluorene in aqueous solution at alkaline pH, Carcinogenesis, 1:459.

Laval, J., O'Connor, T.R., and Boiteux, S., 1988, Repair of imidazole ring-opened purines by E. coli and mammalian cells, this volume.

Margison, G.P. and Pegg, A.E., 1981, Enzymatic release of 7-methylguanine from methylated DNA by rodent liver extracts, Proc. Natl. Acad. Sci. USA., 78:861.

O'Connor, T., Boiteux, S., and Laval, J., 1988, Repair of imidazole ring-opened purines in DNA: Overproduction of the formamidopyrimidine-DNA glycosylase of E. coli with plasmids containing the fpg[+] gene, in: "Annali dell'Instituto Superiore di Sanita", Bignami and Essigman, eds., Roma, 25, 27-32.

Singer, B., and Grunberger, D., 1983, Molecular Biology of mutagens and carcinogens, Plenum, New York.

Tang, M.S., and Lieberman, M., 1983, Quantification of adducts formed in DNA treated with N-acetoxy-2-acetylaminofluorene, or N-hydroxy-2-aminofluorene: Comparison of trifluoroacetic acid and enzymatic digestions, Carcinogenesis, 4:1001.

INVESTIGATION OF SEQUENCE SPECIFICITY IN DNA ALKYLATION AND REPAIR USING OLIGODEOXYNUCLEOTIDE SUBSTRATES

Anthony E. Pegg and M. Eileen Dolan

Departments of Physiology and Pharmacology, Milton S. Hershey Medical Center, Pennsylvania State University, Hershey, Pennsylvania 17033, USA

INTRODUCTION

Alkylating carcinogens such as N-methyl-N-nitrosourea and dimethylnitrosamine and their ethyl analogs form at least 13 different alkylation adducts in DNA. Although all of these adducts appear to be repaired in vivo, in the sense that their content declines more rapidly than can be accounted for by cell turnover (Pegg, 1977,1983; Singer and Kusmierek, 1982; Singer, 1984,1986; Saffhill, et al., 1985), only a few of the proteins responsible for such repair have been fully characterized from mammalian cells. These include two glycosylase enzymes. One of these removes 3-methyladenine, 7-methylguanine and 3-methylguanine from DNA and is usually referred to as 3-alkyladenine DNA glycosylase (Margison and Pegg, 1981; Gallagher and Brent, 1984; Male, et al.,1985,1987). This glycosylase clearly plays a major role in the loss of the N-alkyl purines, which make up the bulk of the DNA base adducts.

The other, termed formamidopyrimidine-DNA glycosylase, has been detected and assayed in crude extracts of rat and hamster liver (Margison and Pegg, 1981). It appears to carry out the same reaction as the formamidopyrimidine-DNA glycosylase which has been purified to homogeneity and cloned from E.coli (Boiteaux, et al., 1987). This enzyme removes the residues formed from 7-methylguanine in DNA at an alkaline pH. Such treatment leads to the opening of the imidazole ring forming the two rotameric forms of 2,6-diamino-4-hydroxy 5-(N-methylformamido)pyrimidine. This adduct is clearly non-coding (Boiteaux and Laval, 1983) and presents a block to DNA replication (O'Connor, et al., 1988). However, the extent to which 7-methylguanine in DNA is actually converted to the ring-opened form is unclear since a relatively high pH is needed for this reaction. Reports of the presence of the ring-opened adduct have been limited (Beranek, et al., 1983; Kadlubar, et al., 1984) and the possibility of artifactual generation of this adduct during isolation and analysis of the DNA has not been entirely ruled out.

DNA Repair Mechanisms and Their Biological Implications in Mammalian Cells
Edited by M.W. Lambert and J. Laval
Plenum Press, New York

The third protein which repairs DNA damaged by alkylating agents is O^6-alkylguanine-DNA alkyltransferase (Lindahl, 1982; Lindahl and Sedgwick, 1988; Pegg, 1983; Yarosh, 1985; Pegg and Dolan, 1987). This protein catalyzes the transfer of the alkyl group to a cysteine acceptor site within its polypeptide sequence. The alkylcysteine formed at the acceptor site is not converted back to cysteine and the protein therefore acts stoichiometrically rather than catalytically. The number of O^6-alkylguanine lesions that can be repaired rapidly by the cell is thus limited to the number of molecules of the alkyltransferase protein (Pegg and Hui, 1978; Pegg, 1978,1983; Pegg, et al.,1984). When the supply of the protein is exhausted, further repair depends on the de novo synthesis of the alkyltransferase.

Further information on the biochemistry underlying the repair of alkylated DNA and on the properties and regulation of the relevant enzymes is clearly needed for a better understanding of the consequences for mammalian cells of the exposure to alkylating agents. In the present communication, our studies using short oligodeoxynucleotides containing alkylated bases for analysis of the properties of the repair proteins are described.

INVESTIGATIONS OF O^6-ALKYLGUANINE-DNA-ALKYLTRANSFERASE

Tests with a number of different oligodeoxynucleotides of different lengths which contained O^6-methylguanine at an internal site indicated that both the mammalian and bacterial alkyltransferase proteins were able to act to demethylate such substrates (Scicchitano, et al., 1986; Pegg, et al., 1987; Graves, et al., 1987; Dolan, et al., 1988a-c). Even tetramers were acted upon, although the rate for repair was much slower with these than with dodecamers (Scicchitano, et al., 1986). It is, in fact, not surprising that such small oligodeoxynucleotides are substrates for the reaction since even the free base O^6-methylguanine can be demethylated albeit at a very slow rate (Dolan, et al.,1985; Yarosh, et al., 1986). However, a comparison of the rates of reaction indicates that self complementary dodecamers containing O^6-methylguanine are demethylated by the alkyltransferase protein at rates comparable to substrates made by methylating double stranded calf thymus DNA. These oligodeoxynucleotide substrates are therefore useful for: (a) the study of factors affecting the alkyltransferase reaction; (b) the development of a very sensitive assay of alkyltransferase activity; and (c) the investigation of sequence specificity in repair of O^6-methylguanine.

Several assay procedures have been used to investigate the repair of such low molecular weight substrates containing O^6-methylguanine by the alkyltransferase protein. These include: the use of unlabeled substrates and the analysis of the residual O^6-alkylguanine using HPLC separation of the base liberated by dilute acid hydrolysis and fluorescence detection (Scicchitano, et al., 1986); the use of oligodeoxynucleotides, which are labeled at the 5' end with ^{32}P by reaction with polynucleotide kinase and (γ-^{32}P)ATP (Scicchitano, et al., 1986; Graves, et al., 1987; Dolan, et al., 1988c); and the preparation of ^3H-methylated oligodeoxynucleotides by reaction with [^3H]N-methyl-N-nitrosourea. These can then be used to assay the alkyltransferase activity by standard techniques (Dolan, et al., 1988a).

When the ^{32}P-labeled oligodeoxynucleotides were used as substrates, the assay mixture also contained 100 mM Na-phosphate, pH 7.8, 2 mM EDTA, 25 mM spermidine, 20 mM dithiothreitol and the alkyltransferase protein. After incubation for various times up to 90 minutes, the reaction mixtures were chilled to 0^0C, diluted to 0.8 ml and the protein precipitated by addition of 0.24 ml of 1 N perchloric acid. The precipitate was removed by centrifugation at 15,000 g for 10 minutes and the supernatant neutralized by addition of 0.36 ml of 1 M Tris. The samples were then analyzed by HPLC and the extent of conversion of the methylated oligodeoxynucleotide substrate to its unmethylated oligodeoxynucleotide was determined. HPLC analysis of the oligodeoxynucleotides was carried out on a Beckman C_{18} 5μ Ultrasphere reverse phase column (4.6 x 250 mm) at 45^0C using a flow rate of 2 ml/minute and a linear gradient of increasing acetonitrile at 0.5% per minute over 40 minutes in 50 mM Na-phosphate, pH 6.3. The separation between the methylated and the unmethylated form was better for some of the dodecamers tested than for others with 5'-[^{32}P]dCGCGAATTCm^{6}GCG-3' being particularly good but that adequate resolution was achieved even in the latter case (Dolan, et al., 1988a).

The reaction of the alkyltransferase protein with the oligodeoxynucleotide substrates was found to follow second order kinetics. The second order rate constants for a number of substrates are shown in Table 1. These show that the rates of repair of dodecamers containing O^6-methylguanine are at least 100 times that of the tetramers and more than a million times faster than the reaction with the free base. At least a part of this difference is likely to be due to the fact that double stranded substrates are preferred by the alkyltransferase protein (Lindahl, et al., 1982; Pegg, et al.,1983; Pegg and Dolan, 1987). Indeed, this is clearly indicated by the results found using a non self-complementary dodecamer as a substrate and testing the rate at which it is repaired by the alkyltransferase in the presence or absence of the complementary sequence. Addition of the complement increased the rate of repair by about 10 times (unpublished observations). Oligodeoxynucleotides containing O^6-ethylguanine were also substrates for repair by the bacterial or the mammalian alkyltransferase but the rate was considerably lower with the ethyl derivative as previously reported for DNA substrates prepared by the reaction of DNA with alkylnitrosoureas (Sedgwick and Lindahl, 1982; Pegg, et al., 1983,1984). It should be pointed out that the results of these studies may be confused by the presence of a range of other alkylation products in the DNA which might affect the rate of repair of the O^6-adduct by the alkyltransferase, particularly since these additional alkylation products are formed in different proportions by ethylating and methylating agents. More accurate comparisons of the relative rates of repair are therefore provided by the oligodeoxynucleotide substrates which contain only a single alkylated base. However, even in these experiments, the E.coli alkyltransferase showed a greater preference for the methylated oligodeoxynucleotide substrate than did the mammalian protein (Table 1) which is consistent with published results with alkylated DNA substrates.

Table 1. Second-order rate constants for demethylation of various oligodeoxynucleotides.

Source of alkyltransferase	Rate Constant ($M^{-1} \cdot h^{-1}$) E. coli	HT29
5'-dCGCGAATTCm^{6}GCGGG-3'	7.9×10^9	7.6×10^8
5'-dCGCCAATTGm^{6}GCG-3'	4.4×10^9	3.7×10^8
5'-dCGCm^{6}GAGCTCGCG-3'	2.5×10^9	1.6×10^9
5'-dCGCe^{6}GAGCTCGCG-3'	4.0×10^7	8.9×10^7
5'-dTm^{6}GCA-3'	8.9×10^7	2.8×10^7
O^6-methylguanine	8.0×10^2	5.0×10^2

The use of ^{32}P-labeled oligodeoxynucleotide substrates of very high specific activity provides a potential method for the assay of very small amounts of alkyltransferase activity. Such an assay might be of value in tests on biopsy samples and of extracts from cells available in very limited quantities. In order to establish the validity of this assay, the activities of alkyltransferase present in a variety of cultured mammalian cells and tissues were compared using the assay procedures with either ^{3}H-methylated DNA or 5'-[^{32}P]dCG-Cm^{6}GAGCTCGCG-3' as substrates Dolan, et al., 1988c). There was good agreement between the two methods except in a few samples in which the oligodeoxynucleotide substrate was extensively degraded by nucleases present in the extracts. This assay using 5'-[^{32}P]dCGCGAATTCm^{6}GCG-3' as a substrate could be used to quantitate as little as 0.1 fmol of the alkyltransferase protein in the cell extracts. It should, therefore, be useful for the determination of the content of alkyltransferase in samples in which the activity is very low and/or the amount of material available for assay is limited. It does however require that the second rate constant for the reaction be known in order to calculate the amount of alkyltransferase present. Another minor disadvantage is that precautions must be taken to ensure that degradation of the substrate by nucleases is minimized. The assay might also be improved by using methods other than HPLC for the separation of the reaction products. The use of immunoprecipitation of the methylated oligodeoxynucleotides (Pegg, et al.,1983; Adamkiewicz, et al., 1985) or the cleavage of the demethylated form by restriction enzymes (Wu, et al., 1987) could provide such improvements.

Comparison of the rates of repair by the mammalian and bacterial alkyltransferases of double stranded dodecamers containing O^6-methylguanine in various sequences indicates that there are significant differences in these rates (Dolan, et al., 1988a; Dolan and Pegg, unpublished observations).

Although a full analysis has not yet been carried out, it appears that there is a significant effect of the 5' flanking base and that the relative rates of repair when this base is varied are 4.0:2.4:2.1:1.0 for adenine:cytosine:thymine and guanine, respectively. It is likely that other features of the DNA sequence and of DNA conformation are also extremely important in determination of the rates of repair. These features can be investigated most conveniently by using sets of oligodeoxynucleotides substrates of defined sequences and the techniques described here.

There is other evidence for sequence specificity in the repair of O^6-methylguanine in DNA by the *E. coli* alkyltransferase (Topal, *et al.*, 1986; Topal, 1988). A bacteriophage *fl/pBR* 322 chimera containing O^6-methylguanine opposite T at multiple positions in the ampicillinase gene was synthesized and passaged through *E. coli.* Analysis of the mutations observed after this passage showed great variation in the frequency at which certain sequences were mutated. This is likely to be due to the differences in repair (Topal, *et al.*, 1986; Topal, 1988). Analysis of the sites at which repair was lacking gave a consensus sequence having considerable similarity to the sequence present in the *H-ras* oncogene about the 12th codon that is known to be mutated by N-methyl-N-nitrosourea (Sukumar, *et al.*, 1983; Zarbl, *et al.*, 1985; Topal, *et al.*, 1986; Topal, 1988).

INVESTIGATIONS OF FORMAMIDOPYRIMIDINE-DNA GLYCOSYLASE

There is currently no information on the possible sequence specificity of the mammalian glycosylase enzymes which remove ring opened 7-methylguanine and 3-methyladenine from DNA. However, we have observed recently that the crude mammalian and the purified *E. coli* formamidopyrimidine-DNA glycosylases (the latter was generously provided by Dr. J. Laval) do act on dodecamers containing ring-opened 7-methylguanine and that the repair of these substrates occurs at rates comparable to those seen with DNA of high molecular weight. Therefore, it will be possible to examine this question with the same techniques as those described above for the alkyltransferase.

REPAIR OF O^4-METHYLTHYMINE

At present, the mechanism of repair of O^4-methylthymine in mammalian cells is not understood. It appears that this product is removed from the DNA of rat liver and lung. In rat liver, O^4-methylthymine which was produced by the administration of a 20 mg/kg dose of 1,2-dimethylhydrazine was lost from DNA with a half life of about 20 hours (Richardson, *et al.*, 1985). However, the relatively poor capacity of the repair system is indicated by the fact that the O^4-methylthymine: O^6-methylguanine ratio rose from less than 0.01 to 1 in DNA in rat liver during chronic treatment with 1,2-dimethylhydrazine (Richardson, *et al.*,1985). Similarly, in rats treated with 4-(N-methyl-N-nitrosamino)-1-(3-pyridyl)-1-butanone, O^4-methylthymine accumulated in the lung to a level of about 5% of O^6-methylguanine but then declined when the carcinogen treatment was stopped with a half life of about 18 hours

(Belinsky, _et al._, 1986). The fact that O^4-methylthymine declined with this relatively rapid rate over a period in which O^6-methylguanine was quite stable because of the exhaustion of the alkyltransferase capacity indicates that the repair of O^4-methylthymine is unlikely to be carried out by this protein. Further support for this conclusion arises from the study of the persistence of ethylated bases in rat DNA after chronic treatment with 40 ppm of diethylnitrosamine. Under these conditions, O^4-ethylthymine accumulated in the hepatocytes to levels that were at least 50 times greater than the steady state level of O^6-ethylguanine (Swenberg, _et al._, 1984; Dyroff, _et al._, 1986). [The initial ratio of of O^4-ethylthymine: O^6-ethylguanine formed by diethylnitrosamine is about 0.25 (Pegg, 1983; Singer, 1984, 1986)]. The pronounced accumulation of O^4-ethylthymine in rat liver DNA is consistent with studies showing that this adduct is lost very slowly from hepatic DNA with a half life of about 11 days (Richardson, _et al._, 1985). Similarly, the rapid removal of O^6-ethylguanine from the hepatic DNA is consistent with the fact that ethyl groups are removed by the mammalian alkyltransferase at a physiologically significant rate (Pegg, _et al._,1984; Pegg and Dolan, 1987). Again, it is clear that the alkyltransferase protein is not acting on O^4-ethylthymine under these conditions.

These results suggest that the mammalian alkyltransferase protein does not act on O^4-methylthymine in DNA although the _E. coli_ alkyltransferase protein clearly does repair this product (McCarthy, _et al._, 1984; Lindahl and Sedgwick, 1988). Direct attempts to demonstrate that the mammalian alkyltransferase acts on O^4-methylthymine (Dolan, _et al._, 1984; Domoradzki, _et al._, 1984; Dolan and Pegg, 1985) and elsewhere (Brent, 1985; Yarosh, _et al._, 1985; Brent, _et al._, 1988) have been entirely unsuccessful. Thus, no repair of O^4-methylthymine occurred when ^3H-methylated poly d(T) combined with poly d(A) was used as a substrate for partially purified rat liver alkyltransferase (Dolan, _et al._, 1984) or a crude extract of transformed human fibroblasts (Domoradzki, _et al._, 1984). These experiments suffer from the drawback that the synthetic polymer used may not be a good substrate for the enzyme activity being looked for. Therefore, DNA was methylated by reaction with [^3H]N-methyl-N-nitrosourea to provide a substrate containing both O^6-methylguanine (7.5% of the total methylation) and O^4-methylthymine (0.06% of the total methylation). This substrate suffers from the disadvantage that there is 150 times more O^6-methylguanine than O^4-methylthymine; but, even when an excess of rat liver alkyltransferase protein was used and all of the O^6-methylguanine was removed, there was no removal of O^4-methylthymine (Dolan and Pegg, 1985). Similarly, no loss of O^4-ethylthymine was observed when a substrate DNA that was ethylated by reaction with [^3H]N-ethyl-N-nitrosourea was incubated with partially purified rat liver alkyltransferase. This substrate contained O^4-ethylthymine equivalent to 2.5% of the total ethylation and O^6-ethylguanine equivalent to 7.8% of the total. Sufficient alkyltransferase was added to remove all of the O^6-ethylguanine in 1 hour and the incubation was continued for up to 8 hour without any loss of the O^4-ethylthymine (Dolan, _et al._, 1984; Brent, _et al._, 1988).

These studies suggest strongly that the rat liver alkyltransferase protein which removes alkyl groups from the O^6 position of guanine, does not act on the O^4 position of thymine. It remained possible, however, that the rate of attack on the thymine derivative was so much slower that the reaction could not be detected in these experiments. A second possibility was that there was a separate alkyltransferase protein with specificity similar to that of the E. coli protein and that this protein is present in such low amounts that it was not revealed because of the limit of sensitivity of these assays. These questions were addressed by using a dodecadeoxynucleotide containing O^4-methylthymine as a substrate for the detection of alkyltransferase activity. The self complementary dodecadeoxynucleotide 5'-dCGCAAGCT-m^4TGCG-3' (kindly provided by Drs. P. F. Swann and B. F. L. Li) was labeled at the 5' end by reaction with [γ-^{32}P]ATP catalyzed by polynucleotide kinase. This substrate was then incubated with either the alkyltransferase protein from E. coli or the alkyltransferase partially purified from rat liver or crude extracts from rat liver or mammalian cells. The reaction mixture was deproteinized and the dodecadeoxynucleotide separated from the demethylated form ([^{32}P]5'-pdCGCAAGCT-TGCG-3') by reversed-phase HPLC on a Ultrasphere C_{18} column eluted at 42^0C with a linear gradient of 0.35% methanol per minute in 50 mM sodium phosphate, pH 6.3. The results are summarized in Table 2. When the E. coli alkyltransferase, which is known to be able to act on O^4-methylthymine, was used as a substrate, there was a time dependent product of the demethylated dodecamer. However, none of the mammalian extracts used led to any detectable demethylated product. The crude rat liver extract did lead to some degradation of the oligodeoxynucleotide, which is presumably due to nucleases in the extract, but a major part of the substrate was present unchanged at the end of a 2 hour incubation and was no sign of any loss of the methyl group. The human cell extracts did not degrade or demethylate the substrate at all over the incubation period used. The limit of detection by this assay at the specific activity of the ^{32}P used was about 1 fmol and the extracts contained more than 100 units of the alkyltransferase activity (where 1 unit is defined as the amount of protein which removes 100 fmol of O^6-methylguanine in a reaction which goes to completion in about 15 minutes). Therefore, the rate of reaction of the mammalian alkyltransferase with O^4-methylthymine was at least 400 times slower than with O^6-methylguanine. Comparisons with the rate of demethylation

Table 2. Repair of O^4-methylthymine in oligodeoxynucleotide substrate.

Source of enzyme	O^4-methylthymine removed (fmol)
E. coli alkyltransferase	154
HeLa cell extract	<1
HT 29 cell extract	<1
Rat liver alkyltransferase	<1

of the dodecamer substrate by the E. coli alkyltransferase when equal amounts of the bacterial and the mammalian activities were added indicates that the rate is at least 1000 times greater with the bacterial protein.

These results show clearly that the mammalian alkyltransferase does not repair O^4-methylthymine in DNA. Since no repair of the dodecamer substrate containing O^4-methylthymine occurred with the crude cell extracts, they also imply strongly that there is no other alkyltransferase protein which carries out this reaction. This does make the assumption that such an alkyltransferase would react with the oligodeoxynucleotide substrate tested. These results are not in agreement with those of Becker and Montesano (1975) and Hall and Karran (1986) suggesting that an activity repairing O^4-methylthymine was present in mammalian cell extracts in very limited amounts; but these workers were unable to characterize any reaction products. The possibility that the loss of O^4-alkylthymine observed *in vivo* is due solely to excision repair has not been ruled out. Samson, *et al.* (1988), have recently demonstrated that such excision repair does remove O^6-alkylguanine and O^4-alkylthymine from DNA in *E. coli.*

ALKYLATION OF OLIGODEOXYNUCLEOTIDES BY N-METHYL-N-NITROSOUREA

In order to investigate the possible sequence specificity in the methylation of DNA at the O^6 and N7 position of guanine by N-methyl-N-nitrosourea, a series of self complementary dodecadeoxynucleotides were synthesized containing guanine residues in different sequences (5'-TATACGCGTATA-3', 5'-TATACCGGTATA-3', 5'-TATAGGCCTATA-3', 5'-TATACATGTATA-3' and 5'-TATACTAGTATA-3'). These oligodeoxynucleotides were then alkylated by reaction with [^{3}H]N-methyl-N-nitrosourea in a reaction mixture which contained 1 mM of the dodecamers, 0.15 mM (290 µCi) [^{3}H]N-methyl-N-nitrosourea, 50 mM Tris-HCl, pH 7.4, in a total volume of 0.2 ml. After incubation for 2 hours at 37^{0}C, the reaction mixture was passed through two successive Sephadex G-25 spin columns to remove residual unbound

Table 3. Extent of guanine methylation at the O^6 position by N-methyl-N-nitrosourea.

Oligodeoxynucleotide	O^6-methylguanine formed (pmol)
5'-dTATACGCGTATA-3'	10.1
5'-dTATACCGGTATA-3'	19.0
5'-dTATAGGCCTATA-3'	29.7
5'-dTATACATGTATA-3'	6.7
5'-dTATACTAGTATA-3'	13.8

radioactivity, and aliquots were acid hydrolyzed and analyzed to determine the extent of modified purine bases by reverse phase HPLC. The O^6-methylguanine, 7-methylguanine and 3-methyladenine that were produced were separated by HPLC using a Micromeritics Microsil C_{18} 5 micron column preceded by a Brown Lee RP-18 precolumn with 7.5% methanol in 0.05 M ammonium formate, pH 4.5, at a flow rate of 1.0 ml/minute and a temperature of 35^0C.

As shown in Table 3, there were considerable differences in the formation of O^6-methylguanine in these oligodeoxynucleotides, although the O^6-methylguanine/N7-methylguanine ratio remained relatively constant (Dolan and Pegg, 1988a). For example, a direct comparison between the methylation at guanine with adenine or thymine as the 5' flanking base which was made with two dodecamers, 5'-TATACATGTATA-3' and 5'-TATACTAGTATA-3' indicated that the level of formation of O^6-methylguanine was 2.1 times greater when the guanine residue was preceded 5' by an adenine rather than thymine. It is also obvious from the results in Table 1 that methylation of guanine was greater when the 5' flanking base was guanine rather than cytosine. Since the oligodeoxynucleotides used for this comparison contain two guanine residues and it was not possible to analyze the alkylation at each position separately, it was necessary to make a crude comparison by assuming that the alkylation of the first guanine in the sequence 5'-CGG-3' was equal to that of the guanine in the sequence 5'-dTATACGCGTATA-3'. When this calculation was done and the methylation of the first guanine (5.0 pmol) was subtracted from the total (19.0 pmol), it was apparent that the methylation occurs to a much greater extent at the second guanine which has a 5' flanking guanine. Although the validity of this calculation might be questioned, the results for the methylation of the sequence 5'-TATAGGCCTATA-3' are in agreement with it. A similar calculation can be made for the second guanine in this sequence by assuming the alkylation of the first guanine is the same as that for guanine in the sequence 5'-TATACTAGTATA-3' and subtracting this value (13.8 pmol) from the total (29.7 pmol). This indicates that the extent of methylation of guanine which has a 5' flanking guanine was 15.9 pmol which agrees well with the previous estimate of 14 pmol. These results demonstrate that the formation of both 7-methylguanine and O^6-methylguanine is favored when the guanine alkylated was flanked on the 5' side by a purine (Dolan, et al., 1988a).

Other groups have also provided evidence that there may be sequence specificity in the formation of 7-alkylguanine in DNA by various alkylating agents. Such specificity was seen in the alkylation of synthetic polydeoxynucleotides by, N-methyl-N-nitrosourea or by 1,3,bis(2-chloroethyl)-1-nitrosourea; in both cases, runs of guanines were apparently preferentially alkylated at the N7 position (Briscoe and Cotter, 1984,1985; Briscoe and Duarte, 1988). DNA sequencing techniques were used to provide more direct evidence that formation of 7-alkylguanine occurred predominantly in regions consisting of runs of contiguous guanines after treatment with chloroethylating agents (Hartley, et al., 1986), nitrogen mustards (Mattes, et al., 1986; Kohn, et al., 1987), triazines (Hartley, et al., 1988), and N-methyl-N-nitrosourea (Wurdeman and Gold, 1988).

The mechanistic basis for the preferential alkylation of guanine in sequences in which it is on the 3' side of a purine, particularly guanine, is not yet understood. Explanations for neighboring base effects of the specificity of alkylation which involve the negative molecular electrostatic potential induced by the neighboring base have been suggested by Furois-Corbin and Pullman (1985) and by Kohn, et al.(1987). An alternative hypothesis put forward by Buckley (1987) supposes a mechanism which involves the initial formation from the imidourea derivative of a tetrahydral intermediate at the 5' dG. Subsequent intramolecular displacement reaction of the 3' dG with this tetrahydral intermediate then generates the alkylated guanines at this site (Buckley, 1987). Irrespective of the mechanism by which the effect is brought about, it is clear that the sequence specificity of DNA damage by these alkylating agents must be taken into account in understanding their effects. Obviously, this information must be combined with equivalent data on the effect of sequence specificity on the repair of various sequences.

It should be emphasized that, although most of the discussion above centers on the possible role of the 5' flanking base pair on the alkylation of guanine, other features of the sequences including the 3' base and residues further away from the site of alkylation may play a significant role in imparting sequence specificity. It is also extremely probable that the DNA conformation may affect the extent of alkylation and that sequence features imparting various conformations would therefore be expected to greatly influence the extent of alkylation. Finally, although only 7-alkylguanine and O^6-methylguanine have been examined for sequence specificity at present, it is likely that this will extend to all of the other alkylation products.

SEQUENCE SPECIFICITY IN MUTAGENESIS

The results described above indicate that there is sequence specificity both in DNA alkylation by N-methyl-N-nitrosourea, and in the rate of repair of various sequences by the alkyltransferase protein. A crude quantitation of the possible consequences of this specificity can be obtained by dividing the amount of O^6-methylguanine formed in a particular sequence by its rate of repair. Although it is highly probable that other features of the sequence surrounding the methylated guanine are also important, the most obvious factor seems to be the 5' flanking base. Such a calculation based only on the 5' flanking base indicates that the relative risk for mutation associated with the 5' flanking base is in the ratio of about 7:2:1.4:1 for guanine:adenine:thymine:cytosine, respectively.

This is in agreement with recent findings on the influence of neighboring base sequence on the mutations caused by N-methyl-N'-nitro-N-nitrosoguanidine, ethyl methanesulfonate, or N-ethyl-N-nitrosourea in the lacI gene of E. coli (Glickman, et al., 1987; Burns, et al., 1986, 1987,1988) and the E. coli gpt gene (Richardson, et al., 1987a,b). In all cases, the majority of the mutations observed were G-C->A-T transitions. These mutations in the lacI gene occurred predominantly in the sequence 5'-purine-G-3' (Burns, et al., 1986, 1988) when N-methyl-N'-nitro-N-nitrosoguanidine and

N-ethyl-N-nitrosourea as alkylating agents but ethyl methanesulfonate did not show this specificity (Burns, et al., 1986). On average, these G-C -> A-T transition mutations produced by N-methyl-N-nitro-N-nitrosoguanidine were 9 and 5 times more likely to occur when the 5' flanking base was guanine or adenine, respectively (Glickman, et al., 1987; Burns, et al., 1988). Similarly, in studies in which the E. coli gpt gene was used as a target, the majority of the G-C -> A-T transitions induced by N-methyl-N-nitro-N-nitrosoguanidine, N-methyl-N-nitrosourea, or N-ethyl-N-nitrosourea were present in the middle guanine of the sequence 5'-GG(A,T)-3' (Richardson, et al., 1987a,b).

When N-ethyl-N-nitrosourea was used there was also a significant incidence (17% in the lacI gene and 21% in the gpt gene) of A-T -> G-C transitions (Richardson, 1987a; Burns, et al., 1988). This is consistent with the miscoding potential of the O^4-alkylthymine adduct since N-ethyl-N-nitrosourea forms a much higher proportion of O^4-alkylthymine than the methylating agents. These mutations occurred in sequences 5'-purine-T-3' again suggesting that formation and/or repair favors the persistence of the adduct at such sites (Burns, et al., 1988).

ACKNOWLEDGEMENTS

This research was supported by grant CA-18137 from the National Cancer Institute, National Institutes of Health, Bethesda, MD.

REFERENCES

Adamkiewicz, J., Eberle, G., Huh, N., Nehls, P., and Rajewsky, M.F. Quantitation and visualization of alkyl deoxynucleosides in the DNA of mammalian cells by monoclonal antibodies. Environ. Health Perspect., 62:49-55, 1985.

Becker, R. A., and Montesano, R. Repair of O^4-methyldeoxythymidine residues in DNA by mammalian liver extracts. Carcinogenesis, 6:313-317, 1985.

Belinsky, S. A., White, C. M., Boucheron, J. A., Richardson, F.C., Swenberg, J. A., and Anderson, M. Accumulation and persistence of DNA adducts in respiratory tissue of rats following multiple administrations of the tobacco specific carcinogen 4-(N-methyl-N-nitrosamino)-1-(3-pyridyl)-1-butanone. Cancer Res., 46:1280-1284, 1986.

Beranek, D.T., Weis, C.C., Evans, F.E., Chetsanga, C.J., and Kadlubar, F. F. Identification of N^5-methyl-N^5-formyl-2,5,6-triamino-4-hydroxypyrimidine as a major adduct in rat liver DNA after treatment with the carcinogens N,N-dimethyl-nitrosamine or 1,2-dimethylhydrazine. Biochem. Biophys. Res. Commun., 110:625-631, 1983.

Boiteaux, S. and Laval, J. Imidazole open ring 7-methylguanine: an inhibitor of DNA synthesis. Biochem. Biophys. Res. Commun., 110:552558, 1983.

Boiteaux, S., O'Connor, T.R., and Laval, J. Formamidopyrimidine-DNA glycosylase of Escherichia coli: cloning and sequencing of the fpg structural gene and overproduction of the protein. EMBO J., 6:3177-3183, 1987.

Brent, T. P. Isolation and purification of O^6-alkylguanine DNA alkyltransferase from human leukemic cells. Prevention of chloroethylnitrosourea-induced cross-links by purified enzyme. Pharmac. Ther., 31:121-140, 1985.

Brent, T.P., Dolan, M.E., Fraenkel-Conrat, H., Hall, J., Karran, P., Laval, F., Margison, G.P., Montesano, R., Pegg, A.E., Potter, P. M., Singer, B., Swenberg, J. A., and Yarosh, D.B. Repair of O-alkylpyrimidines in mammalian cells: A present consensus. Proc. Nati. Acad. Sci. USA, 85:1759-1762, 1988.

Briscoe, W.T., and Cotter, L.E. DNA sequence has an effect of the extent and kinds of alkylation of DNA by a potent carcinogen. Chem. Biol. Interactions, 56:321-331, 1985.

Briscoe, W.T., and Cotter, L.E. DNA sequence has an effect on the extent and kinds of alkylation of DNA by a potent carcinogen. Chem. Biol. Interactions, 56:321-331, 1985.

Briscoe, W.T., and Duarte, S.P. Preferential alkylation by 1, 3-bis(2-chloroethyl)-1-nitrosourea (BCNU) of guanines with guanines as neighboring bases in DNA. Biochem. Pharm., 37:1061-1066, 1988.

Buckley, N. A regioselective mechanism for mutagenesis and oncogenesis caused by alkylnitrosourea sequence-specific DNA alkylation. J. Am. Chem. Soc., 109:7918-7920, 1987.

Burns, P.A., Allen, F.L., and Glickman, B.W. DNA sequence analysis of mutagenicity and site specificity of ethylmethanesulfonate in UvrA- and UvrB- strains of Escherichia coli. Genetics, 113:811-819, 1986.

Burns, P.A., Gordon, A.J.E., and Glickman, B.W. Influence of neighboring base sequence on N-methyl-N-nitrosoguanidine mutagenesis in the lacI gene of Escherichia coli. J. Mol. Biol., 194:385-390, 1987.

Burns, P.A., Gordon, A.J.E., Kunsmann, K., and Glickman, B. W. Influence of neighboring base sequence on the distribution and repair of N-ethyl-N-nitrosourea-induced lesions in Escherichia coli. Cancer Res., 48:4455-4458, 1988.

Dolan, M.E., and Pegg, A.E. Extent of formation of O^4-methylthymine in calf thymus DNA methylated by N-methyl-N-nitrosourea and lack of repair of this product by rat liver O^6-alkylguanine-DNA alkyltransferase. Carcinogenesis, 6:1611-1614, 1985.

Dolan, M.E., Scicchitano, D., Singer, B., and Pegg, A.E. Comparison of repair of methylated pyrimidines in poly(dT) by extracts from rat liver and Escherichia coli. Biochem. Biophys. Res. Commun., 123:324-330, 1984.

Dolan, M.E., Morimoto, K., and Pegg, A.E. Reduction of O^6-alkylguanine-DNA-alkyltransferase activity in HeLa cells treated with O^6-alkylguanines. Cancer Res., 45:6413-6417, 1985.

Dolan, M.E., Oplinger, M., and Pegg, A.E. Sequence specificity of guanine alkylation and repair. Carcinogenesis, 9:2139-2143, 1988a.

Dolan, M.E., Oplinger, M., and Pegg, A.E. Use of a dodecadeoxynucleotide to study repair of the O^4-methylthymine lesion. Mutation Res., 193:131-137, 1988b.

Dolan, M.E., Scicchitano, D., and Pegg, A.E. Use of oligodeoxynucleotides containing O^6-alkylguanine for the assay of O^6-alkylguanine-DNA-alkyltransferase activity. Cancer Res., 48:1184-1188, 1988c.

Domoradzki, J., Pegg, A.E., Dolan, M.E., Maher, V.M., and McCormick, J.J. Correlation between O^6-methylguanine-DNA-methyltransferase activity and resistance of human cells to the cytotoxic and mutagenic effect of N-methyl-N'-nitro-N-nitrosoguanidine. Carcinogenesis, 5:1641-1647, 1984.

Dyroff, M.C., Richardson, F.C., Popp, J.A., Bedell, M.A., and Swenberg, J.A. Correlation of O^4-ethyldeoxythymidine accumulation, hepatic initiation and hepatocellular carcinoma induction in rats continuously administered diethylnitrosamine. Carcinogenesis, 7:241-246, 1986.

Furois-Corbin, S., and Pullman, B. Specificity in carcinogen-DNA interaction; a theoretical exploration of the factors involved in the effect of neighboring bases on N-methyl-N-nitrosourea alkylation of DNA. Chem. Biol. Interactions, 54:9-13, 1985.

Gallagher, P., and Brent, T.P. Further purification and characterization of human 3-methyladenine-DNA glycosylase. Biochim. Biophys. Acta, 782:394-401, 1984.

Glickman, B.W., Horsfall, M.J., Gordon, A.J.E., and Burns, P. A. Nearest neighbor affects G:C to A:T transitions induced by alkylating agents. Environ. Health Perspect., 76:29-32, 1987.

Graves, R.J., Li, B.F.L., and Swann, P.F. Repair of synthetic oligonucleotides containing O^6-methylguanine, O^4-ethylguanine, and O^4-methylthymine, by O^6-alkylguanine-DNA-alkyltransferase. In: H. Bartsch, I.K. O'Neill, R. Schulte-Hermann (Eds.), Relevance of N-Nitroso Compounds to Human Cancer: Exposures and Mechanisms (IARC Scientific Publications No. 84), pp. 41-43. Lyon: IARC, 1987.

Hall, J., and Karran, P. O-Methylated pyrimidines: important lesions in cytotoxicity and mutagenicity in mammalian cells. In: B. Myrnes, H. Krokan (Eds.), Repair of DNA lesions introduced by N-nitroso compounds, pp. 73-88. Oslo: Norwegian University Press, 1986.

Hartley, J.A., Gibson, N.W., Kohn, K.W., and Mattes, W.B. DNA sequence selectivity of guanine-N7 alkylation by three antitumor chloroethylating agents. Cancer Res., 46:1943-1947, 1986.

Hartley, J.A., Mattes, W.B., Vaughan, K., and Gibson, N.W. DNA sequence specificity of guanine N7-alkylations for a series of structurally related triazenes. 9:669-674, 1988.

Kadlubar, F.F., Beranek, D.T., Weis, C.C., Evans, F.E., Cox, R., and Irving, C.C. Characterization of the purine ring-opened 7-methylguanine and its persistence in rat bladder epithelial DNA after treatment with the carcinogen N-methyl-nitrosourea. Carcinogenesis, 5:587-592, 1984.

Kohn, K., Hartley, J.A., and Mattes, W.B. Mechanisms of DNA sequence selective alkylation of guanine-N7 position by nitrogen mustards. Nucleic Acids Res., 15:10531-10549, 1987.

Lindahl, T. DNA repair enzymes. Annu. Rev. Biochem., 51:61-87, 1982.

Lindahl, T., and Sedgwick, B. Regulation and expression of the adaptive response to alkylating agents. Annu. Rev. Biochem., 57:133-157, 1988.

Lindahl, T., Demple, B., and Robins, P. Suicide inactivation of the E. coli O^6-methylguanine-DNA methyltransferase. EMBO J., 1:1359-1363, 1982.

Male, R., Helland, D.E., and Kleppe, K. Purification and characterization of 3-methyladenine-DNA glycosylase from calf thymus. J. Biol. Chem., 260:1623-1629, 1985.

Male, R., Haukanes, B.I., Helland, D.E., and Kleppe, K. Substrate specificity of 3-methyladenine-DNA glycosylase from calf thymus. Eur. J. Biochem., 165:13-19, 1987.

Margison, G.P., and Pegg, A.E. Enzymatic release of 7-methylguanine from methylated DNA by rodent liver extracts. Proc. Natl. Acad. Sci. USA, 78:861-865, 1981.

Mattes, W.B., Hartley, J.A., and Kohn, K.W. DNA sequence of guanine-N7 alkylation by nitrogen mustards. 14:2971-2987, 1986.

McCarthy, T.V., Karran, P., and Lindahl, T. Inducible repair of O-alkylated pyrimidines in Escherichia coli. EMBO J., 3:545-550, 1984.

O'Connor, T.R., Boiteaux, S., and Laval, J. Ring-opened 7-methylguanine residues in DNA are a block to in vitro DNA synthesis. Nucleic Acids Res., 16:5879-5894, 1988.

Pegg, A.E. Formation and metabolism of alkylated nucleosides: Possible role in carcinogenesis by nitroso compounds and alkylating agents. Adv. Cancer Res., 25:195-269, 1977.

Pegg, A.E. Dimethylnitrosamine inhibits enzymatic removal of O^6-methylguanine from DNA. Nature, 274:182-184, 1978.

Pegg, A.E. Alkylation and subsequent repair of DNA after exposure to dimethylnitrosamine and related carcinogens. Rev. Biochem. Toxicol., 5:83-133, 1983.

Pegg, A.E., and Dolan, M.E. Properties and assay of mammalian O^6-alkylguanine-DNA alkyltransferase. Pharmac. Ther., 34:167-179, 1987.

Pegg, A.E., and Hui, G. Formation and subsequent removal of O^6-methylguanine from DNA in rat liver and kidney after small doses of dimethylnitrosamine. Biochem. J., 173:-7739-748,1978.

Pegg, A.E., Wiest, L., Foote, R.S., Mitra, S., and Perry, W. Purification and properties of O^6-methylguanine-DNA-transmethylase from rat liver. J. Biol. Chem., 258: 2327-2333, 1983.

Pegg, A.E., Scicchitano, D., and Dolan, M.E. Comparison of the rates of repair of O^6-alkylguanines in DNA by rat liver and bacterial O^6-alkylguanine-DNA alkyltransferase. Cancer Res., 44:3806-3811, 1984.

Pegg, A.E., Scicchitano, D., Morimoto, K., and Dolan, M.E. Specificity of O^6-alkylguanine-DNA alkyltransferase. In: H. Bartsch, I.K. O'Neill, R. Schulte-Hermann (Eds.), The relevance of N-nitroso compounds to human cancer: exposures and mechanisms, IARC Scientific Publication No. 84, pp. 30-34. Lyon: IARC, 1987.

Richardson, F.C., Dyroff, M.C., Boucheron, J.A., and Swenberg, J.A. Differential repair of O^4-alkylthymidine following exposure to methylating and ethylating hepatocarcinogens. Carcinogenesis, 6:625-629, 1985.

Richardson, K.K., Crosby, R.M., Richardson, F.C., and Slopek, T.R. DNA base changes induced following in vivo exposure of unadapted, adapted or Ada-Escherichia coli to N-methyl-N'-nitro-N-nitrosoguanidine. Mol. Gen. Genet., 209:526-532, 1987a.

Richardson, K.K., Richardson, F.C., Crosby, R.M., Swenberg, J.A., and Skopek, T.R. DNA base changes and alkylation following in vivo exposure of Escherichia coli to N-methyl-N-nitrosourea or N-ethyl-N-nitrosourea. Proc. Natl. Acad. Sci. USA, 84:344-348, 1987b.

Saffhill, R., Margison, G.P., and O'Connor, P.J. Mechanisms of carcinogenesis induced by alkylating agents. Biochim. Biophys. Acta., 823:111-145, 1985.

Samson, L., Thomale, J., and Rajewsky. M.F. Alternative pathways for the in vivo repair of O^6-alkylguanine and O^4-alkylthymine in Escherichia coli: the adaptive response and nucleotide excision repair. EMBO J., 7:2261-2267, 1988.

Scicchitano, D., Jones, R.A., Kuzmich, S., Gaffney, B., Lasko, D.D., Essigmann, J.M., and Pegg, A.E. Repair of oligodeoxynucleotides containing O^6-methylguanine by O^6-alkylguanine-DNA-alkyltransferase. Carcinogenesis, 7:1383-1386, 1986.

Sedgwick, B., and Lindahl, T.A common mechanism for repair of O^6-methylguanine and O^6-ethylguanine in DNA. J. Mol. Biol., 154:169-174, 1982.

Singer, B. Alkylation of the O^6 of guanine is only one of many chemical events that may initiate carcinogenesis. Cancer Invest., 2:233-238, 1984.

Singer, B. O-Alkyl pyrimidines in mutagenesis and carcinogenesis: occurrence and significance. Cancer Res., 46:4879-4885, 1986.

Singer, B., and Kusmierek, J.T. Chemical mutagenesis. Annu. Rev. Biochem., 52:655-693, 1982.

Sukumar, S., Notario, V., Martin-Zanca, D., and Barbacid, M. Induction of mammary carcinomas in rats by nitrosomethylurea involves malignant activation of H-ras-1 locus by single point mutations. Nature, 306:658-661, 1983.

Swenberg, J.A., Dyroff, M.C., Bedell, M.A., Popp, J.A., Huh, N., Kirstein, U., and Rajewsky, M.F. O^4-Ethyldeoxythymidine, but not O^6-ethyldeoxyguanosine, accumulates in hepatocyte DNA of rats exposed continuously to diethylnitrosamine. Proc. Natl. Acad. Sci. USA, 81:1692-1695, 1984.

Topal, M.D. DNA repair, oncogenes and carcinogenesis. Carcinogenesis, 9:691-696, 1988.

Topal, M.D., Eadie, J.S., and Conrad, M. O^6-Methylguanine mutation and repair is nonuniform. Selection for DNA most interactive with O^6-methylguanine. J. Biol. Chem., 261:9879-9885, 1986.

Wu, R.S., Hurst-Calderone, S., and Kohn, K.W. Measurement of O^6-alkylguanine-DNA alkyltransferase activity in human cells and tumor tissues by restriction endonuclease inhibition. Cancer Res., 47:6229-6235, 1987.

Wurdeman, R.L., and Gold, B. The effect of DNA sequence, ionic strength, and cationic DNA affinity binders on the methylation of DNA by N-methyl-N-nitrosourea. Chem. Res. Toxical., 1:146-147, 1988.

Yarosh, D.B. The role of O^6-methylguanine-DNA methyltransferase in cell survival, mutagenesis and carcinogenesis. Mutation Res., 145:1-16, 1985.

Yarosh, D.B., Fornace, A.J., and Day, R. S. O^6-alkyltransferase fails to repair O^4-methylthymine and methyl phosphotriesters in DNA as efficiently as does the alkyltransferase from Escherichia coli. Carcinogenesis, 6:949-953, 1985.

Yarosh, D.B., Hurst-Calderone, S., Babich, M.A., and Day, R. S. Inactivation of O^6-methylguanine-DNA methyltransferase and sensitization of human tumor cells to killing by chloroethylnitrosourea by O^6-methylguanine as a free base. Cancer Res., 46:1663-1668, 1986.

Zarbl, H., Sukumar, S., Arthur, A.V., Martin-Zanca, D., and Barbacid, M. Direct mutagenesis in Ha-ras-1 oncogenes by N-nitroso-N-methylurea during initiation of mammary carcinogenesis in rats. Nature, 315:382-385, 1985.

TOWARDS A ROLE FOR PROMUTAGENIC LESIONS IN CARCINOGENESIS[1]

Peter J. O'Connor

CRC Section of Carcinogenesis, Paterson Institute for Cancer Research, Christie Hospital and Holt Radium Institute, Manchester, M20 9BX, UK

INTRODUCTION

The reactions of alkylating agents with DNA are now widely considered to be major determinants in the toxic, mutagenic, carcinogenic and other biological properties of these compounds. Until recently, it has not been possible to begin to confirm the relevance of such observations for human disease, but techniques now permit preliminary studies of these potentially initiating events in human tissues.

DNA repair mechanisms represent one of the cell's first lines of defence against the deleterious effects of DNA lesions. Repair deficient cells within a tissue which is a target for carcinogenesis may be at special risk. If these natural protective functions are ultimately to be exploited for the benefit of man, either in terms of cancer prevention or protection of tissues from the effects of chemotherapy, then it is not only necessary to understand the mechanisms which control them, but also to determine which of the repair functions are of greatest importance in protecting against the effects of damage resulting from exposure to environmental alkylating agents.

The following account illustrates progress in these areas of research.

---

[1]This paper was presented as a tribute to Dr. Roy Saffhill who died suddenly and unexpectedly on 30 December, 1987. Over the last fifteen years Roy devoted the greater part of his research effort to understanding mechanisms in carcinogenesis induced by chemicals, particularly in relation to their ability to affect the fidelity of DNA replication. He played a large part in the early development of radioimmunoassay procedures for the detection of alkylation damage in DNA. The success of the collaborative study involving laboratories in Lyon, Essen, Beijing and Manchester (Umbenhauer, et al., 1985), which showed that these procedures could detect human exposure to environmental alkylating agents, was a source of great personal satisfaction to him.

DNA Repair Mechanisms and Their Biological Implications in Mammalian Cells
Edited by M.W. Lambert and J. Laval
Plenum Press, New York

ALKYLATION OF HUMAN TISSUES

The modification of mammalian tissues by environmental alkylating agents results in the dose-dependent formation of promutagenic lesions in DNA as determined from animal models (Saffhill, et al., 1985). More recently, with the development of sensitive radioimmunoassay procedures (Adamkiewicz, et al., 1982; Wild, et al., 1983), low levels of O^6-methylguanine (O^6-MeG) have been observed in the DNA extracted from tissues of patients from a district of North China who were undergoing surgery for cancer of the oesophagus and related problems (Umbenhauer, et al., 1985).

Subsequent collections of human material (Saffhill, et al., 1988a) from a region of South East Asia where, like the district in North China, there is a high incidence of oesophageal cancer, have also revealed similar levels of O^6-MeG in tissue DNA (Figure 1). As with the Chinese group, alkylation was detected in the majority of the DNA samples, indicating that the causative agent might be widespread in the environment. In contrast, the data for a group of patients from the Manchester area (Saffhill, et al., 1988a) showed a skewed distribution with O^6-MeG being undetectable in many of the samples, which is more suggestive of the effect of an exposure due to lifestyle or medication (Figure 1). The samples from this group were from the mucosa of the stomach and colon and included tissue from patients presenting with benign conditions, as well as from the tumor and uninvolved tissue adjacent to the tumour in the cancer patients. A limited analysis of the data has been made in an attempt to gain some clues as to the origins of the environmental alkylating agent (Hall, et al., unpublished data). These patients (n = 35, after elimination of samples for which case notes were incomplete) proved to be a group comprised mainly of non-smokers and non-drinkers, so that for this group at least, there was no evidence of a relationship between the methylation of tissue DNA and smoking or alcohol consumption. Similarly, there was no evidence for a correlation between the extent of methylation and the various categories of drug treatment (e.g., antihypertensive, antibiotic, tranquillizer, antidepressant, antacid, etc.) prescribed prior to entry into the hospital for surgery. There was, however, an indication of a correlation ($P = 0.08$) between the level of DNA methylation and the premedicative treatment preceding surgery when this included tranquillizers. If the levels of DNA alkylation in the different tissues were compared, no significant differences were observed between stomach tissue from patients with benign conditions, uninvolved tissues or those of tumor origin. Again, there was no significant difference between the benign and uninvolved colonic tissues. However, when the DNA extracted from the colonic tumors was examined, the level of alkylation was either very low or undetectable (0.0026 ± 0.0026 vs 0.083 ± 0.034 µmole O^6-MeG per mole adenine for tumor and uninvolved tissue, respectively; $P = 0.02$). Since these tumors are generally exposed to the lumen of the gut, physical protection seems unlikely, and these results possibly indicate a loss of capacity for metabolic activation of carcinogens by the tumor cells. Although the numbers of samples in this study are too small to provide definitive indications, they do confirm that alkylation events can be detected in the DNA of human tissues and further implicate exposure to an environmental alkylating species as the causative agent. They suggest that the exposure was of recent origin and that the agent may require metabolism to generate the alkylating species. In this context, it should be noted that drugs which are not themselves alkylating agents (e.g., Isoniazid) can be metabolised to an alkylating intermediate that will react with tissue DNA (Saffhill, et al., 1988b). In considering environmental exposure, therefore, attention should not be restricted solely to alkylating agents or their more obvious precursors (e.g., the nitrosation of amines, Bartsch and Montesano, 1984).

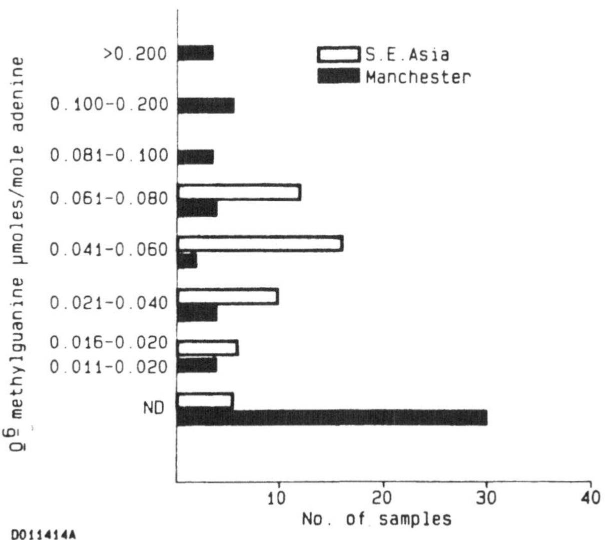

Figure 1. Presence of O^6-MeG in the DNA of human tissues. Samples from South East Asia were from the oesophagus and stomach; those from the Manchester area were from the stomach and bowel. Lower limit of detection was 0.01 μmoles O^6-MeG/mole adenine.

REPAIR DEFICIENT CELLS IN MAMMALIAN TISSUES

In order to gain further insight into these potentially initiating events in man, animals have been used to study the formation and repair of O^6-MeG in the DNA of different organs after exposure to the environmental alkylating agent N-nitrosodimethylamine (NDMA). Tissues vary greatly in their capacity to metabolize nitrosamines so that many remain virtually unaffected, and this partially explains the tissue specific induction of tumors after treatment with the simple alkylating nitrosamines (O'Connor, et al., 1979). However, data for the amount of alkylation present in DNA extracted from whole tissue homogenates, which hitherto have been used as a guide to the extent of tissue damage, merely represent a mean average value and do not provide information about the extent of DNA damage in specific, target cells. Such data can now be obtained by procedures using antibodies to O^6-MeG. A rabbit polyclonal anti-O^6-MeG (Saffhill, et al., 1988b; O'Connor, et al., 1988) now permits the detection of O^6-MeG in the DNA of hepatic nuclei of rats treated with NDMA down to doses as low as 100 µg/kg, a treatment which produces a tissue average alkylation value of ~5 µmole O^6-MeG per mole guanine for extracted hepatic DNA. Under these conditions, relatively few of the centrilobular hepatic nuclei are alkylated so that in the affected cells themselves the actual level of alkylation must be very much higher.

A similar situation may also be found in the cells of organs which do not metabolize nitrosamines so readily as liver and, therefore, overall sustain a much lower level of alkylation (see O'Connor, et al., 1979). Thus, in rats treated with NDMA (40 mg/kg, a dose which is toxic to liver) the epithelial cells lining some of the bronchioles are all very heavily alkylated when examined 5 hours after treatment. Throughout the rest of the lung, alkylated nuclei in the alveoli and stroma are relatively rare. After an interval of 12 days, many of these epithelial nuclei are still highly positive while others in the same region of the bronchiole no longer contain detectable O^6-MeG in their DNA. Removal Of the O^6-MeG is presumably due to repair reactions, since under these conditions, turnover of epithelial cells can be excluded by the absence of cells staining positively for BUdR (i.e., in animals which have been exposed to BUdR for 18 hours via the drinking water; paraffin wax sections were stained using anti-BUdR, Dako Ltd, High Wycombe, UK, as the primary antibody). Thus, although the overall extent of alkylation in lung is relatively low in comparison with the liver, most of the damage is confined to a limited population of cells, and some of these evidently have a much reduced capacity for repair. These observations clearly illustrate that there is a category of cells in which a high metabolic proficiency is coupled with a repair deficiency. Together, these represent genetically controlled, double risk factors that could predispose to malignant transformation and may eventually prove to be a relatively common phenomenon in tissues which have a high risk for cancer development in response to exposure to alkylating carcinogens.

The clarity of the observations in the above systems stimulated an analysis of a model tumor system to determine whether cells containing promutagenic lesions in their DNA could be followed from the time of treatment through the latent period, to correlate with the eventual emergence of tumors. The system selected was as follows: weanling rats which had been maintained on a protein-free diet for 3 days were given a dose of NDMA (30 mg/kg i.p.) which should result in the development of renal tumors of mesenchymal origin in 100% of the animals (Driver, et al., 1987). Protein deprivation reduces the hepatic capacity for metabolism of the nitrosamine (Swann, et al., 1980), thereby, reducing the "first pass" effect and permitting more of the agent to circulate to other organs. Under these conditions, tissues such as kidney, which have a lesser ability to metabolize NDMA, sustain several times more alkylation than occurs in the kidney of normal animals. In earlier studies using this system for the production of renal tumors (Nicoll, et al., 1975), a retention of O^6-MeG had been observed for up to four days in the DNA extracted from the entire kidneys compared with the liver of animals treated in this way.

Sections of liver taken from rats maintained on a normal diet were stained for the presence of O^6-MeG (Fan, et al., in preparation). Five hours after treatment with NDMA, positive nuclei were present throughout the greater part of the liver lobule and around the radical blood vessels, while only the hepatocytes lying in the vicinity of the portal triads were O^6-MeG negative. In the protein deprived animals, however, the O^6-MeG positive nuclei were restricted to a relatively narrow zone close to the central veins, thereby confirming earlier biochemical observations indicating the lower capacity for NDMA metabolism in these animals (Swann, et al., 1980). In sections of kidney taken at the same time, many cells in the cortex were stained positively but below the cortico-medullary boundary, within the medulla, no positive cells were found. In the renal cortex of the protein deprived rats, the nuclei of the tubular epithelium were prominently stained and were the dominant feature of the cortex up to 2 days after treatment. Thereafter, O^6-MeG was lost from the DNA of these nuclei by a combination of repair and cell turnover (as determined by BUdR incorporation; see above), but it was not until six days after treatment that the O^6-MeG positive nuclei of the mesenchymal elements became more obvious (Figure 2).

At around this time (i.e., 6-8 days after NDMA treatment), cell turnover was readily demonstrated (see above) and was most abundant in the central region of the cortex where the mesenchymal tumors tend to arise.

Again the cells of the tubular epithelium featured most prominently, but occasionally cell divisions could be observed among the mesenchymal target cell population. The system, therefore, appears to have the necessary prerequisites for carcinogenesis induced by alkylating agents in as much as DNA lesions are persistent in the target cell population up to the time of DNA replication. Even up to ten weeks after the initial treatment with NDMA, repair deficient mesenchymal cells are still readily distinguishable in the cortex at a time when the preneoplastic lesions are beginning to emerge (Figure 2). Some O^6-MeG positive mesenchymal nuclei can even be identified lying within the preneoplastic lesions themselves, although these particular cells clearly cannot be the progenitors of the preneoplasia (Fan, et al., in preparation). Further studies are in progress in an attempt to examine whether a closer association can be established between the persistence of these promutagenic lesions in the DNA of target cells and the emergence of the malignant phenotype. At this stage, however, the study already provides unambiguous evidence for the presence of a population of repair-deficient cells within a tissue and suggests that such cells may be at high risk for transformation to malignancy. Observations of this kind highlight the need to understand mechanisms controlling the expression of DNA repair genes.

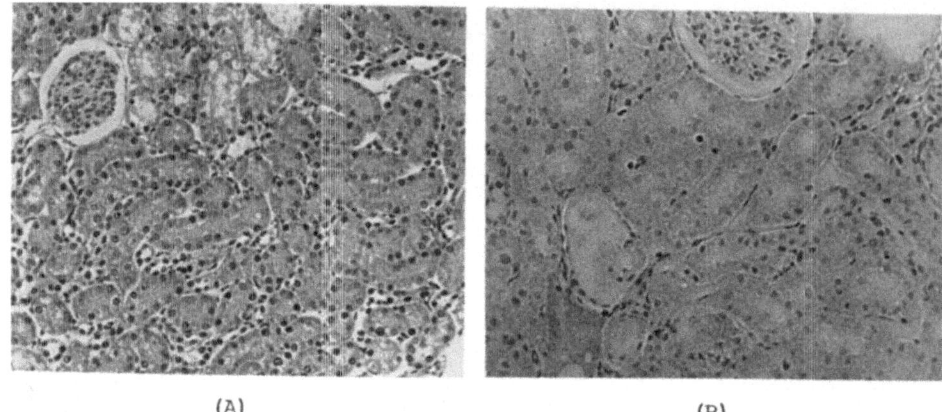

(A) (B)

Figure 2. Immunochemical staining of kidney sections taken 6 days (A) or 10 weeks (B) after treating protein depleted rats with NDMA i.p. (30 mg/kg). Animals were fed a sucrose-corn starch diet for 3 days prior to the nitrosamine treatment. A, Cortex showing residual O^6-MeG positive nuclei, still mainly in the tubular epithelium. B, O^6-MeG positive nuclei are now restricted to the mesenchymal, target cell population. (Paraffin wax sections, 3 μm from 70% ethanol fixed tissues were stained using a rabbit polyclonal anti-O^6-MeG. Positive cells were identified using a rabbit PAP complex and 3,3-diaminobenzidine. The counter stains were eosin, A, and Harris's Hematoxylin, A and B; phase contrast, x 260).

INDUCTION OF $\underline{O}^6$-METHYLGUANINE REPAIR IN MAMMALIAN TISSUES

In mammalian cells, the principle that the repair of O^6-MeG can be induced by a variety of agents is well established (see Saffhill, et al., 1985). Although synthesis of new protein is involved (e.g., Schmerold and Wiestler, 1986), the increase in repair capacity is only 3-6 fold compared to 100-fold increase encountered with E. coli exposed to N-methyl-N'-nitro-N-nitrosoguanidine (Saffhill, et al., 1985). Agents inducing this response in animals have been either genotoxic or in some cases, physiological stimuli, but all are capable of inducing DNA synthesis, although DNA synthesis itself is not a prerequisite (Saffhill, et al., 1985). The response to genotoxic agents has generally been restricted to rat liver, but in the case of X-rays (Margison, et al., 1986) and poly I:C (O'Connor and Bailey, unpublished data), both can induce the response in several tissues, and X-rays are effective in mice and hamsters. Ethionine, which only binds minimally to DNA, and agents such as pronethalol, which do not bind to DNA at all, have proved inactive (Saffhill, et al., 1985). In keeping with earlier short-term observations (Pegg and Perry,

1981), phenobarbital administered in the drinking water (0.05%) for 21 weeks did not alter the amount of O^6-MeG-DNA-alkyltransferase activity present in cell-free extracts of liver, assayed using [^{3}H]-methylated substrate DNA (O'Connor, et al., 1988). Even when the repair activity of cell free extracts was increased 2-3 fold by treating the animals with 2-acetylaminofluorene (Saffhill, et al., 1985), exposure to phenobarbital still failed to maintain the increased level of O^6-MeG-alkyltransferase activity (O'Connor, et al., 1988). However, the repair capacity of liver can also be assayed by administration of a low dose of NDMA (2 mg/kg) and subsequently following changes in the levels of O^6-MeG in hepatic DNA. When this was done, it was evident that the treatment with phenobarbital had increased the repair of O^6-MeG compared with the control animals. Over the course of 21 weeks, the altered repair capacity showed a biphasic response which was presumably limited to those cells with a capacity to activate NDMA (Figure 3). The increased repair was maximal at 3 weeks and was specific for O^6-MeG since i) the ratios of O^6-MeG to 7-methylguanine in hepatic DNA followed a similar time course, changing from 0.089 in the control rats to 0.029 in rats treated with ^{14}C-labelled NDMA and exposed to phenobarbital for 3 weeks, and ii) the repair of O^4-methylthymine (O^4-MeT) was not affected significantly (Figure 3). These changes in the O^6-MeG content of hepatic DNA in response to 3 weeks of phenobarbital treatment could also be followed by immunohistochemical staining of liver sections (see above). When

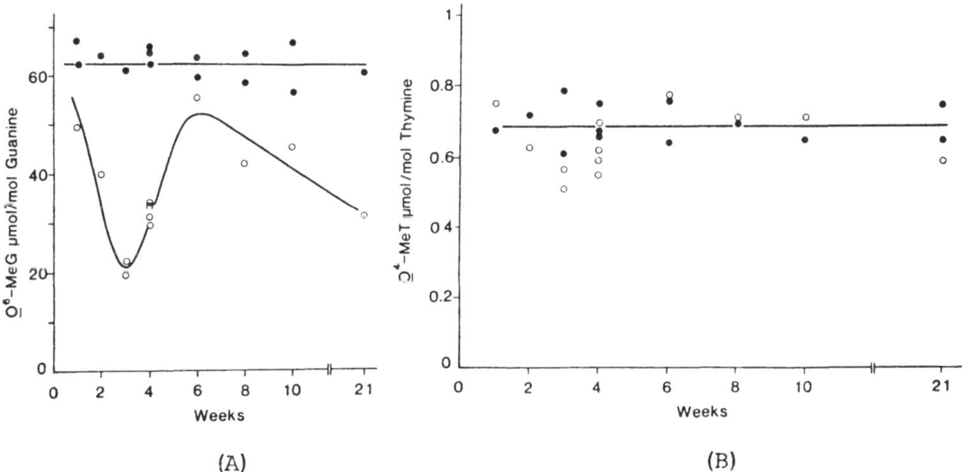

Figure 3. Amounts of O^6-MeG (A) or O^4-MeT (B) in the liver DNA of rats exposed continuously to phenobarbital in the drinking water (O), or in the liver DNA of unexposed controls (●), 5 hours after a single dose of [^{14}C]-NDMA (2 mg/kg). Amounts of alkylated bases were determined by HPLC analysis of enzyme DNA hydrolysates and radioimmunoassay (figures taken from O'Connor, et al., 1988).

animals were sampled at 3, 9, 12, and 24 hours after administration of NDMA (2 mg/kg), fewer O^6-MeG positive nuclei were observed in the livers of phenobarbital animals when sufficient time for repair had elapsed (e.g., at 9 hours after NDMA treatment). The importance of these observations may lie, therefore, not so much in the ability of a promoting agent to induce this response, but that phenobarbital may act via non-DNA targets, possibly at membrane receptor sites (Evarts, et al., 1985; Schwartz, et al., 1987), to induce O^6-MeG repair without the need to elicit the hazardous events associated with genotoxic damage.

ASSIGNING BIOLOGICAL ROLES TO SPECIFIC DNA ALKYLATION PRODUCTS

The simultaneous introduction of 13 different lesions into DNA by the alkylating agents has made studies of the correlation of specific biological effects with individual DNA lesions a precarious exercise. Recently, much more definitive conclusions have been possible from studies in which specific DNA repair genes have been transfected into repair deficient mammalian cells in order to observe the effects of the removal of specific lesions (see Table 1). Such cells are essentially the mammalian equivalent of bacterial wild type and mutant since they differ from the control, repair deficient cells only by the presence of the repair gene DNA sequence. For example, in V79 cells in which the full coding sequence of the ada gene is expressed, the O^6-MeG-alkyltransferase activity of cell-free extracts is approximately 1.5 pmoles/mg protein in comparison with approximately 2 fmoles/mg protein in the control cells. If these cells are treated with N-nitrosomethylurea (NMU), O^6-MeG is eliminated from the host DNA very rapidly (Brennand and Margison, 1986a). Although it is evident that the nuclear membrane can be penetrated readily, as yet there is no information on the kinetics of nuclear transport of the repair protein, but studies on the accessibility of chromatin are in progress.

In general (Table 1), these studies support earlier correlations with respect to the effects of O^6-MeG in DNA, particularly in relation to mutagenesis. However, the role of O^4-MeT cannot be dissociated from that of O^6-MeG because the same catalytic site in the ada protein is responsible for the repair of both lesions (Demple, et al., 1985), but the relative abundance of these two products in DNA (1:100; see above) suggests that of the two, O^6-MeG is making the greater contribution. In these cell lines at least, O^6-MeG (O^4-MeT) appears to be a toxic lesion since protection is also afforded by the truncated ada gene from which the coding sequence for the alkylphosphotriester repair function has been removed (Brennand and Margison, 1986b). The presence of an efficient repair system (either the full or the truncated ada protein) can also protect against the toxicity of the chemotherapeutic, chloroethylating agents (Brennand and Margison, 1986a,b), presumably by binding to O^6-chlorethylguanine, thereby precluding its rearrangement and the formation of the potentially lethal crosslink (see this volume).

The role of this repair system in protecting against the clastogenic effects of O^6-MeG (O^4-MeT) and possibly alkylphosphotriesters is also patent when cells are treated with

Table 1. Protection in Repair Deficient Cell Lines by Prokaryotic DNA Repair Gene Products

Endpoint	Cell Line[a]	Agent	Repair Capacity[b]	Protective Effect	Reference
Toxicity	V79	HN2	$\underline{O}^6$-AG/$\underline{O}^4$-AT/AP	-	Brennand & Margison 1986a
		MMS		-	Brennand & Margison 1986a
		NMU		++	Brennand & Margison 1986a
		CLZ		+++	Brennand & Margison 1986a
		MZ		+++	Brennand & Margison 1986a
	FDCP Mix	HN2		-	Jelinek et al, 1988
		MMS		+	Jelinek et al, 1988
		NMU		+++	Jelinek et al, 1988
		MZ		+++	Jelinek et al, 1988
		TCNU		+++	Jelinek et al, 1988
	V79	NMU	$\underline{O}^6$-AG/$\underline{O}^4$-AT	++	Brennand & Margison 1986b
		CLZ		++	Brennand & Margison 1986b
Mutations Forward TG[R]	V79	NMU	$\underline{O}^6$-AG/$\underline{O}^4$-AT/AP	+++	Brennand & Margison 1986, Fox & Margison 1988
Reverse TG[S]	V79	NMU		+++	Fox & Margison 1988
		NEU		++	Fox & Margison 1988
SCE	V79	MMS	$\underline{O}^6$-AG/$\underline{O}^4$-AT/AP	-	White et al, 1986
		NMU		+++	White et al, 1986
		NBU		-	White et al, 1986
		MZ		+++	White et al, unpublished
	FDCP Mix	NMU		+++	Jelinek et al unpublished
Chromatid Aberrations	V79	NMU	$\underline{O}^6$-AG/$\underline{O}^4$-AT/AP	+++	White et al, 1986
Micronuclei	V79	NMU	$\underline{O}^6$-AG/$\underline{O}^4$-AT/AP	++	White et al, 1986

a Cell lines originally deficient in endogenous DNA-alkyltransferase activity ($\leqslant$ 4 fmoles per mg protein)

b Transfected with (and expressing) the full ada coding sequence $\underline{O}^6$-AG/$\underline{O}^4$-AT/AP or the truncated ada sequence $\underline{O}^6$-AG/$\underline{O}^4$-AT (DNA-alkyltransferase activity 0.3-2.2 pmoles/mg protein).

CLZ, chlorozotocin; HN2, nitrogen mustard; MMS, methyl methanesulphonate; MZ, mitozolomide; NBU, N-nitrosobutylurea; NEU, N-nitrosoethylurea; NMU, N-nitrosomethylurea; TCNU, taurylchloroethylnitrosourea; $\underline{O}^6$-AG, $\underline{O}^6$-alkylguanine; $\underline{O}^4$-AT, $\underline{O}^4$-alkylthymine; AP, alkylphosphotriester.

NMU. However, protection is only marginal when they are treated with methyl methanesulfonate, an agent which alkylates the O^6-position of guanine to a much lesser extent (approximately 20-30 fold) than NMU. One interpretation of these results would be that there is more than one route for the initiation and formation of SCE events. On the other hand, the lack of protection afforded after treatment with N-nitrosobutylurea might be explained if O^6-butylguanine is repaired principly via nucleotide excision mechanisms.

Taken overall, these results and those of others (Samson, et al., 1986; Kataoka, et al., 1986) and the lack of protection afforded against toxicity induced by agents that predominantly alkylate the N7-position of guanine (e.g., HN2, MMS; Table 1) indicate that the principle promutagenic lesions in DNA constitute a major and broad spectrum hazard for the cell.

ACKNOWLEDGEMENTS

I am grateful to Dr. G. P Margison and Dr. D. P. Cooper for critical comments and to Miss Sarah J. Morrissey for preparation of the manuscript. This work was supported by the Cancer Research Campaign, UK.

REEERENCES

Adamkiewicz, J., Drosdziok, W., Eberhardt, W., Langen-berg, H., and Rajewski, M. F., 1982. High-affinity monoclonal antibodies specific for DNA components structurally modified by alkylating agents, In: Banbury report No. 13; Indicators of Genetic Exposure, p.37, B.A. Bridges, D. E. Butterworth and I. B. Weinstein, eds. Cold Sring Harbor Laboratory, New York.

Bartsch, H. and Montesano, R. 1984. Relevance of nitrosamines to human cancer. Carcinogenesis, 5: 1381.

Brennand, J. and Margison, G. P. 1986a. Reduction of the toxicity and mutagenicity of alkylating agents in mammalian cells harbouring the Escherichia coli alkyltransferase gene. Proc. Natl. Acad. Sci. USA, 83: 6292.

Brennand, J. and Margison, G. P. 1986b. Expression in mammalian cells of a truncated Escherichia coli gene coding for $\underline{O}^6$-alkylguanine alkyltransferase reduced the toxic effects of alkylating agents. Carcinogenesis, 7: 2081.

Demple, B., Sedgwick, B., Robins, P., Totty, N., Waterfield, M.D., and Lindahl, T. 1985. Active site and complete sequence of the suicidal methyltransferase that counters alkylation mutagenesis. Proc. Natl. Acad. Sci. USA, 82: 2688.

Driver, H. E., White, I.N.H., and Butler, W.H. 1987. Dose response relationships in chemical carcinogensis: renal mesenchymal tumors induced in the rat by single dose dimethylnitrosamine. Br. J. Exp. Path. 68: 133.

Evarts, R. P., Marsden, E. R., and Thorgiersson, S. S. 1985. Modulation of asialoglycoprotein receptor levels in rat liver by phenobarbital treatment. Carcinogenesis, 6: 1767.

Fan, C.Y., Butler, W. H., and O'Connor, P. J. Data in preparation.

Fox, M. and Margison, G. P., 1988. Expression of an E. coli $\underline{O}^6$-alkylguanine alkyltransferase gene in Chinese hamster cells protects against N-methyl- and N-ethyl-nitrosourea-induced reverse mutation at the hypoxanthine phosphoribosyl transferase locus. Mutagenesis, 3: 409.

Hall, C.N., Saffhill, R., Badawi, A. G. and O Connor, P.J. Data in preparation.

Jelinek, J., Keible, K., Dexter, T.M., and Margison, G. P. 1988. Transfection of murine multipotent haemopoietic stem cells with an E. coli DNA alkyltransferase gene confers resistance to the toxic effects of alkylating agents. Carcinogenesis, 9: 81.

Kataoka, M., Mali, J., and Karran, P. 1986. Complementation of sensitivity to alkylating agents in E. coli and Chinese hamster ovary cells by expression of a cloned bacterial repair gene. EMBO Journal, 5:3195.

Margison, G. P., Butler, J., and Hoey, B. 1985. $\underline{O}^6$-Methylguanine methyltransferase activity is increased in rat tissues by ionizing radiation. Carcinogenesis, 6:1699.

Nicoll, J.W., Swann, P.F., and Pegg, A.E. 1975. Effect of dimethylnitrosamine on persistence of methylated guanines in rat liver and kidney DNA. Nature, 254:261.

O'Connor, P. J., Fida, S., Fan, C.Y., Bromley, M. and Saffhill, R. 1988. Phenobarbital: a non-genotoxic agent which induces the repair of $\underline{O}^6$-methylguanine from hepatic DNA. Carcinogenesis, 9:2033.

O'Connor ,P.J., Saffhill, R., and Margison, G.P. 1979. N-Nitroso compounds: biochemical mechanisms of action, In: Environmental Carcinogenesis. p.73. P. Emmelot and E. Kriek, eds., Elsevier/North Holland Biomedical Press, Amsterdam.

Saffhill, R., Badawi, A.F., and Hall, C.N. 1988a. Detection of $\underline{O}^6$-methylguanine in human DNA, In: Methods for Detecting DNA Damaging Agents in Humans: Applications for Cancer Epidemiology and Prevention, p.301. H. Bartsch, K. Hemminki, and I.K. O'Neill, eds., IARC Sci. Pub. No. 89, Lyon.

Saffhill, R., Fida, S., Bromley, M., and O'Connor, P.J. 1988b. Promutagenic alkyl lesions are induced in the tissue DNA of animals treated with Isoniazid, Human Toxicol. 7: 311.

Saffhill, R., Margison, G. P., and O'Connor, P. J. 1985. Mechanism of carcinogenesis induced by alkylating agents. Biochim. Biophys. Acta, 823:111.

Schmerold, I. and Wiestler, O.D. 1986. Induction of rat liver $\underline{O}^6$-alkylguanine-DNA alkyltransferase following whole body X-irradiation. Cancer Res. 46:245.

Scbwartz, M., Peres, P., Buchmaun, A., Friedberg, T., Waxman, D. J., and Kunz, W. 1987. Phenobarbital induction of cytochrome p450 in normal and preneoplastic liver: comparison of enzyme and mRNA expression as detected by immunohistochemistry and in situ hybridization. Carcinogenesis, 8:1355.

Swann, P. F., Kaufman, D. G., Magee, P.N., and Mace, R. 1980. Induction of kidney tumors by a single dose of dimethylnitrosamine: Dose reponse and influence of diet and benzo(a)pyrene treatment. Br. J. Cancer, 41:285.

Umbenhauer, D., Wild, C. P., Montesano, R., Saffhill, R., Boyle, J.M., Huh, N., Kirstein, N., Thomale, J., Rajewski, M.F., and Lu, S.H., 1985. $\underline{O}^6$-methyldeoxyguanosine in oesophageal DNA among individuals at high risk of oesophageal cancer. Int. J. Cancer, 36:661.

White, G.R.M., Ockey, C. H., Brennand, J., and Margison, G. P. 1986. Chinese Hamster cells harbouring the Escherichia coli $\underline{O}^6$-alkylguanine alkyltransferase gene are less susceptible to sister chromatid exchange induction and chromosome damage by methylating agents. Carcinogenesis, 7:2077.

Wild, C.P., Smart, G., Saffhill, R., and Boyle, J.M. 1983. Radioimunoassay of $\underline{O}^6$-methyldeoxyguanosine in DNA of cells alkylated in vitro and in vivo. Carcinogenesis, 4: 1605.

ACQUISITION OF RESISTANCE TO ALKYLATING AGENTS BY EXPRESSION OF METHYLTRANSFERASE GENE IN REPAIR-DEFICIENT HUMAN CELLS

Mutsuo Sekiguchi[1], Hiroshi Hayakawa[1], Ken-ichi Kodama[1], Kanji Ishizaki[2] and Mituo Ikenaga[2]

[1]Department of Biochemistry, Faculty of Medicine, Kyushu University, Fukuoka 812, and [2]Radiation Biology Center, Kyoto University, Kyoto 606, Japan

INTRODUCTION

Alkylating agents are potent mutagens and carcinogens and sometimes cause cell death. These effects of alkylating agents are mainly attributed to the formation of various alkylated bases in DNA. More than ten kinds of methylated bases are produced by the treatment of cells with a simple methylating agent, such as N-methyl-N-nitrosourea (MNU) and N-methyl-N'-nitro-N-nitrosoguanidine (MNNG).

To counteract such effects, many biological systems possess elaborate mechanisms to repair alkylated lesions in DNA. There are a variety of enzymes that recognize and repair such lesions, and DNA methyltransferase is one of the most notable. It transfers methyl groups from O^6-methylguanine and/or other methylated moieties of the DNA to its own molecule, thereby repairing DNA lesions by a single step reaction. The enzyme is found in many organisms, from bacteria to human cells.

Some human tumor cell lines show the <u>Mer</u>$^-$ (<u>Mex</u>$^-$) phenotype characterized by defects in repair of O^6-methylguanine (Yarosh, <u>et al.</u>, 1983). Since the <u>Mer</u>$^-$ cells are deficient in O^6-methylguanine-DNA methyltransferase activity, we examined whether introduction of the cloned bacterial gene encoding methyltransferase would make <u>Mer</u>$^-$ cells resistant to alkylating agents. Here we describe the result of these experiments and approaches which can be used for cloning of the human repair genes.

METHYLTRANSFERASE GENES OF <u>E. COLI</u>

The <u>ada</u> gene of <u>Escherichia coli</u> encodes a 39,000-dalton methyltransferase. The gene has been cloned from <u>E. coli</u> strains K12 and B, and the entire nucleotide sequence of the <u>ada</u> coding and control regions has been determined (Demple, <u>et</u>

DNA Repair Mechanisms and Their Biological Implications in Mammalian Cells
Edited by M.W. Lambert and J. Laval
Plenum Press, New York

al., 1985;Nakabeppu, et al., 1985). The enzyme was overproduced and purified to physical homogeneity (Nakabeppu, et al., 1985).

The Ada protein catalyzes transfer of the methyl groups from alkylated DNA to its own protein molecule. Analysis of the purified Ada protein revealed that it carries two distinct methyltransferase activities, one to transfer a methyl group from methylphosphotriester and the other to transfer a methyl group from O^6-methylguanine. These two activities reside on the N-terminal and the C-terminal halves of the protein, and can be separated by proteolytic cleavage (Yoshikai, et al., 1988). A specific cysteine residue (Cys^{69}) close to the amino-terminus of the Ada protein could accept the methyl group from one of the two stereoisomers of methylphosphotriesters in alkylated DNA while a cysteine residue (Cys^{321}) which is present near the carboxyterminus could be an acceptor site from O^6-methylguanine (Teo, et al.,1986). Takano, et al. (1988), recently constructed mutant forms of the ada gene in which each one of the codons for cysteine is replaced by that for alanine. The results obtained with such mutant genes support the notion that Cys^{69} and Cys^{321} are the methyl acceptor sites from methylphosphotriester and O^6-methylguanine, respectively, and further revealed that Cys^{321} accepts a methyl group also from a minor methylated base, O^4-methylthymine.

Evidence for the presence of a second methyltransferase enzyme in E. coli was recently reported (Potter, et al., 1987). This gene, named ogt, codes for a protein with a molecular weight of 19,000, which resembles the C-terminal half of Ada protein. The enzyme repairs O^6-methylguanine in the DNA.

An amount of the Ada protein increases when cells are exposed to a low level of alkylating agents whereas an amount of the Ogt protein is unchanged by such treatments. On induction of the ada gene expression, genes such as alkA, alkB, and aidB which are involved in repair of alkylation damage are also induced. Since no such induction occurs in ada^- cells, it was suggested that the Ada protein, which itself is a methyltransferase, acts as a positive regulator both for its own synthesis and for the expression of other genes belonging to the ada regulon. Support for this idea was obtained by findings in in vitro transcription experiments (Teo, et al., 1986; Nakabeppu and Sekiguchi, 1986).

EXPRESSION OF THE ADA GENE IN HUMAN CELLS

Yarosh, et al. (1983), reported that about a fifth of human tumor cell lines show the Mer^- (Mex^-) phenotype, which is characterized by the inability to support the growth of MNNG-treated adenovirus and also by defects in the repair of O^6-methylguanine produced in cellular DNA. The Mer^- cells are deficient in methyltransferase activity and are much more sensitive than repair-proficient Mer^+ cells to killing by alkylating agents. Thus, experiments were performed to observe whether introduction of the cloned ada gene would make Mer^- cells resistant to alkylating agents.

A DNA fragment carrying the ada gene was inserted into pSV2neo vector. The ada gene was flanked by an SV40 promoter sequence and a poly(A) site on the 5' and 3' ends, respectively

(Ishizaki, et al., 1986). The resulting plasmid, pSV2ada-neo, was introduced into Mer^- HeLa MR cells with the aid of calcium phosphate precipitation, and neomycin-resistant transformants were selected. After cultivation for 10 days, the cells were treated with 50 μg/ml of 1-(4-amino-2-methyl-5-pyrimidyl)-methyl-3-(2-chloroethyl)-3-nitrosourea hydrochloride (ACNU) and resistant cells were obtained. ACNU was used for selection of Mer^+ cells because fewer spontaneous resistant cells appeared on exposure to ACNU than to MNNG.

All of the 8 ACNU-resistant clones examined showed significant methyltransferase activity, of varying degrees. The results of Southern hybridization of high-molecular-weight genomic DNA and RNA dot blot hybridization indicated that the ada gene is integrated into the chromosomal DNA and effectively transcribed. The methyltransferase activities in these clones roughly correlated with copy numbers of integrated ada gene and the amount of transcript (Clone 5-1 was estimated to carry 30-40 ada genes per genome).

The E. coli Ada protein has two functional domains; the N-terminal domain accepts a methyl group from methylphosphotriester while the C-terminal domain is responsible for repair of O^6-methylguanine and a minor methylated base, O^4-methylthymine. To clarify the biological role of O^6-methylguanine in human cells, we constructed a truncated ada gene that produces only the C-terminal half of the protein and introduced it into the Mer^- cells (Ishizaki, et al., 1987). pSVada5'd-neo carried a truncated ada gene which had lost the 5' region coding for the N-terminal domain and when it was introduced into HeLa Mer^- cells, the isolated ACNU-resistant cells showed significantly high levels of O^6-methylguanine-DNA methyltransferase activity. When the activity was assayed with [^{3}H]MNU-treated poly(dA-dT)·poly(dA-dT) as the substrate, no activity was detected, thereby confirming that it is defective in methylphosphotriester-DNA methyltransferase activity.

SDS-polyacrylamide gel electrophoresis (Figure 1) revealed that the molecular weight of methyltransferase in clone 5-1 with the intact ada gene is exactly the same as that of E. coli Ada protein (about 39 kDa). This value is completely different from that of wild type HeLa cells, which is around 25 kDa. The methyltransferase in clone 5'dD, which carries the truncated ada gene, was found to contain a peak at 28 kDA. The peak around 20 kDa observed in both clone 5-1 and 5'dD represents proteolytic cleavage products of the methyltransferase proteins.

Figure 2 shows survival curves of clone 5-1 and 5'dD after treatment with MNNG, compared with those of HeLa S3 (Mer^+) and HeLa MR (Mer^-) cells. Clearly, both clones were much more resistant to MNNG than were the recipient HeLa MR cells. Similar results were obtained with Mer^- cells of humans and Chinese hamsters to which the cloned ada gene was introduced (Samson, et al., 1986; Brennand and Margison, 1986; Kataoka, et al., 1986). Thus, bacterial methyltransferase seems to be able to repair alkylated bases, in mammalian cells.

THE ADA GENE ON THE AUTONOMOUSLY REPLICATING VECTOR

pSV2 cannot replicate in human cells, thus for maintenance

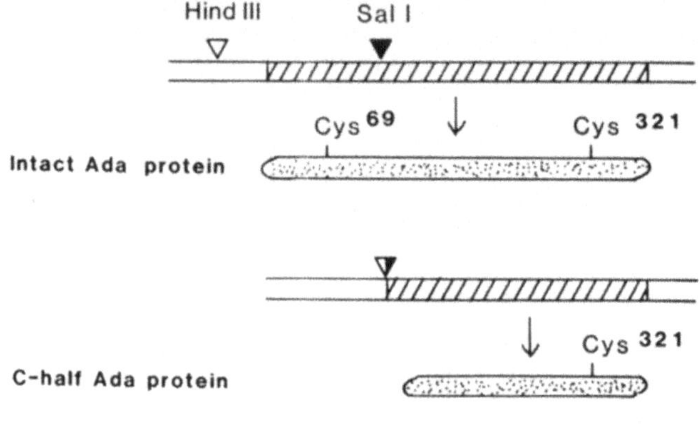

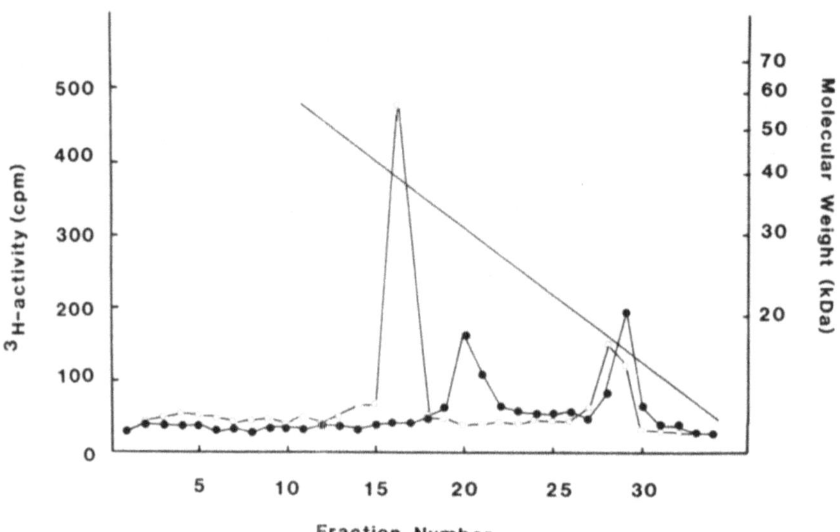

Figure 1. Expression of the intact and the truncated ada gene in HeLa MR cells. Extracts of HeLa MR cells carrying the intact ada gene (clone 5-1) and those carrying the 5'-half of the ada gene (clone 5'dD) were incubated with [^{3}H]MNU-treated calf thymus DNA. The distribution of [^{3}H] radioactivity after SDS polyacrylamide gel electrophoresis is shown. The straight line indicates the molecular weight for each fraction. - ● -, clone 5'dD; - O -, clone 5-1.

it must be integrated into the chromosome. To overcome this difficulty, we developed the p500 vector that carries EBNA-1 and oriP, derived from the EB virus, together with the hygromycin B resistance maker gene. This can replicate autonomously in EB-transformed lymphoblasts as well as in fibroblast cell lines such as HeLa (H. Hayakawa, et al., unpublished result).

To examine the feasibility of this vector system, we placed the cloned ada gene downstream from the RSVVLTR promoter on the vector. Two EB-transformed lymphoblast cell lines were used; FA99 is an HSC99 line with the normal level of O^6-methylguanine-DNA methyltransferase activity (Duckworth-Rysiecki, et al., 1983) and TK6 is a methyltransferase-deficient cell line (K. Tatsui, personal communication). TK6 cells were transfected by p500 or $p500^{\Delta ada}$ carrying the 5'-truncated ada. Cells were cultivated for five days in medium containing various concentrations of MNNG and the number of cells was counted using an electronic cell counter. The result shown in Figure 3 indicates that TK6 cells with $p500^{\Delta ada}$ are significantly more resistant to MNNG, as compared to TK6 cells carrying the vector.

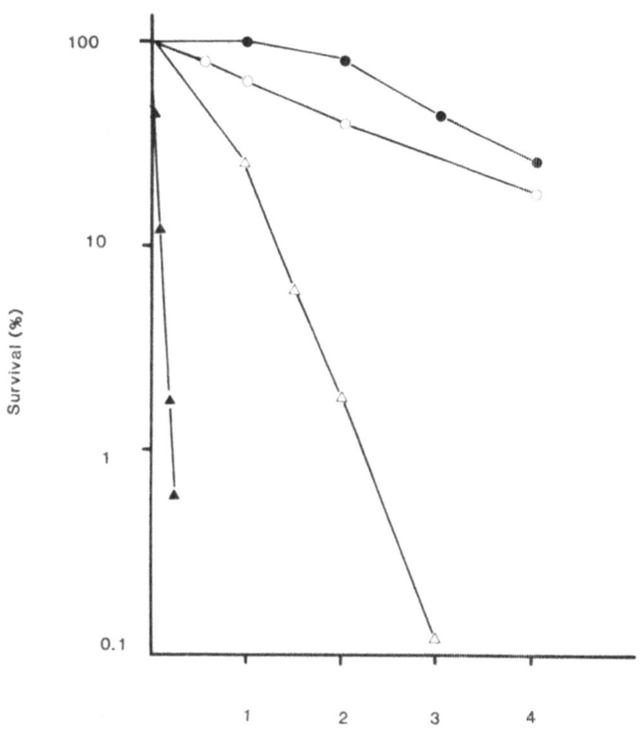

Figure 2. Survivals of HeLa MR, HeLa S3, clone 5-1 and clone 5'dD cells after treatment with MNNG for 1 hour at 37^0C. - ● -, clone 5'dD; - O -, clone 5-1; -Δ -, HeLa S3; -▲-, HeLa MR.

We also examined the ACNU sensitivity of methyltransferase-deficient HeLa MR cell, with or without $p500^{\Delta ada}$. We found that the cloned ada gene on the plasmid vector is effective for increasing survival rates of Mer^- cells.

Thus, the p500 vector may be useful for cloning the methyltransferase gene from mammalian cells. We inserted cDNA of human cells in the cloning site of p500 vector and introduced the resulting cDNA library clones into HeLa MR cells.

After successive selections with ACNU and hygromycin B, we obtained cells resistant to ACNU as well as to MNNG. Some of the lines possessed significantly high levels of methyltransferase activity (H. Hayakawa, et al., unpublished result).

CLONING OF FOREIGN GENES IN E. COLI

It is assumed that E. coli ada⁻ mutant cells deficient in O^6-methylguanine-DNA methyltransferase activity may become resistant to alkylating agents, when foreign DNAs carrying coding sequences for methyltransferase are introduced and properly expressed. Once this is established, cloning of DNA repair genes of various organisms will be facilitated.

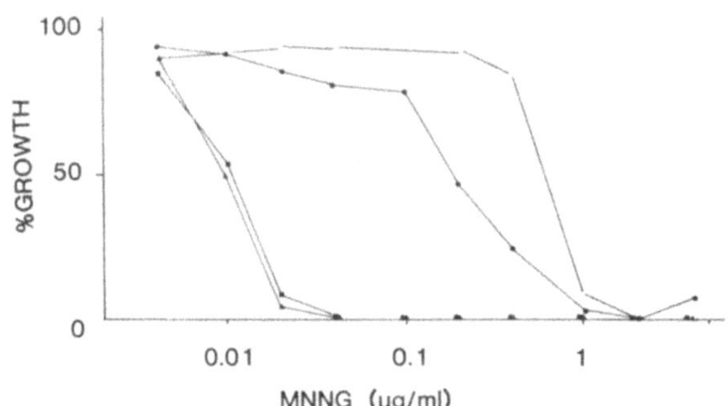

Figure 3. Growth inhibition following treatment of cell lines with MNNG. Lymphoblasts were harvested by centrifugation, resuspended in serum-free RPMI 1640 and seeded in 24 well plates (2-5 x 10^5 cells/0.5 ml/well). 0.5 ml of serum-free RPMI 1640 containing various concentrations of MNNG was added and the suspension was incubated at 37^0C for 1 hour. After the treatment, 0.2 ml of FCS was added and the culture was incubated for five days. - O -, FA99; - ● -, TK6/p500$^{\Delta ada}$; - ▲ -,TK6/p500; - ■ -, TK6.

Bacillus subtilis possesses two molecular species of O^6-methylguanine-DNA methyltransferase, one being constitutive and the other inducible (Morohoshi and Munakata, 1987). We used this organism as the source of DNA. DNA of B. subtilis strain 168T was partially digested with a mixture of three restriction enzymes, AluI, DpnI and RsaI, each recognizing different tetranucleotide sequences, and the resulting DNA fragments were inserted into a cloning vector as to place downstream from the lac promoter. The recombinant plasmids were introduced to E. coli ada$^-$ cells and MNNG-resistant transformants were isolated. Among 3 x 10^4 transformants examined, 10 clones showed a distinct methyltransferase activity (K. Kodama, et al., 1989).

In this manner, we cloned the B. subtilis gene, dat, that encodes a constitutive O^6-methylguanine-DNA methyltransferase. Based on the nucleotide sequence of the gene, it was deduced that the protein comprises 165 amino acids and that the molecular weight is 18,779. The presumptive amino acid sequence of Dat protein is homologous to the sequence of E. coli Ogt

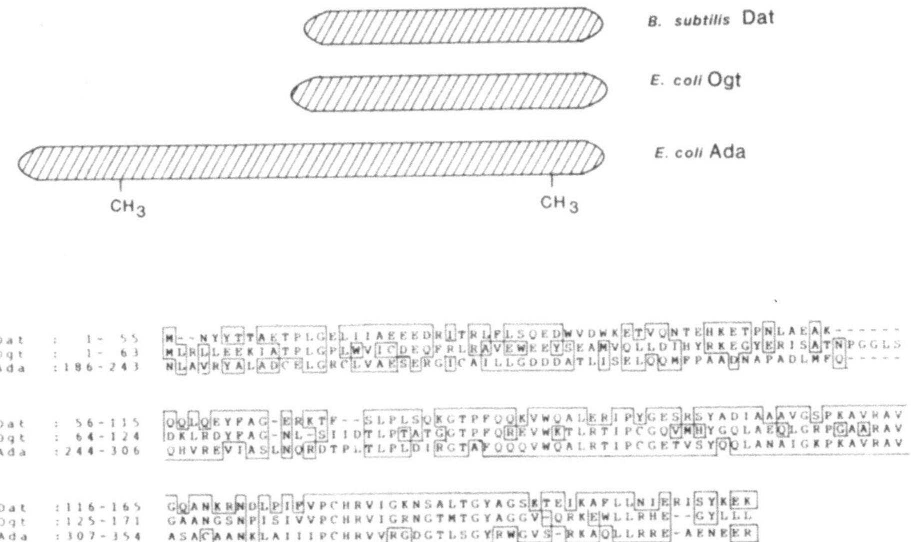

Figure 4. Comparison of the predicted amino acid sequences of the B. subtilis Dat protein, the E. coli Ogt protein and the C-terminal half of the E. coli Ada protein. The homologous regions are boxed. The numbers indicate the position from the N-terminal of each protein. ▲ denotes the methyl acceptor site of Ada protein.

protein and the C-terminal half of Ada protein, both of which carry O^6-methylguanine-DNA methyltransferase activity. The pentaamino acid sequence, Pro-Cys-His-Arg-Val, the cysteine residue of which is the methyl acceptor site in Ada protein, were conserved in the three methyltransferase proteins (Figure 4).

In principle, this method is applicable for cloning of cDNA for methyltransferase from higher organisms. The sequence homology evidenced in the present experiments provides a useful tool for screening the gene or cDNA for methyltransferase.

ACKNOWLEDGEMENTS

We extend special thanks to Dr. F. Morohoshi for pertinent advice and to M. Ohara for comments. We also thank Drs. Y. Nakabeppu, C. Fujio, T. Tsujimura, H. Yawata, Y. Zhang and G. Koike, who participated in some of the experiments.

REFERENCES

Brennand, J. and Margison, G.P., 1986. Reduction of the toxicity and mutagenecity of alkylating agents in mammalian cells harboring the Escherichia coli alkyltransferase gene. Proc. Natl. Acad. Sci. USA, 83:6292-6296.

Demple, B., Sedgwick, B., Robins, P., Totty, N., Waterfield, M.D. and Lindahl, T., 1985. Active site and complete sequence of the suicidal methyltransferase that counters alkylation mutagenesis. Proc. Natl. Acad. Sci. USA, 82: 2688-2692.

Duckworth-Rysiecki, G., Cornish, K., Clarke, C.A. and Buchwald, M., 1983. Identification of two complementation groups in Fanconi anemia. Somat. Cell Mol. Genet., 11:35-41.

Ishizaki, K., Tsujimura, T., Fujio, C., Yangpei, Z., Yawata, H., Nakabeppu, Y., Sekiguchi, M. and Ikenaga, M., 1987. Expression of the truncated E. coli O^6-methylguanine methyltransferase gene in repair-deficient human cells and restoration of cellular resistance to alkylating agents. Mutation Res., 184:121-128.

Ishizaki, K., Tsujimura, T., Yawata, H., Fujio, C., Nakabeppu, Y., Sekiguchi, M. and Ikenaga, M., 1986. Transfer of the E. coli O^6-methylguanine methyltransferase gene into repair-deficient human cells and restoration of cellular resistance to N-methyl-N'nitro-N-nitrosoguanidine. Mutation Res., 166:135-141.

Kataoka, H., Hall, J. and Karren, P., 1986. Complementation of sensitivity to alkylating agents in Escherichia coli and Chinese hamster ovary cells by expression of a cloned bacterial DNA repair gene. EMBO J., 5:3195-3200.

Kodama, K., Nakabeppu, Y. and Sekigucji, M., 1989 Cloning and expression of the Bacillus subtilis methyltransferase gene in Escherichia coli ada^- cells, Mutation Res., 218:153-163.

Morohoshi, F. and Munakata, N., 1987. Multiple species of Bacillus subtilis DNA alkyltransferase involved in the adaptive response to simple alkylating agents. J. Bacteriol., 169:587-592.

Nakabeppu, Y., Kondo, H., Kawabata, S., Iwanaga, S. and Sekiguchi, M., 1985. Purification and structure of the intact Ada regultaory protein of Escherichia coli K12, O^6-methylguanine-DNA methyltransferase. J. Biol. Chem., 260:7281-7288.

Nakabeppu, Y. and Sekiguchi, M., 1986. Regulatory mechanisms for induction of synthesis of repair enzymes in response to alkylating agents: Ada protein acts as a transcriptional regulator. Proc. Natl. Acad. Sci. USA, 83:6297-6301.

Potter, P.M., Wilkinson, M.C., Fitton, J., Carr, F.J. Brennand, J., Cooper, D.P. and Margison, G.P., 1987. Characterization and nucleotide sequence of ogt, the O^6-alkylguanine-DNA-alkyltransferase gene of E. coli. Nucleic Acids Res., 15:9177-9193.

Samson L., Derfler, B. and Waldstein, E.A., 1986. Suppression of human DNA alkylation-repair defects by Escherchia coli DNA repair gene. Proc. Natl. Acad. Sci. USA, 83:5607-5610.

Takano, K., Nakabeppu, Y. and Sekiguchi, M., 1988. Functional sites of the Ada regulatory protein of Escherichia coli: Analysis by amino acid substitutions. J. Mol. Biol., 201: 261-271.

Teo, I., Sedgwick, B., Kilpatrick, M.W., McCarthy, T.V. and Lindahl, T., 1986. The intracellular signal for induction of resistance to alkylating agents in E. coli. Cell, 45: 315-324.

Yarosh, D.B., Foote, R.S., Mitra, S. and Day, R.S. III, 1983. Repair of O^6-methylguanine in DNA by demethylation is lacking in Mer^- human tumor cell strains. Carcinogenesis 4:199-205.

Yoshikai, T., Nakabeppu, Y. and Sekiguchi, M., 1988. Proteolytic cleavage of Ada protein that carries methyltransferase and transcriptional regulator activities. J. Biol. Chem., 263:19174-19180..

THE MER MINUS PHENOTYPE, PATIENT RESPONSE TO NITROSOUREAS, AND PROTOONCOGENE ACTIVATION IN HUMAN GLIOBLASTOMAS

Rufus S. Day, III[1], Junji Miyakoshi[1], Kelly Dobler[1], Joan Allalunis-Turner[2], John D.S. McKean[3], Kenneth Petruk[3], Peter B.R. Allen[3], Keith N. Aronyk[3], Bryce Weir[3], Debbie Huyser-Wierenga[4], Dorcas Fulton[4], and Raul C. Urtasun[4]

[1]Molecular Genetics and Carcinogenesis Laboratory, Department of Medicine; [2]Radiobiology Laboratory, Department of Radiation Oncology, and [4]Department of Radiation Oncology, Cross Cancer Institute, 11560 University Avenue, Edmonton, Alberta T6G 1Z2, Canada

[3]Division of Neurosurgery, Faculty of Medicine, University of Alberta, Edmonton, Alberta T6G 2G3, Canada

INTRODUCTION

It is about 10 years since the identification of the first Mer$^-$ line, A172 (Day and Ziolkowski, 1979), a human astrocytoma line produced by Giard, et al.(1973). To review, Mer$^-$ lines are human cell lines defined by their relative inability to support the growth of adenovirus 5 that has been treated with N-methyl-N -nitro-N-nitrosoguanidine (MNNG) prior to infection of cell monolayers (Day, et al., 1980a,b). Such lines lack, without exception, the ability to repair m6Gua produced in their DNA by certain methylating agents (Day, et al., 1980a; Day, et al., 1984), and are thus Mex$^-$ by the definition of Sklar and Strauss (1981). Mer$^-$ strains produced from human tumors are highly sensitive with respect to Mer$^+$ cells as assessed by several endpoints to agents that react with the O^6 of guanine: sister chromatid exchange (Day, et al., 1980a), mutation induction (Baker, et al., 1979, 1980, Domoradski, et al., 1984), cell killing by MNNG or nitroso-ureas (Day, et al., 1980a,b; Erickson, 1980a,b; Scudiero, et al., 1984 a,b; Gibson, et al., 1986). The results obtained in cell culture are clear-cut; for example, with MNNG as the damaging agent, the inactivation slopes of survival curves of Mer$^-$ cells are up to 50 fold steeper than are those of Mer$^+$ cells (Scudiero, et al., 1984a). Several groups have inserted parts or all of the *E. coli* *ada* gene into Mex$^-$ cells and have provided evidence that such differential sensitivity is likely due to differential repair of m6Gua (Brennand and Margison,

DNA Repair Mechanisms and Their Biological Implications in Mammalian Cells
Edited by M.W. Lambert and J. Laval
Plenum Press, New York

1986, Ishizaki, et al., 1986; Kataoka, et al., 1986; Samson, et al., 1986; Fox, et al., 1987), a point which has been discussed previously, and for which there is substantial evidence (Day, et al., 1987). No matter how persuading the cell culture evidence, there is little evidence to demonstrate that Mer$^-$ cells occur in tumors; i.e., that some fraction of human tumors is composed of Mer$^-$ cells.

We are attempting to demonstrate: 1) whether tumors are composed of Mer$^-$ cells, and 2) an involvement of activation of proto-oncogenes or inactivation of tumor repressor genes in converting cells from Mer$^+$ to Mer$^-$. This paper reviews our work in these areas.

HUMAN MALIGNANT GLIOMAS AS A SUBJECT FOR RESEARCH

Human Brain Tumors. The term malignant glioma refers to astrocytomas with anaplastic foci (AAF; were termed astrocytoma, grade III) and glioblastomas (were termed glioblastoma multiforme or astrocytoma grade IV), but not to astrocytoma grades I or II or to oligodendrogliomas. We are approaching human malignant gliomas in many aspects of our research because:

1. They are a good source of Mer$^-$ cell lines. To date 12 of 37 lines produced from malignant gliomas are Mer$^-$ (Day and Ziolkowski, 1979; Day, et al., 1979; Day, et al., 1980a,b; Scudiero, et al., 1984a; Sariban, et al., 1987; Day, unpublished results). Transformed cell lines can be produced from 20-30% of the glioblastoma biopsies procured (Ponten, 1975; and this paper). Ponten (1975) pointed out that astrocytoma grades I and II, like normal brain tissue, never give rise to permanent lines. Only malignant gliomas give rise to permanent lines. Therefore, a permanent line must have arisen from tumor tissue. Of course, tumor tissue that does not produce a permanent line is not necessarily normal. The work of James, et al.(1988), who found loss of heterozygosity for loci on chromosome 10 in 28 of 29 glioblastoma biopsies, shows that their biopsy material is largely tumor material.

2. Brain tumors are difficult to treat successfully (see, for example Wilson, 1976), so that research inroads may more likely be clinically useful than for other tumors.

3. Brain tumors are one of the few human tumors which are treated with single agent chemotherapy rather than the multi-agent regimens proscribed for many other tumors. The single agent is usually a nitrosourea, to which Mer$^-$ cells are sensitive in vitro. Initial post-diagnostic treatment at the CCI entails surgery followed by radiation therapy. CCNU is given upon relapse, and thus treatment with CCNU is isolated in time from other treatments, and the response of the tumor to CCNU may be followed by CT scan with few complications.

4. There is a growing literature on the genomic changes in human glioblastomas as detected at 1) the cytological level in short and long term cell cultures of astrocytoma/ glioblastoma biopsies (see for example, Bigner, et al., 1988); the level of proto-oncogene activation as observed in biopsies and longer term cultures (Kinzler, et al., 1987; Wong, et al.,

1987); and the level of loss of heterozygosity (James, et al., 1988).

Specific questions that we are approaching are:

1. Does patient response (or resistance) to CCNU [3-cyclohexyl-1-(2- chloroethyl)-1-nitrosourea, a nitrosourea to which Mer$^-$ lines are sensitive, Erickson, et al., 1980a] correlate with the Mer phenotype of the cell line developed from the surgically removed biopsy? Do short term (3 - 5 day) tests, in which cells from disaggregated biopsies are treated with either radiation or BCNU and are assayed for survival, reflect the behavior of the line produced?

2. Do plasmids containing activated oncogenes convert Mer$^+$ cells to Mer$^-$?

3. Are there patterns of activation of proto-oncogenes or inactivation of tumor suppressor genes that suggest their involvement in producing the Mer$^-$ phenotype?

4. Are there nucleic acid sequences from Mer$^-$ brain tumor cells that are not contained in Mer$^+$ cells or vice versa?

PATIENT RESPONSE AND THE MER MINUS PHENOTYPE

In the two years of this collaborative effort, 37 biopsies have been obtained from brain tumor patients seen both at the Cross Cancer Institute/University of Alberta Hospital system and at the Royal Alexandra Hospital, both in Edmonton (Table 1). These were cultured by three methods:

1) Biopsies were washed with PBS, cut with crossed scalpels into fragments of volume less than 1 mm^3, which were placed in 60 mm culture dishes. To ensure attachment to the dish, coverslips with stopcock grease (on diagonally opposing corners) were pushed gently on top of the fragments. The cultures were supplemented with either F12, Dulbecco's Modified Eagle Medium (DMEM), or Biorich medium [all with 10% fetal calf serum (FCS)].

2) Washed biopsies were shaken in a 50 ml tube with 20 ml 0.5% trypsin plus 33 units/ml collagenase in PBS at 37 degrees. The small clumps and single cells were harvested twice, after 20 minutes and 40 minutes of treatment, washed 2x with PBS and plated in 60 mm dishes with medium as above.

3) This is the method of Freeman and Hoffman (1986). Spongostan was cut into 1 cm x 1 cm x 0.5 cm pieces, one which was placed in each 35 mm dish with one of its 1 cm x 1 cm sides contacting the plate surface. Medium (as above) was added such that 1 mm of the Spongostan was not submerged (but medium was maintained in this portion by capillary action), and 1 to 4 biopsy fragments were placed on the protruding Spongostan. In all methods, plates were kept at 35-37 degrees at 85% humidity, and received medium changes weekly. We have prepared 8 transformed lines from 30 biopsies, 7 by method 1 and 1 by method 2 which was introduced at biopsy 16.

Table 1. Patients of the brain tumor study.

#	Age Sex	Diagnosis	Mer*	Response to Therapy (through 1 Aug 88)
1	13F	Oligodendroglioma		No CCNU (RT only; no recurrence)
2	58M	Glioblastoma	+	No CCNU (died during treatment with RT)
3	55M	Oligodendroglioma		No CCNU (RT only; no recurrence)
4	51M	AAF		2 courses CCNU; adverse change on CT after 1st course.
5	30F	Grade I-II Astrocytoma		No CCNU (RT only; no recurrence)
6	67M	Glioblastoma	-	No RT; 4 courses CCNU; resp. to 1st 2 cycles; died 28 Feb 1988.
7	47M	Glioblastoma	+	CCNU
8	41M	Benign Meningioma		No CCNU
9	58F	Meningioma		No CCNU
10	49M	Glioblastoma	-	No CCNU (RT only; no recurrence)
11	66M	Glioblastoma	+	No CCNU
12	53M	Grade III Astrocytoma		No CCNU (recurrence in March 88; pt. refused CCNU)
13	21M	Recurrent Astrocytoma		4 Cycles CCNU; enhancing area unchanged or improved on CT; died July 88.
14	36M	Glioblastoma		CCNU - 1 course; no response; repeat resection and RT in Fall 87
15	41M	AAF		No CCNU.
16	63M	Glioblastoma	+	No CCNU (RT only; pt. moved to Oregon April 88; no recurrence)
17	68M	Glioblastoma		No CCNU (died during RT)
18	54M	Glioblastoma		RT + IUdR (no recurrence)
19	17M	Ganglioglioma		No CCNU
20	46M	Grade II Astrocytoma		No CCNU (RT only; no recurrence)
21	40M	Glioblastoma		RT + IUdR (no recurrence)
22	24M	Germ Cell Tumor		No CCNU (RT only; no recurrence)
23	58M	Glioblastoma		CCNU (1st course given July 88)
24	29M	AAF		No CCNU (RT only; no recurrence)
25	51M	Metastatic Adenoma		No CCNU
26	35M	Grade II Astrocytoma		No CCNU (RT only; no recurrence)
27	67M	Glioblastoma	-	No CCNU (? recurrence; may start CCNU)
28	70M	Glioblastoma		No CCNU (refused treatment of any kind)
29	51M	Glioblastoma		No CCNU (RT only; no recurrence)
30	35M	AAF		No CCNU (RT + IUdR only; no recurrence)
31	61F	Glioblastoma		No CCNU (no treatment with RT or CCNU poor prognostic factors + pt. choice)
32	47M	Glioblastoma		No CCNU (finished RT 1 month ago)
33	37M	Grade II Astrocytoma		No CCNU (currently on RT)
34	36F	Glioblastoma		No CCNU (currently on RT)
35	25M	Grade II - AAF		No CCNU (currently on RT)
36	30M	Astro/glial hyperp.		No CCNU
37	17M	Grade II Astrocytoma		No CCNU (currently on RT)

*Mer phenotypes reported for transformed cell lines only

Note: tumors other than malignant gliomas were included during preliminary work.

Using method 1, two patterns of outgrowth of a transformed line were observed. Biopsies from patients 2, 6, 16, and 27 grew directly into transformed lines on all plates (12 to 16 plates prepared per biopsy) with little or no apparent growth of non-transformed cells. Biopsies from patients 7, 10, 11, and 12 grew into strains on 1 or 2 of the 12 to 16 original plates, and appeared after the outgrowth of non-transformed cells. No other biopsy has yet produced a transformed line. Method 2, introduced on receipt of the biopsy from patient 16, was successful in producing a transformed line from patient 16, but from no other biopsy. Method 3 was used for biopsies 1-4, and is being reintroduced as a replacement for method 2. Biopsy material from several grade IV astrocytoma (glioblastoma) patients has been in culture for more than 1 year, and has not yet produced transformed lines. Strains produced from patients 2, 6, etc., were given the names M002, M006, etc.

Of the eight transformed lines, seven are from grade IV astrocytomas (4 Mer^+: M007, MO11, and M016; 3 Mer^-: M006, MO1O, and M027), and one from a grade III astrocytoma (Mer^+).

As yet limited patient response data is available. Six patients have received CCNU: patients 4, 6, 7, 13, 14, and 23. No response data is available for patients 7 or 23. Patient 14 was resistant as judged by CT scan. Patients 4, 6, and 13 were sensitive, showing tumor regression or no progression. Patient 6 was unique in that radiation treatment was refused; the patient received oral CCNU only. In patient 6 the response to CCNU correlated positively with the Mer^- phenotype of the cell culture derived from the tumor.

We have compared the MTT assay (Mosmann, 1983) as a short term test of cytotoxicity with the clonogenic assay used previously (Scudiero, et al., 1984b). Our data indicate that cellular sensitivity to BCNU of a disaggregated biopsy, as assessed by the MTT assay, is a good indicator of BCNU sensitivity of the line produced from that biopsy, as assessed by a clonogenic assay.

PLASMIDS WITH ACTIVATED PROTOONCOGENES HAVE NOT YET PRODUCED THE MER^- PHENOTYPE

We have prepared three plasmids bearing activated oncogenes.

1) The src gene was selected because it appeared that Rous sarcoma virus converted a human osteosarcoma cell line, HOS, a clone of TE85, to a Mer^- strain RHOS (Yarosh, et al., 1984b). The following experimentation shows why we now suspect that this finding is erroneous. We obtained the plasmid pMS484c (Jakobovits, et al., 1984) as a kind gift from H. Varmus, and cloned its 2.8 kb Bam HI fragment containing both v-src-RSV and the viral 3' LTR into pRSV deltalinker Neo (our construct) to obtain pRSVsrcNeoI (Figure 1). This construct contains both an RSV-LTR driven v-src gene and an RSV-LTR driven neo gene. pRSVsrcNeo was transfected into HOS cells by polybrene mediated transfection (Aubin, et al.,1988), and colonies stable to selection by 200 ug/ml G418 in F12 medium

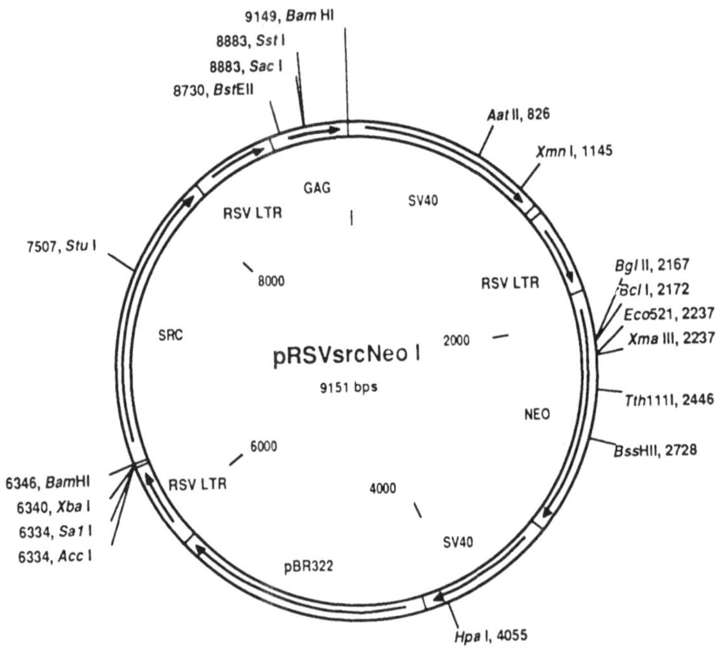

Figure 1. Plasmid pRSVsrcNeo I

were selected, cloned, grown into cell strains, and tested for their Mer phenotype. In summary, 4 of 78 clones prepared in this manner were Mer$^-$. Mer$^-$ clones were found only in uncloned HOS cells; no Mer$^-$ clones were observed when HOS clone 8 (selected as a clone with flat cells) was used as a host for transfections. Furthermore, when selected transfected clones were grown in the absence of G418 and tested for expression of their <u>src</u> gene by Northern and dot blotting, no correlation between expression of the <u>src</u> gene and Mer phenotype was observed. In Figure 2, HOS CW3 and HOS AC12 are Mer$^-$; the remaining strains are Mer$^+$. Only HOS D, HOS CA, HOS CS, and HOS CV show expression of the <u>src</u> gene, detected by a <u>src</u> gene probe. Control blots using the gamma actin gene as a probe (Figure 3) show roughly equal expression of this gene. In a similar protocol, we observed no Mer$^-$ transformants due to transfection of HOS-8 by c-hu-<u>myc</u> gene (exons II and III) placed downstream from an RSV promoter. We did not analyse the strains by blots.

2) We examined the expression of the SV40 genes in Mer$^+$ and Mer$^-$ cells that were produced by SV40 transformation of Mer$^+$ human fibroblasts. The supposition was that cells expressing greater levels of a SV40 mRNA or any amount of a given mRNA species would more likely be Mer$^-$ than those cells producing lesser amounts of SV40 mRNA. However, Figure 4 shows that there is no obvious relationship between SV40 mRNA production and Mer phenotype SV40. Strains IMR90-830, IMR90-890, SV80, XP12T703, GM638, WI38VA13, W18VA2, and CRL1584 are Mer$^-$ strains; AT5BIVA2, W98VA1, WI26VA4, and GM637 are Mer$^+$.

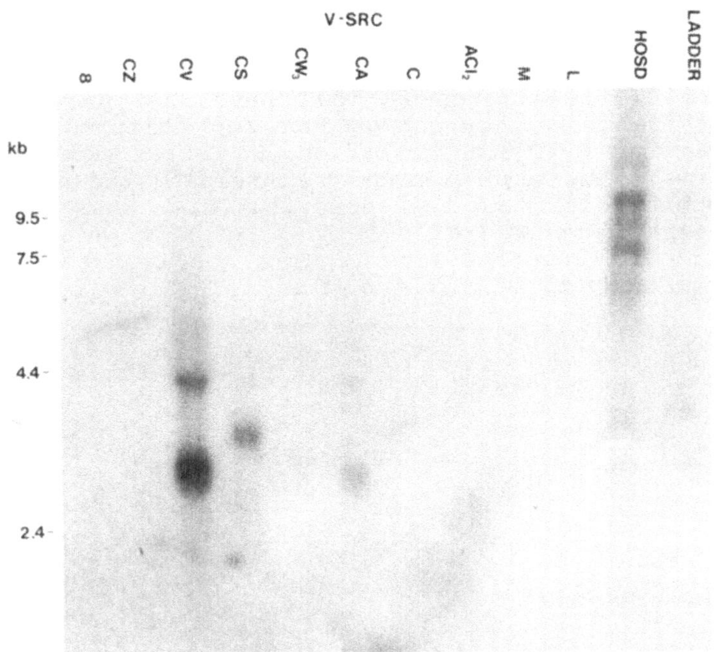

Figure 2. src Specific mRNA Production by pRSVsrcNeo I. Transformed HOS Clones. HOS clones D, L, ACl2, M, C, CA, CW3, CS, CV, CZ were selected with G418 and passaged without G418 for 10 to 50 generations prior to RNA extraction. HOS8 is a clone of HOS selected without transfection and in the absence of G418.

A Southern blot, not shown, shows that there is no specific integration site associated with the production of Mer⁻ cells. We can not conclude that SV40 insertional inactivation of the Mer gene is impossible. Such inactivation could occur because SV40 transformants occur at approximately 0.0003 per cell infected. If, as is common, an input multiplicity of 100 SV40 genomes per cell were used, and if there are a million genes per cell, three "target genes" of which, when inactivated give rise to transformation, and if each infecting SV40 genome were to insert randomly, 0.0003 transformants per cell infected would be generated. According to this idea, two of the three target genes when inactivated would be supposed to generate the Mer⁻ phenotype (in a dominant fashion) to account for the fact that 2/3 of the transformants are Mer⁻. On the other hand, SV40 is known to transform human cells by a "two step" mechanism. An SV40 transformed "focus" goes through a growth crisis (at approximately the same doubling as non-infected parent fibroblasts do) from which a permanent cell line may or may not arise. Perhaps it is this second, immortalizing, event that generates the Mer⁻ phenotype.

3) The adenovirus transformed human embryonic kidney (HEK) cell line 293 is another Mer$^-$ cell line. All primary HEK strains (3) and one HEK line (Flow 4000) that we have tested are Mer$^+$. We presume, but have not proved, that adenovirus transformation can produce Mer$^-$ cultures from Mer$^+$ cultures as does SV40 transformation. L. Babiss (Rockefeller) kindly provided us with a plasmid containing the adenovirus XhoC fragment containing the adenovirus E1A, E1B, and 55 kD terminal protein genes (pMK 0-15.5) into which we inserted a <u>neo</u> gene to produce pMKO-15neoI. This was transfected into A549 cells, known to support the growth of adenoviruses well.

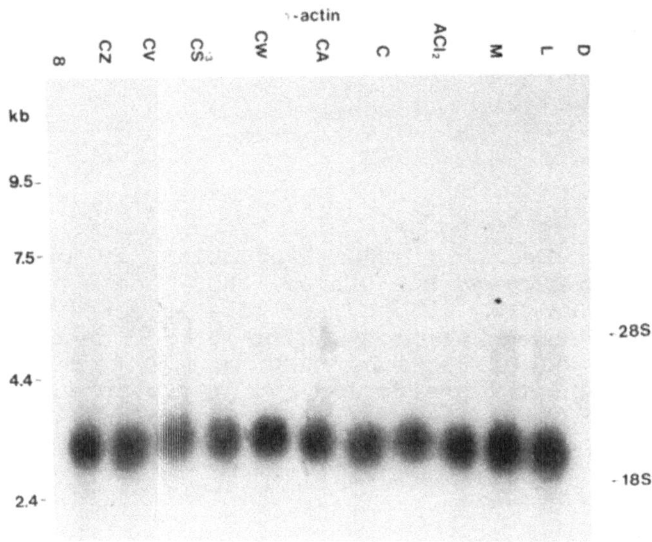

Figure 3. Gamma actin - Specific mRNA Production by the HOS Clones in Figure 2.

Selected clones, maintained in G418 at 350 ug/ml in F12, were tested for their ability to express the E1A product. This was done by assaying the plaque forming activity of the adenovirus 5 deletion mutant Ad5d1312, shown to require the adenovirus early products for growth (Jones and Shenk, 1979) and kindly supplied to us by T. Shenk. Of 10 clones tested, 2 promoted plaquing by Ad5d1312, showing that these are producing functional E1A protein. Initial tests show that these strains are Mer$^+$. We are passaging these cells in increasing G418 concentrations in attempts to enhance E1A expression and possibly produce a Mer$^-$ cell.

STUDIES OF PROTOONCOGENES IN LINES FROM MALIGNANT GLIOMAS

Because the Mer$^-$ phenotype accompanies transformation, we believe that either the activation of a gene involved in transformation (or the loss of function of that gene) may pro-

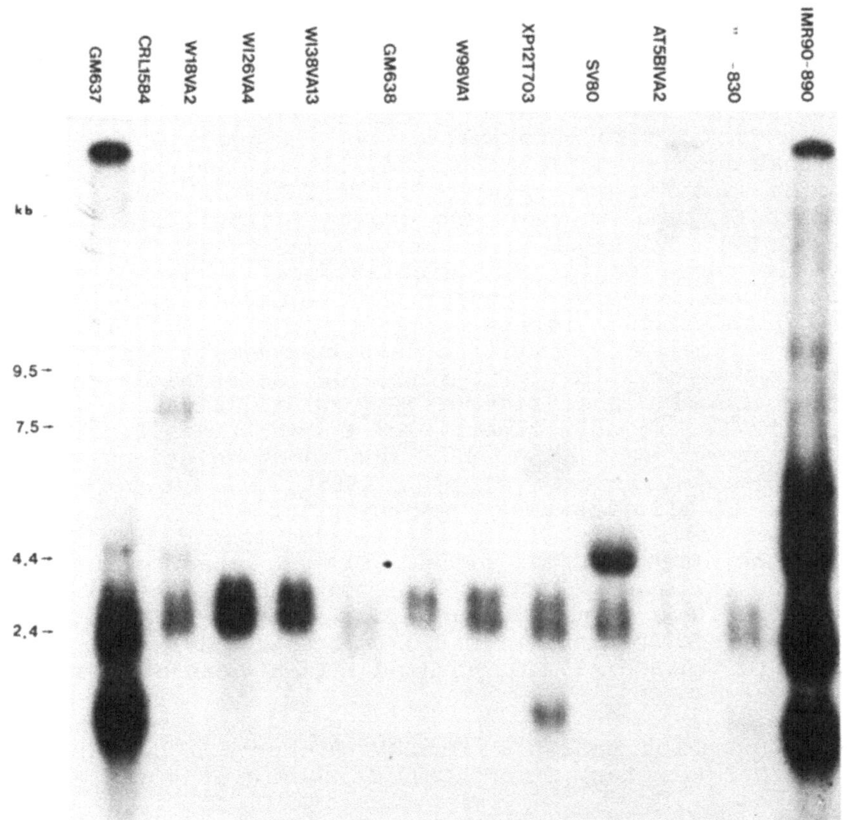

Figure 4. SV40-Specific mRNA production by SV40 Transformed Mer$^+$ and Mer$^-$ Cell Lines. Strains GM637, W126VA4, W98TA1, and AT5BIVA2 are Mer$^+$; CRL1584, W18VA2, WI38VA13, GM638, XP12T703, SV80, IMR90-830, and IMR90-890 are Mer$^-$. Probe was entire SV40 genome from pBRSV. (Southern blots showed a similar pattern).

duce transformed Mer$^-$ cells from nontransformed Mer$^+$ cells. The fact that the Mer$^-$ phenotype is most often dominant in Mer$^+$/Mer$^-$ hybrids (Yarosh, et al., 1984b; Day and Dobler, unpublished; but see Ayres, et al., 1982), argues that the Mer$^+$ phenotype is inactivated by activation of a gene during transformation, possibly by a transforming gene or oncogene.

We have done a preliminary study of proto-oncogene activation in 5 Mer^+ and 5 Mer^- cell lines from human glioblastomas: The Mer^+ strains are: SAN, MIL, U118MG, GRE, and T98; the Mer^- strains include A172, CLA, RIC, U87MG, and P4. The origins of all but P4, which was from Drs. Kornblith and Smith (NIH) are published (Day and Ziolkowski, 1979; Day, et al, 1980a,b; Scudiero, et al., 1984 a,b).

EGFR gene. The first proto-oncogene reported to be altered in human glioblastomas was the epidermal growth factor receptor (EGFR) gene (Libermann, et al., 1985). The EGFR gene is a 26 exon, 110 Kb gene specifying 5.1 and 9.5 kb transcripts that encode a 1126 amino acid, 175 kd product. EGFR has a tyrosine specific kinase activity, an ATP binding site, and binds both EGF and transforming growth factor alpha (Haley, et al., 1987). The gene is on chromosome 7, which is often present in three copies in glioblastomas (Bigner, et al., 1988). It was amplified and/or overexpressed in 4 of 10 primary glioblastomas (Libermann, et al., 1985). More recently 24 of 63 gliomas were found to have both amplification and elevated mRNA expression, while none of the remaining glioblastomas showed either sign of EGFR gene activation (Wong, et al., 1987). In cell lines, 2 of 6 cell lines from glioblastomas showed amplification of and short deletions within the EGFR gene (Yamazaki, et al., 1988). In our work, two strains of 17 glioblastomas show an altered EGFR allele by Southern analysis (see Figure 5) using plasmid pE7 (exons 2-21) as the source of the probe. CLA appeared to suffered a deletion, while A172 contained an insertion. Other differences in the 10 kb region may be RFLPs because no consistent differences were observed with EcoRI or PstI cut genomic DNAs as were with CLA and A172. The Hind III 5.3 kb RFLP was ob-

Table 2. Dot Blot analysis of mRNA Production by Mer^+ and Mer^- Malignant Glioma Cell Lines for Selected Genes.

Cell line/ Strain	(Mer)	Her-2/ NEU	EGFR	c-Ha-ras	c-myc	v-src	c-raf	c-sis	gli	γ-actin
A172	(-)	++	+++	+++	+	+	++	+/++	+	++
U87	(-)	-	+/++	+++	+/++	+	++	+/++	+	++
RIC	(-)	+	+	+++	+	+	++	+	+	+
P4	(-)	++	+++	+++	+	+	++	+	+	++++
CLA	(-)	+	+++	+++	+	-/+	++	+/++	+	+++
T98	(+)	+	+	+++	+/++	+/++	++	+/++	+	-/+
SAN	(+)	++++	-/+	+++	+	++	++	+/++	+	+
118MG	(+)	+	++	+++	+	+/++	++	+/++	+	+
GRE	(+)	+	++	+++	+	+	++	+/++	+	+++
MIL	(+)	+	++	+++	++	++	++	+/++	+	+++
CCL2	(+)	+	NT	+++	++++	+	++	NT	NT	++
HOS8	(+)	+	NT	+++	-/+	-/+	++	NT	NT	++
KD	(+)	++	+++	+++	-/+	NT	++	+/++	+	++
A431	(+/-)	+++	++++++	NT	NT	NT	NT	NT	+	NT

(pluses are approximately linearly related to mRNA production; NT, not tested)

served: none of 6 analyzable Mer^- glioblastoma strains contained the 5.3 kb band, whereas 6 analyzable Mer^+ strains did. One Mer^+ glioblastoma line was observed not to produce little or no EGFR mRNA, whereas 10 Mer^+ and Mer^- lines did (See Table 2).

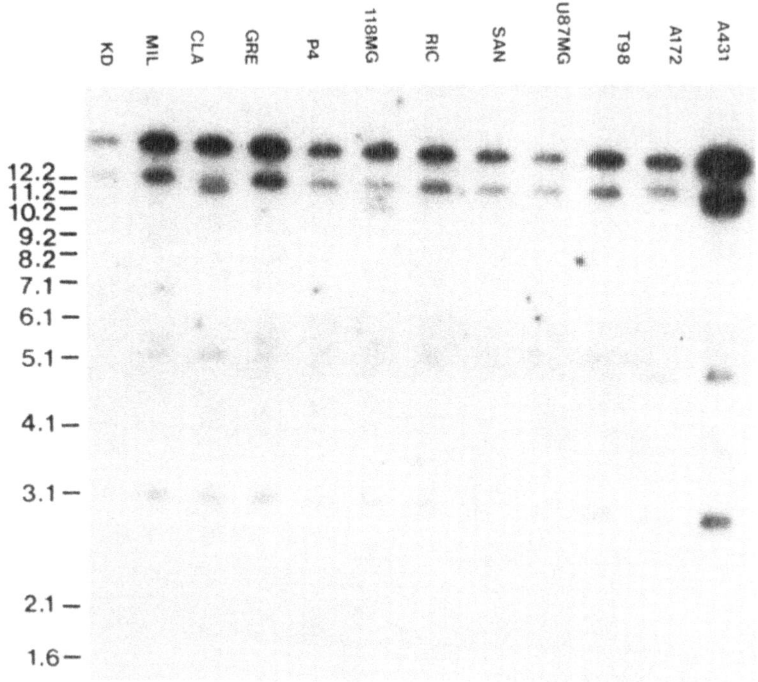

Figure 5. EGFR gene in Human Glioblastoma Cell Lines. Genomic DNAs were digested with HindIII, electrophoresed, blotted, and probed with the EGFR insert in plasmid pE7. A431 is known to have an amplified EGFR gene and KD is a normal fibroblast cell strain.

C-sis gene. Another proto-oncogene expressed in human glioblastomas is the platelet-derived growth factor beta chain (PDGFB) gene (Nister, et al., 1988), whose product is structually similar to the v-sis product of simian sarcoma virus (Waterfield, et al., 1983), a virus that causes gliomas in primates (Deinhardt, 1980). In these studies, all of 21 lines tested expressed PDGF A chain mRNA, 16 of 21 expressed the PDGF B chain mRNA and PDGF receptor mRNA was expressed in 15-16 of the 21 lines. A mechanism involving autocrine stimulation of malignant growth is suspected, but not proved (Nister, et al., 1988). In our studies no extra chromosomal material was seen in dot blot and Southern analysis of EcoRI, HindIII, and PstI cut genomic DNA from 34 transformed and 2 non-transformed strains. The Hind III RFLP pattern was analysed with the pSM-1 probe: 22 strains had a 30 kb band only, 5 strains had a 14 plus a 6 kb band, and 9 strains contained all these bands. There was no correlation of the RFLP pattern with Mer phenotype.

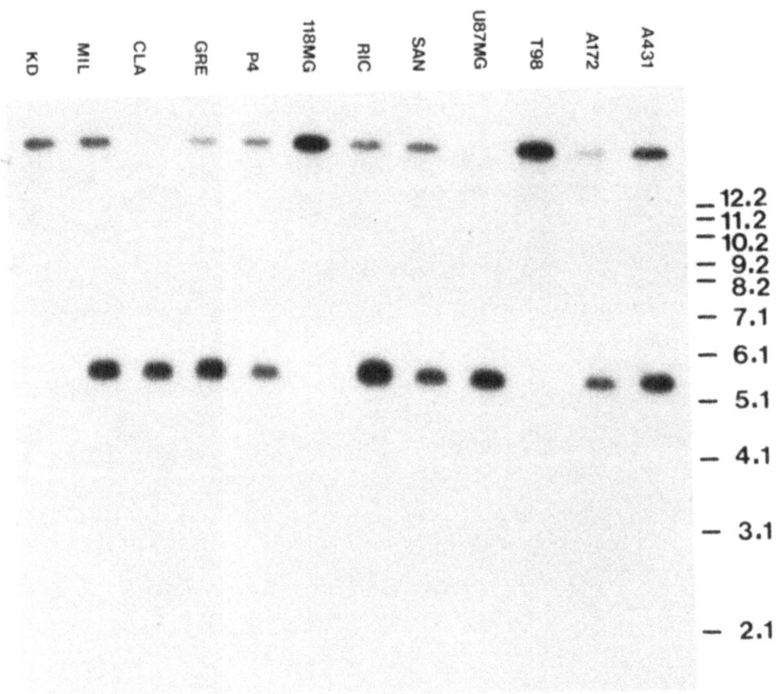

Figure 6. gli gene in Human Glioblastoma Cell Lines. Genomic DNAs were digested with EcoRI, electrophoresed, blotted, and probed with the gli insert in plasmid pKK36Pl.

gli gene. The gli gene was isolated from a glioblastoma cell line with double minute chromosomes using a library of sequences isolated by selecting for amplified DNA. The gene is amplified and overexpressed in the cell line, D-259MG, used to isolate it and in 1 of 63 primary glioblastomas (Wong, et al., 1987; Kinzler, et al., 1987). To our knowledge there is no literature published on studies of the gli gene in cell lines. In our work (see Figure 6) an apparently new restriction length polymorphism (RFLP) was detected with a gli gene probe (kindly supplied by B. Vogelstein) by Southern analysis of Eco RI cleaved genomic DNAs isolated from 39 strains. 19 strains had only a 15-16 kb band; 3 had only a 5.3 kb band, while 17 had both. Patterns in 10 tumor strains were consistent with extra chromosome 12 material in the region of gli. There was no correlation between either the polymorphism or the extra gli material and Mer phenotype.

ROS1 gene. In a study of 45 lines produced from many solid tumor types, the ROS1 gene on chromosome 6 was found to be expressed to moderate or high levels almost solely by astrocytomas and glioblastomas (Birchmeier, et al., 1987). In one glioblastoma, U118MG, the ROS1 gene was altered in a way suggesting activation. It was found to be truncated, likely due to an intrachromosomal deletion event, and gave a product lacking the extracellular receptor site, but retaining tyrosine kinase activity and the transmembrane region (Wigler, personal communication). We have not yet studied the ROS1 gene, although published data on ROS1 gene mRNA production (Birchmeier, et al., 1987) does not appear to correlate with the Mer$^-$ phenotype of the same strains as determined in our laboratory.

Interferon (IFN) genes. A possibly telling hint about Mer$^-$ glioma lines is their sensitivity to growth inhibition by interferons (IFNs) alpha and beta (Cook, et al., 1983; Yarosh, et al., 1985). IFN alpha and beta genes have not yet been shown to be altered in brain tumors. However, the p arm of chromosome 9 in glioblastomas frequently is a breakpoint for translocations and deletions (Bigner, et al., 1988), the location the genes for IFN alpha and beta (Shows, et al., 1982). Furthermore, the c-ets-1 gene on chromosome 6 is translocated to the interferon region in chromosome 9 in human acute monocytic leukemia (Diaz, et al., 1986). The finding that some patients with acute lymphoblastic leukemia or non-Hodgkin lymphoma have a homozygous deletion of the IFN-alpha and -beta genes (Diaz, et al., 1988) and the possibility that IFNs may act as tumor suppressor genes (see, for example, Reznitzky, et al., 1986) supports the hypothesis that deletion of IFN genes in brain tumor cell strains may constitute an oncogenic event. We are currently testing this possibility. We have found that the IFN-beta gene is deleted from several of our human brain tumor lines, using as probe the IFN-beta gene of Mark, et al., (1984).

Protooncogene expression. Northern and RNA dot blotting was used to assess the expression of mRNAs of selected protooncogenes in five Mer$^+$ and five Mer$^-$ strains. The results are shown in Table 2. Although there are significant differences among the strains, there is no correlation of expression of any one gene with Mer phenotype.

CONCLUSIONS

We have begun a patient study in order to assess whether patient tumors behave as if they are composed of Mer$^-$ cells, defective in the repair of O^6-methylguanine. This study is promising in that a tumor from one patient both responded to CCNU chemotherapy and gave rise to a Mer$^-$ line. In addition, to determine possible relationships between gene activation and Mer phenotype, we have begun a study of gene expression and structure in Mer$^+$ and Mer$^-$ lines either produced by SV40 transformation or from human brain tumor biopsies. This study has led to interesting findings about brain tumors, but has not led us to understand the basis for the Mer$^-$ phenotype. A study of the mRNAs differentially expressed by Mer$^+$ and Mer$^-$ strains may lead us further.

REFERENCES

Aubin, R., Weinfeld, M., and Paterson, M.C., 1988, Factors influencing efficiency and reproducibility of polybrene-assisted gene transfer. Somatic Cell and Mol. Genet., 14:155.

Ayres, K., Sklar, R., Larson, K., Lindgren, V., and Strauss, B., 1982, Regulation of the capacity for O^6-methylguanine removal from DNA in human lymphoblastoid cells studied by cell hybridization. Mol. Cell. Biol., 2:904.

Baker, R.M., Van Voorhis, W.C., and Spencer, L.A., 1979, HeLa cell variants that differ in sensitivity to monofunctional alkylating agents, with independence of cytotoxic and mutagenic responses, Proc. Natl. Acad. Sci. USA, 76:5249.

Baker, R.M., Zuerndorfer, G., and Mandel, R., 1980, Enhanced susceptibility of a xeroderma pigmentosum cell line to mutagenesis by MNNG and EMS, Environ. Mutagenesis, 2:269.

Bigner, S.H., Mark, J., Burger, P.C., Mahaley, M.S., Jr., Bullard, D.E., Muhlbaier, L.H., and Bigner, D.D., 1988, Specific chromosomal abnormalities in malignant human gliomas. Cancer Res., 88:405.

Birchmeier, C., Sharma, S., and Wigler, M., 1987, Expression and rearrangement of the ROS1 gene in human glioblastoma cells. Proc. Natl. Acad. Sci. USA, 84:9270.

Brennand, J., and Margison, G.P., 1986, Reduction of the toxicity and mutagenicity of alkylating agents in mammalian cells harboring the Escherichia coli alkyltransferase gene. Proc. Natl. Acad. Sci. USA, 83:6292.

Cook, A.W., Carter, W.A., Nidzgorski, R., and Akhtar, L., 1983, Human brain tumor-derived cell lines: growth rate reduced by human fibroblast interferon. Science, 219:881.

Day, R. S., III and Ziolkowski, C.H.J., 1979, Human brain tumour cell strains with deficient host-cell reactivation of N-methyl-N'-nitro-N-nitrosoguanidine-damaged adenovirus 5, Nature, 279:797.

Day, R.S., III, Ziolkowski, C.H.J., Scudiero, D.A., Meyer, S. A., Mattern, M.R., 1980a, Human tumor cell strains defective in the repair of alkylation damage, Carcinogenesis, 1:21.

Day, R.S., III, Yarosh, D.B., and Ziolkowski, C.H.J., 1984, Relationship of methyl purines produced by MNNG in adenovirus 5 DNA to viral inactivation in repair-deficient (Mer-) human tumor cell strains. Mutation Res., 131: 45.

Day, R.S., III, Ziolkowski, C.H.J., Scudiero, D.A., Meyer, S. A., Lubiniecki, A.S., Girardi, A.J., Galloway, S.M., and Bynum, G.D., 1980b, Defective repair of alkylated DNA by human tumour and SV40-transformed human cell strains, Nature, 288:724.

Day, R.S., III, Babich, M.A., Yarosh, D.B., and Scudiero, D. A., 1987, The role of O^6-methylguanine in human cell killing, sister-chromatid exchange induction and mutagenesis: a review. J. Cell. Sci. Suppl., 6:333.

Deinhardt, F., 1980, Biology of primate retroviruses, in: "Viral Oncology," G. Klein, ed., Raven Press, New York, p. 357.

Diaz, M.O., Le Beau, M.M., Pitha, P., and Rowley, J.D., 1986, Interferon and c-ets-1 genes in the translocation

(9;11)(p22;q23) in human acute monocytic leukemia, Science, 231:265.

Diaz, M.O., Zieman, S., Le Beau, M.M., Pitha, P., Smith, S. D., Chilcote, R., and Rowley, J.D., 1988, Homozygous deletion of the alpha- and beta-1-interferon genes in human leukemia and derived cell lines. Proc. Natl. Acad. Sci. USA, 85:5259.

Domoradzki, J., Pegg, A.E., Dolan, M.E., Maher, V.M., and McCormick, J.J., 1984, Correlation between O^6-methylguanine-DNA-methyltransferase activity and resistance of human cells to the cytotoxic and mutagenic effect of N-methyl-N'-nitro-N-nitrosoguanidine, Carcinogenesis, 5: 1641.

Erickson, L.C., Laurent, G., Sharkey, N.A., and Kohn, K.W., 1980a, DNA cross-linking and monoadduct removal in nitrosourea-treated human tumour cells. Nature, 288:727.

Erickson, L.C., Bradley, M.O., Ducore, J.M., Ewig, R.A.G., and Kohn, K.W., 1980b, DNA crosslinking and cytotoxicity in normal and transformed human cells treated with antitumor nitrosoureas. Proc. Natl. Acad. Sci. USA, 77:467.

Fox, M., Brennand, J., and Margison, G.P., 1987, Protection of Chinese hamster cells against the cytotoxic and mutagenic effects of alkylating agents by transfection of the Escherichia coli alkyltransferase gene and a truncated derivative. Mutagenesis, 2:491.

Freeman, A.E., and Hoffman, R.M., 1986, In vivo-like growth of human tumors in vitro. Proc. Natl. Acad. Sci. USA, 83: 2694.

Giard, D.J., Aaronson, S.A., Todaro, G.J., Arnstein, R., Kersey, J.H., Dosik, H., and Parks, W.P., 1973, In vitro cultivation of human tumors: establishment of cell lines derived from a series of solid tumors, J. Natl. Cancer Inst., 51:1417.

Gibson, N.W., Hartley, J.A., Strong, J.M., and Kohn, K.W., 1986, 2-chloroethyl(methylsulfonyl)methanesulfonate(NSC-338947) a more selective DNA alkylating agent than the chloroethylnitrosoureas. Cancer Res., 46:553.

Haley, J., Whittle, N., Bennett, P., Kinchington, D., Ullrich, A., and Waterfield, M., 1987, The human EGF receptor gene: structure of the 110 kb locus and identification of sequences regulating its transcription. Oncogene Res., 1:375.

Ishizaki, K., Tsujimura, T., Fujio, C., Yangpei, Z., Yawata, H., Nakabeppu, Y., Sekiguchi, M., and Ikenaga, M., 1986, Expression of the truncated E. coli O^6-methylguanine methyltransferase gene in repair-deficient human cells and restoration of cellular resistance to alkylating agents. Mutation Res., 184:121.

Jakobovits, E.B., Majors, J.E., and Varmus, H.E., 1984, Hormonal regulation of the Rous sarcoma virus src gene via a heterologous promoter defines a threshold dose for cellular transcription. Cell, 38:757.

James, C.D., Carlbom, E., Dumanski, J.P., Hansen, M., Nordenskjold, M., Collins, V.P. and Cavenee, W.K., 1988, Clonal genomic alterations in glioma malignancy stages. Cancer Res., 48:5546.

Jones, N., and Shenk, T., 1979, Isolation of adenovirus type 5 host range deletion mutants defective for transformation of rat embryo cells. Cell, 17:683.

Kataoka, H., Hall, J., and Karran, P., 1986, Complementation of sensitivity to alkylating agents in Escherichia coli and Chinese hamster cells by expression of a cloned bacterial repair gene. EMBO J., 5:3195.

Kinzler, K.W., Bigner, S.H., Bigner, D.D., Trent, J.M., Law, M.L., O'Brien, S.J., Wong, A.J., and Vogelstein, B., 1987, Identification of an amplified, highly expressed gene in a human glioma. Science, 236:70.

Libermann, T.A., Nusbaum, H.R., Razon, N., Kris, R., Lax, I., Soreq, H., Whittle, N., Waterfield, M.D., Ullrich, A., and Schlessinger, J., 1985, Amplification, enchanced expression, and possible rearrangement of EGF receptor gene in primary human tumors of glial origin. Nature, 313:144.

Mark, D.F., Lu, S.D., Creasey, A.A., Yamamoto, R., and Lin, L.S., 1984, Site-specific mutagenesis of the human fibroblast interferon gene. Proc. Natl. Acad. Sci. USA, 81:5662.

Mosmann, T.J., 1983, Rapid colorimetric assay for cellular growth and survival: application to proliferation and cytotoxicity assays. J. Immun. Methods, 65:55.

Nister, M., Libermann, T.A., Betsholtz, C., Petterson, M., Clesson-Welsh, L., Heldin, C.H., Schlessinger, J., and Westermark, B., 1988, Expression of messenger RNAs for platelet-derived growth factor and transforming growth factor-alpha and their receptors in human malignant cell lines. Cancer Res., 48:3910.

Ponten, J., 1975, Neoplastic human glia cells in culture. in: "Human Tumor Cells in Vitro", J. Fogh, ed., Plenum Press, New York. p 175.

Resnitzky, D., Yarden, A., Zipori, D., and Kimchi, A., 1986, Autocrine beta-related interferon controls c-myc suppression and growth arrest during hematopoietic cell differentiation. Cell, 46:31.

Samson, L., Derfler, B., and Waldstein, E.A., 1986, Suppression of human DNA alkylation-repair defects by Escherichia coli DNA-repair genes. Proc. Natl. Acad. Sci. USA, 83: 5607.

Sariban, E., Kohn, K.W., Zlotogorski, C., Laurent, G., D'Incalci, M., Day, R.S., III, Smith, B.H., Kornblith, P.L., and Grickson, L.C., 1987, DNA cross-linking responses of human malignant glioma cell strains to chloroethylnitrosoureas, cisplatin, and diaziquone. Cancer Res., 47:3988.

Scudiero, D.A., Meyer, S.A., Clatterbuck, B.E., Mattern, M. R., Ziolkowski, C.H.J., and Day, R.S., III, 1984a, Relationship of DNA repair phenotypes of human fibroblast and tumor strains to killing by N-methyl-N'-nitro-N-nitrosoguanidine, Cancer Research, 44:961.

Scudiero, D.A., Meyer, S.A., Clatterbuck, B.E., Mattern, M. R., Ziolkowski, C.H.J., and Day, R.S., III, 1984b, Sensitivity of human cell strains having different abilities to repair O^6-methylguanine in DNA to inactivation by alkylating agents including chloroethylnitrosoureas. Cancer Res., 44:2467.

Shows, T.B., Sakaguchi, A.Y., Naylor, S.L., Goeddel, D.V., and Lawn, R.M., 1982, Clustering of leucocyte and interferon genes on human chromosome 9. Science, 218:373.

Sklar, R., and Strauss, B., 1981, Removal of O^6-methylguanine from DNA of normal and xeroderma pigmentosum-derived lymphoblastoid cell lines, Nature, 289:417.

Watatani, M., Ikenaga, M., Hatanaka, T., Kinuta, M., Takai, S., Mori, T. and Kondo, S., 1985, Analysis of N-methyl

N'-nitro-N-nitrosoguanidine (MNNG)-induced DNA damage in tumor cells strains from Japanese patients and demonstration of MNNG hypersensitivity of Mer- xenografts in athymic nude mice. Carcinogenesis, 6:549.

Waterfield, M.D., Scrace, G.T., Whittle, N., Stroobant, P., Johnsson, A., Wasteson, A., Westermark, B., Heldin, C.H., Huang, J.S., and Deuel, T.F., 1983, Platelet-derived growth factor is structurally related to the putative transforming protein p28 sis of simian sarcoma virus. Nature, 304:35.

Wilson, C.B., 1976, Chemotherapy of brain tumors, in: "Advances in Neurology, Vol. 15, Neoplasia in the Central Nervous System," R.A. Thompson and J.R. Green, eds., Raven Press, New York. p. 361.

Wong, A.J., Bigner, S.H., Bigner, D.D., Kinzler, K.W., Hamilton, S.R., and Vogelstein, B., 1987, Increased expression of the epidermal growth factor receptor gene in malignant gliomas is invariably associated with gene amplification. Proc. Natl. Acad. Sci. U.S.A., 84:6899.

Yamazaki, H., Fukui, Y., Ueyama, Y., Tamaoki, N., Kawamoto, T., Taniguchi, S., and Shibuya, M., 1988, Amplification of the structurally and functionally altered epidermal growth factor receptor gene (c-erbB) in human brain tumors. Mol. Cell. Biol., 8:1816.

Yarosh, D.B., Rice, M., Day, R.S., III, Foote, R.S., and Mitra, S., 1984a, O^6-methylguanine-DNA methyltransferase in human cells. Mutat. Res., 131:27.

Yarosh, D.B., Scudiero, D.A., Ziolkowski, C.H.J., Rhim, J.S., and Day, R.S., III. , 1984b, Hybrids between human tumor cells differing in repair of MNNG-produced DNA damage. Carcinogenesis, 5:627.

Yarosh, D.B., Scudiero, D.A., Yagi, T., and Day, R.S., III, 1985, Human tumor cell strains both unable to repair O^6-methylguanine and hypersensitive to killing by human alpha and beta interferons. Carcinogenesis, 6:883.

THE MOLECULAR BASIS OF ALKYLATING AGENT RESISTANCE IN MAMMALIAN CELLS

Peter Karran[1], Janet Hall[1], Hiroko Kataoka[1], Claire Stephenson[1], Michael Green[2], Jill Lowe[2], Corrine Petit-Frere[2]

[1]Imperial Cancer Research Fund, Clare Hall Laboratories, South Mimms, UK

[2]MRC Cell Mutation Unit, Sussex University Falmer, Brighton, UK

INTRODUCTION

The Mex$^-$(or Mer$^-$) (Sklar and Strauss, 1981; Day, et al., 1980) phenotype of human cells results in a sensitivity to certain alkylating agents. Foremost among the compounds to which Mex$^-$ cells are sensitive are the alkylnitrosamines and alkylnitrosamides which can form covalent adducts with cellular DNA at the O^6-position of guanine. A particularly important class of such compounds is the cross-linking nitrosoureas which are widely used in chemotherapy. The molecular basis of the Mex$^-$ phenotype is not fully characterised, although it appears to be associated with the transformed state in as far as normal human cells are Mex$^+$ but may become Mex$^-$ following transformation *in vitro*. Thus, the phenotype is important from the practical viewpoint of chemotherapy, as well as a more fundamental one of understanding the events which occur during cellular transformation.

The sensitivity of Mex cells to methylating agents correlates with the inability of these cells to remove O^6-MeGua from their DNA (Sklar and Strauss, 1981); a defect which is a direct consequence of the absence of a specific repair enzyme, O^6-MeGua-DNA methyltransferase (Harris, et al., 1983). This enzyme removes the methyl group from the O^6-position in a well-characterised suicidal methyl transfer reaction. Unlike its bacterial counterpart, the *E. coli ada*$^+$ gene product, it does not demethylate methylphosphotriesters (MePTEs) in DNA, nor does it act on the minor methylation product O^4-methylthymine (Dolan, et al., 1984).

In order to address the question of the role of O^6-MeGua in the cytotoxicity of methylating agents, we have examined the effects of expressing the Ada protein in Chinese hamster ovary (CHO) cells. These cells do not express a methyltrans-

DNA Repair Mechanisms and Their Biological Implications in Mammalian Cells
Edited by M.W. Lambert and J. Laval
Plenum Press, New York

ferase and exhibit a sensitivity to monofunctional alkylating agents which is comparable to that of Mex$^-$ human cells. By expressing Ada proteins with altered functions, we have directly examined the relation between O^6-MeGua and cell killing.

An intriguing property of Mex$^-$ cells is their ability to acquire a resistance to methylating agents which is not dependent on the excision of O^6-MeGua from DNA or the expression of a methyltransferase function. Such resistant lines exhibit the high susceptibility to methylating agent-induced mutation which is characteristic of Mex$^-$ cells but are able to survive treatment with much higher concentrations of agents such as N-methyl-N'-nitro-N-nitrosoguanidine (MNNG) and N-methyl-N-nitrosourea (MNU) (Goldmacher, et al., 1986; Goth-Goldstein, 1987). These data suggest that the resistant cells are able to 'tolerate' O^6-MeGua in their DNA by some mechanism (Roberts, et al., 1971) but that they nevertheless remain susceptible to errors during DNA replication. We have tested this hypothesis directly using the human SV40-transformed fibroblast cell line MRC5V1 which is Mex$^-$. By studying resistant variants of this line, we have defined some of the properties of the 'tolerance' pathway.

RESULTS

CHO cells were transfected with the plasmid pHJ2 which contains the full coding sequence of the E. coli Ada protein and the E. coli gpt^+ gene both under the control of the SV40 early promoter (Kataoka, et al., 1986). Cells which expressed the Gpt protein were selected by resistance to mycophenolic acid and surviving cells were further selected for resistance

Table 1. Methyltransferase Activities In Transfected Or Resistant Cell Lines

	Methyltransferase Activity (1 unit[a]/mg protein)	
Cell Line	O^6-MeGua	MePTE
CHO	<0.01	<0.01
CHOCNU3	0.5	0.5
CH0623	<0.01	0.5
CHO7.1	0.12	<0.01
CHO7.6	0.12	<0.01
MRC5V1	<0.05	ND
M1	<0.05	ND
M2	<0.05	ND
MRC5V1a	0.54	ND
MRC5V1b	0.24	ND
MRC5V1c	0.53	ND
M2a	0.88	ND

[a]1 unit of methyltransferase removes 1 pmole methyl groups from its substrate under standard assay conditions.

to the alkylating agent chloroethylnitrosourea (CNU). Five mycophenolic acid-resistant/CNU-resistant clones were obtained all of which expressed levels of O^6-MeGua-DNA methyltransferase activity comparable to those of Mex^+ cell lines. The value for one of these lines, CHOCNU3 is shown in Table 1. The sensitivity of the cell line CHOCNU3 to MNNG was determined (Figure 1a). Expression of the Ada protein additionally conferred protection against the cytotoxicity of CNU and, to a lesser degree, methylmethanesulfonate (data not shown). The degrees of resistance observed were closely similar to those seen in Mex^+ human fibroblast cell lines (Scudiero, et al., 1984). Resistance to other DNA damaging agents, such as ultraviolet light, was unaltered in CHOCNU3 cells (data not shown).

The Ada protein comprises two active domains; one of which repairs O^6-MeGua while the other acts on MePTEs in DNA. We constructed a recombinant plasmid, pHJ24, in which the introduction of a frameshift mutation into the Ada protein coding sequences renders the O^6-MeGua repair domain non-functional (Kataoka, et al., 1986). Following transfection of pHJ24 into CHO cells, we obtained cell lines which expressed only the MePTE repair function of the ada^+ gene (Table 1). The resistance to alkylating agents of one of these cell lines, CHO623, was examined. Expression of the MePTE repair domain of the Ada protein conferred little or no protection against the cytotoxicity of MMNG except at relatively high doses (Figure 1a).

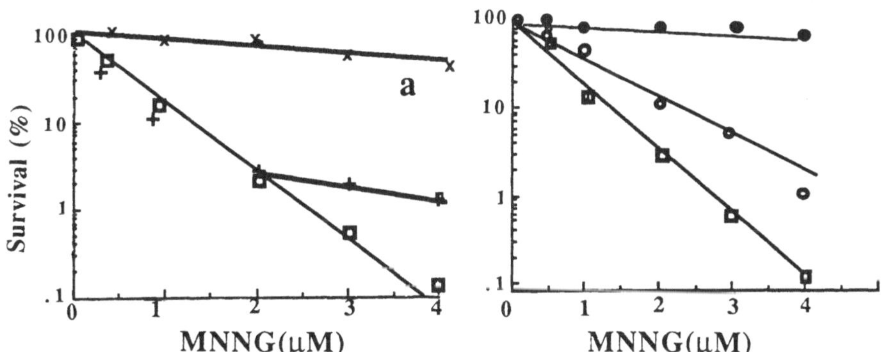

Figure 1. Survival of CHO cells expressing different methyltransferase functions.

(a) Cells were treated for 60 minutes at 37^0 at the concentrations of MNNG shown. CHO (□), CHOCNU3 (X), CHO623 (+).
(b) CHO (□), CHO7.1 (●), CHO7.6 (O).

CHO623 cells were not more resistant to CNU or ultraviolet light (data not shown). Taken together, the data from the cell lines CHOCNU3 and CHO623 suggest that protection against the cytotoxic effects of MNNG is conferred by the ability to repair O^6-MeGua in DNA. We tested this possibility using the plasmid pHJ3 which encodes a truncated form of the Ada protein from which the MePTE repair domain has been removed (Hall, et al., 1988). Two independent transfectant CHO cell lines which expressed the truncated Ada protein, CHO7.1 and CHO7.6 were isolated. Both cell lines expressed identical levels of O^6-MeGua methyltransferase activity, although somewhat less than that seen in CHOCNU3. Despite their lower level of Ada protein, CHO7.1 and CHO7.6 were both resistant to killing by MNNG (Figure 1b). However, CHO7.1 was better protected against MNNG killing than CHO7.6 even though their methyltransferase levels, as determined in cell-free extracts, were equivalent. Thus, while the ability to repair O^6-MeGua in DNA confers resistance to killing by MNNG, the relation between the degree of protection conferred and the level of methyltransferase as determined in cell-free extracts is not quantitative. In the particular case of CHO7.1 and CHO7.6, this may reflect the stability of the truncated Ada protein in CHO cells. However, it appears that the relation between methyltransferase expression in rodent cells and resistance to MNNG is, at best, only semi-quantitative (Yagi, et al., 1984). It is possible that,

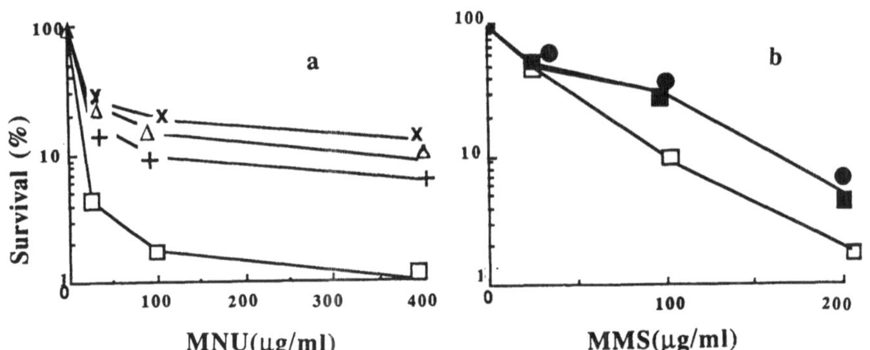

Figure 2. Survival of MRC5V1 cells and resistant derivatives.

(a) Cells were treated with MNU at the concentrations shown for 15 minutes at 37^0 in 0.1 M citrate buffer, pH5.0. MRC5V1 (□), M2 (+), MRC5V1a (△), M2a (x). For clarity, we have shown only a single example of each type of cell line. Data for the cell line M1 are essentially identical to M2. Data for MRC5V1b and MRC5V1c are closely similar to MRC5V1a. Data for M1a and M2a are closely similar.

(b) Cells were treated at the MMS concentrations shown for 60minutes in PBS. MRC5V1 (□), M1 (■), M2 (●).

for rodent cells at least, factors other than the level of methyltransferase activity may determine resistance to methylating agents.

Table 2. MNU-Induced Lethal Hits In MRC5V1 And Its MNU-Resistant Derivatives

	NUMBER OF LETHAL HITS[1]					
	Observed				Expected	
MNU (μg/ml)	MRC5V1	M2[1]	MRC5V1a-c[1]	M2a[1]	Model 1 M2a	Model 2 M2a
25	2.76	1.58	1.31	1.26	0.75	0.13
100	4.42	2.27	2.01	1.61	1.06	-0.08
400	5.47	2.79	3.27	2.29	1.67	0.59

The number of lethal hits was calculated from the zero term of the Poisson distribution. On Model 1, the tolerance mechanism and the methyltransferase are postulated to operate on the same potentially cytotoxic DNA lesion. Each is postulated to remove or tolerate the same proportion of the lesion independently of the other. On Model 2, two toxic lesions of approximately equal frequency are assumed; O^6-methylguanine repaired by methyltransferase action and the second lesion by a mechanism active only in M2.

[1]The data are presented as average values: M2 = Average for M1 & M2; M2a = Average for M2a,M2b & M2c; MRC5V1a-c = Average for MRC5V1a, MRC5V1b & MRC5V1c.

A second pathway which confers resistance to methylating agents may be expressed in mammalian cells. In this case, protection does not involve the active removal of O^6-MeGua lesions from DNA. In order to study the properties of this pathway, we isolated MNU-resistant variants of the SV40-transformed human fibroblast cell line MRC5V1. Two cell lines, designated M1 and M2, were isolated following multiple treatments with MNU. Figure 2a shows that both M1 and M2 exhibit considerable resistance to MNU and a more modest increase in resistance to MMS (Figure 2b).

In this case, the resistance is not accompanied by expression of a methyltransferase function and both M1 and M2 are Mex$^-$ as is the MRC5V1 parental line (Table 1). Transfection of the plasmid pHJ2 into MRC5V1 cells and expression of the Ada protein also conferred resistance to MNU (Figure 2, Table 1). The level of protection observed in the transfectant cell lines MRC5V1a, b and c, is closely similar to that seen in M1 and M2. When the MNU-resistant line M2 was transfected with pHJ2, we observed that the Ada protein-expressing derivatives M2a, b, and c, did not exhibit any significant further enhancement in their resistance to MNU (Figure 2a, Table 1). It appears that M1 and M2 have become resistant to MNU through some mechanism which allows them to tolerate O^6-MeGua in their DNA since the ability to remove this lesion, which is conferred by the Ada protein, does not result in enhanced resistance in M2a, b, or c. This proposition is presented more formally in Table 2 in which we compare the hypotheses that the potentially lethal methylation product, to which M1 and M2 are resistant, is the same as, or different to, the lesion acted on by the transfected <u>ada⁺</u> gene product. It is clear from Table 2 that the data fit best to a model in which the enhanced resistance of M1 and M2 results from an increased ability to tolerate O^6-MeGua lesions without excising them from DNA.

There are differences in the specificities of the tolerance mechanism and the methyltransferase. The cell lines, M1 and M2, are not resistant to the chloroethylating agent mitozolomide and, in fact, are somewhat more sensitive to this compound (data not shown). Expression of the Ada protein in M2 restores resistance to mitozolomide, indicating that M2 cells are not able to tolerate chloroethyl adducts and that resistance to this agent requires their removal from DNA.

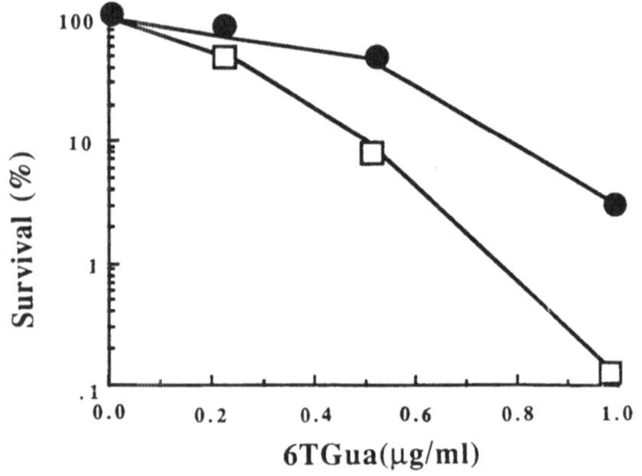

Figure 3. Survival of MRC5V1 cells and resistant derivatives. Cells were plated in the presence of 6-thioguanine at the concentrations shown. The 6-thioguanine was not subsequently removed. MRC5V1 (□), M2 (●). (The data for M1 are essentially identical to those for M2).

In order to investigate possible involvement of mismatch correction in the observed tolerance process in M1 and M2, we tested their sensitivity to a number of base analogues. No differences were seen in the sensitivity of M1 and M2 to the base analogues 2-aminopurine and diaminopurine (data not shown). However, both M1 and M2 exhibited an increased resistance to 6-thioguanine (Figure 3).

DISCUSSION

The means by which the expression of methyltransferase is controlled in mammalian cells is still unknown, although there is evidence to suggest that the phenotype is unstable and that Mex^+ cell lines may arise from a Mex^- population in the presence (Morten and Margison, 1988; Satoh, et al., 1987; Hori, et al., 1988) or the absence (Huh and Rajewsky,1988) of a selective pressure. The data presented here indicate that the ability to demethylate O^6-MeGua in DNA confers on the cell a considerable degree of protection against alkylation damage. Nevertheless, it appears that mammalian cells do not rely completely on this method of protection, and that resistance may be acquired via the tolerance mechanism investigated here.

Although we are unable to draw firm conclusions about the precise mechanism by which tolerance is brought about, it seems that the following conclusions can be drawn: 1) Tolerance acts on O^6-MeGua in DNA; 2) The ability to tolerate O^6-MeGua does not confer tolerance to O^6-chloroethylGua in DNA nor to the base analogues 2-aminopurine and diaminopurine; and 3) Tolerance does, however, appear to extend to the altered purine 6-thioguanine which exerts its effect via incorporation into DNA.

In view of the considerable bias exhibited by the mammalian mismatch correction system towards correction in favour of Gua bases in DNA (Brown and Jiricny, 1988), it is tempting to speculate that the tolerance to O^6-MeGua may parallel the acquisition of resistance to MNNG of *E. coli* *dam* mutants. In that case, *dam* mutants may become resistant to MNNG by the introduction of a second mutation which inactivates the mismatch correction system (Karran and Marinus, 1982). It is possible that an alteration in mismatch correction has occured in M1 and M2 which results in an enhanced ability to withstand the cytotoxic potential of O^6-MeGua in DNA. The alteration may confer protection against 6-thioguanine but not against base analogues in general.

ACKNOWLEDGEMENTS

We thank Derval Byrne for her help with the manuscript. This work was in part supported by E.C. contract B16.042UK(H).

REFERENCES

Arita,I., Tatsumi, K., Tachibana, A., Toyoda, M., Takebe, H., 1988, *Mutat.Res.*, 208:167-172.

Brown,T.C., and Jiricny, J., 1988, *Cell*, 54:705-711.

Day,R.S., Ziolkowski, C.H.J., Scudiero, D.A., Myer, S., Lubiniecki,A.S., Girardi, A.J., Galloway, S.M. and Bynum, G.D., 1980, *Nature*, 288:724-727.

Dolan,M.E., Scicchitano, D., Singer, B., and Pegg A.E. (1984). *Biochem. Biophys. Res. Comun.*,123:324-330.

Goldmacher,V.S., Cuzick, R.A., and Thilly, W.G., 1986, *J. Biol. Chem.*, 261:2462-2471.

Goth-Goldstein,R., (1987). *Carcinogenesis*, 8:1449-1453.

Hall,J., Kataoka, H., Stephenson, C., and Karran, P., 1988, *Carcinogenesis*, 9:1587-1593.

Harris,A.L., Karran, P. and Lindahl, T., 1983, Cancer Res., 43:3247-3252.
Huh, N. and Rajewsky, M.F., 1988, Int. J. Cancer, 41:762-766.
Karran,P. and Marinus, M.G., 1982, Nature, 296:868-869.
Kataoka,H., Hall, J., and Karran, P., 1986, EMBO J.,5:3195-3200.
Morten, J.E.N. and Margison, G.P.,1988, Carcinogenesis, 9: 45-49.
Roberts,J.J., Pascoe, J.M., Smith, B.A., and Crathorn, A.R., 1971, Chem. Biol. Interactions, 3:49-68.
Satoh, M.S., Huh, N., Hori, Y., Thomale, J., and Rajewsky, M.F., 1987, Gann, 78:1094-1099.
Scudiero, D.A., Meyer, S.A., Clatterbuck, B.E., Mattern, M.R., Ziolkowski, C.H.J., and Day, R.S., 1984, Cancer Res., 44: 961-969.
Sklar, R. and Strauss, B., 1981, Nature, 289:417-420.
Yagi,T., Yarosh, D.B., and Day, R.S., 1984, Carcinogenesis, 5:593- 600.

DIFFERENTIAL REPAIR OF O^4-METHYLTHYMINE AND O^6-METHYLGUANINE IN RAT AND HAMSTER LIVER

Janet Hall, Henriette Brésil, Ghyslaine Martel-Planche, Mireille Serres*, Christopher P. Wild, Ruggero Montesano

Unit of Mechanisms of Carcinogenesis, International Agency for Research on Cancer, Lyon, France

ABSTRACT

After treatment with a single dose of the methylating agent dimethylnitrosamine (DMN), differential rates of repair of O^6-methylguanine (O^6-MedG) and O^4-methylthymine (O^4-MeT) were found in rat and hamster liver DNA. In the rat, both methylated lesions were actively repaired ($t^1/_2$ 19 and 30 hours, respectively), while in the hamster, the level of O^6-MedG remained stable between 6 and 48 hours after treatment, during which time the level of O^4-MeT was reduced by 42%. A comparison of the levels of two DNA repair enzymes after the carcinogen treatment also revealed species differences. While in both species increases in the methylpurine-DNA glycosylase level were seen, the inactivation and subsequent time course of recovery of the O^6-methylguanine-DNA methyltransferase was significantly different. In the hamster, no active O^6-methylguanine-DNA methyltransferase was detected in liver extracts for up to 96 hours after treatment, compared with the rat liver extracts, where active enzyme was detectable from 24 hours and by 96 hours an increase in enzyme levels compared to control values was observed. This lack of active O^6-methylguanine-DNA methyltransferase explains the apparent stability of O^6-MedG in hamster liver DNA over the period 6 to 48 hours and indicates that it is not this DNA repair enzyme that is responsible for the observed removal of O^4-MeT from this tissue.

INTRODUCTION

Alkylation of DNA at the oxygen atoms of guanine and

*Present address: Laboratoire de Biologie, Ecole Normale Supérieure, Lyon, France

DNA Repair Mechanisms and Their Biological Implications in Mammalian Cells
Edited by M.W. Lambert and J. Laval
Plenum Press, New York

thymine appears to be of particular importance in the induction of mutations and tumors by carcinogens such as nitrosamines. The majority of such studies have concentrated on the measurement of O^6-alkylguanine, which is formed in higher concentrations than the pyrimidine derivatives. However, O^4-alkylthymine has been shown to exhibit miscoding properties during *in vitro* DNA synthesis (Abbott and Saffhill, 1979), and by analogy with O^6-alkylguanine could thus significantly contribute both to mutagenesis and carcinogenesis. Indeed, in the induction of hepatocellular carcinomas by chronic diethylnitrosamine administration in rats, O^4-ethylthymidine has been implicated as being relevant since it accumulates in liver DNA to 50 times the O^6-ethylguanine concentration (Swenberg, *et al*., 1984). The mechanism of repair of O^6-methylguanine in DNA has been shown to be the same in *E. coli* (reviewed Lindahl, 1982), rodent (Craddock, *et al*., 1982; Pegg, *et al*., 1983) and human tissue (Pegg, *et al*., 1982). However, the substrate specificity of the purified enzymes from bacterial and mammalian sources do not appear to be the same. In *E. coli* the 39 kd *ada*$^+$ gene product demethylates O^6-MedG, O^4-MeT, and particular isomers of methylphosphotriesters in DNA (Olsson and Lindahl, 1980; McCarthy and Lindahl, 1985), while the 24.5 kd mammalian enzyme apparently only repairs O^6-MedG (Yarosh, 1985). The repair mechanism for O^4-MeT remains unclear (see Brent, *et al*., 1988), however, *in vivo* removal of this adduct from rat liver has been clearly demonstrated (Richardson, *et al*., 1985; Belinsky, *et al*., 1986). In this paper, we describe the rates of disappearance of O^4-MeT and O^6-MedG in liver DNA of rats and hamsters treated with a single dose of DMN and the rates of recovery of O^6-methylguanine-DNA methyltransferase and methylpurine-DNA glycosylase. The differential repair of these two DNA adducts observed in the liver of these two species suggests a model system for the identification of the repair enzyme responsible for the removal of O^4-MeT.

MATERIALS AND METHODS

Chemicals

Radiolabelled compounds were obtained from Amersham or New England Nuclear, as indicated in the text. Enzymes used for DNA extraction and hydrolysis were purchased from Boehringer, Mannheim, FRG. DMN was obtained from Merck, Munich, FRG, and deoxyribonucleoside standards were from Sigma Chemical Company, St Louis, Mo., USA. 2'-Deoxycoformycin was obtained from the NCI, Bethesda, MD, USA.

Animal experiments

Male outbred BDIV rats (eight weeks old), bred in the animal house of the International Agency for Research on Cancer, and male Syrian Golden hamsters (eight weeks old) from TNO Central Institute for the Breeding of Laboratory Animals, Zeist, The Netherlands, were treated with a single dose of DMN (in water) at 20 mg/kg and 25 mg/kg, respectively. The livers were taken at various times up to 96 hours after this exposure and immediately frozen and stored at -80^{0}C until DNA was extracted or enzyme extracts were prepared.

DNA extraction and chromatography

DNA was isolated from liver samples, enzymatically hydrolyzed to deoxynucleosides and subsequently the methylated and normal deoxynucleosides were separated on an Aminex A7 cation exchange column. These methods have been described in detail previously (Umbenhauer, et al., 1985). The co-chromatography of O^4-MeT with deoxyadenosine and deoxycytosine necessitated a second chromatographic step to purify O^4-MeT. This was carried out on a reverse phase Lichrosorb RP18 column eluted isocratically with 15% methanol at a flow rate of 1 ml/minute. Under these conditions O^4-MeT, dA and dC had retention times of 26, 10, and 4 minutes respectively. Fractions containing O^6-MedG (from Aminex A7) and O^4-MeT (from reverse phase) were pooled, dried and reconstituted with 400 µl of PBS containing 1% foetal bovine serum, 3 mM sodium azide and 1 µM 2'-deoxycoformycin, for radioimmunoassay (RIA). Parent deoxynucleosides were quantitated by absorbance at 260 nm.

Antibodies and RIA analysis

The antibody used to quantitate O^6-MedG was monoclonal antibody C4 (Wild, et al., 1983), while a polyclonal antisera was prepared against O^4-MeT-BSA in New Zealand male albino rabbits by the immunization procedure described for other nucleoside-protein conjugates (Müller and Rajewsky, 1980). The specificity of this antisera has been described (Wild, et al., 1987). RIAs were performed by a standard methodology (Wild, et al., 1983) using tritiated tracers of O^6-MedG (Amersham, 29 Ci/mmole) and O^4-MedT (16 Ci/mmole). The latter compound was purified from a reaction of diazomethane with [^{3}H]-thymidine (30 Ci/mmole) and was kindly provided by Dr. R. Saffhill (Manchester, UK). The amounts of deoxynucleoside required to give 50% inhibition in the respective RIAs were 0.2 pmoles O^6-MedG and 1.0 pmole O^4-MedT.

Preparation of tissue extracts

Tissues were homogenized in 2 mls of extraction buffer (50 mM TRIS, pH 7.5, 1 mM EDTA, 10 mM dithiothreitol) and sonicated for three periods of ten seconds. Cellular debris was removed by centrifugation at 17,000 g for 30 minutes and the supernatant fraction used as the crude cell extract. All operations were performed at 4^0C. Protein concentrations were determined by the Bradford assay.

Enzyme assays

The substrate for the O^6-methylguanine-DNA methyltransferase assay was *M. luteus* DNA treated with [^{3}H]-MNU (28 Ci/mmol, Amersham) and partially depurinated by heating at neutral pH (Karran, et al., 1979). The substrate for the methylpurine-DNA glycosylase was calf thymus DNA treated with [^{3}H]-DMS (4500 mCi/mmol, New England Nuclear) (Harris, et al., 1983). The O^6-methylguanine-DNA methyltransferase and the methylpurine-DNA glycosylase were assayed as described by Harris, et al.(1983).

RESULTS

Repair of methylation adducts in vivo

The results presented in Figure 1 show the repair of O^6-MedG and O^4-MeT in hamster and rat liver after doses of DMN which induce similar initial levels of these two adducts in each species (i.e., 25 mg/kg for hamsters and 20 mg/kg for rats). Active repair of O^6-MedG and O^4-MeT was found in rat liver with $t^1/_2$ of 19 hours and 30 hours, respectively. However, in he hamster, the level of O^6-MedG was stable between 6 hours and 48 hours, while the level of O^4-MeT was reduced by 42% over this period (mean of two experiments).

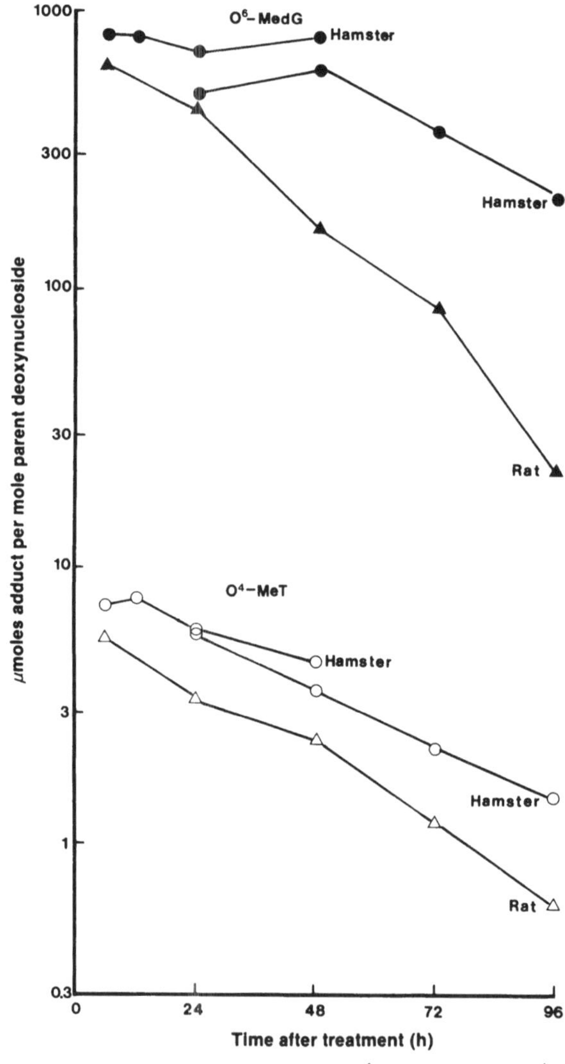

Figure 1. Rate of Disappearance of O^6-MedG and O^4-MeT from Rat and Hamster Liver DNA.

At later time points, the kinetics of removal of O^6-MedG and O^4-MeT in the rat liver remained the same as that seen up to 48 hours. However, in hamster liver there was a sharp decline in O^6-MedG levels , which could principally be an effect of toxicity.

DNA repair enzyme activities in liver extracts

Extracts prepared from control livers from both rat and hamster were found to contain comparable O^6-methylguanine-DNA transferase and methylpurine-DNA glycosylase activity. In the hamster liver extracts, higher levels of the methylpurine-DNA glycosylase were detected 24 hours after the DMN treatment compared to control values, while in the rat liver extracts this increase was less marked and observed only at later time points (Table 1). In both species a depletion of O^6-methylguanine-DNA methyltransferase activity was found

Table 1. DNA Repair Enzyme Levels in Hamster and Rat Liver after Treatment with a Single Dose of DMN

	Hamster (DMN 25mg/kg)						
Enzyme	Time after Treatment (Hours)						
	Control	6	12	24	48	72	96
Methylpurine-DNA glycosylase[1]	0.686 (8)	0.755 (3)	0.615 (5)	1.455 (8)	1.458 (9)	1.502 (3)	1.471 (3)
O^6-methylguanine-DNA methyl-transferase[1]	0.191 (8)	<0.01 (3)	<0.01 (5)	<0.01 (8)	<0.01 (9)	<0.01 (3)	<0.01 (3)

	Rat (DMN 20mg/kg)					
Enzyme	Time after Treatment (Hours)					
	Control	6	24	48	72	96
Methylpurine-DNA glycosyl-ase[1]	0.579 (1)	0.693 (3)	0.772 (2)	0.687 (5)	0.71 (4)	1.029 (3)
O^6-methylguanine-DNA methyl-transferase[1]	0.167 (1)[2]	<0.01 (3)	0.012 (3)	0.033 (5)	0.191 (4)	0.448 (3)

[1]pmoles/mg protein
[2]This value falls within the range previously observed for rats of this species and age in this laboratory.
( ) no. of animals in each group

immediately after DMN treatment and in the hamster no active O^6-methylguanine-DNA methyltransferase was measurable up to 96 hours after this exposure. In contrast, in the rat liver extracts, active enzyme was detectable from 24 hours after the 20 mg/kg DMN treatment and its activity at 96 hours was higher than in control livers.

DISCUSSION

The results in vivo for the DNA methylation adduct levels in the rat liver after a single dose of DMN are comparable to those previously observed (Richardson, et al., 1985; Belinsky, et al., 1986) and clearly show that both O^6-MedG and O^4-MeT are lost rapidly from liver DNA ($t^1/_2$ 19 and 30 hours respectively). These values fall within the range of reported values compiled by Brent, et al. (1988). In contrast, in the hamster, the level of O^6-MedG appeared stable over the period 6 to 48 hours, while the level of O^4-MeT was reduced by 42%. It would not seem likely that cell division and cell death play a major role in establishing the $t^1/_2$ over this initial period and, if it did, it should equally affect the various DNA adducts, and this is not the case (see Figure 1). However, it probably explains the apparent rapid elimination of both methylated adducts observed from 48 hours, this dose of DMN having previously been shown to be toxic to the liver of the hamster (Margison, et al., 1976).

These results point to two qualitative and quantitative differences between rat and hamster livers in their capacity to repair the DNA alkylation adducts, O^6-MedG and O^4-MeT. Firstly, rat liver repairs O^6-MedG more efficiently than hamster liver. This has been previously observed (Stumpf, et al., 1979) and is seen both after high and low doses of DMN. Secondly, while the constitutive levels of O^6-methylguanine-DNA methyltransferase are similar and in both species a depletion in the level of active enzyme immediately after DMN treatment was observed, the subsequent time course of recovery of O^6-methylguanine-DNA methyltransferase to control levels is markedly different in the two species, the recovery being much slower in the hamster liver than in the rat liver (Table 1). With the dose of DMN used here (20 mg/kg and 25 mg/kg for the rat and hamster respectively) substantial levels of adduct are produced (approximately 1 modification per 103 unmodified deoxynucleosides). With such high levels of O^6-MedG being formed in the DNA, any newly synthesized protein could be exhausted immediately following its synthesis. Thus, it cannot be excluded that, in the hamster, the repair protein is beginning to be synthesised at low levels during the period 48 to 96 hours and could contribute to the observed elimination of O^6-MedG from cellular DNA over this time period.

These phenomena, of inactivation and subsequent recovery of O^6-methylguanine-DNA-methyltransferase following carcinogen treatment, have previously been described both in cultured cells (e.g., Yarosh, et al., 1984) and animals (e.g., Montesano, et al., 1983; Becker and Montesano, 1986) and have been shown to be dependent on RNA and protein synthesis and the dose and frequency of carcinogen exposure.

The repair of O-alkylpyrimidines in mammalian cells has

been the subject of much recent discussion (reviewed in Brent, _et al_., 1988) and the apparent consensus of results, _in vivo_ and _in vitro_, is that the mode of O-alkylpyrimidine repair in mammalian cells differs from that of _E. coli_. Thus, it would appear that, at least in the hamster liver, it is not the O^6-methylguanine-DNA methyltransferase that repairs O^4-MeT in DNA in a manner analogous to the _E. coli_ _ada_$^+$ gene encoded O^6-methylguanine-DNA-methyltransferase. A similar suggestion of a different repair protein for these two adducts in rats has been made by Belinsky, _et al_.(1986), based on the observation of differential rates of removal of O^6-MedG and O^4-MeT in rat liver, lung and nasal mucosa following multiple doses of 4-(N-methyl-N-nitrosamine-1-(3-pyridyl)-1-butanone.

An increase in the methylpurine-DNA glycosylase, which removes the major cell killing lesions, 3-methyladenine and 3-methylguanine, was also observed in both the hamster and rat after DMN treatment. This enzyme, although recognizing several methylated purines in DNA (Harris, _et al_., 1983; Gallagher and Brent, 1984), has not previously been shown to repair O-alkylpyrimidines (Hall and Karran, 1986). Becker and Montesano (1985) have demonstrated the repair of O^4-MeT by mammalian liver extracts _in vitro_ and shown that a DNA glycosylase is probably not involved. Hall and Karran (1986) using Raji cell extracts, which contained repair activity for O^4-MeT, detected no methylated adduct released as an ethanol soluble base or nucleoside. Exposure of the hamster to the alkylating agent DMN provides a means of manipulating the levels of O^6-methylguanine-DNA methyltransferase and suggests a model system in which the enzymatic repair activity for O^4-MeT may be further studied.

ACKNOWLEDGEMENTS

This study was partially supported by US NIEHS Grant No. 5 U01 ES04281-02 and CEC Contract No. EVAV 0040-F (CD).

REFERENCES

Abbott, P.S., and Saffhill R., 1979, DNA synthesis with methylated poly(dA-dT) templates, possible role of O^4-methylthymine as a promutagenic base, _Nucleic Acid Res._, 4:761-769.

Becker, R.A., and Montesano, R., 1985, Repair of O^4-methyldeoxythymidine residues in DNA by mammalian liver extracts, _Carcinogenesis_, 6:313-317.

Becker, R.A., and Montesano, R., 1986, In vitro studies of O^6-methylguanine-DNA-methyltransferase activity: mammalian liver and rat reproductive tissues, in: Repair of DNA lesions introduced by N-nitroso compounds, B. Myrnes and H. Krokan, eds, Norwegian University Press, Oslo, pp 101-111.

Belinsky, S.A., White, C.M., Boucheron, J.A., Richardson, F.C., Swenberg, J.A., and Anderson, M., 1986, Accumulation and persistence of DNA adducts in respiratory tissue of rats following multiple administrations of the tobacco specific carcinogen 4-(N-Methyl-N-nitrosamine)-1-(3-pyridyl)-1-butanone, _Cancer Res._, 46:1280-1284.

Brent, T.P., Dolan, M.E., Fraenkel-Conrat, H., Hall, J.,

Karran, P., Laval, F., Margison, G.P., Montesano, R., Pegg, A.E., Potter, P.M., Singer, B., Swenberg, J.A., and Yarosh, D.B., 1988, Repair of O-alkylpyrimidines in mammalian cells: A present consensus, Proc. Natl. Acad Sci, USA., 85:1759-1762.

Craddock, V.M. , Henderson, A.R., and Gash, S., 1982, Nature of the constitutive and induced mammalian O^6-methylguanine DNA repair enzyme, Biochem. Biophys. Res. Commun., 107:546,553.

Gallagher, P.E., and Brent, T.P., 1984, Further purification and characterisation of human 3-methyladenine DNA glycosylase. Evidence for broad specificity, Biochem. Biophys. Acta., 782:394-401.

Hall, J., and Karran, P., 1986, O-methylated pyrimidines - important lesions in cytotoxicity and mutagenicity in mammalian cells, in: 'Repair of DNA lesions introduced by N-Nitroso compounds,', B. Myrnes and H. Krokan eds, Norwegian University Press, Oslo, pp 73-88.

Harris, A.L., Karran, P., and Lindahl, T., 1983, O^6-methylguanine-DNA-methyltransferase of human lymphoid cells: structural and kinetic properties and absence in repair deficient cells, Cancer Res., 43:3247-3252.

Karran, P., Lindahl, T., and Griffin, B., (1979), Adaptive response to alkylating agents involves alteration in situ of O^6-methylguanine residues in DNA, Nature, 280:76-78.

Lindahl, T., 1982, DNA repair enzymes, Ann. Rev. Biochem., 51:61-87.

Margison, G.P., Margison, J.M., and Montesano, R., 1976, Methylated purines in the deoxyribonucleic acid of various Syrian-Golden-Hamster tissues after administration of a hepatocarcinogenic dose of dimethyl-nitrosamine, Biochem. J., 157:627-634.

McCarthy, T.M.V., and Lindahl, T., 1985, Methyl phosphotriester in alkylated DNA are repaired by the Ada regulatory protein of E. coli, Nucl. Acids Res., 13:2683-2698.

Montesano, R., Brésil, H., Planche-Martel, G., Margison, G.P., and Pegg A.E., 1983, Stability and capacity of dimethylnitrosamine-induced O^6-methylguanine repair system in rat liver, Cancer Res., 43:5808-5814.

Müller, R., and Rajewsky, M.F., 1980, Immunological quantification by high affinity antibodies of O^6-ethyldeoxyguanosine in DNA exposed to N-ethyl-N-nitrosourea, Cancer Res., 40:887-896.

Olsson, M., and Lindahl, T., 1980, Repair alkylated DNA in E. coli. Methyl group transfer from O^6-methylguanine to a protein cystein residue, J. Biol. Chem., 255:10569-10571.

Pegg, A.E., Roberfroid, M., Van Bahr, C., Foote, R.S., Mitra, S., Brésil, H. Likhachev, A. , and Montesano, R., 1982, Removal of O^6-methylguanine from DNA by human liver fractions, Proc. Natl. Acad. Sci. USA., 79:5162-5165.

Pegg, A.E., Wiest, L., Foote, R.S., Mitra, S., and Perry, W., 1983, Purification and properties of O^6-methylguanine-DNA-methyltransferase from rat liver, J. Biol Chem., 258:2327-2333.

Richardson, F.C., Dyroff, M.C., Boucheran, J.A., and Swenberg, J.A., 1985, Differential repair of O^4-alkylthymidine following exposure to methylating and ethylating hepatocarcinogens, Carcinogenesis, 6:625-629.

Stumpf, R. , Margison, G.P., Montesano, R., and Pegg, A.E., 1979, Formation and loss of alkylated purines from DNA of hamster liver after administration of dimethylnitrosamine, Cancer Res., 39:50-54.

Swenberg, J.A., Dyroff, M.C., Bedell, M.A., Popp, J.A., Huh, N., Kirstein, U., and Rajewsky, M.F., 1984, O^4-Ethyl deoxythymidine, but not O^6-ethyldeoxyguanosine accumulates in hepatocyte DNA of rats exposed continuously to diethylnitrosamine, Proc. Natl. Acad. Sci. USA, 81:1692-1695.

Umbenhauer, D.R., Wild, C.P., Montesano, R., Saffhill, R., Boyle, J.M. , Huh, N., Kirsten, U, Thomale, J., Rajewsky, M.F., and Lu, S.H., 1985, O^6-methyldeoxyguanosine in oesophogeal DNA among individuals at high risk of oesophageal cancer, Int. J. Cancer, 36:661-665.

Wild, C.P., Smart, G., Saffhill, R., and Boyle, J.M. , 1983, Radioimmunoassay of O^6-methyldeoxyguanosine in DNA of cells alkylated in vitro and in vivo, Carcinogenesis, 4:1605-1609.

Wild, C.P., Lu, S.H., and Montesano, R., 1987, Radio-immunoassay used to detect DNA alkylation adducts in tissues from populations at high risk for oesophageal and stomach cancer, in: "The Relevance of N-Nitroso Compounds to Human Cancer Exposures and Mechanisms, H. Bartsch, I. O'Neill, and R. Schulte-Hermann, eds, IARC Scientific Publications, 84:538-543.

Yarosh, D.B., 1985, The role of O^6-methylguanine-DNA methyltransferase in cell survival, mutagenesis and carcinogenesis, Mutation Res. , 145:1-16.

Yarosh, D.B., Rice, M., Day, R.S., Foote, R.S., and Mitra, S., 1984, O^6-methylguanine-DNA methyltransferase in human cells. Mutation Res., 131: 27-36.

FORMATION OF 1,N^6-ETHENODEOXYADENOSINE AND 3,N^4-ETHENODEOXYCYTIDINE IN DNA FROM SEVERAL ORGANS OF RATS EXPOSED TO VINYL CHLORIDE

Alain Barbin, Françoise Ciroussel and Helmut Bartsch

International Agency for Research on Cancer, 150 cours Albert-Thomas, 69372 Lyon Cedex 08, France

ABSTRACT

Seven-day old (group I) and 28-day old (group II) BDVI rats were exposed for two weeks to 500 ppm vinyl chloride (VC) in air [7 hours per day, for 7 days (group I) or 5 days (group II) per week]. DNA from several organs was analysed for the formation of three VC adducts: 7-(2-oxoethyl)-guanine (oxetG), 1,N^6-ethenodeoxyadenosine (εAdR) and 3,N^4-ethenodeoxycytidine (εCdR). oxetG was measured as 7-(2-hydroxy-2-[^{3}H]-ethyl)guanine by a post-labelling/HPLC procedure. εAdR and εCdR were dosed from enzymatic DNA hydrolysates separated by reversed-phase HPLC: a competitive radioimmunoassay (RIA) in the presence of specific murine monoclonal antibodies (Mab;obtained in collaboration with M.F. Rajewsky and G. Eberle) was used. Both ethenonucleosides were detected in the DNA from the liver, lung and brain of group I rats, at levels (fmoles/mg DNA) ranging from 62 to 133 for εAdR and 162 to 394 for εCdR. Molar ratios of oxetG/εCdR and oxetG/εAdR in DNA were about 30 and 80, respectively, in these three organs. In contrast, εAdR and εCdR were not detected (detection limit, 25 fmoles/mg DNA) in the kidney of group I nor in the liver from group II rats. These findings are discussed in relation to the organotropism of VC-induced carcinogenesis.

INTRODUCTION

VC has been implicated in the etiology of occupational cancers, in particular of hepatic angiosarcomas (Creech and Johnson, 1974; Forman, <u>et al.</u>, 1985) and it is generally accepted that the carcinogenic and mutagenic effects of VC are due to its reactive metabolite, chloroethylene oxide (review articles: Bartsch and Montesano, 1975; Barbin and Bartsch, 1986; Bolt, 1986). Four adducts of VC (chloroethylene oxide) with nucleic acid bases are presently known (Figure 1). Following short treatments of rodents with [^{14}C]-VC, oxetG was

DNA Repair Mechanisms and Their Biological Implications in Mammalian Cells
Edited by M.W. Lambert and J. Laval
Plenum Press, New York

the major adduct observed in DNA (Osterman-Golkar, et al., 1977; Laib, et al., 1981), and low amounts of N^2, 3-ethenoguanine (εG) were detected in rat liver DNA (Laib, 1986). Under similar experimental conditions, [^{14}C]-labelled 1,N^6-ethenoadenine and 3,N^4-ethenocytosine moieties were easily detected in RNA but not in DNA of rat liver (Laib, et al., 1981). In contrast, Green and Hathway (1978) presented limited mass-spectral evidence, but no quantitative data, for the formation of εCdR and, tentatively, εAdR in liver DNA of rats after a two-year exposure to VC. Recently, a sensitive immunoanalytical method using Mab was developed for dosing εAdR and εCdR (Eberle, et al., 1989). With this method, Eberle, et al., could demonstrate the formation of, and quantitate both ethenoadducts in lung and liver DNA from infant Sprague-Dawley rats exposed to VC.

The detection of ethenobases in DNA may be an important step towards the understanding of VC-induced mutagenesis and carcinogenesis. Indeed, in vitro replication or transcription fidelity assays using synthetic templates suggest that the three ethenobases, but not oxetG may be potential promutagenic lesions of VC (Barbin and Bartsch, 1986; Singer, et al., 1987; Bolt, 1988). Therefore, determination of the kinetics of formation/repair of ethenobases in vivo should permit to further elucidate how VC exerts its genotoxic effects. As a first step towards this aim, we exposed young BDVI rats to VC and analysed the formation of εAdR and εCdR in the DNA of several organs known to be sensitive to VC-induced carcinogenesis. In addition, we determined the levels of oxetG.

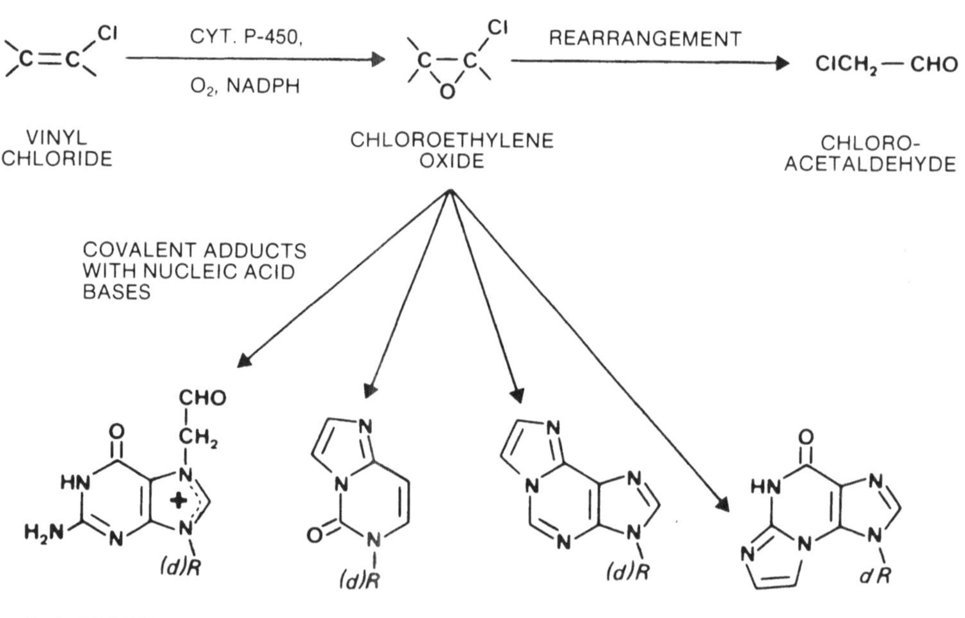

Figure 1. Vinyl chloride metabolites and nucleic acid adducts.

MATERIALS AND METHODS

Chemicals

VC (3% in nitrogen) was obtained from Airgaz, Salaise, France. Nucleosides and bases were purchased from P.L. Biochemicals (St. Goar, FRG) or Sigma (St. Louis, MO, USA). εCdR, oxetG, [^{3}H]-εAdR and [^{3}H]-εCdR were prepared as previously described (Eberle, et al., 1989). [^{3}H]-sodium borohydride ([^{3}H]-$NaBH_4$, 20 Ci/mmole) was obtained from CEA, Gif- sur--Yvette, France. Cetrimonium bromide (hexadecyltrimethylammonium bromide) was obtained from Merck (Darmstadt, FRG) and heptafluorobutyric acid from Pierce Chemical Co. (Rockford, IL, USA). Enzymes used for DNA preparation or hydrolysis were purchased from Boehringer-Mannheim, Mannheim, FRG.

Monoclonal antibodies (Mab)

The production and characterization of murine Mab which specifically recognize εAdR or εCdR have been described in detail elsewhere (Eberle, et al., 1989). Mab EM-A-1 (anti-εAdR) and EM-C-1 (anti-εCdR) were used in this study.

Animals

BDVI rats were bred in this laboratory and were given commercial rat chow (Biscuits Extra Labo from Société Piétremont, Provins, France) and water ad libitum.

VC exposure

A dynamically operating inhalation exposure system was used according to the original design from Barrow and Steinhagen (1982), with some modifications. It consisted of an acrylic plastic chamber (68 x 52 x 52 cm) divided into two compartments which could accommodate up to four normal rat cages. Chamber inlet and outlet were fabricated from 4 cm O.D., high density polyethylene tubing. VC was trapped downstream of the chamber in a system which consisted of four, 0.25 m O.D. x 1.23 m high density polyethylene cylinders connected in series. The two first cylinders were filled with Siliporite NK 30 (CECA SA, Velizy-Villacoublay, France) and the two last ones with activated charcoal (CECA). A rotary vane vacuum pump fitted with a T connection and a ball valve at intake was used to draw the test atmosphere through the chamber and the traps. Chamber airflow was adjusted by means of the ball valve to correspond to about 15 chamber volume changes per hour. Under these conditions, the chamber was operated at a sub-atmosphere pressure, thus system leakage was minimized. Diluted VC (3% in nitrogen) was metered from a pressurized cylinder fitted with a fine-adjustment valve and a calibrated flowmeter, and was mixed with ambiant air inside the chamber inlet pipe. Samples of the chamber atmosphere were taken through sampling ports and the concentration of VC was determined by gas chromatography on Porapack QS (80-100 mesh, from Waters, Milford, MA, USA), with flame ionization detection.

During exposure, rats were housed in four plastic and brass cages placed in the inhalation chamber and had access to food and water ad libitum. Two groups of rats were exposed to 500 ppm VC in air. Group I consisted of 39 seven-day old male and female rats, originating from 6 litters, which were exposed for 14 consecutive days, for 7 hours per day. These infant rats were kept with their mothers during exposure. Group II consisted of 4 male 28-day old rats which were exposed for 2 weeks, 5 days per week, 7 hours per day. Animals from both groups were killed at 12 hours following the end of the last exposure; their organs were removed and kept at -80°C until further processing.

DNA purification

Nucleic acids were extracted from rat tissues as described by Krieg, et al.(1983). RNA was removed by treatment with 100 µg/ml bovine pancreas RNase (preheated at 80°C for 30 minutes) in 1 mM EDTA, 10 mM Tris-HCl buffer (pH 7.4), for 1 hour at 37°C.

Determination of εAdR and εCdR

εAdR and εCdR were analyzed by competitive RIA/HPLC as described previously by Eberle, et al. (1989), with some modifications. Purified DNA was digested enzymatically: following a 30 minute incubation at 37°C with DNase I (50 µg/mg DNA) in a 0.05 M Tris-HCl buffer (pH 6.75) containing 0.01 M $MgCl_2$, snake venom phosphodiesterase (15 µg/mg DNA) and alkaline phosphatase (10 U/mg DNA) were added, and incubation was continued for two further hours. Nucleosides and bases generated by DNA hydrolysis were separated by reversed-phase HPLC on 5 µm Nucleosil C_{18} (10 mm I.D. x 25 cm prepacked column from Société Francaise Chromato Colonne, Gagny,France). A mobile phase gradient, made up from eluant A (0.025 M ammonium formate, pH 5.5) and eluant B (methanol: water, 80:20, v/v), was used. The gradient program was: Step 1, 0% eluant B (isocratic, 10 minutes); Step 2, 0 to 2% eluant B (linear, 5 minutes); Step 3, 2 to 9% eluant B (linear, 6 minutes); Step 4, 9 to 13% eluant B (linear, 6 minutes); Step 5, 13 to 25% eluant B (linear, 20 minutes); Step 6, 25% eluant B (isocratic, 15 minutes). Elution was carried out at room temperature, at a flow rate of 3 ml/minute, and was monitored by UV absorbance at 254 nm. About 1.5-2 mg of hydrolysed DNA could be separated per each chromatographic run. Fractions containing isolated 2'-deoxyribonucleosides were pooled from several runs for quantitative determinations. Amounts of unmodified 2'-deoxyribonucleosides were determined from absorbance at 260 nm. The amounts of εAdR and εCdR in the respective fractions were measured by competition RIA as previously described (Eberle, et al., 1989), using Mab EM-A-1 or Mab EM-C-1, respectively, and $[^3H]$-εAdR and $[^3H]$-εCdR as tracers. The degree of inhibition of tracer-antibody binding in the RIA was read from a standard curve and the values were converted to fmoles of εAdR or εCdR.

Determination of oxetG

oxetG was measured as 7-(2-hydroxy-2-$[^3H]$-ethyl)guanine, after reductive tritiation of the cetrimonium salt of DNA with $[^3H]$-$NaBH_4$ in isopropanol, as described by Scherer, et al., (1986). DNA was then depurinated by mild acid hydrolysis in 0.1 N hydrochloric acid (1 hour at 70°C). Purine bases were separated by reversed-phase HPLC on Nucleosil C_{18} (5 µm; 4.5

mm I.D. x 25 cm column), using a non-linear gradient made up from 3% to 80% methanol in 0.1% heptafluorobutyric acid, at a flow rate of 1 ml/minute at room temperature. 7-(2-Hydroxy-2-[^{3}H]-ethyl)guanine, which eluted at 20.4 minutes, was dosed by liquid scintillation counting. Amount of non-modified guanine (elution time, 19.7 minutes) was determined by UV absorbance at 260 nm.

RESULTS

A typical HPLC separation of standard nucleosides and bases, including VC-DNA adducts, is shown in Figure 2. Under the conditions used, εCdR and εAdR were well resolved from natural nucleosides and from other VC-DNA adducts, and eluted at 57.8 and 60.2 minutes, respectively.

Using an HPLC pre-separation step and a competitive RIA for quantitation, εCdR and εAdR were analysed in enzymatic DNA hydrolysates from several organs of rats treated with VC. Both adducts were detected in the liver, lung and brain of group I animals, at levels ranging respectively from 0.16 to 0.39 and from 0.06 to 0.13 pmoles per mg DNA (Table I). Values from

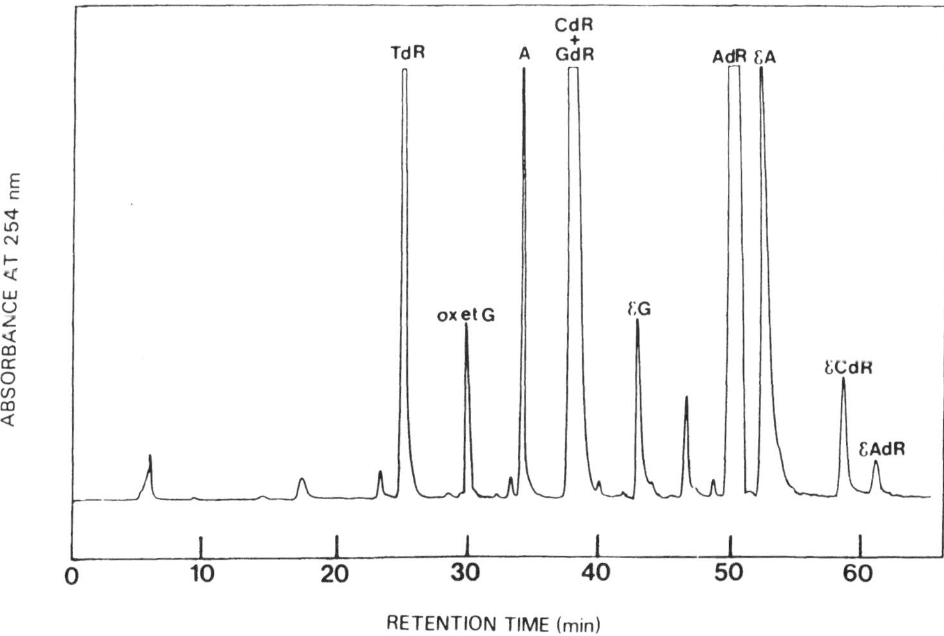

Figure 2. Reversed-phase HPLC of VC-DNA adducts (standards). TdR, thymidine; oxetG, 7-(2-oxoethyl)guanine; A, adenine; CdR, 2'-deoxycytidine; GdR, 2'-deoxyguanosine; εG, N^2,3-ethenoguanine; AdR, 2'-deoxyadenosine; εA, 1,N^6-ethenoadenine; εCdR, 3,N^4-etheno-2'-deoxycytidine; εAdR, 1,N^6-etheno-2'-deoxyadenosine.

duplicate analyses (Table I) were very close, indicating a good reproducibility of the measurements. In contrast to the data shown in Table I, εCdR and εAdR could not be detected in kidney DNA from young rats (group I) nor in liver DNA from adult animals (group II) exposed to VC, at a detection limit of 25 fmoles of adduct per mg DNA. These two ethenonucleosides were also undetectable in the liver DNA from untreated young BDVI rats (same detection limit as above).

For comparison, oxetG, the major VC-DNA adduct, was also analyzed in the DNA of the liver, lung and brain from group I rats (Table I), using a post-labelling procedure. Levels of oxetG were found to range between 4 and 8 pmoles per mg DNA. Total DNA alkylation, defined as the sum of the three adducts, oxetG, εCdR and εAdR, decreased in the following order: liver, lung and brain. However, constant molar ratios between the three adducts were observed irrespective of the organ. Values (mean ± SD) were 28 ± 6 for the ratio oxetG/εCdR and 83 ± 5 for the ratio oxetG/εAdR.

Table I. DNA adducts measured in young BDVI rats exposed to vinyl chloride[1].

Organ[2]	pmoles/mg DNA of		
	oxetG	εCdR	εAdR
Liver	8.16	0.415[3] 0.373	0.133[3] 0.133
Lung	6.51	0.196[3] 0.178	0.107[3] 0.091
Brain	4.18	0.173[3] 0.151	0.062

[1]500 ppm VC in air from days 7 to 21 after birth, 7 hours per day; sacrifice at 12 hours following end of exposure.
[2]Organs were pooled for DNA analyses.
[3]Values from two analyses are listed.

DISCUSSION

Mab of high affinity and specificity for εCdR and εAdR had been previously developed and characterized (Eberle, et al., 1989). Using these antibodies in competitive RIA, in combination with an HPLC pre-separation step of DNA hydrolysates, Eberle, et al. (1989), were able to show the formation of εCdR and εAdR in the DNA of the liver and lung of 11-day old Sprague-Dawley rats exposed for 10 days (7 hours on days 1-9, 24 hours on day 10) to 2000 ppm VC and sacrificed immediately following the end of exposure. The following molar ratios of ethenoadducts/unmodified bases in DNA were reported (Eberle, et al., 1989): εCdR/CdR, 1.6×10^{-7} (liver),

3.3×10^{-7} (lung); εAdR/AdR, 5.0×10^{-8} (liver), 1.3×10^{-7} (lung).

Using the same Mab and a very similar methodology, we report here the formation of these two ethenoadducts in the DNA of several organs of 7-day old BDVI rats exposed for 14 days (7 hours per day) to 500 ppm VC and killed at 12 hours following the end of exposure. In addition to liver and lung DNA, εCdR and εAdR were also found in brain DNA. The molar ratios of ethenoadducts/unmodified bases in DNA were as follows: εCdR/CdR, 4.9×10^{-7} (liver), 2.3×10^{-7} (lung) and 2.1×10^{-7} (brain); εAdR/AdR, 1.3×10^{-7} (liver), 0.97×10^{-7} (lung), 0.61×10^{-7} (brain).

These data allow a comparison of the formation of ethenoadducts in Sprague-Dawley and BDVI rats, following VC exposure. Treatment schedules were quite similar in both strains, except that Sprague-Dawley rats were exposed to higher concentrations of VC in air; nevertheless, the effective biological dose of VC in these two experiments was probably similar because the pharmacokinetics of VC in rats is saturated at about 250 ppm of VC in air (Bolt, et al., 1977). Thus, comparing the molar ratios listed above, we observe similar levels of either εCdR or εAdR in the lung DNA of both Sprague-Dawley and BDVI rats. In contrast, levels of ethenoadducts in liver DNA are three-fold lower in Sprague-Dawley than in BDVI rats. This observation, which needs to be confirmed by further analyses, suggests that the livers from these two rat strains may differ in their capacity to activate VC and/or to repair εCdR and εAdR.

In parallel to the analysis of ethenoadducts, we determined also oxetG in the DNA from VC-exposed rats, using a post-labelling procedure (adapted from Scherer, et al., 1986). Molar ratios between oxetG and εCdR or εAdR moieties were found to be similar in the DNA of the three organs, liver, lung and brain: values were about 30 and 80 for oxetG/εCdR and oxetG/εAdR molar ratios, respectively. In a previous study, the third ethenoadduct, εG, was measured in liver DNA from VC-treated infant rats, and detected at a molar ratio oxetG/εG of about 100 (Laib, 1986).

Since εCdR and εAdR are putative promutagenic lesions of VC (see Introduction), it is tempting to compare their formation in DNA from different organs of rats, as determined in the present work and by Eberle, et al. (1989), to the target sites for VC-induced tumorigenesis (IARC, 1979, 1987; Maltoni, et al., 1984). Unfortunately, this comparison is presently biased by the absence of carcinogenicity data on BDVI rats. The principal target organ appears to be the liver which in four rat strains is sensitive to tumour induction by VC (Drew, et al., 1983; Feron, et al., 1981; Groth, et al.; 1981; Hong, et al., 1981; Maltoni, et al., 1984). In our study on BDVI rats, the liver had the highest level of DNA akylation. Furthermore, the detection of εCdR and εAdR in liver DNA from young rats but not from adult animals, following a VC-treatment, reflects well the higher sensitivity of the former to VC-induced carcinogenesis (Maltoni, et al., 1984). Maltoni, et al.(1984), also showed perinatal exposure to VC to be essential for the development of lung adenomas in Sprague-Dawley rats. Our data, which demonstrate the formation of εCdR and εAdR in lung DNA of VC-treated infant BDVI rats, are in accord with the lung of infant rats being sensitive to tumour development by VC. We also detected the formation of ethenoadducts in the brain DNA of young BDVI rats, following exposure to VC. The brain is another established target site for VC-induced carcinogenesis in rats (Sprague-Dawley and Wistar strains; Maltoni, et al., 1984). In experiments on VC-induced carcinogenesis, a significant increase of kidney tumours has been observed in one rat strain only, and following lifetime exposures (Sprague-Dawley rats; Maltoni, et al., 1984). Thus, the lack of detection of ethenoadducts (<25 fmoles/mg DNA) in kidney DNA from infant BDVI rats exposed to VC probably results from a low capacity of the kidney of these young animals to activate VC. Our data, although limited, therefore suggest that the levels of ethenoadducts in DNA may be indicative of the organs which are at risk for tumor development, following exposure to VC.

ACKNOWLEDGEMENTS

We wish to thank Ms. Y. Granjard for typing the manuscript. This work was supported in part by a contract with the "Groupe de Recherche sur les Hépatites, Cirrhoses et Cancers du Foie" (INSERM, Lyon, France) and ATOCHEM (Paris).

REFERENCES

Barbin, A., and Bartsch, H., 1986, Mutagenic and promutagenic properties of DNA adducts formed by vinyl choride metabolites, in: "The Role of Cyclic Nucleic Acid Adducts in Carcinogenesis and Mutagenesis," B. Singer and H. Bartsch, eds., (IARC Scientific Publications No. 70), International Agency for Research on Cancer, Lyon, France, 345-358.

Barrow, C.S., and Steinhagen, W.H., 1982, Design, construction and operation of a simple inhalation exposure system, Fund. Appl. Toxicol., 2:33-37.

Bartsch, H., and Montesano, R., 1975, Mutagenic and carcinogenic effects of vinyl chloride, Mutat. Res., 32:93-114.

Bolt, H.M., 1986, Metabolic activation of vinyl chloride, formation of nucleic acid adducts and relevance to carcinogenesis, in: "The Role of Cyclic Nucleic Acid Adducts in Carcinogenesis and Mutagenesis," B. Singer and H. Bartsch, eds., (IARC Scientific Publications No. 70), International Agency for Research on Cancer, Lyon, France, 261-268.

Bolt, H.M., 1988, Roles of etheno-DNA adducts in tumorigenicity of olefins. CRC Critical Reviews in Toxicology, 18:299-309.

Bolt, H.M., Laib, R.J., Kappus, H., and Buchter, A., 1977, Pharmacokinetics of vinyl chloride in the rat, Toxicology, 7:179-188.

Creech, J.L., Jr., and Johnson, M., 1974, Angiosarcoma of the liver in the manufacture of polyvinyl chloride, J.Occup. Med., 16:150-151.

Drew, R.T., Boorman, G.A., Haseman, J.K., McConnell, E.E., Busey, W.M., and Moor, J.A., 1983, The effect of age and exposure duration on cancer induction by a known carcinogen in rats, mice, and hamsters, Toxicol. Appl. Pharmacol., 68:120-130.

Eberle, G., Barbin, A., Laib, R.J., Ciroussel, F., Thomale, J., Bartsch, H., and Rajewsky, M.F., 1989, 1,N^6-Ethenodeoxyadenosine and 3,N^4-ethenodeoxycytidine detected by monoclonal antibodies in lung and liver DNA of rats exposed to vinyl chloride, Carcinogenesis, 10:209-212.

Feron, V.J., Hendriksen, C.F.M., Speek, A.J., Til, H.P., and Spit, B.J., 1981, Lifespan oral toxicity study of vinyl chloride in rats, Food Cosmet. Toxicol., 19:317-333.

Forman, D., Bennet, B., Stafford, J., and Doll, R., 1985, Exposure to vinyl chloride and angiosarcoma of the liver: a report of the register of cases, Br.J. Ind. Med., 42: 750-753.

Green, T., and Hathway, D.E., 1978, Interactions of vinyl chloride with rat liver DNA in vivo, Chem. Biol. Interact., 22:211-224.

Groth, D.H., Coate,W.B., Ulland, B.M., and Hornung, R.W., 1981, Effects of aging on the induction of angiosarcoma, Environ. Health Perspect., 41:53-57.

Hong, C.B., Winston, J.M., Thornburg, L.P., Lee, C.C., and Woods, J.S., 1981, Follow-up study on the carcinogenicity of vinyl chloride and vinylidene chloride in rats and mice: tumor incidence and mortality subsequent to exposure, J. Toxicol. Environ. Health, 7:909-924.

IARC, 1979, "IARC Monographs on the Evaluation of the Carcinogenic Risk of Chemicals to Humans: Some Monomers, Plastics and Synthetic Elastomers,and Acrolein," International Agency for Research on Cancer, Lyon, France, 19:377-438.

IARC, 1987, "IARC Monographs on the Evaluation of the Carcinogenic Risk of Chemicals to Humans; Supplement 7: Overall evaluations of carcinogenicity:an updating of IARC Monographs Volumes 1 to 42", International Agency for Research on Cancer, Lyon, France, 373-376.

Krieg, P., Amtmann, E., and Sauer, G., 1983, The simultaneous extraction of high molecular weight DNA and of RNA from solid tumors, Anal. Biochem., 134:288-294.

Laib, R.J., 1986, The Role of cyclic base adducts in vinyl chloride induced carcinogenesis: Studies on nucleic acid alkylation in vivo, in: "The Role of Cyclic Nucleic Acid Adducts in Carcinogenesis and Mutagenesis," B. Singer and H. Bartsch, eds., (IARC Scientific Publications No. 70), International Agency for Research on Cancer, Lyon, France, 101-108.

Laib, R.J., Gwinner, L.M., and Bolt, H.M., 1981, DNA alkylation by vinyl chloride metabolites: etheno derivatives or 7-alkylation of guanine?, Chem. Biol. Interact., 37:219-231.

Maltoni, C., Lefemine, G., Ciliberti, A., Cotti, G., and Caretti, D., 1984, Experimental research on vinyl chloride carcinogenesis, in: "Archives of Research on Industrial Carcinogenesis," C. Maltoni, and M.A. Mehlman, eds., Princeton Scientific Publishers, Inc. , Princeton, N.J., USA, vol. 2.

Osterman-Golkar, S., Hultmark, D., Segerbeck, D., Calleman, C.J., Gothe, R., Ehrenberg, L., and Wachtmeister, C.A., 1977, Alkylation of DNA and proteins in mice exposed to vinyl chloride, Biochem. Biophys. Res. Commun., 76: 259-266.

Scherer, E., Winterwerp, H., and Emmelot, P., 1986, Modification of DNA and metabolism of ethyl carbamate in vivo: formation of 7-(2-oxoethyl)-guanine and its sensitive determination by reductive tritiation using ^{3}H sodium borohydride, in: The Role of Cyclic Nucleic Acid Adducts in Carcinogenesis and Mutagenesis," B. Singer and H. Bartsch, eds., (IARC Scientific Publications No. 70), International Agency for Research on Cancer, Lyon, France, 109-125.

Singer, B., Spengler, S.J., Chavez, F., and Kusmierek, J.T., 1987, The vinyl chloride-derived nucleoside, N^2,3-ethenoguanosine, is a highly efficient mutagen in transcription, Carcinogenesis, 8:745-747.

THE ROLE OF O^6-ALKYLGUANINE IN CELL KILLING AND SISTER CHROMATID EXCHANGE INDUCED BY ALKYLATING AGENTS

Angelo Abbondandolo[1,2], Gabriele Aquilina[3], Margherita Bignami[3], Stefania Bonatti[1,4], Roberto Cosani[1], Eugenia Dogliotti[3], Guido Frosina[1] and Andrea Zijno[3]

[1]Laboratorio di Mutagenesi, IST, Genova; [2]Cattedra di Genetica, Universita' di Genova; [3]Istituto Superiore di Sanita', Roma; [4]Istituto di Mutagenesi e Differenziamento, CNR, Pisa, Italy

INTRODUCTION

When DNA is exposed to alkylating agents, reactions occur at as many as fifteen different sites on the bases and at the phosphodiester backbone (Beranek, et.al., 1980). The relative proportion of the alkylation products depends mainly on the nucleophilicity of the reactive sites and on the chemical properties of the alkylating agents (Lawley, 1974).

The biological consequences of DNA alkylation in mammalian cells include cell killing, neoplastic transformation and a wide variety of genetic changes. To assign each of these biological effects to one or more specific DNA alkylation products is an exceedingly difficult task. It is, therefore, not surprising that our knowledge on this aspect is largely incomplete. Indeed, the correlation between defined modifications and genetic effects has only been shown for point mutations. Sound experimental evidence accumulated over the years indicates that O^6-alkylguanine (O^6alkG) and O^4-alkylthymine (O^4alkT) are critical lesions for the induction of point mutations, the role of the latter being probably less important, owing to its relatively low frequency in DNA.

It is interesting that O^6alkG is repaired in all organisms by a non-catalitic process mediated by suicidal methyltransferases. It has been stressed several times (e.g., Demple, 1988) that the apparent prodigality of cells, prepared to waste a whole protein to repair a single lesion, is probably justified by the need to repair that lesion rapidly. Implicit in this view is the assumption that the lesion to be repaired has to be particularly harmful to the cell. As mentioned, the causative role of O^6alkG in mutation induction is a relative-ly old acquisition. In more recent times, evidence is accumulating pointing to O^6alkG as a key lesion that, at

DNA Repair Mechanisms and Their Biological Implications in Mammalian Cells
Edited by M.W. Lambert and J. Laval
Plenum Press, New York

least in eukaryotic cells, may be involved in other biological effects as well.

In this article we focus on the possible role of O^6alkG in the induction of two effects of alkylating agents, sister chromatid exchange (SCE) and cell killing. The role of O^6alkG in the induction of other forms of chromosomal damage, such as chromosome aberrations and micronuclei, will not be considered here since it has been the subject of a recent paper (Bonatti and Abbondandolo, 1989) to which the reader is referred. A role of O^6alkG in SCE induction was first proposed by Carrano and coworkers (1978) and by Wolff (1978). The possibility that O^6alkG is a lethal lesion did not receive serious consideration until recently, probably because separate lesions had previously been shown to be responsible for mutation and cell death in bacteria (Karran, et al., 1982). The evidence for and against O^6alkG being involved in SCE induction and cell killing comes from different experimental approaches that will be presented in the separate sections of this paper.

EARLY STUDIES: THE "ADAPTIVE RESPONSE"

The existance of an adaptive response to alkylating agents in bacteria (Samson and Cairns, 1977) had been known for a short time when Samson and Schwartz (1980) reported that an adaptive treatment rendered Chinese hamster ovary (CHO) cells resistant to cell killing and SCE induction by N-methyl-N'-nitro-N-nitrosoguanidine (MNNG). Other reports soon followed, indicating the existance of an adaptive response in mammalian cells, analogous to that in bacteria.

Studies on mammalian cell adaptation, however, did not lead to a consistent pattern of results and left many questions unresolved (Frosina and Abbondandolo, 1985). Indeed, some of the cell lines in which adaptation to alkylating agents was described contain little or no O^6-methylguanine-DNA-methyltransferase (O^6-MT) activity (Warren, et al., 1979; Goth-Goldstein, 1980; Bignami, et al., 1987) and, more importantly, this activity is not increased under adaptive conditions (Mariani, et al., 1988). These studies therefore, with few exceptions (Laval and Laval, 1984), could not attribute specific biological effects to defined DNA lesions. Certainly, the approach based on the adaptive response did not establish a role of O^6alkG in cell killing and SCE induction.

CORRELATION STUDIES BASED ON MOLECULAR DOSIMETRY DATA

A typical approach to the identification of DNA lesions responsible for specific biological effects has been to compare the effects of alkylating agents with different reactivities towards specific sites in DNA (Vogel and Natarajan, 1981; van Zeeland, 1988). This approach implies the measurement of the "molecular" or "DNA dose" at many reactive sites in DNA. The approach has some limitations, because (1) it does not allow distinction between lesions that are formed in similar proportions by each alkylating agent, and (2) it does not identify a lesion unless it is the only (or major) lesion responsible for the effect.

The majority of data obtained in molecular dosimetry studies fail to indicate a correlation between O^6alkG and SCE (Connel and Medcalf, 1982; Heflich, et al., 1982; Morris, et al., 1983; Natarajan, et al., 1984; de Kok, et al., 1985). In one study, correlation with O^6alkG was reported for SCE induced by ethylmethanesulfonate (EMS), MNU and N-nitroso-N-ethylurea (ENU), but not by methylmethanesulfonate (MMS). It was proposed that alkylation at O^6 of guanine was relevant to SCE induced by the oxygen reactive agents, while other alkylated bases, such as 3-methyladenine and 3-methylthymine could account for the MMS-induced SCE (Swenson, et al., 1980). Other proposed alkylated sites relevant for SCE induction include 3-methyladenine and 7-methylguanine (Connel and Medcalf,1982), 3-ethyladenine, 3-ethylguanine and ethylated phosphodiesters (Heflich, et al., 1982; Morris, et al., 1983), and O^2-ethylcytosine and 3-ethylguanine (de Kok, et al., 1985).

As described above, the molecular dosimetric approach has not led to the identification of the DNA lesion responsible for SCE induction. Moreover, no correlation of specific DNA lesions with cytotoxicity was found in any of the mentioned studies, as well as in other studies where mutation and other end points were considered (e.g., Newbold, et al., 1980). The reason, in our opinion, is that more than one lesion is involved in each of the two effects under consideration, a situation in which the dosimetric approach becomes ineffective, as illustrated in more detail elsewhere (Bonatti and Abbondandolo, 1989). In conclusion, a role of O^6alkG in cell killing and SCE induction can neither be proved nor disproved by a dosimetric approach.

STUDIES ON CELLS WITH DIFFERENT CAPACITIES TO REPAIR O^6-ALKYLGUANINE

(a) Mer^+ and Mer^- Cells

Useful information on the biological role of O^6alkG has come from the study of cells with or without capacity to repair this lesion. Day and coworkers (1980a) classified cell strains derived from human tumors according to their ability to support the growth of MNNG-damaged adenovirus as Mer^+ and Mer^-, Mer standing for methylation repair. Mer^- cells seem identical with Mex lymphoblastoid cell lines (Sklar and Strauss, 1981), Mex standing for methyl excision. Mer and Mex cells are generally defective in O^6-methylguanine (O^6meG) repair (Day, et al., 1980b; Sklar and Strauss, 1981) and devoid of the associated repair protein, O^6-MT (Harris, et al., 1983; Yarosh, et al., 1983; Domoradzki, et al., 1984).

All Mer^- cells examined are hypersensitive to killing by MNNG (Day, et al., 1980a; 1980b; 1984; Scudiero, et al.,1984). Mer^- cells were also more sensitive than Mer^+ cells to SCE induction by MNNG (Day, et al., 1980b). In rodent cells, lack of O^6meG removal correlated well with elevated SCE frequency induced by MNNG, but with some exceptions (Yagi, et al., 1984).

In general, therefore, these data are consistent with a role of O^6meG (or of a lesion induced by agents that produce O^6meG and are repaired by mechanisms that repair O^6meG) in cell killing and SCE induction. A word of warning against this conclusion has been introduced by the finding that lack of O^6meG repair and sensitivity to MNNG can be separated (Ishida and Takahashi, 1987; Goth-Goldstein, 1987). In fact, this finding does not necessarily rule out O^6meG as a lethal lesion, but rather indicates that resistance to killing by MNNG is controlled by other factors in addition to the capacity to repair O^6meG. Acquisition of MNNG-resistance by a Mer^- HeLa cell line was not accompained by the acquisition of O^6-MT activity or resistance to SCE induction (Samson and Linn, 1987). In this case, therefore, sensitivity to killing, but not to SCE induction, was separated from lack of O^6meG repair.

(b) DNA Transfection Experiments

Recently, a number of studies have been reported in which the ada gene of Escherichia coli, encoding the O^6-MT protein, was transfected into and expressed by mammalian cells devoid of O^6-MT activity. In a few studies, human DNA was transfected into O^6-MT-deficient cells with the aim of transfering human genes involved in the repair of O^6alkG. One advantage of this approach is that the effect of O^6alkG repair can be studied in cells of identical or very similar genetic and biochemical background.

In all cases reported, expression of the bacterial O^6-MT gene protected the mammalian cells against both cell killing and SCE induction by methylating or chloroethylating agents which react extensively with oxygen atoms (Brennand and Margison, 1986a; Ishizaki, et al., 1986; Kataoka, et al., 1986; Samson, et al., 1986; White, et al., 1986). In the study by Samson and coworkers (1986), the whole ada-alkB operon of E. coli was transfected into Mer^- HeLa cells; in all other cases, the transfected plasmids contained only the ada gene sequence. Brennand and Margison (1986b) constructed a plasmid containing the region of the ada gene responsible for the repair of O^6alkG and O^4alkT, but not the region responsible for the repair of alkylphosphotriesters. V79 cells transfected with the truncated gene acquired resistance to oxygen reactive alkylating agents. Transfer of a truncated gene containing only the alkylphosphotriester repair activity, on the contrary, did not result in the acquisition of resistance to MNNG killing by CHO cells (Kataoka, et al., 1986).

Transfection of human DNA from liver cells (Ding, et al., 1985) or a Mer^+ tumour cell line (Yarosh, et al., 1986) resulted in the appearance of O^6-MT activity in O^6alkG repair-deficient cells. Acquisition of resistance by the transfected cells was observed only in the latter study. Transfection of CHO cells with DNA from human diploid fibroblasts rendered the Chinese hamster cells resistant to MNNG killing, but no increased removal of O^6meG was observed (Kaina, et al., 1987).

We (Bignami, et al., 1987) have tested the effect of MNNG and MNU on CHO cells transfected with human liver DNA by Ding and coworkers (1985). The results concerning cell killing and SCE induction are illustrated in Figure 1. They show that CHO cells with increased O^6-MT activity (MT^+) were protected against the killing effect of MNNG and MNU and the induction of SCE by MNNG.

The transfected (MT$^+$) CHO cells were further selected for resistance to higher levels of MNNG and two resulting clones were again tested for their response to methylating agents (Aquilina, et al., 1988). As shown in Figure 2 (a-c), resistance to SCE induction paralleled the resistance to killing in the tested clones. Figure 2 (d) shows that clone 13 and clone B have similar O^6-MT levels, significantly higher than the activity of the original MT$^+$ cells. These data indicate that expression of O^6-MT activity protects CHO cells against cell killing and SCE induction, but further protection can be acquired through other mechanisms that are not associated with O^{6}meG repair, as exemplified by the behaviour of clone B.

Taken together, the data reported in subsections (a) and (b) strongly favour a role for O^{6}alkG in cell killing and SCE induction. Moreover, there is good evidence (Kaina, et al., 1987; Aquilina, et al., 1988) for the existance of other, so far unidentified, mechanisms of resistance.

O^6-METHYLGUANINE-DNA-METHYLTRANSFERASE INACTIVATION BY O^6-METHYLGUANINE AS A FREE BASE

O^{6}meG in double-stranded DNA is the best substrate for the E. coli O^6-MT (Demple, 1988). However, slow methyl transfer from free O^{6}meG also occurs in vitro (Yarosh, et al., 1986) and leads to inactivation of O^6-MT because of its suicidal reaction. In addition, O^{6}meG may be incorporated into tRNA, thus becoming a better substrate for O^6-MT (Karran, 1985). These observations have been exploited to deplete mammalian cells of the active O^6-MT protein.

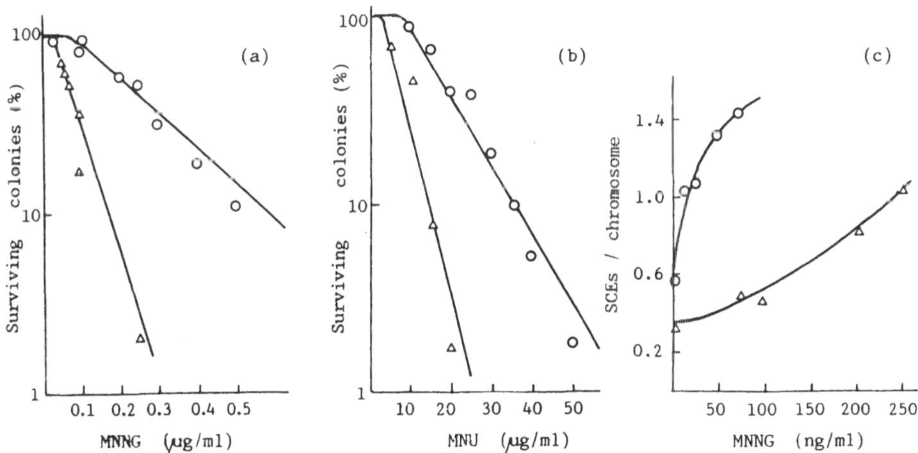

Figure 1. The induction of cell killing by MNNG (a) and MNU (b) and of SCE by MNNG (c) after treatment for 30 minutes at 37^{0}C of O^6-MT-deficient (Δ) and O^6-MT-proficient (o) CHO cells. Redrawn after Bignami, et al., 1987.

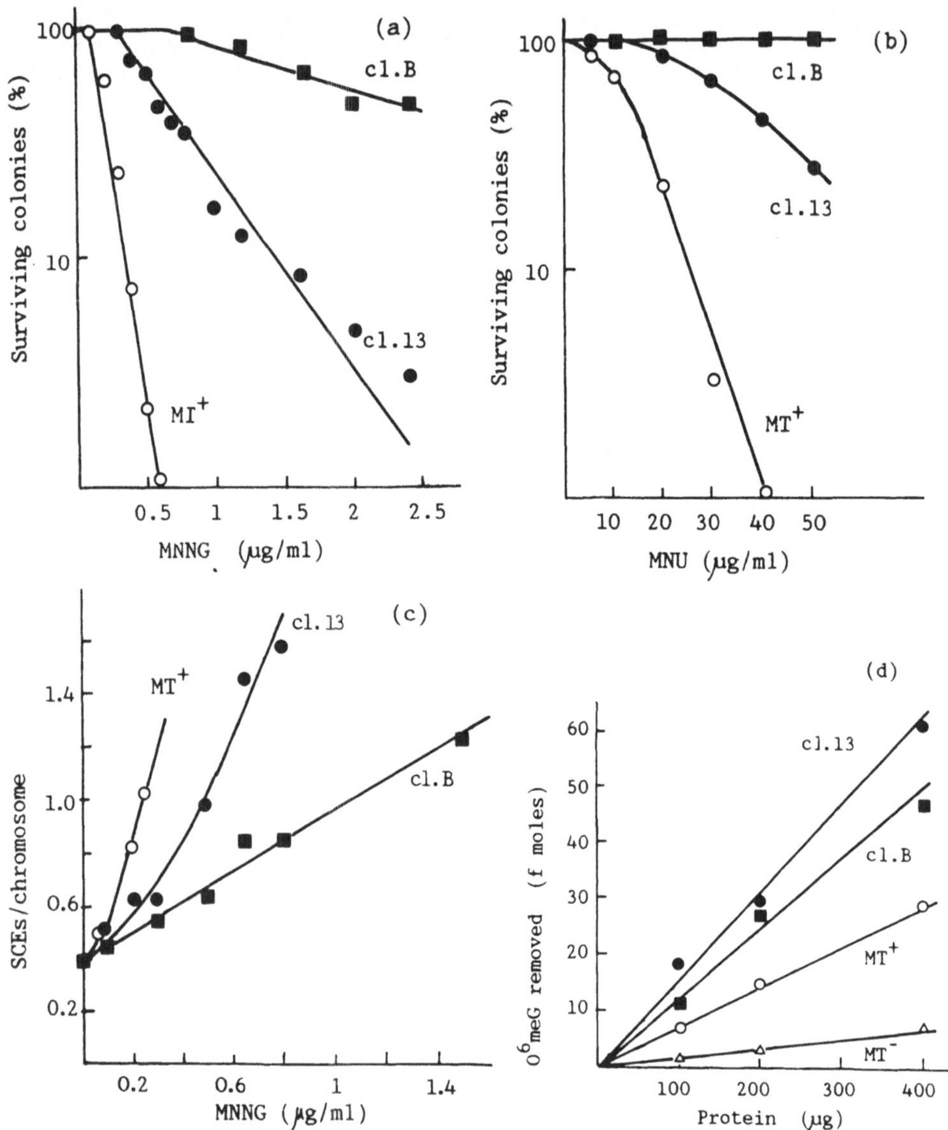

Figure 2. The induction of cell killing by MNNG (a) and MNU (b) and of SCE by MNNG (c) after treatment for 30 minutes at 37^{0}C of O^6-MT-proficient (o), clone 13 (●) and clone B (■) CHO cells. O^{6}mG-DNA-methyltransferase activity is shown in (d); symbols as in (a-c) and (Δ), O^6-MT-deficient CHO cells. (a-c) redrawn after Aquilina, et al., 1988.

Treatment of human cells with constitutive levels of O^6-MT with O^{6}meG as a free base was found to deplete the cells of their transferase activity, usually by 60-80% (Domoradzki, et al., 1985; Dolan, et al., 1985; 1986; Karran, 1985; Karran and Williams, 1985; Yarosh, et al., 1986). The transferase-depleted cells showed increased sensitivity to chloroethylating agents (Dolan, et al., 1986; Yarosh, et al., 1986), but with 1-(2-chloroethyl)-1-nitrosourea the opposite result was also reported (Karran and Williams, 1985). Moreover, sensitization to MNNG-killing did not occur (Karran and Williams, 1985; Domoradzki, et al., 1985), although the expected increase in mutagenic response was observed (Domoradzki, et al., 1985). Sensitization to chloroethylating agents does not necessarily imply that O^6alkG is a lethal lesion, since chloroethylation at the O^6 position of guanine is the first of a two-step reaction leading to interstrand cross-links, considered to be the cytotoxic lesions (Kohn, 1977).

In conclusion, methyltransfgrase inactivation has not provided positive evidence for a cytotoxic role of O^6alkG. However, these experiments have not ruled out a cytotoxic role for O^6alkG, as discussed extensively by Kataoka and coworkers (1986).

CONCLUDING REMARKS

The assignment of biological effects of alkylating agents to specific DNA lesions has largely relied upon indirect approaches. These include molecular dosimetry studies, treatments to stimulate ("adaptive" treatments) or inactivate (treatments with O^6meG) repair functions, and studies on cells with different repair capacities, either naturally occurring, or obtained by DNA transfer.

A direct approach, based on construction of DNA molecules containing a single lesion in a known position and their transfer into untreated cells has been used to study the role of O^6meG in gene mutation (Loechler, et al., 1984; Dogliotti, et al., 1987), but not in other biological effects. Our conclusions on the possible role of O^6alkG in cell killing and chromosomal effects must therefore rely on circumstantial evidence.

As shown in this article, positive evidence for a role of O^6meG in cell killing and SCE induction is mainly provided by the comparison of such effects in cells with or without O^6meG repair capacity. Other approaches produced in general inconclusive results. Circumstantial evidence for a role of O^6meG in micronucleus induction was also provided by studies with O^6meG repair proficient or deficient cells, as discussed elsewhere (Bonatti and Abbondandolo, 1989).

ACKNOWLEDGEMENTS

We thank Dr. A. Minks for critical reading of the manuscript and Miss T. Tavilla for skillful typing. This paper is based on work supported by the Commission of European Communities (Contracts n. 530 ENV Is and n. EV4V-0036-I(A)) and by CNR Special Projects "Medicina Preventiva e Riabilitativa" (Contracts n. 85.00811.56 and n. 86.02068.56),"Oncologia" (Contracts n. 86.00620.44 and n. 87.01493.44) and "Progetto Strategico Mutagenesi" (Contract n. 87.02346.74). Part of reported results have been published previously (Bignami, et al., 1987; Aquilina, et al., 1988).

REFERENCES

Aquilina, G., Frosina, G., Zijno, A., Di Muccio, A., Dogliotti, E., Abbondandolo, A., and Bignami, M., 1988, Isolation of clones displaying enhanced resistance to methylating agents in O^6-methyl-guanine-DNA-methyltransferase-proficient CHO cells, Carcinogenesis, 9:1217.

Beranek, D.T., Weis, C.C., and Swenson, D.H., 1980, A comprehensive quantitative analysis of methylated and ethylated DNA using high pressure liquid chromatography, Carcinogenesis, 1:595.

Bignami, M., Terlizzese, M., Zijno, A., Calcagnile, A., Frosina, G., Abbondandolo, A., and Dogliotti, E., 1987, Cytotoxicity, mutations and SCEs induced by methylating agents are reduced in CHO cells expressing an active mammalian O^6-methylguanine-DNA-methyltransferase gene, Carcinogenesis, 8:1417.

Bonatti, S., and Abbondandolo, A., 1989, The search for the molecular lesions responsible for the induction of chromosomal damage by alkylation agents, in: M. Bignami, E. Dogliotti and J. Essigman, eds., Annali Istituto Superiore di Sanita, Vol. 25, Rome, Italy.

Brennand, J., and Margison, G.P., 1986a, Reduction of the toxicity and mutagenicity of alkylating agents in mamnalian cells harboring the Escherichia coli alkyltransferase gene, Proc. Natl. Acad. Sci. U.S.A., 83:6292.

Brennand, J., and Margison, G.P., 1986b, Expression in mammalian cells of a truncated Escherichia coli gene coding for O^6-alkylguanine alkyltransferase reduces the toxic effects of alkylating agents, Carcinogenesis, 7: 2081.

Carrano, A.V., Thompson, L.H., Lindl, P.A., and Minkler, J.L., 1978. Sister chromatid exchange as an indicator of mutagenesis, Nature, 271:551.

Connel, J.R., and Medcalf, A.S.C., 1982, The induction of SCE and chromosomal aberrations with relation to specific base methylation of DNA in Chinese hamster cells by N-methyl-N-nitrosourea and dimethylsulphate, Carcinogenesis, 3:385.

Day, R.S., III, Yarosh, D.B., and Ziolkowski, C.H.J., 1984, Relationship of methyl purines produced by MNNG in adenovirus 5 DNA to viral inactivation in repair-deficient (Mer-) human tumor cell strains, Mutation Res., 131: 45.

Day, R.S., III, Ziolkowski, C.H.J., Scudiero, D.A., Meyer, S. A., and Mattern, M.R., 1980a, Human Tumor cell strains defective in the repair of alkylation damage, Carcinogenesis, 1:21.

Day, R.S., III, Ziolkowski, C.H.J., Scudiero, D.A., Meyer, S.A., Lubiniecki, A.S., Girardi, A.J., Galloway, S.M., and Bynum, G.D., 1980b, Defective repair of alkylated DNA by human tumor and SV40-transformed human cell strains, Nature, 288:724.

de Kok, A.J., van Zeeland, A.A., Simons, J.W.I.M., and Den Engelse, L., 1985, Genetic and molecular mechanisms of the in vitro transformation of Syrian hamster embryo cells by the carcinogen N-ethyl-N-nitrosourea. II. Correlation of morphological transformation, enhanced fibrinolytic activity, gene mutations, chromosomal alterations and lethality to specific carcinogen-induced DNA lesions, Carcinogenesis, 6:1571.

Demple, B., 1988, Self-methylation by suicide DNA repair enzymes, in: "Protein Methylation", W.K. Paik and S. Kim, eds., CRC Press, Boca Raton, U.S.A.

Ding, R., Gosh, K., Eastman, A., and Bresnick, E., 1985. DNA-mediated transfer and expression of a human DNA repair gene that demethylates O^6-methylguanine, Molec. Cell. Biol., 5:3293.

Dogliotti, E., Ellison, K.S., Basu, A.K., Bignami, M., and Essigman, J.M., 1987, Construction of shuttle vectors for studying the genetic effects of defined chemical carcinogen-DNA base adducts in mammalian cells, in: "Gene Transfer Vectors for Mamnalian Cells", Current Communications in Molecular Biology, Cold Spring Harbor Laboratory, Cold Spring Harbor, USA.

Dolan, M.E., Young, G.S., and Pegg, A.E., 1986, Effect of O^6-alkylguanine pretreatment on the sensitivity of human colon tumor cells to the cytotoxic effects of chloroethylating agents, Cancer Res., 46:4500.

Dolan, M.E., Morimoto, K., and Pegg, A.E., 1985, Reduction of O^6-alkyl-guanine-DNA-alkyltransferase activity in HeLa cells treated with O^6 -alkylguanines, Cancer Res., 45: 6413.

Domoradzki, J., Pegg, A.E., Dolan, M.E., Maher, V.M., and McCormick, J.J., 1984, Correlation between O^6-methylguanine-DNA-methyltransferase activity and resistance of human cells to the cytotoxic and mutagenic effect of N-methyl-N'-nitro-N-nitrosoguanidine, Carcinogenesis, 5:1641.

Domoradzki, J., Pegg, A.E., Dolan, M.E., Maher, V.M., and McCormick, J.J., 1985, Depletion of O^6-methylguanine-DNA-methyltransferase in human fibroblasts increases the mutagenic response to N-methyl-N'-nitro-N-nitrosoguanidine, Carcinogenesis, 6:1823.

Frosina, G., and Abbondandolo, A., 1985, The current evidence for an adaptive response to alkylating agents in mammalian cells, with special reference to experiments with in vitro cell cultures, Mutation Res., 154:85.

Goth-Goldstein, R., 1980, Inability of Chinese hamster ovary cells to excise O^6-alkylguanine, Cancer Res., 40:2623.

Goth-Goldstein, R., 1987, MNNG-induced partial phenotypic reversion of Mer^- cells, Carcinogenesis, 8:1449.

Harris, A.L., Karran, P., and Lindahl, T., 1983, O^6-methylguanine-DNA-methyltransferase of human lymphoid cells: structural and kinetic properties and absence in repair-deficient cells, Cancer Res., 43:3247.

Heflich, R.H., Beranek, D.T., Kodell, R.L., and Morris, S.M., 1982, Induction of mutations and sister-chromatid exchanges in Chinese hamster ovary cells by ethylating agents. Relation to specific DNA adducts, Mutation Res., 106:147.

Ishida, R., and Takahashi, T., 1987, N-methyl-N'-nitro-N-nitrosoguanidine-resistant HeLa S3 cells still have little O^6-methylguanine-DNA-methyltransferase activity and are hypermutable by alkylating agents, Carcinogenesis, 8: 1 109.

Ishizaki, K., Tsujimura, T., Yawata, H., Fujio, C., Nakabeppu, Y., Sekiguchi, M., and Ikenaga, M., 1986, Transfer of the E. coli O^6-methylguanine methyltransferase gene into repair-deficient human cells and restoration of cellular resistance to N-methyl-N'- nitro-N-nitrosoguanidine, Mutation Res., 166:135.

Kaina, B., van Zeeland, A.A., Backendorf, C., Thielmann, H.W., and van de Putte, P., 1987, Transfer of human genes conferring resistance to methylating mutagens, but not

to UV irradiation and cross-linking agents, into Chinese hamster ovary cells, Molec. Cell. Biol., 7:2024.
Karran, P., 1985, Possible depletion of a DNA repair enzyme in human lymphoma cells by subversive repair, Proc. Natl. Acad. Sci. U.S.A., 82:5285.
Karran, P., Hjelmgren, T., and Lindahl, T., 1982, Induction of a DNA glycosylase for N-methylated purines in part of the adaptive response to alkylating agents, Nature, 296: 770.
Karran, P., and Williams, S.A., 1985, The cytotoxic and mutagenic effects of alkylating agents on human lymphoid cells are caused by different DNA lesions, Carcinogenesis, 6:789.
Kataoka, H., Hall, J., and Karran, P., 1986, Complementation of sensitivity to alkylating agents in Escherichia coli and Chinese hamster ovary cells by expression of a cloned bacterial DNA repair gene, EMBO J., 5:3195.
Kohn, K.W., 1977, Interstrand cross-linking of DNA by 1,3-bis-(2- chloroethyl)-1-nitrosourea and other 1-(2-haloethyl)-1-nitrosoureas, Cancer Res., 37 : 1450.
Laval, F., and Laval, J., 1984, Adaptive response in mammalian cells: Cross reactivity of different pretreatments on cytotoxicity as contrasted to mutagenicity, Proc. Natl. Acad. Sci. U.S.A., 81:1062.
Lawley, P.D., 1974, Alkylation of nucleic acids and mutagenesis, in:"Molecular and Environmental Aspects of Mutagenesis", L. Prakash, F. Sherman, M.W. Miller, C.W.Lawrence, and H.W. Taber, eds., C.C. Thomas, Springfield, U.S.A.
Loechler, E.L., Green, C.L., and Essigman, J.M., 1984, In vivo mutagenesis by O^6-methylguanine built into a unique site in a viral genome, Proc. Natl. Acad. Sci. U.S.A. , 81: 6271.
Mariani, L., Bertini, R., Fiorio, R., Gervasi, P., and Citti, L., 1988, O^6-methylguanine-DNA-methyltransferase activity in V79 Chinese hamster cells, Mutation Res., 208:73.
Morris, S.M., Beranek, D.T., and Heflich, R.H., 1983, The relation between sister-chromatid exchange induction and the formation of specific methylated DNA adducts in Chinese hamster ovary cells, Mutation Res., 121:261.
Natarajan, A.T., Simons, J.W.I.M., Vogel, E.W., and van Zeeland, A.A., 1984, Relationship between cell killing, chromosomal aberrations, sister-chromatid exchanges and point mutations induced by monofunctional alkylating agents in Chinese hamster cells, Mutation Res., 128:31.
Newbold, R.F., Warren, W., Medcalf, A.S.C., and Amos, J., 1980, Mutagenicity of carcinogenic methylating agents is associated with a specific DNA modification, Nature, 283:596.
Samson, L., and Cairns, J., 1977, A new pathway for DNA repair in Escherichia coli, Nature, 267:281.
Samson, L. and Linn, S., 1987. DNA alkylation repair and the induction of cell death and sister chromatid exchange in human cells, Carcinogenesis, 8:227.
Samson, L., and Schwartz, J.L., 1980, Evidence for an adaptive DNA repair pathway in CHO and human skin fibroblast cell lines, Nature, 287:861.
Samson, L., Derfler, B., and Waldstein, E.A., 1986, Suppression of human DNA alkylation-repair defects by Escherichia coli DNA repair genes, Proc. Natl. Acad. Sci. U.S.A., 83:5607.
Scudiero, D.A., Meyer, S.A., Clatterbuck, B.E., Mattern, M.R.,

Ziolkowski, C.H.J., and Day, R.S., III, 1984, Relationship of DNA repair phenotypes of human fibroblast and tumor strains to killing by N-methyl-N"-nitro-N-nitrosoguanidine, Cancer Res., 44:961.
Sklar, R., and Strauss, B., 1981, Removal of O^6-methylguanine from DNA of normal and Xeroderma pigmentosum derived lymphoblastoid lines, Nature, 289:417.
Swenson, D.H., Harbach, P.R., and Trzos, R.J., 1980, The relationship between alkylation of specific DNA bases and induction of sister chromatid exchange, Carcinogenesis, 1:931.
Van Zeeland, A.A., 1988, Molecular dosimetry of alkylating agents: quantitative comparison of genetic effects on the basis of DNA adduct formation, Mutagenesis, 3:179.
Vogel, E.W., and Natarajan, A.T., 1981, The relation between reaction kinetics and mutagenic action of monofunctional alkylating agents in higher eukaryotic systems. Interspecies comparison, in: "Chemical Mutagens", A. Hollaender and F.J. de Serres, eds., Vol. 7, Plenum, New York, U.S.A.
Warren, W., Crathorn, A.R., and Shooter, K.V., 1979, The stability of methylated purines and of methylphosphotriesters in the DNA of V79 cells after treatment with N-methyl-N-nitrosourea, Biochim. Biophys. Acta, 563:82.
White, G.R.M., Ockey, C.H., Brennand, J., and Margison, G.P., 1986, Chinese hamster cells harbouring the Escherichia coli O^6-alkyl-transferase gene are less susceptible to sister chromatid exchange induction and chromosome damage by methylating agents, Carcinogenesis, 7:2077.
Wolff, S., 1978. Relation between DNA repair, chromosome aberrations, and sister chromatid exchanges, in: P.C. Hanawalt, E.C. Friedberg, and C.F. Fox, eds., DNA Repair Mechanisms, ICN-UCLA Symposium on Molecular and Cellular Biology, Vol. 9, Academic Press, New York, U.S.A.
Yagi, T., Yarosh, D.B., and Day, R.S., III, 1984, Comparison of repair of O^6-methylguanine produced by N-methyl-N'-nitro-N-nitrosoguanidine in mouse and human cells, Carcinogenesis, 5:593.
Yarosh, D.B., Barnes, D., and Erickson, L.C., 1986, Transfection of DNA from a chloroethylnitrosourea-resistant tumor cell line (MER+) to a sensitive tumor cell line (MER-) results in a tumor cell line resistant to MNNG and CNU that has increased O^6-methylguanine-DNA-methyltransferase levels and reduced levels of DNA interstrand crosslinking, Carcinogenesis, 711603.
Yarosh, D.B., Foote, R.S., Mitra, S., and Day, R.S., III, 1983, Repair of O^6-methylguanine in DNA by demethylation is lacking in Mer human tumor cell strains, Carcinogenesis, 4:199.
Yarosh, D.B., Hurst-Calderone, S., Babich, M.A., and Day, R.S., III, 1986, Inactivation of O^6-methylguanine-DNA-methyltransferase and sensitization of human tumor cells to killing by chloroethylnitrosourea by O^6-methylguanine as a free base, Cancer Res., 46:1663.

INDUCIBLE REPAIR IN MAMMALIAN CELLS TREATED WITH ALKYLATING AGENTS

Patricia Lefebvre and Francoise Laval

Groupe Radiochimie de l'ADN, Institut Gustave Roussy, 94805 Villejuif, France

INTRODUCTION

When E. coli cells are exposed to low doses of an alkylating agent, they become more resistant to the toxic and mutagenic effects of this compound (Samson and Cairns,1977). This adaptive response is under the control of the ada gene. The O^6-methylguanine-DNA-methyltransferase (transferase) which repairs the promutagenic lesion O^6-methylguanine, is the product of the ada gene. It regulates its own synthesis and the expression of the alk A gene (Teo, et al., 1986) which encodes DNA glycosylase II: this enzyme removes several alkylated bases and especially the toxic lesions 3-methyladenine and 3-methylguanine from alkylated DNA (Karran, et al., 1982).

As alkylating agents are efficient mutagens and carcinogens (Saffhill, et al., 1985), many attempts have been made to know whether an adaptive response exists in mammalian cells. Pretreatment with MNNG decreases the number of SCE in CHO cells and in a SV40 transformed human cell lines and reduces the cell killing in CHO cells treated with MNNG or MNU (Samson and Shwartz, 1980). Pretreatment with MNNG or MNU reduces the mutation frequency in V79 cells challenged with the same agent (Kaina, 1982). We have shown that the number of transferase molecules (Laval and Laval, 1984) and the 3-methyladenine glycosylase activity (Laval, 1985) could be enhanced in a rat hepatoma cell line (H4 cells) pretreated with repeated non-toxic doses of MNNG. This pretreament increases also the cell resistance to the toxic and mutagenic effects of MNNG. However, negative results have been reported (reviewed in Frosina and Abbondandolo, 1985), suggesting that this "adaptive response" varies from cell line to cell line.

Therefore, the aim of our study was to know whether the increased resistance to alkylating agents described in several mammalian cell lines was due to a process ressembling, in some respects, the adaptive response described in bacteria, or was due to a different process.

DNA Repair Mechanisms and Their Biological Implications in Mammalian Cells
Edited by M.W. Lambert and J. Laval
Plenum Press, New York

RESULTS

In previous experiments (Laval and Laval, 1984), we have shown that H4 cells (rat hepatoma cells) pretreated during 72 hours with non-toxic doses of either MMS, MNU or MNNG were more resistant to the toxic effect of these compounds. Cross-reactivity was observed for survival, but not for mutagenesis, as resistance to mutagenicity was only observed in cells pretreated with MNNG. This increased resistance to mutagenicity in MNNG-pretreated cells correlated with a faster and more complete removal of the O^6-methylguanine residues from the cellular DNA compared to control cells, when challenged with a high MNNG dose. This faster removal was due to an increase of the number of transferase molecules per cell: about 54,000 and 132,000 molecules were measured in control and pretreated cells, respectively. Measurement of the amount of alkylated residues formed in the cellular DNA during the pretreament suggested that induced mutagenesis resistance was related to the amount of O^6-methylguanine residues produced during the pretreatment.

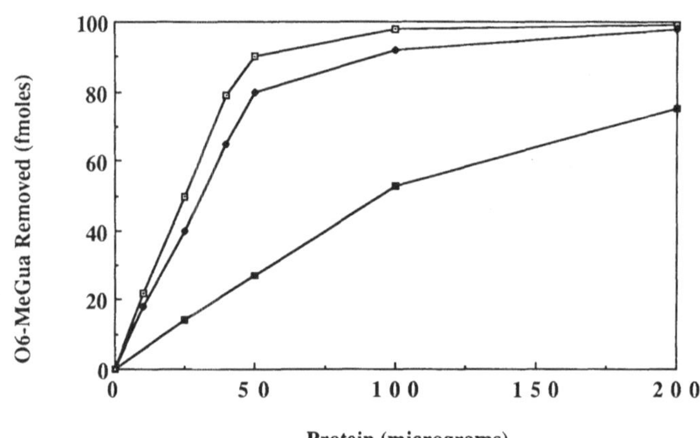

Figure 1. Removal of O^6-methylguanine residues from alkylated DNA by H4 cell extracts. [^{3}H]-MNU-treated DNA was incubated with extracts of control cells (■--■) or of cells treated with N-methyl-9-hydroxy-ellipticin (2.5 µM) (●--●) or cis-platinum II (5 µM) (□--□). The transferase activity was measured 48 hours after the treatment.

The influence of different pretreatments was studied in order to determine whether this increased resistance to alkylating agents is an adaptive response, as described in bacteria, and especially whether it is related to the formation of alkykated bases in the cellular DNA during the pretreatment, or is due to a different process.

H4 cells were treated with a single low toxic dose of either chemical compounds or physical agents, then the transferase activity was determined by incubating cell extracts with [^{3}H]-MNU-DNA and measuring the disappearance of O^6-methylguanine from the substrate (Boiteux and Laval, 1985). Figure 1 shows the transferase activity 48 hours after a single dose of cis-platinum II (5 µM) or of N-methyl-9-hydroxy-ellipticin (ellipticin) (2.5 µM). The number of transferase molecules per cell, calculated from the linear part of the curves, is 54,000, 207,000 and 252,000 in control, cis-platinum II and ellipticin treated cells, respectively. The transferase activity is increased after other treatments, as shown in Figure 2, which represents the number of transferase molecules per cell in control cells, or 48 hours after treatment with MNNG (10 µM), MMS (1 mM), γ-rays (300 rads), bleomycin (25 µg/ml), cis-platinum II (5 µM) and UV light (10 J/m^2). All these treatments were delivered at equitoxic doses.

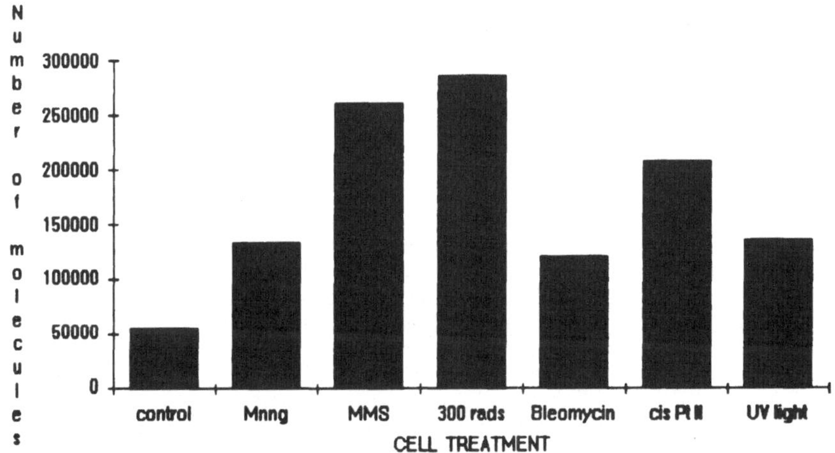

Figure 2. Number of transferase molecules per H4 cell treated with different agents. The transferase activity was measured 48 hours after treatment, using increasing amounts of cell extracts. The number of molecules was calculated from the linear part of the curves.

This increase in transferase activity is time-dependent. In the case of H4 cells, it is maximal 48 hours after the treatment (about three cell cycles), then it begins to decrease and reaches the control value after 120 hours. This increase is also dose-dependent: in γ-irradiated cells, for example, the number of transferase molecules per cell increases from 54,000 (control cells) to 95,500, 161,600 and 285,000 in cells irradiated with 50, 100 and 300 rads, respectively. This process requires <u>de novo</u> protein synthesis, since incubating the cells with cycloheximide (1 µg/ml) after the different treatments abolishes the response. Polyacrylamide gel electrophoresis of cellular proteins shows that the newly synthesized transferase has the same molecular weight as the constitutive one (data not shown).

These treatments also increase the 3-methyladenine glycosylase activity in H4 cells. Cell extracts were incubated with [^{3}H]-DMS-treated DNA, then the released 3-methyladenine residues were separated by HPLC and counted (Laval, 1985). Results show that extracts from treated cells release about 2 to 3-fold more 3-methyladenine residues compared to extracts of control cells (data not shown).

Therefore, treatment with a single dose of different agents induces at least two different repair activities in H4 cells. This induction is not due to a block of DNA replication in the treated cells as, (1) the doses delivered to the cells do not inhibit DNA synthesis and (2) incubating the cells with hydroxyurea (1 mM) does not result in an increased transferase activity. Furthermore, the enhancement of transferase activity observed in irradiated stationary cultures (Lefebvre and Laval, 1986) rules out a regulation of this protein due to cell proliferation.

The various pretreatments which enhance the transferase activity produce different types of damage in the cellular DNA and therefore, there is no direct correlation between the lesions produced during the treatment and the enhanced repair proteins. A common damage resulting from the different treatments is the formation of DNA single strand breaks, either directly, or during the enzymatic repair of the lesions. The role of DNA single strand breaks in the induction of repair is also suggested by the fact that incubating the cells in the presence of poly (ADP-ribose) synthesis inhibitors (Sims, _et al._, 1982), after the treatments, results in a higher enhancement of the transferase activity. Table 1 shows that the inhibitors of poly (ADP-ribose) synthesis alone do not modify the number of transferase molecules per cell, but that they do enhance the increase in this number produced by MNNG when they are applied together with MNNG.

Table 1. Influence of poly (ADP-ribose) synthesis inhibitors on the number of transferase molecules.

Cell treatment	Number of transferase molecules/cell
None	54,000
3-AB (2 mM)	55,000
3-MB (2.5 mM)	55,500
MNNG (10 μM)	134,500
MNNG (10 μM) + 3-AB (2 mM)	286,200
MNNG (10 μM) + 3-MB (2.5 mM)	253,800

H4 cells were incubated for 1 hour with MNNG and then grown either in control medium or in the presence of 3-amino benzamide(3-AB) or of 3-methoxybenzamide (3-MB). The transferase activity was measured 48 hours after the MNNG treatment.

The enhanced repair activities are biologically active in the cells. When H4 cells are pretreated with a single dose of γ-rays (300 rads) or of cis-platinum II (5 μM) then challenged with radioactive MNU, they remove the base damage (O^6-methylguanine and N^3-methyladenine) faster and to a greater extent than the control cultures (Table 2). They also become more resistant to the toxic and mutagenic effects of MNNG: the D_0 dose for MNNG-treated cells is 8.5 μM and 12.6 μM in control and pre-irradiated cultures, respectively, and the number of mutants (6-thioguanine resistant cells) decreases from $38/10^5$ survivors to $8/10^5$ survivors after a 10 μM MMNG dose in control and γ-irradiated cells, respectively.

The question arises whether this inducible response is a specific property of H4 cells or whether it exists in other cell lines. We have measured the number of transferase molecules in cell lines of different origin: in three tumorigenic cell lines, H4 cells (rat hepatoma), LICH cells (human hepatoma), Ro cells (rat rhabdomyosarcoma), two non-tumorigenic cell lines, 3T3 cells (mouse embryo), MRC 5 cells (normal human fibroblasts) and in CHO cells which have a low constitutive level of transferase activity. Results are summarized in Table 3. They show that the different pretreatments enhance the transferase activity in the tumorigenic cell lines tested but not in the normal cells and that the different treatments have no effect in cells defective in this type of repair.

Table 2. Removal of methylpurines by control or γ-irradiated cells.

Time	μmoles methylpurines per mole DNA P	
After treatment	Control cells	Irradiated cells
N3-meAdenine		
0	0.75	0.75
4 hours	0.56	0.32
8 hours	0.34	0.12
24 hours	0.05	0
O^6-meGuanine		
0	0.98	0.98
4 hours	0.50	0.10
8 hours	0.40	0
24 hours	0.30	0

Cells were incubated with [^{3}H]-MNU 48 hours after irradiation (300 rads) and the amount of O^6-methylguanine and N^3-methyladenine residues in the cellular DNA was measured by HPLC after various incubation times (Laval and Laval, 1984).

Table 3. Number of transferase molecules in various cell lines.

CELL LINE	CELL TREATMENT			
	None	γ-rays	cis-PtII	Ellipticin
H4 cells	54,000	285,000	207,000	252,000
LICH cell	135,000	247,600	230,000	514,000
Ro cells	147,000	441,500	162,000	356,000
3T3 cells	72,000	85,200	86,000	85,000
MRC5 cells	36,000	41,000	38,600	38,000
CHO cells	<200	<200	<200	<200

The cells were treated with γ-rays (300 rads), cis platinum II (5 μM), or N-methyl-9-hydroxy-ellipticin (2.5 μg/ml) for 1 hour. The transferase activity was measured in increasing amounts of extracts of each cell line, and the number of molecules was calculated from data of the linear part of the curves.

DISCUSSION

The results described in this paper show that at least two DNA repair activities, the O^6-methylguanine-DNA-methyltransferase and to a lesser extent the N^3-methylguanine glycosylase activities, can be increased in some cell strains. This increase occurs after treatment with a single dose of chemical or physical agents. It should be recalled that an increased transferase activity has been observed in different tissues of rats which had been treated with either ionizing radiations (Margison, et al., 1985; Schmerold and Wiestler, 1986) or chemical compounds (Chu, et al., 1981).

It has been shown that O^6-methylguanine residues are more efficiently removed during the G_1 phase of the cell cycle (Schuster, et al., 1985). However, the increased activity measured in the pretreated cells does not seem to be related to cell cycle modifications, since the different treatments delivered to the cells neither modify the doubling time nor the number of cells in S phase (Frosina and Laval, 1987). Furthermore, an increased transferase activity was also measured in γ-irradiated stationary cultures (Lefebvre and Laval, 1986).

The mechanisms underlying the enhancement of the repair activities is not established. The common damage produced by the different treatments is the DNA single strand break made either directly by the treatments or during the repair of the lesions. Therefore, there is no direct correlation between the initial type of damage produced in the cellular DNA and the induced repair activities. It has been shown that environmental changes or genotoxic stress are able to induce DNA amplification in mammalian cells (Stark and Wahl, 1984). MNNG-treatment, for example, induces amplification of SV40 sequences in CO60 cells, and this amplification is increased by the poly (ADP-ribose) synthesis inhibitor, 3-aminobenzamide (Burkle, et al., 1987). Implication of such mechanisms in the induction of the transferase activity can not be ruled out.

This induced repair of alkylated damage is different from an "adaptive response": (1) it is induced by a single treatment, whereas adaptation is induced by low doses delivered repeatedly during a long period of time, and (2) it is observed after treatment with different agents and there is no apparent correlation between the damage produced by these treatments and the induced repair activities, whereas, adaptation is observed when cells are pretreated and challenged with the same agent. In previous experiments (Laval and Laval, 1984), we observed an increased resistance of H4 cells to alkylating agents after pretreatment with repeated low doses of these compounds. As this resistance also occurs after a single dose of either alkylating drugs or other agents, it is probably due to a process different from an adaptive response.

These induced repair activities, which render the cells more resistant to the toxic and mutagenic effects of alkylating drugs, seem to occur preferentially in tumor cell lines. It is well known in cancer therapy, that some tumors develop resistance to radio or chemotherapy during the course of the treatments. Therefore, this inducible repair might be one of the processes which modify the efficiency of the treatments.

ACKNOWLEDGEMENTS

The technical assistance of Mrs. M. Letourneur was greatly appreciated. This work was supported by grants from Institut National de la Santé et de la Recherche Médicale and from Association pour la Recherche sur le Cancer, Villejuif.

REFERENCES

Boiteux, S. and F. Laval, 1985. Repair of O^6-methylguanine, by mammalian cell extracts, in alkylated DNA ad poly-($dG-m^5$ dC)·poly) ($dG-m^5dC$) in B and Z forms, Carcinogenesis, 6:805.

Burkle, A., T. Meyer, H. Hiltz and H. Zur Hausen, 1987. Enhancement of MNNG-induced DNA amplification in a Simian transformed chinese hamster cell line by 3-aminobenzamide, Cancer Res., 47:3632.

Chu, H.Y., A.W. Craig and P.J. O'Connor, 1981. Repair of O^6-methylguanine in rat liver is enhanced by pretreatment with single or multiple doses of aflatoxin B, Br. J. Cancer, 43:850.

Frosina, G. and A. Abbondandolo, 1985. The current evidence for an adaptive response to alkylating agents in mammalian cells with special reference to experiments with in vitro cell cultures, Mutation Res., 154:85.

Frosina, G. and F. Laval, 1987. The O^6-methylguanine-DNA-methyl transferase activity of rat hepatoma cells is increased after a single exposure to alkylating agents, Carcinogenesis, 8:91.

Kaina, B., 1982. Enhanced survival and reduced mutation and aberration fraquencies induced in V79 Chinese hamster cells pre-exposed to low levels of methylating agents, Mutation Res., 93:195.

Karran, P., T. Hjelmgreen and T. Lindahl, 1982. Induction of a glycosylase for N-methylated purines is part of the adaptive response to alkylating agents, Nature, 296:770.

Laval, F., 1985. Repair of methylated bases in mammalian cells during adaptive response to alkylating agents, Biochimie, 67:361.

Laval, F. and J. Laval, 1984. Adaptive response in mammalian cells: cross-reactivity of different pretreatments on cytotoxicity as contrasted to mutagenicity, Proc. Natl. Acad. Sci. USA, 81:1062.

Lefebvre, P. and F. Laval, 1986. Enhancement of O^6-methylguanine-DNA-methyltransferase activity induced by various treatments in mammalian cells, Cancer Res., 46:5701.

Margison, G.P., J. Butler and B. Hoey, 1985. O^6-methylguanine methyltransferase activity is increased in rat tissues by ionising radiations, Carcinogenesis, 6:1699.

Saffhill, R., G.P. Margison and P.J. O'Connor, 1985. Mechanisms of carcinogenesis induced by alkylating agents, Biochim. Biophys. Acta, 823:111.

Samson L., and J. Cairns, 1977. A new pathway for DNA repair in Escherichia coli, Nature, 267:281.

Samson, L. and J.L. Schwartz, 1980. Evidence for an adaptive DNA repair pathway in CHO and human skin fibroblast cell lines, Nature, 237:861.

Schmerold, I. and O.D. Wiestler, 1986. Induction of rat liver O^6-alkylguanine-DNA-alkyltransferase following whole body X-irradiation, Cancer Res., 46:245.

Schuster, C., G. Rode and H.M. Rabes, 1985. O^6-methylguanine repair of methylated DNA in vitro: cell cycle dependence of rat liver methyltransferase activity, J. Cancer Res. Clin. Oncol., 110:98.

Sims, J.L., W. Sikorski, D.M. Catino, S.J. Berger and N.A. Berger, 1982. Poly(adenosine diphosphoribose)polymerase inhibitors stimulate unscheduled DNA synthesis in normal human lymphocytes, Biochemistry, 21:1813.

Stark, G.R. and G.M. Wahl, 1984. Gene amplification, Ann. Rev. Biochem., 53:447.

Teo, I., B. Sedgwick, M.W. Kilpatrick, T.V. MacCarthy and T. Lindahl, 1986. The intracellular signal for induction of resistance to alkylating agents in E.coli, Cell, 45:315.

AN UPDATE OF THE MAMMALIAN UV RESPONSE: GENE REGULATION AND INDUCTION OF A PROTECTIVE FUNCTION

Bernd Kaina, Bernd Stein, Axel Schonthal, Hans Jobst Rahmsdorf, Helmut Ponta and Peter Herrlich

Kernforschungszentrum Karlsruhe, Institut fur Genetik und Toxikologie, PO Box 3640, D-7500 Karlsruhe 1, FRG

INTRODUCTION

Bacteria and yeast react to DNA damaging agents by synthesizing more of several gene products (Walker, 1985; Moustacchi, 1987; Friedberg, et al., 1988). Among the induced products there are DNA repair functions and proteins required for mutagenesis. For mammalian cells, formally a similar response has been detected. New gene products are synthesized (Herrlich, et al., 1986). No DNA repair induction has yet been characterized although it is likely that it exists since viral and shuttle vector probes have revealed enhanced reactivation and mutagenesis (Radman, 1980; Cornelis, et al., 1982; Defais, et al., 1983; Sarasin, 1985; Dion and Hamelin, 1987; Herrlich, 1988a; Protic, et al., 1988). Also, low doses of mutagens may, under certain conditions, modify subsequent mutagenic and clastogenic responses (Samson and Schwartz, 1980; Kaina, 1982, 1983; Rieger, et al., 1982; Olivieri, et al., 1984; Laval and Laval, 1984). Fluctuation analyses supplied support for the idea that proneness to mutation is passed on to daughter cells (Maher, et al., 1988). As another parallel between bacteria and mammalian cells DNA damage causes reinitiation of replication (Kogoma, et al., 1979) which in mammalian cells is observed as gene amplification (Lavi, 1981; Brown, et al., 1983; Schimke, 1984) or recruitment of quiescent cells into the cell cycle (Cohn, et al., 1984). Molecular genetic techniques have made both the mechanisms of signal transduction and gene functions experimentally accessible. Here, current ideas on damage-induced gene expression and its phenotypic consequences are presented.

WHICH GENES ARE INDUCED?

An increase in transcription following DNA damage can be detected in cultured cells for a rather large number of genes. The estimate is in the order of 100. Only a few have been identified. Table 1 shows the genes and the inducing DNA

DNA Repair Mechanisms and Their Biological Implications in Mammalian Cells
Edited by M.W. Lambert and J. Laval
Plenum Press, New York

damaging agents so far tested. We note that several genes are addressed by more than one DNA damaging agent, while others respond in a more restricted fashion.

TOWARDS AN UNDERSTANDING OF PRIMARY EVENTS

The gene-inducing agents cause different types of DNA damage. Is DNA damage a necessary intermediate in the gene induction process? For the induction of gene expression by UV the need for DNA damage has been proven by two types of experiments:

1) the relevant action spectrum for UV induced collagenase, metallothionein, HIV-1, FOS and EPIF resembles a DNA absorption spectrum (Stein, et al., 1988; B. Stein, unpublished). 2) UV induced expression of genes in fibroblasts from patients with putative UV DNA repair deficiency (Xeroderma pigmentosum, Cockayne's syndrome) requires less dose than the same expression in fibroblasts from a healthy individual (Schorpp, et al., 1984; Stein, et al., 1988; B. Stein, unpublished).

We suggest that other agents such as MNNG (Kleinberger, et al., 1988) or alpha irradiation (Herrlich, et al., 1986) also activate genes through DNA damage. The types of DNA damage produced are quite diverse. For one class of genes, damage specific signals are apparently relevant in that these genes are induced by only one or the other agent. For other genes a rather general structural alteration of the DNA helix or a common consequence of DNA damage could be postulated as the origin of signal transduction to transcription. A common consequence could be an arrest of replication or of transcription (suggested by Dr. Mullenders, Leiden) in a unit other than the UV induced reporter gene. For instance, UV causes arrests of both replication and transcription (Sauerbier and Hercules, 1978; Kaufmann, et al., 1980). An arrest of replication does not seem to be necessary since UV induces at least six genes (FOS, JUN, collagenase, metallothionein: MTIIA, HIV-1, SV40) in serum-starved non-cycling cells. The UV doses required for gene induction are sufficiently high to interfere with larger transcriptional units. The transcriptional arrest has been used for size-mapping of transcriptional units in pro- and eukaryotes (Scherzinger, et al., 1972; Herrlich, et al., 1974; Hirsch-Kauffman, et al., 1975; Sauerbier and Hercules, 1978). It is, however, unlikely that it is the transcriptional arrest which triggers signal transduction to genes since transfection of UV irradiated random DNA substitutes for cellular UV irradiation in the induction of genes (S. Mai and S. Lavi, preliminary results).

Most transcriptional regulations in mammalian cells examined so far, involve positive controls, that is, in order to induce the expression of a gene, a transcription factor is activated which selects the responsive gene. For instance, a steroid hormone binds to a specific receptor protein conferring to the protein the ability to interact with a specific cis-acting sequence element in appropriate hormone-responsive promoters (Chandler, et al., 1983; Hynes, et al., 1983; Ponta, et al., 1985; Yamamoto, 1985). Likewise, phorbol esters activate a number of transcription factors that operate at

TABLE 1. DNA damage inducible gene functions in mammalian cells.

gene ∨ / inducing agent >	UV	NQO	MNNG	MMC	DMBA	Phorbol ester	EPIF
human collagenase	+[1]		+[2]	+[3]		+[1]	+[1]
human plasminogen activator	+[4]					+[5]	+[6]
c-FOS	+[7,8]					+[9]	-[7]
c-JUN	+(fig.3)					+[10,11,12]	
ß-actin			+[13]				
metallothionein (MTIIA)	+[14,15]					+[14]	+[14]
ornithine decarboxylase	+[16]					+[17]	
p53	+[18]	+[18]					
Invariant chain	+[19]			+[19]			
HIV-1	+[20,21]	+[20]		+[20,21]		+[22,23]	+[2]
SV40 early prom./enh.	+[24]	+[24]	+[13,24]	+[24]	+[13]	+[25]	
intracist.A particle, LTR			+[13]				
sprI, II;other cDNA's	+[26,27]	+[26,27]				+[26,27]	
cDNA clones of Fornace et al. (1988b)	+ (several groups with different inducibility)						
EPIF	+[6,28]		+[2]				
gene amplification	+[29,30]					+[31]	

+ inducible; - not inducible; UV=Ultraviolet radiation; NQO=Nitroquinoline oxide; MNNG=N-Methyl-N'-nitro-N-nitroso-guanidine; MMC=Mitomycin C; DMBA=7,12-Dimethyl-benz(a)anthracene; EPIF=Extracellular protein synthesis inducing factor.

1) Angel,<u>et al.</u>,1987a; 2) M. Krämer and B. Stein, unpublished; 3) H.J. Rahmsdorf, unpublished; 4) Miskin and Ben-Ishai,1981; 5) Waller and Schleuning,1985; 6) Rotem,<u>et al.</u>,1987, the identity of the described factor with EPIF has not been demonstrated; 7) Angel,<u>et al.</u>,1985; 8) Büscher,<u>et al.</u>,1988; 9) Greenberg and Ziff,1988; 10) Schönthal,<u>et al.</u>,1988b; 11) Herrlich,<u>et al.</u>,1989; 12) Lamph,<u>et al.</u>,1989; 13) Kleinberger, <u>et al.</u>,1988; 14) Angel, <u>et al.</u>,1986; 15) Fornace,<u>et al.</u>,1988; 16) Verma,<u>et al.</u>,1979, irradiation of mice with UVB; 17) Rose-John,<u>et al.</u>,1987; 18) Maltzman and Czyzyk,1984; 19) Rahmsdorf,<u>et al.</u>,1983; 20) Stein,<u>et al.</u>,1988; 21) Valerie, <u>et al.</u>,1988; 22) Dinter, <u>et al.</u>,1987; 23) Kaufman, <u>et al.</u>,1987; 24) Vanetti,1988; 25) Imbra and Karin,1986; 26) Kartasova, <u>et al.</u>,1987; 27) Kartasova and van de Putte,1988; 28) Schorpp,<u>et. al.</u>,1984; 29) Lavi,1981; 30) Lücke-Huhle and Herrlich,1989; 31) Varshavsky,1981.

defined cis-acting sequences within phorbol ester responsive genes (Sen and Baltimore, 1986; Angel, <u>et al.</u>, 1987b; Lee, <u>et</u> al. , 1987; Karin and Herrlich, 1988; Baeuerle and Baltimore,

1988). It is expected that UV also leads to activation of transcription factors. Since UV induced FOS transcription occurs within less than 5 minutes and does not seem to require prior protein synthesis, it is suggested that UV activates a preformed transcription factor.

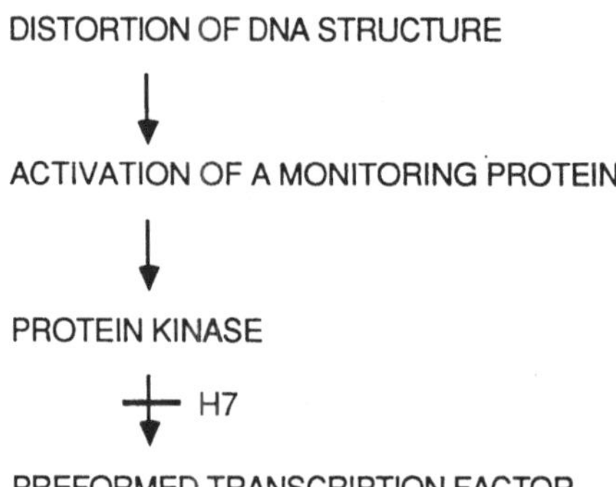

Figure 1. Hypothetical primary events in the UV induced expression of genes (see also Büscher, et al., 1988).

Thus, at least for FOS, UV induces the post-translational modification of a transcription factor (Figure 1). The nature of the modification and the connection to the site of DNA damage are as yet unknown. Since an inhibitor of protein kinases (H7) blocks the UV induced expression of genes we suggest that a nuclear or cytoplasmic protein kinase participates. The cis-acting sequence elements which form the targets of the transcription factors, have been identified in several genes. These sequences will lead to the identification of the transcription factors. For the FOS gene the UV responsive element is located between ~320 and ~299 (dyad symmetry element, Büscher, et al., 1988). It is identical to the element required for growth factor and phorbol ester dependent regulation (Treisman, 1985; Fisch, et al., 1987; Büscher, et al., 1988). The putative target of the UV induced modification reaction is the serum responsive factor (Treisman,1986; Prywes and Roeder, 1986). The UV responsive elements of the collagenase gene and of HIV-1 are also identical with the enhancer elements activated by phorbol esters: -73/-65 in the collagenase promoter and -121/-76 of HIV-1 (Angel, et al., 1987b; Dinter, et al., 1987).

FOS AS KEY INTERMEDIATE IN THE UV INDUCTION OF GENES

The induction of FOS expression by UV appears to form a decisive intermediate in the induction by UV of other genes. This has been tested by experiments using antisense sequences. Cells can be deprived of Fos protein by the introduction of antisense Fos RNA. By this technique mRNA-antisense RNA hybrid molecules form in the cytoplasm. These structures can not be translated and are subjected to degradation. Since Fos protein turns over with a half life of 30 minutes, the cells soon loose all Fos protein. Such cells can no longer transmit the UV induced signal to the responsive genes. This statement is supported by experiments with as yet only two UV responsive genes: collagenase and HIV-1 (Stein, et al., unpublished). The UV induced expression from promoters carrying either the collagenase or HIV-1 UV responsive elements, are severely reduced in Fos-deprived cells. Thus either Fos protein must be present as a preformed receiving structure for the signal generated by UV, or new synthesis of Fos protein must occur.

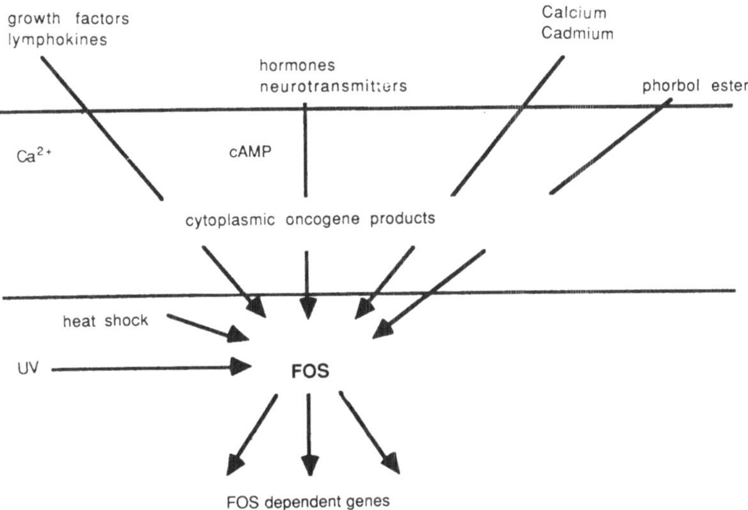

Figure 2. The central role of FOS in signal transduction.

FOS is not a UV specific intermediate. Rather Fos is part of physiologic signal transduction pathways, such as the signal transfer from a growth factor receptor to genes. Cells deprived of Fos protein and unable to make new Fos, are also deficient in growth factor and phorbol ester induced collagenase expression (Schönthal, et al., 1988a,b), and, in appropriate situations, the transforming ability of oncogenes is obliterated (reviewed by Herrlich and Ponta, 1989). Phorbol esters bypass the need for a growth factor and growth factor receptor by activating protein kinase C (reviewed by Nishizuka, 1986). The exact location where UV feeds into the physiological signal transduction pathway, has as yet not been defined.

The activation of the Fos gene by many different inducing agents (Figure 2) and the pleiotropic nature of the phenotypes whose expression seem to require Fos protein, suggests that Fos is involved in the regulation of many promoters. Fos protein is a nuclear protein and could itself act as a transcription factor. Fos could either participate in binding to the cis-acting DNA sequences (the enhancer sequences of collagenase and HIV-1) or Fos protein could influence the activity of other transcription factors, e.g., the activity of the collagenase enhancer (TRE) binding factor which may recognize a series of promoters carrying the same cis-acting element, or the activity of the factor binding to the HIV-1 enhancer which may stand for another class of genes. In fact using antibodies to Fos protein, Fos associated factors have been detected (Curran and Teich, 1982; Franza, et al.,1988; Rauscher, et al., 1988) one of which proved later to be identical to the transcription factor Jun which binds to the collagenase TRE (Angel, et al., 1987b; Bohmann, et al., 1987; Angel, et al., 1988). The specific binding can be detected in gel retardation and DNase I footprinting experiments. The DNA-protein complexes formed contain Jun and also Fos protein. The complexes formed at the HIV-1 enhancer differ from those at the collagenase TRE with regard to both migration in the gel and competition experiments (Stein, et al., unpublished). Also the HIV-1 enhancer-protein complexes carry antigenic determinants of Fos protein. The Fos protein appears to be instrumental to binding since Fos antibodies disturb complex formation at both enhancers. The existence of Fos protein in two different transcription complexes can also be deduced from competition experiments. Competing concentrations one magnitude higher than normally used destroy complex formation at the other sequence. As shown in Figure 3, a 250 fold excess of competing oligonucleotide carrying the collagenase TRE sequence interferes with band formation at the radioactive oligonucleotide comprising the HIV-1 enhancer sequence. A 250 fold excess of an oligonucleotide carrying a non-binding mutant sequence of the HIV-1 enhancer, does not disturb. The specificity is retained at even higher concentrations of competing oligonucleotide (not shown). We interpret these results to indicate a requirement for Fos protein in the binding of specific protein to the HIV-1 enhancer. The Fos protein is removed from the reaction by the collagenase TRE, which serves as a Jun-Fos binding sequence.

How are the transcription factors activated? One part of the answer is an activation of the transcription factor genes. Both FOS and JUN transcription are co-induced by phorbol ester (Herrlich, et al., 1989) as well as UV treatment (Figure 4). The increased levels of Fos and Jun proteins then cooperatively could activate collagenase-like promoters. For the major HIV-1 enhancer binding factor NFkB posttranslational activation by phorbol esters depends on the removal of an inhibitor (Baeuerle and Baltimore, 1988). The increase in binding activity in extracts from UV treated cells (Herrlich, et al., 1989) may be based on the same type of mechanism. The inhibitor-free NFkB seems to be associated with FOS. FOS transcription and translation is highly transient and does not correlate with the much longer kinetics of transcription factor activation (Herrlich, et al., 1989). This suggests the existence of another posttranslational modification. Such modifications may be agent-specific and could provide for the necessary molecular specificity as different Fos inducing agents cause agent-specific phenotypic consequences (Herrlich, et al., 1989).

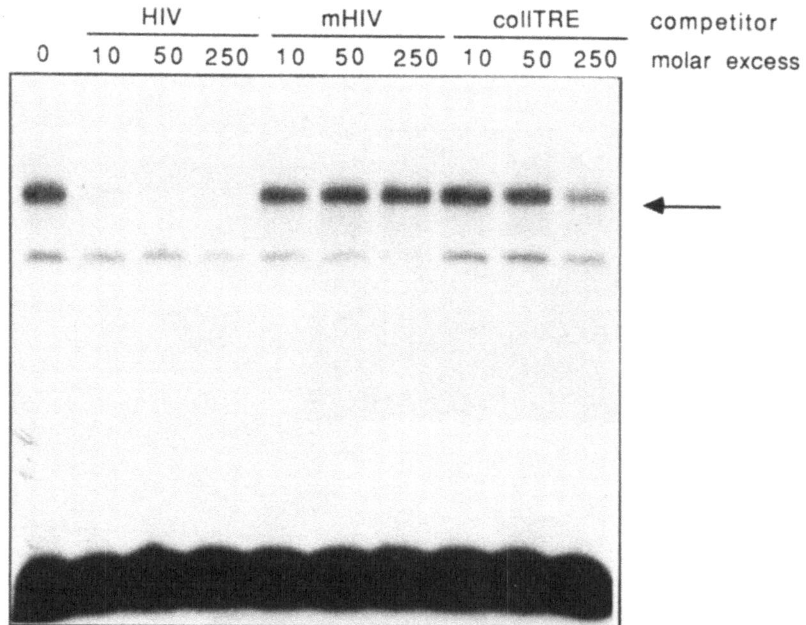

Figure 3."Indirect" competition for Fos protein. Gel retardation analysis of DNA protein complexes formed at the HIV-1 enhancer sequence (UV responsive element, sequence shown). Competing oligonucleotides: HIV = same sequence as radioactive probe. mHIV = point mutant: 5'TGCTCACTTTCCAGCCG3', coll TRE = UV responsive element of the collagenase promoter: 5'TGATGAGTCAGCCG3'. Whole cell extracts were prepared from logarithmically growing HeLa thymidine kinase deficient cells (Angel, et al., 1987a).

FUNCTIONAL CONSEQUENCES OF UV INDUCED GENE EXPRESSION

In bacteria, the SOS response includes DNA repair, recombination and mutagenesis functions as well as reinitiation of replication. To what extent does the mammalian UV response resemble the SOS response? For several genes that have been identified as UV induced (e.g., collagenase), such functional context can not be deduced in any straightforward manner. Other genes, however, match the expected scheme. These will now be discussed.

Subsequent to DNA damage, integrated SV40 sequences in rodent cells are heavily amplified (Lavi, 1981; Lücke and Herrlich, 1987). Similarly to the activation of transcription factors described above, a cellular DNA binding protein is activated: the minimal origin "early domain" binding protein (Lücke, et al., 1989; Lücke and Herrlich, 1989). This activation seems to be the limiting factor in the replication of SV40 in rodent cells. Its activity is elevated in primate

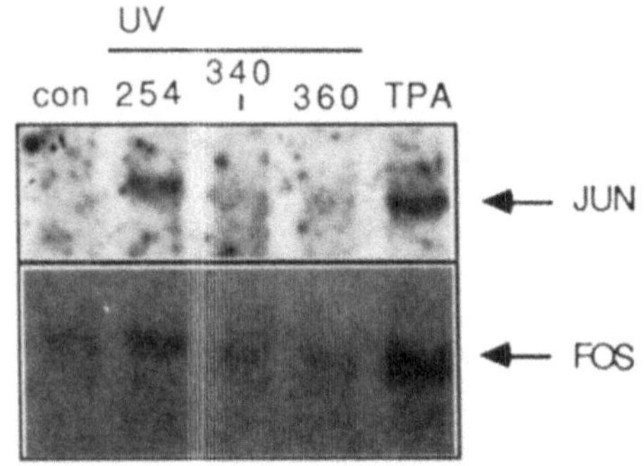

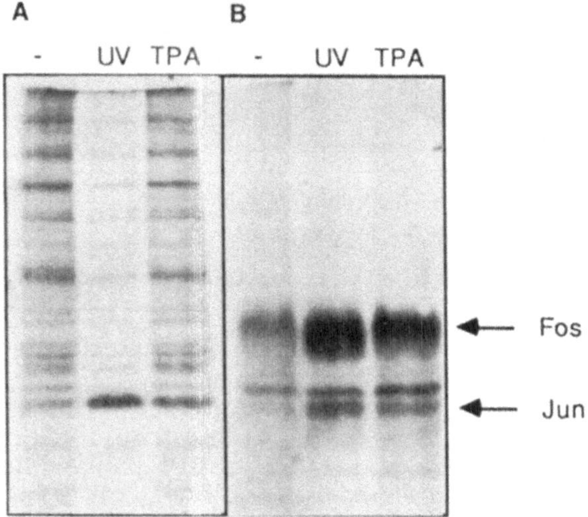

Figure 4. UV and phorbol ester induced expression of JUN and FOS. Upper panel: Northern RNA transfer hybridization of RNAs isolated from control cells (HeLa-thymidine kinase deficient, Angel, et al., 1987a, con), irradiated cells (254, 340 or 360 nm, 45 J/m^2, 2000 J/m^2, 2000 J/m^2, respectively) and from phorbol ester treated cells (TPA, 20 ng/ml), 45 minutes after treatment. Lower panel: Cells were incubated for 3 hours with culture medium lacking methionine. They were irradiated with UV (254 nm, 45 J/m^2), treated with phorbol ester (TPA, 20 ng/ml), or not treated (-) and incubated for 2 hours with 250 µCi S-35 methionine/ml. The cells were disrupted with detergents and specific proteins precipitated by antibodies to Jun (A, polyclonal serum produced in rabbits against a TrpE-jun fusion construct, Angel, et al., 1988) and Fos protein (B, polyclonal serum produced in rabbits against a β-gal-Fos fusion construct, Verrier, et al., 1986). The Fos antibodies coprecipitate Jun protein.

cells which are permissive for SV40 (Traut and Fanning, 1988). Although the mechanisms involved in SV40 amplification are not identical to the bacterial SOS system, both the induced replication of SV40 and the induction of HIV-1 resemble formally the proviral activation of Lambda bacteriophage.

We and others have shown earlier that UV induces the release of an extracellular factor (EPIF) which causes in non-irradiated cells the activation of a subset of the UV inducible promoters (Table 1; Schorpp, et al., 1984; Rotem, et al., 1987). Several cell lines can be induced to EPIF secretion and several cell types serve as target cells. For instance, UV treated epidermal cells release EPIF which in lymphoid cells increases HIV-1 expression (Stein, unpublished). This observation may be important for the activation of latent viruses in humans (Herrlich, 1988b).

Whether mammalian cells activate an error-prone replication mechanism as bacteria do is still debated. Some observations could be taken as indication for the existance of error-prone replication. As a physiological example, pre B cells and perhaps activated peripheral lymphocytes in an early stage introduce point mutations into the IgH gene and its neighborhood (Meyer, et al., 1985). The "mutator" can be switched off by fusion with a terminal cell: plasmocytoma (M. Wabl, personal communication). That UV may induce such an activity is suggested by the non-targeted mutations in the vicinity of UV DNA damage (Seidman, et al., 1987) and by the point mutations introduced into the genes conferring resistance to thioguanine and diphtheria toxin, by the UV induced extracellular protein factor EPIF. Human diploid fibroblasts treated with an EPIF containing ammonium sulfate fraction of conditioned medium from UV treated cells, for only 6 hours, accumulate about four times more point mutations in these genes than control cells (Herrlich, et al., 1987; Maher, et al., 1988). These data are compatible with the assumption that the UV response encodes an error-prone mechanism.

A classical instrument in establishing the existance of damage inducible repair and mutagenesis functions, is the application of split doses. When a radiation dose leads to a lower survival or mutation rate of cells as compared to the same dose given in more than one portion, the result is usually taken as an indication for the induction of cellular functions involved in DNA repair or mutagenesis. When inhibitors of protein synthesis given between the split doses abolish the effect, this may imply, that the increased survival or mutagenesis is due to the synthesis of new proteins. Contradictory results have in fact been reported (Domon and Rauth, 1973; Thilly and Heidelberger, 1973; Warren and Stich, 1975; Chang, et al., 1978; Stone-Wolff and Rossman, 1981). Although it is not proven that such functions would be FOS dependent, it may be worth while to consider how the data known for FOS would affect the interpretation of split dose experiments: (I) The induction of FOS by UV is transient, the RNA and protein disappearing from the treated cells with half lifes of 8 minutes and 30 minutes, respectively. The half life of FOS dependent gene products (e.g. repair or mutagenesis functions) is of course unknown; (II) Immediately after UV treatment, inducibility of FOS becomes refractory in that a second dose of UV at 3 or 24 hours does not lead to new induction of FOS (Büscher, et al., 1988); (III) UV in the presence of cycloheximide causes enormous overinduction of FOS RNA both by preventing effective repression of transcription and by stabilizing FOS mRNA. With cycloheximide permitting some 5% residual protein synthesis, the net induction of FOS and of possibly dependent repair and mutagenesis functions may be significant. These three parameters may make the outcome of split dose experiments rather unpredictable. Caution, therefore, is appropriate in their interpretation.

Recently, we have obtained first evidence for a UV-induced function that is engaged in cell survival:metallothionein. The induction of metallothionein by UV (and by other DNA damaging agents, see Table 1) has led to the idea that metallothionein may exert protection against genotoxic stress. Metallothionein confers resistance to heavy metals (Beach and Palmiter, 1981) and, as an isolated protein, scavenges free oxygen and hydroxyl radicals (Thornalley and Vasak, 1985). Exposing cultivated cells chronically to high doses of cadmium, causes an increase of intracellular metallothionein and slight radiation tolerance (Bakka, et al., 1982). Cadmium exposure, however, could have side effects causing resistance.

Table 2. Effect of gamma-rays and various chemical mutagens on survival of CHO cells and CHO-cells transfected with pBPV-MTII-A.

	Kl-2	Kl-2MT	Bcll	BcllMT
Gamma-rays (Gy)	6.2	5.1	2.2	2.2
Bleomycin (µg/ml)	78	84	40	43
N-Methyl-N'-nitro-N-nitrosoguanidine (MNNG) (µM)	1.2	2.0*	1.6	8.7
N-Methyl-N-nitrosourea (MNU) (mM)	0.35	0.8*	0.45	1.8
Methyl methanesulfonate(mM)	0.7	0.7	0.7	0.7
N-Hydroxy-N-chloroethyl-nitrosourea (µM)	48	25	26	26
Mitomycin C (µg/ml)	3.0	2.6	1.2	1.7
Cis-platinum (µM)	8	16	9	16

CHO Kl-2, parental strain; Bcll, Xray hypersensitive mutant derived from CHO Kl-2 (Jeggo, et al., 1982); Kl-2MT and BcllMT denote the corresponding pBPV-MTII-A transfectants. Data represent the dose reducing survival by 90% (D_{10}). Treatment with chemicals was performed for 60 minutes.

*Survival curve with plateau at 5-10% survival.

To examine the role of metallothionein in the absence of any other change, we generated overexpressing rodent cells by transfecting a BPV-MTIIA construct. The total cellular content of metallothionein was increased dramatically by the transfection. Less than 4% of radioactive cysteine incorporated into protein was in metallothionein in the control cells and 97.7% in the overexpressing cells (in the absence of cadmium). While metallothionein did not help against the lethal effects of ionizing radiation, metallothionein conferred strong protection from the lethal effects of alkylation.

Expression of metallothionein-mediated alkylation resistance depended on the agent used (Table 2). The highest degree of resistance of rodent cells overexpressing metallothionein was observed for MNNG and MNU. Protection from killing by these mutagens probably occurred at the level of damage handling because the degree of total DNA alkylation was not affected in the transfectants. Protection from alkylation-induced cytotoxicity is, in our knowledge, the first physiologic function ascribed to metallothionein in addition to its protecting cells from heavy metal intoxication.

CONCLUSIONS

Using cloned UV induced genes as a handle we have defined one of the critical intermediate functions between DNA damage and transcriptional activation of effector genes: the FOS gene. The gap between DNA damage and FOS still needs to be filled. From inhibitor studies it seems that a protein kinase is involved. Putative effector genes are numerous. With respect to SOS function, the bacterial prototype example teaches us: Repair genes need to be cloned first before we will be able to test their regulation and function. Functions have, however, been detected for the induction of latent viruses and of mutagenesis. The role of metallothionein described here suggests the existence of inducible repair or damage tolerance mechanisms for which metallothionein appears to be a cofactor or regulatory element.

REFERENCES

Angel, P., Rahmsdorf, H.J., Pöting, A., and Herrlich, P., 1985, C-fos mRNA levels in primary human fibroblasts after arrest in various stages of the cell cycle. Cancer Cells, 3:315.

Angel, P., Poting, A., Mallick, U., Rahmsdorf, H.J., Schorpp, M., and Herrlich, P., 1986, Induction of metallothionein and other mRNA species by carcinogens and tumor promoters in primary human skin fibroblasts. Mol. Cell. Biol., 6:1760.

Angel, P., Baumann, I., Stein, B., Delius, H., Rahmsdorf, H.J., and Herrlich, P., 1987a, 12-O-tetradecanoyl-phorbol-13-acetate induction of the human collagenase gene is mediated by an inducible enhancer element located in the 5'-flanking region, Mol. Cell. Biol., 7:2256.

Angel,P., Imagawa, M., Chiu, R, Stein, B., Imbra, R.J., Rahmsdorf, H.J., Jonat, C., Herrlich, P., and Karin, M., 1987b, Phorbol ester inducible genes contain a common cis element recognized by a TPA-modulated trans-acting factor, Cell, 49:729.

Angel, P., Allegretto, E., Okino, S., Hattori, K., Boyle, W. J., Hunter, T., and Karin, M., 1988, The c-jun proto-oncogene encodes a sequence specific DNA-binding protein similar or identical to the transcriptional activator AP-1, Nature, 332:166.

Baeuerle, P.A. and Baltimore, D., 1988, IkB: a specific inhibitor of the NF-kB transcription factor. Science, 242:540.

Bakka, A., Johnson, A.S., Endresen, C., and Rugstad, H.E., 1982, Radioresistance in cells with high contents of metallothionein, Experienta, 32:381.

Beach, L.R. and Palmiter,R.D., 1981, Amplification of the metallothionein-I gene in cadmium-resistant mouse cells. Proc. Natl.Acad.Sci.USA, 78:2110.

Bohmann, D., Bos, T.J., Admon,A., Nishimura, T., Vogt, P.K. and Tjian, R., 1987, Human proto-oncogene c-jun encodes a DNA binding protein with structural and functional properties of transcription factor AP-1. Science, 238: 1386.

Brown, P.C., Tlsty, T.D., and Schimke, R.T., 1983, Enhancement of methotrexate resistance and dihydrofolate reductase gene amplification by treatment of mouse 3T6 cells with hydroxyurea. Mol. Cell. Biol. 3:1097.

Büscher, M., Rahmsdorf, H.J., Litfin, M., Karin, M., and Herrlich, P., 1988, Activation of the c-fos gene by UV and phorbol ester: different signal transduction pathways converge to the same enhancer element, Oncogene, 3:301.

Chandler, V.L., Maler, B.A., and Yamamoto, K.R., 1983, DNA sequences bound specifically by glucocorticoid receptor in vitro render a heterologous promoter hormone responsive in vivo, Cell, 33:489.

Chang, C.C., Ambrosio, S.M., Schultz, R., Trosko, J.E., and Setlow, R.B., 1978, Modification of UV-induced mutation frequencies in Chinese hamster cells by dose fractionation, cycloheximide and caffeine treatments, Mutation Res., 52:231.

Cohn, S.M., Krawisz, B.R., Dresler, S.L., and Lieberman, M. W., 1984, Induction of replicative DNA synthesis in quiescent human fibroblasts by DNA damaging agents, Proc. Natl. Acad. Sci. USA, 81:4828.

Cornelis, J.J., Su, Z.Z., and Rommelaere, J., 1982, Direct and indirect effects of ultraviolet light on the mutagenesis of parvovirus H-1 in human cells, Embo J., 1:693.

Curran, T., and Teich, N.M., 1982, Candidate product of the FBJ murine osteosarcoma virus oncogene: Characterization of a 55,000 Dalton phosphoprotein, J. Vir.. 42:114.

Defais, M.J., Hanawalt, P.C., and Sarasin, A.R., 1983, Viral probes for DNA repair, Adv. Radiat. Biol., 10:1.

Dinter, H., Chiu, R., Imagawa, M., Karin, M., and Jones, K.A, 1987, In vitro activation of the HIV-1 enhancer in extracts from cells treated with a phorbol ester tumor promoter, EMBO J., 6:4067.

Dion, M. and Hamelin, C., 1987, Relationship between enhanced reactivation and mutagenesis of UV-irradiated human cytomegalovirus in normal human cells, EMBO J., 6:397.

Domon, M., and Rauth, A.M., 1973, Cell cycle specific recovery from fractionated exposures of ultraviolet light, Rad. Res., 55:81.

Fisch, T.M., Prywes, R., and Roeder, G., 1987, C-fos sequences necessary for basal expression and induction by epidermal growth factor, 12-O-tetradecanoyl phorbol-13-acetate, and the calcium ionophore, Mol. Cell. Biol., 7:3490.

Fornace, Jr. A.J., Schalch, H., and Alamo, Jr. I. , 1988a, Coordinate induction of metallothioneins I and II in rodent cells by UV irradiation, Mol. Cell. Biol., 8:4716

Fornace, Jr. A.J., Alamo, Jr. L., and Hollander, M.C., 1988b, DNA damage-inducible transcripts in mammalian cells, Proc. Natl. Acad. Sci. USA, 85:8800.

Franza, Jr. B.R., Rauscher, III. F.J., Josephs, S.F., and Curran, T., 1988, The fos complex and fos-related antigens recognize sequence elements that contain AP-1 binding sites. Science, 239:1150.

Friedberg, E.C., Burtscher, H.J., Cooper, A.J., Couto, L.B., Harosh, I., Kalainov, D., Lambert, C., Naumovski, L., Robinson, G.W., Siede, W., Song, J.M., and Weiss, W.A., 1988, Rad genes and rad proteins for nucleotide excision repair in the yeast Saccharomyces cerevisiae: Recent progress, in: "Mechanisms and Consequences of DNA Damage Processing," pp. 185, E. Friedberg and P. Hanawalt, eds., Alan R. Liss,Inc., New York.

Greenberg, M.E., and Ziff E.B., 1984, Stimulation of 3T3 cells induces transcription of the c-fos proto-oncogene. Nature, 311:433.

Herrlich, P., Rahmsdorf, H.J., Pai, S.H., and Schweiger, M., 1974, Translational control induced by bacteriophage T7, Proc. Natl. Acad. Sci. USA., 71:1088.

Herrlich, P., Mallick, U., Ponta, H., and Rahmsdorf, H.J., 1984, Genetic changes in mammalian cells reminiscent of an SOS response, Human Genet., 67:360.

Herrlich, P., Angel, P., Rahmsdorf H.J., Mallick, U., Pöting, A., Hieber, L., Lücke-Huhle, Ch., and Schorpp, M., 1986, The mammalian genetic stress response, in: "Advances in Enzyme Regulation," G. Weber, ed., Pergamon Press, 25, pp. 485-504.

Herrlich, P., Imagawa, M., Maher, V., Sato, K., McCormick, J.J., Angel, P., Karin, M., Baumann, I., Lücke-Huhle, Ch., and Rahmsdorf, H.J., 1987, The molecular basis for the UV response: cis and trans acting elements responsible for gene induction, in: "Accomplishments in Oncology", zur Hausen, H., Schlehofer, J., eds., J.B. Lippincott, Philadelphia.

Herrlich, P. and Ponta, H., 1989, Nuclear oncogenes transform extracellular stimuli into changes in the genetic program, TIG, 5:112.

Herrlich, P., 1988a, Endogenous cellular contributions to mammalian mutagenesis, in: "DNA replication and mutagenesis",R.E. Moses, W.C. Summers, eds., American Society for Microbiology, Washington, pp.457-464.

Herrlich, P., 1988b, The problems of latency in human disease: Molecular actions of tumor promoters and carcinogens, in: "Accomplishments in Cancer Research , 1987", J.G. Fortner, J.E. Rhoads, eds., pp. 213-230. Lippincott, Philadelphia.

Herrlich, P., Ponta, H., Stein, B., Gebel, S., König, H., Schönthal, A., Büscher, M., and Rahmsdorf, H.J., 1989, The role of fos in gene regulation, in: "Molecular Mechanisms and Consequences of Activation of Hormone and Growth Factor Receptors." C.E. Sekeris, ed., in press.

Hirsch-Kauffmann, M., Schweiger, M., Herrlich, P., Ponta, H., Rahmsdorf, H.J., Pai, S.H., and Wittmann, H.G., 1975, Transcriptional units for ribosomal proteins of Escher

ichia coli, Eur. J. Biochem., 52:469.
Hynes, N.E., van Ooyen, A.J.J., Kennedy, N., Herrlich, P., Ponta, H., and Groner, B., 1983, Subfragment of the long terminal repeat cause glucocorticoid responsive expression of mouse mammary tumor virus and of an adjacent gene. Proc. Natl. Acad. Sci. USA, 80:3637.
Imbra, R.J., and Karin, M., 1986, Phorbol ester induces the transcriptional stimulatory activity of the SV40 enhancer. Nature, 323:555.
Jeggo, P.A., Kemp, L.M., Holliday, R., 1982, The application of the microbial "tooth-pick" technique to somatic cell genetics, and its use in the isolation of X-ray sensitive mutants of the Chinese hamster ovary cells, Biochimie, 64:713.
Kaina, B., 1982, Enhanced survival and reduced mutation and aberration frequencies induced in V79 Chinese hamster cells pre-exposed to low levels of methylating agents, Mutation Res., 93:195.
Kaina, B., 1983, Studies on adaptation of V79 Chinese hamster cells to low doses of methylating agents. Carcinogenesis, 4:1437.
Karin, M., and Herrlich, P., 1988, Cis- and trans-acting genetic elements responsible for induction of specific genes by tumor promoters, serum factors and stress, in: "Genes and Signal Transduction in Multistage Carcinogenesis", N.H. Colburn, ed., pp. 415-440, Marcel Dekker, Inc., New York.
Kartasova, T., Cornelissen, B.J.C., Belt, P., and van de Putte, P., 1987, Effects of UV, 4-NQO and TPA on gene expression in cultured human epidermal keratinocytes, Nucl. Acids Res., 15:5945.
Kartasova, T., and van de Putte, P,. 1988, Isolation, characterization, and UV-stimulated expression of two families of genes encoding polypeptides of related structure in human epidermal keratinocytes. Mol. Cell Biol., 8:2195.
Kaufmann, W.K., Cleaver, J.E., and Painter, R.B., 1980, Ultraviolet radiation inhibits replicon initiation in S phase human cells, Biochim. et Biophys. Acta, 608:191.
Kaufman, J.D., Valandra, G., Roderiquez, G., Bushar, G., Giri, C., and Norcross, A., 1987, Phorbol ester enhances human immunodeficiency virus-promoted gene expression and acts on a repeated 10-base-pair functional enhancer element, Mol. Cell. Biol., 7:3759.
Kleinberger, T., Flint, Y.B., Blank, M., Etkin, S., and Lavi, S., 1988, Carcinogen-induced trans activation of gene expression. Mol. Cell. Biol., 8:1366.
Kogoma, T., Torrey, T.A., and Connaughton, M.J., 1979, Induction of UV-resistant DNA replication in Escherichia coli: Induced stable DNA replication as an SOS function. Molec. Gen. Genet., 176:1.
Lamph, W.W., Wamsley, P., Sassone-Corsi, P., and Verma, I.M., 1988, Induction of proto-oncogene JUN/AP-1 by serum and TPA. Nature, 334:629.
Laval, F., and Laval, J., 1984, Adaptive response in mammalian cells: Crossreactivity of different pretreatments on cytotoxicity as contrasted to mutagenicity. Proc. Natl. Acad. Sci. USA, 81:1062.
Lavi, S., 1981, Carcinogen-mediated amplification of viral DNA sequences in simian virus 40-transformed Chinese hamster embryo cells, Proc. Natl. Acad. Sci. USA, 78:6144.
Lee, W., Mitchell, P., and Tjian, R., 1987, Purified trans-

cription factor AP-1 interacts with TPA-inducible enhancer elements, Cell, 49:741.

Lücke-Huhle, C., and Herrlich, P., 1987, Alpha radiation induced amplification of integrated SV40 sequences is mediated by a trans-acting mechanism, Int. J. Cancer, 39:94.

Lücke-Huhle, C., Gloss, B., and Herrlich, P., 1989, Radiation-induced gene amplification in rodent and human cells. Acta Biologica Hungarica, 40: in press.

Lücke-Huhle, C., Mai, S. , and Herrlich, P., 1989, UV induced "early domain" binding factor as the limiting component in UV induced SV40 amplification in rodent cells, in press.

Maher, V.M., Sato, K., Kateley-Kohler, S., Thomas, H., Michaud, S., McCormick, J.J., Kraemer, M., Rahmsdorf, H.J., and Herrlich, P., 1988, Evidence of inducible error-prone mechanisms in diploid human fibroblasts, in: "DNA Replication and Mutagenesis", R. Moses and W.C. Summers, eds., pp. 465-471, ASM, Washington D.C.

Maltzman, W., and Czyzyk, L., 1984, UV irradiation stimulates levels of p53 cellular tumor antigen in nontransformed mouse cells. Mol. Cell. Biol., 4:1689.

Meyer, J., Jäck, H.M., Ellis, N., and Wabl, M.R., 1986, High rate of somatic point mutations in vitro in and near the variable-region segment of an immunoglobulin heavy chain gene, Proc. Natl. Acad. Sci. USA, 83:6950.

Miskin, R., and Ben-Ishai, R., 1981, Induction of plasminogen activator by UV light in normal and xeroderma pigmentosum fibroblasts, Proc. Natl. Acad. Sci. USA, 78:6236.

Moustacchi, E., 1987, DNA Repair in Yeast: Genetic control and biological consequences, Adv. Radiat. Biol., 13:1.

Nishizuka, Y., 1986, Studies and perspectives of protein kinase C, Science, 233:305.

Olivieri, G., Bodycote, J., and Wolff, S., 1984, Adaptive response of human lymphocytes to low concentrations of radioactive thymidine, Science, 223:594.

Ponta, H., Kennedy, N., Skroch, P., Hynes, N., Groner, B., 1985, Hormonal response region in the mouse mammary tumor virus long terminal repeat can be dissociated from the proviral promoter and has enhancer properties, Proc. Natl. Acad. Sci. USA, 82:1020.

Protic, M., Roilides, E., Levine, A.S., and Dixon, K., 1988, Enhancement of DNA repair capacity of mammalian cells by carcinogen treatment. Somatic Cell and Molec. Genetics, 14:351.

Prywes, R., and Roeder, R.G., 1986, Inducible binding of a factor to the c-Fos enhancer, Cell., 47:777.

Radman, M., 1980, Is there SOS induction in mammalian cells? Photochem. Photobiol., 32:832.

Rahmsdorf, H.J., Koch, N., Mallick, U., and Herrlich, P., 1983, Regulation of MHC class II invariant chain expression: Induction of synthesis in human and murine plasmocytoma cells by arresting replication. EMBO J., 2:811.

Rauscher, III F.J., Cohen, D.R., Curran, T., Bos, T.J., Vogt, P.K., Bohman, D., Tjian, R., and Franza, Jr. B.R., 1988, Fos-associated protein p39 is the product of the jun proto-oncogene, Science, 240:1010.

Rieger, R., Michaelis, A., and Nicoloff, H., 1982, Inducible repair processes in plant root tip meristems? "Below-additivity effects" of unequally fractionated clastogen concentrations. Biol. Zbl., 101:125.

Rose-John, S., Rincke, G., and Marks, F., 1987, The induction of ornithine decarboxylase by the tumor promoter TPA is controlled at the post-transcriptional level in murine Swiss 3T3 fibroblasts, Biochem. Biophys. Res. Comm., 147:219.

Rotem, N., Axelrod, J.H., and Miskin, R., 1987, Induction of urokinase-type plasminogen activator by UV light in human fetal fibroblasts is mediated through a UV-induced secreted protein. Mol. Cell. Biol., 7:622.

Samson, L., and Schwartz, J.L., 1980, Evidence for an adaptive DNA repair pathway in CHO and human skin fibroblast cell lines, Nature, 287:861.

Sarasin, A., 1985, SOS response in mammalian cells, Cancer Invest., 2:163.

Sauerbier, W., and Hercules, K., 1978, Gene and transcription unit mapping by radiation effects. Ann. Rev. Genet., 12:329.

Scherzinger, E., Herrlich, P., Schweiger, M., and Schuster, H., 1972, The early region of the DNA of bacteriophage T7. European J. Biochem., 25:341.

Schimke, R.T., 1984, Gene amplification in cultured animal cells, Cell, 37:705.

Schönthal, A., Herrlich, P., Rahmsdorf, H.J., and Ponta, H., 1988a, Requirement for fos gene expression in the transcriptional activation of collagenase by other oncogenes and phorbol esters, Cell, 54:325.

Schönthal, A., Gebel, S., Stein, B., Ponta, H., Rahmsdorf, H. J., and Herrlich, P., 1988b, Nuclear oncoproteins determine the genetic program in response to external stimuli, Cold Spring Harbor Symposia on Quantitative Biology, 53: 779.

Schorpp, M., Mallick, U., Rahmsdorf, H.J., and Herrlich, P., 1984, UV induced extracellular factor from human fibroblasts communicates the UV response to non-irradiated cells, Cell, 37:861.

Seidman, M.M., Bredberg, A., Seetharam, S., and Kraemer, K. H., 1987, Multiple point mutations in a shuttle vector propagated in human cells: Evidence for an error-prone DNA polymerase activity, Proc. Natl. Acad. Sci. USA, 84:4944.

Sen, R., and Baltimore, D., 1986, Inducibility of κ immunoglobulin enhancer-binding protein NF-kB by a posttranslational mechanism, Cell, 47:921.

Stein, B., Rahmsdorf, H.J., Schönthal, A., Büscher, M., Ponta, H., and Herrlich, P., 1988, The UV induced signal transduction pathway to specific genes, in: "Mechanisms and Consequences of DNA damage Processing." E. Friedberg and P. Hanawalt, eds., UCLA Symposia on Molecular and Cellular Biology, New Series, Vol. 83:557, Alan R. Liss, Inc., New York.

Stone-Wolff, D.S., and Rossman, T.G., 1981, Effects of inhibitors of de novo protein synthesis on UV-mutagenesis in Chinese hamster cells, Mutation Res., 82:147.

Thilly, W.G., and Heidelberger, C., 1973, Cytotoxicity and mutagenicity of ultraviolet irradiation as a function of the interval between split doses in cultured Chinese hamster cells, Mutation Res., 17:287.

Thornalley, P.J., and Vasak, M., 1985, Possible role for metallothionein in protection against radiation-induced oxidative stress. Kinetics and mechanism of its reaction with superoxide and hydroxyl radicals. Biochem. Biophys. Acta., 827:36.

Traut, W., and Fanning, E., 1988, Sequence-specific interactions between a cellular DNA-binding protein and the simian virus 40 origin of DNA replication, Mol. Cell. Biol., 8:903.

Treisman, R., 1985, Transient accumulation of c-fos RNA following serum stimulation requires a conserved 5' element and c-fos 3' sequences, Cell, 42:889.

Treisman, R., 1986, Identification of a protein-binding site that mediates transcriptional response of the c-fos gene to serum factors. Cell, 46:567.

Valerie, K., Delers, A., Bruck, C., Thiriart, C., Rosenberg, H., Debouck, C., and Rosenberg, M., 1988, Activation of human immunodeficiency virus type 1 by DNA damage in human cells. Nature, 333:78.

Vanetti, M., 1988, Der unterschiedliche Beitrag der Motive im SV40 Enhancer zur Induktion mit UV Strahlung, Dipl.-Thesis, University of Karlsruhe.

Varshavsky, A., 1981, Phorbol ester dramatically increases incidence of methotrexate-resistant mouse cells: possible mechanisms and relevance to tumor promotion, Cell, 25:561.

Verrier, B., Müller, D., Bravo, R., and Müller, R., 1986, Wounding a fibroblast monolayer results in the rapid induction of the c-fos proto-oncogene, EMBO J., 5:913.

Verma, A.K., Lowe, N.J., and Boutwell, R.K., 1979, Induction of mouse epidermal ornithine decarboxylase activity and DNA synthesis by ultraviolet light, Cancer Research, 39: 1035.

Walker, G.C., 1985, Inducible DNA repair systems, Ann. Rev. Biochem., 54:425.

Waller, E.K., and Schleuning, W.D., 1985, Induction of fibrinolytic activity in HeLa cells by phorbol myristate acetate. J. Biol. Chem., 260:6354.

Warren, P.M., and Stich, H.F., 1975, Reduced DNA repair capacity and increased cytotoxicity following split doses of the mutagen 4-nitroquinoline-I-oxide in cultured human cells. Mutation Res., 28:285.

Yamamoto, K.R., 1985, Steroid receptor regulated transcription of specific genes and gene networks, Ann. Rev. Genet., 19:209.

THE TRANSPORTATION OF AP ENDONUCLEASE FROM CYTOPLASM TO NUCLEUS IN MAMMALIAN CELLS

Walter G. Verly, Daniel Maréchal, and Régis César

Biochimie, Faculté des Sciences, Université de Liège, Sart Tilman B6, B-4000 LIEGE I, Belgium

We have shown (Thibodeau and Verly, 1980) that more than 95% of the rat-liver AP endonuclease is located in chromatin. However, after the cell nuclei have been discarded, a significant activity remains in microsomes although the cell sap is practically devoid of it.

We devised a quantitative assay of the microsomal AP endonuclease. To Triton-dissociated microsomes, ϕX174 RF-I DNA is added. After a brief heating that denatures DNA but leaves AP sites intact and addition of ethidium bromide, the remaining RF-I molecules are measured at pH 11.8 by fluorescence. The average number, n, of strand breaks per molecule is calculated from the fractional decrease of RF-I molecules. The microsomal preparation has no action on RF-I molecules that do not contain AP sites; on the other hand, AP site-containing RF-I molecules are transformed into RF-II. The RF-I molecules contain an average of 1 AP site per molecule and the AP site concentration in the assay is much lower than the K_m of the microsomal AP endonuclease, so that the reaction rate is first order to the substrate concentration. If, indeed, we plot ln (S_0/S) versus the amount of enzyme, where S_0 and S are the average numbers, per RF molecule, of intact AP sites at 0 time and after the 15 minutes of standard incubation, a straight line is obtained passing through the origin. We thus have a quantitative assay for the microsomal AP endonuclease.

Microsomes from rat liver were exposed to different ionic strengths before high-speed centrifugation, followed by enzyme assays on the pellet and the supernatant fraction. With a high enough ionic strength, all the AP endonuclease activity can be released in the supernatant fraction. This means two things: first, the enzyme is not an intrinsic membrane protein, but a peripheral one; second, the enzyme is totally located on the outside of the microsomes. In rat liver, the greater part of the microsomes are derived from the endoplasmic reticulum, and the outside of the microsome corresponds to the cytosolic side of the membranes of the endoplasmic reticulum.

DNA Repair Mechanisms and Their Biological Implications in Mammalian Cells
Edited by M.W. Lambert and J. Laval
Plenum Press, New York

Using an exposure to 0.8 M NaCl followed by centrifugation, AP endonuclease-free microsomes and free microsomal AP endonuclease were prepared. We then mixed a constant amount of AP endonuclease-free microsomes with different amounts of free microsomal AP endonuclease in 0.15 M KCl and, after 1 hour incubation at 37^{0}C, separated, by centrifugation, microsomes and supernatant. Bound and free enzymes were measured and the results plotted in Scatchard coordinates. A straight line was obtained showing that all the microsomal binding sites had the same affinity for the AP endonuclease. We concluded that the outside surface of the microsomes and thus, very likely, the cytosolic side of the endoplasmic reticulum membranes, present receptors for the AP endonuclease.

It is known that transportation of nuclear proteins from the cytoplasm, where they are synthesized, into the nucleus depends on a karyophilic signal (Dingwall, 1985; Feldherr, 1985). The protein accumulates at the nuclear pores and this step is signal-dependent, it subsequently migrates into the nucleus and this step is ATP-dependent (Newmeyer and Forbes, 1988; Richardson, et al., 1988). For SV-40 T antigen, the karyophilic signal is the heptapeptide Pro-Lys-Lys-Lys-Arg-Lys-Val (Kalderon, et al., 1984); replacement of the third aminoacid Lys by Asn suffices to interfere with the nuclear migration of the protein (Lanford and Butel, 1984).

We have mixed a constant amount of AP endonuclease-free microsomes, a constant amount of free microsomal AP endonuclease, and different amounts of the SV-40 normal or mutated heptapeptide. The normal heptapeptide displaces the AP endonuclease from its microsomal receptors; a 50 nM concentration displaces half of the AP endonuclease. On the other hand, the mutated heptapeptide does not displace the enzyme.

It thus seems that SV-40 T antigen and AP endonuclease might use the same receptors on the endoplasmic reticulum. It is likely that, when the aminoacid sequence of the AP endonuclease will be known, a karyophilic signal very similar if not identical with that of SV-40 T antigen will be found.

Our hypothesis for the transportation of karyophilic proteins within the nucleus is the following (Figure 1). Unoccupated receptors for the karyophilic signal are free to move on the cytosolic side of the endoplasmic reticulum and the external nuclear membrane. Attachment of the signal of a karyophilic protein to the receptor induces a conformational change so that the protein-receptor complex is blocked by the nuclear pore proteins. ATP is then needed for the passage into the nucleus. In other words, to the two steps already proposed by other authors, concentration at the nuclear pores and translocation into the nucleus (Newmeyer and Forbes, 1988: Richardson, et al., 1988), we add a third one, which is the first in the sequence of events: fixation to receptors on the endoplasmic reticulum. The endoplasmic reticulum would serve as a large net thrown in the cytosol to collect karyophilic proteins in order to concentrate them at the nuclear pores.

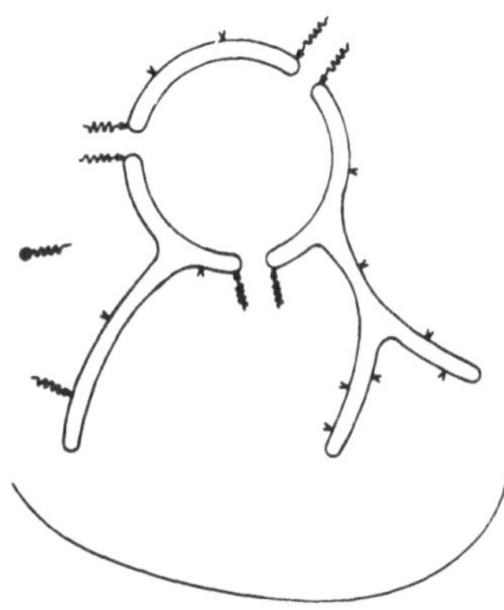

Figure 1. Transportation into the nucleus of a karyophilic protein. The karyophilic protein (zigzag line) is synthesized on a ribosome (black dot) in the cytosol. The receptors for the signal of the karyophilic protein (open angle), when unoccupied, move freely on the cytosolic side of the endoplasmic reticulum and external nuclear membrane; they have no affinity for the nuclear pore proteins. Attachment of the signal of a karyophilic protein to the receptor induces a conformational change in the receptor which adheres to the nuclear pore proteins when it meets them. The protein-receptor complex is blocked near the nuclear pores where it accumulates.

REFERENCES

Dingwall,C., 1985, The accumulation of proteins in the nucleus, TIBS, 10:64-66

Feldherr,C.M., 1985, The uptake and accumulation of proteins in the cell nucleus, BioEssays, 3:52-55.

Kalderon,D., Roberts,B.L., Richardson,W.D., and Smith,A.E., 1984, A short amino acid sequence able to specify nuclear location, Cell, 39:499-509.

Lanford,R.E. and Butel,J.S., 1984, Construction and characterization of an SV40 mutant defective in nuclear transport of T antigen, Cell, 37:801-813.

Newmeyer,D.D. and Forbes,D.J., 1988, Nuclear import can be separated into distinct steps in vitro: nuclear pore binding and translocation, Cell, 52:641-653.

Richardson,W.D., Mills,A.D., Dilworth,S.M., Laskey,R.A. and Dingwall,C., 1988, Nuclear protein migration involves two steps: rapid binding at the nuclear envelope followed by slower translocation through nuclear pores, Cell, 52:655-664.

Thibodeau,L. and Verly,W.G., 1980, Cellular localization of the AP (apurinic/apyrimidinic) endodeoxyribonucleases in rat liver, Eur.J.Biochem., 107:555-563.

DNA DAMAGE AND OXYGEN SPECIES

J. Rueff, A. Laires, A. Bras, H. Borba, T. Chaveca, J. Gaspar, A. Rodrigues, L. Cristovâo and M. Monteiro

Department of Genetics, Faculty of Medical Sciences UNL, R. Junqueira 96, P-1300 Lisbon, Portugal

INTRODUCTION

Various genes are involved in cell protection against oxidative stress (e.g., SOD, catalase, GSH peroxidase) and some genes recognize and repair oxidative damage to DNA (both base and sugar damage). Yet, the nature and relative rates of the reactions involving active oxygen species make it difficult to ascertain, in each particular situation, what reactants may act as the predominant genotoxicant and what DNA repair mechanisms may ensue.

The broad term "oxygen stress" is used here to refer mainly to processes where partially reduced oxygen products may be involved.

Partially reduced oxygen products encompass the species listed below which can result from consecutive univalent reductions of oxygen:

$$O_2 \rightarrow O_2^{\cdot -} \rightarrow H_2O_2 \rightarrow HO^{\cdot} + H_2O \rightarrow H_2O$$

Superoxide radical ($O_2^{\cdot -}$) is mostly reactive in a hydrophobic environment but can not, as such, freely cross biological membranes. Its protonated form ($HO_2^{\cdot}$) seems, however, to be able to cross membranes as effectively as H_2O_2. Although only a minor amount of $O_2^{\cdot -}$ will exist in the form of $HO_2^{\cdot}$ at neutral pH, that amount should be considerably higher in the proximity of membranes. Another process for permeation of $O_2^{\cdot -}$ is the existence of an "anion channel", the known example being the erythrocyte membrane (Halliwell and Gutteridge, 1986).

Unlike the superoxide, H_2O_2 readily crosses biological membranes. Cytotoxicity and damage to biomolecules by H_2O_2 and $O_2^{\cdot -}$ are thought to be due to the formation of the highly reactive $HO^{\cdot}$. Transition metals (iron in particular) are cata-

DNA Repair Mechanisms and Their Biological Implications in Mammalian Cells
Edited by M.W. Lambert and J. Laval
Plenum Press, New York

lysts of the Haber-Weiss reaction whose net result is the conversion of $O_2^{\cdot-}$ and H_2O_2 to $HO^{\cdot}$. Very reactive free radicals like $HO^{\cdot}$ interact with biomolecules under diffusion control, this implying that they will react in the vicinity of the site of formation (Pryor, 1986; Slater and Cheeseman, 1988).

Both base and sugar damage may result from reaction of oxygen radicals with DNA. Thymine glycol, methyl-tartronyl-urea, urea and 5-hydroxy-methyluracil are amongst the products which may be formed from thymine. Single-stranded DNA breaks may occur as a result of sugar fragmentation (Ames, 1986; Imlay and Linn, 1988).

Mammalian cells and bacteria possess enzymes that act on oxidatively damaged bases. Two such well characterized repair enzymes are thymine glycol-DNA glycosylase (in *E. coli* the product of *nth* gene) and formamidopyrimidine-DNA glycosylase (Breimer and Lindahl, 1985).

Besides selective recognition activities, the activation of poly-ADP-ribose polymerase by DNA strand breaks might seemingly be a first step response to DNA damage by oxygen radicals (Schraufstatter, *et al.*, 1986; Cleaver and Morgan, 1985). Other inducible responses exist in mammalian cells, though, which are triggered by oxidative stress. These include the induction of SOD (Fridovich, 1983) and probably also the SOS-like response leading to induction of the c-*fos* gene (P. Herrlich, personal communication).

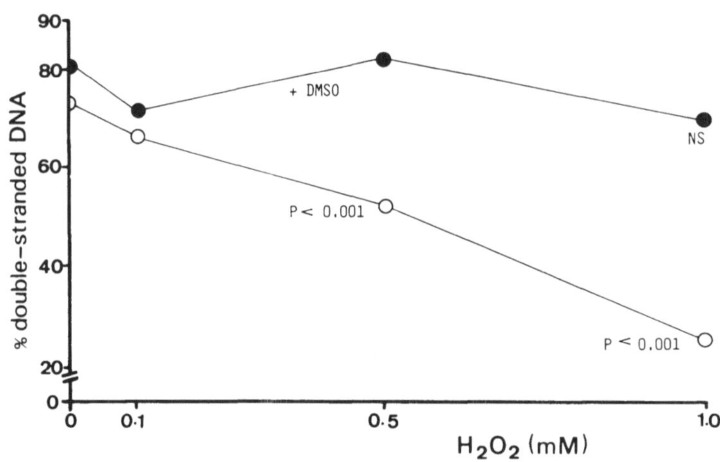

Figure 1. Influence of H_2O_2 on production of DNA strand breaks as determined by fluorimetric analysis of DNA unwinding (FADU). Human leukocytes from normal volunteers non-smokers and without recent exposition to drugs and radiation, suspended in salt solution (2×10^6/ml), were incubated with different doses of H_2O_2 for 10 minutes at 37^0C. Results in the absence (open circles) or in the presence (closed circles) of 150 mM DMSO are the mean of at least 4 independent assays.

GENETIC DAMAGE BY OXYGEN RADICALS

When human leukocytes are incubated with H_2O_2 for 10 minutes it can be shown that the percentage of double-stranded DNA steadily decreases in a dose-dependent way (Figure 1). The assays were carried out at a viable cell concentration of 2 x 10^6/ml and the production of breaks analysed essentially as described by Birnboim and Jevcak (1981). When the cells were treated with H_2O_2 in the presence of DMSO, a hydroxyl radical scavenger, the production of DNA strand breaks was markedly inhibited (Figure 1). This suggests, as already demonstrated in V-79 cells (Bradley and Erickson, 1981), that DNA damage by H_2O_2 is mainly mediated by $OH^{\cdot}$. DNA strand breakage in leukocytes by $O_2^{\cdot-}$ has also been demonstrated and seems to occur through mechanisms partly different from the damage pathway of H_2O_2 (Birnboim and Kanabus-Kaminska, 1985).

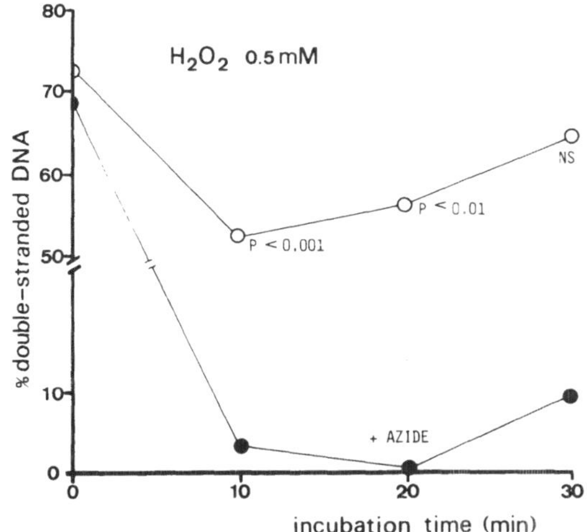

Figure 2. The effect of incubation time with 0.5 mM H_2O_2 on development of DNA strand breaks as determined by FADU. Results in the absence (open circles) or in the presence (closed circles) of 3.96 mM sodium azide are the mean of at least 4 independent assays.

Recovery of DNA damage by H_2O_2 seems to occur in a short delay after challenge (Figure 2). This could be associated with induction of poly (ADP-ribose) polymerase (Schraufstatter, et al., 1986). Besides the protective role of nuclear defense/repair mechanisms, cytoplasmic enzymes also play a major role in DNA protection against oxydative damage. An example of the extent of protection afforded by catalase can be inferred when challenging the cells in the presence of azide, an inhibitor of catalase (Figure 2).

Oxygen free radicals could also produce genetic damage through indirect mechanisms, namely by initiating lipid peroxidation in biological membranes. One end product of unsaturated lipid peroxidation, malondialdehyde (MDA) is a well known mutagen (Bird, et al., 1982; Marnett, et al., 1984). In the conditions used for analysis of DNA strand breaks, production of MDA (Ohkawa, et al., 1979) could be detected at a significant level at 30 minutes incubation with H_2O_2 (Figure 3). MDA can cross-link a variety of biomolecules and modify adenine and cytosine which can result in a decrease of template activity (Nair, et al., 1984).

OXYGEN RADICALS AS INTERMEDIATE SPECIES IN DNA DAMAGE

Numerous genotoxins produce DNA damage through direct or metabolic-mediated reaction with nucleophilic centers in DNA. Some genotoxins, however, seem to produce genetic damage via generation of oxygen radicals; nitro compounds (Biaglow and Varnes, 1982) and compounds containing the quinone nucleus (Chesis, et al., 1984) are seemingly amongst these group of compounds. Oxygen radicals may represent for some chemicals the main genotoxic pathway, or may add to other pathways leading to DNA damage. A network of different pathways may act as an "amplification cascade" for some chemicals.

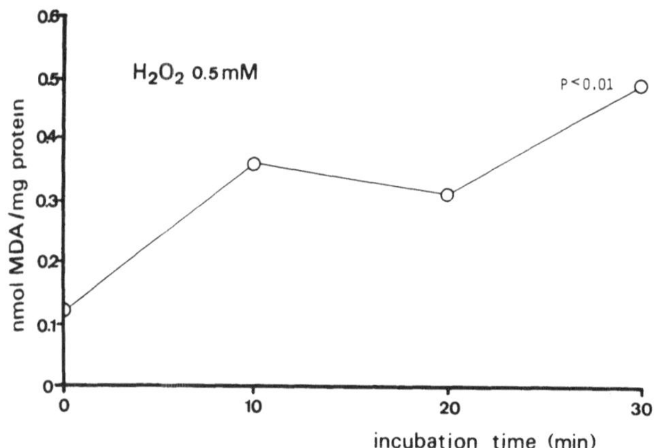

Figure 3. The effect of the incubation time with 0.5 mM H_2O_2 on thiobarbituric acid-reactive products. Tetramethoxipropane was used as standard. Results are the mean of 3 independent assays.

The widely distributed plant flavonol, quercetin, spontaneously produces $O_2^{\cdot-}$. This production, as monitored by reduction of nitro-blue-tetrazolium (NBT), is pH-dependent and SOD markedly decreases the reduction of NBT by quercetin (Figure 4).

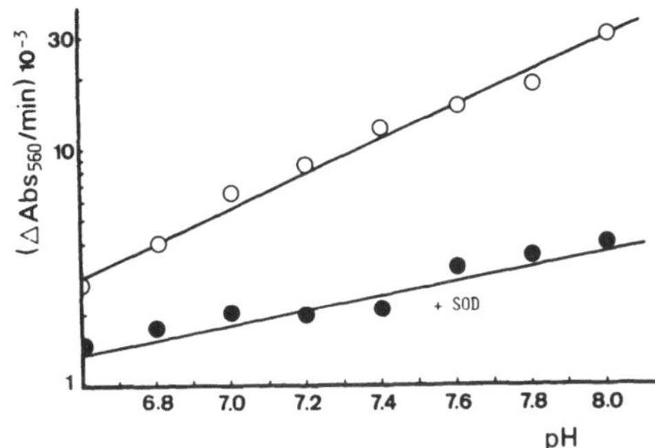

Figure 4. Initial velocities of the reduction of NBT in 0.01 M phosphate buffer by quercetin (20 µg/ml) at different pH values in the presence (closed circles) and absence (open circles) of SOD (120 units/ml).

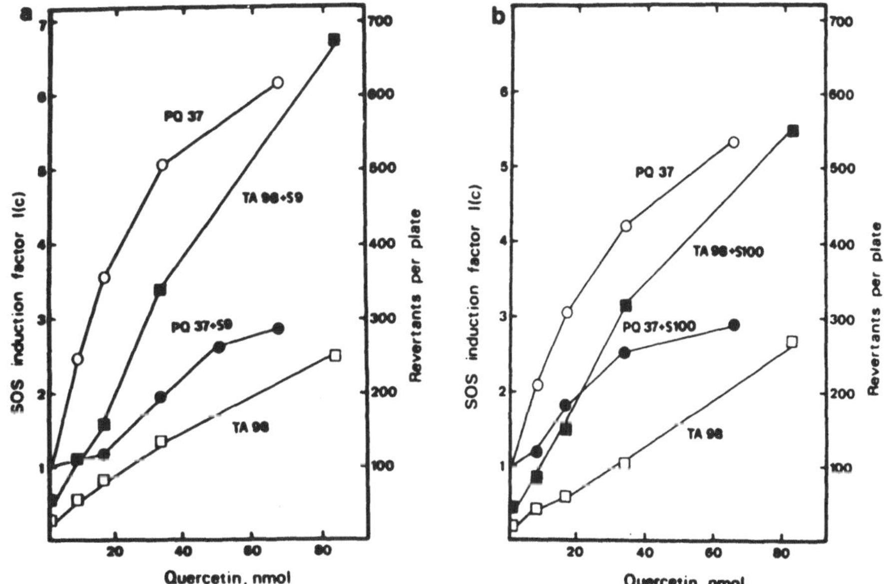

Figure 5. Dose-response curves of quercetin in the Ames assay (strain TA98) and the SOS Chromotest (strain PQ37) both in the absence of rat-liver enzymes or in the presence of S9 (a) or S100 (b). Circles show the SOS induction factor I(c), which is at concentration c: I(c) = R(c)/R(0). R is the ratio of beta-galactosidase activity/alkaline phosphatase activity. (Reproduced from Rueff, et al., 1986, by courtesy of IRL Press).

The responses of quercetin in different genotoxicity assays seem to suggest that quercetin exerts DNA-damage via more then one mechanism (Rueff, et al., 1986; Llagostera, et al., 1987). When assessing the activity of quercetin in the induction of SOS functions in the SOS chromotest (Quillardet and Hofnung, 1985) or by reverse mutation in the Ames test (Maron and Ames, 1983) it can be shown that the patterns of response of quercetin with the rat liver microsomal fraction (S9) or the rat liver cytosolic fraction (S100) are similar in both assays (Figure 5). The S100 fraction decreases the SOS inducing activity of quercetin, whereas the mutagenic activity is increased. Since the S100 fraction contains SOD, it has been postulated that it should act in the Ames assay by protecting the autooxidative degradation of quercetin (Ochiai, et al., 1984).

The degradation of quercetin can be followed spectrophotometrically and SOD protects against autooxidation (Figure 6). HPLC analysis further confirmed the protection afforded by SOD, and additionally showed that a stable product is

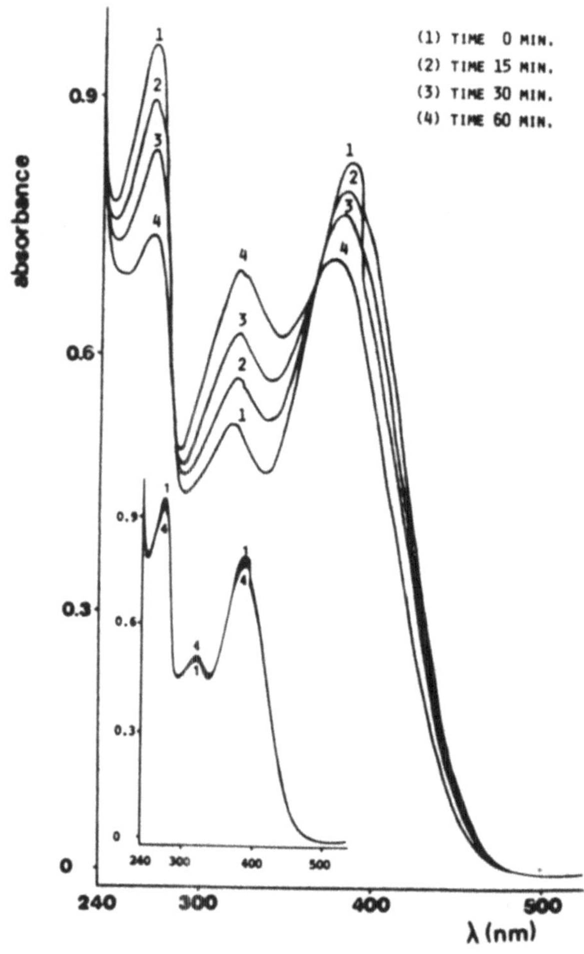

Figure 6. UV/VIS spectra of 20 µg/ml quercetin in 0.01 M phosphate buffer, pH 8.0, at different incubation times in the absence or in the presence (inset) of 40 units/ml SOD.

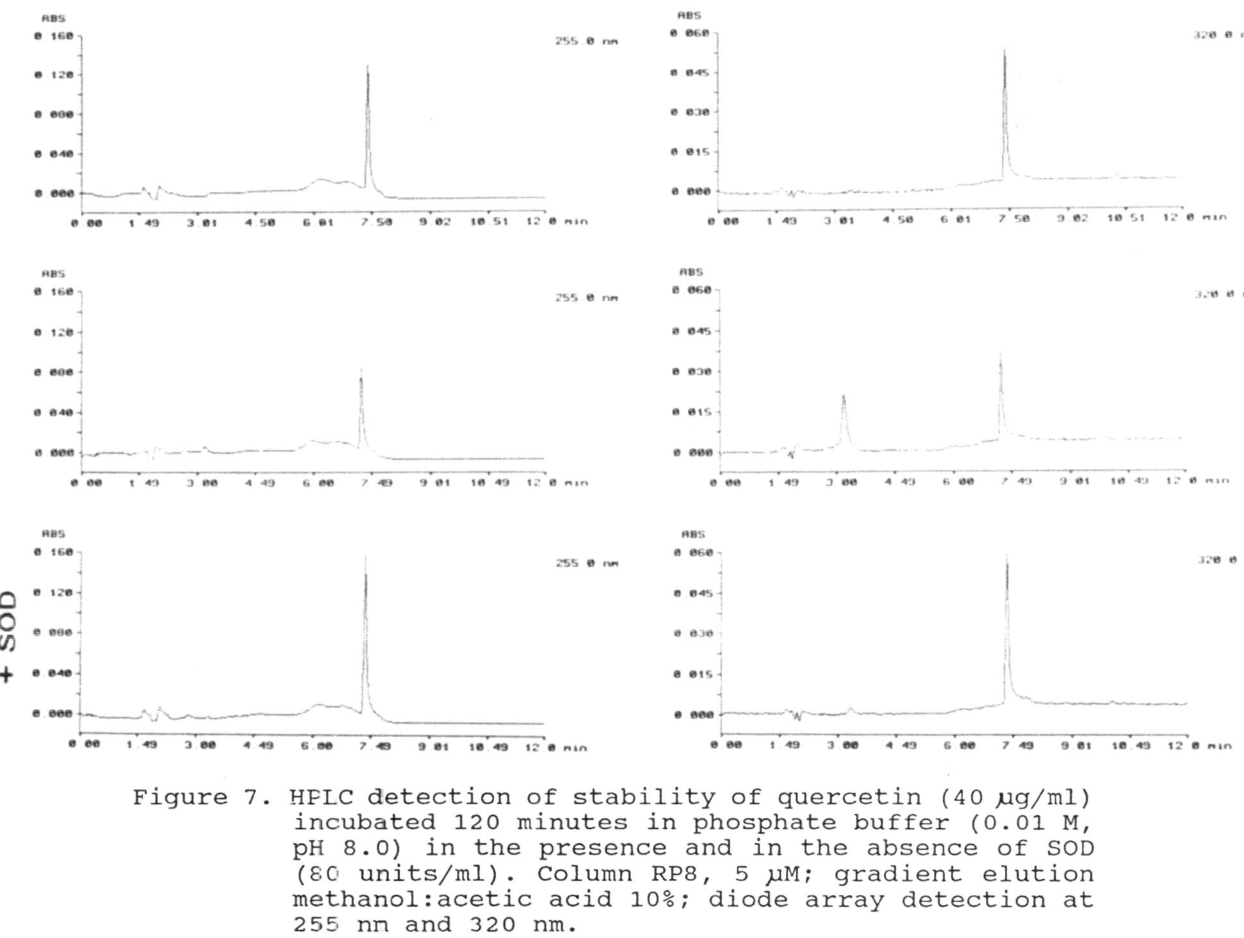

Figure 7. HPLC detection of stability of quercetin (40 µg/ml) incubated 120 minutes in phosphate buffer (0.01 M, pH 8.0) in the presence and in the absence of SOD (80 units/ml). Column RP8, 5 µM; gradient elution methanol:acetic acid 10%; diode array detection at 255 nm and 320 nm.

formed in the absence of SOD (Figure 7). Either this stable reaction product, or another intermediate with lower half-life, could behave as the major inducer of SOS functions. It is interesting to note in this connection, that in mammalian cells quercetin also exhibits a higher DNA-damaging activity when cells are treated in the absence of rat liver enzymes (Meltz and MacGregor, 1981; Carver, et al., 1983). Higher intracellular levels of SOD do not seem, however, to modify

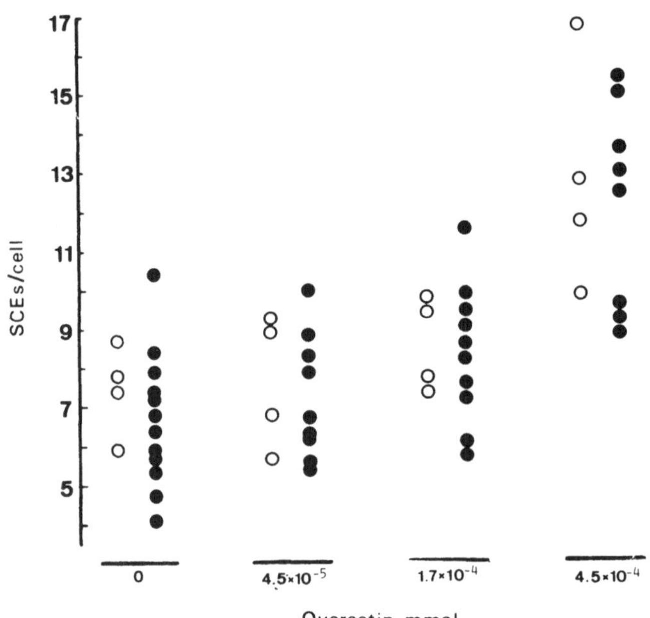

Figure 8. Induction of SCEs by quercetin in euploid (open circles) and trisomic 21 (closed circles) lymphocytes from different individuals. No statistically significant differences could be found between the two groups for each dose tested (Mann-Whitney U Test). Quercetin at the dose of 4.5×10^{-4} per assay significantly induced SCEs in lymphocytes from each of the individuals irrespective of the presence or absence of +21. Each point is the mean of at least 30 metaphases scored.

the genotoxic activity of quercetin. The SOD-1 (CuZnSOD) gene is located in human cells in chromosome 21 (21q22.1), and a gene dosage effect occurs for SOD-1. When induction of sister-chromatid exchanges (SCEs) is assessed in trisomic 21 cells, as compared to euploid cells, no differences could be found between the two groups (Figure 8). The concept of "site specificity" (for review see Halliwell and Gutteridge, 1986) should be evoked in this context.

CONCLUDING REMARKS

The mechanisms of genetic damage mediated by oxygen radicals certainly represent a major topic of research interest in biology and medicine for its implications in processes like aging and cancer. "Oxygen stress" in carcinogenesis may be relevant in the initiation, promotion and progression stages. Numerous physiological processes contribute to an endogenous production of oxygen radicals, and these may in turn modulate the activity of genotoxins/carcinogens.

Interesting to note is that evolution seems to have equipped cells with an array of stress-defense genes coordinately expressed. In bacteria both heat-shock proteins and enzymatic defenses against oxidative stress are expressed under positive control of the oxyR gene. The oxyR gene seems to act as a positive regulator of the regulon encoding those stress-inducible genes (Quillardet and Hofnung, 1988). In mammalian cells both c-*fos* and c-*myc* genes are induced by oxygen stress (Muehlematter, *et al.*, 1988), and c-*fos* is also heat and UV-inducible (P. Herrlich, personal communication). A coordinated expression of stress-inducible genes in mammalian cells can not be excluded.

ACKOWLEDGEMENTS

Our current research is supported by FLAD, CEC (EV4V-0069) and the Calouste Gulbenkian Foundation.

REFERENCES

Ames, B.N., 1986. Carcinogens and Anticarcinogens, in:"Antimutagenesis and Anticarcinogenesis Mechanisms", D.M. Shankel, P.E. Hartman, T. Kada and A. Hollaender, eds, Plenum Press, New York and London.

Biaglow, J.E. and M.E. Varnes, 1982. The metabolic activation of carcinogenic nitro compounds to oxygen-reactives intermediates, in:" Free Radicals and Cancer", R.A. Floyd, ed., Marcel Dekker, Inc. New York and Basel.

Bird, R.P., H. Draper and P.K. Basrur, 1982. Effect of malonaldehyde on cultured mammalian cells. Production of micronuclei and chromosomal aberrations, *Mutation Res.*, 101:237.

Birnboinm, H.C. and J.J. Jevcak, 1981. Fluorometric method for rapid detection of strand breaks in human white cells produced by low doses of radiation, *Cancer Res.*, 41:1889.

Birnboim, H.C. and M. Kanabus-Kaminska, 1985. The production of DNA strand breaks in human leukocytes by superoxide anion may involve a metabolic process, *Proc. Natl. Acad. Sci. USA*, 82:6820.

Bradley, M.O., C. Erickson, 1981. Comparison of the effects of hydrogen peroxide and X-Ray irradiation on toxicity, mutation, and DNA damage/repair in mammalian cells (V-79), *Biochim. Biophys. Acta*, 654:135.

Breimer, L.H.B. and T. Lindahl, 1985. Enzymatic excision of DNA bases damaged by exposure to ionizing radiation or oxidizing agents, *Mutation Res.*, 150: 85.

Carver, J.H., A.V. Carrano, J. MacGregor, 1983. Genetic effects of flavonols quercetin, kaempferol and galangin on Chinese hamster ovary cells in vitro, Mutation Res., 113:45.

Chesis, P.L., D.E. Levin, M.T. Smith, L. Ernster, and B.N. Ames, 1984. Mutagenicity of quinones: Pathways of metabolic activation and detoxification, Proc. Natl. Acad. Sci. USA, 81:1696.

Cleaver, J.E. and W.F. Morgan, 1985. Poly (ADP-ribose) synthesis is involved in the toxic effects of alkylating agents but does not regulate DNA repair, Mutation Res., 150:69.

Fridovich, I., 1983. Superoxide radical: an endogenous toxicant, Ann. Rev. Pharmacol. Toxicol., 23:239.

Halliwell, B. and J.M.C. Gutteridge, 1986. Oxygen free radicals and iron in relation to Biology and Medicine: Some problems and concepts. Arch. Biochem. Biophys., 246: 501.

Imlay, J.A. and S. Linn, 1988. DNA Damage and Oxygen Radical Toxicity, Science, 240:1302.

Llagostera, M., S. Garrido, J. Barbe, R. Guerrero and J. Rueff, 1987. Influence of S9 mix in the induction of SOS system by quercetin, Mutation Res., 191:1.

Marnett, L.J., H.K. Hurd, M.C. Hollstein, D.E. Levin, H. Esterbauer and B.N. Ames, 1984. Naturally occurring carbonyl compounds are mutagens in Salmonella tester strain TA 104, Mutation Res., 148:25.

Maron, D.M. and B.N. Ames, 1983. Revised methods for the Salmonella mutagenicity test, Mutation Res., 113:173.

Meltz, M.L. and J. MacGregor, 1981. Activity of the plant flavonol quercetin in the Mouse lymphoma L5178Y TK+/- mutation, DNA single stand break, and BALB/c 3T3 chemical transformation assay, Mutation Res., 88:317.

Muehlematter, D., R. Larsson and P. Cerutti, 1988. Active oxygen induced DNA strand breakage and poly ADP-ribosylation in promotable JB6 mouse epidermal cells, Carcinogenesis, 9:239.

Nair, V., G.A. Turner and R.J. Offerman, 1984. Novel adducts from the modification of nucleic acid bases by malondialdehyde, J. Am. Chem. Soc., 106:3370.

Ochiai, M., M. Nagao, K. Wakabayashi and T. Sugimura, 1984. Superoxide dismutase acts as an enhancing factor for quercetin mutagenesis in rat liver cytosol by preventing its decomposition, Mutation Res., 129:19.

Ohkawa, H., N. Ohishi and K. Yagi, 1979. Assay for peroxides in animal tissues by thiobarbituric acid reaction, Anal. Biochem., 95:351.

Pryor, W.A., 1986. Oxy-radical and related species: Their formation, lifetimes, and reactions. Ann. Rev. Physiol., 48:657.

Quillardet, P. and M. Hofnung, 1988. The screening, diagnosis and evaluation of genotoxic agents with batteries of bacterial tests, Mutation Res., 205:107.

Quillardet, P. and M. Hofnung, 1985. The SOS chromotest, a colorimetric bacterial assay for genotoxins: procedures, Mutation Res., 147:65.

Rueff, J., A. Laires, H. Borba, T. Chaveca, M.I. Gomes and M. Halpern, 1986. Genetic toxicology of flavonoids: the role of metabolic conditions in the induction of reverse mutation, SOS functions and sister chromatid exchanges, Mutagenesis, 1:179.

Schraufstatter, I.U., D.B. Hinshaw, P. Hyslop, R. Spragg and C.G. Cochrane, 1986. DNA strand-breaks activate polyaden-

osine diphosphate-ribose polymerase and lead to depletion of nicotinamide adenine dinucleotide, J. Clin. Invest., 77:1312.
Slater, T.F. and K.H. Cheeseman, 1988. Free radical mechanisms of tissue injury and mechanisms of protection, in: "Reactive oxygen species in Chemistry, Biology, and Medicine", A. Quintanilha, ed., Plenum Press, New York and London.

MOLECULAR STUDIES OF MUTAGENESIS USING PLASMID VECTORS IN XERODERMA PIGMENTOSUM CELLS

Kenneth H. Kraemer, Saraswathy Seetharam, Douglas E. Brash, Anders Bredberg[1], Miroslava Protic'-Sabljic' and Michael M. Seidman

National Cancer Institute,Bethesda,20892,and Otsuka Pharmaceutical Company, Rockville, MD 20850, USA

ABSTRACT

A plasmid shuttle vector, pZ189, containing a marker gene, supF, was used to measure DNA repair and mutagenesis in human cells. Plasmids were treated with ultraviolet radiation (UV) in vitro and transfected into xeroderma pigmentosum (XP) and repair proficient human cells. Replicated episomal plasmids were harvested and assayed for survival and mutations by transforming indicator bacteria. Plasmids with mutated supF gene yield white or light blue colonies after transformation while those with wild type supF result in blue colonies. After passage of the UV damaged plasmid through SV40 transformed XP cells of complementation group A (XP-A) or D (XP-D), plasmid survival was markedly reduced in comparison to plasmids harvested from normal cells. The plasmid mutation frequency was greater with the XP than with the normal cells. Removal of cyclobutane dimers by photolyase treatment prior to transfection indicated that dimers are the major mutagenic lesions. The major UV photoproduct, the TT dimer, was only weakly mutagenic in these human cells. Most mutations occured at cytosines that were 3' to a cytosine or a thymine. There was a restricted spectrum of UV induced mutations with XP-A and XP-D cells: Plasmids recovered from normal cells had single, tandem, and multiple base substitution mutations in the supF gene. With XP-A and XP-D, fewer plasmids had multiple base substitutions. In addition, the XP-D cells had fewer plasmids with tandem mutations. Unique mutagenic hotspots were present with each XP line.

---

[1]Present address: Dept of Medical Microbiology, University of Lund, Malmo General Hospital, Malmo, Sweden

DNA Repair Mechanisms and Their Biological Implications in Mammalian Cells
Edited by M.W. Lambert and J. Laval
Plenum Press, New York

INTRODUCTION

Xeroderma pigmentosum (XP) is a rare, autosomal recessive disease with sun sensitivity, and a markedly increased frequency of ultraviolet radiation (UV) induced skin cancer (Kraemer and Slor, 1985, Kraemer, 1987, Kraemer, et al., 1987a, Cleaver and Kraemer, 1989). Cells from XP patients are hypersensitive to killing by UV, are hypermutable, and exhibit defective DNA excision repair (reviewed in Kraemer and Slor, 1985, Kraemer, 1987, Cleaver and Kraemer, 1989).

We employed a plasmid shuttle vector, pZ189, (Figure 1) (Seidman, et al., 1985) in a host cell reactivation assay (Kraemer, et al., 1987b) to measure DNA repair and mutagenesis in XP and normal human cells. This chimeric plasmid contains pBR sequences which permit replication in bacteria and selection for ampicillin resistance and contains Simian virus 40 (SV40) sequences that permit replication in mammalian cells. The mutagenic target is an approximately 150 base pair (bp) bacterial tyrosine suppressor transfer RNA gene, supF, which is extremely sensitive to mutagenic inactivation (Kraemer and Seidman, 1989).

Plasmids were damaged in vitro and transfected into XP and repair proficient human cells. The plasmid utilizes the human host cell enzymes for repair, replication and mutagenesis. After 2-3 days, replicated episomal plasmids were harvested and assayed by transformation of indicator bacteria.

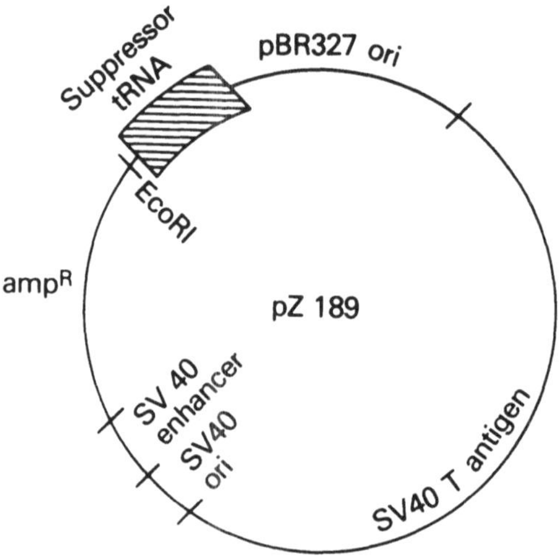

Figure 1. Shuttle vector plasmid, pZ189, used for mutagenesis studies. The plasmid contains SV40 sequences that permit replication in mammalian cells and plasmid sequences that permit replication in bacteria. The suppressor tRNA gene serves as a marker for mutations. The construction and use of this plasmid in described in Seidman, et al.,(1985).

The number of ampicillin resistant colonies reflects plasmid survival. Since the bacteria strain contains a suppressible (amber) mutation in its beta galactosidase gene, transformation with plasmids having a functioning supF gene confers the ability to produce active beta galactosidase. On agar with indicator dye, these bacteria form blue colonies. Plasmids with a mutated supF gene yield white or light blue colonies after transformation. Thus, the proportion of white or light blue colonies reflects the mutation frequency. Mutated plasmids are purified and the DNA sequence is determined.

We performed studies with this plasmid to examine DNA repair and UV induced mutations in XP and normal cells (Bredberg, et al., 1986, Seetharam, et al., 1987, Seidman, et al., 1987). We examined the survival and the mutations introduced into UV treated pZ189 by XP cells and the effect of selective removal of cyclobutane dimers (Brash, et al., 1987, Kraemer, et al., 1988).

RESULTS AND DISCUSSION

Plasmid Survival. UV treatment of pZ189 followed by passage through XP-D cells and transformation of indicator bacteria is shown in Figure 2. With the normal cell line at doses up to 500 Jm^{-2}, there was no reduction in the number of colonies recovered. In marked contrast, with the XP-D line the relative number of bacterial colonies was about 10% of control after 500 Jm^{-2}. This is similar to the sensitivity of these XP-D cells to killing by UV (Protic-Sabljic, et al., 1986a) and reflects the DNA repair defect.

In order to determine the relative contribution of different photoproducts to plasmid killing, cyclobutane dimers were selectively removed from UV treated plasmid by treatment with photolyase (a gift from Dr. A. Sancar) before transfection, as described in Protic'-Sabljic' and Kraemer, 1986a, Protic'-Sabljic', et al., 1986b, and Brash et al., 1987). Figure 2 shows the result of passage of this dimer free plasmid through the XP-D cells. Survival of the plasmid increased substantially, corresponding to removal of about 60% of the lethal hits. This indicates that the major portion of plasmid killing (about 60%) in the repair deficient XP-D cells can be ascribed to cyclobutane dimers. Since plasmid survival did not increase fully to normal levels, the remaining non-dimer photoproducts must also be lethal in the XP-D cells. This also implies that the XP-D cells are not able to repair non-dimer photoproducts. Similar results have been reported previously with XP-A cells (Brash et al., 1987) and with monkey kidney cells (Protic'-Sabljic', et al., 1986b).

Plasmid Mutagenesis. The UV treated plasmid was passed through the XP-D cells and used to transform bacteria which were then plated on agar containing indicator dye. Blue colonies indicate a functional supF gene without mutation while white or light blue colonies indicate mutations (Seidman, et al., 1985, Kraemer and Seidman, 1989). The spontaneous mutation frequency in the absence of UV was about 0.06% (Seetharam, et al., 1987). The frequency of mutant colonies as a function of UV dose to the plasmid is shown in Figure 3. There was a dose dependent increase in mutant colonies recovered with a maximum of about 8% at UV dose of 500 J/m^2. This plasmid mutation frequency is much greater than that with the normal cells and is similar to the plasmid hypermutability found with XP-A cells (Brash, et al., 1987,

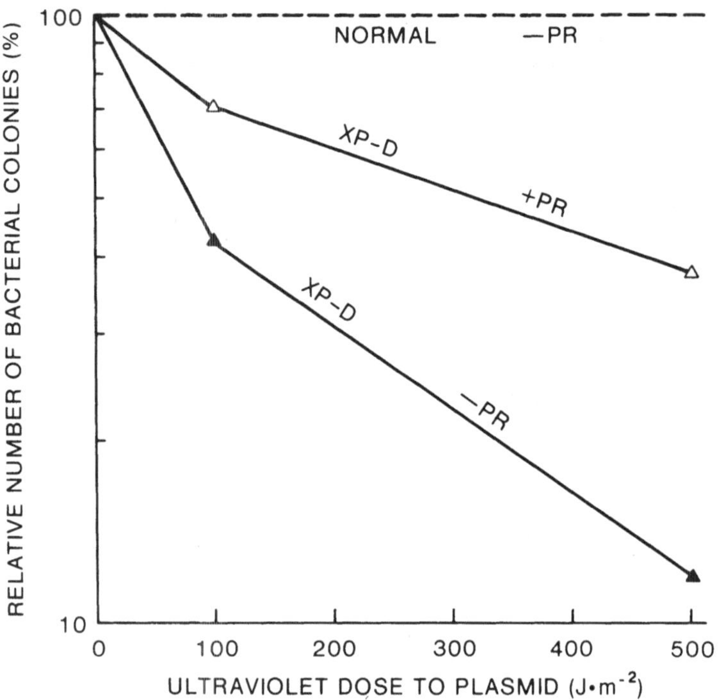

Figure 2. Survival of UV-treated pZ189 in xeroderma pigmentosum group D cells with and without photoreactivation. Plasmid pZ189 was treated with 254 nm UV, and transfected into the XP-D cell line, XP6BE(SV40) (Protic'-Sabljic', et al., 1986a). After 24 hours, replicated plasmids were harvested and used to transform indicator bacteria. The relative number of ampicillin resistant bacterial colonies obtained is plotted vs the UV dose to the plasmid. A portion of the UV damaged pZ189 was treated with photolyase (a gift from Dr. A. Sancar) before transfection. Solid triangles: without photoreactivation. Open triangles: with photoreactivation. Methods are as described in Protic'-Sabljic' and Kraemer (1986), Bredberg, et al., (1986), Seetharam, et al., (1987), Protic-Sabljic', et al., (1986b) and Brash, et al., (1987). The data for the normal line is obtained from Bredberg, et al., (1986).

Kraemer, et al., 1988). These results with the plasmid vector are consistent with earlier studies demonstrating that primary strains of XP-A and XP-D cells are hypermutable to UV (Maher, et al., 1979, Glover, et al., 1979).

The relative contribution of dimer and non-dimer photoproducts to mutations in the XP-D cell was estimated by selectively removing the cyclobutane dimers by photolyase treatment. As shown in Figure 2, the mutation frequency fell dramatically after photoreactivation. This indicates that 70% or more of the mutations were related to the presence of cyclobutane dimers in the plasmid. At 100 Jm^{-2} UV treatment about 30% of the mutations remained after photoreactivation. This indicates that non-dimer photoproducts also are linked to generation of mutations. These findings with the XP-D cell are similar to those previously reported with the XP-A line (Brash, et al., 1987, Kraemer, et al., 1988).

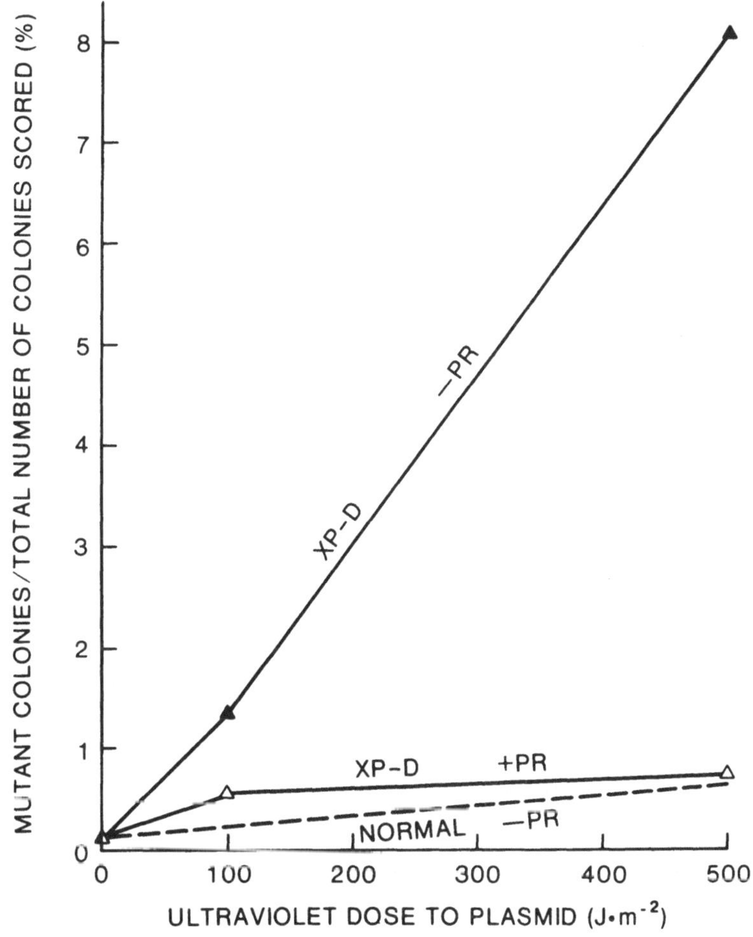

Figure 3. Mutations of UV-treated pZ189 in xeroderma pigmentosum group D cells with and without photoreactivation. Experimental details are described in the legend to Figure 2. Mutant colonies were scored as light blue or white colonies on agar containing the dye X-gal and the inducer of beta-galactosidase, IPTG. The mutant colonies result from inactivation of the marker supF gene in the plasmid during passage through the XP cells. The data for the normal line is obtained from Bredberg, et al. (1986).

DNA sequence analysis. Mutant plasmids recovered from white or light blue colonies obtained after passage through XP-D cells were sequenced and compared to those recovered from XP-A and normal cells. Table 1 shows the classification of the base substitution mutations recovered. With the normal cell line, point mutations consisted of single base substitutions, tandem (adjacent) base substitutions and plasmids with multiple (3 or more) base substitutions within the <u>supF</u> gene. With the XP lines there was a restricted number of changes found. Plasmids with multiple base substitutions were significantly fewer with the XP-D and XP-A lines while plasmids with tandem mutations were significantly fewer with the XP-D line.

Table 1. Mutations in UV treated pZ189 replicated in XP-D, XP-A or Normal human cells[1].

	Number of plasmids with base changes (%)		
	XP-D	XP-A	Normal
Independent plasmids sequenced	69 (100)	61 (100)	89 (100)
Base substitutions:			
Single base	59 (86)	47 (77)	48 (54)
Tandem bases	4 (6)	12 (20)	16 (18)
Multiple bases	6 (9)	1 (2)	24 (27)

[1]Modified from Seetharam, <u>et al</u>.(1987).

The types of single or tandem mutations found with the plasmid passed through the XP-D cells was compared to that with the XP-A and normal (Table 2). All 6 base substitution mutations possible with double stranded DNA (2 types of

Table 2. Types of single or tandem base substitutions in UV-treated pZ189 replicated in XP-D, XP-A or normal human cells[1].

	Number of changes (%)		
	XP-D	XP-A	Normal
Transitions	59 (88)	67 (94)	61 (75)
G:C to A:T	57 (85)	66 (93)	59 (73)
A:T to G:C	2 (3)	1 (1)	2 (2)
Transversions	8 (12)	4 (6)	20 (25)
G:C to T:A	5 (7)	0	8 (10)
G:C to C:G	1 (1)	1 (1)	5 (6)
A:T to T:A	0	3 (4)	6 (8)
A:T to C:G	2 (3)	0	1 (1)
Total	67 (100)	71 (100)	81 (100)

[1]Modified from Seetharam, <u>et al.</u>(1987).

transitions and 4 transversions) were found with the normal line. In marked contrast, the types of mutations were limited in plasmids passed through the XP-A and XP-D cells. There were significantly fewer transversions overall with the XP-A and fewer A:T to T:A transversions with the XP-D. It is remarkable that XP, a disorder with cellular hypermutability and increased cancer frequency, has fewer types of mutations. This suggests that the mutations present with both the XP and the normal cells may be associated with skin cancer in XP and normal patients (Bredberg, et al., 1986).

With all three cell strains, the major mutagenic class found following UV treatment of the plasmid was the G:C to A:T transition. This indicates that the major mutagenic lesion contains cytosine and not thymine. Thus the major UV photoproduct, the TT cyclobutane dimer, cannot be the major mutagenic lesion. Similar observations have been made in prokaryotes more than 25 years ago by Drake ((1963) and Howard and Tessman (1964). An explanation for this apparent failure to mutate at TT dimers was afforded by studies of purified polymerases. Boiteux and Laval (1982), Strauss, et al.(1983) reported that polymerases when reaching non-instructional lesions have a tendency to insert adenines. This "A rule" is of particular advantage when encountering UV photoproducts that involve thymine. The major cyclobutane dimer is the TT cyclobutane dimer but TC and CT dimers are also found (Brash, et al., 1987, Kraemer, et al., 1988). The A rule would predict that adenine would be inserted opposite the thymine in photoproducts. Since this is the correct base, the T would rarely be mutagenic. Further, TC and CC photoproducts would lead to mutations only at the C. This behavior is seen with the XP cells (Figure 4).

The location of the sites of single and tandem base substitution mutations in the supF gene with the XP-D, normal, and XP-A cells is shown in Figure 4. More than 80% of the single or tandem base substitutions in the supF sequence (base pairs 99-183) inactivate tRNA activity and are thus detectable (Kraemer and Seidman, 1989). There is a non-random distribution of mutations with all three cell lines. Most mutations occur at TC or CC sequences. A hotspot for mutations is present at base pair 156 in all 3 cell lines. This represents a G:C to A:T transition at a 5'TC3' site (bp 157-156). Unique hotspots were present at bp 159 with the XP-D line and at base pairs 168 and 169 with the XP-A cells. These different hotspots indicate that cells with different DNA repair capacities handle the same UV damaged DNA in markedly different fashions.

Photoreactivation of mutations. The location of the mutations obtained after passage of the photoreactivated plasmid through the XP-A and XP-D cells is shown in Figure 5. The same hotspots are seen as in the plasmid treated with UV without photoreactivation (Figure 4). Since the mutation frequency fell by 70% - 90% following removal of cyclobutane dimers with XP-D photolyase (Figure 3), dimers are the major mutagenic lesions. The finding of the same hotspots after photoreactivation implies that the lesions that remain, i.e. non-dimer photoproducts such as (6-4) photoproducts (Brash, et al., 1987), are also mutagenic at these same sites. These hotspots occur at 5'TC and 5'CC bases, the dipyrimidines which are the most likely to form 6-4 lesions (Brash, et al., 1987, Kraemer, et al., 1988).

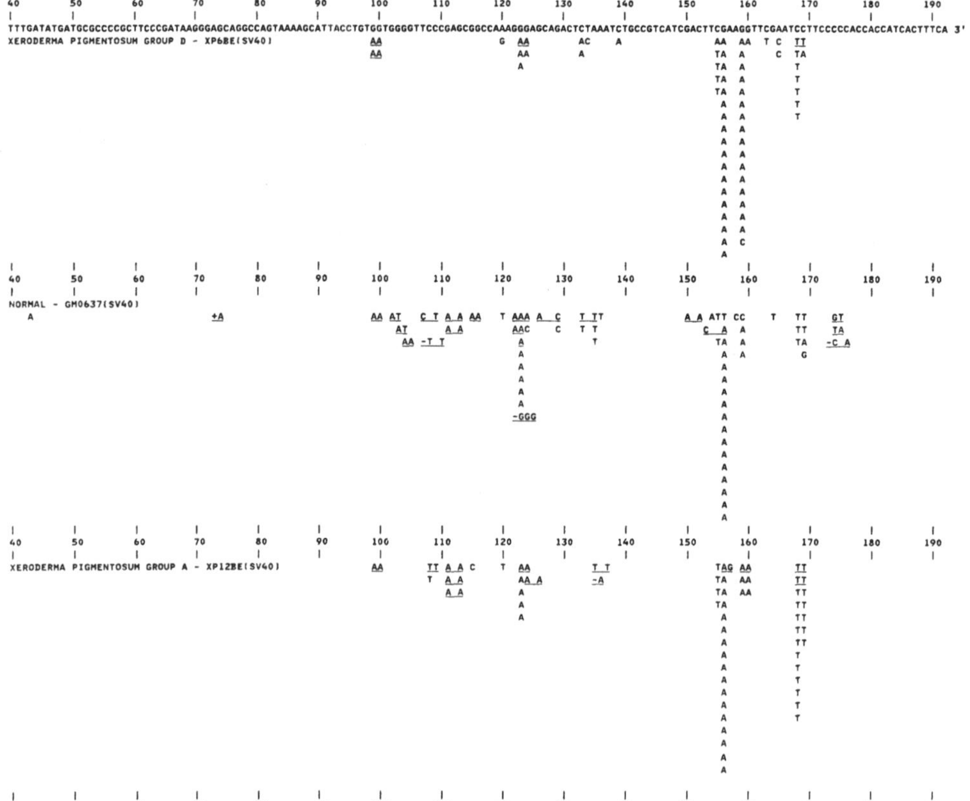

Figure 4. Distribution of single and tandem base substitution mutations in the <u>supF</u> gene of pZ189 after UV treatment and passage through XP-A, XP-D and normal cells. A portion (153 bases) of the DNA sequence of one strand of the <u>supF</u> gene is shown. This contains part of the promoter region (base pairs 24-58), the pre-tRNA sequence (base pairs 59-98) and the suppressor tRNA sequence (base pairs 99-183). Base substitutions are indicated below the altered base as a change in this strand. Each letter represents the change found in an independent plasmid sequenced. Tandem or closely spaced base substitutions, deletions (-), or insertions (+) are indicated by underlining. The mutations obtained with the XP-D, normal and XP-C lines are shown separately. Data modified from Bredberg, <u>et al.</u> (1986) and Seetharam, <u>et al</u>. (1987).

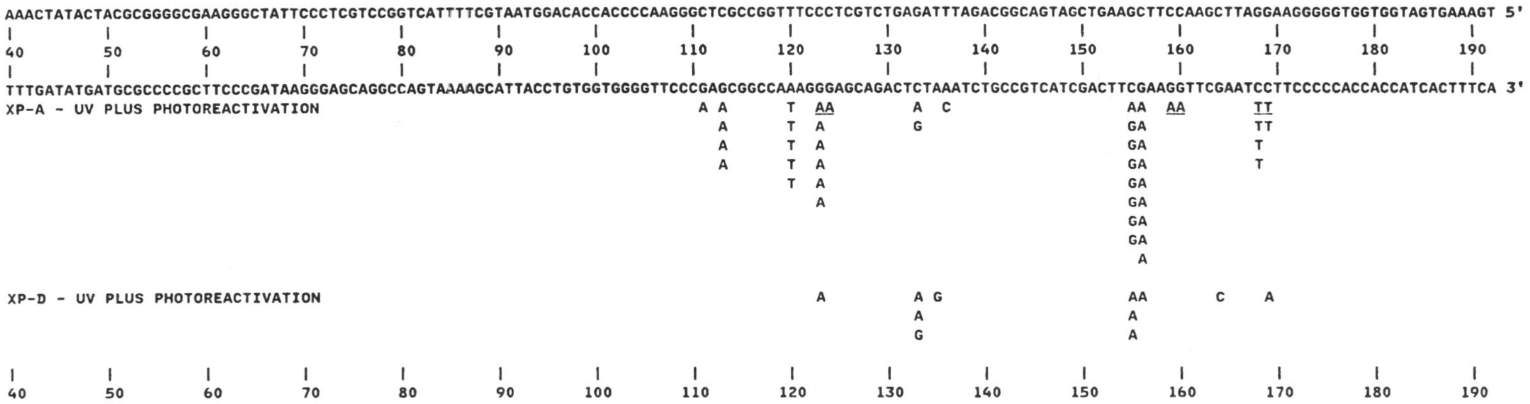

```
AAACTATACTACGCGGGGCGAAGGGCTATTCCCTCGTCCGGTCATTTTCGTAATGGACACCACCCCAAGGGCTCGCCGGTTTCCCTCGTCTGAGATTTAGACGGCAGTAGCTGAAGCTTCCAAGCTTAGGAAGGGGGTGGTGGTAGTGAAAGT 5'
|         |         |         |         |         |         |         |         |         |         |         |         |         |         |         |
40        50        60        70        80        90        100       110       120       130       140       150       160       170       180       190
|         |         |         |         |         |         |         |         |         |         |         |         |         |         |         |
TTTGATATGATGCGCCCCGCTTCCCGATAAGGGAGCAGGCCAGTAAAAGCATTACCTGTGGTGGGGTTCCCGAGCGGCCAAAGGGAGCAGACTCTAAATCTGCCGTCATCGACTTCGAAGGTTCGAATCCTTCCCCCACCACCATCACTTTCA 3'
XP-A - UV PLUS PHOTOREACTIVATION                                       A A      T  AA        A  C                  AA  AA       TT
                                                                         A      T  A         G                     GA           TT
                                                                         A      T  A                               GA           T
                                                                         A      T  A                               GA           T
                                                                                T  A                               GA
                                                                                   A                               GA
                                                                                                                   GA
                                                                                                                   GA
                                                                                                                    A

XP-D - UV PLUS PHOTOREACTIVATION                                                   A         A G                   AA       C    A
                                                                                             A                     A
                                                                                             G                     A

|         |         |         |         |         |         |         |         |         |         |         |         |         |         |         |
40        50        60        70        80        90        100       110       120       130       140       150       160       170       180       190
```

Figure 5. Distribution of single and tandem base substitution mutations in the supF gene of pZ189 after UV treatment, photoreactivation, and passage through XP-A and XP-D cells. See legend to Figure 4 for interpretation of symbols. The data for the XP-A line is modified from Brash, et.al.(1987).

With the XP-A line more transversions were recovered following photoreactivation. In particular, transversion hotspots were present at base pairs 120 and 155. This suggests that transversions are more likely to be formed at non-dimer photoproducts than at dimers.

REFERENCES

Boiteux, S. and Laval J. (1982): Coding properties of poly-(deoxycytidylic acid) templates containing uracil or apyrimidinic sites: In vitro modulation of mutgenesis by deoxyribonucleic acid repair enzymes. Biochemistry, 21: 6746-6751.

Brash, D.E., Seetharam, S., Kraemer, K.H., Seidman, M.M. and Bredberg, A. (1987): Photoproduct frequency is not the major determinant of ultraviolet mutation hotspots or coldspots in human cells. Proc. Natl. Acad. Sci. USA, 84:3782-3786.

Bredberg, A., Kraemer, K.H., and Seidman, M.M. (1986): Restricted mutational spectrum in an UV-treated shuttle vector propagated in xeroderma pigmentosum cells. Proc. Natl. Acad. Sci. USA, 83:8273-8277.

Cleaver, J. and Kraemer, K.H. (1989): Xeroderma pigmentosum. In Scriver, C.R., Beaudet, A.L., Sly, W.S., and Valle, D. (Eds.) The Metabolic Basis of Inherited Disease,Sixth Edition. New York, McGraw Hill, pp. 2949-2971.

Drake, J.W. (1963): Properties of ultraviolet-induced rIII mutants of bacteriophage T4. J. Mol. Biol., 6:268-283.

Glover, T.W., Chang, C., Trosko, J. and Li, S.S. (1979): Ultraviolet light induction of diphtheria toxin-resistant mutants in normal and xeroderma pigmentosum fibroblasts. Proc Natl Acad Sci USA, 76:3982-3986.

Howard, B.D. and Tessman, I. (1964): Identification of the altered bases in mutated single stranded DNA. III Mutagenesis by ultraviolet light. J. Mol. Biol., 9:372-375.

Kraemer, K.H., and Slor, H. (1985): Xeroderma pigmentosum. Clinics in Dermatol., 3:33-69.

Kraemer, K.H. (1987): Heritable diseases with increased sensitivity to cellular injury. In Fitzpatrick, T.B., Eisen, A.Z., Wolff, K., Freedberg, I.M., and Austen, K.F. (Eds.): Dermatology in General Medicine. New York, McGraw Hill, 1987, pp 1791-1811.

Kraemer, K.H., Lee, M.M., and Scotto, J. (1987a): Xeroderma pigmentosum: Cutaneous, ocular and neurologic abnormalities in 830 published cases. Arch. Dermatol., 123:241-250.

Kraemer, K.H., Protic'-Sabljic', M., Bredberg, A., and Seidman, M.M. (1987b): Plasmid vectors for study of DNA repair and mutagenesis. Curr. Prob. Dermatol., 17:166-181.

Kraemer, K.H., Seetharam, S., Protic'-Sabljic', M., Brash, D.E., Bredberg, A., and Seidman, M.M. (1988): Defective DNA repair and mutagenesis by dimer and non-dimer photoproducts in xeroderma pigmentosum measured with plasmid vectors. In Friedberg, E. and Hanawalt, P. eds. Mechanisms and Consequences of DNA Damage Processing, UCLA Symposia on Molecular and Cellular Biology, New Series, Vol 83. Alan R. Liss, Inc, New York, N.Y., pp. 325-335.

Kraemer, K.H., and Seidman, M.M. (1989): Use of supF, and Escherichia coli tyrosine suppressor tRNA gene, as a mutagenic target in shuttle vector plasmids. Mutat. Res., 220:61-72.

Maher, V.M., Dorney, D.J., Mendrala, A.L., Konze-Thomas, B. and McCormick, J.J. (1979): DNA excision repair processes in human cells can eliminate the cytotoxic and mutagenic consequences of ultraviolet irrradiation. Mutat. Res, 62:311.

Protic'-Sabljic', M. and Kraemer, K. H.(1986): Reduced repair of non-dimer photoproducts in a gene transfected into xeroderma pigmentosum cells. Photochem. Photobiol., 43: 509-513.

Protic'-Sabljic', M., Seetharam, S., Seidman, M.M., and Kraemer, K.H. (1986a): An SV40-transformed xeroderma pigmentosum group D cell line: Establishment, ultraviolet sensitivity, transfection efficiency and plasmid mutation induction. Mutat Res., 166:287-294.

Protic'-Sabljic', M., Tuteja, N., Munsen, P., Hauser, J., Kraemer, K.H., and Dixon, K. (1986b): UV-induced cyclobutane pyrimidine dimers are mutagenic in mammalian cells. Mol. Cell. Biol., 6:3349-3356.

Schaaper, R.M., Kunkel, T.A., and Loeb, L. (1983): Infidelity of DNA synthesis associated with bypass of apurinic sites. Proc. Natl. Acad. Sci. USA, 80:487-491.

Seidman, M.M., Dixon, K., Razzaque, A., Zagursky,R.J., & Berman, M.L. (1985): A shuttle vector plasmid for studying carcinogen-induced point mutations in mammalian cells. Gene, 38:233-237.

Seidman, M.M., Bredberg, A., Seetharam, S. and Kraemer, K.H. (1987): Multiple point mutations in a shuttle vector propagated in human cells: Evidence for an error-prone polymerase activity. Proc. Natl. Acad. Sci..USA., 84: 4944-4948.

Seetharam, S., Protic'-Sabljic', M., Seidman, M.M. and Kraemer, K.H. (1987): Abnormal ultraviolet mutagenic spectrum in DNA replicated in cultured fibroblasts from a patient with the skin cancer-prone disease, xeroderma pigmentosum. J. Clin. Invest., 80:1613-1617.

Strauss, B.S., Rabkin, S., Sagher, S. and Moore, P. (1982): The role of DNA polymerase in base substitution mutagenesis on non-instructional templates. Biochimie, 64:829-838.

MOLECULAR ANALYSIS OF MUTATIONS AT THE HPRT LOCUS: UV MUTATION SPECTRA IN NORMAL AND UV-SENSITIVE V79 CHINESE HAMSTER CELLS

H. Vrieling[1,2], M.Z. Zdzienicka[1], J.W.I.M. Simons[1], P.H.M. Lohman[1] and A.A. van Zeeland[1,2]

[1]Department of Radiation Genetics and Chemical Mutagenesis, State University of Leiden, Wassenaarseweg 72, 2333 AL Leiden, The Netherlands

[2]J.A. Cohen Institute, Interuniversity Research Institute for Radiopathology and Radiation Protection, Leiden, The Netherlands

SUMMARY

A method has been developed which enables fast determination of the nature of point mutations in the HPRT gene in mammalian cells. It is based on the Polymerase Chain Reaction procedure (PCR), which enables *in vitro* amplification of specific DNA sequences up to a million-fold. Total cytoplasmic RNA was used to synthesize HPRT cDNA which contained the entire HPRT coding region. This HPRT cDNA could then be amplified, cloned and sequenced. In this way mutations in HPRT exon sequences, which are spread over 34 and 44 kb of DNA for mouse and man respectively, can be determined in a fast and efficient way.

The influence of DNA repair processes on UV-mutagenesis was investigated in normal V79 Chinese hamster cells and its UV-sensitive (UV^s) derivative V-H1. The nature of point mutations in HPRT exon sequences was determined for 19 mutants from V79 cells and for 17 mutants from V-H1 cells. All mutations in V79 were single and tandem double base pair substitutions with no preference for a particular type of change, although the GC => CG transversion was not detected. Mutations in V-H1 cells consisted of single and tandem double base pair changes, which were all GC => AT transitions, and frame shift mutations. The assumption was made that the UV-induced mutations were caused by photoproducts at dipyrimidine sequences. In V79 cells 11 out of 17 mutations were then caused by photoproducts in the non-transcribed strand of the HPRT gene. However, in V-H1 cells, which are completely deficient in the removal of pyrimidine dimers from the HPRT gene, 10 out of 11 mutations were caused by photoproducts in the transcribed strand of the HPRT gene. This strand specificity in mutation-

DNA Repair Mechanisms and Their Biological Implications in Mammalian Cells
Edited by M.W. Lambert and J. Laval
Plenum Press, New York

induction in V-H1 cells may be caused by differences in the fidelity of replication between the leading and lagging strand, when photoproduct containing DNA is copied. Possible preferential removal of photoproducts from the transcribed strand of the HPRT gene may explain why in V79 no strand specificity in mutation induction was detected.

INTRODUCTION

The molecular analysis of point mutations in mammalian genes has been a complicated and tedious matter for many years. Large deletions in a particular gene can be detected with Southern blot analysis, because they cause changes in the restriction profiles. A new method was developed which allows rapid determination of point mutations in the coding region of the HPRT gene. It is based on the Polymerase Chain Reaction (PCR) procedure as described by Saiki, et al.(1985), which allows in vitro amplification of specific genomic DNA segments. The PCR procedure has already been used for the molecular analysis of various specific DNA sequences. It has enabled diagnoses to be performed of genetic disorders such as sickle cell anemia (Saiki, et al., 1985, 1986) and and ß-thalassemia (Chehab, et al., 1987; Wong, et al., 1987), the detection of proviral sequences of the human immunodeficiency virus (HIV-1) in DNA isolated from peripheral blood cells of seropositive persons (Ou, et al., 1988), and the analysis of activated ras oncogenes (Verlaan de Vries, et al., 1986; Bos, et al., 1987; McMahon, et al., 1987). In order to be able to detect the majority of point mutations at the HPRT locus giving rise to an HPRT deficient phenotype, cDNA instead of genomic DNA is used as target sequence for amplification. This has the advantage that the presence of point mutations in all 9 exons can be investigated following one single amplification experiment. The assumption is made that most HPRT mutations give rise to a change in the structure of the HPRT protein and therefore will be located in the coding region of the HPRT gene.

Ultraviolet light (254 nm) introduces various types of lesions into the DNA, from which the cyclobutane pyrimidine dimer and the (6-4) pyrimidine-pyrimidone photoproduct are implicated as playing the major roles in cell killing and mutagenesis. V-H1 is a UV^s cell line which was isolated from V79 Chinese hamster cells (Zdzienicka and Simons, 1987). V-H1 is about 10x more sensitive than normal V79 cells to the cell killing effects of UV and displays a 7-fold higher UV-induced mutation frequency at the HPRT locus (Zdzienicka, et al., 1988). The molecular nature of UV-induced HPRT mutations was determined for 19 HPRT mutants from V79 and 17 mutants from UV^s V-H1 Chinese hamster cells. Large differences in mutation spectra between the two cell lines were found.

MOLECULAR ANALYSIS OF MUTATIONS AT THE HPRT LOCUS.

Most mammalian genes have a complex structure of intron and exon sequences and a relatively large size. These features have prevented large scale molecular analysis of mutations in mammalian genes. The development of the Polymerase Chain Reaction (PCR) procedure (Saiki, et al., 1985), which allows in vitro amplification of specific DNA sequences, has made it possible to determine the nature of mutations in a specific DNA fragment in a very rapid way. Two oligonucleotide primers, which flank the DNA segment to be amplified and which can hybridize to opposite strands, are used in repeated cycles of denaturation, primer annealing and extension of the primers with a DNA polymerase. This procedure results in specific amplification of the target DNA sequence up to a million-fold.

We have adapted the PCR method in such a way (Figure 1) that it allows rapid detection of point mutations in the coding region of the HPRT gene (Vrieling, et al., 1988b). Instead of starting from chromosomal DNA, where the HPRT exons are scattered over 34-40 kb of DNA, we used HPRT cDNA, where all exons are linked together, as a target for amplification. Through the use of primers which contained mismatches with the wild type sequence, restriction sites were introduced in the amplified DNA. These sites were used for cloning of the amplified HPRT cDNA into M13 sequencing vectors when no natural occurring restriction sites were available. The original protocols used the Klenow fragment of E. coli DNA polymerase I to extend the annealed primers. This required the addition of fresh enzyme during every cycle, because the Klenow fragment was inactivated by the heat denaturation step. Recently a thermostable DNA polymerase purified from the thermophilic bacterium Thermus aquaticus was introduced into the PCR reaction, which overcame the aforementioned problem. The use of higher temperatures for annealing and extension with this enzyme resulted in increased specificity and yield. We observed that, when primers with large differences in melting temperatures (Tm) were used, not only a double-stranded but also a single-stranded amplified HPRT fragment was generated. This single-stranded DNA fragment probably resulted from partial denaturation of one of the primer template complexes during the extension step. The use of an extended version of the primer with the lower Tm resulted in the production of double-stranded HPRT fragment only.

UV-INDUCED MUTATIONS IN NORMAL AND UV-SENSITIVE CHINESE HAMSTER CELLS.

Short wavelength (254 nm) UV irradiation of mammalian cells results in the formation of various adducts in the DNA. The major UV-induced lesion is the cyclobutane pyrimidine dimer which is formed between two adjacent pyrimidines. Dimers play a significant role in mutagenesis and probably also in cell killing. Pyrimidine dimers in genomic DNA have been shown to occur randomly (Bohr, et al., 1985) indicating that there is no significant bias between large domains of chromatin. Recently however, it has been shown that dimer distribution in core DNA of nucleosomes is not uniform (Gale, et al.,1987), but that a 10.3-base periodicity exists which may be related to the inherent structure of the nucleosome and reflect the nature of histone-DNA interactions. The order of preference for dimer formation at possible dipyrimidine sites was shown to be TT > TC = CT > CC (Setlow and Carrier, 1966). Another UV-induced lesion that may play an important role in the mutagenic effects of ultraviolet light is the pyrimidine-pyrimidone (6-4) photoproduct (Lippke, et al., 1981; Brash and Haseltine, 1982). Like pyrimidine dimers it is also formed

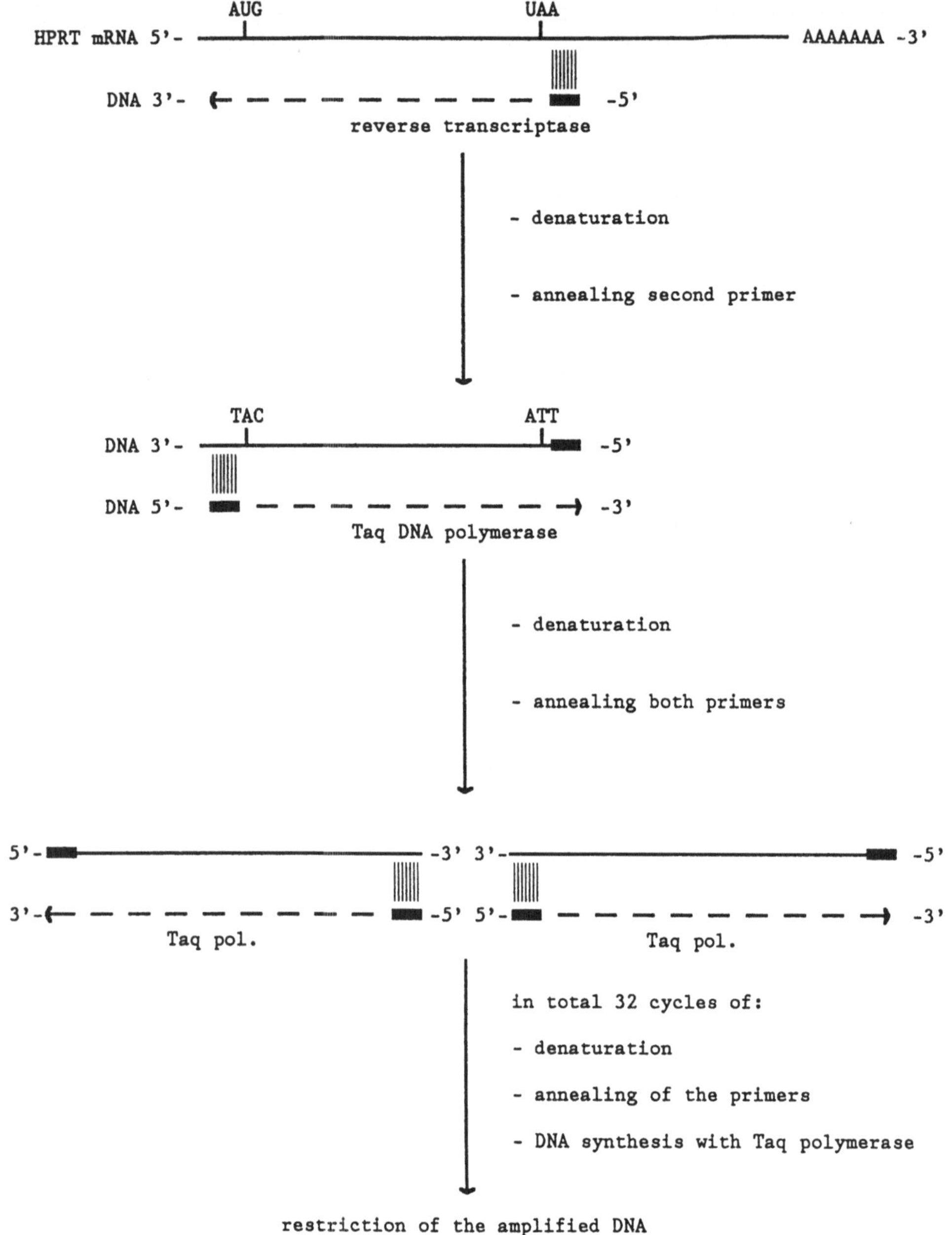

Figure 1. PCR procedure for the amplification of the entire HPRT coding region starting from total cytoplasmic RNA. Solid blocks represent HPRT specific oligonucleotide primers (20-mers). cDNA is synthesized using reverse transcriptase and a primer which can anneal to HPRT mRNA just 3' of the UAA stop codon. Following denaturation, another primer, which can anneal to the newly synthesized DNA strand just 5' of the AUG start codon, is added and a double-stranded DNA fragment is synthesized using thermostable Taq DNA polymerase. This DNA fragment is then amplified with the PCR procedure.

at dipyrimidine sites, almost exclusively at TC and CC and at an overall frequency which is about 5-10-fold lower than the frequency of pyrimidine dimers (Brash, et al., 1987; Bourre, et al., 1987).

Persisting DNA damage in mammalian cells can interfere with DNA replication and transcription. However, DNA repair processes exist which can remove UV-induced adducts from the DNA. The molecular nature of UV-induced HPRT mutations was determined for 19 HPRT mutants from normal V79 and for 17 HPRT mutants from UV^s V-H1 Chinese hamster cells (Vrieling, et al., in press) in order to investigate the influence of DNA repair on mutation induction (Table 1).

Table 1. Types of UV-induced mutations at the HPRT locus in normal V79 and UV-sensitive V-H1 cells.

	V79	V-H1
Single base changes	16	7
Tandem double mutations	2	4
Frame shift mutations	0	3
Splice mutations	1	3
Total	19	17

Large differences in mutation spectra between the two cell lines existed (Table 2). Mutations in V79 consisted only of single and tandem double base pair changes and involved all of the possible base substitutions except the GC => CG transversion, with no preference for a particular type of change. In contrast to this, all base pair changes in the single and tandem base pair substitutions of the V-H1 mutants were GC => AT transitions. Furthermore also frame shift mutations were detected in the V-H1 mutants. In 3 of the mutants from V79 and in 1 mutant from V-H1, we detected HPRT cDNA which lacked one or more exons. Probably, these mutants were mutated in the 3' splice site of the intron adjacent to the first missing exon. Other investigators, using the supF gene on the replicating shuttle vector pZ189 (Hauser, et al., 1986; Bredberg, et al., 1987) or the endogenous APRT gene (Drobetsky, et al., 1987) in primate and Chinese hamster ovary cells respectively, determined UV-mutation spectra which differ significantly from our spectrum in V79 cells. GC => AT transitions were the predominant type of base substitutions with both of these genes accounting for about 70% of all base changes. Differences may exist, not only between endogenous and exogenous DNA sequences but also between different chromosomal genes, in the rate and efficiency with which cellular processes like DNA repair act upon DNA lesions in these sequences. The V-H1 cells have been shown to be unable to remove pyrimidine dimers from their genome, also not from the active HPRT gene, whereas removal of (6-4) photoproducts was 50-80% of that found in normal cells depending on the repair time. Probably in the repair deficient background pyrimidine dimers were the major cause for mutations giving only rise to GC = AT transitions. Preliminary data show extensive removal of dimers from the active HPRT gene in normal V79 cells, as has also been shown for the DHFR gene in CHO cells (Bohr, et al., 1985, 1986; Mellon, et al., 1986). In these repair proficient cells a large fraction of the mutations consisted of transversions, which were probably primarily caused by (6-4) photoproducts, as has already been proposed by Protic-Sabljic, et al.(1986). Recently, even a strand specificity in dimer removal has been shown to exist for the DHFR gene in both cultured human and CHO cells (Mellon, et al., 1987). Pyrimidine dimers were removed at a higher rate (human cells) and to a larger extent (CHO cells) from the transcribed strand of the gene. We expected in V79 cells to find a preference for mutation induction in the non-transcribed strand of the HPRT gene, due to possible preferential DNA repair of that strand. Assuming that UV-induced mutations were caused by photoproducts at dipyrimidine sites, it is not yet clear whether in V79 a significant difference in mutation induction between the two strands exists.

Table 2. Types of single and tandem double base pair substitutions in UV-induced HPRT mutants from V79 and V-H1 Chinese hamster cells.

	V79	V-H1
Transitions	6	15
GC > AT	3	15
AT > GC	3	0
Transversions	14	0
GC > TA	6	0
GC > CG	0	0
AT > TA	4	0
AT > CG	4	0
Total	20	15

In V-H1 cells, however, which do not show any preferential removal of dimers from the HPRT gene, there was a predominance in mutation induction due to photoproducts in the transcribed strand. This result seems to indicate that indeed in V79 cells preferential repair of the transcribed strand of the HPRT gene occurred, but that through some other mechanism lesions in the non-transcribed strand are less mutagenic than in the transcribed strand. Recently data on *in vitro* SV40 DNA replication were presented which implicate DNA polymerase δ and DNA polymerase α as the leading and lagging strand polymerase, respectively (Prelich and Stillman, 1988). A difference in fidelity between these polymerases when bypassing DNA lesions could be the cause for these unexpected results concerning strand specific mutation induction.

CONCLUDING REMARKS

Investigation of mutation spectra will be useful for many purposes. It may point out which lesions are mutagenic and what kind of influence chromosomal position and local DNA structure have on mutation induction. Furthermore, it can provide more insight in cellular processes such as (preferential) DNA repair and DNA replication. Since mutation spectra can also be determined in vivo by isolating HPRT deficient T-lymphocytes from human blood (Albertini, et al., 1982; Morley, et al., 1983), epidemiological studies may be performed, which will possibly allow correlation of these spectra to occupational and environmental exposure of individuals to mutagenic agents. The mutagenic properties of indirect acting mutagens which have to be metabolized in order to become mutagenic can be monitored following the isolation of HPRT mutant lymphocytes from mouse blood (Jones, et al., 1985).

At the moment the HPRT gene is the most suitable mammalian gene to be used in mutation studies, since it is X-linked and molecular studies are not restricted to those performed in only a small number of special (heterozygous) cell lines. However, mutation spectra also have to be determined in other mammalian genes. If necessary, this must be done in heterozygous cell lines, and with shuttle vector systems in order to study the generality of mutation spectra obtained at the HPRT locus, the effect of different chromosomal positions and the influence of cellular processes, such as DNA repair, on mutation induction. For the same purposes, methods have to be developed which will enable the determination of mutation spectra in silent genes and non-coding DNA sequences, where no selection for a mutant phenotype is possible.

We thank N.A. Groen and M.L. Van Rooijen for excellent technical assistance. This research was sponsored by the Queen Wilhelmina Fund of the Netherlands (contract no. IKW 85-64), Euratom (contract no. BIO-E-407-81NL) and Royal Dutch Shell.

REFERENCES

Albertini, R.J., K.S. Castle and W.R. Borcherding (1982) T cell cloning to detect the mutant 6-thioguanine resistant lymphocytes present in human peripheral blood, Proc. Natl. Acad. Sci. (U.S.A.), 79, 6617-6621.

Bohr, V.A., C.A. Smith, D.S. Okumoto and P.C. Hanawalt (1985) DNA repair in an active gene: removal of pyrimidine dimers from the DHFR gene of CHO cells is much more efficient than in the genome overall, Cell, 40, 359-369.

Bohr, V.A., D.S. Okumoto, L. Ho, P.C. Hanawalt (1986) Characterization of a DNA repair domain containing the dihydrofolate reductase gene in Chinese hamster ovary cells, J. Biol. Chem., 261, 16666-16672.

Bos, J.L., E.R. Fearon, S.R. Hamilton, M. Verlaan-de Vries,

J.H. van Boom, A.J. van der Eb and B. Vogelstein (1987) Prevalence of ras gene mutations in human colorectal cancers, Nature, 327, 293-297.

Bourre, F., G. Renault and A. Sarasin (1987) Sequence effect on alkali-sensitive sites in UV-irradiated SV40 DNA, Nucleic Acids Res., 15, 8861-8875.

Brash, D.E. and W.A. Haseltine (1982) UV-induced mutation hotspots occur at damage hotspots, Nature, 298, 189-192.

Brash, D.E., S. Seetharam, K.H. Kraemer, M.M. Seidman and A. Bredberg (1987) Photoproduct frequency is not the major determinant of UV base substitution hot spots or cold spots in human cells, Proc. Natl. Acad. Sci. (U.S.A.), 84, 3782-3786.

Bredberg, A., K.H. Kraemer and M.M. Seidman (1986) Restricted ultraviolet mutational spectrum in a shuttle vector propagated in xeroderma pigmentosum cells, Proc. Natl. Acad. Sci. (U.S.A.), 83, 8273-8277.

Chehab, F.F., M. Doherty, S. Cai, Y.W. Kai and E.M. Rubin (1987) Detection of sickle cell anaemia and thalassaemias, Nature, 329, 293-294.

Drobetsky, E.A., A.J. Grosovsky and B.W. Glickman (1987) The specificity of UV induced mutations at an endogenous locus in mammalian cells, Proc. Natl. Acad. Sci. (U.S.A.), 84, 9103-9107.

Gale, J.M., K.A. Nissen and M.J. Smerdon (1987) UV-induced formation of pyrimidine dimers in nucleosome core DNA is strongly modulated with a period of 10.3 bases, Proc. Natl. Acad. Sci. (U.S.A.), 84, 6644-6648.

Hauser, J. , M.M. Seidman, K. Sidur and K. Dixon (1986) Sequence specificity of point mutations induced during passage of a UV-irradiated shuttle vector plasmid in monkey cells, Mol. Cell. Biol., 6, 277-285.

Jones, I.M., Burkhart-Schultz, K. and Carrano, A.V. (1985) A method to quantify spontaneous and in vivo induced thioguanine-resistant mouse lymphocytes, Mutation Res., 147, 97-105.

Lippke, J.A., L.K. Gordon, D.E. Brash and W.A. Haseltine (1981) Distribution of UV light-induced damage in a defined sequence of human DNA: Detection of alkaline-sensitive lesions at pyrimidine nucleoside-cytidine sequences, Proc. Natl. Acad. Sci. (U.S.A.), 78, 3388-3392.

McMahon, G., E. Davis and G.N. Wogan (1987) Characterization of c-Ki-ras oncogene alleles by direct sequencing of enzymatically amplified DNA from carcinogen-induced tumors, Proc. Natl. Acad. Sci. (U.S.A.), 84, 4974-4978.

Mellon, I., V.A. Bohr, C.A. Smith and P.C. Hanawalt (1986) Preferential DNA repair of an active gene in human cells, Proc. Natl. Acad. Sci. (U.S.A.), 83, 8878-8882.

Mellon, I., G. Spivak and P.C. Hanawalt (1987) Selective removal of transcription-blocking DNA damage from the transcribed strand of the mammalian DHFR gene, Cell, 51, 241-249.

Morley, A.A., K.J. Trainor, R. Seshadri and R.B. Ryall (1983) Measurements of in vivo mutations in human lymphocytes, Nature (London), 302, 155-156.

Ou, C.Y., S. Kwok, S.W. Mitchell, D.H. Mack, J.J. Sninsky, J.W. Krebs, P. Feorino, D. Warfield and G. Schochetman (1988) DNA amplification for direct detection of HIV-1 in DNA of peripheral blood mononuclear cells, Science, 239, 295-297.

Prelich, G. and B. Stillman (1988) Coordinated leading and lagging strand synthesis during SV40 DNA replication *in vitro* requires PCNA, *Cell*, 53, 117-126.

Protic-Sabljic, M., N. Tuteja, P.J. Munson, J. Hauser, K.H. Kraemer and K. Dixon (1986) UV light-induced cyclobutane pyrimidine dimers are mutagenic in mammalian cells, *Mol. Cell. Biol.*, 6, 3349-3356.

Saiki, R.K., S. Scharf, F. Faloona, K.B. Mullis, G.T. Horn, H.A. Erlich and N. Arnheim (1985) Enzymatic amplification of ß-globin genomic sequences and restriction site analysis for diagnosis of sickle cell anemia, *Science*, 230, 1350-1354.

Saiki, R.K., T.L. Bugawan, G.T. Horn, K.B. Mullis and H.A. Erlich (1986) Analysis of enzymatically amplified ß-globin and HLA-DQ DNA with allele-specific oligonucleotide probes, *Nature*, 324, 163-166.

Setlow, R.B. and W.L. Carrier (1966) Pyrimidine dimers in ultraviolet-irradiated DNA's, *J. Mol. Biol.*, 17, 237-254.

Verlaan-de Vries, M., M.E. Bogaard, H. van den Elst, J.H. van Boom, A.J. van der Eb and J.L. Bos (1986) A dot-blot screening procedure for mutated ras oncogenes using synthetic oligodeoxynucleotides, *Gene*, 50, 313-320.

Vrieling, H., J.W.I.M. Simons and A.A. van Zeeland (1988) Nucleotide sequence determination of point mutations at the mouse HPRT locus using in vitro amplification of HPRT mRNA sequences, *Mutation Res.*, 198, 107-113.

Vrieling, H., M.L. van Rooijen, N.A. Groen, M.Z. Zdzienicka, J.W.I.M. Simons, P.H.M. Lohman and A.A. van Zeeland (1989) DNA strand specificity for UV-induced mutations in mammalian cells, *Mol. Cell. Biol.*, 9, 1277-1283.

Wong, C., C.E. Dowling, R.K. Saiki, R.G. Higuchi, H.A. Erlich and H.H. Kazazian (1987) Characterization of ß-thalassaemia mutations using direct genomic sequencing of amplified single copy DNA, *Nature*, 330, 384-386.

Zdzienicka, M.Z. and J.W.I.M. Simons (1987) Mutagen-sensitive cell lines are obtained with a high frequency in V79 Chinese hamster cells, *Mutation Res.*, 178, 235-244.

Zdzienicka, M.Z., G.P. van der Schans, A. Westerveld, A.A. van Zeeland and J.W.I.M. Simons (1988) Phenotypic heterogeneity within the first complementation group of UV-sensitive mutants of Chinese hamster cell lines, *Mutation Res.*, 193, 31-41.

COMPARISON OF THE MUTATIONAL SPECIFICITIES EXHIBITED BY BPDE IN *ESCHERICHIA COLI* AND CHO CELLS

Mary Mazur, Alasdair J.E. Gordon, Cecilia Bernelot-Moens, and Barry W. Glickman

Department of Biology
York University
Toronto, Ontario, Canada M3J 1P3

INTRODUCTION

BPDE is presumably the ultimate reactive metabolite of benzo[a]pyrene (B[a]P), a well known carcinogenic environmental pollutant. It has become evident that most chemical carcinogens are active only after metabolism to an ultimate carcinogenic and mutagenic form. These ultimate carcinogens tend to be electron deficient and thus react with nuclephilic sites which are abundant in DNA. It has been established that B[a]P is metabolized by the mixed function oxygenases to a variety of products, including the four enantiomeric forms of BPDE (*anti* or *syn*; (+) or (-)) (Fahl, 1982). Interestingly, metabolism of B[a]P in mammalian cells produces primarily the (+) *anti*-BPDE isomer (Yang, *et al.*, 1978). However, not all DNA reactions with these ultimate carcinogens are of equal biological importance. It has been shown at the *hprt* locus in CHO cells that the respective mutagenic efficiency of BPDE is (+)*anti* > >(-)*anti* = (+/-)*syn*, and remarkably the reverse is seen at the *gpt* locus in TA100 bacteria (-)*anti* = (+/-)*syn* > (+)*anti* (Stevens, *et al.*, 1985). It is also known that the (+)*anti* enantiomer is more than 60 fold more active as a tumor initiator in CD-1 and Sencar mice (Pelling, *et al.*, 1984).

Our lab has been studying the mutagenic and site specificity of (+/-)*anti*-BPDE in mammalian CHO cells at the *aprt* locus (Mazur and Glickman, 1988), and at the *lacI* gene in *E. coli* (Gordon, *et al.*, in preparation). This paper examines the spectra obtained in the two systems. This comparison provides insights into the potential role played by DNA repair systems in the different biological systems.

METHODS

A. Isolation and cloning of APRT mutants

Independent cultures of the hemizygous CHO cell line D422

DNA Repair Mechanisms and Their Biological Implications in Mammalian Cells
Edited by M.W. Lambert and J. Laval
Plenum Press, New York

were exposed to a dose of 0.7 μM anti-BPDE [(+/-)7-a,8-b-dihydroxy-9-b-epoxy-7,8,9,10-tetrahydrobenzo[a]pyrene, National Cancer Institutes chemical repository, lot CSL-85-008-09, diluted in Aldrich ethanol] for 1 hour at 37^0C. Five days were allowed for expression of the mutant phenotype. Cells were then plated in selective media containing the purine analog 8-azaadenine (0.4 mM). The details of the cloning methodology are presented elsewhere (Mazur and Glickman, 1988).

B. Isolation and cloning of lacI mutants

Induced $lacI^-$ mutations were selected in E. coli strain EE125 [F' pro-lac; ara thi rfa (lac pro)/pKM101] obtained from Dr. E. Eisenstadt (Harvard University, U.S.A.). EE125 carries an uncharacterised mutation rendering it permeable to high molecular weight compounds. The details of the genetic and cloning methodologies are presented in detail elsewhere (Schaaper, et al., 1985).

Bacteria were grown in minimal medium to late log phase and then were washed and resuspended in Vonner salt solution. (+/-)trans-BPDE exposure (50 μg/ml, same source and lot number) was for 20 minutes at room temperature. After washing, $lacI^-$ mutants were selected by plating the treated cells directly onto plates containing the sugar phenyl-β-D-galactoside (P-gal; Research Organics, Inc.) as the sole carbon source, ensuring independence of mutations. Appropriate dilutions of treated cells were also plated on LB and supplemented minimal media plates to determine survival. P-gal is a substrate for β-galactosidase but does not induce synthesis of the enzyme. Therefore, when supplied as the sole carbon source, only cells that constitutively synthesize β-galactosidase form colonies.

RESULTS

Examination of the sequence alteration in 117 $LacI^-$ mutants and comparison of these results with our previous findings for 21 mutants in APRT (Mazur and Glickman, 1988) allows examination of the mutagenic and site specificities manifested in the two systems.

Mutagenic specificity of BPDE-induced mutation: Base substitution comprises the major class of mutation recovered after BPDE treatment, accounting for 19/21 and 60/117 mutations, respectively, for CHO and E. coli. The base substitution spectra are similar in CHO and E. coli. Mutation occurs predominantly at G:C base pairs (17/19 and 58/60) consistent with guanines being the preferred site of BPDE damage. Transversions predominate with the G:C —> T:A transversion accounting for 13/19 and 39/60 of the base substitutions. The distribution among the types of base substitutions are presented in Table 1.

Site Specificity of BPDE-induced base substitution: Consideration of the site specificity of the predominant base substitution, the G:C —> T:A transversion, reveals an intriguing difference between the two systems. Both systems display a

Table 1. Mutagenic specificity of BPDE in CHO and E. Coli.

Substitution	CHO	E. coli
G:C —> T:A	13	39
G:C —> A:T	1	15
G:C —> C:G	3	4
A:T —> T:A	2	2

characteristic sequence motif for the site of the transversion which is presented in Table 2. In CHO cells the transversion occurs five times more frequently at 5'-purine-G-3' (5'-R-G-3') sites than at 5'-pyrimidine-G-3' sites (5'-Y-G-3'). Specifically, 5'-G-G-3' sites account for 10 of 13 mutations. Half of these transversions (7/13) occurred in runs of Gs flanked by A residues (Mazur and Glickman, 1988). Conversely, in E. coli G:C —> T:A transversions are recovered four times more frequently at 5'-Y-G-3' sites. In E. coli this influence of the 5' flanking base on occurrence of mutation was not seen for the G:C —> A:T transition (8:7 for 5'-Y versus 5'-R sites). The smaller numbers in the other categories do not allow such analysis. This 5' influence on the G:C —> T:A transversion is not restricted to the pKM101 background, the same site specificity is recovered in isogenic E. coli that do not harbour this plasmid (data not shown).

Table 2. Site specificity of BPDE-induced base substitution.

Class of mutation	CHO	E. coli
G:C —> T:A	13	39
5'-R-G-3'	11	8
5'-Y-G-3'	2	31
G:C —> C:G	3	4
5'-R-G-3'	3	1
5'-Y-G-3'	0	3
G:C —> A:T	1	15
5'-R-G-3'	1	7
5'-Y-G-3'	0	8

DISCUSSION

The principal BPDE adduct to DNA is at the N-2 exocyclic position of guanine, although modification at the N-7 position of guanine is a minor adduct. When BPDE binds to the N2 position of guanine, a stable adduct is formed. However, BPDE binding at the N7 position of guanine creates an unstable glycosidic bond, which results in the rapid loss of the modified base from the DNA (depurination), and the formation of an apurinic (AP) site (Sage and Haseltine, 1984). For both lesions mechanisms can be envisaged that will lead to the G:C —> T:A transversion: (1) preferential incorporation of deoxyadenosine residues opposite non-coding AP sites by DNA polymerases subsequent to N7 damage (Foster, et al., 1983), and (2) purine-purine mispairing between syn-BPDE-G and the imino tautomer of adenine subsequent to N2 damage (Eisenstadt, et al., 1982). Therefore, in this scheme an N7 modified guanine would have a similar mutagenic consequence as modification at the N2 position.

Base substitution was the major class of mutation recovered after BPDE treatment in CHO and E. coli. There was an overwhelming preference for mutation at G:C base pairs, with the predominant sequence alteration observed in our study being the G:C —> T:A transversion. While the mutagenic specificity for base substitutions in the two systems is similar, the site specificity for the predominant G:C —> T:A transversion is strikingly dissimilar. We would like to direct attention to this intriguing difference, and address the possibility that the biologically relevant BPDE-DNA adducts may differ between the two systems.

A recent study enables the prediction of alkali-labile sites (which can be equated with sites of N7-guanine adducts) within DNA of known sequence modified by diol-epoxides (Lobanenkov, et al., 1987). It was shown that a G flanked 5' by a pyrimidine was more susceptible to damage resulting in alkali-labile sites (83 of 92 potential sites) than when it was flanked 5' by a purine (13 of 112 sites). Internal guanines within poly(dG) sequences were found to be very poor targets for damage leading to alkali-labile sites. Therefore, potential sites of BPDE damage (i.e., G residues) can be divided into two general groups based upon such a classification: guanines that will most likely suffer N2 damage (5'-R-G-3' sites; stable lesions) and those that will most likely suffer N7 damage (5'-Y-G-3' sites; AP sites). In Table 2 such an analysis is performed. Strikingly the majority of G:C —> T:A transversions in baceria are likely to result from the N7 modification of guanine (5'-Y-G-3'). Precisely the opposite specificity is seen in mammalian CHO cells where the site specificity for the G:C —> T:A transversion (5'-R-G-3') is consistent with N2 modified guanine as the major mutagenic premutational lesion. That 10 of 13 CHO G:C —> T:A transversions occurred at 5'-G-G-3' sites strongly argues that these mutations are due to damage at the N2 position.

This pattern of preferred sites of N7-directed events in bacteria and N2-directed events in CHO cells is also consistent with the different biological potencies of the BPDE isomers. (-)anti-BPDE is most mutagenic in bacteria and produces a high proportion of N7-dG adducts upon reaction with DNA. Whereas, the (+)anti-BPDE is most mutagenic and carcinogenic in mammalian cells and stereospecifically binds the N2 position of dG. Both (+) and (-)anti-BPDE react primarily with the N2 position of guanine. However, the extent of that binding to N2 differs for the isomers, (+)anti-BPDE (in vitro) is approximately 91% and (-)anti-BPDE is approximately 45% (Osborne, et al., 1981). The N7-dG adduct is a minor product exclusively of the weakly carcinogenic (-)anti-isomer, where it constitutes 31% of binding seen in vitro with the (-) isomer, compared to 2% seen with the (+) isomer.

The various sensitivities may be a manifestation of differences in repair abilities of the systems to the various adducts. The probable reason that the (+)anti isomer is most potent in mammalian cells is because they are proficient in repairing AP sites, but relatively less efficient in repairing the N2-dG adduct. Support is provided by studies of transformation efficiencies of NIH 3T3 cells with DNA modified by anti-BPDE or apurinic sites, which indicate that depurination does not appear to be a major mechanism for mutation in oncogenic transformation studies (Vousden, et al., 1987). The reverse is probably true for bacteria, suggesting differences in processing specificity by the polymerases between the two systems.

The greater carcinogenic potency of the (+)anti isomer may be accounted for by the complementary nature of the preferred damage (N2 position of guanine) and the resultant mutation (G:C —> T:A transversion) at specific sequences. Over half of these transversions (7/13) are localised to a hotspot sequence 5'-A-G-$(G)_n$-3'. There are seven such possible sites within the coding sequence of APRT; mutations were found at five of them. Two other transversions were located at 5'-A-G-A-3' sites. This specificity was not seen in the bacterial lacI gene. BPDE activates via alteration of the 61st codon more frequently than at the 12th codon of the human c-Ha-ras1 proto-oncogene (Vousden, et al., 1987). Surprisingly, the 61st codon includes a run of Gs flanked by As, the same target hotspot region found in our study. This provides an explanation for the selective activation by BPDE of that proto-oncogene. It would be of interest to compare the sequence targeting by the (+) and (-) isomers, since our data suggest each isomer is responsible for a particular site specificity of the G:C —> T:A transversion.

Analysis of the mutational specificity of BPDE in the CHO and E. coli systems has revealed a different site specificity for the G:C —> T:A transversion in the two systems. This difference in site specificity can be interpreted as a reflection of the mutationally relevant lesions in the two systems; the bacterial specificity is consistent with damage at the N7 position of guanine, the eukaryotic specificity is consistent with damage at the N2 position of guanine. If the initial damage distribution in the two systems is similar (the same racemic BPDE mixture was used for the two studies), the different resultant specificities may be attributable to the different repair capacities of the two systems.

REFERENCES

Eisenstadt, E., Warren, A.J. Porter, J., Atkins, D., and Miller, J.H., 1982. Carcinogenic epoxides of benzo[a]-pyrene and cyclopenta[cd]pyrene induce base substitutions via specific transversions, Proc. Natl. Acad. Sci., U.S.A., 79:1945-1949.

Fahl, W.E., 1982. The kinetics of beno[a]pyrene anti-7,8-dihydrodiol-9,10-epoxide formation from benzo[a]pyrene and

regulatory membrane effects, Arch. Biochem. Biophys., 216:581-592.
Foster, P.L., Eisenstadt, E., and Miller, J.H., 1983. Base substitution mutations induced by metabolically activated aflatoxin B1, Proc. Natl. Acad. Sci., U.S.A., 50:2695-2698.
Lobanenkov, V.V., Plumb, M., Goodwin, G.H., and Grover, P.L., 1986. The effect of neighbouring bases on G-specific DNA cleavage mediated by treatment with the anti-diol epoxide of benzo[a]pyrene in vitro, Carcinogenesis, 7:1689-1695.
Mazur, M., and Glickman, B.W., 1988. Sequence specificity of mutations induced by benzo[a]pyrene-7,8-diol-9,10-epoxide at endogenous aprt gene in CHO cells, Som. Cell. Molec. Genet., 14:393-400.
Osborne, M.R., Jacobs, S., Harvey, R.G., and Brookes, P., 1981. Minor products from the reaction of (+) and (-) benzo[a]-pyrene-anti-diolepoxide with DNA, Carcinogenesis, 2:553-558.
Pelling, J.C., Slaga, T.J., and DiGiovanni, J., 1984. Formation and persistence of DNA, RNA, and protein adducts in mouse skin exposed to pure optical enantiomers of 7β,8α-dihydroxy-9α,10α-epoxy-7,8,9,10-tetrahydrobenzo[a]pyrene in vivo, Cancer Res., 44:1081-1086.
Sage, E., and Haseltine, W.A., 1984. High ratio of alkali-sensitive lesions to total DNA modification induced by benzo[a]pyrene diol epoxide, J. Biol. Chem., 259:11098-11102.
Schaaper, R.M., Danforth, B.N., and Glickman, B.W., 1985. Rapid repeated cloning of mutant lac repressor genes, Gene, 39: 181-189.
Stevens, C.W., Bouck, N., Burgess, J.A., and Fahl, W.E., 1985. Benzo[a]pyrene diolepoxides: different mutagenic efficiency in human and bacterial cells, Mutat. Res., 152:5-14.
Vousden, K.H., Bos, J.L., Marshall, C.L., and Philips, D.H., 1986. Mutations activating human c-Ha-ras1 protooncogene (HRAS1) induced by chemical carcinogens and depurination, Proc. Natl. Acad. Sci., U.S.A., 83:1222-1226.
Yang, S.K., Roller, P.P., and Gelbion, H.V., 1978. Benzo[a]-pyrene metabolism: Mechanism in the formation of epoxides, phenols, dihydrodiols, and the 7,8-diol-9,10-epoxides in "Carcinogenesis, Polycyclic Aromatic Hydrocarbons, Vol. III" Jones, P.W., and Freudenthal, R.I., (eds.), Raven Press, New York, pp. 285-301.

DEOXYRIBONUCLEOTIDE POOLS, DNA REPAIR AND MUTAGENESIS

Andrew R. Collins and Diane T. Black

University of Aberdeen, Department of Biochemistry
Marischal College, Aberdeen, AB9 1AS Scotland

ABSTRACT

The DNA resynthesis step of excision repair makes only a small demand on the pool of DNA precursors in the cell. However, DNA repair is sensitive to drugs that act via the dNTP pool; hydroxyurea, which depletes certain of the dNTPs by inhibiting ribonucleotide reductase, slows the resynthesis step and causes incomplete repair sites to accumulate as easily measurable DNA breaks. The dependence of DNA repair on nucleotide synthesis is examined in a mutant cell line, Ade$^-$C, a purine auxotroph. When deprived of hypoxanthine as a purine source, Ade$^-$C cells cease DNA replication, and although DNA repair continues, it is abnormal. Purine starvation increases both the cytotoxicity and the mutagenicity of ultraviolet light. We find 4-5 times more forward or back mutations in Ade$^-$C cells irradiated in the purine-starved state compared with irradiated but unstarved cells.

INTRODUCTION

DNA repair, in almost all its guises, involves a certain amount of DNA synthesis. Two or three nucleotides may be inserted when a strand break caused by ionizing radiation is resealed; at the other extreme, removal of bulky lesions by the nucleotide excision pathway is followed by synthesis of a patch upwards of 30 nucleotides in length. Figure 1 summarises the pathways by which DNA precursors are provided. It shows the key role played by the enzyme ribonucleotide reductase and how the salvage pathways feed into the scheme, thymidine providing pyrimidines and hypoxanthine purines.

An average experimental dose of DNA damaging agent inflicts a very small number of lesions relative to the total amount of DNA in the cell, and so the demands of repair synthesis for DNA precursors are exiguous. However, it is still an intriguing question how these demands are met. The total cellular pool of dNTPs in the cell is not large, and it has been suggested that replicative DNA synthesis is supplied, not from this pool, but by conversion of rNDPs to dNDPs and then to dNTPs at the replication fork, giving a very high local

DNA Repair Mechanisms and Their Biological Implications in Mammalian Cells
Edited by M.W. Lambert and J. Laval
Plenum Press, New York

concentration (Reddy and Pardee, 1980; Wickremasinghe and Hoffbrand, 1983). An enzyme complex including ribonucleotide reductase and DNA polymerase is probably part of the structure to which DNA loops are attached, the nuclear matrix, and the current model for replication is of DNA being reeled through these fixed sites on the matrix (Pardoll, et al., 1980).

Little is known about the organization of DNA repair at this level. There is some evidence that repair occurs preferentially at the matrix (McCready and Cook, 1984; and see Mullenders, et al., this vol.). K_mS of DNA polymerase involved in repair (inversely reflecting affinity for dNTPs) are lower than the K_ms of the same enzyme in replication (Dresler, et al., 1982) which suggests that polymerase can synthesize repair patches efficiently even when substrates are at a low concentration; so perhaps repair does utilize the dilute cellular pool of dNTPs. The differing K_ms may arise from the organization of repair and replicative enzymes in distinct macromolecular complexes.

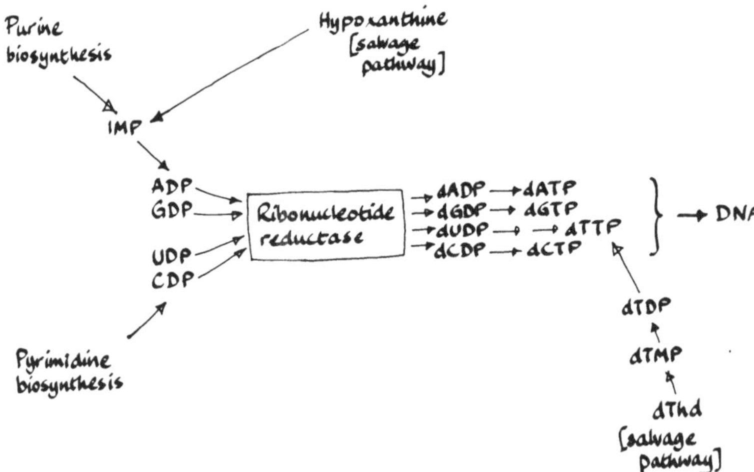

Figure 1. DNA precursor metabolism; de novo biosynthesis and salvage pathways.

For several years we have been investigating the dependence of DNA repair on the pool of deoxyribonucleotides. We have looked at effects of drugs such as hydroxyurea that block dNTP production; we have recently measured the concentration of dNTPs in hydroxyurea treated cells; and we have studied a mutant cell line which is defective in purine biosynthesis. As well as the immediate effect of a perturbed dNTP pool on repair, there is the question of long-term consequences. In the case of DNA replication, there is now much evidence that an unbalanced pool, in vitro or in whole cells, can adversely affect the fidelity of synthesis, introducing mutations (Kunkel, et al., 1982; Meuth, 1981). Is this true also for DNA repair in cells with artificially manipulated dNTP pools? There is a wider significance in this question, since many of the cells in the body are in a non-dividing state and have extremely small dNTP pools; yet they receive DNA damage and, in some cases at least, repair it. If this repair is inaccurate, mutations may result.

EFFECTS OF HYDROXYUREA ON dNTP POOLS AND ON DNA REPAIR

Hydroxyurea is a commonly used inhibitor of replicative DNA synthesis which acts on ribonucleotide reductase. We recently measured the concentrations of dNTPs after adding hydroxyurea to exponentially growing, asynchronous cultures of the cell lines most used in our DNA repair studies (Collins and Oates, 1987). One of these is the CHO cell line K1. Figure 2 shows the rapid depletion of the pools of dATP and dGTP within 30 minutes of adding the drug. The slow increase in the dTTP pool is typical of the behavior of this nucleotide after

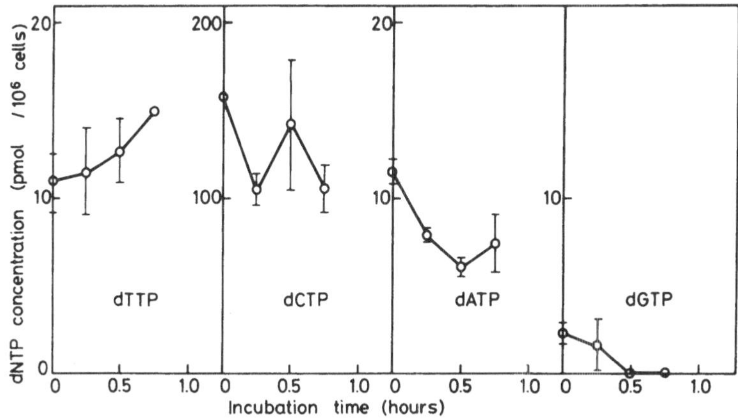

Figure 2. dNTP concentrations in exponentially growing CHO cells; effect of adding hydroxyurea (10 mM) at time zero. dNTP pools were measured by HPLC, using essentially the method of Garrett and Santi (1979). Bars represent the standard errors of the mean, or the range of duplicate determinations. From Collins and Oates (1987) with permission of FEBS.

addition of hydroxyurea (Nicander and Reichard, 1985). In hamster cells the dCTP pool is very large (an order of magnitude larger than the others), and no significant change is seen with hydroxyurea. In human cells, dCTP is present at about the same concentration as the other dNTPs, and it falls in parallel with dATP and dGTP after adding the inhibitor (Collins and Oates, 1987).

There are two ways of looking at the depletion of dNTPs, depending on the relationship between the pools and replication. If the dNTP pools are the actual source of precursors for replication, the depletion represents a continuing activity by DNA polymerase until it has exhausted the available pool of the limiting precursor. It does seem to be true that the fall in dNTPs is initially rapid but then slows or stops when the nucleotide present at the lowest concentration (dGTP) has disappeared, as if replication only then ceases. According to the alternative view of replication, in which ribonucleotides are channelled through the enzyme complex at the replication fork, the pools of dNTPs are irrelevant to replication (perhaps representing leakage from the replication complex); the falls in the pools after hydroxyurea treatment then simply reflect a turnover which is independent of replication and which is seen when the input to the pools is blocked.

Even when dGTP is undetectable by our technique, there might be enough present for the cell to carry out repair of DNA, which after all has very modest needs. But if repair is affected by nucleotide depletion, and incision occurs without ligation, incomplete repair sites will be expected to accumulate as breaks in the DNA. This is what is seen when CHO cells are irradiated with ultraviolet and incubated with hydroxyurea (Figure 3).

Of course, it is well known that, in contrast to the rapid cessation of replicative DNA synthesis on adding hydroxyurea, repair synthesis, as indicated by the "unscheduled"

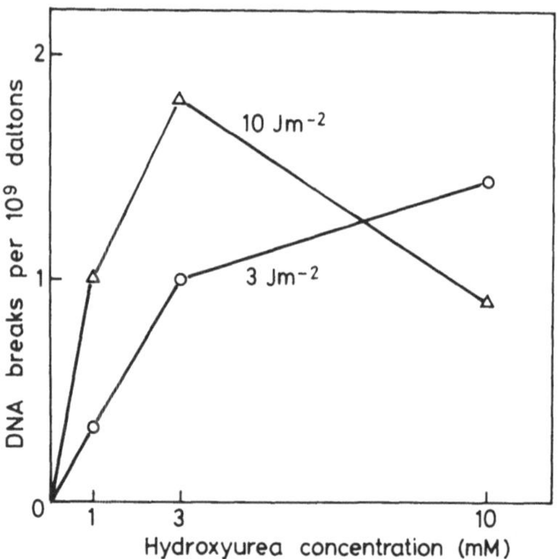

Figure 3. DNA breaks accumulated by CHO cells incubated with hydroxyurea after ultraviolet irradiation. Cells (labelled with [^{3}H]dThd) were preincubated for 30 minutes with hydroxyurea at the concentrations shown, irradiated with ultraviolet at 3 Jm^{-2} (○) or 10 Jm^{-2} (Δ) and incubated for a further 30 minutes with the inhibitor. At the end of this time accumulated DNA breaks were measured by an alkaline unwinding/hydroxyapatite chromatography assay (Squires, et al., 1982). Redrawn from Collins and Oates, 1987.

incorporation of [^{3}H]thymidine, is not reduced by hydroxyurea. This is the basis of the common use of hydroxyurea to suppress replication and facilitate detection of repair synthesis. It seems like a paradox if, as Figure 3 suggests, repair sites are not completed. What probably happens is that the reduced precursor supply has the dual effect of slowing polymerisation and holding the repair sites open for an abnormal length of time; but slow polymerisation over this extended period does achieve substantial repair synthesis.

Cells that are not dividing tend to have very small dNTP pools (Snyder, 1984) and repair sites in such cells are found to stay open for a relatively long time. It was shown several years ago (Yew and Johnson, 1979) that DNA repair in unstimulated lymphocytes was speeded up when deoxyribonucleosides were provided in the growth medium, presumably replenishing the diminished dNTP pools. Another example (Squires and Johnson, 1983) is seen in cells from Cockayne syndrome patients; repair sites are open for longer than in normal cells, suggesting a possible defect in precursor metabolism or utilization in these cells.

DNA REPAIR IN THE PURINE AUXOTROPH HAMSTER CELL LINE ADE$^-$C

To study further the relationship between DNA repair and DNA precursor pools, we have used the mutant CHO cell line Ade$^-$C, which is defective in the enzyme glycinamide ribonucleotide synthetase on the pathway of de novo purine synthesis. Ade$^-$C cells grow normally if purines are supplied in the medium in the form of hypoxanthine which is taken up by the salvage pathway (Figure 1). We looked first of all at replication (Figure 4). Within two or three hours of transferring Ade$^-$C cells to medium without hypoxanthine, replicative DNA synthesis virtually ceases. What little remains is accounted for by a few cells that are for some reason slow to respond to starvation; after pulse-labelling with [^{3}H] thymidine, these appear as heavily labelled cells in autoradiographs, amounting to 3% of the total. Now, if hypoxanthine-starved Ade$^-$C cells are irradiated with ultraviolet, repair synthesis is seen as a stimulation of the near-background level of [^{3}H] thymidine incorporation (solid triangles in Figure 4). We see this stimulation also after treatment with chemical DNA damaging agents. So it seems that repair is less stringent than replication in its demands on the dNTP pool; it clearly does not cease when the purine supply is disrupted. In this respect, the situation is similar to that of normal cells incubated with hydroxyurea; replication stops while repair synthesis continues.

If repair synthesis is occurring in purine-starved Ade$^-$C cells, we would certainly expect to detect the earlier step of incision. Incision is measured as the accummulation of DNA breaks (incomplete repair sites) when cells are incubated with hydroxyurea and cytosine arabinoside. Whether incision is measured immediately after ultraviolet irradiation (Figure 5a), or four hours later (Figure 5b), purine-depleted Ade$^-$C cells show a healthy activity compared with the unstarved cells. The lower levels of incision at the later time are to be expected if by then a significant fraction of the damage has been dealt with. In hypoxanthine-starved S phase cells, in which purine nucleotide pools should be particularly rapidly depleted by ongoing replication, there is still a normal rate of incision (Figure 5c).

Another aspect of repair that might be affected by purine starvation is the final step, ligation. If repair sites remained open for longer than usual, DNA breaks would accumulate even in the absence of hydroxyurea and cytosine arabinoside. No such accumulation was seen in control experiments (not shown).

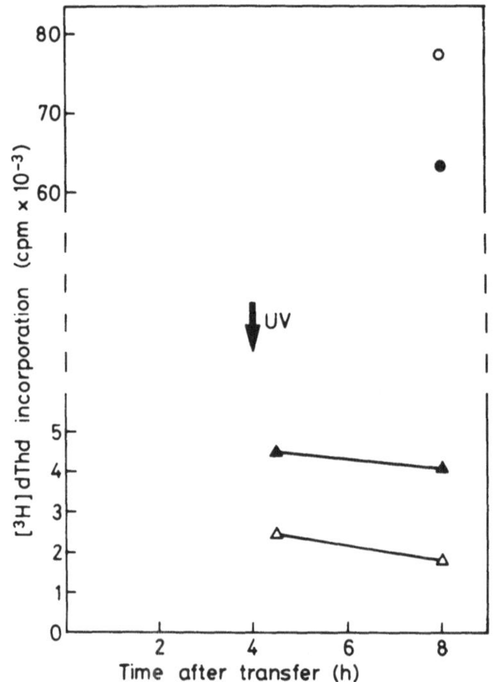

Figure 4. DNA synthesis after transfer of Ade⁻C cells to medium without hypoxanthine. At intervals after transfer, samples of cells were incubated with [^{3}H]dThd for 1 hour (▲). Other samples remained in complete medium (○). Solid symbols represent cells irradiated with ultraviolet (10 Jm^{-2}). ^{3}H incorporated into DNA was measured.

The only defect in repair that we have been able to find in starved Ade⁻C cells is an apparent inability to remove the damage induced by ultraviolet light. We used a modified, extra sensitive assay for cyclobutane pyrimidine dimers measured as UV endonuclease-sensitive sites. The assay detects dimers after very low ultraviolet doses in the range 0.5 - 4 Jm^{-2}. We found no sign of removal of dimers in the hypoxanthine-starved cells, whereas unstarved cells removed about a quarter of the dimers in 5 hours following irradiation with 1 or 2 Jm^{-2} of ultraviolet (Collins, et al., 1988). We have yet to confirm this finding with an alternative assay. If it is a real effect, it implies an aberrant form of repair with DNA incision, repair synthesis and ligation occurring in a

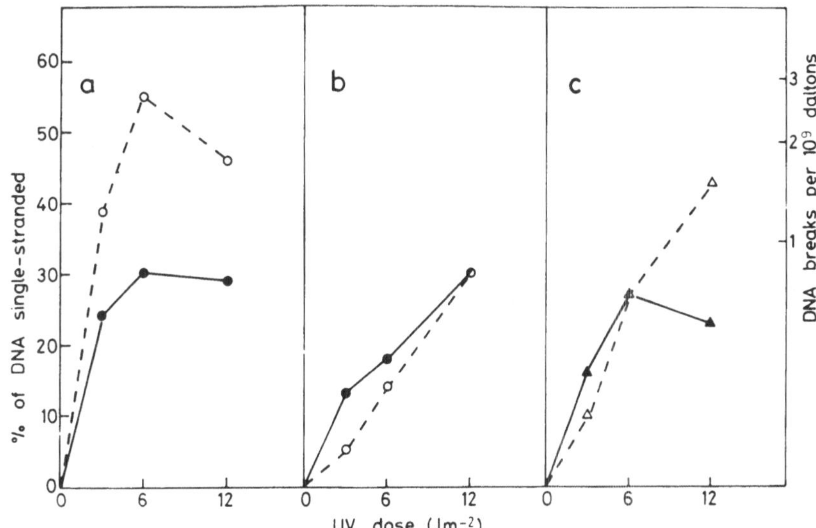

Figure 5. Incision in ultraviolet irradiated Ade C cells (a,b). Cells (labelled the previous day with [^{3}H]dThd) were incubated for 4 hours with (●) or without (O) hypoxanthine before irradiation with the doses of ultraviolet shown. In (a), cells were then incubated with hydroxyurea and cytosine arabinoside for 40 minutes to accumulate incomplete repair sites which were measured by an alkaline unwinding assay. In (b), cells were incubated for 3.5 hours after irradiation before adding inhibitors. In (c), cells were pulse-labelled with [^{3}H]dThd just before transfer to hypoxanthine-free medium, so that cells still in S phase were selectively examined for incision 4 hours later (△); (▲) indicates cells maintained in medium with hypoxanthine. (From Collins, et al., 1988, with permission of Elsevier Science Publishers B.V.).

dose-dependent manner, but on the wrong strand, perhaps, or down- or up-stream from the dimers. This has been suggested as an explanation of the low level of dimer removal associated with a relatively high level of repair synthesis in xeroderma pigmentosum cells of complementation group D (Paterson, et al., 1987). Alternatively, a minor lesion, not detected in the ultraviolet endonuclease assay, might be repaired normally, accounting for the incision and resynthesis response, while dimers are not repaired.

Whatever the nature of this aberrant repair, it has serious consequences for the cells. There is a significant increase in cell killing in starved, ultraviolet-irradiated cells; after 5 Jm^{-2}, the relative plating efficiency was 23%, compared with 42% for unstarved cells. There is a pronounced enhancement of the mutagenic effect of ultraviolet irradiation in purine-starved cells, too. We have examined both forward mutations (to resistance to the ATPase inhibitor ouabain) and reversion to the hypoxanthine-independent state. Whereas 12 ouabain-resistant colonies arose per 10^6 surviving cells after ultraviolet irradiation of unstarved cultures with 5 Jm^{-2}, 52 such colonies per 10^6 surviving cells were seen in purine-starved cultures (Collins, et al., 1988). No ouabain-resistant colonies occurred in over 3 million surviving cells of unirradiated cultures, whether unstarved or starved.

In terms of reversion, too, aberrant repair induces a high level of mutation. Five times more revertants occurred in starved, irradiated populations than in controls. Some of the revertants have been isolated and studied in detail to elucidate the nature of the mutations (Black, Oates and Collins, in preparation). In one revertant line, OB1, GARS activity is detectable though still very low compared with wild type cells, and specific chromosome rearrangements lead us to hypothesize that amplification of the defective gene may have occurred to provide sufficient purine biosynthetic capacity for the cells to survive hypoxanthine starvation.

POOL IMBALANCES IN REPLICATION AND REPAIR; SUMMARY

In various experimental cell systems, dNTP pools have been artificially perturbed, either by adding high concentrations of certain deoxyribonucleosides to the medium, or by the use of mutants with altered pools. Peterson, et al. (1978) found an increase in mutations induced by a methylating agent when thymidine and deoxycytidine were added to the medium, and an altered ratio of dCTP/dTTP resulted in more mutations in CHO cells treated with an ethylating agent (Meuth, 1981). In both cases, a probable explanation is that an imbalance of DNA precursors makes misincorporation more likely when an alkylated base is encountered during replication. In this context, it is interesting that we found no ouabain resistant mutants in the purine-starved but unirradiated Ade^-C cultures. The clear implication is that purine starvation alone is not mutagenic, perhaps because replication is so effectively suppressed that there is no chance for misincorporation to occur even though the pools of dNTPs are grossly disturbed.

Evidence of mutagenesis resulting specifically from misincorporation during repair rather than replication is not easy to obtain. In the Ade^-C system, we are able to examine effects of DNA precursor pool disturbances in the virtual absence of replication, and we clearly demonstrate that mutations can result from repair carried out with an unbalanced pool. Whether there is misincorporation of bases into repair patches, or whether aberrant repair induces chromosomal alterations that show as mutations, remains to be established.

In conclusion, it should be remembered that many "resting" cells in the organism have pools of deoxyribonucleotides very different in concentrations and proportions from those in proliferating cells, so repair of DNA damage by these cells might well be associated with high levels of mutation and potential carcinogenesis.

ACKNOWLEDGEMENTS

This work was supported by the Cancer Research Campaign, of which ARC is the Michael Sobell Research Fellow. We thank Dr. R.T. Johnson and Professor H.M. Keir for their support.

REFERENCES

Collins, A.R., Black, D.T., and Waldren, C.A., 1988, Aberrant DNA repair and enhanced mutagenesis following mutagen treatment of Chinese hamster Ade⁻C cells in a state of purine deprivation, Mutat. Res., 193:145-155.

Collins, A., and Oates, D.J., 1987, Hydroxyurea: effects on deoxyribonucleotide pool sizes correlated with effects on DNA repair in mammalian cells, Eur. J. Biochem., 169: 299-305.

Dresler, S.L., Roberts, J.D., and Lieberman, M.W., 1982, Characterization of deoxyribonucleic acid repair synthesis in permeable human fibroblasts, Biochemistry, 21:2557-2564.

Garrett, C., and Santi, D.V., 1979, A rapid and sensitive high pressure liquid chromatography assay for deoxyribonucleoside triphosphates in cell extracts, Analyt. Biochem., 99:268-273.

Kunkel, T.A., Silber, J.R., and Loeb, L.A., 1982, The mutagenic effect of deoxynucleotide substrate imbalances during DNA synthesis with mammalian DNA polymerases, Mutat. Res., 94:413-419.

McCready, S.J., and Cook, P.R., 1984, Lesions induced in DNA by ultraviolet light are repaired at the nuclear cage, J. Cell Science, 70:189-196.

Meuth, M., 1981, Role of deoxynucleoside triphosphate pools in the cytotoxic and mutagenic effects of DNA alkylating agents, Somat. Cell Genet., 7:89-102.

Nicander, B., and Reichard, P., 1985, Relations between synthesis of deoxyribonucleotides and DNA replication in 3T6 fibroblasts, J. Biol. Chem., 260:5376-5381.

Pardoll, D.M., Vogelstein, B., and Coffey, D.S., 1980, A fixed site of DNA replication in eukaryotic cells, Cell, 19: 527-536.

Paterson, M.C., Middlestadt, M.V., MacFarlane, S.J., Gentner, N.E., Weinfeld, M., and Eker, A.P.M., 1987, Molecular evidence for cleavage of intradimer phosphodiester linkage as a novel step in excision repair of cyclobutyl pyrimidine photodimers in cultured human cells, J. Cell Sci., Suppl. 6:161-176.

Peterson, A.R., Landolph, J.R., Peterson, H., and Heidelberger, C., 1978, Mutagenesis of Chinese hamster cells is facilitated by thymidine and deoxycytidine, Nature, (London), 276:508-510.

Reddy, G.P.V. , and Pardee, A.B., 1980, Multienzyme complex for metabolic channeling in mammalian DNA replication, Proc. Natl. Acad. Sci. USA, 77:3312-3316.

Snyder, R.D., 1984, Deoxyribonucleoside triphosphate pools in human diploid fibroblasts and their modulation by hydroxyurea and deoxynucleosides, Biochem. Pharmacol., 33:1515-1518.

Squires, S., and Johnson, R.T., 1983, U.V. induces long-lived DNA breaks in Cockayne's syndrome and cells from an immunodeficient individual (46BR): defects and disturbance in post incision steps of excision repair, Carcinogenesis, 4:565-572.

Squires, S., Johnson, R.T., and Collins, A.R.S., 1982, Initial rates of DNA incision in UV-irradiated human cells; differences between normal, xeroderma pigmentosum and tumor cells, Mutat. Res., 95:389-404.

Wickremasinghe, R.G., and Hoffbrand, A.V., i983, Inhibition by aphidicolin and dideoxythymidine triphosphate of a multienzyme complex of DNA synthesis from human cells, FEBS Lett., 159 : 175-179.

Yew, F.H., and Johnson, R.T., 1979, Ultraviolet-induced DNA excision repair in human B and T lymphocytes, II. Effect of inhibitors and DNA precursors, Biochim. Biophys. Acta., 562:240-251.

DUPLEX-DUPLEX HOMOLOGOUS RECOMBINATION CATALYSED BY A HUMAN NUCLEAR EXTRACT. INVOLVEMENT IN DOUBLE-STRAND BREAK REPAIR

Bernard Lopez and Jacques Coppey

Institut Curie - Biologie
75231 Paris, France.

INTRODUCTION

Homologous recombination is implicated in many basic biological processes, such as molecular evolution of multigenic families (Baltimore, 1981), diversification of immunoglobulin genes (Kourilsky, 1983; Radman, 1983; Reynaud, et al., 1987) induction of sister-chromatid exchanges, and DNA repair (Friedberg, 1985).

Strand transfer, i.e., invasion of a duplex DNA by a homologous single-stranded DNA, is a central step in homologous recombination. The best characterized strand transferase is Rec A from E. coli (reviewed by Cox and Lehman, 1987). Strand transferases have also been purified from the lower eukaryotes, Ustilago maydis (Rec 1) (Kmiec and Holoman, 1982) and yeast Saccharomyces cerevisiae (Kolodner, et al., 1987), the first being ATP dependent and the second ATP independent. Similar activities have been described in human nuclear extracts, which appear to exhibit either ATP dependency (Kenne and Ljungquist, 1984; Cassuto, et al., 1987; Ganea, et al., 1987; Fishel, et al., 1988) or ATP independency (Hsieh, et al., 1986; Lopez, et al., 1987). In addition, Rec A, Rec 1, and one human strand transferase were found to possess Z-DNA binding activity (Blaho and Wells, 1986; Kmiec, et al., 1985; Fishel, et al., 1988).

The situation seems as yet unclear in human cells. Moreover, recombination between two homologous duplex DNA molecules remains poorly documented in mammalian cells. This prompted us to devise a system allowing us to study recombination between two homologous duplex DNA molecules, catalysed by human cell extracts. One advantage of our system is its potential ability to study this process at both the phenotypic level of expression of the recombined sequences and the molecular level in elucidating the structure of the recombining DNA's. These experiments were conducted with nuclear extracts from transformed human cell lines of different origins (Lopez, et al., 1987).

DNA Repair Mechanisms and Their Biological Implications in Mammalian Cells
Edited by M.W. Lambert and J. Laval
Plenum Press, New York

I. RECOMBINATION BETWEEN HOMOLOGOUS SUBSTRATES

Phenotypic analysis

The strategy of these experiments is based upon the restoration by recombination of a functional gene (lac Z' gene) in the replicative form (RF) of a M13 mp8 bacteriophage DNA. After incubation of substrate DNA molecules in nuclear extract, recombination was monitored by transfection of indicator *E. coli* JM109 bacteria (recA) with the products of the reaction (Figure 1A).

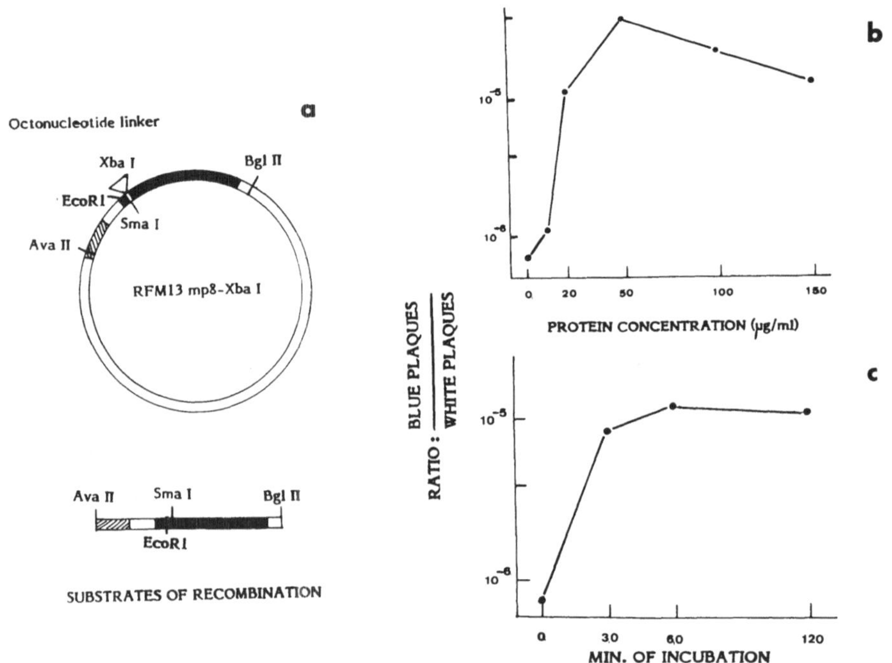

Figure 1. a) Substrates used: one substrate is the RF of M13 mp8 bacteriophage containing a XbaI linker (8 nucleotides) in the SmaI site. The linker insertion destroys the SmaI cleavage site and disrupts the reading frame of the lac Z' gene. Transfection of JM109 bacteria (rec A) with such modified RF gives white plaques in the presence of IPTG and X-gal, whereas transfection with wild type M13 mp8 RF gives blue plaques (Yannish-Perron, et al., 1985). The other substrate is the one kb AvaII-BglII fragment isolated from wild type RF. This fragment contains a functional lac Z' sequence but is not self-replicating when introduced in bacteria (no plaques). Recombination between the two substrates may give rise to RF's containing a restored functional lac Z' gene producing blue plaques when transfected in JM109 bacteria. Black: lac Z' gene; hatched box: lac i'. The ratio of blue to white plaques is plotted as a function of: b) protein extract concentration, and c) time incubation with 50 µg/ml of nuclear extract (a minimum of 10 blue plaques scored per point).

The extent of recombination (given by the proportion of blue plaques) was enhanced from 7 x 10^{-7} to 5 x 10^{-5} depending on the protein concentration and the time of incubation (Figure 1b and c). About 10% of the blue plaques exhibited blue sectoring and the progeny from these plaques gave both blue and white plaques. This result indicates that the corresponding RF's were heteroduplexes. Restriction analysis of the progeny of 20 recombined clones (blue plaques) shows the reappearance of the SmaI site (see legend to Figure 1). Moreover, these results exclude the possibility of tandem or random integration of the wild type donor fragment in the recombined RF.

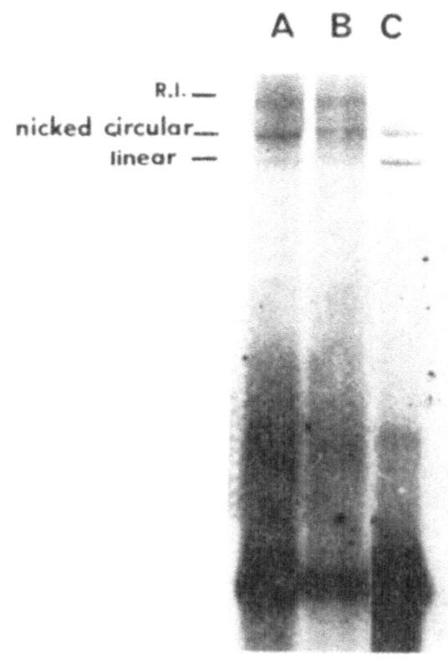

Figure 2. Analysis by gel electrophoresis of the recombination products. Labelled wild type donor fragment and unlabelled RF were incubated in the nuclear extract, then extracted and analyzed by gel electrophoresis (agarose 1%) and autoradiography. A trace amount of labelled RF was added just before electrophoresis as an internal size marker. A) 50 µg/ml of extract; B) 10 µg/ml of extract; C) no extract. The proportion of radioactivity in the upper band (RI) relative to the total radioactivity is 2.2% (lane A) and 0.32% (lane B).

Separate incubation of the two DNA substrates in nuclear extract, followed by purification of the two DNA's and coincubation in hybridization conditions prior to transfection did not give rise to an increased frequency of blue plaques. We conclude that the catalysis of recombination by nuclear factors does not result simply from single-strand exonucleolytic activity on both substrates in the regions of homology followed by heteroduplex formation between such randomly gapped molecules.

Molecular analysis

After incubation of labelled AvaII-BglII fragment and cold RF in the extract, the DNA was analyzed by gel electrophoresis and autoradiography (Figure 2).

A slowly migrating band is seen after incubation in the extract (Figure 2, lanes A and B). This band was expected to represent recombination intermediates (RI). The formation of such a band was independent of the addition of ATP and dNTP's (Lopez, et al., 1987). The DNA from this band was recovered by electroelution. Transfection of JM109 bacteria with a fraction of this electroeluted RI DNA gave 50% blue plaques of normal size, which indicates a great enrichment in recombined molecules. Electron microscopic examination of this electroeluted DNA revealed the presence of a significant proportion of double-stranded circular molecules, having the size of M13 RF, with a tail of 1 kb (Figure 3).

A 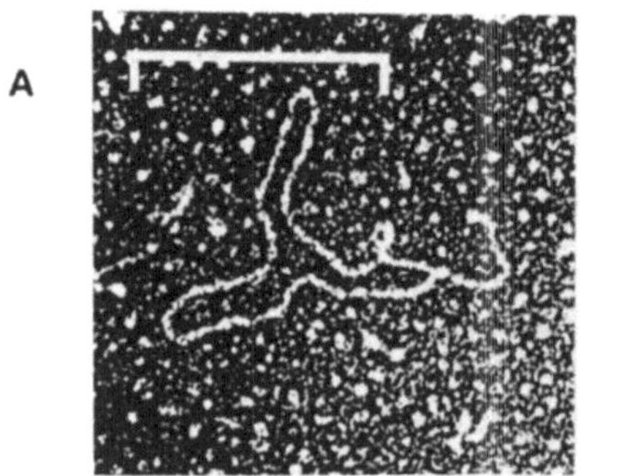

B 

Figure 3. A) Electron micrograph of one typical molecule present in the electroeluted RI DNA. Scale bar: 0.5 µm. (Photograph by Dr. R. Rousset). B) Probable structure of an RI molecule. The heavy lines correspond to the fragment and the thin line to the RF. The black squares indicate the mismatched region.

These results show the existence of a fair correlation between the recombined phenotype (blue plaques) and the RI structure (Figure 3b).

To further define the structure of the RI, we have performed a restriction analysis of the DNA (labelled fragment + cold RF) after incubation with the extract (Figure 4).

The digestion patterns compared to those predicted according to the position of the tail (Figure 4C) allow the deduction that the tail can be located at the AvaII or BglI sites. This result implies that initiation (hence termination) of recombination can take place at either site. If this recombination process exhibits polarity, as seen for all strand transferases so far studied, the present results could indicate that either strand of the fragment can invade the RF.

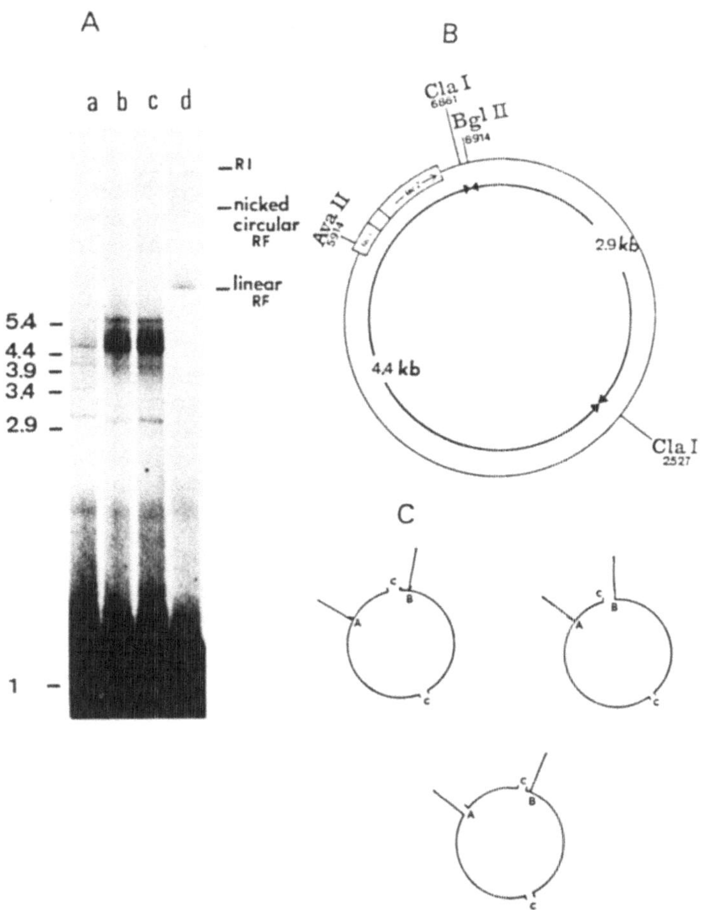

Figure 4. A) Restriction patterns of DNA incubated with human nuclear extract. a: AvaII/ClaI double digestion; b: BglII/ClaI double digestion; c: ClaI digestion; d: no digestion. B) Restriction map of M13 mp8RF. C) Possible patterns of RI digestion.

Sequencing of 20 recombined clones (blue plaques from electroeluted RI DNA) spanning from the primer annealing site upstream from the AvaII site (400 nt/clone) did not reveal any sequence changes compared to the parental M13 mp8 sequence. This analysis includes the lac i' sequence which is silent (no selection bias). For 20 independent recombination events, the process appears to be error free in human nuclear extract, despite nuclease activities present in the extract.

The absence of any alteration all along the sequence analysed suggests the existence of protective factors in the extract. To test this, we have performed a DNase I protection assay. The results show a transient protection (between 1 and 30 minutes) covering the potentially recombining DNA sequences. The extent of protection is a function of protein extract concentration and requires homologous sequences. The protection is optimal for equivalent numbers of either substrate molecules. The size of the protected region indicates that several protein molecules act all along the recombining sequences. Furthermore the fair correlation between i) the time dependence of homologous recombination (Figure 1c) and that of DNase I protection, ii) the proportion of protected DNA ($\sim 10^{-3}$) and that of RI DNA formed (see legend to Figure 2), could indicate that the protective factors of the extract are proteins implicated in the recombination process. Moreover the time course of DNase I protection can indicate that the search of homology takes place very rapidly (< 1 minute incubation), the overall reaction consuming a longer time (> 30 minutes). In order to characterize the enzymatic activities governing this type of recombination, stepwise purification of the nuclear extract will be required.

We may presently conclude that human nuclear extracts contain activities promoting recombination between two homologous duplex DNA's which results in the replacement of an altered sequence by its functional counterpart, and that the overall recombination proceeds via single-strand exchange according to three steps: i) unwinding of at least one of the duplexes; ii) single-strand exchange which can be reciprocal and synaptic pairing which creates heteroduplexes; iii) resolution of the cross-junction. Furthermore the process seems to be error-free.

In E. coli, an increase of recombination of λ phages is accompanied by a decrease of mutagenesis (Blanco and Devoret, 1973), and in yeast, meiotic gene conversion is error-free for a total of 189 alleles analysed (Fogel and Mortimer, 1970). Recombination seems therefore to be error-free in prokaryotes and in eukaryotes including mammalian cells, as shown here by a molecular approach.

More generally, in the theory of concerted evolution, gene conversion plays a central role in molecular evolution and sequence homogenization of repeated sequence families (ref. in Baltimore, 1981; Hess, et al., 1984). Moreover, gene conversion has been proposed to account for polymorphism and diversity in the immunoglobulin gene system (Kourilsky, 1983; Radman, 1983; Reynaud, et al., 1987). These phenomena are critically dependent on the fidelity of recombination. On the other hand, the promising applications of gene targeting (Smithies, et al., 1985; Thomas, et al., 1986; Thomas and Cappechi, 1986; Nandi, et al., 1988) basically rely on the fidelity of homologous recombination.

II. REPAIR OF DOUBLE-STRAND BREAK

The DNA double-strand break (dsb) is a critical damage induced by genotoxic agents such as ionizing radiation. On the other hand, dsb's stimulate recombination in mammalian cells (Subramani and Seaton, 1988). In order to study dsb repair in human nuclear extract, we have used the strategy described in Figure 5.

The presence of purified fragment leads to an increased (up to 15 times) reactivation rate as a function of protein extract concentration, time of incubation or relative amount of either substrate, provided that i) the dsb be located in the region homologous with the fragment, ii) the fragment be intact in the region corresponding to that of the dsb in the RF (Lopez and Coppey, 1987). The reaction is ATP dependent and partially dNTP's independent. This last requirement shows that limited polymerization is involved.

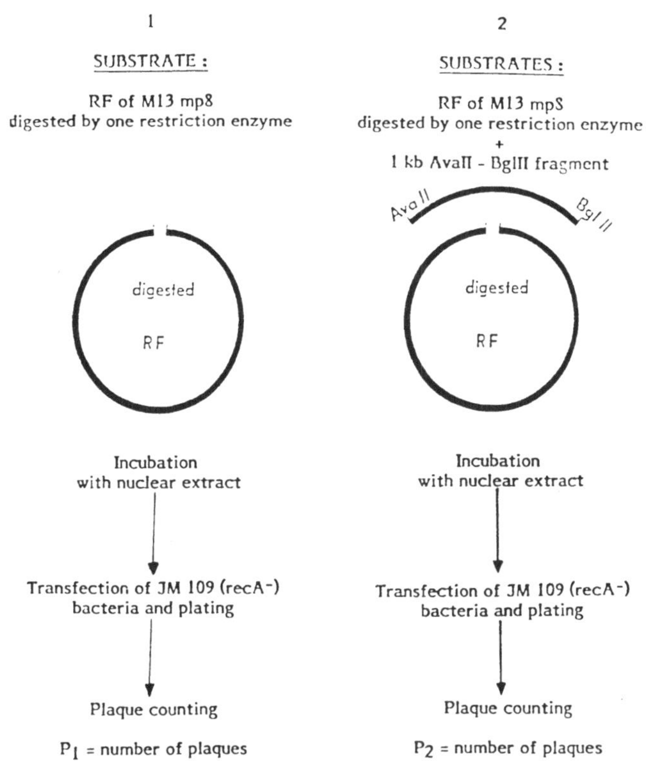

Figure 5. Strategy employed to study dsb repair. (1) A break is introduced in the RF by a restriction endonuclease: all RF's are cut at the same site. (2) An intact homologous fragment covering the dsb is added in some experiments.

Combined with restriction analysis of repaired molecules, these data demonstrate that the repair of a dsb in the presence of an intact fragment occurs via a pathway of homologous recombination initiated at the location of the dsb.

Influence of the structure of the break.

For carrying out this analysis, RF was cut by restriction

endonucleases generating different types of extremities. Recombination is involved for repairing dsb's with 5' protruding or blunt ends, but not for repairing dsb's with 3'protruding ends (Table 1).

Two possible mechanisms can account for the lack of recombinational dsb repair on 3' protruding ends: such structures could be repaired mainly by ligation but also by a recombinational pathway involving factor(s) lost during the partial purification of extracts. This would imply that the enzymatic apparatus acting on 5' protruding or blunt ends is distinct from that acting on 3' protruding ends. These results seem to indicate that the exonuclease and/or recombinase associated with the recombinational repair present a polarity.

Interestingly, dephosphorylation of the ends at the cut prevents ligation (P1 value decreased), but is without effect on recombination (P2 value unchanged) (Lopez and Coppey, 1987). In addition, we have showm that a homology disruption (8 bp heterology), located at 7 bp from the extremity on one side of the break prevents repair by recombination (Lopez and Coppey, 1987). More recently, we have demonstrated that a sequence homology (comprised between 15 and 27 bp) is required on both sides of the break to allow initiation of recombinational repair (unpublished data).

Table 1. Effect of the structure of the cut on the reactivation rate.

Enzymatic treatment	Structure at the cut	P_1*	P_2*	Reactivation rate
EcoR1	5' protruding	275	1375	5
BamH1	5' protruding	234	1172	5
HindIII	5' protruding	251	1258	5
PstI	3' protruding	228	63	0.3
BglI	3' protruding	216	172	0.8
PstI then T_4 polymerase + dNTPs	blunt ends	76	836	11
BglI then T_4 polymerase + dNTPs	blunt ends	72	648	9

* The P_1 and P_2 values are the sum from 4 independent experiments. DNA manipulations were performed using standard procedures.

Mammalian cells contain highly reiterated sequences distributed throughout the genome. It has been suggested that the minimal length of homology required for homologous recombination would increase with genome complexity to avoid deleterious recombination events (Thomas, 1966). Indeed the results of intrachromosomal recombination in mammalian cells (minimal homology of 95 bp) (Liskay, et al., 1987), compared to those with bacteriophage T4 (50 bp) (Singer, et al., 1982) and E. coli (20-74 bp) (Watt, et al., 1985) are rather consistent with such a hypothesis, but not those of extrachromosomal (14 bp or 25 bp) (Rubnitz and Subramani, 1984; Ayares, et al., 1985) or of in vitro recombination (present data). In other words, homology requirements would be larger for intrachromosomal recombination with packaged DNA's than for naked DNA's. This could indicate that the human recombination machinery itself requires short streches of homology to initiate homologous recombination on free DNA's. Chromatin structure may reduce the simultaneous accessibility to the recombination apparatus of sufficientlylong stretches on both DNAs. In turn, this can prevent undesirable recombination. If this is true, we can predict that actively transcribed sequences are more prone to recombination than the inactive ones.

CONCLUSION

Dsb repair in human nuclear extracts takes place preferentially via recombination with a homologous DNA sequence. A multienzymatic complex appears to be involved in the successive steps of this process (summarized in Figure 6):

1. Initiation at the location of the break with an associated exonucleolytic degradation.
2. Strand exchange with the intact DNA sequence.
3. Polymerization to fill in the single-stranded regions.
4. Resolution of the cross-junctions.

Figure 6. Two characteristic steps of recombination in the region of a dsb in nuclear extracts.

It is interesting to note that we have observed the filling of a 30 bp gap by recombinational repair. All the data obtained in our system are in general agreement with the model of Szostak, et al. (1983) established from genetic data in yeast.

ACKNOWLEDGEMENTS

This work was supported by grants from INSERM, France,

no. 852017 and 87460 and from the Commission des Communautés Européennes no. B16-151F. We thank Miss Malot for typing the text.

REFERENCES

Ayares, D., Chekuri, L., Song, K.Y. and Kucherlapati, R., 1986, Sequence homology requirements for intermolecular recombination in cells, Proc. Natl. Acad. Sci. USA, 83: 5199.

Baltimore, D., 1981, Gene conversion: some implications for immunoglobulin genes, Cell, 24:592.

Blaho, J.A. and Wells, R.D., 1986, Left-handed Z-DNA binding to the RecA protein of E. coli, J. BiOl. Chem., 262:6082.

Blanco, M. and Devoret, R., 1973, Repair mechanisms involved in prophage reactivation and UV reactivation of UV-irradiated phage , Mutation Res., 17:293.

Cassuto, E., Lightfoot, L.A. and Howard-Flanders, P., 1987, Partial purification of an activity from human cells that promotes homologous pairing and the formation of heteroduplex DNA in the presence of ATP, Mol. Gen. Genet., 208: 10.

Cox, M.M. and Lehman, I.R., 1987, Enzymes of general recombination, Ann. Rev. Biochem., 56:229.

Fishel, R.A., Detmer, K. and Rich, A., 1988, Identification of homologous pairing and strand-exchange activity from a human tumor cell line based on Z-affinity chromatography, Proc. Natl. Acad. Sci. USA, 85:36.

Fogel, S. and Mortimer, R.K., 1970, Fidelity of meiotic gene conversion in yeast, Mol. Gen. Genet., 109:177.

Friedberg, E.C., 1985, DNA Repair, W.H. Freeman and Company, N.Y.

Ganea, D., Moore, P., Chekuri, L.A. and Kucherlapati R.S., 1987, Characterization of an ATP-dependent DNA strand transferase from human cells, Mol. Cell. Biol., 7:3124.

Hess, J.F., Schmid, C.W. and Shen, C.K.J., 1984, A gradient of sequence divergence in the human adult α-globin duplication units, Science, 226:67.

Holliday, R., 1964, A mechanism for gene conversion in fungi, Genet. Res., 5:282.

Hsieh, P., Meyn, M.S. and Camerini-Otero, R.D., 1986, Partial purification and characterization of a recombinase from human cells, Cell, 44:885.

Kenne, K. and Ljungquist, S., 1984, A DNA-recombinogenic activity in human cells, Nucleic Acids Res., 12:3057.

Kmiec, E. and Holloman, W.K., 1982, Homologous pairing of DNA molecules promoted by a protein from Ustilago, Cell, 29: 367.

Kmiec, E., Angelidess, K.J. and Holloman, W.K., 1985, Left-handed DNA and the synaptic pairing reaction promoted by Ustilago Rec 1 protein, Cell, 40:139.

Kolodner, R., Evans, D.H. and Morrisson, P.T., 1987, Purification and characterization of an activity from Saccharomyces cerevisiae that catalyzes homologous pairing and strand exchange, Proc. Natl. Acad. Sci. USA, 84:5560.

Kourilsky, P., 1983, Genetic exchanges between partially homologous nucleotide sequence: possible implication for multigene families, Biochimie, 65:85.

Liskay, R.M., Letsou, A. and Stachelek, J.J., 1987, Homology requirement for efficient gene conversion between duplicated chromosomal sequences in mammalian cells, Genetics, 115:161.

Lopez, B., Rousset, S. and Coppey, J., 1987, Homologous recombination intermediates between two duplex DNA's catalysed by human cell extracts, Nucleic Acids Res., 15:5643.

Lopez, B. and Coppey, J., 1987, Promotion of double-strand break repair by human nuclear extracts preferentially involves recombination with intact homologous DNA, Nucleic Acids Res., 15:6813.

Nandi, A.K., Roginski, R.S., Gregg, R.G., Smithies, O. and Skoultchi, A.I., 1988, Regulated expression of genes inserted at the human chromosomal α-globin locus by homologous recombination, Proc. Natl. Acad. Sci. USA, 85:3845.

Radman, M., 1983, Diversification and conservation of genes by mismatch repair: a case for immunoglobulin genes, in Cellular Response to DNA Damage, eds. E. Friedberg, and B. Bridges, A.R. Liss, N.Y., p. 287.

Reynaud, C.A., Anquez, V., Grimal, H. and Weill, J.C., 1987, A hyperconversion mechanism generates the chicken light chain preimmune repertoire, Cell, 48:379.

Rubnitz, J. and Subramani, S., 1984, The minimum amount of homology required for homologous recombination in mammalian cells, Mol. Cell. Biol., 4:2253.

Singer, B.S., Gold, L., Gauss, P. and Doherty, D.M., 1982, Determination of the amount of homology required for recombination in bacteriophage T4, Cell, 31:25.

Smithies, O., Gregg, R.G., Boggs, S.S., Koralewshi, M.A. and Kucherlapati, R.S., 1985, Insertion of DNA sequences into the human chromosomal -globin locus by homologous recombination, Nature, (Lond.), 317:230.

Subramani, S. and Seaton, B.L., 1988, Homologous recombination in mitotically dividing mammalian cells, in Genetic Recombination, eds. R.S. Kucherlapati and G. Smith, ASM, Washington, D.C., p.549.

Szostak, J.W., Orr-Weaver, T.L., Rothstein, R.J. and Stahl, F.W., 1983, The double strand break repair model for recombination, Cell, 33:25.

Thomas, C.A., 1966, Recombination of DNA molecules, Progress Nucleic Acids Res. Mol. Biol., 5:315.

Thomas, K.R. and Capecchi, M.R., 1986, Introduction of homologous DNA sequences into mammalian cells induces mutations in the cognate gene, Nature (Lond.), 324:134.

Thomas, K.R., Folger, K.R. and Capecchi, M.R., 1986, High frequency targeting of genes to specific sites in the mammalian genome, Cell, 44:419.

Watt, V.M., Ingles, C.J., Urdea, M.S. and Rutter, W.J., 1985, Homology requirements for recombination in Escherichia coli, Proc. Natl. Acad. Sci. USA, 82:4768.

Yannish-Perron, C., Viera, J. and Messing, J., 1985, Improved M13 phage cloning vectors and host strains: nucleotide sequences of the M13 mp18 and pUC19 vectors, Gene, 33: 103.

PROTEINS FROM YEAST AND HUMAN CELLS SPECIFIC FOR MODEL HOLLIDAY JUNCTIONS IN DNA

Stephen C. West, Kieran M. Elborough, Carol A. Parsons, and Steven M. Picksley

Imperial Cancer Research Fund
Clare Hall Laboratories
South Mimms, Herts EN6 3LD, U.K.

INTRODUCTION

Genetic recombination involves the exchange of genetic material between chromosomes to produce new assortments of alleles. As such, it affects one of the most fundamental and important components of heredity, the genome itself. Genetic rearrangements can be favourable or unfavourable, and certain forms of cancer have been linked to gene translocations. To understand the molecular basis of recombination, we have directed our efforts to try to determine how simple organisms recombine their DNA. In bacteria and lower eukaryotes, the enzymes involved in genetic recombination also play a role in the repair of DNA following irradiation or chemical damage. This overlap between recombination and repair is indicative of a need for recombinational repair, a process which ensures that the integrity of the chromosomal material is maintained.

The enzymatic process of genetic recombination requires the breakage and reunion of homologous DNA molecules. Biochemical studies have shown that the process may be sub-divided into three stages: (i) synapsis, in which homologous DNA sequences are brought together, (ii) strand exchange, in which single-strands of DNA from each parent unite to form a heteroduplex joint, and (iii) resolution, in which the crossover is cut, enabling separation of the recombinant DNA molecules. All three aspects of the process may now be studied *in vitro*. Steps (i) and (ii) are catalyzed by the RecA protein of *E. coli*, and step (iii) may be studied using endonucleases from bacteriophages T4 or T7 or by using a recently purified yeast endonuclease.

A central intermediate in the process of genetic recombination is a structure in which recombining helices are linked by two single-stranded crossovers that form a Holliday junction (Holliday, 1964). This junction is capable of branch migration along the DNA, thereby making heteroduplex DNA symmetrically in the joined DNA molecules. Although Holliday

DNA Repair Mechanisms and Their Biological Implications in Mammalian Cells
Edited by M.W. Lambert and J. Laval
Plenum Press, New York

structures have been identified from bacteria (Thompson, *et al.*, 1975; Benbow, *et al.*, 1975), yeast (Bell and Byers, 1979), and human cells (Wolgemuth and Hsu, 1980), little is known of the geometric structure of the junctions or the way in which they are resolved. Two bacteriophage-encoded nucleases are capable of cleaving Holliday junctions *in vitro* (Mizuuchi, *et al.*, 1982; Lilley and Kemper, 1984; deMassy, *et al.*, 1984), and studies with these enzymes have helped define some general properties which may be expected of cellular proteins that resolve Holliday junctions during genetic recombination. Paramount to the process of resolution is recognition of the junction within many kilobases of DNA. Since Holliday junctions are capable of branch migration, the basis of recognition is likely to be a structural feature of the junction itself rather than the flanking DNA sequences. Once bound by the nuclease, cleavage occurs by the introduction of nicks into two homologous arms of duplex DNA close to the junction. The DNA sequences close to the junction may, at this stage, exert some influence over the sites of cleavage.

RESULTS AND DISCUSSION

Characterisation of an endonuclease from Saccharomyces cerevisiae that resolves model Holliday junctions in DNA

A nuclease that cleaves artificial Holliday junctions has been partially purified from mitotically grown *S. cerevisiae* (West and Korner, 1985; West, *et al.*, 1987). The nuclease cleaves model Holliday junctions *in vitro* by the introduction of nicks in regions of duplex DNA adjacent to the crossover point. Using supercoiled plasmids that extrude inverted repeat sequences to form cruciform junctions, we observed that cleavage occurred by the introduction of nicks at positions symmetrically related across the junction. Figure 1 shows the cleavage sites for the three unrelated plasmids pColIR215 (Lilley, 1981), pIRbke8 and pIRekb/ColL (Sullivan and Lilley, 1986). In all cases, the major sites of resolution were targeted towards the extruded cruciform arms, i.e., the regions of DNA sharing sequence homology (Parsons and West, 1988).

The native molecular weight of the yeast nuclease, determined by gel filtration, is approximately 200 kDa. It is routinely purified from fermentor batches of cells, and at this stage we observe seven major bands on silver-stained SDS polyacrylamide gels ranging from 75 to 200 kDa. The enzyme has been detected in a range of strains, both wild-type and mutant. Extracts from a series of haploid X-ray sensitive yeast mutants, which include rad50, rad51, rad52, rad54, rad55, rad56 and rad57, have been assayed for the presence of this junction-specific nuclease. The autoradiograph presented in Figure 2 shows that crude protein extracts prepared from the various mutants contain the activity.

The effect of DNA sequence changes upon the site of cleavage of a cruciform structure has recently been investigated (Parsons, *et al.*, 1989). Eight derivatives of the plasmid pIRbke8 were employed, each with a single base change at, or close to, the site of cleavage by the yeast endonucle-

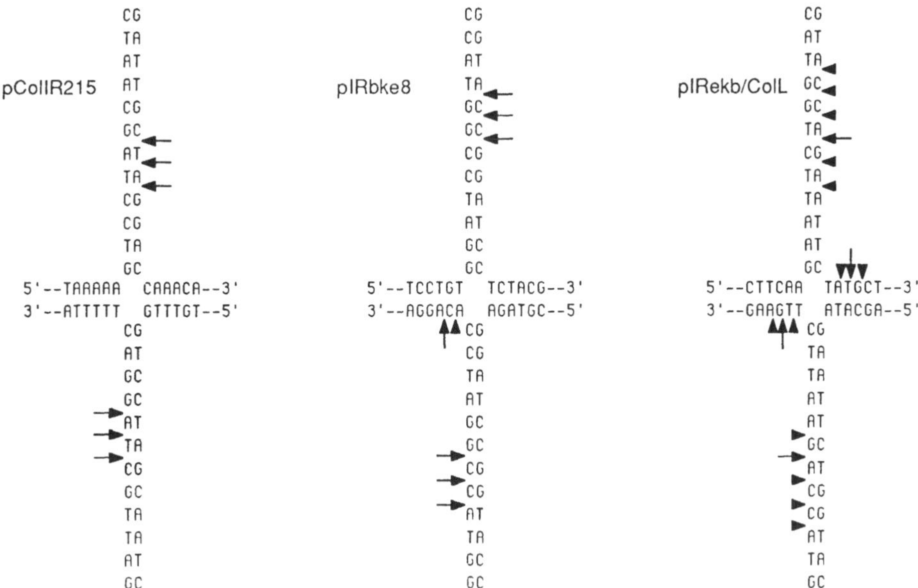

Figure 1. Cruciform-specific cleavage of three unrelated plasmid DNAs by the yeast endonuclease <u>in vitro</u>. The large and small arrows indicate sites of major and minor cutting, respectively (West, <u>et al.</u>, 1987; Parsons and West, 1988). The vertical arms represent a portion of the inverted repeat sequences that are extruded to form the cruciform junctions.

ase. Since all changes made were symmetric within the inverted repeat sequence, perfect homology between the two arms was retained while the DNA sequence was altered. In all cases, the cleavage pattern was similar to that of pIRbke8, with the major sites of cutting occurring 6-8 nucleotides from the base of the junction. We, therefore, concluded that this cleavage pattern occurred irrespective of the base sequence at this site.

Studies of Holliday junctions that utilize cruciforms as model substrates are limited by structural constraints. For example, whereas true Holliday junctions produced during homologous genetic recombination are free to branch migrate along the DNA, cruciform junctions are immobile. Secondly, cruciforms contain only one pair of homologous arms as opposed to the two pairs present in a Holliday junction. This asymmetry has been exploited to study the effects of homologous or heterologous sequences flanking the junction upon cleavage by the yeast endonuclease. It was found that a relationship exists between homology and the ability of the enzyme to promote symmetrical cleavage (Parsons and West, 1988; Parsons, <u>et al.</u>, 1989). Using DNA substrates that contain heterologous arm sequences, we showed that cleavage occurred asymmetrically across the junction. Moreover, cleavage within the heterologous arms was dependent upon DNA sequence since sequence changes or base methylation affected the cleavage pattern (Figure 3). This relationship between homology and symmetrical cleavage by the yeast nuclease has not been observed with the

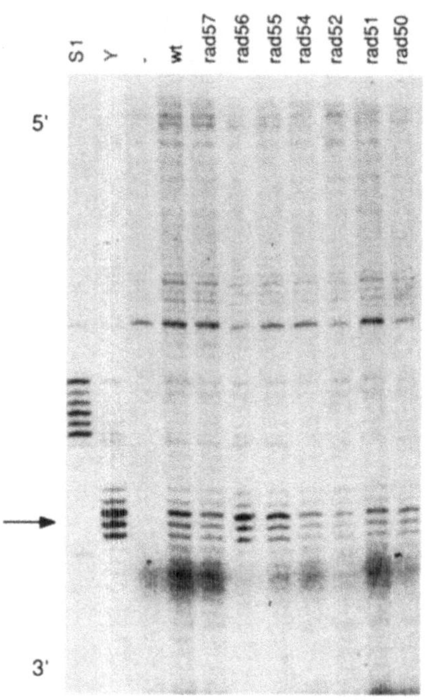

Figure 2. Specific cleavage of pColIR215 DNA by protein extracts prepared from the RAD50 epistasis group. The plasmid pColIR215 was treated with S1 nuclease, purified cruciform-specific nuclease (Y), no protein (-), or protein extracts prepared from 1 L cultures of the following strains: BJ2168 (wild-type with respect to the cruciform-specific activity), LP1365-4C (rad57), LP1400-2D (rad56), LP1363-8D (rad55), XL4-22C (rad54), S95 (rad52), LP1452-10D (rad51) and LP1398 (rad50). The DNA was then treated with EcoRI and 3'-end labelled with the Klenow fragment of DNA polymerase I. Fragments were denatured and run on a 5% polyacrylamide sequencing gels as described previously and the DNA visualised by autoradiography (West, et al., 1987). The presence of the junction-specific activity is indicated by the series of bands on the 3'-side of the junction (arrowed). The temperature sensitive strain LP13638D was grown at 23^{0}C.

T4 or T7 nucleases and may reflect a more specialized role within the cell.

The contrasting results with homologous or heterologous arms can therefore be summarised. In the case of homologous arm sequences, symmetrical cleavage is independent of base sequence and the state of methylation. With heterologous arm sequences, the sites of cleavage are asymmetric and are affected by changes to the base sequence or by base modification. Homologous and heterologous sequences are cleaved with the same efficiency by the yeast endonuclease, indicating that the junction itself is the target for binding. However, subsequent to binding, sequence context may then play a role in determining the sites of cleavage relative to the junction.

(1) Base sequence

5' TA AT 3'
AATTTTCACCTAAACTACAGAGAGTGCTGATCTTCTTTCTAC CCACTCATTTCCACTATTTTCTGTGCTCTGTGATCATTGTGC
TTAAAAGTGGATTTGATGTCTCTCACGACTAGAAGAAAGATG GGTGAGTAAAGGTGATAAAAGACACGAGACACTAGTAACACG
3' TA AT 5'

5' TA AT 3'
TGCCGGCCACGATGCGTCCGGCGTAGAG GATCTTCTTTCTAC CCACTCATTTCCACTATTTTCTGTGCTCTGT GATCCACAGGA
ACGGCCGGTGCTACGCAGGCCGCATCTCCTAG AAGAAAGATG GGTGAGTAAAGGTGATAAAAGACACGAGACACTAG GTGTCCT
3' MboI TA AT MboI 5'

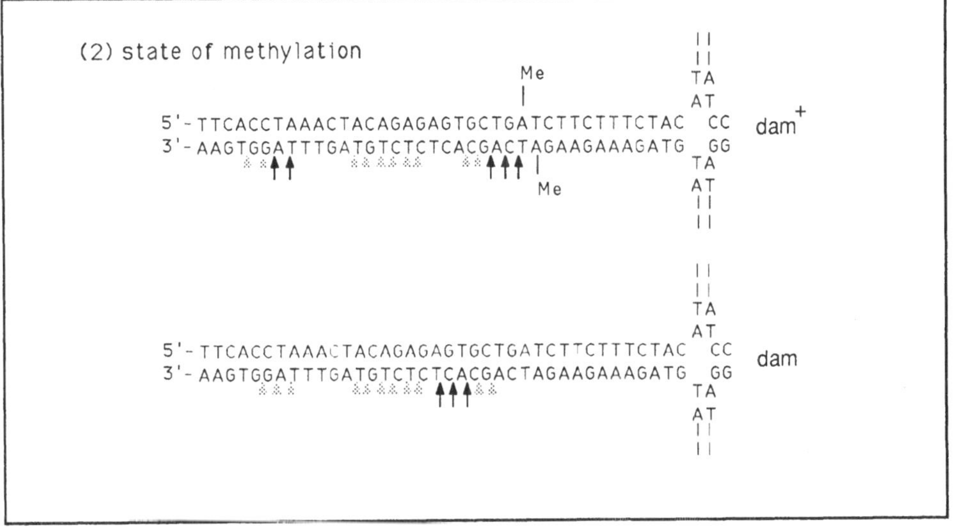

Figure 3. Factors that affect cleavage within heterologous arm sequences. (1) Changes to base sequence affect cleavage pattern. The sites of cleavage of the plasmid pXG540 were determined. Using MboI restriction sites, the region containing the cruciform was then subcloned such that the left arm sequences were replaced, whereas, the right arm sequences remained unaltered. The sites of cleavage were again determined, and found to be different in both arms (Parsons and West, 1988). (2) Changes in the state of methylation effect cleavage pattern. The cleavage sites on pXG540 DNA grown in wild-type or dam mutant E. coli were determined and found to be different according to the state of methylation at the GATC sequence close to the junction (Parsons, et al., 1989).

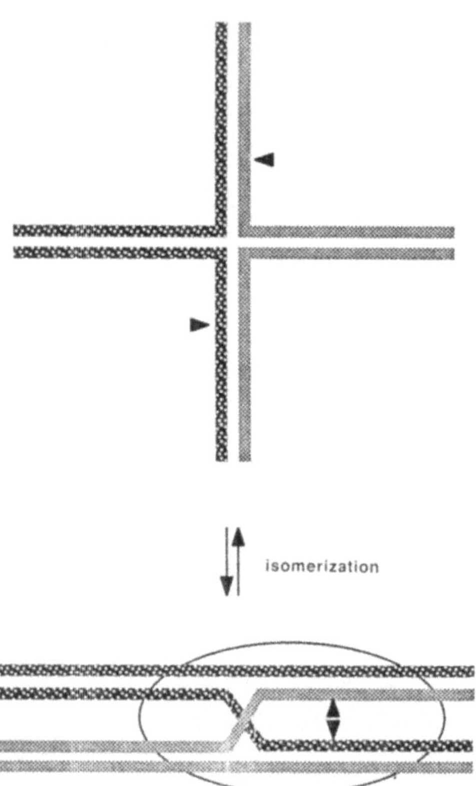

Figure 4. Symmetrical cleavage across the junction requires alignment of homologous arm sequences, presumably within the active site of the nuclease. Cleavage close to the site of the junction occurs by the introduction of nicks into strands of like polarity. For simplicity, this diagram shows the protein as a monomer (shaded area). However, the symmetry inherent in a Holliday junction may require that a resolving protein function as a dimer or tetramer.

When the arms to which the enzyme binds are homologous, symmetrical cleavage results from nicking of the two duplexes within identical sites located at a defined distance from the base of the junction (Figure 4). Cleavage at symmetrical sites several nucleotides from the base of the junction is necessary for strand realignment and the formation of duplexes containing single stranded nicks. These may subsequently be repaired by DNA ligase. Previous results show that nicking of one strand occurs rarely, indicating that cleavage within both duplex helices occurs in a concerted reaction (West, et al., 1987). When the arms are heterologous, identical sequences are not available for juxtaposition and concerted symmetrical cleavage. In this case, the nuclease may favor cleavage within sequences that show similarities or short homologies, resulting in asymmetric cleavage.

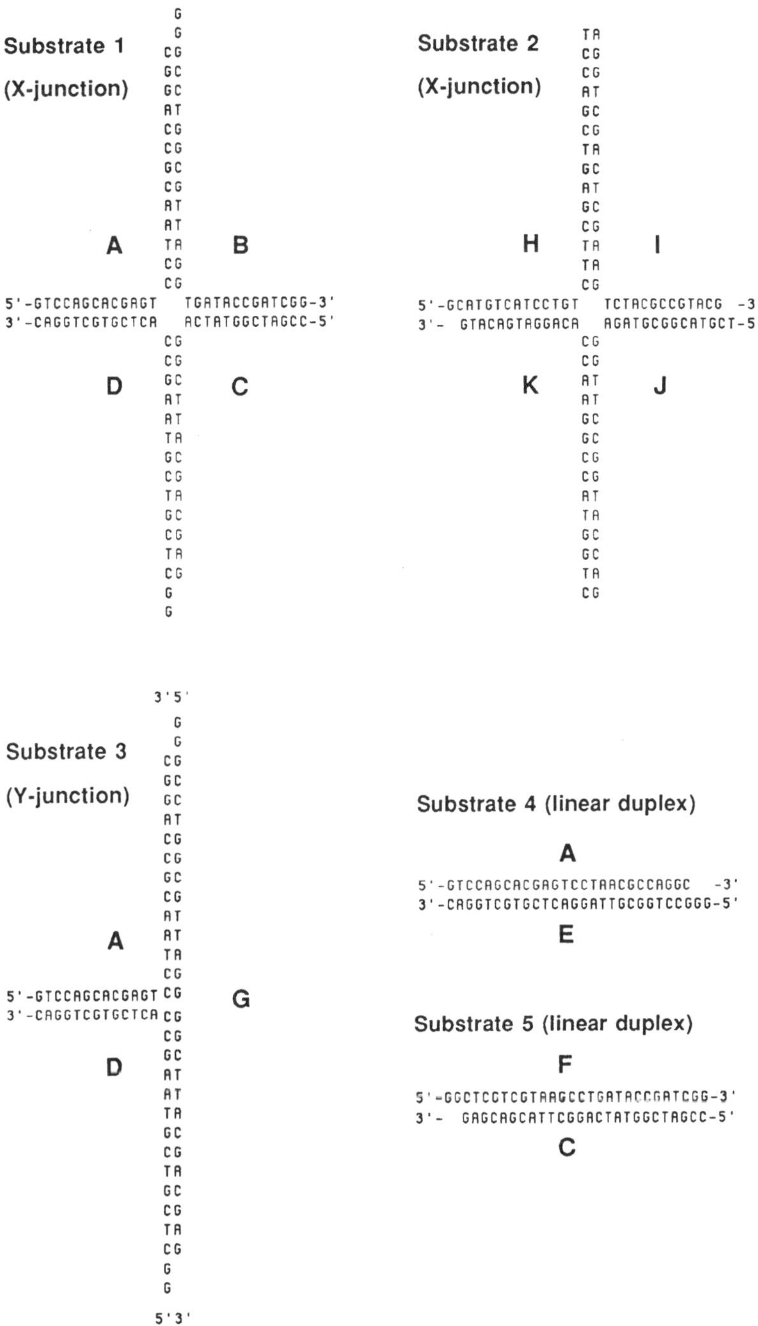

Figure 5. DNA substrates used in the gel retardation assay. Substrates 1-5 were produced by annealing equimolar amounts of oligonucleotides A-J as described by Kallenbach, et al., (1983). They were then purified by preparative gel electrophoresis, and 3'-recessed termini were labelled using the Klenow fragment of DNA polymerase I.

Specific binding of synthetic Holliday junctions by a protein from human cell extracts

Efforts to detect a nuclease from human cells that specifically cleaves Holliday junctions have so far proven unsuccessful. However, in an alternative approach, we have used gel electrophoretic binding assays to probe extracts for proteins that bind synthetic Holliday junctions. Structures analogous to Holliday junctions were produced by annealing equimolar amounts of various synthetic oligonucleotides to produce substrates 1-5 shown in Figure 5. Annealed structures were purified by preparative gel electrophoresis and ^{32}P-labelled at recessed 3'- termini using the Klenow fragment of DNA polymerase I.

The assay used to detect binding to X-junction DNA is based on the observation that DNA-protein complexes migrate slower through low ionic strength polyacrylamide gels than unbound DNA (Fried and Crothers, 1981; Garner and Revzin, 1981). To reduce the effect of non-specific binding, a simple copolymer [poly(dI-dC)· poly(dI-dC)] was included as competitor in each assay. Cell-free extracts prepared from human lymphoblasts were preincubated with an excess of competitor DNA prior to the addition of ^{32}P-labelled X-junction. After further incubation, reactions were loaded directly onto 4% polyacrylamide gels and electrophoresed. The results are shown in Figure 6. In the absence of extract, the X-junction DNA migrated as a discrete band (lane a). However, when X-junction DNA was incubated with whole cell extract, one distinct band of lower mobility was observed (lane b). The relative intensity of this band was a function of extract concentration (Elborough and West, 1988).

Since the phage nucleases that cleave X-junctions are also able to cleave Y-junctions in DNA (Jensch and Kemper, 1986; deMassy, et al., 1987), we also tested to see whether Y-junction DNA (substrate 3, Figure 5) could be retarded by the extract. The results presented in Figure 6, lanes c and d, show the Y-junction in the absence and presence of extract. Again, the DNA was bound by the cell-free extract to produce one distinct band of lower mobility. This band was of a similar mobility to that observed when X-junction DNA was used as a probe.

Several lines of evidence suggested that the observed binding was specific for DNA structure rather than DNA sequence: (1) linear duplexes (substrates 4 and 5, Figure 5) containing the DNA sequences present in the X-junction were not bound by the extract (Figure 6, lanes e-h) ; (2) a second X-junction (substrate 2, Figure 5) constructed from DNA sequences unrelated to those used in substrate 1, was also bound by the extract; (3) binding to one X-junction was competed by addition of a second X-junction but not by the same concentration of linear duplex DNA or single-stranded DNA; and (4) protein-DNA complexes were observed even at probe:non-specific competitor DNA ratios of 1:10000.

Several cultured human lymphoblast cell lines, both normal and those containing inheritable DNA repair defects, have been assayed for activity. These include HeLa, GM1953 and GM0621 (from phenotypically normal individuals), GM2249 (xeroderma pigmentosum complementation group C), and GM3403 (Bloom's syndrome). In all extracts, specific protein-DNA binding activities have been observed. A similar junction-specific protein was recently identified and partially purified from rat liver nuclei (Bianchi, 1988).

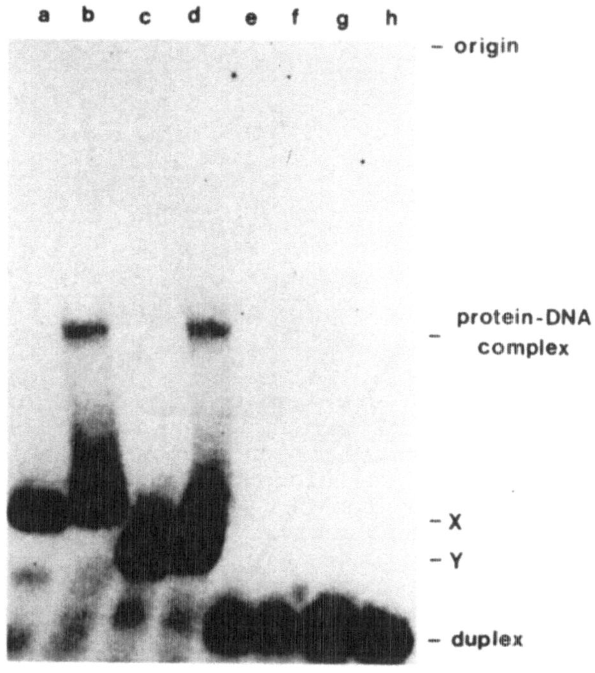

Figure 6. Specific binding to X- and Y-structures in DNA by whole cell extract from GM3403 lymphoblasts. Cell-free extract was incubated with [poly(dI-dC)·-poly(dI-dC)] prior to addition of labelled X-junction DNA. Following further incubation, samples were loaded directly onto low ionic strength polyacrylamide gels and electrophoresed as described (Elborough and West, 1988). Lanes a and b, X-junction DNA (Figure 5, substrate 1) ; Lanes c and d, Y-junction DNA (substrate 3) ; Lanes e and f, linear duplex DNA (substrate 4) ; Lanes g and h, linear duplex DNA (substrate 5). Protein extract was included in the reactions of lanes b, d, f and h.

The junction-specific protein from HeLa cells is currently being purified. The activity is known to bind, and elute from single-stranded DNA-cellulose and DEAE-Biogel. Fractions from the DEAE column are shown in Figure 7. At the present time, we have little knowledge of the role this protein may play in nucleic acid metabolism.

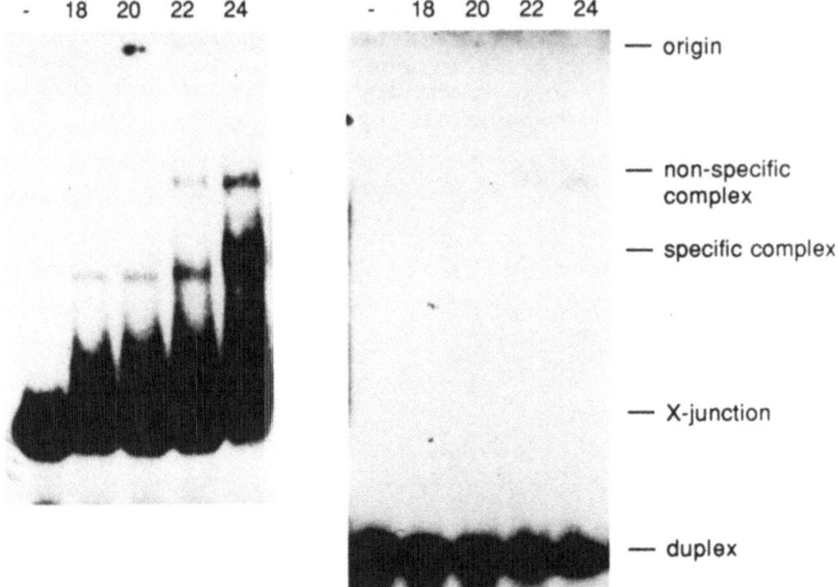

Figure 7. Cell extracts were prepared from 30 L of HeLa cells as described (Elborough and West, 1988). Proteins were fractionated by chromatography using single-stranded DNA-cellulose followed by DEAE-Biogel. Fractions from the DEAE column were assayed for X-junction binding activity, and this photograph shows a portion of the elution profile. (Left) Gel retardation assay using X-junction DNA as a probe; (Right) the same fractions assayed using linear duplex DNA as a probe.

ACKNOWLEDGEMENTS

We are grateful to Dr. L. Prakash and Dr. R. Mortimer for providing the yeast RAD50 epistasis group, and to Dr. D. Lilley for supplying the plasmids used in these studies.

REFERENCES

Bell, L., and Byers, B., 1979, Occurrence of crossed strand-exchange forms in yeast during meiosis, Proc. Natl. Acad. Sci. USA, 76;3445.

Benbow, R.M., Zuccarelli, A.J., and Sinsheimer, R.L., 1975, Recombinant DNA molecules of ϕX174, Proc. Natl. Acad. Sci. USA, 72;235.

Bianchi, M.E., 1988, Interaction of a protein from rat liver nuclei with cruciform DNA, EMBO J., 7:843.

deMassy, B., Studier, F.W., Dorgai, L., Appelbaum, E. and Weisberg, R.A., 1984, Enzymes and sites of genetic recombination: Studies with gene 3 endonuclease of phage T7 and with site affinity mutants of phage lambda, Cold Spring Harb. Symp. Quant. Biol., 49:715.

deMassy, B., Weisberg, R.A. and Studier, F.W., 1987, Gene 3 endonuclease of bacteriophage T7 resolves conformationally branched structures in double-stranded DNA, J. Mol. Biol., 193:359.

Elborough, K.M., and West, S.C., 1988, Specific binding of cruciform DNA structures by a protein from human extracts, Nucl. Acid Res., 16:3603.

Fried, M., and Crothers, D.M., 1981, Equilibria and kinetics of lac repressor-operator interactions by polyacrylamide gel electrophoresis, Nucl. Acid. Res., 9:6505.

Garner, M.M., and Revzin, A., 1981, A gel electrophoresis method for quantifying the binding of proteins to specific DNA regions: application to components of the Escherichia coli lactose operon regulatory system, Nucl. Acid. Res., 9:3047.

Holliday, R., 1964, A mechanism for gene conversion in fungi, Genet. Res. Camb. , 5;282.

Jensch, F., and Kemper, B., 1986, Endonuclease VII resolves Y-junctions in branched DNA in vitro, EMBO J., 5:181.

Kallenbach, N.R., Ma, R., and Seeman, N., 1983, An immobile nucleic acid junction constructed from oligonucleotides, Nature, 305:829.

Lilley, D.M.J., 1981, Hairpin loop formation by inverted repeats in supercoiled DNA is a local and transmissible property, Nucl. Acid Res., 9:1271.

Lilley, D.M.J., and Kemper, B., 1984, Cruciform-resolvase interactions in supercoiled DNA, Cell, 36:413.

Mizuuchi, K., Kemper, B., Hays, J. and Weisberg, R.A., 1982, T4 endonuclease VII cleaves Holliday structures, Cell, 29:357.

Parsons, C.A., and West, S.C., 1988, Resolution of model Holliday junctions by yeast endonuclease is dependent upon homologous DNA sequences, Cell, 52:621.

Parsons, C.A., Murchie, A.I.H., Lilley, D.M.J., and West, S.C., 1989, Resolution of model Holliday junctions by yeast endonuclease: Effect of DNA structure and sequence, submitted for publication.

Thompson, B.J., Escarmis, C., Parker, B., Slater, W.C., Doniger, J., Tessman, I., and Warner, W.C., 1975, Figure-8 configuration of dimers of S13 and ϕX174 replicative form DNA, J. Mol. Biol., 91;409.

Sullivan, K.M., and Lilley, D.M.J., 1986, A dominant influence of flanking sequences on a local structural transition in DNA, Cell, 47:817.

West, S.C., and Korner, A., 1985, Cleavage of cruciform DNA structures by an activity from Saccharomyces cerevisiae, Proc. Natl. Acad. Sci. USA, 82:6445.

West, S.C., Parsons, C.A., and Picksley, S.M., 1987, Purification and properties of a nuclease from Saccharomyces cerevisiae that cleaves DNA at cruciform junctions, J. Biol. Chem., 262:12752.

Wolgemuth, D.J., and Hsu, M.T., 1980, Visualization of genetic recombination intermediates of human adenovirus type 2 DNA from infected HeLa cells, Nature, 287;168.

CHARACTERIZATION OF A STRAND TRANSFERASE ACTIVITY FROM HUMAN CELLS

Era Cassuto[1] and Paul Howard-Flanders[2]

[1]INRA-CNRS, Génétique Microbienne, Domaine de Vilvert, 78350 Jouy en Josas, France

[2]Department of Molecular Biophysics and Biochemistry, Yale University, New Haven, CT 06511, USA

INTRODUCTION

In procaryotes and in lower eucaryotes, mutants deficient in homologous recombination display a high sensitivity to a variety of DNA-damaging agents. In these organisms, recombination is clearly associated with DNA repair. In E. coli, for example, the recA gene controls homologous recombination, but its expression is also required for numerous processes of recovery from DNA-damage. So far, none of the repair mutants identified in mammalian cells have been shown to be deficient in recombination. This is not altogether surprising, since attempts to merely detect recombination activities in higher eucaryotes have only recently become successful. The development of shuttle vectors and drug resistance markers has allowed the demonstration of recombination between both chromosomal and extrachromosomal genes. Most studies have monitored, in whole cells, the reconstruction of a functional gene from pairs of plasmids or viruses carrying mutant or truncated genes. The same type of investigation has recently been applied to cell extracts and revealed that, as in yeast, the introduction of double-strand breaks in plasmid DNA substrates markedly increased the frequency of exchanges. All the above studies, however, were designed for the detection of complete recombinants rather than intermediates, and provide little insight into the mechanism of the recombination process.

Several groups have attempted to detect recombination enzyme activities in crude extracts of mammalian cells. The studies on RecA protein, the recombination enzyme of E. coli, serve as a paradigm for the search for similar enzymes in eucaryotic cells. The bacterial enzyme promotes homologous pairing and strand exchange between suitable DNA substrates. Both 3-strand and 4-strand reactions have been investigated. The substrates for the 3-strand reaction consist of circular

DNA Repair Mechanisms and Their Biological Implications in Mammalian Cells
Edited by M.W. Lambert and J. Laval
Plenum Press, New York

single-stranded and homologous full-length linear duplex molecules which are converted to nicked circular heteroduplexes and linear single-stranded species (Cox and Lehman, 1981a). This reaction requires single-strand binding protein (SSB) to overcome the effects of secondary structure in the singlestranded DNA (ssDNA) (Cox and Lehman, 1981b; Cox and Lehman, 1982; West, et al., 1982). In the 4-strand reaction, which has no requirement for SSB, reciprocal exchange takes place between gapped circular duplexes and full length linear duplexes with a 3' end complementary to the single-strand in the gap (West, et al., 1982a). The products are nicked circular and linear heteroduplexes, of which the linear has a 5' single-stranded tail. Both reactions are ATP-dependent and both allow strand exchange to proceed to completion without topological hindrance. In contrast, hindered 3-strand reactions, such as the formation of D-loops from covalent circular duplexes and single-stranded fragments, do not lead to complete strand exchange but provide a convenient assay for homologous pairing (Shibata, et al., 1979).

RecA protein is required in stoichiometric amounts for efficient reaction. It binds to ssDNA or gapped DNA to form complexes which are ready to pair with duplex molecules at any position. When the 3' end of the duplex is paired with the ssDNA, strand exchange is initiated and proceeds in the 5' to 3' direction with respect to the ssDNA (West, et al., 1982a; Cox and Lehman, 1981b; Kahn, et al., 1981). Proteins that catalyze pairing and strand exchange have been purified from Ustilago maydis (Kmiec and Holloman, 1982) and Saccharomyces cerevisiae(Kolodner, et al., 1987). The former catalyzes ATP-dependent pairing and strand exchange in the 3' to 5' direction (Kmiec and Holloman, 1983), whereas the yeast enzyme promotes strand exchange with the same polarity as the bacterial enzyme, albeit in an ATP-independent manner.

The search for RecA-like activities in mammalian cells has led to conflicting results. Homologous pairing has been detected in extracts from Bloom syndrome fibroblasts (Kenne and Lindquist, 1984), EJ carcinoma cells (Ganea, et al., 1987) and human B cells (Hsieh, et al., 1986). The latter activity does not require a nucleotide cofactor. ATP-dependant pairing and strand exchange activities were partially purified from HeLa cells (Cassuto, et al., 1987) and human T lymphoblast cells (Fishel, et al., 1988), the latter promoting branch migration in the 3' to 5' direction. In this paper, we report on the further purification and characterization of the strand transferase activity from HeLa cells and show that it promotes strand exchange in 3-strand and 4-strand reactions, in an ATP-dependent manner, and with the same polarity as the bacterial RecA protein.

METHODS

Preparation of extracts

8 liters of HeLa S3 cells were supplemented with either 40 ng/ml TPA (12-0-Tetradecanoyl Phorbol-13-Acetate) or 1 µg/ml Mitomycin C 18 hours before harvesting. The cells were centrifuged at 3000 x g for 10 minutes, washed twice with 0.15 M NaCl, 10 mM sodium phosphate, pH 7.5, and resuspended in 5

mM potassium phosphate, pH 7, 2 mM $MgCl_2$, 0.5 mM DTT, 0.1 mM EDTA, 1 mM β-mercaptoethanol, 1 mM PMSF (cell extraction buffer). The cells were swollen at 0^0C, broken with a Dounce homogenizer and centrifuged at 200 x g for 10 minutes. The nuclei were washed once with cell extraction buffer, without magnesium, and resuspended in the same buffer. EDTA was added to 4 mM. The nuclei were lysed with 2 M NaCl, 50 mM Tris-HCl, pH 7.5, 10 mM β-mercaptoethanol, and 1 mM PMSF. The lysed nuclei were spun at 25K for 30 minutes in a type 35 Beckman rotor. The supernatant fraction was dialyzed versus 20 mM Tris-HCl, pH 7.5, 2 mM DTT, 0.1 mM EDTA, 10% glycerol, 1 mM PMSF, 0.5 M NaCl and passed over a DEAE-Biogel A column equilibrated with the same buffer. Ammonium sulfate (0.313 g/ml) was added slowly with stirring. The resulting precipitate was collected by centrifugation, resuspended in 20 mM Tris-HCl, pH 7.5, 2 mM DTT, 0.1 mM EDTA, 10% glycerol, 1 mM PMSF (buffer A) and dialyzed against the same buffer. The yield of protein at this stage was close to 120 mg.

Purification steps

The protein solution (fraction I) was applied to a DEAE-Sepharose column (1 ml/10 mg protein) equilibrated with buffer A and eluted with a gradient from 0.05 to 1 M NaCl in buffer A. The fractions were assayed for ssDNA-dependent ATPase activity and also for topoisomerase and endo- and exo-nuclease activities on ss and double-stranded (ds) DNA. The active fractions, which eluted between 0.2 and 0.35 M NaCl (fraction II, 20 mg protein), were pooled, dialyzed against buffer A and applied to a CM-Sepharose column developed with a gradient from 0.05 to 0.7 M NaCl in buffer A. The fractions were assayed for strand transfer activity as well as ATPase activity. The active fractions eluted between 0.2 and 0.25 M NaCl (fraction III, 2 mg protein), were pooled, dialyzed against buffer A and loaded onto a 0.3 ml ssDNA-agarose column developed with a gradient from 0.05 to 0.4 M NaCl in buffer A. The fractions were assayed for strand transfer activity. The pooled active fractions (fraction IV, 0.3 mg protein) were dialyzed against buffer A, supplemented with BSA (100 μg/ml), and stored at -80^0C. The activity was stable under these conditions for 2 weeks. Fraction IV was found to be free of topoisomerase activity and of endo- and exo-nuclease activity on ss and dsDNA.

Assay of exonuclease activity

Mixtures contained in 100 μl: 17 μM ^{32}P linear ϕX174 single-strand or duplex DNA, 25 mM Tris-HCl, pH 7.5, 10 mM $MgCl_2$, 2 mM DTT, 2 mM ATP and 50 μg protein fraction. After 60 minutes incubation at 37^0C, the mixtures were supplemented with 0.4 ml salmon sperm DNA at 2 mg/ml and 0.5 ml 1.75% perchloric acid, centrifuged, and the supernatant was counted for ^{32}P radioactivity in liquid fluor. This method was sensitive enough to detect the release of 0.05% of the input radioactivity, or less than 3 nucleotides per ϕX174 DNA strand.

Assay of endonuclease activity

The endonucleolytic activity of the protein fractions was measured on ϕX174 ds linear and ss circular DNA as follows:

mixtures containing 17 µM linear ϕX174 duplex DNA, 25 mM Tris-HCl, pH 7.5, 10 mM $MgCl_2$, 2 mM DTT, 2 mM ATP and 500 µg/ml protein fraction were incubated at 37^{0}C for 60 minutes. The linear dsDNA was precipipated with ethanol, denatured in 0.1 N NaOH, 1 mM EDTA, 8% sucrose, and 0.05% bromophenol blue, and applied to a neutral 1% agarose gel. Its mobility was compared to that of untreated linear dsDNA similarly denatured. For ssDNA the alkali denaturation was omitted.

ATPase assay

Reaction mixtures (50 µl) contained 20 mM Tris-HCl, pH 7.5, 10 mM $MgCl_2$, 0.1 mg/ml BSA, 1 mM α^{32}P ATP, 1 to 5 µl of protein fraction, and either no DNA, 1 µg of native calf thymus DNA or 1 µg of ϕX174 viral DNA. The mixtures were incubated at 37^{0}C for 60 minutes and the reaction stopped with EDTA. 1 to 5 µl of each fraction were spotted onto Cellulose 300 polyethyleneimine thin layer plates, developed in 1 M formic acid, 0.5 M LiCl and air dried. The plates were then covered with cellophane and exposed to X-ray films. For each fraction, the amount of ^{32}P label found at the position of ADP in the presence of ssDNA was compared visually to the levels observed in the presence of dsDNA or in the absence of DNA. Although not suitable for quantitative measurements, this assay allowed us to test very rapidly a large number of fractions for ssDNA-dependent ATPase activity.

DNA substrates

Gapped DNA was prepared by annealing of a purified unlabeled PstI-AvaII ϕX174 fragment to unlabeled ϕX174 viral DNA (West, et al., 1982a), to produce a gap of 340 nucleotides. The gapped DNA was purified by gel electrophoresis and electroelution.

Strand exchange assays

Strand exchange assay mixtures (20 µl) contained 25 mM Tris-HCl, pH 7.5, 2 mM DTT, 10 mM $MgCl_2$, and 2mM ATP. For the 3-strand reaction, 60 ng of ^{32}P ϕX174 linear ss-DNA were reacted with 30 ng of ϕX174 viral DNA. To study reciprocal (4-strand) exchange, reaction mixtures (20 µl) contained 60 ng gapped DNA, and 70 ng of uniformly labeled ^{32}P form III DNA obtained by digestion of form I ϕX174 DNA with either PstI, AvaII, or SstII.

RESULTS

3-strand reaction

The DNA substrates and products of this reaction are illustrated in Figure 1, as it was described for RecA protein. A complete strand transfer yields a nicked heteroduplex and releases the (+) strand of the linear duplex.

The results of incubation of the same substrates with fraction IV are shown in Figure 2. The appearance of the ^{32}P label present in the linear duplex at the position of form II is indicative of the formation of a nicked heteroduplex (lane e). This species was not observed if ϕX174 viral DNA was

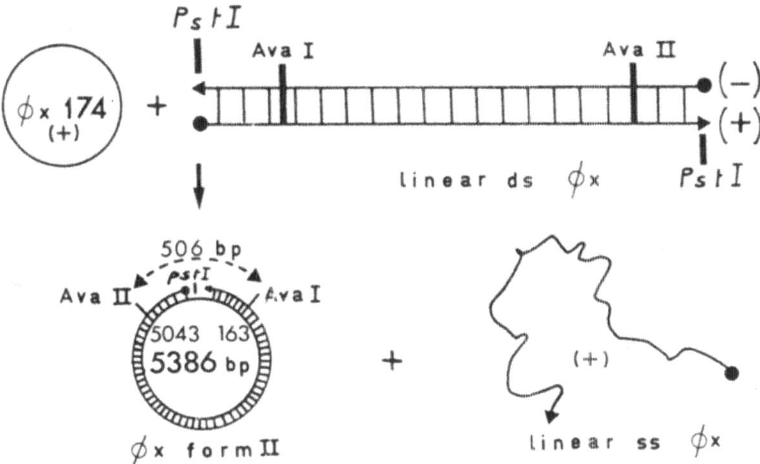

Figure 1. Strand transfer (3-strand reaction) promoted by RecA protein.

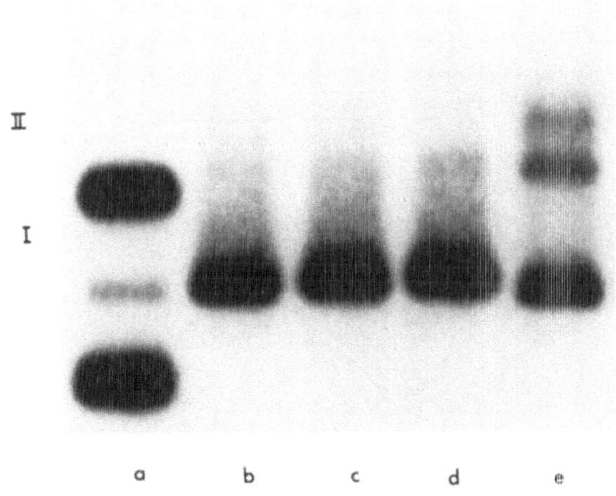

Figure 2. Heteroduplex formation (3-strand reaction) promoted by the human strand transferase.

replaced by M13 viral DNA and could not therefore be due to the circularization of the linear duplex by contaminating ligase (lane c). Nor was it observed in the absence of Mg++ or ATP (lanes b and d). However we did not detect the release of the ^{32}P labeled linear (+) strand.

In order to determine whether a full-length heteroduplex was formed in the reaction, we measured the restoration of 2 restriction sites, AvaI and AvaII, flanking the PstI site. Treatment of ϕX174 form II with AvaI and AvaII yields 2 fragments of 4880 and 506 bp, which in this case can only be produced if strand exchange has proceeded through both the AvaI and AvaII sites. We introduced the ^{32}P label in the viral DNA only, so that any new band detected by autoradiography after treatment with AvaI and AvaII would necessarily originate from digestion of a heteroduplex product. As anticipated the radioactivity in the form II product was quantitatively recovered in 2 bands located at the expected positions (about 5000 and 500 bp).

4-strand reaction

The combinations of DNA substrates used in this experiment are illustrated in Figure 3. The gapped DNA was made from ϕX174 DNA as described in Methods and contained a defined 340 nucleotide gap in the (-) strand, between the AvaII and PstI sites.

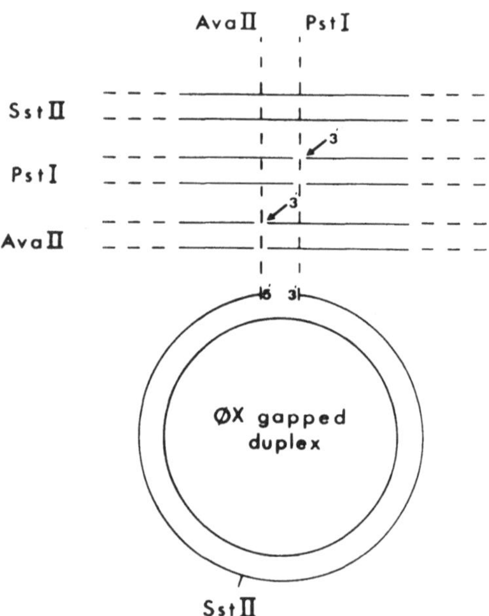

Figure 3. Substrates used in the 4-strand reaction. The arrow indicates the position of the 3' end of the (-) linear strand with respect to the gap.

The gapped DNA was incubated with fraction IV and ^{32}P labeled AvaII, PstI or SstII linearized ϕX174 DNA. The (-) strand of the linear species had, respectively, a 3' free end, a 5' free end, and no free end complementary to the gap. The appearance of the ^{32}P label at the position of form II is indicative of the formation of full length heteroduplex DNA resulting from an exchange of strands between the gapped and the linear duplex. The other product of the reciprocal exchange, a heteroduplex gapped linear molecule with a 5' tail, cannot be seen since it comigrates with the unreacted linear duplex. Figure 4 shows the results obtained with the various linear species. A substantial amount of ^{32}P-labeled form II DNA was seen with the AvaII linear duplex (lane d). This band was not detected in the absence of gapped DNA (lane C), and could not therefore be due to contamination by ligase. Nor was it detected in the absence of ATP (lane g) or with the SstII linear duplex (lane f). Only trace amounts of form II heteroduplex could be seen in the reaction with the PstI linear duplex (lane e).

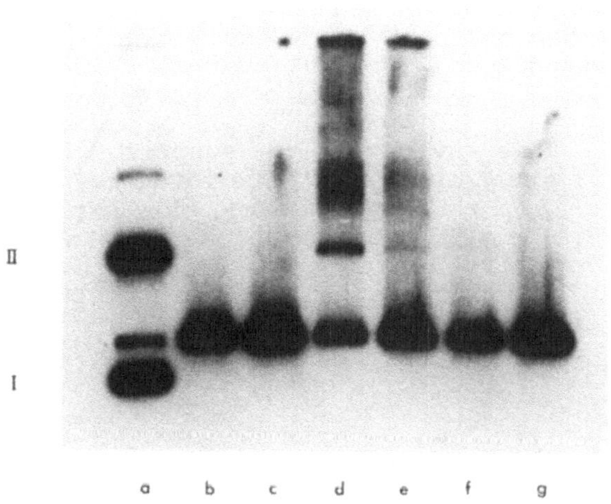

Figure 4. Heteroduplex formation (4-strand reaction) promoted by the human strand transferase.

DISCUSSION

We have described the partial purification of an activity for homologous pairing and 3- and 4-strand exchange reactions. Crude extracts were prepared from exponentially growing HeLa cells treated with Mitomycin C or TPA 18 hours before harvesting. In pilot tests, this treatment was found to increase the activity several fold. Purification by column chromatography was based on assays for ssDNA-dependent ATPase and strand transfer between circular ssDNA and linear dsDNA. The strand transfer (3-strand) assay provides no information on directionality. Because it requires very little substrate construction, it was used in the present work to follow the strand exchange activity in the last purification steps, when interference by nucleases and topoisomerases was below detection level. The 4-strand assay, somewhat more difficult to perform, allows the determination of the polarity of branch migration. In addition, it may be more relevant to recombination in living cells, which for the most part takes place between duplex molecules. The reaction mediated by fraction IV produced substantial yields of nicked circular heteroduplex with the AvaII linear duplex that had a 3' end complementary to the single-strand in the gap, but only traces with the PstI linear duplex with a 5' complementary end. No heteroduplex product was found in reactions where both ends of the linear duplex were at a distance from the single-strand region (SstII linear duplex), or in the absence of ATP. These results show that the human activity generates heteroduplexes with the same polarity as the bacterial RecA protein, in the 5' to 3' direction with respect to the single-strand in the gap.

It is at present difficult to compare the activity described above to other mammalian activities which have been tested for homologous pairing rather than strand exchange (Ganea, et al., 1987; Kenne and Lindquist, 1984) promote strand exchange in the opposite direction to that of the bacterial enzyme (Fishel, et al., 1988), or have no cofactor requirement (Hsieh, et al., 1986). In some cases, the difference may be due to the type of assay used and further tests will be needed to sort out what may only appear to be conflicting results.

There is some evidence to support the notion that recombinational strand exchanges are involved in recovery from damage in mammalian cells. However, the experiments are as difficult to perform as the results are to interpret, in the absence of a genetic framework. It is also important to remember that direct extrapolation of molecular mechanisms demonstrated in E.coli may be a thankless exercise.

REFERENCES

Cassuto E, Lightfoot LA, and Howard-Flanders P, (1987) Partial purification of an activity from human cells that promotes homologous pairing and the formation of heteroduplex DNA in the presence of ATP. Mol Gen Genet. 208:10-14.

Cox MM and Lehman IR (1981A). RecA protein of Escherichia coli promotes branch migration, a kinetically distinct phase of DNA strand exchange. Proc Natl Acad Sci USA. 78:3433-3437.

Cox MM and Lehman IR (1981B). Directionality and polarity in RecA protein-promoted branch migration. Proc Natl Acad Sci USA. 78:6018-6022.

Cox MM and Lehman IR (1982). RecA protein-promoted DNA strand exchange. J Biol Chem. 257:8523-8532.

Fishel RA, Detmer K and Rich A (1988). Identification of homologous pairing and strand-exchange activity from a human tumor cell line based on Z-DNA affinity chromatography. Proc Natl Acad Sci USA. 85:36-40.

Ganea D, Moore P, Chekuri L and Kucherlapati R (1987). Characterization of an ATP dependent DNA strand transferase from Human Cells. Mol Cell Biol. 7:3124- 3130.

Hsieh P, Meyn MS and Camerini-Otero D (1986). Partial purification and characterization of a recombinase from human cells. Cell. 44:885-894.

Kahn R, Cunningham RP, DasGupta C and Radding CM (1981). Polarity of heteroduplex formation promoted by Escherichia coli RecA protein. Proc Natl Acad Sci USA. 78:4786-4790.

Kenne K and Lindquist S (1984). A DNA-recombinogenic activity in human cells. Nucl Acids Res. 12:3057-3068.

Kmiec E and Holloman WK (1982). Homologous pairing of DNA molecules promoted by a protein from Ustilago. Cell. 29:367-374.

Kmiec E and Holloman WK (1983). Heteroduplex formation and polarity during strand transfer promoted by Ustilago Recl protein. Cell. 33:857-864.

Kolodner R, Evans DH and Morrison PT (1987). Purification and characterization of an activity from Saccharomyces cerevisiae that catalyzes homologous pairing and strand exchange. Proc Natl Acad Sci USA. 84:5560-5564.

Shibata T, Cunningham RP, DasGupta C and Radding CM (1979). Homologous pairing in genetic recombination: complexes of RecA protein and DNA. Proc Natl Acad Sci USA. 76:5100-5104.

West SC, Cassuto E and Howard-Flanders P (1981). Heteroduplex formation by RecA protein: Polarity of strand exchanges. Proc Natl Acad Sci USA. 78:6149-6153.

West SC, Cassuto E and Howard-Flanders P (1982A). Postreplication repair in E. coli: strand exchange reactions of gapped DNA by RecA protein. Mol Gen Gen. 187:209-217.

West SC, Cassuto E and Howard-Flanders P (1982). Role of SSB protein in RecA promoted branch migration reactions. Mol Gen Genet. 186:333-339.

MISMATCH REPAIR IN MAMMALIAN CELLS: APPROACHES TO THE IN VITRO STUDY OF DNA MISMATCH CORRECTION REACTIONS

William C. Summers and Peter M. Glazer

Radiobiology Laboratories
Yale University School of Medicine
New Haven, CT. 06510 U.S.A.

INTRODUCTION

Mismatches in duplex DNA can occur as the result of replication infidelity, damage to individual bases in DNA, and the formation of joint molecules in the process of genetic recombination. In one sense, the correction of mismatches can be considered as a subclass of the general repair reactions that exist to restore DNA to its intact duplex state. Several reactions which repair mismatched DNA are known to exist in mammalian cells: the editing by DNA polymerases is a specific reaction to detect and remove terminal mismatches; excision repair of damaged bases is another reaction which detects some types of base mismatches.

However, what is the evidence that non-terminal mismatches occuring between normal, physiologically occurring bases (A,G,C and T) can be recognized and corrected in mammalian cells? First there are data from the genetic analysis of simpler eucaryotes that suggest, by analogy, that mismatch correction might occur in mammalian cells. These data (reviewed in Rossignol, et al. 1988) come from non-Mendelian segregation of markers during meiosis (termed gene-conversion) that implies the existence of mismatched heteroduplex DNA. This mismatched heteroduplex DNA is thought to be the product of the recombination and "crossing-over" believed to be required for chromosome pairing at meiosis (Meselson and Radding, 1975). More direct experimental evidence for mismatch repair in mammalian cells comes from analysis of the fate of synthetic heteroduplex DNA introduced into mammalian cells by DNA transfection or microinjection. Thus, Miller, Cooke and Fried (1975) analyzed the results of transfection with trans-heteroduplexes of polyoma DNA and concluded that mammalian cells are remarkably efficient at correcting mismatches. Similar results and conclusions were obtained by Glazer (1987) who used shuttle vector DNA and a sensitive bacterial assay for detecting the repaired molecules. These results could not rigorously exclude the possibility of recombination between replicated DNA, however. In an attempt

DNA Repair Mechanisms and Their Biological Implications in Mammalian Cells
Edited by M.W. Lambert and J. Laval
Plenum Press, New York

to circumvent this difficulty, Folger, Thomas and Cappecchi (1985) injected non-replicating trans-heteroduplexes of non-viral origin; they, too observed a very efficient repair of mismatches. This work, too, could not exclude the possibility of post-replication recombination (after integration) between two single mutant forms of the marker used.

More recently, Hare and Taylor (1985) and Brown and Jiricny (1987, 1988) have used single heteroduplex molecules of SV40 and studied the bursts of virus from single infective centers. Pure bursts indicate repair of the heteroduplex while mixed bursts are interpreted as arising from segregation of unrepaired heteroduplex to give progeny of both genotypes. Both of these studies showed that mismatch correction is very efficient in mammalian cells. Brown and Jiricny (1988) determined the DNA sequence at an unselected site of the mismatch and were able to measure the efficiency of correction as well as the preference for correction, i.e., which strand or base was used as template for the "correction". These results demonstrated, for the first time, that some specificity exists for the mismatch repair reaction. With over 95 percent of the G/T mismatches being repaired to G/C rather than A/T, it is hard to argue that random processes such as nick-translation or excision repair generated by other lesions are responsible for the correction of the mismatches. It appears that for G/T mismatches, at least, the repair is provoked by the lesion and the reaction specifically selects the G-containing strand as the template for correction.

If mismatch correction reactions occur in mammalian cells, it should be possible to determine the enzymatic steps which are involved by studying the reaction _in vitro_. As a first step toward this goal, we have searched for conditions in which the mismatch repair reaction will occur in cell-free extracts of mammalian cells. We found (Glazer, _et al_. 1987) that whole-cell-extracts of the sort used to carry out transcription reactions were able to correct trans-heteroduplex DNA with G/T and C/A mismatches. As will be described below, these substrates are less than optimal and on-going studies employ improved procedures for construction of single heteroduplex substrates.

USE OF _E. COLI_ TO ASSAY HETERODUPLEX CORRECTION

We reasoned that we needed a sensitive and simple assay for the formation of corrected DNA molecules after the _in vitro_ repair reaction. In the absence of _a priori_ information about the efficiency of the _in vitro_ reaction, we set up a bioassay for the functional _supF_ gene (since our mismatches were constructed in that DNA sequence, see Glazer, Sarkar and Summers, 1986). Since we did not want the unprocessed heteroduplexes to be repaired by _E. coli_, we first constructed bacterial hosts that were deficient in the genes (_mut_H,U) that control the methyl-directed mismatch repair pathway in _E. coli_. This strain showed little reduction in the overall processing of the heteroduplex (see Table 1). A combination of _mut_H,U and _rec_A, however, reduced the background in this bioassay by about ten-fold.

Since we wished to test the role of methylated DNA in

Table 1. Processing of trans-heteroduplex in E. coli strains.

Strain	Relevant markers	Percent Blue[1]
SY204	wild-type	2.0 (166/8350)
EG826	ssb-1	0.86 (17/1970)
SY302	rec A56	0.90 (24/2661)
SY208	mut H3, mut U4	2.1 (118/5730)
SY209	rec A56, mut H3, mut U4	0.17(37/22235)

The trans-heteroduplexes were mixtures of the following two duplexes:

```
--c----------C---         ---T----------t--
                    and
--A----------a---         ---g----------G--
```

where the upper case letters designate the wild-type base at each mismatch. The mismatches in the supF gene were separated by 61 base pairs.
[1]Data adapted from Glazer, et al., 1987.

directing mismatch repair in mammalian cell extracts, as suggested by the results of Hare and Taylor (1985), we also

Table 2. Processing of hemi-methylated heteroduplex in E. coli strains.

Strain	Heteroduplex[1]	Percent WT correction (Blue)/(Blue+ White)
SY204 (wt)	- / +M	99
	-M / +	24
EG826 (ssb)	- / +M	50
	-M / +	47
SY302 (recA)	- / +M	83
	-M / +	9
SY208 (mutH,U)	- / +M	51
	-M / +	74
SY209 (mutH,U, recA)	- / +M	43
	-M / +	46

[1] Mismatched heteroduplexes were single mismatches constructed by annealing p3AC DNA (Sarkar, Dasgupta and Summers, 1984) prepared from E. coli SY204 (wt) or from E. coli GM272 (dam, dcm, hsdS) to obtain DNA devoid of methylated bases. (-) and (+) designates the mutant (T) and wild-type (G) strand, respectively and (M) indicates that the strand has wild-type methylation.

tested for the processing of hemimethylated DNA in these E. coli strains (Table 2). For the G/T heteroduplex with one or the other strand methylated by the E. coli dam methylase, as expected, the mutH,U defect abolished the role of methylation in determining the direction of correction (Wagner and Meselson, 1976; Meselson, 1988). A new finding was that the Ssb function is also needed in the methyl-directed mismatch correction reaction. Fischel and Kolodner (1983) also reported evidence for methyl-independent mismatch correction in E. coli.

PROCESSING OF HETERODUPLEXES IN MAMMALIAN CELLS AND CELL-FREE EXTRACTS

To test if the trans-heteroduplexes in the supF gene (see Table 3) in the shuttle vector p3AC (Sarkar, Dasgupta and Summers, 1984) could be corrected in mammalian cells, we transfected these heteroduplexes into COS-1 monkey cells. We found that about 5 percent of the molecules processed in the mammalian cells gave blue plaques on the bacterial indicator (indicative of correction to the wild-type sequence at both sites in the trans-heteroduplex). The data of Brown and Jiricny (1988) allow calculation (see below) of an upper limit on the expected frequency of blue plaques of about 22 percent. From such a crude analysis, we conclude that at least one-fourth of the molecules were repaired in this experiment.

These substrate molecules were incubated in the whole-cell-extract system (Glazer, et al. 1987) under different conditions (Table 4). About a 10-fold increase in the frequency of blue plaques was observed with the complete reaction. This frequency (1.2 percent), when compared to the calculated upper limit (22 percent), suggests that about 5 percent of the molecules are repaired in this reaction.

We also assayed the overall incorporation of dCTP into DNA as a measure of the DNA polymerase activities and nick-translation reactions in these extracts (Table 4). These results show a discordance between the degree of inhibition

Table 3. Processing of heteroduplex DNA in monkey cells.

DNA	Replication in COS-1 Cells	Percent Blue
Trans-heteroduplex[1]	Yes	4.9 (62/1263)
Trans-heteroduplex	No	2.0 (166/8350)
Mix of homoduplexes	Yes	0.0 (0/1800)
Mix of homoduplexes	No	0.0 (0/50000)

DNA was introduced into COS-1 monkey cells where it replicated prior to being extracted and assayed in E. coli SY204. In the control experiments the heteroduplex DNA was introduced directly into the bacteria without processing in the mammalian cells.

[1]See Table 1 for description of the structure of the trans-heteroduplex DNA.

of incorporation and mismatch correction, e.g., in the presence of dideoxynucleotides. This finding suggests that at least some component of the correction reaction is not the result of the random nick-translation reaction but is likely more specific for the mismatched regions.

SIMPLE ANALYSIS OF EXPECTED CORRECTION OF TRANS-HETERODUPLEXES

If a given heteroduplex is repaired with equal probability to one of the two bases, and if all repair events are independent, then the frequency of the wild-type molecules will be one-fourth of the total of all the repaired molecules. However, the results of Brown and Jiricny (1988) show that the first of these two assumptions is incorrect. Further, it is likely that for closely linked markers, co-repair is frequent (e.g., in E. coli, repair tracts seem to be of the order of kilobases); thus, the second simplifying assumption is likely to be incorrect as well. To help in the analysis of our results:

Let: F(i) = fraction of single heteroduplex (type i) that is unrepaired
P = probability of repair to the wild-type base
W = observed frequency of colonies (or plaques) which are phenotypically wild-type (i.e., blue)

Then: $$W(i) = 0.5F(i) + P[1 - F(i)]$$

This is because the unrepaired heteroduplexes segregate and, if replication is unbiased, yield 50 percent wild-type progeny molecules. As an example, we can take the particular mismatches we used and the values of F and P from Brown and Jiricny (1988). For G/T mismatches F(g/t) = 0.04, P(g/t) = 0.92 and for A/C mismatches F(a/c) = 0.22 and P(a/c) = 0.50.

Table 4. Repair of trans-heteroduplexes in Hela cell extracts[1].

Reaction conditions	Percent blue	Relative Correction	Relative dCTP incorporation
Complete	1.2	100	100
+ Dideoxynucleotides	0.44	26	63
- ATP, creatine phosphate and creatine kinase	0.42	25	3
+ Aphidicolin	0.33	16	2
No extract	0.17	0	0

[1]Data modified from Glazer, et al., 1987.

It is then a simple combinatorial calculation to sum the contributions of the nine classes of trans-heteroduplexes (three outcomes at the G/T site: unrepaired, wild-type, and mutant, together with the corresponding three outcomes at the A/C site). For the trans-heteroduplex designated 4/8, W(4/8) = 0.32; for the reciprocal trans-heteroduplex designated 8/4, W(8/4) = 0.11. Since we used an equimolar mixture of these two substrates, W(average) = 0.215. In these calculations it is important to keep in mind that unrepaired duplexes give mixed colonies that appear phenotypically wild-type.

PREPARATION OF SUBSTRATES WITH SINGLE MISMATCHES

It is clear from the analyses above that trans-heteroduplex DNA is less than ideal as a substrate for mismatch correction reactions. Recently we have employed a different approach to prepare large quantities of pure single mismatched heteroduplexes in covalently closed circular DNA. In principle, it should be possible to anneal an oligonucleotide with a single mismatched nucleotide to a single-standed circular DNA (e.g., from a preparation of SS phage DNA) and extend the DNA with a DNA polymerase using the oligonucleotide as a primer to make the DNA completely duplex. In practice this approach is complicated by lack of processivity of the DNA polymerase and the 5'-3' exonuclease activity of some DNA polymerases: either the duplex may be incomplete or the mismatch in the primer may be removed.

The replication system encoded by bacteriophage T4 has been shown by Alberts and colleagues (Nossal and Alberts, 1983) to be highly efficient and processive. In the absence of gene 32 protein the strand displacement does not occur and the synthesis stops when the replication apparatus encounters the 5' end of the primer oligonucleotide. We have used this system to prepare large quantities of covalently closed circular DNA containing a single mismatched base pair as follows: a synthetic oligonucleotide, phosphorylated at the 5' end, containing a single mismatch in a unique Eco RI restriction site is annealed to a preparation of single-stranded M13 phage DNA. This primer-template complex is extended with T4 DNA polymerase in the presence of the purified proteins encoded by T4 genes 44, 45, and 62. These accessory proteins increase the processivity of the polymerase by preventing its dissociation from the primer-template complex (Nossal and Alberts, 1983). The completely duplex DNA is covalently closed by the action of T4 DNA ligase. This reaction is essentially quantitative and from 50 ug of SS DNA we can obtain nearly 100 µg of closed circular DNA. This product is then treated with Eco RI to digest any molecules which have an intact site for this restriction enzyme created by low level synthesis past the mismatch site. The remaining covalently closed DNA, all of which contains the mismatch at the Eco RI site, is purified by equilibrium density gradient centrifugation in the presence of ethidium bromide.

The ease of preparation of this homogeneous substrate in large quantities should greatly facilitate the *in vitro* assay of the mismatch repair reactions as well as the subsequent dissection of these pathways into their individual components.

ACKNOWLEDGEMENTS

This work was supported by USPHS grant PO-1-CA 39238. The authors thank Brooks Low and Efim Golub for advice about the genetic manipulation of E. coli, and Maureen Munn for assistance with the T4 replication system.

REFERENCES

Brown, T.C. and Jiricny, J. 1987. A specific mismatch repair event protects mammalian cells from loss of 5-methylcytosine. Cell, 50:945-950.

Brown, T.C. and Jiricny, J. 1988. Different base/base mispairs are corrected with different efficiencies and specificies in monkey kidney cells. Cell, 705-711.

Fischel, R.A., and Kolodner, R. 1983. Gene conversion in E. coli: The identification of two repair pathways for mismatched nucleotides. U.C.L.A. Symp. Mol. Cell. Biol. 11:309-326.

Folger, K., Thomas, K., and Capecchi, M.R. 1985. Efficient correction of mismatched bases in plasmid heteroduplexes injected into cultured mammalian cell nuclei. Mol. Cell. Biol. 5:70-74.

Glazer, P.M. 1987. Mutagenesis and DNA Repair in Mammalian Cells. Ph.D. Thesis, Yale University. New Haven. pp 106.

Glazer, P.M., Sarkar, S.N., and Summers, W.C. 1986. Detection and analysis of UV-induced mutations in mammalian cell DNA using a Lambda phage shuttle vector. Proc. Natl. Acad. Sci. U.S.A. 83:1041-1044.

Glazer, P.M., Sarkar, S.N., Chisholm, G.E., and Summers, W.C. 1987. Mismatch repair activity detected in cell-free extracts of human cells. Molec. Cell. Biol., 7:218-224.

Hare, J.T. and Taylor, J.H. 1985. One role of DNA methylation in vertebrate cells is strand discrimination in mismatch repair. Proc. Natl. Acad. Sci. U.S.A. 82:7350-7354.

Meselson, M. 1988. Methyl-directed repair of DNA mismatches. pp 91-114 in: The Recombination of Genetic Material, ed. K.B. Low, Academic Press, San Diego. pp 506.

Meselson, M.S. and Radding, C.M. 1975. A general model for genetic recombination. Proc. Natl. Acad. Sci. U.S.A. 72:358-361.

Nossal, N.G., and Alberts, B.M. 1983. Mechanism of DNA replication catalyzed by purified T4 replication proteins. pp 71-81 in: Bacteriophage T4, eds. C.K. Mathews, E.M. Kutter, G. Mosig, and P.B. Berget. American Soc. Microbiol. Washington,D.C.

Rossignol, J-L., Nicolas, A., Hamza, H., and Kalogeropoulos. 1988. Recombination and gene conversion in Ascobolus. pp 24-72 in: The Recombination of Genetic Material, ed. K.B. Low, Academic Press, San Diego. pp 506.

Sarkar, S.N., Dasgupta, U.B., and Summers, W.C. 1984. Error-prone mutagenesis detected in mammalian cells by a shuttle vector containing the supF gene of E. coli. Molec. Cell. Biol. 4:2227-2230.

Wagner, R. and Meselson, M. 1976. Repair tracts in mismatch DNA heteroduplexes. Proc. Natl. Acad. Sci. U.S.A. 73: 4135-4139.

MISMATCH REPAIR PATTERNS IN SIMIAN CELLS CORRELATE WITH THE SPECIFICITY OF A MISMATCH BINDING PROTEIN ISOLATED FROM SIMIAN AND HELA CELLS

J. Jiricny, T.C. Brown, N. Corman, B.B. Rudkin

Friedrich Miescher-Institut
P.O. Box 2543, CH-4002 Basel
Switzerland

INTRODUCTION

Mismatches are formed in DNA during recombination of homologous but nonidentical sequences, as errors of DNA replication, and during the spontaneous hydrolytic deamination of 5-methylcytosine (Modrich, 1987). The efficiency and specificity of mismatch correction thus influences the outcome of genetic events such as gene conversion (Holliday, 1974; White, et al., 1985; Kourilsky, 1986), homogenization of repeated sequence families (Dover, 1986), generation of antibody diversity (Baltimore, 1974; Kunkel, et al., 1986), and DNA replication fidelity (Hare and Taylor, 1985; Reyland and Loeb, 1987). In addition, specific repair of G/T mismatches to G/C may stabilize patterns of 5-methylcytosine distribution (Brown and Jiricny, 1987).

Mismatch repair studies in procaryotic systems suggest that repair patterns differ according to the circumstances of mismatch formation. In E. coli, mispairs formed as errors of DNA replication are corrected in favor of the parental strand (Radman and Wagner, 1986), mismatched heteroduplexes formed during recombination are corrected randomly (Fishel, et al., 1983; Fishel, et al., 1986), and G/T mispairs arising through 5-methylcytosine deamination are restored to G/C pairs (Lieb, 1985; Lieb, et al., 1986; Jones, et al., 1987; Zell and Fritz, 1987). These events are mediated by different, albeit overlapping, sets of gene products (Modrich, 1987).

We were interested in studying whether the correction of mismatches in mammalian cells also involves several distinct repair pathways. We describe a way of introducing specific mispairs into the genome of Simian Virus 40 (SV40) and of determining the fate of the mispaired bases in simian cells (Brown and Jiricny, 1987). We show that the G/T, but no other mismatch, is addressed by a specific pathway, which counteracts the effects of spontaneous deamination of 5-methyl-

DNA Repair Mechanisms and Their Biological Implications in Mammalian Cells
Edited by M.W. Lambert and J. Laval
Plenum Press, New York

cytosine. This hypothesis is substantiated by the identification of a mammalian protein that binds solely to DNA duplexes containing a G/T mispair.

RESULTS

Preparation of mismatch-containing SV40 DNA

We constructed SV40 DNA with specific mispairs by replacing a 21 bp sequence between the BstXI and TaqI restriction sites (Figure 1) with mismatched synthetic 12 bp duplexes (Brown and Jiricny, 1987). DNA from wild type SV40 (strain 776) was digested with BstXI and then with TaqI. After each

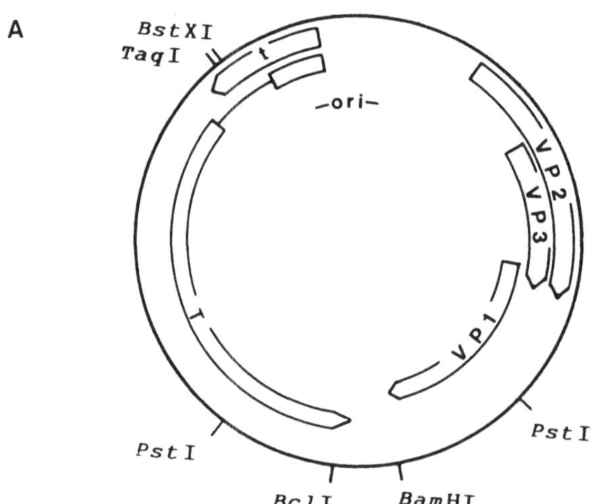

B

CAATCGAAGCAGTAGCAATCAACCCACACAAGTGGATC
GTTAGCTTCGTCATCGTTAGTTGGGTGTGTTCACCGAG
TaqI BstXI

Figure 1. Construction of SV40 with mismatched base pairs. (A) Map of the SV40 genome showing restriction sites relevant to this study. (B) Sequences surrounding the BstXI and TaqI restriction sites. Arrows indicate the sites of cleavage.

digestion linear DNA was isolated to exclude partially-digested circular molecules and to eliminate the 21 bp fragment between these restriction sites. Cleavage with these two enzymes produced linear DNA with noncompatible sticky ends (Figure 1). We then ligated aliquots of this DNA to the duplexes shown in Figure 2. Each synthetic oligonucleotide of a duplex pair contained a different restriction enzyme recognition sequence. Duplexes were mismatched at the 5' terminal bases of the two different restriction sites. Single-stranded ends complementary to the sticky ends of the viral DNA ensured efficient ligation in a defined orientation. We isolated circular, form II viral DNA and used it to transfect host CV-1 simian cells. Transfection produced plaques, each plaque corresponding to a productive infection initiated by a single viral DNA molecule. Duplexes were designed so that correction of the mispair in the host cell would create distinct restriction sites. Noncorrection, or viral DNA replication before repair occurred, would generate a mixture of viral DNA molecules, some with one restriction site and some with the other. Repair patterns could therefore be determined by diagnostic restriction analysis of DNA derived from individual plaques.

Analysis of progeny SV40 DNA

We transfected CV-1 cells with SV40 DNA modified to contain the duplexes shown in Figure 2. The results of the restriction analyses of the progeny viral DNAs are summarized in Tables 1 and 2.

```
                 BamHI                                BclI
G/T1   CGTGATCGGATCCCACAA      G/T2   CGGGGATCTGATCAACAA
          ACTAGTCTAGGG                   CCCTAGGCTAGT
          BclI                           BamHI

            ClaI
G/T3   CGGATCGATTCGAGACAA
          CTAGCTGAGCTC
                XhoI

               BglII                     BamHI
A/C1   CGCGATCAGATCTCACAA      A/C2   CGGGATCCGATCACACAA
          GCTAGCCTAGAG                   CCTAGACTAGTG
          PvuI                                 BclI

               BclI                      BamHI
C/T1   CGCGATCTGATCACACAA      C/T2   CGGGATCCGATCTCACAA
          GCTAGCCTAGTG                   CCTAGTCTAGAG
          PvuI                                 BglII

               BglII                     PvuI
A/G1   CGGGATCAGATCTCACAA      A/G2   CGCGATCGGATCACACAA
          CCTAGGCTAGAG                   GCTAGACTAGTG
          BamHI                                BclI

               SalI                      BamHI
G/G    CGGGATCGTCGACCACAA      C/C    CGGGATCCTCGACCACAA
          CCTAGGAGCTGG                   CCTAGCAGCTGG
          BamHI                                SalI

               ClaI                      BglII
A/A    CGAGATCATCGATCACAA      T/T    CGAGATCTTCGATCACAA
          TCTAGAAGCTAG                   TCTAGTAGCTAG
          BglII                                ClaI
```

Figure 2. Heteroduplexes ligated into SV40 DNA between BstXI and TaqI. Each strand of the heteroduplex contains a restriction enzyme recognition sequence as indicated.

The correction of the G/T mispair contained in the duplexes G/T1, G/T2 and G/T3 depended neither on the orientation of the mispair within the SV40 genome, nor on the sequence context. As shown in Table 1, this mismatch was corrected with extremely high efficiency, and almost exclusively to a G/C.

Table 1. Efficiencies and directionalities of correction of heteromispairs G/T, A/C, C/T and A/G. Asterisk denotes heteroduplexes that were transfected without purification of the closed circular SV40 DNA.

Mismatch	Corr. to G/C	Corr. to A/T	Uncorr.
G/T1	92	4	4
G/T2*	85	13	2
G/T3*	94	5	1
A/C1	52	33	15
A/C2	33	40	27
C/T1	64	14	22
C/T2	54	8	38
A/G1	22	11	67
A/G2	31	23	46

Restriction analysis of DNA derived from plaques obtained following transfection of the CV-1 cells with SV40 containing the duplexes A/C1 and A/C2 indicated that, as for G/T mispairs, correction patterns observed for A/C mispairs were about the same regardless of the orientation of the mismatched bases in the viral genome. The correction efficiency was much lower, however, and there appeared to be no repair bias. The repair of the two remaining heteromispairs C/T and A/G was also inefficient, 72 and 39% respectively, and a slight bias in favour of G/C was apparent.

Analysis of the correction efficiency of the homomispairs G/G, C/C, A/A and T/T, summarized in Table 2, shows that these were also repaired with different efficiencies (92, 66, 64 and 39%, respectively). The directionality of repair varied, the bias favouring G/C over A/T.

Detection of a G/T mismatch binding protein

In an effort to determine whether the efficient, biased correction of the G/T mispairs is mediated by a specific repair pathway, we attempted to establish whether mammalian cell extracts contained mismatch binding activity. According to the rationale of these experiments, detection of one or more proteins binding exclusively to G/T-mismatched duplexes would attest to a repair pathway specific for this mispair.

Binding substrates were prepared by annealing the synthetic 5'-^{32}P-labeled 34-mers 5'-AATTCCCGGGGATCCGTC**R**ACCTGCAGCCAAGCT-3' (**R**= **G** or **A**) to the respective unlabeled complementary strand 5'-AGCTTGGCTGCAGGT**Y**GACGGATCCCCGGGAATT-3' (**Y**= **T** or **C**) to yield homoduplexes **G/C** and **A/T**, and heteroduplexes **G/T** and **A/C**. HeLa whole cell extracts, prepared by the method

Table 2. Correction efficiency of homoduplexes G/G, C/C, A/A and T/T.

Mispair	Corrected	Uncorrected
G/G	92	8
C/C	66	34
A/A	64	36
T/T	39	61

of Manley (Manley, 1984), were incubated with the labeled oligonucleotide duplexes according to the procedure of Fried and Crothers (1983). The binding reactions were allowed to proceed at room temperature for 30 minutes. Protein-bound duplexes were resolved from unbound duplexes using 6% non-denaturing polyacrylamide gels. Figure 3 shows the autoradiograph of the dried gel. This result indicates that the HeLa cell extract contains at least two factors that bind selectively to the duplex containing the G/T mispair. Similar results have been obtained using CV-1 cell extracts (results not shown).

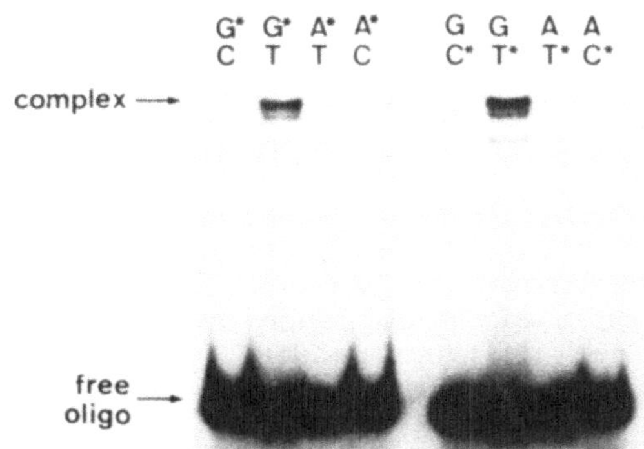

Figure 3. Binding of protein factors contained in HeLa whole cell extracts to labeled oligonucleotide duplexes G/C, G/T, A/T and A/C. The duplexes were labeled either in the top strand or in the bottom strand. The asterisk denotes the labeled strand.

DISCUSSION

Spontaneous deamination of cytosine is thought to occur at a rate of 100 per mammalian cell genome per day (Lindahl, 1982). 5-Methylcytosine deaminates about 2.5 times more rapidly than cytosine at neutral pH (Wang, <u>et al.</u>, 1982). Cells in which 5% of cytosines are methylated would therefore accumulate 12 G/T mispairs per day, or 2000 per year. The loss of 5-methylcytosine could seriously affect cell behavior because these bases are crucial DNA residues implicated in gene regulation (Doerfler, 1983; Bird, 1986; Saluz, et al., 1986), differentiation (Razin, et al., 1986) and tumorigenesis (Holliday, 1979; Riggs and Jones, 1983; Kastan, et al., 1982; Wilson and Jones, 1983; Gama-Soza, et al., 1983). Our results attest to a specific mismatch repair pathway that stabilizes the methylation pattern of mammalian cellular DNA by restoring G/C pairs whenever G/T mispairs arise through the deamination of 5-methylcytosine. Selective correction of G/T mispairs replaces thymine with cytosine rather than 5-methylcytosine. Subsequent methylation of the restored cytosine is accomplished by DNA methyltransferase, the so-called maintenance methylase.

Despite the specificity of G/T mismatch repair in favor of guanine, it is likely that deamination of 5-methylcytosine occasionally leads to its loss. 5-Methylcytosine occurs mainly, if not exclusively, at mCpG dinucleotides (where mC is 5-methylcytosine) (Bird, 1978; Cedar, et al., 1979). The rarity (Bird, 1980) and instability (Barker, et al., 1984) of CpG dinucleotides in mammalian cellular DNA indicates that mCpG dinucleotides sustain high mutation rates (Bird, 1987). Certainly in E. coli, specific correction of G/T mispairs at presumptive sites of cytosine methylation (Lieb, 1985; Lieb, et al., 1986; Jones, et al., 1987; Zell and Fritz, 1987) does not wholly abolish mutations attributable to 5-methylcytosine deamination (Coulondre, et al., 1978). Thus, while our results indicate that specific correction of G/T mispairs protects cells from loss of 5-methylcytosine, protection is probably not complete. In light of the findings presented here, our observation that 5%-10% of G/T mispairs are corrected to A/T pairs is most easily explained by supposing that G/T mispairs are subject to two correction pathways. One, which we believe acts exclusively on G/T, is probably highly specific for correction in favor of G/C. The proteins found to bind selectively to G/T mismatched heteroduplexes may play a role in this specific pathway. A second pathway, which may act on all mispairs upon their occurrence during DNA replication or recombination, may be more random and may account for the majority of repair events in favor of thymine.

We have previously noted that inflexible bias in the correction of G/T mispairs to G/C would be mutagenic, since resolution of mispairs formed by incorporation of guanine opposite thymine during DNA replication would favor fixation of the mutation (Brown and Jiricny, 1987). This difficulty is overcome by supposing that the G/T mismatch repair pathway specific for establishment of G/C does not act on newly replicated DNA, but is supplanted by a pathway correcting mispairs in favor of the parental strand (Hare and Taylor, 1985). According to this hypothesis, mismatch repair patterns in mammalian cells and in E. coli would share at least one point of similarity: correction of G/T mispairs resulting from replication error and from 5-methylcytosine deamination would be addressed by different, though overlapping, repair pathways.

REFERENCES

Baltimore, D., 1974. Is terminal deoxynucleotidyl transferase a somatic mutagen in lymphocytes? Nature. 248:409-411.

Barker, D., M. Schafer and R. White, 1984. Restriction sites containing CpG show a higher frequency of polymorphism in human DNA, Cell, 36:131-138.
Bird, A.P., 1987. CpG islands as gene markers in the vertebrate nucleus, Trends in Genet., 3:342-347.
Bird, A.P., 1986. CpG-rich islands and the function of DNA methylation, Nature, 321:209-213.
Bird, A.P., 1980. DNA methylation and the frequency of CpG in animal cells, Nucl. Acids Res., 8:1499-1504.
Bird, A.P., 1978. Use of restriction enzymes to study eucaryotic DNA methylation. II. The symmetry of methylated sites supports semiconservative copying of the methylation patterns, J. Mol. Biol., 118:49-60.
Brown, T.C. and J. Jiricny, 1987. A specific mismatch repair event protects mammalian cells from loss of 5-methylcytosine, Cell, 50:945-950.
Cedar, H., A. Solange, G. Glaser and A. Razin, 1979. Direct detection of methylated cytosine in DNA by use of the restriction enzyme MspI, Nucl. Acids Res., 6:2125-2132.
Coulondre, C., J.H. Miller, P.J. Farabaugh and W. Gilbert, 1978. Molecular basis of base substitution hotspots in Esherichia coli, Nature, 274:775-780.
Doerfler, W., 1983. DNA methylation and gene activity, Ann. Rev. Biochem., 52:93-124.
Dover, G.A., 1986. Molecular drive in multigene families:how biological novelties arise, spread and are assimilated. Trends in Genet., 2:159-165.
Fishel, R.A., E.C. Siegel and R. Kolodner, 1986. Gene conversion in Escherichia coli. Resolution of heteroallelic mismatched nucleotides by co-repair, J. Mol. Biol., 188: 147-157.
Fishel, R.A., E.C. Siegel and R. Kolodner, 1983. The identification of two repair pathways for mismatched nucleotides, UCLA Symp. Mol. Cell. Biol., 11:309-326.
Fried, M.G. and D.M. Crothers, 1983. CAP and RNA polymerase interactions with the lac promotor: binding stoichiometry and long range effects, Nucleic Acids Res., 11:141-158.
Gama-Soza, M.A., V.A. Slagel, R.W. Trewyn, R. Oxenhandler, K.C. Kuo, C.W. Gehrke and M. Ehrlich, 1983. The 5-methylcytosine content of DNA from human tumors, Nucl. Acids Res., 11:6883-6894.
Hare, J. and H. Taylor, 1985. One role of DNA methylation in vertebrate cells is strand discrimination in mismatch repair. Proc. Natl. Acad. Sci. (USA) 82:7350-7354.
Holliday, R., 1979. A new theory of carcinogenesis, Br. J. Cancer, 40:513-522.
Holliday, R., 1974. Molecular aspects of genetic exchange and gene conversion, Genetics. 78:273-287.
Jones, M., R. Wagner and M. Radman, 1987. Mismatch repair of deaminated 5-methyl-cytosine, J. Mol. Biol., 194:155-159.
Kastan, M.N., B.J. Gowans and M.W. Lieberman, 1982. Methylation of deoxycytidine incorporated by excision-repair synthesis of DNA, Cell, 30:509-516.
Kourilsky, P., 1986. Molecular mechanisms for gene conversion in higher cells, Trends in Genet. 2:60-63.
Kunkel, T.A., K.P. Gopinathan, D.K. Dube, E.T. Snow and L.A. Loeb, 1986. Rearrangements of DNA mediated by terminal transferase. Proc. Natl. Acad. Sci. (USA) 83:1867-1871.
Lieb, M., 1985. Recombination in the lambda repressor gene: evidence that very short patch (VSP) mismatch correction

restores a specific sequence. Mol. Gen. Genet., 199:465-470.

Lieb, M., E., Allen and D. Read, 1986. Very short patch mismatch repair in phage lambda: repair sites and length of repair tracts, Genetics, 114:1041-1060.

Lindahl, T., 1982. DNA repair enzymes, Ann. Rev. Biochem., 51: 61-87.

Manley, J.L., 1984. Transcription of eukaryotic genes in a whole-cell extract. In: Transcription and translation, a practical approach. B.D. Hames & S.J. Higgins (Eds.) IRL Press, Oxford, pp. 71-80.

Modrich, P., 1987. DNA mismatch correction, Ann. Rev. Biochem. 56:435-466.

Radman, M. and R, Wagner, 1986. Mismatch repair in Escherichia coli. Ann. Rev. Genet. 20:523-538.

Razin, A., M. Szyf, T. Kafri, T., M. Roll, H. Giloh, S. Scarpa, D. Carotti and G.L. Cantoni, 1986. Replacement of 5-meth ylcytosine by cytosine: a possible mechanism for transient DNA methylation during differentiation, Proc. Natl. Acad. Sci. (USA) 83:2827-2831.

Reyland, M.E. and L.A. Loeb, 1987. On the fidelity of DNA replication, J. Biol. Chem. 262:10824-10830.

Riggs A.D. and P.A. Jones, 1983. 5-Methylcytosine, gene regulation and cancer, Adv. Cancer Res., 40:1-30.

Saluz, H.-P., J. Jiricny. and J.P. Jost, 1986. Genomic sequencing reveals a positive correlation between the kinetics of strand-specific DNA methylation of the overlapping estradiol/glucocorticoid receptor binding sites and thrate of avian vitellogenin synthesis, Proc. Natl. Acad. Sci.(USA), 83:7167-7171.

Wang, R.Y.-H., K.C. Kuo, C.W. Gehrke, L.-H. Huang, and M. Ehrlich, 1982. Heat- and alkalai-induced deamination of 5-methylcytosine and cytosine residues in DNA, Biochem. Biophys. Acta, 697:371-377.

White, J.H., K. Lusnak, and S. Fogel, 1985. Mismatch-specific post-meiotic segregation frequency in yeast suggests a heteroduplex recombination intermediate, Nature. 315:350-352.

Wilson, V.L. and P.A. Jones, Inhibition of DNA methylation by chemical carcinogens in vitro, Cell, 32:239-246.

Zell, M.J. and H.-J. Fritz, 1987. DNA mismatch repair in Escherichia coli counteracting the hydrolytic deamination of 5-methyl-cytosine residues, EMBO J., 6:1809-1815.

DNA EXCISION REPAIR AT THE NUCLEOSOME LEVEL OF CHROMATIN

Michael J. Smerdon

Biochemistry/Biophysics Program
Washington State University
Pullman, WA 99164-4660

INTRODUCTION

The complexity of enzymatic mechanisms that have evolved for the recognition and removal of DNA lesions in mammalian cells is undoubtedly coupled to the complexity of the very "substrate" itself (i.e., the highly condensed structure of DNA in chromatin). For this reason, over the past decade my laboratory has focused on the role of chromatin structure in the process of excision repair in eukaryotes. In the discussion that follows, I have reviewed our results concerning features of DNA damage formation and excision repair at the nucleosome level of chromatin. It is important to keep in mind that the features discussed are reflective primarily of "bulk" (i.e., transcriptionally inactive) chromatin in mammalian cells. The discussion starts with a brief review of nucleosome structure, and ends with a rather simple, generalized model for excision repair of DNA in bulk chromatin of mammalian cells. Throughout this chapter, I have cited recent reviews, rather than specific papers, pertaining to a given topic. Specific references can be found in these reviews.

BACKGROUND

It is now clear that the primary level of organization of DNA in chromatin is a series of repeating units called nucleosomes (for recent reviews, see Pederson, et al., 1986 and vol. I of Adolph, 1988). In its extended form, the nucleofilament resembles "beads-on-a-string" in the electron microscope with an average thickness of 10 nm. The beads, representing the major component of nucleosomes, are called "core particles" and consist of 146 bp of DNA wound into $1^3/_4$ lefthanded superhelical turns around an octomer of the four core histones H2A, H2B, H3 and H4. The general shape of the core particle is that of a disc 11 nm in diameter and 5.7 nm thick. More detailed structural information has come from X-ray diffraction (Richmond, et al., 1984), neutron diffraction (Bentley, et al., 1984), and chemical crosslinking studies (Shick, et al., 1985). Based on these studies a model

DNA Repair Mechanisms and Their Biological Implications in Mammalian Cells
Edited by M.W. Lambert and J. Laval
Plenum Press, New York

has emerged which details the spatial arrangement of histones in the core particle and the path of the DNA helix around the histone core (see reviews cited above and Morse and Simpson, 1988). Core particles are separated by variable lengths of DNA, known as linker DNA. The average length of these regions is species specific and may vary from 20 to 100 bp. In addition to the four core histones, a fifth histone class, H1, is thought to bind the nucleofilament at positions where the DNA enters and exits the core particle, as well as to linker DNA regions. The binding of histone H1 results in an additional 20 bp of DNA (10 bp on each end) forming a stable complex with the core histones. The resulting particle, which has 166 bp of DNA wound into two complete turns around the core histones, has been termed the chromatosome. The DNA linking chromatosomes, therefore, varies from 0 to 80 bp. This subunit structure appears to be a ubiquitous feature amongst eukaryotes and is present throughout most of the genome in a single cell.

The next level of packaging represents the compaction of the 10 nm nucleofilament into a fiber approximately 30 nm thick in the electron microscope. Histone H1 is required for the formation and/or stabilization of this structure which has been studied in detail by a variety of physical techniques. In contrast to the general agreement on nucleosome structure, the results of these studies have not generated a conclusive structure for the 30 nm fiber. The four major classes of structures that have been proposed are reviewed in Pederson, _et al._ (1986). Higher order levels of folding in the structural hierarchy of chromatin are even less well-defined, although several features of these structures have been elucidated (Nelson, _et al._, 1986; vols. II and III of Adolph, 1988).

There is a large body of evidence indicating that regions of chromatin engaged in replication or transcription are in a more "open" conformation than the compact 30 nm fiber (Pederson, _et al._, 1986; Thoma and Sogo, 1988; Reeves, 1984, 1988). Furthermore, nucleosomes in these regions of chromatin are different from those in bulk chromatin in that they are partially (or totally) deficient in histone H1, contain stoichiometric amounts of certain nonhistone proteins and are preferentially attacked by nucleases (above reviews). It is also clear that histones are subject to one or more modifications of their structure, and many of these modifications occur in regions that most likely interact with DNA (Reeves, 1984; Matthews, 1988). Thus, certain structural features within nucleosomes may vary widely depending on their location in the genome and their degree of modification. These differences may yield differing constraints on the processes of DNA damage and repair. Therefore, the features discussed below reflect _average_ properties of excision repair in the bulk of the DNA in chromatin.

REARRANGEMENT OF NUCLEOSOME STRUCTURE DURING EXCISION REPAIR

A decade has passed since we first reported that rearrangement of nucleosome structure occurs following excision repair of UV-induced photoproducts in DNA of human diploid fibroblasts (Smerdon and Lieberman, 1978). Since that time, several laboratories have observed this phenomenon following repair synthesis induced by a variety of different agents, including bulky chemicals which form adducts preferentially in linker DNA of nucleosomes and methylating agents which may have almost equal access to linker and core DNA (Table 1). Furthermore, this phenomenon has been observed in several different mammalian cell types with varying degrees of repair efficiency, including xeroderma pigmentosum cells and monkey kidney cells (Table 1).

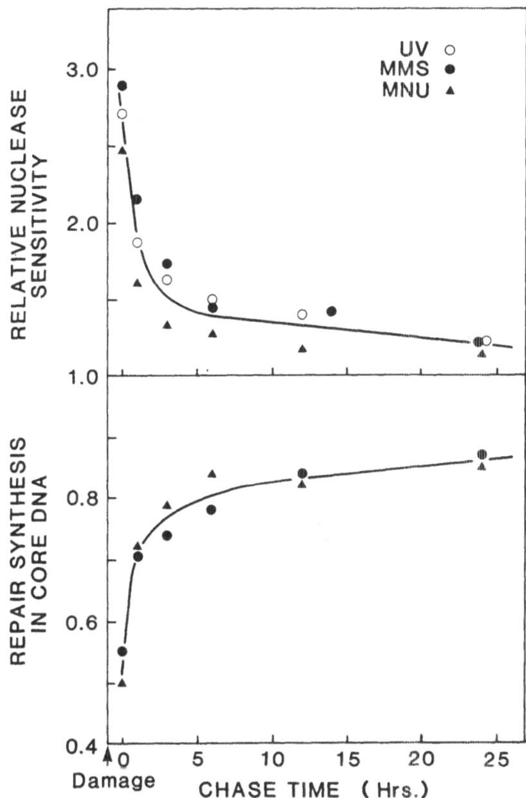

Figure 1. Staphylococcal nuclease sensitivity (top panel) and association with isolated nucleosome core DNA (bottom panel) of repair-incorporated nucleotides, relative to bulk chromatin, following different chase times after a 1 hour pulse-labeling period. Abbreviations for damaging agents used are the same as in Table 1. Data are from Sidik and Smerdon, 1984.

This rearrangement phenomenon was initially observed in two different ways (Figure 1): (1) Newly repaired DNA in chromatin is rapidly digested to small oligonucleotides by staphylococcal nuclease immediately after repair synthesis, and then looses this enhanced nuclease sensitivity during subsequent times; and (2) Newly repaired DNA is initially underrepresented (or absent) in isolated nucleosome core

TABLE I

NUCLEOSOME REARRANGEMENT IN MAMMALIAN CELLS

Damaging Agent[a]	Helix Distortion[b]	Nucleosome Preference[c]	Cell Type	Reference
UV Radiation (254 nm)	++	Random	Normal Human Fibroblasts	Smerdon and Lieberman (1978)
			Xeroderma Pigmentosum	Smerdon *et al.* (1979); Williams and Friedberg (1979)
			Monkey Kidney	Bodell and Cleaver (1981)
			HeLa	Bodell *et al.* (1984)
			Cockayne Syndrome	Cleaver (1982)
			Immunodeficient Human	Smerdon (1986)
NA-AAF	+++	Linker	Normal Human Fibroblasts	Tlsty and Lieberman (1978)
BMBA	+++	Linker	Normal Human Fibroblasts	Oleson *et al.* (1979)
Angelicin	++	Linker	Human Glioblastoma	Zolan *et al.* (1982)
MMS	+	Random[d]	Normal Human Fibroblasts	Sidik and Smerdon (1984)
			Xeroderma Pigmentosum	Sidik and Smerdon (1987)
MNU	+	Linker	Normal Human Fibroblasts	Sidik and Smerdon (1984)

[a]Abbreviations: BMBA, 7-bromomethylbenz(a)anthracene; NA-AAF, N-acetoxy-2-acetylaminofluorene; MMS, methylmethane sulfonate; MNU, methylnitrosourea; UV, ultraviolet.

[b]Relative amount of perturbation associated with the major adducts in each case (see Ciarrocchi and Pedrini, 1982; Wiesehahn and Hearst, 1978; Singer and Kusmierek, 1982).

[c]Region of nucleosome (linker or core) most frequently damaged (on a unit DNA basis) (see also, Niggli and Cerutti, 1982; Lang *et al.*, 1982; Berkowitz and Silk, 1981).

[d]Assumes distribution of MMS damage is similar to dimethyl sulfate (McGhee and Felsenfeld, 1979; Berkowitz and Silk, 1981).

particles and then becomes associated with these regions during subsequent times. The time course for each of these changes (i.e., loss of nuclease sensitivity and increased association with nucleosome cores) is very similar, having both a rapid phase and a slow phase (Smerdon and Lieberman, 1978). Therefore, it was apparent that these two observations were associated with different aspects of the same phenomenon and reflected structural changes at the nucleosome level of chromatin. Because we did not know the cause of these changes in the cell, we chose to refer to this phenomenon as nucleosome rearrangement, rather than use terms such as "nucleosome sliding" or "nucleosome displacement" which imply specific mechanisms. In 1979 we proposed a model for this process which summarized our thinking at that time and was intended to provide some direction for further experimentation on this process (Lieberman, et al., 1979). The model has become known as the "unfolding-refolding" model since it depicts the rearrangement we observe as refolding of newly repaired DNA into a nucleosome structure after an initial unfolding of this DNA for processing by repair enzymes.

Since that time my laboratory has made a considerable effort to understand the underlying mechanism(s) for this process, if not in specific molecular terms, at least in more general terms which would help clarify its relationship to the repair process. Initially, we found that nucleosome rearrangement occurred following repair synthesis regardless of the time after UV damage that repair takes place (Smerdon and Lieberman, 1980). It has been known for many years that repair of UV damage occurs in two distinct phases in human cells (e.g., Kantor and Setlow, 1981). Thus, whether repair synthesis occurred during the early rapid phase of repair or during the late slow phase of repair, nucleosome rearrangement was observed (Figure 2). However, subtle differences were observed in both the degree of nuclease sensitivity (relative to bulk chromatin) and the rate of nucleosome rearrangement following repair during these two phases (Smerdon and Lieberman, 1980). These features are discussed in more detail later.

We also reported that the initial enhanced nuclease sensitivity of newly repaired regions and the subsequent loss of sensitivity of these regions could be observed using DNase I as a probe (Smerdon and Lieberman, 1980). This observation was important for two reasons: (1) It was in keeping with the concept that repair patches are initially in a more "open" conformation, since DNA not tightly packaged into nucleosomes should be more accessible to any nuclease; and (2) It emphasized the fact that "enhanced DNase I sensitivity" is not synonymous with transcriptionally active chromatin (e.g., Reeves, 1984), especially when this enhanced sensitivity is transient. Finally, we also reported that initially newly repaired DNA does not yield the "10.4 base ladder" on denaturing gels following DNase I digestion (Smerdon and Lieberman, 1980). This ladder has become recognized as a footprint for core histone binding to DNA in nucleosomes (Lutter, 1978). Thus, this result appeared to strongly support the notion that during and immediately after repair synthesis, newly repaired DNA is not tightly bound to a surface of core histones in a native nucleosome conformation. Once again, an altered nucleosome structure during excision repair was the most likely explanation for these results.

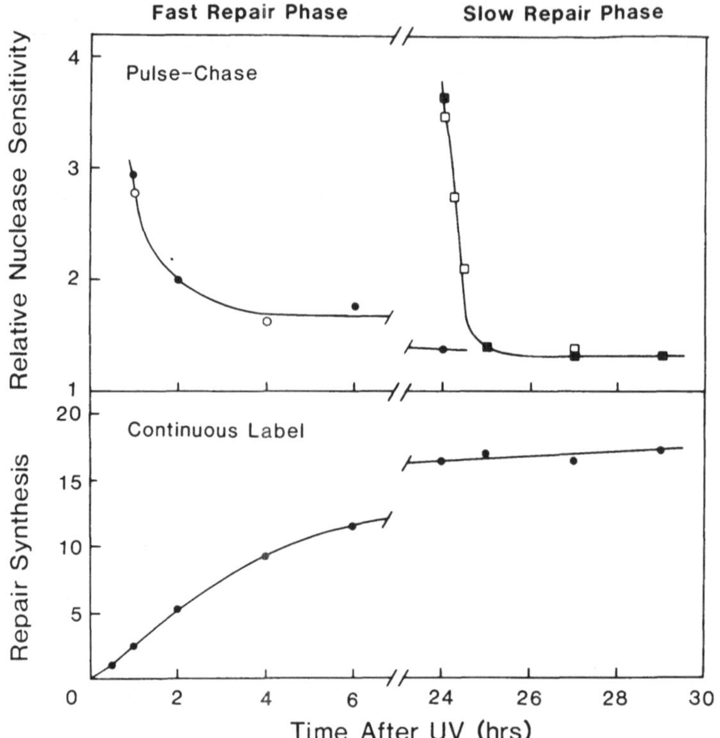

Figure 2. Kinetics of nucleosome rearrangement following repair synthesis occurring at early (fast repair phase) or late (slow repair phase) times after UV irradiation of confluent human cells. Upper panel shows staphylococcal nuclease sensitivity, relative to bulk chromatin, of repair patches inserted during a 1 hour pulse-labeling period (starting at 0 hours or 23 hours after UV irradiation) and following subsequent chase times. These data are from Smerdon and Lieberman (1980). Lower panel shows the corresponding rate of repair incorporation following continuous labeling during this same time period.

It became clear, however, that the <u>timing</u> of repair patch ligation relative to the association of these patches with nucleosome core structures was crucial to our interpretation of the DNase I protection data (see Discussion in Hunting, et al., 1985). Thus, both Lieberman's laboratory and my own examined the relationship between ligation of newly repaired regions and nucleosome formation in these regions (Hunting, et al., 1985; Smerdon, 1986). In both of these studies, it was observed that repair patch ligation preceded the loss of nuclease sensitivity even when ligation was delayed by inhibitors of repair synthesis (Hunting, et al., 1985; Smerdon, 1986) or in a partially ligase-deficient human cell strain (Smerdon, 1986). Furthermore, in one of these studies (Smerdon, 1986) it was shown that most (if not all) of the repair patches associated with isolated nucleosome core particles, shortly after rearrangement, contained a ligated 3' end. Thus, the completion of excision repair in nucleosomes appears to progress from an unligated, nonucleosome structure to a ligated, nonucleosome structure and finally to a ligated, nucleosome structure.

In an entirely different approach toward understanding the features of the nuclease sensitive state(s) of newly repaired regions in chromatin, we examined the ability of these regions to form nucleosome structures _in vitro_ following the induction of nucleosome sliding and/or exchange in nuclei (Watkins and Smerdon, 1985b). Following treatment of intact nuclei (i.e., containing a full complement of histones) or H1-depleted nuclei with increasing salt concentrations or temperature, we found that nucleosome rearrangement _in vitro_ could be divided into two distinct phases (Watkins and Smerdon, 1985a). The first phase involves a reduction in the average nucleosome repeat from 192 bp to 168 bp and occurs at low to moderate salt concentrations (25-175 mM KCl). The second phase, occurring at higher salt concentrations, involves a shift in repeat length to 146 bp. However, following the first salt-induced change, little (or no) additional nucleosome formation was observed in newly repaired DNA, suggesting that these regions were resistant to nucleosome formation during the low salt transition. This resistance did not result from features associated with the DNA alone (e.g., unligated nicks or gaps) since nucleosomes could be formed in these regions to the same extent as in bulk DNA following removal of core histones and subsequent reconstitution of nucleosomes (Watkins and Smerdon, 1985b). These results also support the notion of an altered nucleosome structure in nascent repair patches since they indicate that these regions are resistant to nucleosome formation via sliding of core histones onto these regions. Furthermore, they strongly suggest that a model originally entertained by us was unlikely. In this model, nucleosome rearrangement was envisioned as the result of preferential repair of linker DNA, followed by constitutive sliding of nucleosomes to obtain the loss of nuclease sensitivity and increased association with nucleosome core domains (Smerdon and Lieberman, 1978). Clearly, if this were the case, core histones should be able to slide onto these regions with the same frequency as in linker regions of bulk chromatin. Thus, all of the experimental observations to date are consistent with a mechanism involving transient formation of an altered nucleosome structure during excision repair. This altered structure may involve unfolding of DNA from the core histone surface or induced sliding of core histones in the damaged region. The rapid phase of rearrangement is envisioned as the refolding of these regions into a canonical, nucleosome conformation.

HISTONE MODIFICATION DURING EXCISION REPAIR

From the above discussion, it appears that in a large fraction of mammalian cell chromatin significant structural alterations are required for the multi-step, excision repair process to occur. Therefore, it is not surprising that a number of laboratories have begun to focus on molecular processes that may be involved in modulation of chromatin structure. These processes include chemical modification of nuclear proteins, especially histones. Many laboratories have now reported that ADP-ribosylation of nuclear proteins is stimulated by DNA damage in intact cells (see reviews by Ueda and Hayaishi, 1985 and Matthews, 1988). Interestingly, it has been reported that poly(ADP-ribosylation) of histone H1 yields an extended nucleofilament structure in vitro under conditions where unmodified nucleofilaments fold into the 30 nm fiber (de Murcia, et al., 1986). Another modification that appears to correlate with an open chromatin structure is the reversible acetylation of ε-amino groups of specific lysines on core histones (Reeves, 1984; Matthews, 1988). Indeed, both newly replicating chromatin and transcriptionally active chromatin contain high concentrations of acetylated histones, although the acetylated forms of histones may differ in each case (Matthews, 1988). An indication that this chromatin modification may also play a role in excision repair came from our initial observation that treatment of nonreplicating normal and xeroderma pigmentosum human skin fibroblasts with sodium butyrate, under conditions where the core histones are maximally acetylated, enhances excision repair immediately following UV irradiation (Smerdon, et al., 1982a). Although this short-chained fatty acid can have complex and pleiotropic effects on cultured cells besides the reversible inhibition of histone deacetylase enzymes (Reeves, 1984), the simplest interpretation of our results was that the increased repair efficiency of a fraction of the DNA lesions was caused, in some way, by the increase in "hyperacetylated" chromatin. In another report, it was concluded that sodium butyrate treatment of growing human cells resulted in an artifactual increase in repair synthesis due to changes in nucleotide pools (Williams and Friedberg, 1982). However, since sodium butyrate treatment arrests cells in the G1 phase of the cell cycle (Reeves, 1984), the comparison of nucleotide pool concentrations between butyrate treated and untreated cells may have been confounded by changes in these pools between growing and arrested cells. Alternatively, the above observations may underscore the marked variation in butyrate-induced effects on cellular metabolism and physiology with different cell types (Reeves, 1984). Indeed, we recently reported that repair synthesis in a human lung fibroblast cell strain actually decreased following butyrate treatment under conditions which yield increased repair in human skin fibroblasts (Smerdon, 1986). These apparently contradictory observations may have been resolved by the findings of Dresler (1985) who reported that in permeable human cells, where repair synthesis is dependent on exogeneous nucleotides, pretreatment with sodium butyrate resulted in a stimulation of repair synthesis similar in magnitude to the stimulation we observed (Smerdon, et al., 1982a). Furthermore, Dresler (1985) also observed changes in the nucleotide pools of these cells. Therefore, it appears that both enhanced DNA repair synthesis and changes in nucleotide pools occur in human cells following butyrate treatment, and the extent to which this latter change affects measurements of repair synthesis may vary between cell types.

It is important to note that the increased rate of repair we observed in butyrate treated cells (compared to untreated cells) was short lived, lasting only a few hours after UV

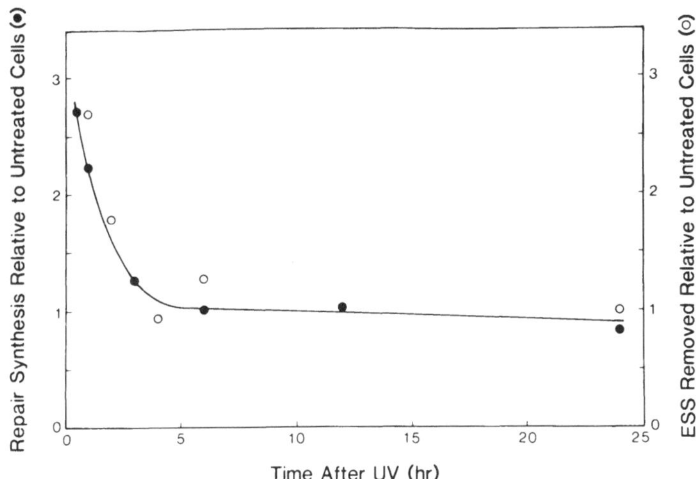

Figure 3. Repair synthesis and endonuclease-sensitive sites (ESS) removed in butyrate-treated normal human fibroblasts, relative to untreated cells, at different times after UV irradiation. For repair synthesis measurements, confluent cells (strain AG1518), prelabeled with [^{14}C]dThd, were either treated with 10 mM sodium butyrate for 48 hours or 10 mM hydroxyurea for 45 minutes (and no butyrate) prior to irradiation with 12 J/m^2 UV. The cells were then labeled continuously with [^{3}H]dThd for the times shown. The relative amount of repair synthesis (●) represents the ratio of the ^{3}H DPM/^{14}C DPM values for butyrate-treated cells and the ^{3}H DPM/^{14}C DPM for cells not treated with butyrate. The values for relative amount of ESS removed (O) were obtained for cells treated, or not treated, with 20 mM sodium butyrate for 48 hours as described in Smerdon, et al. (1982a).

irradiation (Figure 3). During this time period, ~20% of the UV endonuclease sensitive sites (presumably cyclobutyl pyrimidine dimers) were removed (Smerdon, et al., 1982a). Therefore, at most, only 20% of the UV photoproducts were repaired more efficiently in these cells, even though 50-60% of the total core histones were hyperacetylated under the conditions specified for these studies (Smerdon, et al., 1982a and unpublished results). Thus, it was not known if the enhanced repair in these cells, following butyrate treatment, occurred in regions of chromatin containing hyperacetylated histones. Recently, we have examined the levels of repair synthesis in different acetylated species of nucleosomes from butyrate treated human cells (B. Ramanathan and M. Smerdon, submitted for publication). The results of this study indicate that there is an increase in repair synthesis of at least 2-fold associated with core regions of hyperacetylated nucleosomes. Furthermore, this enhanced repair synthesis does not

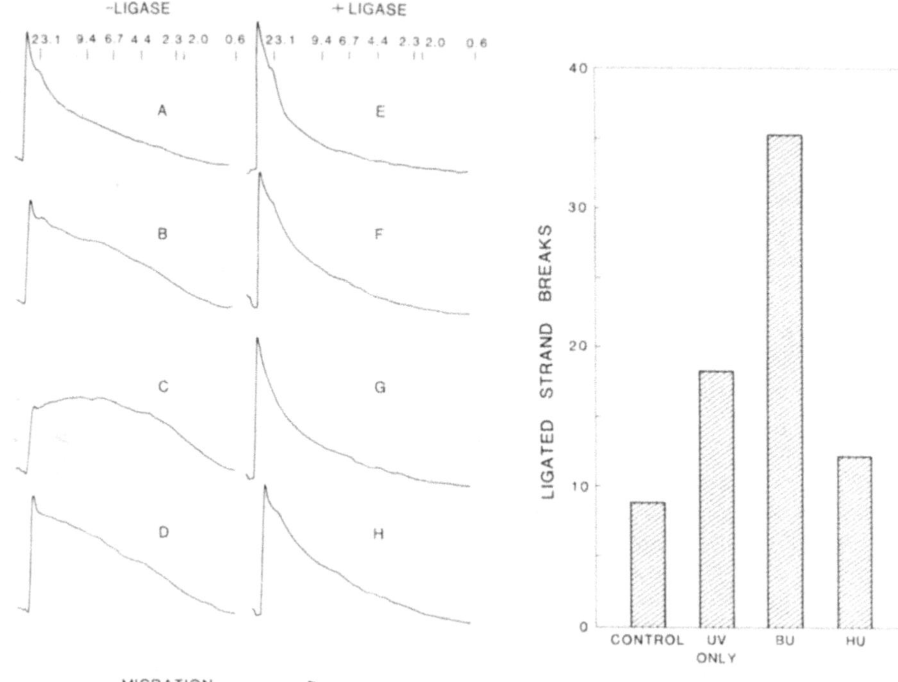

Figure 4. (Left panel) Gel scans of DNA from confluent human cells (strain AG1518) treated with (E-H) or without (A-D) T4 DNA ligase and electrophoresed on an alkaline agarose gel. The DNA samples were isolated from untreated cells (A, E), cells harvested 30 minutes after exposure to 12 J/m^2 UV (B, F), cells pretreated for 48 hours with 20 mM sodium butyrate (BU) prior to the irradiation and incubation steps (C, G), and cells pretreated for 45 minutes with 10 mM hydroxyurea (HU) prior to the irradiation and incubation steps (D, H). Positions of fragments generated by a Hind III digest of DNA are shown at top. (Right panel) Number of ligatable strand breaks per unit DNA determined for the samples used in the left panel by the following equation:

$$\frac{0.6b}{M_r} = \frac{1}{M_{1/2}(-L)} - \frac{1}{M_{1/2}(+L)}$$

where b is the number of strand breaks, M_r is the molecular weight of the intact DNA, and $M_{1/2}(-L)$ and $M_{1/2}(+L)$ represent the median molecular weights determined from the gel scans shown in the left panel for no ligase treatment and with ligase treatment, respectively. Since the value of M_r is not known for these DNA preparations and can not be determined from the gel scans (i.e., many of the DNA fragments were too long to be separated on the gel), it was assumed that this value was the same for each treatment, and we calcualted $0.6b/M_r$, rather than b itself, to obtain the ligated strand breaks per unit mass of DNA. This value allows comparison of the different treatments. For details on the calculation of $M_{1/2}$ and the development of this equation see Sutherland and Shih (1983).

appear to result from increased UV damage in these regions, since the levels of thymine containing pyrimidine dimers (measured by HPLC) was similar in each nucleosome fraction. We conclude, therefore, that in butyrate treated cells, increased excision repair of UV damage does indeed occur in the hyperacetylated chromatin domains.

Since excision repair is a multi-step process, it was of interest to determine which of these steps is affected by butyrate treatment. We found that butyrate treatment had little effect on either the rate or extent of nucleosome rearrangement (i.e., an event associated with completion of repair synthesis) (Smerdon, 1983). More recently, however, several laboratories have reported that the rate of incision may be increased in butyrate treated cells (Dresler, 1985; Smerdon, 1986; Smith, 1986). For example, the number of _ligatable_ single-strand breaks in DNA following a 30 minute incubation after UV irradiation increases ~2 fold in human cells treated with sodium butyrate (Figure 4). This is similar to the enhancement observed in repair synthesis and removal of endonuclease sensitive sites following the same time period (Figure 3). It was also reported that butyrate treatment had no effect on the average size or composition of repair patches inserted after UV irradiation (Dresler, 1985). Thus, at present, it appears that sodium butyrate treatment results in either increased levels of "incision-mediating" proteins or increased accessibility of damaged sites to repair enzymes in a fraction of the total chromatin.

At this point, it should be emphasized that even if postsynthetic modifications of chromatin expose certain DNA lesions to repair enzymes, these observations do not indicate whether such modulation plays an _active_ role in excision repair. To address this question more directly, we have begun to analyze the levels of acetylation in nuclear proteins (from cells not treated with sodium butyrate) during the time in which repair is occurring. Surprisingly, we have observed changes in the acetylation level of the _total_ nuclear protein population following UV irradiation of confluent human cells (Ramanathan and Smerdon, 1986). Following low doses of UV radiation, we observed a small, but measurable, wave of increased acetylation (lasting 1-6 hours) followed by a more pronounced wave of decreased acetylation (lasting 24-72 hours) before returning to control levels. This latter wave of "hypoacetylation" was more pronounced at higher UV doses (>5 J/m^2), while the early wave of "hyperacetylation" was more pronounced at lower UV doses. Furthermore, both the duration and the magnitude of the initial wave was found to be dependent on the presence of hydroxyurea and the age of the cells in culture (Ramanathan and Smerdon, 1986). Analysis of the individual nuclear proteins by electrophoresis indicated that both phases could be observed in the acetylation levels of the core histones. Interestingly, three prominent nonhistone proteins actually showed an increase in acetylation content in the UV irradiated cells during the hypoacetylation phase of the core histones (Ramanathan and Smerdon, 1986). At present, we do not know if these changes reflect actual changes in histone acetylase (or deacetylase) activity, or changes in acetate metabolism (e.g., in production of acetyl CoA).

In related studies, we have also observed that the association of enhanced repair synthesis with hyperacetylated nucleosomes in sodium butyrate treated cells (discussed earlier) is transient, lasting about 12 hours after repair incorporation (B. Ramanathan and M. Smerdon, submitted for publication). This observation may reflect deacetylation of newly repaired chromatin following completion of excision repair in these cells. [It should be noted that histone deacetylase activities can be differentially affected by sodium butyrate (Reeves, 1984)]. Alternatively, these findings (and those discussed earlier) may reflect a general response by nuclear enzymes to UV irradiation, and the changes observed in histone acetylation are not coupled to repair. Clearly, examination of acetylation levels of histones <u>at</u> repair sites in chromatin both during and after repair is required to distinguish between these possibilities.

As mentioned earlier, it appears that different acetylated forms of core histones are present in newly replicated versus transcriptionally active chromatin (Matthews, 1988). Therefore, we have examined the relative levels of different acetylated forms of core histones during the hypoacetylation phase in UV irradiated human cells (B. Ramanathan and M. Smerdon, unpublished results). We have found that this phase involves a decrease in the levels of tri- and tetra-acetylated species of histone H4. Thus, it appears that the prolonged hypoacetylation of histones following UV irradiation (and following the initial hyperacetylation phase) reflects an overall reduction in hyperacetylated chromatin. It is interesting to note that it is this form of chromatin (i.e., where the core histones are in a hyperacetylated state) that is associated with transcriptionally active domains (Reeves, 1984; Matthews, 1988). Thus, it is possible that excision repair of DNA may also require initial "processing" of at least the more compact regions of chromatin by histone modification enzymes to allow effective repair to take place. Such an active role for histone acetylation in repair has been proposed by Perry and Chalkley (1982).

MATURATION OF NEWLY REPAIRED REGIONS FOLLOWING NUCLEOSOME FORMATION

As discussed in the first section, a consistent feature of the nuclease digestion data used to monitor nucleosome rearrangement in newly repaired regions has been the biphasic nature of these curves (Figure 1). The rapid phase, involving the most dramatic loss in nuclease sensitivity, has an associated "half-life" of ~20 minutes. The subsequent slow phase accounts for, at most, 20% of the total change in nuclease sensitivity and takes many hours. A number of possibilities were originally considered to explain this observation, including: (1) The population of repaired regions is heterogeneous in that a small fraction of these regions acquires a nucleosome structure much more slowly than the majority of repair patches; (2) After the rapid phase of rearrangement, the structure of nucleosomes in newly repaired regions differs from that of nucleosomes in bulk chromatin; and (3) Following nucleosome formation (i.e., during the rapid phase) repair patches are nonuniformly distributed in nucleosome subdomains and "weight" the regions that are nuclease sensitive. In regards to the second possibility, newly formed nucleosomes could differ from those in bulk chromatin by lacking histone H1. Since this histone is much less prevalent in transcriptionally active regions of chromatin (Reeves, 1984, 1988) and appears to associate with daughter strands after core histones during replication (DePamphilis and Wassarman, 1980), we investigated the possibility that the slow phase of the nuclease digestion curves reflected the slow reassociation of histone H1 with these regions. These studies were made possible by our observation that the submonomer pattern of DNA fragments generated by staphylococcal nuclease digestion of nuclei (i.e., fragments produced by digestion _into_ core regions) changed when H1 was selectively removed from these nuclei (Smerdon and Lieberman, 1981). In other words, histone H1 leaves a "signature" on the staphylococcal nuclease digestion profile when it is bound to nucleosomes. Therefore, we examined the submonomer pattern for newly repaired DNA which was folded into nucleosome core structures during the rapid phase of rearrangement (Smerdon, _et al._, 1982b). We observed the pattern for H1-containing nucleosomes in these regions, even when only 40% of the total repair patches had undergone the rapid phase of rearrangement. These results indicated that histone H1 rapidly associates with newly formed nucleosomes in repaired regions of chromatin, and the slow phase of the rearrangement time course must be due to some other mechanism.

More recently, we have focused on the distribution of repair patches in nucleosomes during the early and late repair phases in human cells (i.e., the third possibility listed above). We developed a method, using exonuclease III, to "map" the distribution of repair patches in homogeneous sized nucleosome core DNA from human cells (Lan and Smerdon, 1985). The results of this analysis for repair occurring during the early repair phase clearly showed an enhancement of repair patches in the 5' end of nucleosome cores (Figure 5). Furthermore, the data appeared to be sigmoidal in going from the 3' to 5' ends and could be approximated by a distribution in which there is a "gap" in repair synthesis near the center of core DNA regions (Lan and Smerdon, 1985). Using this simple approximation of the distribution, we found that the best fit to the data was achieved by a distribution where enhanced repair synthesis occurred in the first ~60 bases from the 5' end and the first ~30 bases from the 3' end. Thus, this approximation predicted a gap in repair synthesis of some 50 nucleotides near the center of core particles. Interestingly, the boundaries in this distribution correlate well with known structural features observed in isolated nucleosome core particles (see Discussion in Lan and Smerdon, 1985). Furthermore, this distribution accurately predicted the values obtained for the nuclease digestion analysis immediately following the rapid phase of rearrangement.

Since it was clear that repair patches inserted at early times after UV damage are nonuniformly distributed in nucleosome DNA, we wondered if this distribution changed during the prolonged slow phase of the rearrangement curves (Figure 1). To this end, we monitored the repair patch distribution in core DNA from UV irradiated cells that were pulse-labeled for 30 minutes and chased for varying times after the pulse period (Nissen, _et al._, 1986). The results clearly showed that over a 72 hour period, while the cells are at confluence, the distribution of patches becomes random in core DNA. Furthermore, the nonuniform distribution was not restored following replication (Nissen, <u>et al.</u>, 1986). The time course of this randomization process is similar to that of the slow change observed in the nuclease digestion data. Thus, both the apparent enhanced nuclease sensitivity of newly repaired DNA following nucleosome formation and the slow loss of this sensitivity over long time periods could be accounted for by our distribution analyses.

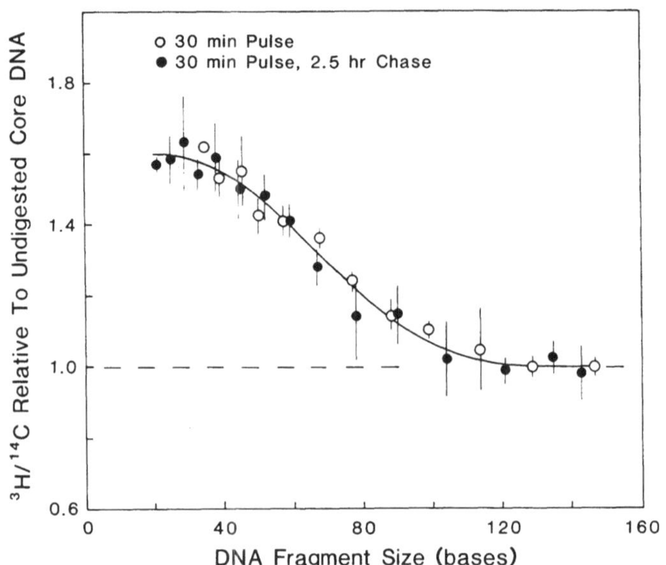

Figure 5. Ratio of ^{3}H and ^{14}C DPM as a function of DNA fragment size for exonuclease III digested 146 bp core DNA from cells harvested either immediately after a 30 minute pulse-labeling period with [^{3}H]dThd (O) or following a 2.5 hour chase period (●). The DNA was uniformly labeled with [^{14}C]dThd during replication. Data represent the mean $\pm$1 SD for either 2 (O) or 3 (●) different experiments. Values are normalized to the ^{3}H/^{14}C value for undigested core DNA in each case to allow direct comparison of the data from the different core DNA preparations. The dashed line denotes a random distribution of ^{3}H label. For details see Lan and Smerdon (1985).

Two new features of the maturation of newly repaired chromatin arose from these studies: (1) The reformation of nucleosome structure in newly repaired regions of chromatin "preserves", at least to some degree, the nonuniform alignment of repair patches in nucleosomes; and (2) There must be a slow repositioning of core histones along the DNA in these regions long after repair takes place. In regards to this latter feature, we examined the _extent_ of repositioning of core histones required to randomize the predicted distribution (Arnold, _et al._, 1987). Our analysis indicated that core histones would have to slide an average distance of ~50 bp in these regions to yield the randomization we measured using the exonuclease III mapping technique. Given the time period over which this randomization occurs, this amount of sliding would represent a linear diffusion rate of 10 to 100 times less than free diffusion of histones along the DNA (Arnold, _et al._, 1987).

An obvious question raised by these studies is: What is the reason for the nonuniform distribution of repair patches in nucleosomes during the early rapid phase of repair? Two possibilities immediately come to mind: (1) The early repair phase involves _preferential repair_ of the 5' and 3' end domains of nucleosome core regions (and presumably linker regions) ; or (2) The distribution of stable UV photoproducts in nucleosome core particles is nonuniform (i.e., _preferential damage_) and gives rise to the nonuniform distribution of repair patches in these regions. The preferential repair model predicts that if UV-induced photoproducts are uniformily distributed in nucleosomes, ~70% of the photoproducts should be removed during the early rapid repair phase in human cells (Lan and Smerdon, 1985). This value is within one standard deviation of the average obtained from ten different studies reported in the literature since 1973 (Figure 6).

A second prediction of the preferential repair model is that the concentration of repair patches should be enhanced in the central region of core DNA following repair synthesis occurring during the late slow repair phase. Recently, we have completed such an analysis of the slow repair phase in confluent human cells (strain AG1518) and found that the distribution of repair patches is, in fact, nearly random in core DNA (Nissen and Smerdon, 1988; and unpublished results). This is the case even at very early times after these regions undergo nucleosome reformation. Thus, although the distribution pattern for repair patches inserted during the late slow repair phase is clearly different from the pattern for repair associated with the early rapid repair phase, it is not the pattern predicted by the simplest form of the preferential repair model stated above. One possible variation of this model, which can account for these results, is that reformation of nucleosomes following repair during the slow phase does not preserve the alignment of patches in core DNA. For example, during the slow repair phase core histones may be _completely_ dissociated from the damaged DNA for repair enzymes to operate on the central core domain and then position randomly along this DNA during reformation of the nucleosome unit. Indeed, the nuclease digestion characteristics of these regions (in the unfolded state) are distinctly different from those associated with repair occurring during the early repair phase (Smerdon and Lieberman, 1987, Figure 2). Clearly, such differences may reflect a more unfolded chromatin conformation occurring during the late repair phase. Alternatively, the size of the repair patch inserted at late times after UV damage may differ from that for early repair.

In regards to preferential damage in nucleosomes (i.e., the second possibility listed above), two different laborator-

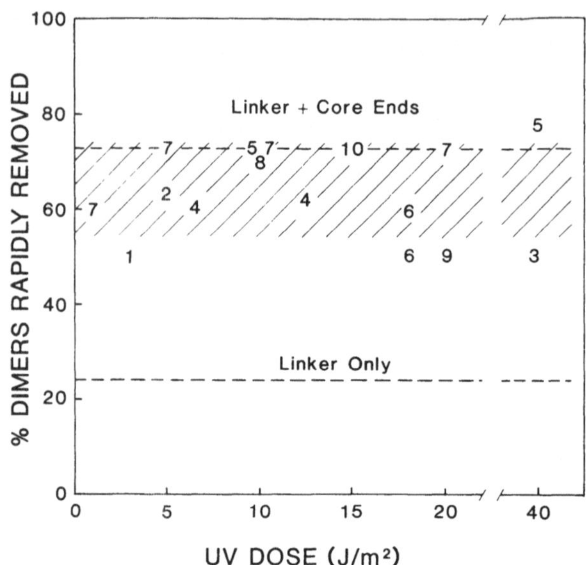

Figure 6. Percent of UV-induced photoproducts removed during the early rapid phase of repair in human cells following different UV doses. Each value was determined from data reported in the literature for the time course of removal of UV photoproducts in human cells. The values, represented as numbers, correspond to the following reports: 1, Zelle and Lohman (1979); 2, van Zeeland, et al.(1981); 3, Konze-Thomas, et al.(1979); 4, Paterson, et al. (1973); 5, Mitchell, et al. (1982); 6, Klocker, et al.(1982); 7, Kantor and Setlow (1981); 8, Clarkson, et al.(1983); 9, Cornelis (1978); and 10, Amacher, et al.(1977). The shaded region represents ±1 SD of the mean of these values.

ies have reported on the distribution of UV-induced cyclobutyl pyrimidine dimers (PD) between linker and core regions of bulk nucleosomes (Williams and Friedberg, 1979; Niggli and Cerutti, 1982). In each case, no difference in the PD level was observed between these two nucleosome subdomains (on a unit DNA basis). However, the results of these studies do not rule out the possibility that PD are distributed nonuniformly _within_ nucleosome core domains. Therefore, we developed a sensitive assay to measure the UV photoproduct yield within nucleosome core regions at the single nucleotide level (Gale, et al., 1987). This assay makes use of the observations made by Haseltine and coworkers (Doetsch, et al., 1985; Chan, et al., 1985) that the 3'—> 5' exonuclease activity of T4 DNA polymerase is quantitatively blocked at the 3' side of both PD and pyrimidine-pyrimidone (6-4) dimers. The results of our study showed that the packaging of DNA by core histones dramatically influences the distribution of PD (Figure 7). This distribution shows a striking 10.3 (±0.1) base periodicity, where regions of enhanced PD formation map to positions

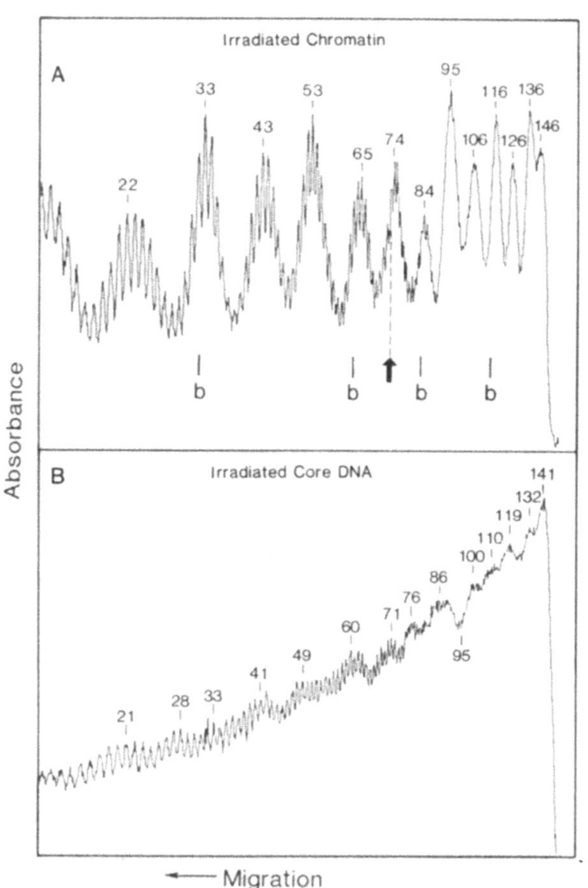

Figure 7. Laser densitometer scans of T4 DNA polymerase digestion profiles of (A) nucleosome core DNA from UV-irradiated calf thymus chromatin and (B) nucleosome core DNA isolated from unirradiated calf thymus chromatin and then irradiated as naked DNA. The values shown in (A) are the positions (in bases) of the ensemble peaks from the 5' end of core DNA. The dyad axis is marked by the solid arrow and the positions of helical bends observed in the crystal structure of isolated nucleosome cores are denoted by b (taken from Richmond, et al., 1984). See Gale and Smerdon (1988a) for details.

along the DNA strands that are farthest from the core histone surface (Gale, et al., 1987). A detailed analysis of this pattern demonstrates that certain characteristics correlate well with known structural features of isolated nucleosome core particles and indicates that these features are preserved in intact chromatin (Gale and Smerdon, 1988a). Thus, the photochemistry of PD formation in DNA is influenced by the folding of DNA into the nucleosome unit.

Currently, we are using this method to determine the

initial distribution and subsequent removal rates of PD from nucleosome subdomains in confluent human cells (K. Jensen and M. Smerdon, unpublished results). We have found that the PD distribution pattern in nucleosome cores of irradiated intact cells (12 J/m^2) is readily detected by this method and is similar to that of irradiated chromatin or isolated mononucleosomes (see also, Gale, et al., 1987). Therefore, it should be possible to determine if there is a "build-up" of PD in the central region of nucleosome core DNA (relative to the yield in the 5' and 3' end domains) following the early rapid phase of repair. Such an enhancement of PD is predicted by the preferential repair model given above.

Finally, we have recently analyzed the PD distribution obtained from irradiated chromatin (e.g., Figure 7A) to determine if this distribution could predict the distribution of repair patches observed by our exonuclease III mapping procedure (M. Smerdon and J. Gale, unpublished results). Assuming average repair patch lengths of 20 to 40 bases, this analysis yields curves which give good approximations of the data shown in Figure 5 except for the very 5' end of the core DNA region. Thus, the initial, nonuniform distribution of PD in nucleosomes of human cells may be a significant factor in determining the nonuniform distribution of repair patches in these regions, and therefore, it may be at the level of PD formation that structural features of the nucleosome play a role in repair distribution.

A MODEL FOR EXCISION REPAIR IN BULK CHROMATIN

As alluded to earlier, a number of physical and biochemical studies have provided evidence for a "break-down" of structural hierarchy in regions of chromatin engaged in transcription and replication (Reeves, 1984, 1988; Pederson, et al., 1986; Thoma and Sogo, 1988). Similarly, observations at both the light microscopy level (Hittleman, 1984, 1986) and the biochemical level (Mathis, et al., 1986; Harris and Boyd, 1987) have suggested that an unfolding of large structural domains in chromatin may occur in cells engaged in excision repair. This is not too surprising since the disruption of nucleosome structure during repair most likely represents a local event which requires relaxation of a considerably larger region of the compact chromatin fiber. Therefore, I have presented a schematic model in Figure 8 which correlates such a change in higher-ordered structure with events observed at the nucleosome level. As can be seen, this model portrays different structural states of the nucleofilament during the repair process. The salient features of this model are as follows: (1) DNA damage by many agents (e.g., UV radiation) results in only minor distortions in nucleosome and higher-order structures and these distortions can not account for the altered nucleosome structure observed following repair synthesis (this aspect of DNA damage was not discussed above, however, see Gale and Smerdon, 1988b); (2) Once a lesion is recognized (e.g., by the repair endonuclease) the chromatin fiber is relaxed and this relaxation may involve postsynthetic modifications of histones and/or nonhistones covering a much larger region of the chromatin fiber; (3) Either during incision or the initiation of repair synthesis, local nucleosome structure (i.e., in the region of the lesion) is dis-

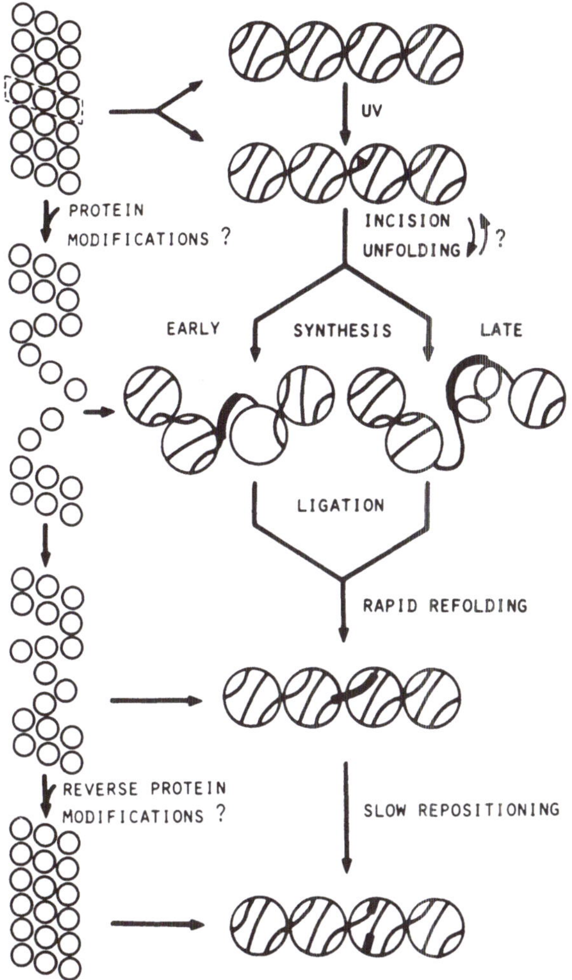

Figure 8. Possible transitions during excision repair of bulk chromatin in mammalian cells. See text for discussion.

rupted via an unfolding mechanism (as depicted in Figure 8), which may involve displacement of histone H1, an induced sliding mechanism, where core histones are "forced" to migrate along the DNA away from the lesion, or perhaps a complete dissociation of core histones from the DNA; (4) This disruption may involve different mechanisms for repair during the early rapid phase and repair during the late slow phase; (5) Following ligation, the DNA is rapidly folded into a nucleosome structure and, at this point, repair patches inserted during early times after damage are heavily weighted toward the 5' end of nucleosome cores; (6) The positioning of core histones in these regions slowly changes during repair at other sites in the chromatin fiber and/or during the refolding of the chromatin fiber into higher-ordered structures; and (7) The reformation of higher-ordered structures may involve reversal of protein modifications. Obviously, more experimentation is required before many of these features can be definitively tested. However, the purpose of presenting this model here is to provide the reader with a conceptual framework in which this rather complex series of events can be discussed. Hopefully, it will stimulate new ideas for the study of DNA repair in mammalian cells, particularly into mechanisms that may facilitate repair through modulation of chromatin structure.

ACKNOWLEDGEMENTS

The recent studies discussed in this report were supported by N.I.H. Grants ES02614, ES04106 and ES03720, and by an N.I.H. Research Career Development Award (ES00110). I would like to thank the members of my laboratory for helpful discussions, and Karen Jensen and Drs. Raymond Reeves, Jeanette Huijzer, Ralph Yount and Michael Griswold for their critical evaluation of this manuscript. I also thank Colleen Fowles, Diane Smerdon and Karen Libey for help in preparation of this manuscript.

REFERENCES

Adolph, K. W. 1988, "Chromosomes and Chromatin", Vols. I, II and III, CRC Press, Baton Rouge, FL.

Amacher, D. E., Elliott, J. A., and Lieberman, M. W., 1977, Differences in removal of acetylaminofluorene and pyrimidine dimers from the DNA of cultured mammalian cells, Proc. Natl. Acad. Sci. USA, 74:1553.

Arnold, G. E., Dunker, A. K., and Smerdon, M. J., 1987, Limited nucleosome migration can completely randomize DNA repair patches in intact human cells, J. Mol. Biol., 196:433.

Bentley, G. A., Lewit-Bentley, A., Finch, J. T., Podjarny, A. D. and Roth, M., 1984, Crystal structure of the nucleosome core particle at 16 Å resolution, J. Mol. Biol., 176:55.

Berkowitz, E. M., and Silk, H., 1981, Methylation of chromosomal DNA by two alkylating agents differing in carcinogenic potential, Cancer Letters, 12:311.

Bodell, W. J., and Cleaver, J. E., 1981, Transient conformation changes in chromatin during excision repair of ultraviolet damage to DNA, Nucleic Acids Res., 9:203.

Bodell, W. J., Cleaver, J. E., and Roti Roti, J. L., 1984, Inhibition by hyperthermia of repair synthesis and chromatin reassembly of ultraviolet-induced damage to DNA, Radiation Res., 100:87.

Chan, G. L., Doetsch, P. W., and Haseltine, W. A., 1985, Cyclobutane pyrimidine dimers and (6-4) photoproducts block polymerization by DNA polymerase I, Biochemistry, 24:5723.

Ciarrocchi, G. and Pedrini, A. M., 1982, Determination of pyrimidine dimer unwinding angle by measurement of DNA electrophoretic mobility, J. Mol. Biol.,155:177.

Clarkson, J. M., and Mitchell, D. L., 1983, The effect of various inhibitors of DNA synthesis on the repair of DNA photoproducts, Biochim. Biophys. Acta, 740:355.

Cleaver, J. E., 1982, Normal reconstruction of DNA supercoiling and chromatin structure in cockayne syndrome cells during repair of damage from ultraviolet light, Am. J. Hum. Genet., 34:566.

Cornelis, J. J., 1978, The influence of inhibitors on dimer removal and repair of single-strand breaks in normal and bromodeoxyuridine substituted DNA of HeLa cells, Biochim. Biophys Acta, 521:134.

de Murcia, G., Huletsky, A., Lamarre, D., Gaudreau, A., Pouyet, J., Daune, M., and Poirier, G. G., 1986, Modulation of chromatin superstructure induced by poly (ADP-ribose) synthesis and degradation, J. Biol. Chem., 261: 7011.

DePamphilis, M. L., and Wassarman, P. M., 1980, Replication of eukaryotic chromosomes: A close-up of the replication fork, Ann. Rev. Biochem. , 49:627.

Doetsch, P. W., Chan, G. L., and Haseltine, W. A., 1985, T4 DNA polymerase (3'-5') exonuclease, an enzyme for the detection and quantitation of stable DNA lesions: the ultraviolet light example, Nucleic Acids Res., 13:3285.

Dresler, S. L., 1985, Stimulation of deoxyribonucleic acid excision repair in human fibroblasts pretreated with sodium butyrate, Biochemistry, 24:6861.

Gale, J. M., Nissen, K. A., and Smerdon, M. J., 1987, UV-induced formation of pyrimidine dimers in nucleosome core DNA is strongly modulated with a period of 10.3 bases, Proc. Natl. Acad. Sci. USA, 84:6644.

Gale, J. M. and Smerdon, M. J. , 1988a, UV photofootprint of nucleosome core DNA in intact chromatin having different structural states, J. Mol. Biol., 204:949.

Gale, J. M. and Smerdon, M. J. , 1988b, UV-induced pyrimidine dimers and trimethylpsoralen cross-links do not alter chromatin folding in vitro, Biochemistry, 27:7197.

Harris, P. V., and Boyd, J. B., 1987, Pyrimidine dimers in Drosophila chromatin become increasingly accessible after irradiation, Mutat. Res., 183:53.

Hittelman W. N., and Pollard, M., 1984, Visualization of chromatin events associated with repair of ultraviolet light-induced damage by premature chromosome condensation, Carcinogenesis, 5:1277.

Hittelman,W. N., 1986, Visualization of chromatin events during DNA excision repair in XP cells: deficiency in localized but not generalized chromatin events, Carcinogenesis, 7:1975.

Hunting, D. J., Dresler, S. L., and Lieberman, M. W., 1985, Multiple conformational states of repair patches in chromatin during DNA excision repair, Biochemistry, 24:3219.

Kantor, G. J., and Setlow, R. B., 1981, Rate and extent of DNA repair in nondividing human diploid fibroblasts, Cancer Res., 41:819.

Klocker, H., Auer, B., Burtscher, H. J., Hirsch-Kauffmann, M., and Schweiger, M., 1982, Repair rate in human fibroblasts measured by thymine dimer excorporation, Mol. Gen. Genet., 188:309.

Konze-Thomas, B. , Levinson, J. W., Maher, V. M., and McCormick, J. J., 1979, Correlation among the rates of dimer excision, DNA repair replication, and recovery of human cells from potentially lethal damage induced by ultraviolet radiation, Biophys. J. , 28:315.

Lan, S. Y., and Smerdon, M. J., 1985, A nonuniform distribution of excision repair synthesis in nucleosome DNA, Biochemistry, 24:7771.

Lang, M. C., de Murcia, G., Mazen, A., Fuchs, R. P. P., Leng,

M., and Daune, M. , 1982, Non-random binding of N-acetoxy-N-2-acetylaminofluorene to chromatin subunits as visualized by immunoelectron microscopy, Chem. Biol. Interact., 41:83.

Lieberman, M. W., Smerdon, M. J., Tlsty, T. D., and Oleson, F. B., 1979, The role of chromatin structure in DNA repair in human cells damaged with chemical carcinogens and ultraviolet radiation, in: "Environmental Carcinogenesis", P. Emmelot and E. Kriek, eds., Elsevier/NorthHolland Biomedical Press, Amsterdam.

Lutter L. C., 1978, Kinetic analysis of deoxyribonuclease I cleavages in the nucleosome core: evidence for a DNA superhelix, J. Mol. Biol., 124:391.

Mathis, G., and Althaus, F. R., 1986, Periodic changes of chromatin organization associated with rearrangement of repair patches accompany DNA excision repair of mammalian cells, J. Biol. Chem., 261:5758.

Matthews, H. R., 1988, Histone modifications and chromatin structure, in: "Chromosomes and Chromatin,, Vol. I, K. W. Adolph, Ed., CRC Press, Baton Rouge, FL.

McGhee, J. D., and Felsenfeld, G., 1979, Reaction of nucleosome DNA with dimethyl sulfate, Proc. Natl. Acad. Sci. USA, 76:2133.

Mitchell, D. L., Nairn, R. S., Alvillar, J. A., and Clarkson, J. M., 1982, Loss of thymine dimers from mamnalian cell DNA. The kinetics for antibody-binding sites are not the same as that for T4 endonuclease V sites, Biochim. Biophys. Acta, 697:270.

Morse, R. H., and Simpson, R. T., 1988, DNA in the nucleosome, Cell, 54:285.

Nelson, W. G., Pienta, K. J., Barrack, E. R., and Coffey, D. S., 1986, The role of the nuclear matrix in the organization and function of DNA, Ann. Rev. Biophys. Biophys. Chem. , 15:457.

Niggli, H. J., and Cerutti, P. A., 1982, Nucleosomal distribution of thymine photodimers following far- and near-ultraviolet irradiation, Biochem. Biophys. Res. Commun., 105:1215.

Nissen, K. A., Lan, S. Y., and Smerdon, M. J., 1986, Stability of nucleosome placement in newly repaired regions of DNA, J. Biol. Chem., 261:8585.

Nissen, K. A., and Smerdon, M. J., 1988, Excision repair in different domains of nucleosome core DNA in human cells, J. Cell. Biochem., Supplement 12A, 295.

Oleson, F. B., Mitchell, B. L., Dipple, A., and Lieberman, M. W.,1979, Distribution of DNA damage in chromatin and its relation to repair in human cells treated with 7-bromomethylbenz(a)anthracene, Nucleic Acids Res., 7:1343.

Paterson, M. C., Lohman, P. H. M., and Sluyter, M. L., 1973, Use of a UV endonuclease from Micrococcus luteus to monitor the progress of DNA repair in UV-irradiated human cells, Mutat. Res., 19:245.

Pederson, D. S., Thoma, F., Simpson, R. T., 1986, Core particle, fiber, and transcriptionally active chromatin structure, Ann. Rev. Cell Biol., 2:117.

Perry, M., and Chalkley, R., 1982, Histone acetylation increases the solubility of chromatin and occurs sequentially over most of the chromatin. Novel model for the biological role of histone acetylation, J. Biol. Chem., 257:7336.

Ramanathan, B., and Smerdon, M. J., 1986, Changes in nuclear protein acetylation in u.v.-damaged human cells, Carcinogenesis, 7:1087.

Reeves, R., 1984, Transcriptionally active chromatin, Biochim. Biophys. Acta, 782:343.

Reeves, R., 1988, Active chromatin structure, in: "Chromosomes and Chromatin', Vol. I, K. W. Adolph, Ed., CRC Press, Baton Rouge, FL.

Richmond, T. J., Finch, J. T., Rushton, B., Rhodes, D., Klug, A., 1984, Structure of the nucleosome core particle at 7 Å resolution, Nature, 311:532.

Shick, V. V., Belyavsky, A. V., Mirzabekov, A. D., 1985, Primary organization of nucleosomes. Interaction of non-histone high mobility group proteins 14 and 17 with nucleosomes, as revealed by DNA-protein crosslinking and immunoaffinity isolation, J. Mol. Biol., 185:329.

Sidik, K., and Smerdon, M. J., 1984, Nuclease sensitivity of repair-incorporated nucleotides in chromatin and nucleosome rearrangement in human cells damaged by methyl methanesulfonate and methylnitrosourea, Carcinogenesis, 5:245.

Sidik, K., and Smerdon, M. J., 1987, Rearrangement of nucleosome structure during excision repair in xeroderma pigmentosum (group A) human fibroblasts, Carcinogenesis, 8:733.

Singer, B., and Kusmierek, J. T., 1982, Chemical mutagenesis, Ann. Rev. Biochem., 52:655.

Smerdon, M. J., 1983, Rearrangements of chromatin structure in newly repaired regions of deoxyribonucleic acid in human cells treated with sodium butyrate or hydroxyurea, Biochemistry, 22:3516.

Smerdon, M. J., 1986, Completion of excision repair in human cells. Relationship between ligation and nucleosome formation, J. Biol. Chem., 261:244.

Smerdon, M. J., Kastan, M. B., and Lieberman, M. W., 1979, Distribution of repair-incorporated nucleotides and nucleosome rearrangement in the chromatin of normal and xeroderma pigmentosum human fibroblasts, Biochemistry, 18:3732.

Smerdon, M. J., Lan, S. Y., Calza, R. E., and Reeves, R., 1982a, Sodium butyrate stimulates DNA repair in UV-irradiated normal and xeroderma pigmentosum human fibroblasts, J. Biol. Chem., 257:13441.

Smerdon, M. J., and Lieberman, M. W., 1978, Nucleosome rearrangement in human chromatin during UV-induced DNA repair synthesis, Proc. Natl. Acad. Sci. USA, 75:4238.

Smerdon, M. J. and Lieberman, M. W., 1980, Distribution within chromatin of deoxyribonucleic acid repair synthesis occurring at different times after ultraviolet radiation, Biochemistry, 19:2992.

Smerdon, M. J. and Lieberman, M. W., 1981, Removal of histone H1 from intact nuclei alters the digestion of nucleosome core DNA by staphylococcal nuclease, J. Biol. Chem., 256:2480.

Smerdon, M. J., Watkins, J. F., and Lieberman, M. W., 1982b, Effect of histone H1 removal on the distribution of ultraviolet-induced deoxyribonucleic acid repair synthesis within chromatin, Biochemistry, 21:3879.

Smith, P. J., 1986, n-Butyrate alters chromatin accessibility to DNA repair enzymes, Carcinogenesis, 7:423.

Sutherland, B. M. and Shih, A. G., 1983, Quantitation of pyrimidine dimer contents of nonradioactive deoxyribonucleic acid by electrophoresis in alkaline gels, Biochemistry, 22:745.

Thoma, F., and Sogo, J. M., 1988, Structures of bulk and transcriptionally active chromatin revealed by electron microscopy. in: "Chromosomes and Chromatin" Vol. I, K. W. Adolph, Ed., CRC Press, Baton Rouge, FL.

Tlsty, T. D., and Lieberman, M. W., 1978, The distribution of DNA repair synthesis in chromatin and its rearrangement following damage with N-acetoxy-2-acetylaminofluorene, Nucleic Acids Res., 5:3261.

Ueda, K., and Hayaishi, O., 1985, ADP-ribosylation, Ann. Rev. Biochem., 54:73.

Van Zeeland, A. A., Smith, C. A., and Hanawalt, P. C. 1981, Sensitive determination of pyrimidine dimers in DNA of UV-irradiated mammalian cells. Introduction of T4 endonuclease V into frozen and thawed cells, Mutat. Res., 82:173.

Watkins, J. F. and Smerdon, M.J. 1985a. Nucleosome rearrangement vitro. Two phases of salt induced nucleosome migration in nuclei. Biochemistry, 24:7279.

Watkins, J.F. and Smerdon, M.J. 1985b. Nucleosome rearrangement in vitro. Formation of nucleosomes in newly repaired regions of DNA. Biochemistry, 24:7288.

Wiesehahn, G. and Hearst, J.E. 1978. DNA unwinding induced by photoaddition of psoralen derivatives and determination of dark-binding equilibrium constants by gel electrophoresis. Proc. Natl. Acad. Sci. USA., 75:2703.

Williams, J.I. and Friedberg, E.C. 1979. Deoxyribonucleic acid excision repair in chromatin after ultraviolet irradiation of human fibroblasts in culture, Biochemistry, 18:3965.

Williams, J.I. and Friedberg, E.C. 1982. Increased levels of unscheduled DNA synthesis in UV-irradiated human fibroblasts pretreated with sodium butyrate. Photochem. Photobiol., 36:423.

Zelle, B. and Lohman, P.H.M. 1979. Repair of UV-endonuclease-susceptible sites in the 7 complementation groups of xeroderma pigmentosum A through G, Mutat. Res., 62:363.

Zolan, M.E., Smith, C.A., Calvin, N.M., Hanawalt, P.C., 1982. Rearrangement of mammalian chromatin structure following excision repair, Nature, 299:462.

MODULATION OF ACTIVITY OF HUMAN CHROMATIN-ASSOCIATED ENDONUCLEASES ON DAMAGED DNA BY NUCLEOSOME STRUCTURE

Muriel W. Lambert and David D. Parrish

Department of Pathology, UMDNJ - New Jersey Medical School, Newark, New Jersey, USA

INTRODUCTION

The excision repair pathway in mammalian cells is a complex process involving a number of enzymatic steps. My laboratory has focused on the initial endonuclease mediated step. Our approach has been to develop a method to isolate biologically active endonucleases from mammalian nuclei and examine them for specificity of action on DNA containing specific types of damage (e.g., apurinic/apyrimidinic sites, interstrand cross-links, pyrimidine dimers, monoadducts, intercalated adducts). Using this methodology, we have isolated a series of nine chromatin-associated DNA endonucleases from the nuclei of normal human cells. We have been able to detect individual selective activities of these endonucleases on a number of different types of damaged DNA, and have found that different ones of them recognize different types of DNA damage. We have also examined the endonucleases in cells from patients with a number of genetic diseases known to be deficient in DNA repair (i.e., xeroderma pigmentosum and Fanconi's anemia). The results have indicated different defects associated with the endonuclease(s), depending upon the genetic disease.

The study of DNA repair mechanisms in mammalian cells is more complex than in bacteria due to the fact that their endonucleases and other DNA repair enzymes have to interact with DNA which is associated with histones and various nonhistone proteins which form chromatin. Chromatin, which is organized into repeating units known as nucleosomes, has been shown to influence both the accessibility of DNA to damage by different DNA damaging agents as well as to repair of this damage (Lan and Smerdon, 1985; Bohr, et al., 1987; Smerdon, 1989). Although a number of approaches have been utilized to study the involvement of chromatin in repair of a number of types of lesions in cellular DNA, little is known regarding the influence of nucleosome structure on the activity of specific, isolated mammalian DNA endonucleases on particular types of DNA adducts. My laboratory has been particularly interested in studying the involvement of nucleosome structure

DNA Repair Mechanisms and Their Biological Implications in Mammalian Cells
Edited by M.W. Lambert and J. Laval
Plenum Press, New York

on the activity of the human endonucleases, which we have isolated, on specific types of damaged DNA. These studies have revealed that interaction of factors closely associated with the endonucleases, or of the endonucleases themselves, with nucleosome structure is of critical importance in regulation of endonuclease activity on damaged nucleosomal substrates. They have also shown that in at least one genetic disease, xeroderma pigmentosum, complementation group A (XPA), this interaction between specific endonucleases and damaged nucleosomes is defective and may represent a general defect in endonucleases from XPA cells involved in repair. This paper will review the methodology employed in isolation of the various endonucleases we have studied, the damaged substrate specificities of these endonucleases, the ability of the endonucleases to bind to damaged DNA and then the important influence of chromatin structure on their activity. Both our results from normal human endonucleases and those from specific endonucleases from XP and Fanconi's anemia cells will be described.

DNA ENDONUCLEASE ISOLATION

The endonucleases we are studying have been extracted from nuclei isolated from normal human lymphoblastoid cells (GM 1989 and GM 3299) and lymphoblastoid cells from patients with XP, complementation group A, XPA (GM 2345 and GM 2250A), both of which have been transformed with Epstein-Barr virus (Institute for Medical Research, Camden, NJ). In addition, we have recently begun to work with Epstein-Barr virus transformed lymphoblastoid cells from patients with Fanconi's anemia, complementation groups A and B (HSC 72 and HSC 230) (gift of Dr. Manual Buckwald). Chromatin-associated and nucleoplasmic proteins were both extracted from cell nuclei and tested for DNA endonuclease and exonuclease activity against calf thymus DNA and ^{3}H-poly d(A-T), respectively (Okorodudu, et al., 1982). Both endonuclease and exonuclease activity were found in the nucleoplasmic proteins, whereas, only endonuclease activity was found in the chromatin-associated proteins (Okorodudu, et al., 1982; Lambert, et al., 1982). The chromatin-associated proteins were passed through a CM-Sephadex column and then separated on an isoelectric focusing column (Okorodudu, et al., 1982; Lambert, et al., 1982). Nine clearly separable endonuclease activities, containing no exonuclease activity, with isoelectric points ranging from 3.8 - 9.8, were obtained (Figure 1). Eight of these endonucleases, pI 3.9 - 9.2, showed some activity on undamaged DNA (Okorodudu, et al., 1982; Lambert, et al., 1982). A similar series of endonucleases has been isolated from XPA nuclei (Okorodudu, et al., 1982) and from nuclei from Fanconi's anemia cells, complementation groups A and B.

In a separate set of experiments we have also found a similar set of nine chromatin-associated DNA endonucleases in the chromatin of Cloudman melanoma cells grown in mice *in vivo*. Their presence in both mouse melanoma cells grown *in vivo* and in human lymphoblastoid cells grown in culture indicates that these are highly conserved mammalian nuclear endonucleases (Lambert, et al., 1982).

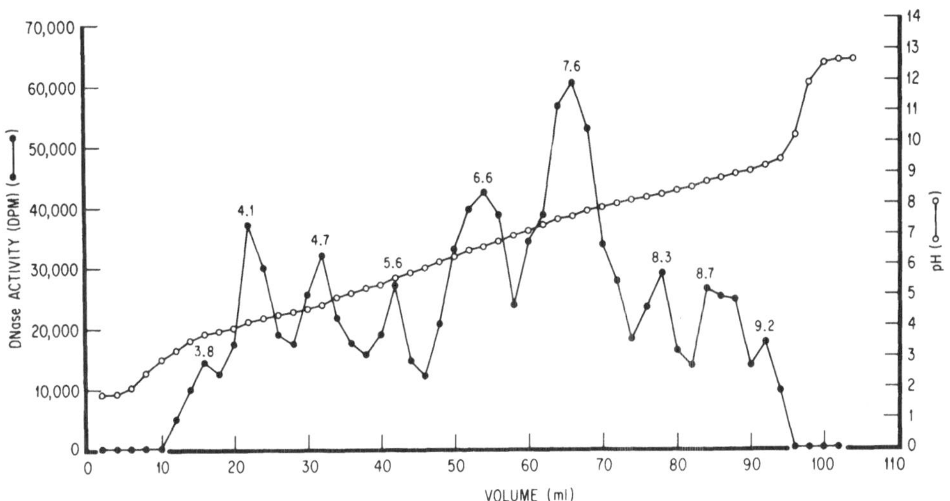

Figure 1. Isoelectric focusing patterns of chromatin-associated DNA endonucleases from normal human lymphoblastoid cells. (● - ●), DNase activity on calf thymus DNA; (O - O), pH gradient. (Okorodudu, et al., 1982).

NORMAL HUMAN DNA ENDONUCLEASE ACTIVITY ON DAMAGED DNA

Each of the nine human chromatin-associated DNA endonucleases we have isolated has been tested on a variety of different types of DNA lesions. These include apurinic/-apyrimidinic (AP) sites, psoralen plus long wavelength ultraviolet radiation (UVA) induced monoadducts and interstrand cross-links, intercalation of various agents, and short wavelength ultraviolet radiation (UVC) induced damage. The incisibilty of DNA containing these various lesions, by each of these nine normal human endonucleases, will be discussed in turn.

AP DNA

AP sites can be produced in cellular DNA by a number of different mechanisms. AP damage is repaired by an excision repair pathway which is initiated by a specific AP endonuclease (Laval and Laval, 1980; Lindahl, 1982). Endonucleases in mammalian cells which specifically recognize AP sites have been isolated and purified from a variety of sources (reviewed in Lindahl, 1982; Sancar and Sancar, 1988). We have examined each of the nine chromatin-associated DNA endonucleases we isolated from cultured human cells for AP endonuclease activity. Circular, supercoiled, duplex PM2 bacteriophage DNA was rendered partially AP by first either alkylating the DNA with methyl methanesulfonate (MMS) and then heating it at 70^{0}C or by heating it at 70^{0}C at pH 5.2 (Lindahl and Nyberg, 1972; Lambert, M.W., et al., 1983). Endonuclease activity on this substrate was determined using a gel electrophoretic assay which measured the conversion of superhelical DNA (form I) to nicked, relaxed, circular DNA (form II) (Lambert, M.W., et al., 1983). We have found that two of the chromatin-associated endonucleases, pIs 9.2 and 9.8, have exceedingly more activity on AP DNA than on undamaged DNA, with lesser activities found in some of the other endonucleases (Figure 2) (Lambert, M.W., et al., 1983). These endonucleases produced single-strand breaks in the AP DNA with the number of breaks produced by endonucleases, pIs 9.2 and 9.8, equal to the number of AP sites per DNA molecule. These endonucleases did not produce any linear, (i.e., form III) DNA. Whether the two endonucleases are different or the same is currently under investigation.

Psoralen Plus UVA Damaged DNA

Repair of DNA interstrand cross-links is known to occur in mammalian cells, however, the mechanism of this repair is poorly understood (Kaye, et al., 1980; Bredberg, 1982; Gruenert and Cleaver, 1985). Although, in mammalian cells, *in vitro* studies have suggested that excision is involved in repair of DNA interstrand cross-links (Kaye, et al., 1980; Bredberg, et al., 1982; Gruenert and Cleaver, 1985), a DNA endonuclease from human cells which specifically recognizes and incises these lesions has not been identified or isolated. We wished to determine whether any of the chromatin-associated endonucleases we have isolated could recognize and incise DNA containing interstrand cross-links. We chose as a damaging agent psoralen plus UVA radiation because, of all the cross-linking agents, it produces the largest proportion of cross-links to other types of adducts and its mechanism of interaction with DNA has been studied extensively. Psoralen first intercalates between adjacent DNA base pairs in the dark and then, upon photoreaction with UVA, forms covalent monoadducts and interstrand cross-links (Ben-Hur and Song, 1984; Cimino, et al., 1985; Vigny, et al., 1985). An advantage of using the psoralen system is that reaction conditions can be adjusted

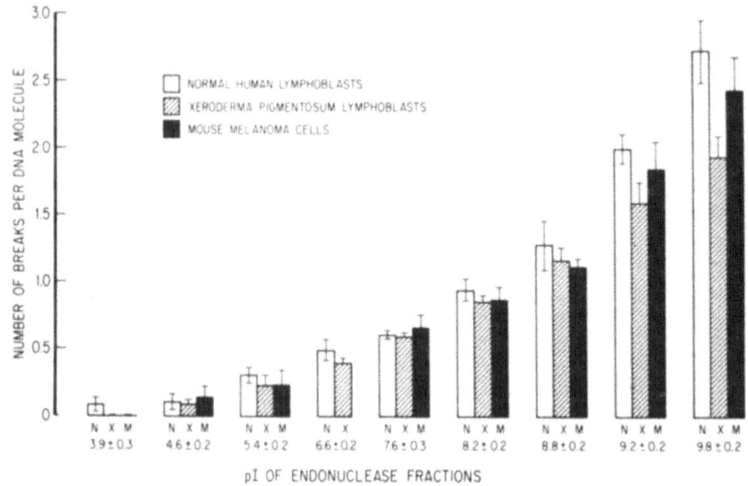

Figure 2. Action of chromatin-associated DNA endonucleases on MMS (12.5 mM) alkylated and heat depurinated PM2 DNA. Vertical lines represent ± S.E.M. (Lambert, M.W., et al., 1983).

so that just one of these three types of lesions produced predominates and each one can be separately studied (Ben-Hur and Song, 1984; Cimino, et al., 1985; Vigny, et al., 1985).

Three different psoralens were utilized, 4,5',8-trimethylpsoralen (TMP) and 8-methoxypsoralen (8-MOP), both of which upon photoactivation produce monoadducts and DNA interstrand cross-links (Ben-Hur and Song, 1984; Cimino, et al., 1985; Vigny, et al., 1985) and angelicin, a closely related furocoumarin derivative, which forms mainly monoadducts (Bredberg, et al., 1982; Cimino, et al., 1985). For both 8-MOP and TMP, photoreaction was carried out using a treatment protocol which involved exposing the psoralen treated DNA to two doses of UVA, an initial one after the 8-MOP or TMP had intercalated into the DNA and a second one after the unbound 8-MOP or TMP had been removed by dialysis (Lambert, et al., 1988). This procedure increases the number of interstrand cross-links formed (Ben-Hur and Elkind, 1973; Bredberg, 1982) and is sufficient to produce cross-links in 99% of the DNA molecules (Lambert, et al., 1988). No cross-links were detected in DNA similarly treated with angelicin plus UVA (Lambert, et al., 1988).

We found that two of the nine chromatin-associated DNA endonucleases, pIs 4.6 and 7.6, were active on both TMP and 8-MOP plus UVA treated DNA (Figure 3A) (Lambert, et al., 1988). The levels of activity between these two endonucleases were similar and both were more than twice as active on psoralen plus UVA treated DNA as on undamaged DNA. PM2 DNA and a plasmid DNA, pWT830/pBR322 (a clone of the entire SV40 and pBR322 genomes), were used as substrate with similar results. The endonuclease, pI 7.6, however, was much more active than the endonuclease, pI 4.6, on angelicin treated DNA; the activity of the latter was only approximately 30% of that observed on 8-MOP plus UVA treated DNA (Figure 3B) (Lambert, et al., 1988). The activity of these two endonucleases was also tested on DNA treated with 8-MOP or TMP in the dark but not exposed to UVA irradiation so that the psoralen only intercalated with DNA and did not form monoadducts or interstrand cross-links. Only the endonuclease, pI 4.6, showed activity on the psoralen-intercalated DNA (Lambert, et al., 1988). This was approximately 20% of the activity that this endonuclease showed on DNA treated with either TMP or 8-MOP photoactivated DNA. This indicates that although intercalation is recognized by this enzyme, it is not the major type of damage recognized in the psoralen plus UVA treated DNA.

These results indicate that the endonuclease, pI 7.6, recognizes the psoralen monoadduct. In support of this is our preliminary finding that a monoclonal antibody for 8-MOP plus UVA monoadducts (Santella, et al., 1985) inhibits the activity of this endonuclease on 8-MOP plus UVA DNA treated with the monoclonal antibody. The activity of the endonuclease, pI 4.6, is only slightly affected. These results also show that the endonuclease, pI 4.6, recognizes the intercalation; whether, as seems likely, it also recognizes the cross-link is under further investigation.

UVC Irradiated DNA

A number of different adducts are produced in DNA by UVC

(254 nm) irradiation, the major one of which is the pyrimidine dimer; pyrimidine-pyrimidine (6-4) lesions and other minor photoproducts are also formed (reviewed in Friedberg, 1984). Mammalian cells repair UVC radiation induced lesions by excision repair (Lehmann and Karran, 1981; Hanawalt, et al., 1982; Teebor and Frenkel, 1983; Rubin, 1988). We examined the

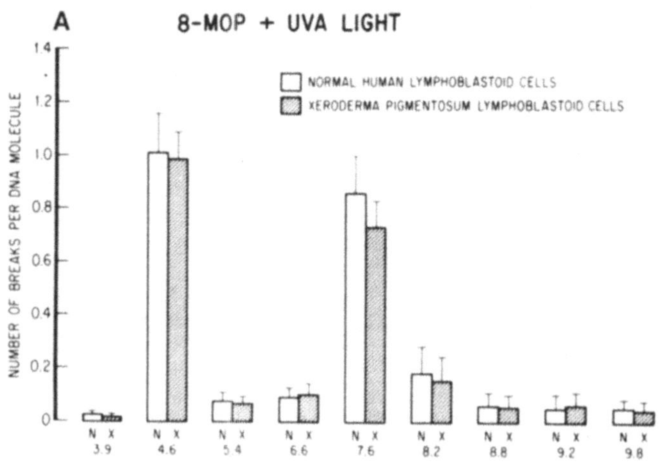

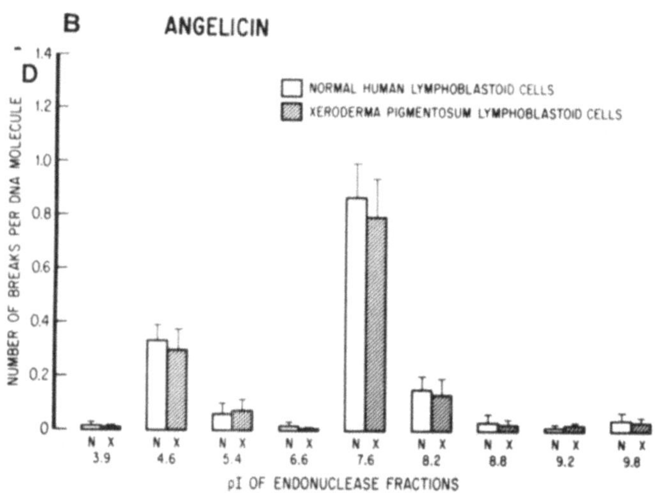

Figure 3. Activity of chromatin-associated DNA endonucleases from normal human and XPA lymphoblastoid cell lines on PM2 DNA treated with (A) 15 µg/ml 8-MOP and (B) 25 µg/ml angelicin plus UVA light. These values have had subtracted from them the enzyme activity on undamaged DNA. Vertical lines represent ± S.E.M. (Lambert, et al., 1988)

activity of each of our nine normal chromatin-associated endonucleases on PM2 or plasmid DNA which had been irradiated with varying dosages of UVC light. One of the endonucleases, pI 7.6, was active on this damaged substrate (Figure 4). Studies in which pyrimidine dimers were removed by treating UVC irradiated DNA with Escherichia coli DNA photolyase (gift of Dr. Aziz Sancar) plus UVA irradiation indicate that the pyrimidine dimer is the major type of UVC induced lesion recognized by this endonuclease (Lambert, et al., in preparation). This is the same endonuclease which incises DNA containing psoralen monoadducts. It has been postulated that a UV endonuclease may be involved in the initial stages of repair of DNA adducts produced by psoralen plus UVA irradiation, thus simulating pyrimidine dimer excision (Song and Tapley, 1979; Kaye, et al., 1980; Gruenert and Cleaver, 1985). The covalent attachment of the psoralen molecule to DNA involves a cyclobutyl linkage just as does the pyrimidine dimer (Song and Tapley, 1979; Ben-Hur and Song, 1984; Vigny, et al., 1985; Cimino, et al., 1985). It is possible that for both the psoralen monoadduct and the pyrimidine dimer the cyclobutyl ring is an important determinant for recognition of the damaged substrate by the enzyme. A DNA endonuclease has been purified from rat liver which incises DNA treated with 260 nm UV irradiation or acetylaminofluorene and this same enzyme also incises DNA treated with psoralen plus UVA irradiation (Tomura and Van Lanker, 1980). Further purification of our enzyme is currently in progress; this will aid in comparing it with other endonucleases reported in the literature.

Adriamycin Treated DNA

There is conflict in the literature regarding the repairability of damage produced by the anthracycline, adriamycin (ADM). This may relate to the assay methods used. ADM did not stimulate unscheduled DNA synthesis (UDS) in rat myocardial cells (Fialkoff, et al., 1979) and human peripheral

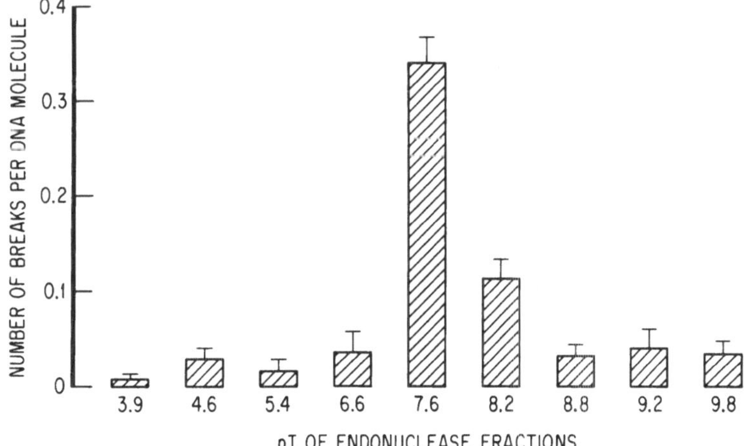

Figure 4. Action of chromatin-associated DNA endonucleases from normal human lymphoblastoid cell lines on plasmid pWT830/pBR322 DNA irradiated with 180 J/m^2 UVC. These values have had subtracted from them the enzyme activity on undamaged DNA. Vertical lines represent ± S.E.M.

leukocytes (Lambert, B., et al., 1983). It is possible that the inability of ADM to induce detectable DNA repair synthesis may be due to inhibitory effects of this drug on DNA polymerase(s) (Lewensohn and Ringborg, 1983), as has been shown for alpha and beta calf thymus DNA polymerases (Tanaka and Yoshida, 1980) and avian myeloblastosis virus DNA polymerase (Matson, et al., 1980). On the other hand, ADM has been shown to induce UDS during spermatogenesis in rabbits (Schmid and Zbinden, 1979). Ross and Smith (1982), using an alkaline elution assay, found that ADM induced effects were repaired slowly in mouse leukemia cells and suggested that the slow rate of repair may be related to the degree of retention of the drug within cells.

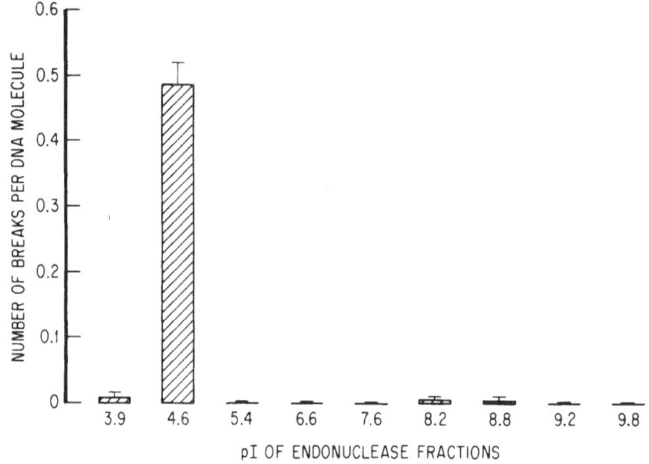

Figure 5. Activity of normal human chromatin-associated DNA endonucleases on PM2 DNA treated with 10 µM adriamycin. These values have had substracted from them the enzyme activity on undamaged DNA. Vertical lines represent ± S.E.M.

The interaction of ADM with DNA involves intercalation between two adjacent nucleotide base pairs in the DNA molecule (Zunio, et al., 1977; Waring, 1981; Pachter, et al., 1982; Eriksson, et al., 1988). The mechanism of repair of damage produced by ADM has not been well studied and specific enzymes that recognize this lesion have not been identified. However, the presence of cellular repair enzymes that recognize DNA damage produced ADM have been hypothesized (Lee and Byfield, 1976; Ross, et al., 1981; Zwelling, et al., 1981). We examined our chromatin-associated endonucleases form normal human cells to determine whether any of them can incise ADM treated PM2 DNA. One endonuclease, pI 4.6, was selectively active on DNA intercalated with ADM (Figure 5). The activity depended upon the concentration of ADM. This endonuclease is also active on DNA intercalated with psoralen.

XPA DNA ENDONUCLEASE ACTIVITY ON DAMAGED DNA

Cells from patients with XPA have been shown to be deficient in repair of damage produced by sunlight or short wavelength (UVC) radiation as well as a variety of chemical agents (reviewed in Lambert and Lambert, 1987; Kraemer, et al., 1987). In vitro studies have shown that XPA cells are defective in repair of both monoadducts and interstrand cross-links in psoralen plus UVA treated DNA (Kaye, et al., 1980; Bredberg, et al., 1982; Gruenert and Cleaver, 1985). A number of studies have shown that the DNA repair defect in XPA cells is in the initial endonuclease mediated recognition step for repair of these types of damage (reviewed in Lambert and Lambert, 1987; Kraemer, et al., 1987).

We examined XPA cells in order to ascertain whether they had endonucleases capable of incising the various types of damaged DNA discussed above. The nine chromatin-associated DNA endonucleases isolated from XPA cell nuclei were each in turn tested on a variety of damaged DNA substrates. Parallel experiments were carried out using DNA endonucleases obtained from normal cells as controls.

We found that XPA cells had, like normal cells, two endonucleases, pIs 9.2 and 9.8, which were active on partially AP DNA (Figure 2) (Lambert, et al., 1983). The two XPA endonucleases were moderately decreased in activity; they were 80% and 70% of normal, respectively. These are similar to results obtained by Kuhnlein, et al. (1976) and Witte and Thielmann (1979). There have been conflicting reports, though, regarding whether AP endonuclease activity in XPA cells is decreased (Moses and Beaudet, 1978; Snyder and Regan, 1982). Cells from patients with XPA are able to repair AP damage produced by monofunctional alkylating agents such as MMS (Cleaver, 1971; Stich, et al., 1973; Snyder and Regan, 1982; Friedberg, et al., 1979). The significance of the decreased AP endonuclease activities is unclear. XPA cells have been shown to have slightly decreased viability compared to normal cells when exposed to alkylating agents such as MMS (Sasaki, et al., 1977; Thielmann and Witte, 1980). It may be that XPA cells have sufficient AP endonuclease levels to handle the doses of MMS utilized in routine in vivo, UDS, assays and that the AP endonuclease defect in XPA cells is only important at high levels of MMS.

Similarly to normal cells, XPA cells had two endonucleases, pIs 4.6 and 7.6, active on DNA treated with either TMP or 8-MOP plus UVA light (Figure 3A) (Lambert, et al., 1988). These two endonucleases had levels of activity similar to those of the corresponding normal endonucleases. The two XPA endonucleases could also incise DNA containing monoadducts produced by angelicin plus UVA (Figure 3B) (Lambert, et al., 1988). Like the corresponding normal endonucleases, the XPA endonuclease, pI 7.6, had over two and one half fold greater activity on the angelicin photoreactivated DNA than did the endonuclease, pI 4.6. Cell lines from XPA patients, however, have been shown to be deficient in repair of both monoadducts and cross-links produced in DNA by psoralen plus UVA irradiation (Kaye, et al., 1980; Bredberg, et al., 1982; Gruenert and Cleaver, 1985). The question thus arose as to where exactly in the initial endonuclease mediated step of the excision repair process the XPA repair defect lies. Our investigation of this question is discussed below.

XPA cells were also examined for endonucleases active on UVC irradiated DNA. Like normal human cells, the chromatin-associated DNA endonuclease, pI 7.6, had activity on this damaged substrate (Lambert, et al., in preparation). The levels of activity were similar to those of the corresponding normal human endonuclease. XPA cells have been shown to be defective in repair of damage produced by UVC irradiation (reviewed in Lambert and Lambert, 1987; Kraemer, et al., 1987). They have been shown to be deficient in removal of both pyrimidine dimers and 6-4 photoproducts (reviewed in Lambert and Lambert, 1987; Kraemer, et al., 1987), and the defect has been shown to occur at the cellular level at the endonuclease-mediated first step of the nucleotide excision repair process (Setlow, 1969; Tanaka, et al., 1975)). However, as for the endonucleases active on psoralen plus UVA irradiated DNA, our results indicate that the defect in XPA cells in repair of UVC radiation damage is not in the ability of the endonuclease to incise damaged naked DNA.

FANCONI'S ANEMIA DNA ENDONUCLEASE ACTIVITY ON DAMAGED DNA

Fanconi's anemia (FA) is another recessively transmitted genetic disease in which there is a defect in DNA repair. Cells from patients with FA are especially hypersensitive to DNA cross-linking agents (Fujiwara, et al., 1977; Ishida and Buchwald, 1982; Sasaki and Tonomura, 1983; Dean, et al., 1988) and have reduced capacity to repair interstrand cross-links and monoadducts produced by these agents (Kaye, et al., 1980; Gruenert and Cleaver, 1985; Papadopoulo, et al., 1987; Averback, et al., 1988). To date, FA patients have been divided into two genetic complementation groups, A and B, (Duckworth-Rysiecki, et al., 1985; Moustacchi, et al., 1987; Digweed, et al., 1988). Cells in complementation group A have a reduced capacity to repair DNA interstrand cross-links as compared to FA cells in complementation group B (Papadopoulo, et al., 1987; Moustacchi, et al., 1989). Cells in complementation group B are much more sensitive to the formation of monoadducts and incision of these adducts is much less in these cells than in cells in group A (Averbeck, et al., 1988; Moustacchi, et al., 1989).

We have examined FA cells in both complementation groups A and B in order to determine whether they have endonucleases which can incise DNA containing monoadducts and interstrand cross-links produced by psoralen plus UVA. Thus far, one cell line from each of the complementation groups has been examined. Our results show that cells from both complementation groups A and B have two endonucleases, pIs 4.6 and 7.6, as do normal human cells, which are active on DNA treated with 8-MOP plus UVA irradiation. However, unlike normal cells, in FA cells from complementation group A, the endonuclease, pI 4.6, is quantitatively reduced in activity and in complementation group B the activity of the endonuclease, pI 7.6, is quantitatively reduced (Table 1). If the endonuclease, pI 4.6, recognizes the cross-link and the endonuclease, pI 7.6, recognizes the monoadduct, then it is possible that the repair defect in FA cells, complementation groups A and B, is due to

Table 1. Activity of DNA endonucleases, pIs 4.6 and 7.6, from normal human cells and Fanconi's anemia (FA) cells, complementation groups A and B, on 8-MOP plus UVA irradiated nucleosomal and non-nucleosomal plasmid DNA.

Complementation Group	Substrate[b]	Average number of breaks/10^5 base pairs[a]: pI of endonuclease 4.6	7.6
Normal	naked	1.7 ± 0.1	1.6 ± 0.1
	core	4.4 ± 0.4	4.2 ± 0.4
	total	2.7 ± 0.1	2.5 ± 0.2
FA, complementation group A	naked	0.5 ± 0.1	1.3 ± 0.1
	core	1.3 ± 0.2	2.9 ± 0.1
	total	0.9 ± 0.1	1.8 ± 0.2
FA, complementation group B	naked	1.6 ± 0.1	0.8 ± 0.1
	core	4.1 ± 0.2	1.9 ± 0.3
	total	2.4 ± 0.2	1.3 ± 0.1

[a]These values have been subtracted from them the enzyme activity on undamaged DNA. 0.35 µg of endonuclease were reacted with 0.1 µg plasmid DNA.
[b]Either naked DNA or core (histone H2A, H2B, H3 and H4) or total (core + H1) nucleosomal DNA were treated with 69 µM 8-MOP plus 2 doses of UVA light.

a deficiency in the endonucleases involved in repairing cross-links and monoadducts, respectively.

MODULATION OF ENDONUCLEASE ACTION BY NUCLEOSOME STRUCTURE

DNA repair processes in mammalian cells are much more complex than in prokaryotes since the repair enzymes in mammalian cells must interact with the nucleosomal structure rather than "naked" DNA or DNA with considerably less organization as regards associated proteins. To date, little information has been obtained on the effect of nucleosome structure on the activity of any specific mammalian DNA endonuclease with specificity for damaged DNA and with a known role in DNA repair. My laboratory has been particularly interested in the role that nucleosome structure has in modulating the action of mammalian DNA endonucleases involved in repair of specific types of DNA damage. We have examined normal human endonucleases and endonucleases from XPA and FA cells. The system we have developed is described below as well as is the progress we have made in this area.

The Nucleosomal System

We have utilized a reconstituted nucleosomal system which consists of a plasmid DNA and histones from either normal or XPA cells (Kaysen, et al., 1986). The plasmid DNA, pWT830/-pBR322, is a clone of the entire SV40 and pBR322 genomes (Kaysen, et al., 1986). The advantages of using this system are that: (1) since the plasmid contains the entire SV40 genome, a comparison of our reconstituted nucleosomal system can be made with the SV40 minichromosome, (2) it permits evaluation of the influence of either core or total histones on enzyme activity, and (3) the source of the histones (i.e., normal vs. XPA) can be varied to determine whether this makes any difference in enzyme activity.

Histones are extracted from both normal and XPA cells and both core (H2A, H2B, H3, H4) and total (core + H1) histones are prepared (Kaysen, et al., 1986). Electrophoretic separation of the histones from normal and XPA cells showed that all five major histone species are present in both cell groups and that there are no quantitative differences between normal and XPA histones (Table 2) (Amari, et al., 1986). Also no significant differences were observed in the binding of either normal or XPA histones to DNA (Table 3) (Amari, et al., 1986).

Nucleosomes are reconstituted by mixing the DNA with either core or total histones and carrying out a gradient dialysis procedure (Kaysen, et al., 1986). Examination of the reconstituted nucleosomal system showed that it gave standard patterns of digestion with micrococcal nuclease and DNAase I (Kaysen, et al., 1986). In addition, using single

Table 2. Electrophoretic separation of histones from normal (N) and xeroderma pigmentosum, complementation group A (XPA) lymphoblastoid cells on 2.5 M urea 15% polyacrylamide gels.

Histone	Percent of total protein[a] (mean ± standard error) N	XPA
H1	28.4 ± 0.7	27.9 ± 0.9
H3	21.1 ± 0.3	21.6 ± 0.8
H2B	18.7 ± 0.9	19.1 ± 0.6
H2A	17.2 ± 0.7	18.3 ± 1.0
H4	14.6 ± 0.8	13.1 ± 1.1

[a]Percentage is based on total protein recovered from the gel and is an average of 5 separate gels representing three different protein extractions. (Amari, et al., 1986).

Table 3. Binding of histones from normal human (N) and xeroderma pigmentosum, complementation group A (XPA) lymphoblastoid cells to DNA.

Ratio of DNA to histones (W/W)	Percent of DNA binding to the filter[a] (mean ± standard error)	
	N	XPA
1:1	24.6 ± 4.2	23.3 ± 5.4
1:2	44.2 ± 2.4	41.6 ± 2.3
1:4	46.9 ± 1.9	42.8 ± 3.6
1:10	44.0 ± 7.9	38.0 ± 8.9

[a]0.13 µg of ^{3}H-thymidine mouse melanoma DNA was reassociated with histones by means of gradient dialysis. Protein-DNA binding was measured by a filter binding assay. Experiments were performed 4-6 times and were from three different protein extractions. (Amari, et al., 1986).

site restriction endonucleases as probes, our studies indicate that the nucleosomes show a non-random arrangement, i.e., are "phased", or positioned, as regards sequences near the origin of replication on the SV40 portion of the plasmid genome (Kaysen, et al., 1987). These studies are in agreement with those on the SV40 minichromosome (Hiwasa, et al., 1981).

NORMAL HUMAN ENDONUCLEASE ACTIVITY ON DAMAGED NUCLEOSOMAL DNA

We have thus far examined whether the presence of nucleosomes (± histone H1) influences the activity of different chromatin-associated endonucleases (i.e., AP, psoralen plus UVA, and UVC) from normal human cells on different damaged DNA substrates. We have found that nucleosome structure enhances the activities of two normal human AP endonucleases, pIs 9.2 and 9.8, on partially AP DNA (Kaysen, et al., 1986). These two endonucleases showed a 2.5 fold increase in activity on AP plasmid DNA reconstituted with core histones as compared to non-nucleosomal AP DNA (Figure 6, B and D). This increase was not observed on undamaged nucleosomal DNA as compared to undamaged non-nucleosomal DNA (Figure 6, A and C); therefore, the presence of nucleosomes did not affect the activity of these endonucleases on undamaged DNA. This increase was also not due to an increase in the number of AP sites since the number of sites on both nucleosomal and non-nucleosomal DNA was approximately 3 per DNA molecule (Kaysen, et al., 1986). Nucleosome assembly was necessary for this increase in endonuclease activity since simple addition of histones to the reaction mixture did not increase enzyme activity on AP DNA. When histone H1 was added to the system, this increase in AP

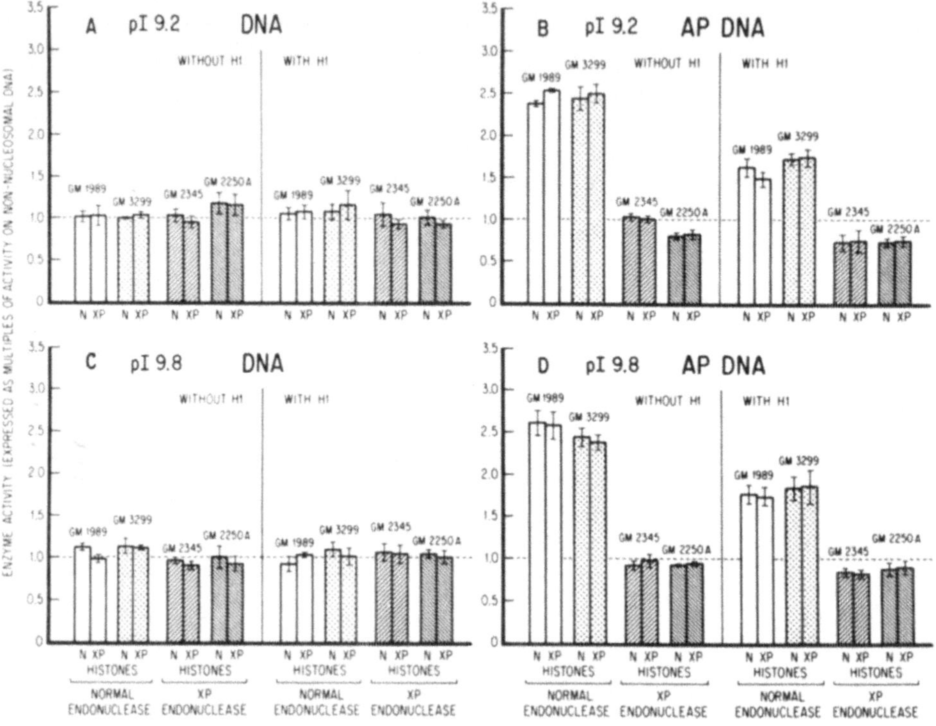

Figure 6. Activity of AP endonucleases, pIs 9.2 and 9.8, from two different normal and XPA lymphoblastoid cell lines, on undamaged (A and C) and partially AP (B and D) nucleosomal DNA. Plasmid pWT830/pBR322 DNA was reconstituted with core histones 1 histone H1 from either normal or XPA lines at a protein/DNA ratio of 1.0. Enzyme activity is expressed in terms of multiples of activity of these endonucleases on non-nucleosomal DNA. Enzyme activity on non-nucleosomal DNA = 1.0 (dotted line). (Kaysen, et al., 1986).

endonuclease activity was reduced approximately 35% (Figure 6, B and D). Whether normal or XPA histones were used in the system made no difference; endonucleases from the two different normal cell lines gave similar results (Figure 6).

We have in addition carried out studies to determine whether nucleosome structure also modulates the activity of the two normal human endonucleases, pIs 4.6 and 7.6, which incise DNA treated with psoralen plus UVA radiation. We utilized nucleosomal DNA (± H1) treated with either 8-MOP or angelicin plus UVA radiation. We found that the presence of nucleosomes (minus histone H1) reduced by approximately 50% the number of 8-MOP adducts bound to the DNA and that when histone H1 was added that total number of adducts was reduced approximately 60% (Parrish and Lambert, 1989). Reduced binding of psoralen to extracted cellular chromatin, compared with its binding to naked DNA, has also been reported by Gia, et al. (1987), although in that study the chromatin substrate was not as well defined as in the present study.

We found that, similar to the activity of the AP endonucleases on AP DNA, the activities of the two normal endonucleases, pIs 4.6 and 7.6, were enhanced by nucleosome structure on DNA treated with 8-MOP or angelicin plus UVA radiation. This increase was approximately 2.5 fold on damaged nucleosomal DNA when histone H1 was absent; when histone H1 was added to the system this increase was reduced approximately 35% (Table 4). This increase in activity also did not correlate with a change in the number of 8-MOP adducts since the number of adducts was reduced on nucleosomal DNA. The presence of nucleosomes did not affect the activity of any of these endonucleases on undamaged DNA.

Table 4. Activity of DNA endonucleases, pIs 4.6 and 7.6, from normal (N) human and XPA cells on 8-MOP or angelicin plus UVA treated nucleosomal and non-nucleosomal plasmid DNA.

		Average number of breaks/10^5 base pairs[a]			
		pI of endonuclease			
Agent[b]	Substrate[c]	4.6		7.6	
		N	XPA	N	XPA
8-MOP + UVA	naked	1.2 ± 0.0	1.2 ± 0.1	1.4 ± 0.0	1.2 ± 0.1
	core	3.1 ± 0.2	1.2 ± 0.0	3.6 ± 0.1	1.4 ± 0.1
	total	1.9 ± 0.1	0.8 ± 0.1	2.4 ± 0.0	0.9 ± 0.1
Angelicin + UVA	naked	0.5 ± 0.0	0.4 ± 0.0	1.2 ± 0.1	1.2 ± 0.0
	core	1.3 ± 0.1	0.4 ± 0.1	3.1 ± 0.2	1.0 ± 0.1
	total	0.7 ± 0.0	0.3 ± 0.0	2.0 ± 0.1	0.2 ± 0.0

[a]These values have subtracted from them the enzyme activity on undamaged DNA. 0.41 ± 0.05 µg of normal (N) or XP enzyme was reacted with 0.1 µg plasmid DNA.
[b]Nucleosomal or non-nucleosomal DNA was treated with 69 µM 8-MOP or 25 µg/ml angelicin plus UVA.
[c]Either naked (non-nucleosomal) DNA or core (histone H2A, H2B, H3 and H4) or total (core + H1) nucleosomal DNA were used as substrate.

Preliminary studies indicate that the activity of the endonuclease, pI 7.6, on UVC irradiated DNA was also enhanced by the presence of nucleosomes (± histone H1). Thus, nucleosome structure enhances the activity of at least three different DNA endonucleases (AP, psoralen plus UVA, and UVC) on damaged DNA, particularly when histone H1 is absent. It may be that the presence of nucleosomes makes the sites of damage more accessible to endonucleolytic incision or that there is associated with the endonucleases an "accessibility factor" which increases accessibility of the various sites of damage on nucleosomal DNA to endonucleolytic attack. It is also possible that a direct interaction occurs between the normal endonuclease and the histones which leads to increased ability of the endonucleases to incise DNA at the sites of damage. Sollner-Webb, et al. (1986) have shown that the apparent affinity of staphylococcal nuclease for chromatin is greater than for protein-free DNA; however, staphylococcal nuclease was less active on chromatin than on the protein free DNA. Kinetic analysis of the two normal endonucleases, pIs 4.6 and 7.6, indicates that their affinity for 8-MOP plus UVA damaged nucleosomal DNA is greater than for damaged naked DNA, but that the maximum turnover number of the enzymes on both substrates is similar (Parrish, et al., in preparation). It is also possible that the histones in the nucleosomes may sequester the endonucleases and, in effect, increase their local effective concentration so as to produce increased activity on damaged nucleosomal DNA. Nucleosomal histones have been shown to sequester the H2A specific protease, but in that case this interaction lead to inhibition of enzyme activity (Davie, et al., 1986; Elia and Moudrianakis, 1988). Investigation into the nature of the interaction of these endonucleases with nucleosomal histones is currently ongoing in my laboratory.

For all of these normal endonucleases (i.e., AP, psoralen plus UVA, and UVC), when histone H1 was added to the nucleosomal system, the increase in endonuclease activity was still present but was reduced by 35%. This could reflect the role of histone H1 in compacting chromatin (McGhee and Felsenfeld, 1980, Igo-Kemenes, et al., 1982; Klingholz and Stratling, 1982; Watanabe, 1984) making it less accessible to endonucleolytic attack. Somewhat similar findings have been reported by Ishimi, et al. (1981) in a study using non-damaged calf-thymus chromatin. They found that depletion of the chromatin of histone H1 caused the nucleosomal DNA to unfold and become sensitive to micrococcal nuclease at a particular site on the DNA which was protected from nuclease attack by histone H1.

XPA ENDONUCLEASE ACTIVITY ON DAMAGED NUCLEOSOMAL DNA

As described above, we have found that the two endonucleases active on AP DNA, the two that are active on psoralen plus UVA irradiated DNA and the one that is active on UVC irradiated DNA are present in XPA cells. The activities of the AP endonucleases are slightly reduced as compared to those of the normal enzymes (Lambert, M.W., et al., 1983), whereas the activities of the psoralen plus UVA and the UVC endonucleases are similar to those of the normal nucleases (Lambert, et al., 1988). Cell culture studies have shown, however, that XPA cells are defective in repair of both monoadducts and cross-links produced in DNA by psoralen plus UVA irradiation (Kaye, et al., 1980; Bredberg, et al, 1982; Gruenert and Cleaver, 1985) and in repair of UVC induced lesions (reviewed in Lambert and Lambert, 1987; Kraemer, et al., 1987). We wished to ascertain whether the repair defect in XPA cells could be at the level of the interaction of the endonucleases with damaged nucleosomal DNA.

We found that, in marked contrast to the normal AP endonucleases, the two XPA AP endonucleases, pIs 9.2 and 9.8, did not show any increased activity on AP core nucleosomal DNA compared with AP naked DNA (Figure 6, B and D) (Kaysen, et al.; 1986). When histone H1 was added to the system, they actually showed a significant decrease in activity (Figure 6, B and D). Similar results were obtained with the two XPA endonucleases, pIs 4.6 and 7.6, active on psoralen plus UVA irradiated DNA (Parrish and Lambert, 1989). These two XPA endonucleases, in contrast to the normal endonucleases, again did not show any increase in activity on 8-MOP or angelicin plus UVA treated core nucleosomal DNA and they actually showed a significant decrease in activity when histone H1 was added to the system (Table 4). Kinetic analysis of the two XPA endonucleases, pIs 4.6 and 7.6, on both 8-MOP plus UVA damaged nucleosomal and non-nucleosomal DNA indicates that the XPA endonucleases have a reduced affinity for the damaged nucleosomal substrate; the maximum turnover numbers of the enzymes on the damaged nucleosomal and non-nucleosomal substrates are similar (Parrish, et al., in preparation). Preliminary studies indicate that the XPA endonuclease, pI 7.6, also does not show increased activity on UVC irradiated DNA when nucleosomes (± histone H1) are present.

The differences between normal and XPA endonucleases on the different types of damaged nucleosomal DNA were not due to different initial levels of enzyme activity, since the concentrations of both normal and XPA endonucleases were adjusted to produce similar numbers of breaks on damaged non-nucleosomal DNA (Kaysen, et al., 1986; Parrish and Lambert, 1989). The differences were also not due to idiosyncratic differences between cell lines, since two different XPA cell lines derived from individuals of different ethnic origins were utilized with the same results.

Our findings indicate, therefore, that a defect exists in the ability of different XPA endonucleases (i.e., AP, psoralen plus UVA, and UVC) to interact with damaged DNA when it is in the form of nucleosomes. These findings further suggest that the defect observed in XPA cells exists in the endonucleases themselves or in a closely associated co-factor. The work of Mortelmans, et al., (1976) and of Kano and Fujiwara (1983), which utilized crude cell extracts, also suggests that XPA cells are defective in a factor which renders the DNA in chromatin accessible to UV endonucleolytic attack by cellular enzymes. The studies of Hittelman (1986) suggest that in XPA cells a defect exists in localized decondensation of chromatin, which he found to be associated with excision repair following exposure to UV irradiation. All of these studies point to a defect in DNA repair in XPA cells which specifically relates to events associated with chromatin.

De Jonge, et al. (1983) have been able to correct the repair defect in UV-irradiated XPA cells by microinjection of calf thymus extracts into these cells. Their work indicates that this "correcting factor" is a protein. Yamaizumi, et al., (1986) microinjected cell extracts from human placenta or from HeLa cells into UVC irradiated XPA cells and were able to restore DNA repair synthesis in these cells. The calf thymus, human placenta and HeLa cell factors have not been analyzed in an isolated nucleosomal system, so the precise level at which they are exerting their effect cannot be ascertained at present. Further purification of these enzymes and factors will be needed before direct comparisons between them can be made.

A system in which soluble extracts of human cells perform repair synthesis on damaged plasmid DNA was used by Wood, et al., (1988). They found that repair replication of closed circular plasmid DNA damaged by UVA activated psoralen or by UVC irradiation was deficient in whole-cell extracts from several XP complementation groups, including XPA. They concluded that their results did not support the hypothesis proposed by Mortelmans, et al. (1976) and Kano and Fujiwara (1983), that XPA cells are deficient in the gene products needed to make UV lesions in chromatin accessible to repair enzymes. Their reaction conditions, however, include a 6 hour incubation of the damaged DNA with the cell extracts. Their cell extracts and reaction conditions were according to the method of Manley, et al. (1980). Hough, et al. (1982) have shown, however, at the ultrastructural level, that incubation of DNA fragments with such soluble whole-cell extracts for only 30 minutes leads to formation of nucleosomes or nucleosome-like structures, indicating that histones are present in these extracts. Therefore, it is very unlikely that the plasmid DNA remained "naked" following the 6 hour incubation with whole cell extracts used by Wood, et al. (1986). Thus an alternative interpretation of their results, in conflict with neither their results, with the results presented here, or with those of Mortelmans, et al. (1976) or of Kano and Fujiwara (1983), is that the XPA endonucleases are defective in their ability to interact with damaged DNA when in the form of chromatin. Moreover, this inability to interact normally with nucleosomes may be a general deficiency in XPA endonucleases active in DNA repair processes.

FANCONI'S ANEMIA ENDONUCLEASE ACTIVITY ON DAMAGED NUCLEOSOMAL DNA

As discussed above, we have found FA, complementation group A, cells defective in an endonuclease, pI 4.6, which appears to recognize psoralen plus UVA irradiation induced DNA interstrand cross-links and FA, complementation group B, cells defective in an endonuclease, pI 7.6, which recognizes psoralen plus UVA induced monoadducts. We wished to ascertain whether nucleosome structure had any influence on the ability of the FA endonucleases to incise psoralen plus UVA damaged DNA. Our studies indicate that in both complementation groups A and B, the two endonucleases, pIs 4.6 and 7.6, show increased activity on 8-MOP plus UVA damaged nucleosomal DNA, similar to normal cells (Table 1). Unlike the XPA endonucleases, the two FA endonucleases are not defective in their ability to interact with psoralen plus UVA irradiated DNA when it is in the form of nucleosomes.

COMPLEMENTATION OF THE XPA REPAIR DEFECT AT THE NUCLEOSOMAL LEVEL

We have carried out experiments to determine whether the two normal human endonucleases, pIs 4.6 and 7.6, can complement the defect in the ability of the corresponding XPA endonucleases to incise nucleosomal DNA damaged with psoralen plus UVA irradiation. Each of the two normal endonucleases was mixed with either of the two corresponding XPA endonucleases and their activity examined on nucleosomal DNA (± histone H1) treated with either 8-MOP or angelicin plus UVA irradiation. We found that the activity of either of the two XPA endonucleases on either type of psoralen damaged nucleosomal DNA was increased to normal or near normal levels when mixed with either of the two corresponding XPA endonucleases (Table 5). This activity was significantly greater than can be accounted for by a simple additive effect. This indicates that there is a correcting factor associated with both normal endonucleases which is capable of correcting the defect in either of the XPA endonucleases. The normal endonuclease, pI 7.6, could similarly complement the ability of the corresponding XPA endonuclease to incise UVC irradiated nucleosomal DNA.

Since we have shown that the XPA AP endonuclease, pI 9.8, also lacks a factor needed for interaction with AP nucleosomal

Table 5. Complementation of the XPA repair defect on 8-MOP plus UVA treated nucleosomal DNA. XPA endonuclease (X), pI 7.6, was mixed with normal endonuclease (N), pI 7.6, and assayed for activity.

Substrate[a]	Endonuclease[b]	Average Number of breaks/10^5 base pairs
Non-nucleosomal DNA	0.2 µg N	0.7 ± 0.0
	0.2 µg N	0.6 ± 0.1
	0.2 µg N + 0.2 µg X	1.3 ± 0.0
	0.4 µg N	1.4 ± 0.0
	0.4 µg X	1.2 ± 0.0
Core Nucleosomal DNA	0.2 µg N	1.7 ± 0.2
	0.2 µg N	0.6 ± 0.1
	0.2 µg N + 0.2 µg X	3.3 ± 0.0
	0.4 µg N	3.5 ± 0.0
	0.4 µg X	1.2 ± 0.1
Total nucleosomal DNA	0.2 µg N	1.2 ± 0.0
	0.2 µg N	0.5 ± 0.1
	0.2 µg N + 2 µg X	2.2 ± 0.0
	0.4 µg N	2.3 ± 0.1
	0.4 µg X	0.8 ± 0.0

[a]Either non-nucleosomal plasmid DNA or plasmid DNA reconstituted with core (H2A, H2B, H3, H4) or total (core + H1) histones was treated with 8-MOP plus UVA irradiation and used as substrate.
[b]Normal (N) and/or XPA (X) endonuclease, pI 7.6, were reacted with 0.1 µg plasmid DNA.

DNA and which is present in the corresponding normal human AP endonuclease, we carried out experiments to determine whether the "correcting factor" present in the normal AP endonuclease could correct the activity of either of the XPA endonucleases, pI 4.6 or 7.6, on psoralen plus UVA damaged nucleosomal DNA. Mixing either of the XPA endonucleases with the normal AP endonuclease, pI 9.8, however, failed to complement their activity on psoralen plus UVA damaged nucleosomal DNA (Parrish and Lambert, 1989). Several explanations can be proposed to account for these findings. It is possible that the factors associated with the psoralen and AP endonucleases are different or exist in modified forms; both factors may interact with chromatin and allow only their respective XPA endonucleases to incise the appropriately damaged DNA. On the other hand, it is possible that the correcting factor associated with all three endonucleases is the same. If this is the case then there are a number of complex hypothetical explanations that can be proposed to account for the lack of complementation (Parrish and Lambert, 1989). Among them is the possibility that some type of enzyme subunit system exists in human cells with at least one component recognizing the specific adduct and a different subunit enhancing its interaction with chromatin. It may be that recognition of the specific damaged site by the appropriate endonuclease must occur before the chromatin factor (or factors) associated with it can exert its effect. According to this hypothesis, therefore, the normal AP endonuclease does not recognize psoralen plus UVA DNA adducts, therefore, the "correcting factor" associated with it cannot be effective.

COMPLEMENTATION OF THE XPA DEFECT AT THE CELLULAR LEVEL

One way of determining whether the DNA endonucleases we have isolated are responsible for the cellular DNA repair defect observed in XPA cells is to introduce the normal endonucleases into damaged XPA cells and observe whether they can correct the repair defect. We accomplished this by introducing the normal human endonucleases into XPA cells via electroporation.

XPA lymphoblastoid cells in culture were treated with 8-MOP plus two doses of UVA radiation. Either of the two normal endonucleases, pIs 4.6 or 7.6, were added to the cell suspension and electroporation was carried out using a high voltage electronic pulse (Tsongalis, et al., 1990). The ability of the cells to carry out excision repair was monitored by measuring levels of unscheduled DNA synthesis (UDS), determined as incorporation of [^{3}H] labeled thymidine into DNA repair patches during a two-hour pulse. Electroporation of 8-MOP plus UVA DNA treated XPA cells with either of the normal endonucleases, pI 4.6 or 7.6, resulted in correction of their DNA repair defect (Table 6). UDS in these XPA cells was over 100% of that in normal cells similarly electroporated but without addition of the enzyme. Electroporation of XPA endonucleases, on the other hand, into damaged XPA cells did not significantly increase UDS above background levels (Table 6). Electroporation by itself did not detectably affect UDS in any cell line tested. Electroporation of the normal endonuclease, pI 7.6, but not the corresponding XPA endonuclease, into XPA cells irradiated with UVC, also corrected the defect in the ability of XPA cells to repair damage produced by UVC (Tsongalis, et al., submitted).

These results indicate that the normal endonucleases we have isolated contain the factor necessary to correct the XPA repair defect. Furthermore electroporation of either of the normal endonucleases into 8-MOP plus UVA treated normal cells increased UDS above that observed in normal cells electroporated without enzyme (Tsongalis, et al., 1990). Moreover, electroporation of either of the XPA endonucleases into damaged normal cells also produced UDS levels above normal (Tsongalis, et al., 1990). These results indicate that introduction of normal enzymes into normal cells increases the efficiency of repair in these cells. The increased UDS seen

Table 6. UDS in XPA cells treated with 8-MOP plus UVA irradiation and electroporated with normal or XPA endonuclease.

Endonuclease electroporated	% of normal UDS[a]	
	pI of endonuclease	
	4.6	7.6
None	10 ± 4	10 ± 4
Normal	118 ± 9	125 ± 7
XPA	20 ± 6	15 ± 5

[a]UDS was monitored by measuring incorporation of [^{3}H] labeled thymidine in DNA repair patches during a two-hour pulse and then visualized via autoradiography. Cells with 40-100 silver grains per nucleus were classified as undergoing UDS.

in damaged normal cells electroporated with either of the XPA endonucleases supports the hypothesis that XPA cells contain endonucleases which can repair 8-MOP plus UVA damage in normal cells but lack a factor necessary to allow these XPA endonucleases to function in their own cells. Normal cells can supply this factor when XPA endonucleases are introduced into them by electroporation (Tsongalis, et al., 1990). Based on their experimental findings in heterokaryons, Gianelli, et al. (1982) have proposed that normal cells have a surplus of the factor needed to complement the DNA repair defect in XPA cells. This hypothesis is supported by our results which, in addition, indicate that in our system this surplus factor allows the XPA endonuclease to repair DNA damage in normal human cells.

ENDONUCLEASE BINDING TO DAMAGED DNA

We have begun, to examine the ability of the human DNA endonucleases we have isolated to bind to damaged DNA. Thus far we have studied the binding of the AP endonuclease, pI 9.8, from both normal and XPA cells to partially AP DNA (Bickley and Lambert, 1988).

We utilized a filter binding assay to determine protein binding to DNA. pWT830/pBR322 plasmid DNA labeled with [^{3}H]-thymidine was used as substrate. The plasmid DNA was linearized by digestion with the single site restriction enzyme, KpnI, and made partially AP by treatment with MMS followed by heating at 70^{0}C (Bickley and Lambert, 1988). Figure 7, A and C, shows that the AP endonuclease, pI 9.8, from normal cells had significantly greater increased binding affinity for AP DNA than for undamaged DNA (Bickley and Lambert, 1988). This increase was seen over a range of protein to DNA ratios. In contrast, no corresponding increase in binding of the XPA AP endonuclease to AP DNA, as compared to binding to undamaged DNA, was observed (Figure 7, B and C). The XPA endonuclease showed only approximately 10% of the increased binding seen with the normal endonuclease (Figure 7C) (Bickley and Lambert,

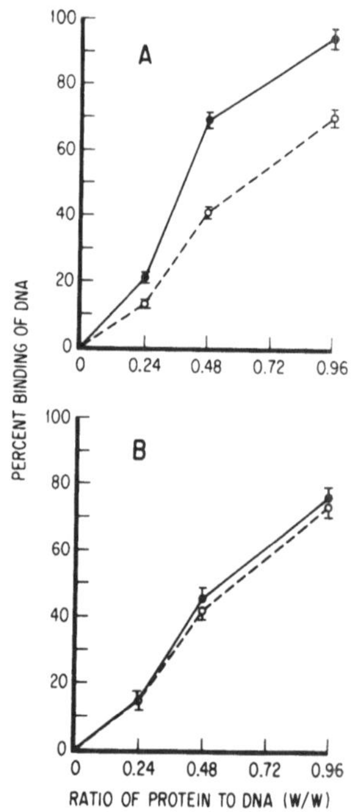

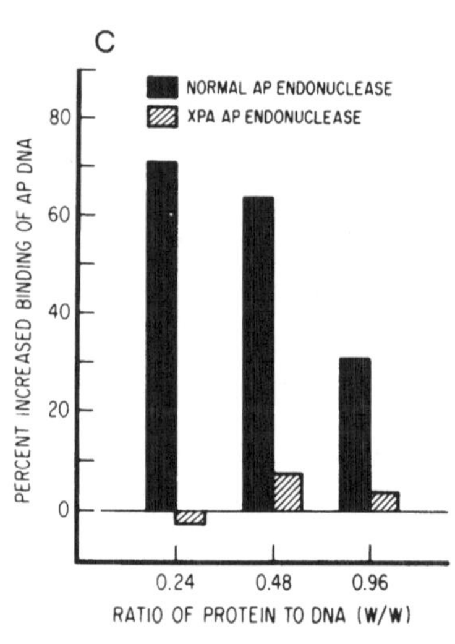

Figure 7. Binding of an AP DNA endonuclease, pI 9.8, from (A) normal human and (B) XPA lymphoblastoid cells to partially AP DNA utilizing a filter binding assay. (O - O), undamaged DNA; (● - ●), AP DNA. (C) Proportional increase in binding, above that seen with undamaged DNA, of the normal versus the XPA AP endonuclease to AP DNA.

1988). Whether this decrease in preferential binding of the XPA endonuclease to AP DNA is related to the decreased activity of this endonuclease on AP DNA is unclear.

To date, no deficiency in binding of a XPA DNA endonuclease to damaged DNA has been reported. However, Kuhnlein, et al., (1983) have shown that a single-stranded DNA binding activity on undamaged DNA is reduced or absent in XPA cells. On the other hand, binding activity to UVC damaged DNA has been detected in XPA cell extracts (Chu and Chang, 1988), and normal levels of a protein which specifically binds to UVC irradiated DNA (Feldberg, 1976), as well as of a partially purified AP DNA binding protein (Kuhnlein, 1985) and all major proteins binding to native, denatured and UV radiation damaged DNA (Lehmann and Kirk-Bell, 1978) have been reported in XPA cells. The role that the deficient binding of the XPA AP endonuclease plays in DNA repair is not clear at present. Further studies to clarify this are in progress.

SUMMARY

We have isolated, in active form, a series of chromatin-associated DNA endonucleases, with specificity for particular types of DNA damage, from the nuclei of normal human lymphoblastoid cells. These same endonucleases are also present in the chromatin of XPA cells. Two of these endonucleases, pIs 9.2 and 9.8, are active on partially AP DNA; two other endonucleases, pIs 4.6 and 7.6, are selectively active on DNA treated with psoralen plus UVA irradiation. The endonuclease, pI 7.6, is much more active than that at pI 4.6 on angelicin plus UVA irradiated DNA, which contains monoadducts and no DNA interstrand cross-links. The activity of the endonuclease, pI 7.6, is inhibited by a monoclonal antibody with specificity for 8-MOP plus UVA irradiation induced monoadducts. The endonuclease, pI 7.6, is also selectively active on UVC treated DNA, where it mainly recognizes the pyrimidine dimer. Our recent studies have shown that the activity of this enzyme is diminished in cells derived from patients with FA, complementation group B, which are defective in repair of psoralen plus UVA induced monoadducts in DNA. The endonuclease, pI 4.6, is selectively active on DNA intercalated with adriamycin and we have recently found it to be defective in cells derived from patients with FA, complementation group A, which are defective in repair of DNA interstrand cross-links. All of these results taken together indicate that the endonuclease, pI 7.6, recognizes psoralen monoadducts and UVC induced pyrimidine dimers, and that the endonuclease, pI 4.6, recognizes intercalation and possibly psoralen plus UVA irradiation induced interstrand cross-links (Table 7).

Our finding of endonucleases in XPA cells that can incise various types of damaged DNA raises the question as to what is the precise nature of the DNA repair defect in XPA cells in the incision step of the excision repair process. My laboratory has addressed this question by examining the action of these endonucleases at the level of the nucleosome. The reconstituted nucleosomal system utilized allows direct analysis of the incision step in the repair process at the molecular level, using well-defined substrates. Using this system we have found that nucleosome structure markedly influences human DNA endonuclease activity. The presence of nucleosomes, minus histone H1, increased the activity of all of the normal endonucleases studied (i.e., pIs 4.6, 7.6, 9.2, and 9.8), on their respectively damaged DNAs by approximately 2.5 fold. When histone H1 was added to the system the increase in endonuclease activity was reduced by approximately 35%. In marked contrast to these results, the corresponding XPA endonucleases did not show this increased activity on damaged nucleosomal DNA, and when histone H1 was present, they actually showed a significant decrease in activity. These results indicate that the defect in XPA cells does not reside in the ability of these endonucleases to act on damaged naked DNA, but rather exists in the ability of the endonucleases or a closely associated factor(s) to interact with damaged DNA when it is in the form of nucleosomes. These findings further emphasize that the interaction of the endonucleases with chromatin is of critical importance and that it is at this level that a defect exists in XPA endonucleases. This defect could either reside in the endonucleases themselves or in a closely associated factor(s). We do not know whether the different endonucleases share a common factor, have different factors or have within the endonucleases themselves, both enzyme and factor associated.

Table 7. Summary of substrate specificity of normal human and XPA DNA endonucleases.

Type of Damage	Endonuclease (pI)[a]			
	4.6	7.6	9.2	9.8
AP sites	-	-	+	+
Psoralen monoadducts	-	+	-	-
Pyrimidine dimers	-	+	-	-
Intercalation	+	-	-	-
Psoralen interstrand cross-Links	+	-	-	-

[a]The predominate type of damage recognized by each endonuclease is marked (+).

Mixing either of the XPA endonucleases, pIs 4.6 or 7.6, with the corresponding normal endonucleases can correct the defect in the ability of the XPA endonucleases to incise psoralen plus UVA damaged nucleosomal DNA. The normal endonuclease, pI 7.6, can also correct the defect in the XPA endonuclease, pI 7.6, to incise UVC irradiated nucleosome DNA. Moreover, introduction of either of the two normal endonucleases, pIs 4.6 or 7.6, into 8-MOP plus UVA damaged XPA cells in culture, or of the normal endonuclease, pI 7.6, into UVC irradiated XPA cells in culture, by electroporation, corrects the XPA repair defect. These results demonstrate that the DNA endonucleases we have isolated from normal human cell chromatin are capable of functioning in DNA repair processes in intact cells in culture. They show that the DNA repair defect in XPA cells resides in the endonucleases we have isolated since this defect can be corrected, at both the nucleosomal and cellular levels, by the corresponding endonucleases from normal cells. The work described here also suggests that this methodology will make possible further elucidation of the XPA repair defect as well as the repair defect in the other XP complementation groups and in other diseases associated with defective DNA repair processes.

ACKNOWLEDGEMENTS

We would like to thank Dr. W. Clark Lambert for critical review of this manuscript and Robert Lockwood for culturing the human cell lines. This work was supported by Grant AM 35148 from the National Institute of Health.

REFERENCES

Adolph, K., 1988. Chromosomes and Chromatin, Volumes 1 and 2, CRC Press, Inc., Boca Raton, Florida.

Amari, N.M.B., W.C. Lambert and M.W. Lambert, 1986. Comparison of histones in normal and xeroderma pigmentosum lymphoblastoid cells, Cell. Biol. Intl. Reports, 10:875-880.

Averbeck, D., D. Papadopoulo and E. Moustacchi, 1988. Repair of 4,5',8-trimethylpsoralen plus light-induced DNA damage in normal and Fanconi's anemia cells, Cancer Res., 48: 2015-2020.

Ben-Hur, E. and M.M. Elkind, 1973. Psoralen plus near ultraviolet light inactivation of cultured Chinese hamster cells and its relation to DNA cross-links, Mutation Res., 18:315-324.

Ben-Hur, E. and P-S. Song, 1984. The photochemistry and photobiology of furocoumarins (psoralens). Adv. Radiat. Biol., 11:131-177.

Bickley, L.R. and M.W. Lambert, 1988. DNA binding of an apurinic/apyrimidinic DNA endonuclease activity from xeroderma pigmentosum cells, Cell Biol. Int. Rep., 12:231-237.

Bohr, V.A., 1987. Differential DNA repair within the genome, Cancer Rev., 7:28-55.

Bredberg, A., 1982. Genetic toxicity of psoralen and ultraviolet radiation in human cells, Acta Dermato-Venereol., 104:1-40.

Bredberg, A., B. Lambert and S. Soderhall, 1982. Induction and repair of psoralen cross-links in DNA of normal human and xeroderma pigmentosum fibroblasts, Mutation Res., 93:221-234.

Cimino, G.D., H.B. Gamper, S.T. Isaacs, and J.E. Hearst, 1985. Psoralens as photoactive probes of nucleic acid structure and function: organic chemistry, photochemistry, Annu. Rev. Biochem., 54:1151-1193.

Chu, G. and E. Chang, 1988. Xeroderma pimentosum group E cells lack a nuclear factor that binds to damaged DNA, Science, 242:564-567.

Cleaver, J.E., 1971. Repair of alkylation damage in ultraviolet sensitive (xeroderma pigmentosum) human cells, Mutation Res., 12:453-462.

Davie, J.R., L. Numerou and G.P. Delcuve, 1986. The nonhistone chromosomal protein, H2A-specific protease, is selectively associated with nucleosomes containing histone H1, J. Biol. Chem., 261:10410-10416.

Dean, S.W., H.R. Sykes, and A.R. Lehmann, 1988. Inactivation by nitrogen mustard of plasmids introduced into normal and Fanconi's anemia cells, Mutation Res., 194:57-63.

De Jonge, A.J.R., W. Vermeulen, B. Klein and J.H.J. Hoeijmakers, 1983. Microinjection of human cell extracts corrects xeroderma pigmentosum defect, EMBO J., 2:637-641.

Digweed, M., S. Zakrzewski-Ludcke, and K. Sperling, 1988. Fanconi's anemia: correlation of genetic complementation group with psoralen/UVA response, Hum. Genet., 78:51-54.

Duckworth-Rysiecki, G., K. Cornish, C.A. Clarke and M.Buchwald, 1985. Identification of two complementation groups in Fanconi's anemia, Somatic Cell and Mol. Genet., 11: 35-41.

Elia, M.C. and E.N. Moudrianakis, 1988. Regulation of H2A-specific proteolysis by histone H3:H4 tetramer, J. Biol. Chem., 263:9958-9964.

Eriksson, M., B. Norden, and S. Eriksson, 1988. Anthracycline-DNA interactions studied with linear dichroism and fluorescence spectroscopy, Biochemistry, 27:8144-8151.

Feldberg, R.S. and L. Grossman, 1976. A DNA binding protein from human placenta specific for ultraviolet damaged DNA, Biochemistry, 15:2402-2408.

Fialkoff, H., M.F. Goodman and M.W. Seraydarian, 1979. Differential effect of adriamycin on DNA replicative and repair synthesis in cultured neonatal rat cardiac cells, Cancer Res., 39:1321-1327.

Friedberg, E.C., V.K. Ehmann and J.I. Williams, 1979. Human diseases associated with defective DNA repair, Adv. Radiat. Biol., 8:85-174.

Friedberg, E.C., 1984. DNA Repair, W.H. Freeman and Company, New York, 40-54.

Fujiwara, Y., M. Tatsumi, and M.S. Sasaki, 1977. Cross-link repair in human cells and its possible defect in Fanconi's anemia cells, J. Mol. Biol., 113:635-649.

Gia, O., G. Palu, M. Palumbo, C. Antonello and S. Marciani Magno, 1987. Photoreaction of psoralen derivatives with structurally organized DNA, Photochem. Photobiol., 45: 87-92.

Giannelli, F., S.A. Pawsey and J.A. Avery, 1982. Differences in patterns of complementation of the more common groups of xeroderma pigmentosum: possible implications, Cell, 29:451-458.

Grunenert, D.C. and J.E. Cleaver, 1985. Repair of psoralen-induced cross-links and mono-adducts in normal and repair-deficient human fibroblasts, Cancer Res., 45:5399-5404.

Hanawalt, P.C., P.K. Cooper, A.K. Ganesan, R.S. Lloyd, C.A. Smith and M.E. Zolan, 1982. Repair responses to DNA damage: Enzymatic pathways in E. coli and human cells, J. Cell. Biochem., 18:271-283.

Hittleman, W.N., 1986. Visualization of chromatin events during DNA excision repair in xeroderma pigmentosum cells: deficiency in localized but not generalized chromatin events, Carcinogenesis, 7:1975-1980.

Hiwasa, T., M. Segana, N. Yamaguchi and K. Oda, 1981. Phasing of nucleosomes in SV40 chromatin reconstituted in vitro, J. Biochem., 89:1375-1389.

Hough, P.V.C., I.A. Mastrangelo, J.S. Wall, J.F. Hainfeld, M.N. Simon and J.L. Manley, 1982. DNA-protein complexes spread on N2-discharged carbon film and characterized by molecular weight and its projected distribution, J. Mol. Biol., 160:375-386.

Igo-Kemenes, T.W., W. Horz and H.G. Zachau, 1982. Chromatin. Annu. Rev. Biochem., 51:89-121.

Ishida, R. and M. Buchwald, 1982. Susceptibility of Fanconi's anemia lymphoblasts to DNA-cross-linking and alkylating agents, Cancer Res., 42:4000-4006.

Ishimi, Y., Y. Ohba, H. Yasuda and M. Yamada, 1981. The interaction of H1 histone with nucleosome core, J. Biochem., 89:1881-1888.

Kano, Y. and Y. Fujiwara, 1983. Defective thymine dimer excision from xeroderma pigmentosum chromatin and its characteristic catalysis by cell-free extracts, Carcinogenesis, 4:1419-1424.

Kaye, J., C.A. Smith and P.C. Hanawalt, 1980. DNA repair in human cells containing photoadducts of 8-methoxypsoralen or angelicin, Cancer Res., 40:696-702.

Kaysen, J.H., N.M.B. Amari and M.W. Lambert, 1986. Enhancement of two apurinic/apyrimidinic endonuclease activities from normal but not xeroderma pigmentosum lymphoblastoid cells by nucleosome structure, Mutation Res., 165:221-231.

Kaysen, J.H., N.M.B. Amari, and M.W. Lambert, 1987. Positioning of nucleosomes reconstituted with xeroderma pigmentosum and normal histones, Cell. Biol. Intl. Reports, 11:95-101.

Klingholz, R. and W.H. Stratling, 1982. Reassociation of histone H1 to H1-depleted polynucleosomes, J. Biol. Chem., 257:13101-13107.

Kraemer, K.H., M.M. Lee and J. Scotto, 1987. Xeroderma Pigmentosum: cutaneous, ocular, and neurological abnormalities in 830 published cases, Arch. Dermatol., 123:241-250.

Kuhnlein, U., S.S. Tsang, O. Lokken, S. Tong, and D. Twa, 1983. Cell lines from xeroderma pigmentosum complementation group A lack a single-stranded-DNA-binding-activity, Bioscience Reports, 3:667-674.

Kuhnlein, U. 1985. Comparison of apurinic DNA-binding protein from an ataxia telangiectasia and a HeLa cell line, J. Biol. Chem., 260:14918-14924.

Kuhnlein, U., E.E. Penhoet and S. Linn, 1976. An altered apurinic DNA endonuclease activity in group A and group D xeroderma pigmentosum fibroblasts, Proc. Natl. Acad. Sci. (U.S.A.), 73:1169-1173.

Lambert, B., M. Sten, S. Soderhall, U. Ringborg and R. Lewenshon, 1983. DNA repair replication, DNA breaks and sister-chromatid exchange in human cells treated with adriamycin in vitro, Mutation Res., 3:171-184.

Lambert, M.W., D. Fenkart and M. Clarke, 1988. Two DNA endonuclease activities from normal human and xeroderma pigmentosum chromatin active on psoralen plus ultraviolet light treated DNA, Mutation Res., 193:65-73.

Lambert, M.W., W.C. Lambert and A.O. Okorodudu, 1983. Nuclear DNA endonuclease activities on partially apurinic/apyrimidinic DNA in normal human and xeroderma pigmentosum lymphoblastoid cells and mouse melanoma cells, Chem.

Biol. Interact., 46:109-120.

Lambert, M.W., D.E. Lee, A.O. Okorodudu, and W.C. Lambert, 1982. Nuclear deoxyribonuclease activities in human lymphoblastoid and mouse melanoma cells: a comparative study, Biochim. Biophys. Acta, 69:192-203.

Lambert, W.C. and M.W. Lambert, 1987. DNA repair deficiency and cancer in xeroderma pigmentosum, Cancer Rev., 7:1-25.

Lan, S.Y. and M.J. Smerdon, 1985. A nonuniform distribution of excision repair synthesis in nucleosomal DNA, Biochemistry, 24:7771-7783.

Laval, J. and F. Laval, 1980. Enzymology of DNA repair, in: Molecular and Cellular Aspects of Carcinogen Screening Tests, IARC Scientific Publication Lyon, 27:55-73.

Lee, Y.C. and J.E. Byfield, 1976. Brief communication: Induction of DNA degradation in vivo by adriamycin, J. Natl. Cancer Inst., 57:221-224.

Lehmann, A.R. and P. Karran, 1981. DNA repair, Int. Rev. Cytol., 72:101-146.

Lehmann, A.R. and S. Kirk-Bell, 1978. DNA - binding proteins in xeroderma pigmentosum fibroblasts, Experimental Cell Research, 114:197-201.

Lewensohn, R. and U. Ringborg, 1983. Inhibition of nitrogen mustard induced DNA repair synthesis by anathracyclines in human peripheral leukocytes, Cancer Lett., 18:305-310

Lindahl, T., 1982. DNA repair enzymes, Annu. Rev. Biochem., 51:61-87.

Lindahl. T. and B. Nyberg, 1972. Rate of depurination of native deoxyribonucleic acid, Biochemistry, 11:3610-3618.

Manley, J.L., A. Fire, A. Cano, P.A. Sharp and M.L. Gefter, 1980. DNA-dependent transcription of adenovirus genes in a soluble whole-cell extract, Proc. Natl. Acad. Sci. (U.S.A.), 77:3855-3859.

Matson, S.W., P.J. Fay and R.A. Bambara, 1980. Mechanism of inhibition of the avian myeloblastosis virus deoxyribonucleic acid polymerase by adriamycin, Biochemistry, 19:2089-2096.

McGhee, J.D and G. Felsenfeld, 1980. Nucleosome structure, Annu. Rev. Biochem., 49:1115-1156.

Mortelmans, K., E.C. Friedberg, H. Slor, G. Thomas, and J.E. Cleaver, 1976. Defective thymine dimer excision by cell-free extracts of xeroderma pigmentosum cells, Proc. Natl. Acad. Sci. (U.S.A.), 73:2757-2761.

Moses, R.E. and A.L. Beaudit, 1978. Apurinic DNA endonuclease activities in repair deficient human cell lines, Nucleic Acid Res., 5:463-473.

Moustacchi, E., D. Papadopoulo, C. Diatloff-Zito and M.Buchwald, 1987. Two complementation groups of Fanconi's anemia differ in their phenotypic response to a DNA-crosslinking treatment, Human Genet., 75:45-47.

Moustacchi, E., D. Papadopoulo, D. Averbeck, D. Fraser and C. Diatloff-Zito, 1989. Processing of photoinduced cross-links and monoadducts in human cell DNA: genetic and molecular features. In: M.W. Lambert and J. Laval (Eds.), DNA Repair Mechanisms and Their Biological Implications in Mammalian Cells, Plenum Press, Inc., New York, in press.

Okorodudu, A.O., W.C. Lambert and M.W. Lambert, 1982. Nuclear deoxyribonuclease activities in normal and xeroderma pigmentosum lymphoblastoid cells, Biochem. Biophys. Res. Commun., 108:576-584.

Pachter, J.A., C.-H. Huang, V.H. Duvernay, Jr., A.W. Prestayko, and S.T. Crooke, 1982. Visometric and fluorometric studies of deoxyribonucleic acid interactions of several new anthracyclins, Biochemistry, 21:1541-1547.

Papadopoulo D., D. Averbeck and E. Moustacchi, 1988. High level of 4,5',8-trimethylpsoralen photoinduced furan-side monoadducts can block cross-link removal in normal human cells, Photochem. Photobiol., 47:321-326.

Parrish, D.D. and M.W. Lambert, 1989. Chromatin-associated DNA endonucleases from xeroderma pigmentosum cells are defective in interaction with damaged nucleosomal DNA, Mutation Res., In press.

Ringborg, U., and B. Lambert, 1978. Inhibition of DNA repair synthesis by adriamycin and daunomycin, J. Supramol. Struct., 2(Suppl.):93.

Ross, W.E. and M.O. Bradley, 1981. DNA double-strand breaks in mammalian cells after exposure to intercalating agents, Biochim. Biophys. Acta, 654:129-134.

Ross, W.E. and M.C. Smith, 1982. Repair of deoxyribonucleic acid lesions caused by adriamycin and ellipticine, Photochem. Photobiol., 31:1931-1935.

Rubin, J.S., 1988. Review: The molecular genetics of the incision step in the DNA excision repair process, Int. J. Radiat. Biol., 3:309-365.

Santella, R.M., N. Dharmaraja, F.P. Gasparro and R.L. Edelson, 1985. Monoclonal antibodies to DNA modified by 8-methoxypsoralen and ultraviolet A light, Nuc. Acids Res., 13: 2533-2544.

Sasaki, M.S., K. Toda and A. Ozawa, 1977. in: M. Seiji and I.A. Bernstein (Eds), Biochemistry of Cutaneous Epidermal Differentiation, Univ. of Tokyo Press, Tokyo, 167.

Sasaki, M.S. and A. Tonomura, 1973. A high susceptibility of Fanconi's anemia to chromosome breakage by DNA-crosslinking agents, Cancer Res., 33:1829-1836.

Schmid, B. and G. Zbinden, 1979. Unscheduled DNA synthesis in male rabbit germ cells induced by methylmethane sulfonate, cyclophosphamide and adriamycin, Arch. Toxicol., 2(suppl.):503-507.

Setlow, R.B., J.D. Regan, J. German, and C.L. Carrier, 1969. Evidence that xeroderma pigmentosum cells do not perform the first step in the repair of ultraviolet damage to their DNA, Proc. Natl. Acad. Sci. USA, 64:1035-1041.

Smerdon, M.J., 1989. DNA excision repair at the nucleosomal level of chromatin. In: DNA Repair Mechanisms and Their Biological Implications in Mammalian Cells. M.W. Lambert and J. Laval, eds. Plenum Press, Inc., New York, in press.

Snyder, R.D., and J.D. Regan, 1982. DNA repair in normal human and xeroderma pigmentosum group A fibroblasts following treatment with various methanesulfonates and demonstration of a long-patch (U.V.-like) repair component, Carcinogenesis, 3:7-14.

Sollner-Webb, B., R.D. Camerini-Otero and G. Felsenfeld, 1976. Chromatin structure as probed by nucleases and proteases: Evidence for the central role of histones H3 and H4, Cell, 9:179-193.

Song, P.-S. and K.J. Tapley, Jr., 1979. Photochemistry and photobiology of psoralens, Photochem. Photobiol., 29: 1177-1197.

Stitch, H.F., R.H.C. San and Y. Kanajoe, 1973. Increased sensitivity of xeroderma pigmentosum cells to some chemical carcinogens and mutagens, Mutation Res., 17:127-137.

Tanaka, K., M. Sekiguchi and Y. Okada, 1975. Restoration of ultraviolet-induced DNA synthesis of xeroderma pigmentosum cells by the contact treatment with bacteriophage T4 endonuclease V (Sendai virus), Proc. Natl. Acad. Sci. (U.S.A.), 72:4071-4075.

Tanaka, M. and S. Yoshida, 1980. Mechanisms of the inhibition of calf thymus DNA polymerase alpha and beta by daunomycin and adriamycin, J. Biochem., 87:911-918.

Teebor, G.W. and K. Frenkel, 1983. The initiation of DNA repair, Adv. Cancer Res., 38:23-59.

Thielmann, H.W. and I. Witte, 1980. Correlation of the colony forming abilities of xeroderma pigmentosum fibroblasts with repair - specific DNA incision reactions, catalyzed by cell free extracts, Arch. Toxicol., 44:197-207.

Tomura, T. and J.L Van Lanker, 1980. The action of a mammalian endonuclease on psoralen-bound DNA, Chem.-Biol.Interact., 31:179-188.

Tsongalis, G.J., W.C. Lambert and M.W. Lambert, 1990. Electroporation of normal human DNA endonucleases into xeroderma pigmentosum cells corrects their DNA repair defect, Carcinogenesis, In press.

Vigny, P., F. Gaboriau, L. Voituriez and J. Cadets, 1985. Chemical Structure of psoralen-nucleic acid photoadducts, Biochimie, 67:317-325.

Waring, M.J., 1981. DNA modification and cancer, Ann. Rev. Biochem., 50:159-192.

Watanabe, F., 1984. Condensation of polynucleosomes by histone H1 binding, FEBS Lett., 170:19-22.

Witte, I., and H.W. Thielmann, 1979. Extracts of xeroderma pigmentosum group A fibroblasts introduce less nicks into methyl methanesulfonate-treated DNA than extracts of normal fibroblasts, Cancer Letters, 6:129-136.

Wood, R.D., P. Robins, and T. Lindahl, 1988. Complementation of the xeroderma pigmentosum DNA repair defect in cell-free extracts, Cell, 53:97-106.

Yamaizumi, M., T. Sugano, H. Asahina, Y. Okada, and T. Uchida, 1986. Microinjection of partially purified protein factor restores DNA damage specifically in group A of xeroderma pigmentosum cells, Proc. Natl. Acad. Sci. (U.S.A.), 83: 1476-1479.

Zunio, F., R. Gambetta, A. DiMarco, A. Velcich, A. Zaccara, F. Quadrifoglio, and V. Crescenzi, 1977. The interaction of adriamycin and its anomer with DNA, Biochem. Biophys. Acta., 476:38-46.

Zwelling, L.A., S. Michaels, L.C. Erickson. R.S. Ungerleider, M. Nichols and K.W. Kohn, 1981. Protein-associated deoxyribonucleic acid strand breaks in L1210 cells treated with the deoxyribonucleic acid intercalating agents 4'-(9-Acridinylamino)methanesulfon-m-anisidide and adriamycin, Biochemistry, 20:6553-6563.

RELATIONSHIPS BETWEEN DNA REPAIR AND TRANSCRIPTION IN DEFINED DNA SEQUENCES IN MAMMALIAN CELLS

Philip Hanawalt, Isabel Mellon, David Scicchitano, and Graciela Spivak

Department of Biological Sciences, Stanford University, Stanford, CA 94305-5020

INTRODUCTION

Certain types of damage to DNA pose blocks to the process of transcription. In particular, the presence of cyclobutane pyrimidine dimers has been shown to result in termination of transcription at the sites of the lesions (Sauerbier and Hercules, 1978). A predictable consequence of this fact is that the persistence of one or more pyrimidine dimers in the transcribed DNA strand of <u>all</u> copies of an essential, active gene will most certainly result in death of the cell. Thus, it might be considered a good strategy for cells to selectively repair the transcription blocking damage in their active genes and, in fact, to focus specifically upon the DNA strands that are being transcribed.

A prediction of the above scenario is that DNA repair may be non-uniform in the genome, with preferential repair in active sequences. Such heterogeneity in repair could render invalid the use of overall genomic repair measurements to predict particular biological responses in mutagen-treated cells. The consequences of unrepaired or misrepaired DNA damage clearly depend upon the precise location of the damage with respect to the relevant domains of the genome. Thus, for example, differences in the repair efficiency in particular protooncogenes might account for profound differences in the carcinogenic responses in different tissues in which those genes are differentially expressed. It is therefore important to understand the rules governing the fine structure heterogeneity of DNA repair in mammalian genomes.

In recent years our laboratory has developed techniques to measure UV-induced pyrimidine dimers and other lesions in defined DNA sequences at the level of the gene. (For reviews see Hanawalt, 1986; Bohr, <u>et al</u>., 1987; Smith, 1987). We have studied the repair of such lesions in mammalian DNA sequences that differ with respect to their state of expression. In the prototype study we discovered that pyrimidine dimers are much more efficiently removed from the transcriptionally active

DNA Repair Mechanisms and Their Biological Implications in Mammalian Cells
Edited by M.W. Lambert and J. Laval
Plenum Press, New York

dihydrofolate reductase (DHFR) gene in Chinese hamster ovary (CHO) cells than from a silent sequence near the gene or in the genome overall (Bohr, et al., 1985). In Swiss mouse 3T3 cells we found that dimers are efficiently removed from the active c-abl protooncogene but not removed from a region containing the inactive c-mos gene (Madhani, et al., 1986a). In repair proficient human cells we noted more rapid removal of dimers from the active DHFR gene than from the silent alpha DNA sequences or the bulk DNA (Mellon, et al., 1986). As a simple model we considered the possibility that the repair enzymes had easier access to the DNA in those genomic regions in which the chromatin structure was in an "open" configuration for transcription than in unexpressed regions. However, if the critical process for cell survival were the removal of transcription-blocking lesions it was also possible that a more direct coupling of DNA repair to transcription might be involved. In order to improve our understanding of DNA repair in active gene sequences we measured repair of dimers in the transcribed and non-transcribed DNA strands in the DHFR gene in CHO and in human cells. In both cases we found very rapid repair of dimers in the transcribed DNA strand, with nearly 80% removal in 4 hours. In the CHO cells hardly any repair occurred in the other strand within 24 hours. In the DHFR gene in human cells, repair in the non-transcribed strand was slower than that in the transcribed strand but nearly complete in 24 hours. In a 5' flanking region of the human DHFR gene, carrying a putative divergent transcript, the selective repair was in the opposite DNA strand relative to the transcribed strand of the DHFR gene thus validating the model (Mellon, et al., 1987). In the present report we discuss several studies designed to further elucidate the factors that determine the relationship between DNA repair and transcription. We have examined strand specificity of DNA repair in the DHFR gene in UV-irradiated CHO cells expressing either the human ERCC-1 DNA repair gene or the T4 bacteriophage denV gene. The denV gene encodes a small enzyme that is both a pyrimidine dimer glycosylase and an AP endonuclease. We have also studied removal of N-methylpurines in the DHFR gene, a process initiated by another small enzyme, a DNA glycosylase that is specific for 7-methylguanine and 3-methyladenine. This analysis of base excision-repair was designed to test the generality of transcription-associated repair in mammalian cells.

METHODOLOGY

Most of the methods that have been applied to measure DNA damage and repair in cells are not sensitive to intragenomic heterogeneity in the distribution of lesions or repair patches. To measure damage in different defined DNA sequences within cells the sequences of interest must either be isolated directly for analysis or some partitioning scheme must be employed to distinguish those DNA fragments containing one or more lesions from those with no lesions, prior to hybridization probing for the sequences of interest. The repair replication event following removal of the damage may also be used as a means for partitioning DNA sequences, but that approach does not provide information about which DNA lesions are being repaired, and it does not provide an initial time point.

An example of the first approach was the purification of the alpha DNA sequences in African green monkey cells for analysis of DNA damage and repair. Alpha DNA is a highly repetitive sequence found in primates and localized near centromeres in the chromosomes. In African green monkey cells these occur in tracts of at least 450 copies in tandem accounting for nearly 20% of the cellular DNA (Madhani, et al., 1986b). The results of our DNA repair analyses on alpha DNA provided the first example of repair heterogeneity in the mammalian genome. For certain chemical adducts (but not pyrimidine dimers) repair was shown to be markedly deficient in the non-transcribed alpha DNA compared to that in the genome overall (Zolan, et al., 1982). More recently, using the second approach, we have developed a versatile technique that can be applied to any DNA sequence for which specific hybridization probes are available, as detailed below.

In our initial experiments a CHO cell line carrying a 50-fold amplification of the DHFR gene was used to enhance the sensitivity of detection of a restriction fragment within the gene when the appropriately restricted DNA was electrophoresed under denaturing conditions. Figure 1 illustrates the essential steps in the procedure. The bacteriophage T4 endonuclease V is used as the specific nicking agent for pyrimidine dimers since it cuts the damaged DNA strands at the site of each dimer in the duplex DNA but does not respond to other lesions. Any restriction fragment containing one or more dimers will be cleaved by the enzyme so the affected strands will not appear in the gel at the position of the intact fragments. The bands, upon transfer to a nylon membrane, are detected by hybridization with ^{32}P-labeled genomic or c-DNA probes. The proportion of fragments free of endonuclease sensitive sites (ESS) in each sample (the zero class) is determined from the ratio of the amount of probe hybridized at the position of full-length fragments for the T4 endonuclease treated and untreated (control) samples. It is important that the DNA to be analyzed not include replicated DNA since that would add to the zero class, so density labeling of the cells with 5-bromodeoxyuridine is used followed by CsCl equilibrium density gradient centrifugation to separate the parental from hybrid daughter DNA. It is also important that the level of damage introduced into the cellular DNA yield a frequency of close to 1 lesion per fragment for optimal sensitivity and quantitation. The methodology for quantifying dimers in defined DNA sequences has been detailed by Bohr and Okumoto (1988.) In principle, any type of damage for which specific strand nicking at lesion sites can be accomplished may be quantified in this way. Thus, Thomas, et al. (1988) have used the UVRABC enzyme complex from E. coli to quantify some other bulky lesions including those produced by cis-platinum, psoralen, and 4-nitroquinoline oxide as well as UV photoproducts. Note that this system would detect other photoproducts in addition to pyrimidine dimers. To investigate the repair of methylated bases, we have developed a quantitative method for examining their removal from specific genes analogous to that devised for pyrimidine dimers (Scicchitano and Hanawalt, submitted). That procedure involves treating the DNA with an appropriate restriction enzyme, but then heating it to release N-methylpurines. During the heating process one portion of each sample is heated with methoxyamine, to reduce the apurinic sites

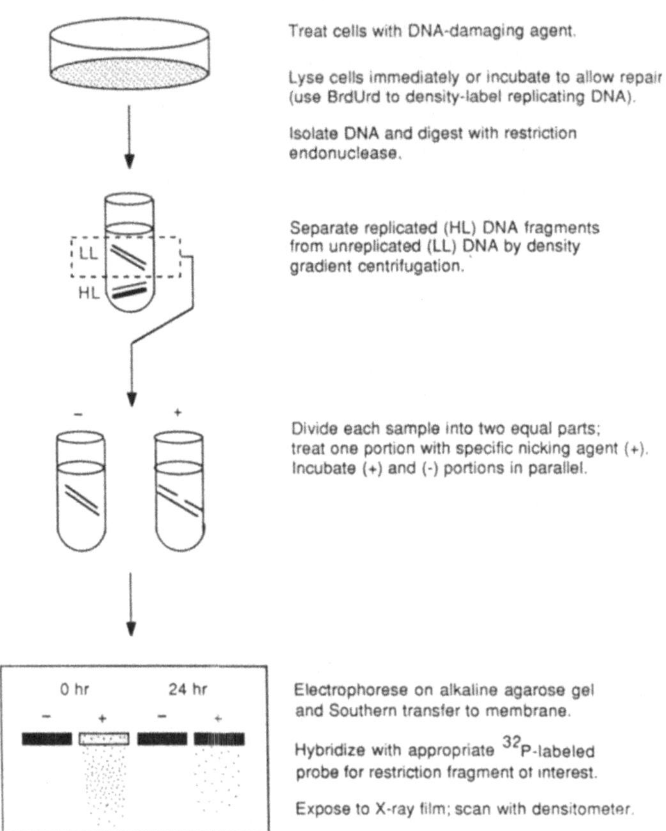

Figure 1. Schematic protocol for assay of lesions in defined DNA sequences.

(AP sites) and protect them from subsequent alkaline degradation. Following alkaline hydrolysis, electrophoresis, transfer to a nylon membrane, and probing for the segment of interest, the ratios of the band intensities of the DNA sample not treated with methoxyamine to its methoxyamine-treated counterpart are calculated to give the percentage of the restriction fragments containing no alkaline labile sites. From this information and the Poisson distribution, the frequency of N-methylpurines is determined at different times for assessment of their removal.

In order to measure repair in the separate DNA strands of a given fragment, labeled RNA probes are prepared to the respective strands as described by Mellon, et al. (1987). The same gel may be probed with the nick translated DNA probe and then with the RNA probes for each strand, respectively. An example of the data obtained for repair of UV-induced pyrimidine dimers in the DHFR gene in CHO cells is given in Figure 2. The dramatic difference between the rate of repair in the transcribed and non-transcribed DNA strands is visually apparent in comparing panels B and C. It is also evident that the repair seen when the nick translated probe is used to quantify both strands (panel A) is almost entirely due to selective repair in the transcribed strand (panel B).

HETEROLOGOUS REPAIR GENES AFFECT STRAND-SPECIFIC REPAIR

As an approach to study the mechanisms by which rodent cells remove dimers preferentially from the transcribed strand of the DHFR gene, and perhaps from other active, housekeeping genes, we examined strand-specific repair in repair deficient CHO cells carrying heterologous repair genes that corrected the deficiency. CHO mutant cell lines have been isolated that are hypersensitive to UV and belong to at least eight complementation groups (Thompson, et al., 1988; also this volume). Some of these mutants have been transfected with DNA sequences that partially or completely restore their resistance to UV. Of these transformants, we have studied two types:

1) the UV-sensitive mutant 43-3B that belongs to complementation group I* (Wood and Burki, 1982) transformed with a human excision-repair gene, ERCC-1 (van Duin, et al., 1986). The colony forming ability of the UV-irradiated transformant cells is significantly increased by the human DNA sequence, but it does not attain the level of UV-resistance of the parental cells, CHO9 (Zdzienicka, et al., 1987).

2) the transformant line I-A1 obtained by introducing the denV gene from phage T4, which codes for the endonuclease V that nicks DNA specifically at cyclobutane pyrimidine dimers (Valerie, et al., 1985), into the UV-sensitive mutant UV5 belonging to complementation group II* (Thompson, et al., 1981). The UV-resistance of the I-A1 cells is intermediate between that of the parental AA8 cells, and that of the UV-sensitive mutant, UV5 (Valerie, et al., 1985).

For the experiments that follow, the cells were grown in minimum essential medium (Eagle) (GIBCO, modified, autoclavable with Earle's salts) containing 10% fetal bovine serum, 2 mM glutamine, non-essential amino acids and antibiotic-antimycotic solution (GIBCO). The transformant cells containing the ERCC-1 gene were maintained in the medium described above with the addition of XHAT (250 µg/ml xanthine, 15 µg/ml hypoxanthine, 10 µg/ml thymidine, 2 µg/ml amethopterin, 25 µg/ml mycophenolic acid); the selection with XHAT was discontinued prior to the ^{3}H-dThd labeling for the repair experiments. The irradiation, lysis, purification of DNA, repair analysis, and construction of hybridization probes were carried out as described (Mellon, et al., 1987).

We examined the removal of ESS from a 14 kb KpnI fragment that encompasses exons I through IV of the DHFR gene by using strand-specific RNA probes derived from the plasmid pZH-4 as

---

*Note that the nomenclature reflects the new designations agreed upon at the Taos DNA repair meeting (Thompson and Bootsma, 1988).

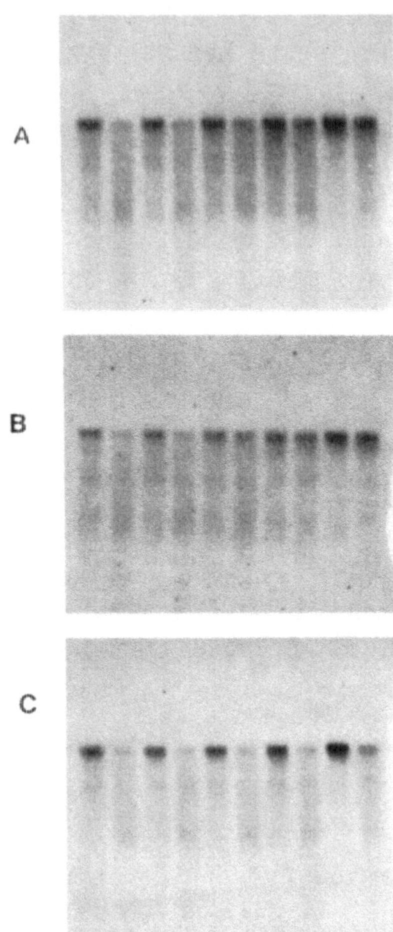

Figure 2. Autoradiograms Illustrating Strand-Specific Repair in the CHO DHFR Gene.

DNA was isolated at the times indicated above the lanes and restricted with KpnI. Samples were not treated (-) or were treated (+) with T4 endonuclease V prior to electrophoresis. The 14 kb KpnI fragment was detected by hybridization with a DNA probe made by nick translation that detects both strands (A), an RNA probe specific for the transcribed strand (B) and an RNA probe specific for the non-transcribed strand (C). Figure reprinted by permission of Cell Press from Mellon, et al., 1987.

described previously (Mellon, *et al.*, 1987). The cells were irradiated with 10 Jm^{-2}, a UV dose that produces in CHO cell DNA an average of 0.086 pyrimidine dimers per kb. The results obtained with the CHO9, 43-3B and *ERCC-1*$^{+}$ cells are shown in Table I. There was no difference in the initial level of damage measured in the transcribed and non-transcribed strands of the *DHFR* gene.

Table 1. Repair Of Complementary Strands In *DHFR* Effect Of Human *ERCC-1* Gene

Repair time (hr)	Percent of ESS removed					
	CHO9		43-3B		ERCC-1^{+}	
	T	NT	T	NT	T	NT
2	15	0	0	0	36	0
4	51	10	0	0	36	10
8	83	6	0	0	51	0
24	78	0	ND	ND	56	0

T: Transcribed strand; NT: Non-transcribed strand; ND: Not done. Two separate experiments were carried out; one included the 2, 4 and 8 hr time points and the other the 24 hr time point. These data have been published (Mellon, *et al.*, 1988).

The *ERCC-1* gene product in the transformed cells restores selective strand repair in the *DHFR* gene, but the kinetics of ESS removal are somewhat slower than in the parental, wild type CHO9 cells. There is essentially no dimer removal from the non-transcribed strand, indicating that the human *ERCC-1* gene does not provide the information necessary to carry out the efficient dimer removal from non-transcribed DNA that is observed in repair proficient human cells. It is of interest that the *ERCC- 1*$^{+}$ cells also behave like the CHO9 wild type cells with respect to relative repair efficiency in the flanking non-transcribed sequences (Bohr, *et al.*, 1988).

The protein encoded by the *ERCC-1* gene is thought to be involved in the incision step of the excision-repair pathway in human cells. *ERCC-1*$^{+}$ hamster cells are able to remove pyrimidine dimers from their DNA, and they show increased resistance to UV; they also completely recover their resistance to 4NQO, ENU and NA-AAF (Zdzienicka, *et al.*, 1987). Based upon our results, we can assume that the *ERCC-1* gene product has a function similar to that of the protein that was rendered non-functional or not produced in the mutant 43-3B, and that this function cannot promote efficient dimer removal from non-transcribed DNA. Both the lower UV-resistance and the slower, less efficient removal of ESS shown by the transformed cells when compared to the parental cells may be explained by the fact that *ERCC-1* is a human enzyme introduced artificially into hamster cells. Thus it may not be working in an optimal intracellular environment. It also may not be able to repair UV-induced lesions other than cyclobutane pyrimidine dimers.

The same approach was used to compare the repair properties of the AA8 and I-A1 cells, and the results are shown in Table 2. The cells containing the denV gene are able to efficiently remove ESS from both strands of the DHFR gene, and this removal is not selective for either strand. The denV gene of phage T4 codes for the enzyme, endonuclease V, that also incises DNA at each cyclobutane pyrimidine dimer, but its properties and mechanism of incision are quite different from those of the ERCC-1 gene product. It is a very small protein (16 kd) that has also been shown to restore repair competence in xeroderma pigmentosum cells (Valerie, et al., 1987). Bohr and Hanawalt (1987) have reported that dimers are removed from both coding and non-coding sequences of the DHFR domain in this same cell line in which we have shown that the I-A1 cells that express denV are able to remove ESS efficiently from both strands of the DHFR gene. This efficient, non-selective removal of ESS from the genome overall, however, does not completely restore the UV-resistance of the CHO cells. One possible explanation would be that the T4 endonuclease attacks all of the dimers in the DNA indiscriminately and very rapidly so that the cell repair machinery does not have adequate time to excise, resynthesize, and/or ligate the DNA repair patches after the initial incisions. Valerie and coworkers have shown that, in the presence of DNA synthesis inhibitors, I-A 1 cells accumulate many more strand breaks than do the wild type AA8 cells (Valerie, et al., 1985). Another explanation could be that other lesions caused by UV-irradiation, (e.g. pyrimidine 6-4 photoproducts), which are not recognized by the T4 endonuclease, are responsible for the reduction in survival. Both possibilities could be experimentally tested.

Table 2. Repair Of Complementary Strands In DHFR Effect Of T4 denV Gene

Repair time (hr)	Percent of ESS removed			
	AA8		I-A1	
	T	NT	T	NT
4	32	7	33	21
8	50	0	38	48

T: Transcribed strand; NT: Non-transcribed strand.

LACK OF STRAND SPECIFIC REPAIR OF N-METHYLPURINES

It was of interest to determine whether the strand specificity in DNA repair would be observed for lesions other than pyrimidine dimers. In particular we wished to know whether the N-methylpurines that are removed from DNA by a 3-methyladenine-DNA glycosylase would be sublect to intragenomic repair heterogeneity and preferential removal from transcribed strands. Whereas the repair of pyrimidine dimers is initiated by the versatile nucleotide excision repair system, the N-methylpurines are repaired by the base excision-repair scheme, initiated by a glycosylase that removes the damaged base. We used the approach outlined in "Methodology".

CHO-B11 cells were exposed to dimethylsulfate (DMS), which produces a number of adducts of which 7-methylguanine and 3-methyladenine constitute about 90% (Hoffman, 1980). Freshly prepared DMS in dimethyl sulfoxide was added to the growth medium to a final concentration of 150 μM DMS for 30 minutes before removal and incubation to permit repair. The results are summarized in Table III. The N-methylpurines within the transcription unit of the DHFR gene were repaired at roughly the same rate as that found in the 3' flanking region that is not transcribed. This result is quite different from the relative repair of dimers in the same fragments. In the case of dimers, the repair rate in the 3' flanking region was similar to that for the bulk DNA while repair in the DHFR gene attained 70% within 24 hours. Also, it is clear that no significant difference in repair of N-methylpurines exists between the transcribed and non-transcribed DNA strands. However, our results for bulk (total cellular) DNA indicate a somewhat slower rate of repair than that in the DHFR domain.

Table 3. Repair Of N-Methylpurines In CHO Cells

Repair time (hr)	Percent of Alkaline Labile Sites Removed				
		DHFR			
	Bulk[1]	Both strands	T	NT	3' Flanking
3	24	39	43	34	30
6	ND	48	54	47	35
10	35	ND	ND	ND	ND
12	ND	80	78	72	70
24	72	100	96	98	96

T: Transcribed strand; NT: Non-transcribed strand; ND: Not done.
[1] Repair in bulk DNA was determined by alkaline sucrose gradient sedimentation.

We do not know the significance of that difference, but it indicates that some regions of the genome must be less accessible to the enzyme than others.

Like the T4 endonuclease V, the 3-methyladenine glycosylase is probably a small enzyme. (The glycosylases from human and rat cells are 25 kd and 24 kd, respectively, so it is likely that the CHO enzyme is of similar size). It is possible that most regions of chromatin are readily accessible to it, and that accounts for the lack of discrimination between active and inactive regions in repair.

SUMMARIZING DISCUSSION

We have shown that pyrimidine dimers are selectively repaired in the transcribed strand of the _DHFR_ gene in wild type CHO cells and in repair deficient mutant cells expressing the human _ERCC-1_ repair gene. However, this selectivity is not seen when the repair defect in CHO mutants is corrected by the _den_V gene, nor is it apparent for another type of lesion, the N-methylpurines that are removed by a glycosylase in the base excision-repair pathway. However, it should be pointed out that the lack of preferential repair of the transcribed DNA strand does not necessarily rule out a transcription coupled repair process; it could be that the efficient repair of lesions in the non-transcribed strand, due to the abundance of a small enzyme, could obscure another dedicated process that achieves similar high repair efficiency in the transcribed DNA strand. Other experimental approaches such as the use of the transcription inhibitor alpha-amanitin are currently under study to elucidate the possible degree of coupling of pyrimidine dimer repair to the actual process of transcription.

One can consider a number of potential levels of excision-repair enzyme (or enzyme complex) accessibility to particular lesions in mammalian DNA. Control at different levels of chromatin condensation may apply for different lesions. Thus, while pyrimidine dimers are efficiently repaired in human cells much of the genome in rodent cells appears to be excluded from repair. It is of interest in that regard that in xeroderma pigmentosum, complementation group C (XP-C), there also appears to be a relatively large portion of the genome that is excluded from repair, when the cells are confluent, as initially reported by Mansbridge and Hanawalt (1983) and discussed in detail by Mullenders and coworkers in this volume. The XP-C cells are UV-sensitive, unlike the rodent cells that are as UV-resistant as normal human cells. Bohr, _et al_.(1986) have suggested that the UV-sensitivity in XP-C may be a consequence of lack of repair in certain essential genes. However, Mullenders, _et al_. (this volume) find that some active genes are proficiently repaired in XP-C and similar results have been obtained by Kantor (to be published). Thus, it is not clear why the XP-C cells are UV-sensitive and which particular regions are excluded from repair in those cells.

In the _DHFR_ gene in CHO and human cells, the selective repair is accounted for by the observed preferential repair in the transcribed DNA strand. One could suppose that the repair complex recognizes the signals that initiate transcription, or more likely that the complex may be physically coupled to the transcription apparatus. In some cases it would appear that the repair is efficient in both strands (_cf_.

Mullenders, et al., this volume) that could be explained by the synthesis of an antisense RNA or simply by the generally "open" state of chromatin in actively expressed regions. There are also other levels of chromatin "openness" to be considered such as the category of sequences that include the inducible genes. These are found in the domains of the genome that replicate early in the cell cycle along with the active genes as a general rule (Goldman, et al.,1984). There may be particular features of the chromatin in these domains that render the DNA sequences more accessible to repair than those sequences in the "closed" chromatin that includes the inactive genes and late replicating regions. In addition, there are other special categories such as the repetitive alpha DNA sequences mentioned earlier, in which the unique chromatin structure may restrict access to DNA by some repair enzymes. It is of some interest in that case that the presence of pyrimidine dimers actually may enhance the accessibility of other "bulky" lesions to repair (Leadon and Hanawalt, 1984), possibly through some chromatin structure perturbation. Finally, there may be some tightly packed chromatin domains in which the DNA is functionally inert, in terminally differentiated cells for example, in which only a small subset of genes may still be expressed. It will be important to systematically learn the rules that operate to regulate accessibility of damaged DNA to repair in these various special circumstances and levels of structural condensation of DNA. In some cases we may be able to make use of the information we obtain on preferential DNA repair to learn from the cells which are the essential regions of the genome worth preserving.

ACKNOWLEDGEMENTS

We thank J. Hoeijmakers for providing the CHO9, 43-3B, and ERCC-1^{+} cells and K. Valerie for the AA8 and I-A1 cells. This work was supported by an Outstanding Investigator Award (CA 44349) from the National Cancer Institute. One of us (David Scicchitano) acknowledges a National Research Training Award (AM 07422).

REFERENCES

Bohr, V.A., Smith, C.A., Okumoto, D.S., and Hanawalt, P.C., 1985, DNA repair in an active gene: Removal of pyrimidine dimers from the DHFR gene of CHO cells is much more efficient than in the genome overall, Cell, 40:359.

Bohr, V.A., Okumoto,D.S., and Hanawalt, P.C., 1986, Survival of UV-irradiated mammalian cells correlates with efficient DNA repair in an essential gene, Proc. Natl. Acad. Sci. USA, 83:3830.

Bohr, V.A., Phillips, D.H., and Hanawalt, P.C., 1987, Heterogeneous DNA damage and repair in the mammalian genome, Cancer Res., 47:6426.

Bohr, V.A., and Hanawalt, P.C., 1987, Enhanced repair of pyrimidine dimers in coding and non-coding genomic sequences in CHO cells expressing a prokaryotic DNA repair gene, Carcinogenesis, 8:1333.

Bohr, V.A., Chu, E.H.Y, Van Duin, M., Hanawalt, P.C., and Okumoto, D.S., 1988, Human repair gene restores normal pattern of preferential DNA repair in repair defective CHO cells, Nucl. Acids Res., 16:7397.

Bohr, V.A., and Okumoto, D.S., 1988, Analysis of pyrimidine dimers in defined genes, in "DNA Repair, A Laboratory Manual of Research Procedures, Volume 3," E.C. Friedberg and P.C. Hanawalt, eds., Marcel Dekker, Inc., New York, New York.

Goldman, M.A., Holmquist, G.P., Gray, M.C., Caston, L.A., and Nag, A., 1984, Replication timing of genes and middle repetitive sequences, Science, 22:686.

Hanawalt, P.C., 1986, Intragenomic heterogeneity in DNA damage processing: potential implications for risk assessment, in "Mechanisms of DNA Damage and Repair", M. Simic, L. Grossman, A. Upton, eds., Plenum Press, New York.

Hoffman, G.R., 1980, Genetic effects of dimethyl sulfate, diethyl sulfate, and related compounds, Mutation Res. 75:63.

Leadon, S.A., and Hanawalt, P.C., 1984, Ultraviolet irradiation of monkey cells enhances the repair of DNA adducts in alpha DNA, Carcinogenesis, 5:1505.

Madhani, H.D., Bohr, V.A., and Hanawalt, P.C., 1986a, Differential DNA repair in transcriptionally active and inactive proto-oncogenes: c-abl and c-mos, Cell, 45:417.

Madhani, H.D., Leadon, S.A., Smith, C.A., and Hanawalt, P.C., 1986b, Alpha DNA in african green monkey cells is organized into extremely long tandem arrays, J. Biol. Chem., 261:2314.

Mansbridge, J.N., and Hanawalt, P.C., 1983, Domain-limited repair of DNA in ultraviolet irradiated fibroblasts from xeroderma pigmentosum complementation Group C, in "Cellular Responses to DNA Damage," E.C. Friedberg and B.R. Bridges, eds., Alan R. Liss, Inc., New York, New York.

Mellon, I., Bohr, V.A., Smith, C.A., and Hanawalt, P.C., 1986, Preferential DNA repair of an active gene in human cells, Proc. Natl. Acad. Sci., USA, 83:8878.

Mellon, I., Spivak, G., and Hanawalt, P.C., 1987, Selective removal of transcription-blocking DNA damage from the transcribed strand of the mammalian DHFR gene, Cell, 51:241.

Mellon, I., Spivak, G., and Hanawalt, P.C., 1988, Strand specificity of DNA repair in CHO cells expressing the human ERCC-1 gene, in "Mechanisms and Consequences of DNA Damage Processing", E.C. Friedberg and P.C. Hanawalt, eds., Alan R. Liss, Inc. , New York, New York.

Sauerbier, W., and Hercules, K., 1978, Gene and transcription unit mapping by radiation effects, Ann. Rev. Genet., 12:329.

Smith, C.A., 1987, DNA Repair in specific sequences in mammalian cells, J. Cell Sci. Suppl., 6:225.

Thomas, D.C., Morton, A.G., Bohr, V.A., and Sancar, A., 1988, General method for quantifying base adducts in specific mammalian genes, Proc. Natl. Acad. Sci., USA, 85:3723.

Thompson, L.H., Busch, D.B., Brookman, K., Mooney, C.L., and Glaser, D.A., 1981, Genetic diversity of UV-sensitive DNA repair mutants of Chinese Hamster Ovary cells, Proc. Natl. Acad. Sci., USA, 78:3734.

Thompson, L.H., and Bootsma, D., 1988, Designation of mammalian complementation groups and repair genes, in "Mechanisms and Consequences of DNA Damage Processing", E.C. Friedberg and P.C. Hanawalt, eds., Alan R. Liss, Inc., New York, New York.

Thompson, L.H., Weber, C.A., and Carrano, A.V., 1988, Human DNA repair genes, in "Mechanisms and Consequences of DNA Damage Processing," E.C. Friedberg and P.C. Hanawalt, eds., Alan R. Liss, Inc. , New York, New York.

Valerie, K., de Riel, J.K., and Henderson, E.E., 1985, Genetic complementation of UV-induced DNA repair in Chinese hamster ovary cells by the denV gene of phage T4, Proc. Natl. Acad. Sci., USA, 82:7656.

Valerie, K., Green, A .P., de Riel, J.K., and Henderson, E.E. 1987, Transient and stable complementation of ultraviolet repair in xeroderma pigmentosum cells by the denV gene of bacteriophage T4, Cancer Res., 47:2967.

Van Duin, M., de Witt, J., Odijk, H., Westerveld, A., Yasuir, A., Koken, M.H.M., Hoeijmakers, J.H.J., and Bootsma, D., 1986, Molecular characterization of the human excision repair gene ERCC-1: cDNA cloning and amino acid homology with the yeast DNA repair gene RAD10, Cell, 44:913.

Wood, R.D., and Burki, H.J., 1982, Repair capability and the cellular age response for killing and mutation induction after UV, Mutation Res., 95:505.

Zdzienicka, M.Z., Roza, L., Westerveld, A., Bootsma, D., and Simons, J.W.I.M., 1987, Biological and biochemical consequences of the human ERCC-1 repair gene after transfection into a repair-deficient CHO cell line, Mutation Res., 183:69.

Zolan, M.E., Cortopassi, G.A., Smith, C.A., and Hanawalt, P.C., 1982, Deficient repair of chemical adducts in alpha DNA of monkey cells, Cell, 28:613.

NON-RANDOM DISTRIBUTION OF UV-INDUCED REPAIR IN HIGHER-ORDER CHROMATIN LOOPS IN HUMAN CELLS AND ITS RELATIONSHIP TO PREFERENTIAL REPAIR OF ACTIVE GENES

L.H.F. Mullenders[1,2], J. Venema[1], L. Mayne[3], A.T. Natarajan[1,2] and A.A. van Zeeland[1,2]

[1]Department of Radiation Genetics and Chemical Mutagenesis, State Univ. of Leiden, Wassenaarseweg 72, 2333 AL Leiden, The Netherlands

[2]J.A. Cohen Institute, Interuniversity Research Inst. for Radiopathology and Radiation Protection, The Netherlands

[3]Sussex Center of Medical Research, University of Sussex, U.K.

SUMMARY

The eukaryotic genome is organized into a series of loops each topologically anchored by a skeletal structure termed scaffold or nuclear matrix. Such an organization is thought to facilitate processes which occur proximal to the nuclear matrix, i.e., replication and transcription. In confluent human fibroblasts exposed to either 5 J/m^2 or 30 J/m^2 no evidence was found for compartmentalization of UV-induced excision repair at the nuclear matrix, i.e., lesions do not require prior attachment to the nuclear matrix to be repaired.

Next we addressed the question whether excision repair occurs randomly within chromatin loops. In cells exposed to 30 J/m^2 repair approached a random distribution. However, at a UV-dose of 5 J/m^2 the distribution of excision repair was inhomogeneous, DNA sequences close to the nuclear matrix being preferentially repaired. The nonrandom distribution of repair was most pronounced directly after irradiation and gradually changed to a more random distribution within 2 hours after treatment. The preferential repair of nuclear matrix associated DNA most likely represents the preferential repair of transcriptionally active DNA located proximal to the nuclear matrix. The observation that pyrimidine dimers are preferentially removed from the active adenosine deaminase gene (ADA) compared to the inactive 754 locus or the genome overall in confluent normal fibroblasts, is in favour of this hypothesis.

DNA Repair Mechanisms and Their Biological Implications in Mammalian Cells
Edited by M.W. Lambert and J. Laval
Plenum Press, New York

Pronounced differences in distribution of excision repair were found among xeroderma pigmentosum complementation group C fibroblasts and Cockayne's syndrome fibroblasts. The residual repair capacity of confluent xeroderma pigmentosum complementation group C cells was found to be very specific for nuclear matrix associated DNA. In contrast, confluent Cockayne's syndrome fibroblasts appeared to be deficient in preferential repair of nuclear matrix associated DNA. This heterogeneity in distribution of repair correlated well with the presence and absence of preferential removal of pyrimidine dimers from the active *ADA* gene in the xeroderma pigmentosum complementation group C and Cockayne's syndrome fibroblasts, respectively. The results suggest that xeroderma pigmentosum group C cells are proficient in repair of active genes, but deficient in repair of chromatin regions outside transcriptionally active DNA. Cockayne's syndrome cells may be deficient in repair of transcriptionally active DNA.

INTRODUCTION

One of the responses of cells to DNA damage is the activation of DNA repair pathways in order to restore the original genetic information. Incomplete repair will give rise to persisting lesions which can impair cellular functions. Unrepaired damage in different regions of the genome may interfere with an appropriate control of expression and may lead to mutations in transcriptionally active DNA affecting both function and survival of cells. It is generally accepted that transcribed chromatin is less compacted than chromatin in a nontranscribable conformation (Weisbrod, 1982). These differences in conformation as well as the folding of eukaryotic DNA into chromatin loops may be an important factor in determining the accessibility of genomic DNA towards damaging agents and enzymes involved in repair of DNA damage. In most eukaryotic cells nuclear DNA is organized in chromatin fibers which are arranged in large loops. In interphase cells, chromatin exists largely in the form of 30 nm fibers by helical coiling of nucleosome filaments into a solenoid. The next level of complexity is the formation of these fibers into loops that are attached at their bases to proteins of the nuclear matrix or scaffold (Berezney and Coffey, 1977; Mirkovitch, *et al.*, 1984). There is considerable evidence that the nuclear matrix plays an important role in the structural organization of chromatin and it is likely that attachment of DNA to the nuclear matrix facilitates both transcription and replication processes. In *Drosophila*, attachment sites tend to be found close to the promotor elements (Mirkovitch, 1986) and transcribed sequences (Small and Vogelstein, 1985). The association of the transcription process itself with the nuclear matrix correlates well with selective attachment of genes during development (Robinson, *et al.*, 1983). Other approaches suggest that the nuclear matrix is involved in DNA replication (Dijkwel, *et al.*, 1979; Vogelstein, *et al.*, 1980) and it is conceivable that the replication complexes are actually fixed to the nuclear matrix (Dijkwel, *et al.*, 1979; Vogelstein, *et al.*, 1980). Each loop might be equivalent to a replicon with the origin of replication attached to the nuclear matrix (Buongiorno-Nardelli, *et al.*, 1982). A model in which attachment to the nuclear matrix is a necessary pre-condition for transcription and replication can be readily extended to include repair. Repair can occur when

DNA loops are reeled through or when lesions become attached to repair complexes fixed within the nuclear matrix.

Analysis of the role of chromatin structure in repair of damage is complicated by an initial nonuniform distribution of lesions within the genome demonstrated for various mutagens (for recent review see Bohr, et al., 1987). For ultraviolet (254 nm) radiation however, most studies show evidence for a uniform distribution of photoadducts (Rahn and Stafford, 1974), and therefore the observed nonuniform distribution of ultraviolet induced excision repair within the genome has been attributed to repair processes operating in a nonrandom manner (Cohn and Lieberman, 1984; Lan and Smerdon, 1985). Interpretations concerning the biological relevance of nonuniform repair within chromatin loops have to consider the nonrandom organization of DNA sequences in loops, transcriptionally active genes being in close proximity to the nuclear matrix. In this study, we have analyzed UV-induced excision repair in confluent human fibroblasts both at the level of chromatin loops, and at the level of defined fragments of genes as described by Bohr, et al.(Bohr, et.al., 1985). Both approaches differ not only with respect to the genomic regions under study, but also with respect to analysis of excision repair. Repair in chromatin loops is detected by radioactive labelling and thus includes all types of lesions under repair, whereas repair at the gene level is quantified for the removal of a single adduct, i.e., pyrimidine dimer. In most of our studies, we used confluent human fibroblasts in order to measure excision repair by incorporation of radioactive precursors. Besides normal human cells we employed two UV-sensitive human cell lines with interesting characteristics: xeroderma pigmentosum fibroblasts belonging to complementation group C (XP-C) and Cockayne's syndrome (CS) fibroblasts. Cells from xeroderma pigmentosum patients are sensitive to lethal and mutagenic effects of UV and in most cases the UV-sensitivity correlates with the relative extent of excision repair defects (Kantor and Hull, 1984). However, nondividing XP-C cells are relatively resistant to lethal action of UV (Kantor and Hull, 1984). XP-C cells are able to recover UV-inhibited RNA synthesis (Mayne and Lehmann, 1982) and the limited amount of excision repair occurs in localized domains (Kantor and Hull, 1984; Mansbridge and Hanawalt, 1983; Mullenders, et al., 1984). Cells from Cockayne's syndrome patients are also hypersensitive to lethal effects of UV, but no defect in the overall repair capacity has been reported (Lehmann, et al., 1979). In spite of the normal overall repair, CS cells are not able to recover UV-inhibited RNA synthesis.

INTRANUCLEAR LOCALIZATION OF UV-INDUCED EXCISION REPAIR

Intact supercoiled DNA loops anchored to the nuclear matrix can be isolated as a rapidly sedimenting complex from nuclei, by extraction with 2 M NaCl and subsequent centrifugation in neutral sucrose gradients (Mullenders, et al., 1983). An alternative way which omits conditions of high ionic strength, is extraction of nuclei with lithium diiodosalicylate (LIS) (Mirkovitch, et al., 1984). The polypeptide composition of DNA-nuclear matrix complexes obtained with both extraction methods, is very similar except that the removal of histones by LIS is less complete (unpublished results).

DNA can be removed from the nuclear matrix by cleaving the DNA with increasing concentrations of DNAse I. The probability of a DNA fragment being released from the nuclear matrix will decrease the closer it is situated to a region bound to the nuclear matrix. Analysis of pulse labelled non-irradiated growing or confluent cells in the presence of the inhibitor hydroxyurea (HU) reveals that nascent DNA is very closely associated with the nuclear matrix (Figure 1). At the single cell level the same conclusions emerge from autoradiographic analysis of DNA halo-matrix structures (Vogelstein, et al., 1980; Mullenders, et al., 1986). Measurement of excision repair was performed with confluent cells UV-irradiated with 5 J/m^2 or 30 J/m^2, and pulse labelled in the presence of HU. In cells exposed to 5 J/m^2 a clear enrichment of repair events at the nuclear matrix was observed, whereas in cells exposed to 30 J/m^2, repair approached a random distribution (Figure 1). The nonrandom distribution of excision repair at 5 J/m^2 gradually changed to a more random distribution within two hours after UV-treatment (Mullenders, et al., 1988). The same results, as shown in Figure 1 for 2 M NaCl extracted nuclei,

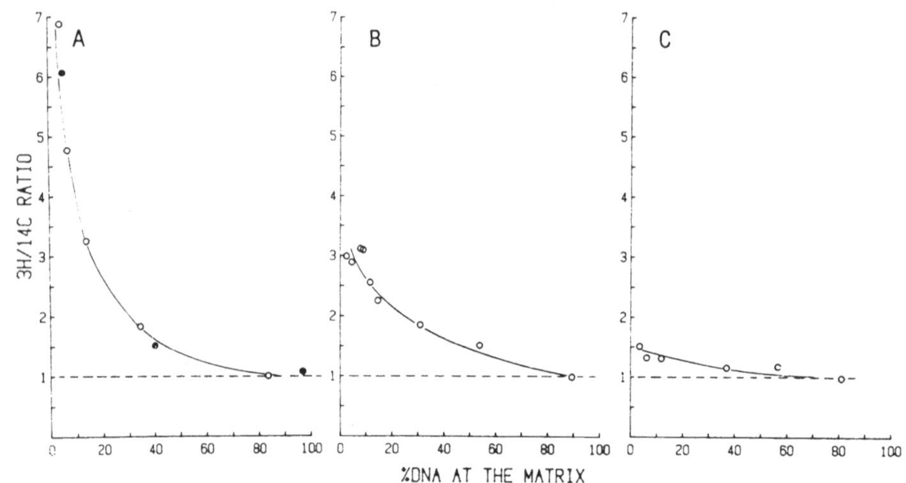

Figure 1. Spatial distribution of pulse labelled DNA in chromatin loops attached to the nuclear matrix. (A) Exponentially growing (●) or confluent (O) human fibroblasts were pulse labelled with [^{3}H]thymidine for 10 minutes in the absence or for 30 minutes in the presence of hydroxyurea, respectively. (B) and (C). Confluent human fibroblasts were UV- (254 nm) irradiated with 5 J/m^2 (B) or 30 J/m^2 (C) and pulse labelled for 10 minutes in the presence of hydroxyurea. Cells were prelabelled with [^{14}C] thymidine. Nuclei were extracted in 2 M NaCl, incubated with DNAse I and analyzed in neutral sucrose gradients. We plotted the relative ^{3}H/^{14}C ratio of the DNA-nuclear matrix complex versus the relative amount of ^{14}C-prelabelled DNA in the nuclear matrix.

were obtained with LIS extracted nuclei. In principle, preferential repair of nuclear associated DNA at 5 J/m^2 could be attributed to transient binding of damaged sites to the nuclear matrix. The results of pulse-chase experiments however, are consistent with the preferential repair of DNA permanently bound to the nuclear matrix (Mullenders, et al., 1988).

Taken together the results suggest that excision repair is not confined to the nuclear matrix as has been found for replication and transcription. It is important to note here, that the vast majority of repair patches labelled during a 10 minute pulse in the presence of HU, are not completed (Mullenders, et al., 1987), ruling out the possibility that excision repair is too fast to be trapped in a transient state of synthesis. At present, we can only speculate about the mechanism underlying the dose dependent association of repair with the nuclear matrix. The pathway involved in repair of nuclear matrix associated DNA, may saturate at a UV-dose significantly lower than the plateau level of 20-30 J/m^2 found for the genome overall (Erixon and Ahnström, 1979). This hypothesis is not unreasonable, if we assume that the preferential repair of nuclear matrix associated DNA represents repair of transcriptionally active DNA. Moreover, evidence has been presented that repair of active DNA is coupled to transcription (Mellon, et al., 1986). Such a pathway may saturate at a lower UV-dose than the repair of the genome overall.

DISTRIBUTION OF EXCISION REPAIR IN NORMAL AND UV-SENSITIVE FIBROBLASTS

During the first two hours following 5 J/m^2 UV-irradiation, normal human fibroblasts preferentially repair nuclear matrix associated DNA (Table 1). The data in Table 1 also show that the residual repair in XP-C cells (15% of normal repair level) is highly specific for matrix associated DNA, whereas the distribution in XP-D cells (25% of normal repair level) resembles that of normal cells. In CS cells the opposite result was observed, i.e., the DNA-nuclear matrix complex became relatively depleted of repair label as more DNA was released from the complex by DNAse I.

The preferential occurrence of repair in nuclear matrix associated DNA in confluent XP-C cells was found in cells labelled for either 20 minutes, 2 hours or 24 hours (Mullenders, et al., 1984). The non-random distribution of repair in normal cells observed initially, after UV-irradiation (Figure 1B), was very similar to the distribution in XP-C cells, but unlike in XP-C changed into a more random distribution within 2 hours after treatment (Mullenders, et al., 1988).

Preferential repair of nuclear matrix associated DNA during a short period after UV-treatment fits in with the concept that functionally important domains in the genome are quickly repaired after treatment. Since repair of UV-induced potentially lethal damage and UV-inhibited RNA synthesis is almost completed within 2 hours (Mayne and Lehmann, 1982; Keyse and Tyrrell, 1987), these domains most likely include transcriptionally active DNA. The different distributions of repair in XP-C and CS cells would then indicate that in XP-C the residual repair is confined to active DNA, whereas CS cells are defective in performing efficient repair of active DNA. The observation that XP-C and CS cells are proficient and deficient, respectively, in the recovery of UV-inhibited RNA synthesis (Mayne and Lehmann, 1982) is consistent with this hypothesis.

REMOVAL OF PYRIMIDINE DIMERS FROM ACTIVE AND INACTIVE GENES

To obtain direct experimental support for the hypothesis that XP-C and CS cells are proficient and deficient, respectively, in repair of active genes, we studied the removal of pyrimidine dimers from active and inactive sequences, using the dimer specific enzyme T4 endonuclease V as described by Bohr, et al.(1985). Dimer removal was determined in an 18 kb DNA fragment located at the 3' end of the active adenosine deaminase gene (ADA) and in a 14 kb fragment of the inactive 754 locus. Dimer removal of the genome overall was calculated from DNA profiles in alkaline sucrose gradients after T4 endonuclease V treatment of the DNA. In order to correlate these results to those of repair studies at the level of DNA-nuclear matrix complexes, we analyzed the position of the ADA gene relative to the nuclear matrix. The ADA gene is located proximal to the nuclear matrix by attachment at the 5' end of the gene as has been found for other genes (Mullenders, et al., 1988).

Table 1. Distribution of excision repair in chromatin loops attached to the nuclear matrix[1]

% ^{14}C-DNA at the matrix	Normal	CS	^{3}H/^{14}C ratio[2] XP-C	XP-D
50	1.05	0.97	1.35	1.05
25	1.31	0.91	1.72	1.15
7.5	1.65	0.55	3.25	1.65

[1]UV-irradiation(254 nm): 5 J/m^2
[2]Ratio of ^{3}H (repair-labelled) to ^{14}C (prelabelled) DNA associated with the nuclear matrix after DNAse I digestion.

Table 2 shows the results of experiments in which confluent normal human fibroblasts were UV-irradiated with 10 J/m^2. Dimer removal from the active ADA gene is 2-3 fold faster than from the genome overall. This result is very similar to preferential repair of dimers from the human DHFR gene as reported by Mellon, et al.(1986) and indicates that preferential repair may be a general feature of active genes. The rate and extent of dimer removal from the inactive 754 locus was similar to that from the genome overall. Confluent XP-C fibroblasts are proficient in removal of dimers from the ADA

Table 2. Removal of dimers from defined DNA fragments and from the genome overall in normal cells[1].

Time of repair (hrs)	% dimers removed ADA	754	Genome overall
4	40%	17%	25%
8	76%	37%	31%
24	92%	71%	69%

[1]UV-irradiation(254 nm): 10 J/m[2]

gene, but are deficient in removal of dimers from the 754 locus (Table 3). In contrast, CS fibroblasts remove dimers to a lesser extent from the ADA gene than normal cells (Table 3). It is obvious from these data, that the heterogeneity in distribution of excision repair in DNA-nuclear matrix complexes correlates well with the heterogeneity in removal of dimers from single genes in various types of cells. In conclusion, the results of two different approaches suggest, that CS cells have a defect in excision of UV-damage from transcriptionally active DNA. XP-C cells may possess a defect in DNA repair associated with chromatin regions outside transcriptionally active DNA.

Table 3. Removal of dimers from defined DNA fragments and the genome overall in XP-C and CS cells.

	% dimers removed			
	XP-C			CS
Time of repair (hrs)	ADA	754	Genome overall	ADA
24	92%	8%	12%	36%

[1]UV-irradiation: 10 J/m[2]

CONCLUDING REMARKS

The organization of the genome into loops has been proposed to have a structural and functional role. The localization of regulatory elements at the base of loops may serve to bring distant sequences close together to create functional complexes for the regulation of DNA synthesis and transcription, and to provide the possibilities for complete unwinding of parental DNA molecules during mitosis (Dijkwel, et al., 1979). The spatial organization of the repair process is different from the latter two processes as it is not confined to the nuclear matrix compartment. The efficiency of repair of UV-induced damage within chromatin of human cells may be based on a sliding model only. In such a model proteins bind non-specifically to DNA and slide along the DNA in search of a specific target site analogous to the model of incision of bulky base damage by the uvr ABC enzyme of *E. coli* (Sancar and Rupp, 1983). Yet it is clear that in UV-irradiated human cells certain domains within the chromatin are more rapidly repaired than the bulk genome. These domains are located proximal to the nuclear matrix and comprise transcriptionally active DNA although other DNA sequences may be included as well. In XP-C and CS cells heterogeneity in distribution of repair sites correlates well with heterogeneity in removal of pyrimidine dimers from the genome, which may indicate that in human cells a major part of repair synthesis occurring initially after treatment can be attributed to repair of dimers. Yet it is possible throughout that at time periods initially following irradiation, repair of 6-4 photoproducts substantially contributes to repair synthesis. In that case, our data suggest that repair of 6-4 photoproducts is subject to the same regime as removal of pyrimidine dimers, i.e., preferentially directed towards repair of damage in transcriptionally active domains. The effective repair of pyrimidine dimers in active genes in confluent XP-C cells closely resembles the rapid removal of dimers from active genes in hamster cells (Bohr, *et al.*, 1985). Yet it is obvious that XP-C cells are considerably more sensitive to the lethal effects of UV than hamster cells. The proficient and deficient repair of 6-4 photoproducts in hamster and XP-C cells, respectively, (Mitchell, *et al.*, 1985) may underlie the differences in UV-sensitivity and may also account for the lack of preferential repair of nuclear matrix associated DNA in hamster cells (Mullenders, *et al.*, 1986). Nevertheless, the effective repair of dimers in active genes is likely to confer a relatively good UV resistance to confluent XP-C cells, since recovery of UV-inhibited RNA synthesis may be the most important factor for survival of UV-irradiated stationary cells.

Finally, our data also suggest that repair pathways involved in processing of pyrimidine dimers in active and inactive chromatin can operate independently: XP-C lost the capacity to repair inactive sequences, whereas CS-cells fail to repair active sequences in a fast and efficient way.

We thank C. Bussmann and A. van Hoffen for excellent technical assistance. This work was supported by the Association of the University of Leiden with Euratom (contract No. BIO-E-407-81-NL) and Medigon (contract No. 900-501-074). L.V.M. is Wellcome Senior Research Fellow (Basic Biomedical Science).

REFERENCES

Berezney, R., and Coffey, D.S., 1977, Nuclear matrix. Isolation and characterization of a framework structure from rat liver nuclei, *J. Cell Biol.*, 73:616.

Bohr, V.A., Smith, C.A., Okumoto, D.S., and Hanawalt, P.C., 1985, DNA repair in an active gene: removal of pyrimidine dimers from the DHFR gene in CHO cells is much more efficient than in the genome overall, *Cell*, 40:359.

Bohr, V.A., Phillips, D.H., and Hanawalt, P.C., 1987, Heterogeneous DNA damage and repair in the mammalian genome, *Cancer Res.*, 47:6426.

Buongiorno-Nardelli, M., Micheli, G., Carri, M.T., and Mavil-

ley, M., 1982, A relationship between replicon size and supercoiled loop domains in the eukaryotic genome, Nature, 298:100.
Cohn, S.M., and Lieberman, M.W., 1984, The distribution of DNA excision repair sites in human diploid fibroblasts following ultraviolet irradiation, J. Biol. Chem., 259:12463.
Dijkwel, P.A., Mullenders, L.H.F., and Wanka, F., 1979, Analysis of the attachment of replicating DNA to a nuclear matrix in mammalian interphase nuclei, Nucleic Acids Res., 6:219.
Erixon, K., and Ahnström, G., 1979, Single-strand breaks in DNA during repair of UV-induced damage in normal human and xeroderma pigmentosum cells as determined by alkaline unwinding and hydroxylapatite chromatography, Mutat. Res., 59:257.
Kantor, G.J., and Hull, D.R., 1984, The rate of removal of pyrimidine dimers in quiscent cultures of normal human and xeroderma pigmentosum cells, Mutat. Res., 132:21.
Keyse, S.M., and Tyrrell, R.M., 1987, Rapidly occurring DNA excision repair events determine the biological expression of UV-induced damage in human cells, Carcinogenesis, 8:1251.
Lan, S.Y., and Smerdon, M.J., 1985, A nonuniform distribution of excision repair synthesis in nucleosomes core DNA, Biochemistry, 24:7771.
Lehmann, A.R., Kirk-Bell, S., and Mayne, L.V., 1979, Abnormal kinetics of DNA synthesis in ultraviolet light-irradiated cells from patients with Cockayne's syndrome, Cancer Res., 39:4237.
Mansbridge, J.N., and Hanawalt, P.C., 1983, Domain-limited repair of DNA in UV-irradiated fibroblasts from xeroderma pigmentosum complementation group C, in: "Cellular responses to DNA damage," E.C. Friedberg and B.R. Bridges, eds., New York.
Mayne, L.V., and Lehmann, A.R., 1982, Failure of RNA synthesis to recover after UV-irradiation: An early defect in cells from individuals with Cockayne's syndrome and xeroderma pigmentosum, Cancer Res., 42:1473.
Mellon, I., Bohr, V.A., Smith, C.A., and Hanawalt, P.C., 1986, Preferential DNA repair of an active gene in human cells, Proc. Natl. Acad. Sci., 83:8878.
Mellon, I., Spivak, G., and Hanawalt, P.C., 1987, Selective removal of transcription-blocking DNA damage from the transcribed strand of the mammalian DHFR gene, Cell, 51:241.
Mirkovitch, J., Mirault, M.E., and Laemmli, U.K., 1984, Organization of the higher-order chromatin loop: specific DNA attachment sites on the nuclear scaffold, Cell, 39:223.
Mirkovitch, J., Spierer, P., and Laemmli, U.K., 1986, Genes and loops in 320.000 basepairs of the Drosophila melanogaster chromosomes, J. Mol. Biol., 190:255.
Mitchell, D.L., Haipek, C.A., and Clarkson, J.M., 1985, 6-4 photoproducts are removed from the DNA of UV-irradiated mammalian cells more efficiently than cyclobutane pyrimidine dimers, Mutat. Res., 143:109.
Mullenders, L.H.F., Zeeland, A.A. van, and Natarajan, A.T., 1983, Comparison of DNA loop size and supercoiled domain size in human cells, Mutat. Res., 112:245.
Mullenders, L.H.F., Kesteren, A.C. van, Bussmann, C.J.M.,

Zeeland, A.A. van, and Natarajan, A.T., 1984, Preferential repair of nuclear matrix associated DNA in xeroderma pigmentosum complementation group C, Mutat. Res., 141:75.
Mullenders, L.H.F., Kesteren, A.C. van, Bussmann, C.J.M, Zeeland, A.A. van, and Natarajan, A.T., 1986, Distribution of UV-induced repair events in higher-order chromatin loops in human and hamster fibroblasts, Carcinogenesis, 7:995.
Mullenders, L.H.F., Zeeland, A.A. van, and Natarajan, A.T., 1987, The localization of ultraviolet-induced excision repair in the nucleus and the distribution of repair events in higher order chromatin loops in mammalian cells, J. Cell Sci., suppl. 6:243.
Mullenders, L.H.F., Kesteren-van Leeuwen, A.C., Zeeland, A.A. van, and Natarajan, A.T., 1988. Nuclear matrix associated DNA is preferentially repaired in normal human fibroblasts exposed to a low dose of ultraviolet light but not in Cockayne's syndrome fibroblasts, Nucleic Acids Res., 16:10607.
Rahn, R., and Stafford, R.S., 1974, Measurements of defects in UV-irradiated DNA by the kinetic formaldehyde method, Nature, 248:52.
Robinson, S.T., Small, D., Idzerda, R., McKnight, G.S., and Vogelstein, B., 1983, The association of transcriptionally active genes with the nuclear matrix of chicken ovioduct, Nucleic Acids Res., 11:5113.
Small, D., and Vogelstein, B., (1985) Anatomy of supercoiled loops in the Drosophila 7F locus, Nucleic Acids Res., 13:7703.
Sancar, A., and Rupp, W.D., 1983, A novel repair enzyme: uvr ABC excision nuclease of Escherichia coli cuts a DNA strand on both sides of the damaged region, Cell, 33:249.
Vogelstein, B., Pardoll, D.M., and Coffey, D.S., 1980, Supercoiled loops and eukaryotic DNA replication, Cell, 22:79.
Weisbrod, S., 1982, Active Chromatin, Nature, 297:289.

ELECTRON MICROSCOPIC VISUALIZATION OF CHROMATIN STRUCTURE AT SITES OF DNA EXCISION REPAIR FOLLOWING ULTRAVIOLET IRRADIATION OF CULTURED HUMAN FIBROBLASTS

B.J. Gowans[1], G. de Murcia[2] and D.J. Hunting[1]

[1]MRC Group in the Radiation Sciences, University of Sherbrooke, QC, Canada and [2]Institut de Biologie Moleculaire et Cellulaire du CNRS, 67084 Strasbourg, Cedex, France

INTRODUCTION

The DNA excision repair process involves transient changes in chromatin structure at the nucleosomal level (Smerdon and Lieberman, 1978; Smerdon, *et al.*, 1978; Tlsty and Lieberman, 1978; Smerdon, *et al.*, 1979; Oleson, *et al.*, 1979; Smerdon and Lieberman, 1980; Bodell and Cleaver, 1981; Bodell, *et al.*, 1982; Zolan, *et al.*, 1982). Although there is no direct evidence for an early disruption step, a late step, termed nucleosome rearrangement, has been identified which results in the restoration of the native nucleosomal structure at the sites of excision repair. This step has the following characteristics: repair incorporated nucleotides are initially very sensitive to digestion by staphylococcal nuclease and DNase I but eventually acquire the same sensitivity as nucleotides in bulk chromatin (Smerdon and Lieberman, 1978; Smerdon, *et al.*, 1978; Tlsty and Lieberman, 1978; Smerdon, *et al.*, 1979; Oleson, *et al.*, 1979; Smerdon and Lieberman, 1980; Bodell and Cleaver, 1981; Bodell, *et al.*, 1982; Zolan, *et al.*, 1982); and repair incorporated nucleotides are initially underrepresented in nucleosomal core-length DNA fragments produced by staphylococcal nuclease digestion but eventually acquire a random distribution (Smerdon and Lieberman, 1978; Smerdon, *et al.*, 1978; Tlsty and Lieberman, 1978; Smerdon, *et al.*, 1979; Oleson, *et al.*, 1979; Smerdon and Lieberman, 1980; Bodell and Cleaver, 1982). Two hypotheses have been proposed on the basis of these data: nucleosome cores either unfold or slide during excision repair, presumably to allow repair enzymes access to damaged DNA (Lieberman, *et al.*, 1979). There is no direct experimental evidence favoring one hypothesis over the other but there are precedents for both processes. For example, nucleosome displacement has been shown to occur, both *in vitro* (Beard, 1978; Watkins and Smerdon, 1985) and *in vivo* (Zolan, *et al.*, 1982), and some type of nucleosomal unfolding must presumably occur at replication forks given the

DNA Repair Mechanisms and Their Biological Implications in Mammalian Cells
Edited by M.W. Lambert and J. Laval
Plenum Press, New York

absence of nucleosomes on newly replicated DNA and the conservation of parental histones at the replication fork. As well, evidence for an unfolded or half-nucleosomal structure has been deduced in transcriptionally active regions (Ryoji and Worcel 1985).

Although conventional approaches, such as the use of nuclease probes, have increased our undertanding of several chromatin related aspects of DNA repair over the last twelve years, the question of the extent of disruption of the nucleosomal structure at and adjacent to sites of excision repair remains unresolved (Smerdon and Lieberman, 1978; Smerdon, et al., 1978; Tlsty and Lieberman, 1978; Smerdon, et al., 1979; Oleson, et al., 1979; Smerdon and Lieberman, 1980; Bodell and Cleaver, 1981; Bodell, et al., 1982; Zolan, et al., 1982; Smerdon, et al., 1982; Smerdon, 1983; Sidik and Smerdon, 1984; Smerdon, 1986; Watkins and Smerdon, 1985; Lan and Smerdon, 1985; Nissen, et al., 1986; Sidik and Smerdon, 1987). We have, therefore, developed a new approach, involving affinity labelling of repair patches with a biotinated nucleotide, psoralen cross-linking of internucleosomal DNA and electron microscopy, in order to visualize the location of nucleosomes at sites of excision repair. Our results show that, following a 15 minute pulse, more than 90% of the sites of UV[1]-induced repair synthesis were located between nucleosomal cores, and nucleosomal structure was not substantially perturbed adjacent to sites of repair.

MATERIALS AND METHODS

Cell Culture and UV Irradiation

Normal human fibroblasts (AG1518; Institute for Medical Research) were grown in monolayers in Dulbecco's MEM medium supplemented with 5% each of fetal and newborn calf serum. Following subculturing, the cells were incubated in medium containing [^{14}C]dThD (50-60 mCi/mmol, 20 nCi/ml) for 1 week, and grown to confluence.

Incorporation of dTTP or BiodUTP into Repair Patches

Permeable cells (Dresler, et al., 1982) were used to permit entry of the deoxyribonucleoside triphosphates, as described previously (Hunting, et al., 1985).

Psoralen Cross-linking

Cross-linking of the DNA of intact or permeable cells was achieved by incubating the cells with 4,5',8-trimethylpsoralen (Sigma Chemical Co.) at 10 to 50 micrograms/ml and irradiating them on ice with 100,000 to 600,000 J/m^2 of 365 nm light (Ultra-Violet Products Inc.).

---

[1]Abbreviatins: UV, ultraviolet; dThd, thymidine; biodUTP, 5-(allylamino)biotin 2'-deoxyuridine triphosphate; dTTP, dCTP, dATP, dGTP, the triphosphates of thymidine, 2'-deoxycytidine, 2'-deoxyadenosine and 2'-deoxyguanosine, respectively.

Purification of DNA by CsTFA Gradients

The permeable cells were washed three times with 10 mM Tris (pH 7.6 at 37^0C), 320 mM sucrose and 0.5% (V/V) Triton X-100, resuspended in 2 mM Tris (pH 7.4 at room temperature), 1 mM EDTA and 1% sodium N-lauroylsarcosine, incubated 1 hour at 37^0C and sheared four times through a 23-gauge needle. The solution was mixed with CsTFA (Pharmacia), the density was adjusted to 1.6 g/cm^2, and centrifuged at 50,000 rpm in a VTi65 rotor (Beckman) for 16 hours. The gradients were fractionated and the peak fractions were precipitated with ethanol.

Measurement of Repair Ligation using Exonuclease III

The basis of this assay is that unligated repair patches should have 3'-OH ends and will therefore be sensitive to exonuclease III. DNA was isolated by addition of proteinase K (0.1 mg/ml) in 200 mM Tris (pH 8.5 at 20^0C), 1% SDS, and 100 mM EDTA to the cell pellet followed by incubation overnight at 37^0C. The samples were extracted three times each with phenol and isoamyl alcohol/chloroform (1:24), precipitated with ethanol and dissolved in digestion buffer (50 mM Tris, pH 8.0 at 37^0C, 5 mM $MgCl_2$, 10 mM mercaptoethanol, 0.5 mg/ml nuclease free bovine serum albumin). All manipulations were designed to minimize shearing of the DNA. Samples were digested at 37^0C with 0.2 units exonuclease III (Bethesda Research Laboratories) per microgram of DNA (Hunting, et al., 1985).

Measurement of DNA Strand Breaks

DNA strand breaks were measured by the alkaline elution method (Kohn, et al., 1976), with modifications described previously (Hunting and Gowans, 1987).

Digestion of Nuclei with Staphylococcal Nuclease

Cells were harvested and nuclei were prepared and digested with staphylococcal nuclease (Worthington Biochemical Corp.) as described previously (Hunting, et al., 1985).

Purification of the DNA-streptavidin-ferritin Complex

Control DNA or biodUMP-DNA was incubated either 3 hours at 37^0C or overnight at room temperature with a 10-fold excess of streptavidin-ferritin conjugate (Bethesda Research Laboratories) in 10 mM Tris (pH 7.4), 100 mM NaCl and 0.2 mM EDTA. The unbound streptavidin-ferritin was removed by gel filtration on a Sepharose 4B column (Pharmacia).

Electron Microscopy

DNA was denatured by incubation overnight in 2 mM Tris (pH 8 at room temperature), 72% (v/v) formamide, 0.4 M glyoxal and 0.2 mM EDTA (Sogo, et al., 1984). Native or denatured DNA was adsorbed onto carbon-coated nickel grids (400 mesh) using the modified Kleinschmidt method (Davis, et al., 1971). Grids were rotary shadowed with platinum at an angle of 7^0.

RESULTS

5-(allylamino)biotin 2'-deoxyuridine triphosphate (biodUTP), an affinity probe developed by Langer, et al. (1981), was synthesized and purified as described previously (Hunting, et al., 1985). The structure of biodUTP is shown in Figure 1.

Effect of BiodUTP on UV-induced Repair Synthesis in Permeable Cells

The results presented in Table 1 demonstrate that biodUTP is capable of serving as a substrate for UV-induced repair synthesis, albeit at slightly lower efficiency than dTTP. Experiments 1 and 2 involved a 20 minute incubation of intact cells following UV irradiation to permit the accumulation of incised sites. In the absence of inhibitors of repair (Experiment 1), ca. 2400 sites per cell accumulated in 20 minutes, whereas, in the presence of 10 mM hydroxyurea and 7.7 μM aphidicolin this number increased to ca. 7500 sites per cell. In Experiment 3, cells were permeabilized and then irradiated such that incision took place in the presence of biodUTP. The results demonstrate both that biodUTP had no inhibitory effect on incision and that when there are not large numbers of incised sites available, biodUTP is as efficient as dTTP as a substrate for repair synthesis.

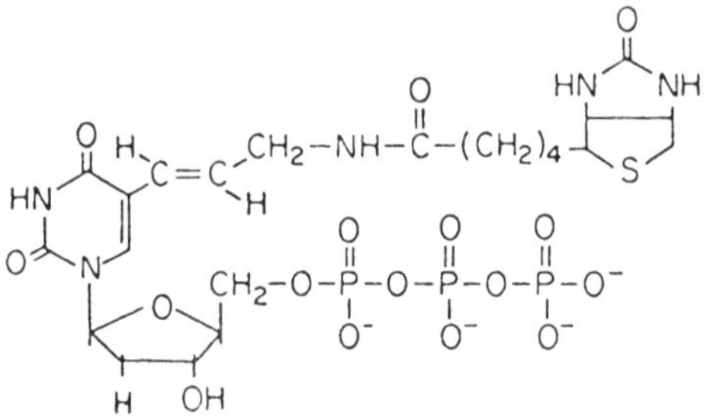

Figure 1. Structure of BiodUTP

Purification of Biotinated DNA

BiodUTP contains an amide linkage which could potentially be sensitive to digestion by proteinase K, an enzyme commonly used in the isolation of DNA from cells. In our earlier work with biotinated DNA, we avoided this potential problem by purifying DNA from cells using CsTFA gradients (Hunting, et al., 1985; Materials and Methods). We have since tested our suspicions by incubating DNA containing repair patches labelled with [^{3}H]dCMP and biodUMP in a buffer containing 1% SDS with or without 100 μg/ml proteinase K at 37^{0}C overnight. Incubation with proteinase K reduced the binding efficiency of the repair patches to avidin-agarose to 60% of control. Therefore, we have continued to avoid treating biotin-containing

Table I: Incorporation of dTTP or BiodUTP into the DNA of permeable human fibroblasts in response to UV radiation[1]

Expt. No.	Experimental Conditions	Rel. Repair Synthesis	Substrate
1	intact cells irradiated, incubated 20 minutes, and then permeabilized	1.0	dTTP
		0.79	BiodUTP
		0.041	dTTP, unir- radiated
2	intact cells irradiated, incubated 20 minutes in presence of hydroxyurea and aphidicolin, and then permeabilized	1.0	dTTP
		0.78	BiodUTP
		0.12	dTTP, unir- radiated
3	cells permeabilized and then irradiated	1.0	dTTP
		0.98	BiodUTP
		0.29	dTTP, unir- radiated

[1]In experiment 1, confluent normal human fibroblasts prelabeled with [^{14}C]dThd were damaged with 10 J/m^2 UV radiation and incubated at 37^{0}C for 20 minutes at which time the number of DNA single-strand breaks (incised and partially polymerized sites) was at a maximum (ca. 2400 breaks per cell as determined by alkaline elution). The cells were then permeabilized, washed to remove endogenous deoxyribonucleotides, and incubated with [^{32}P]dCTP, dATP, dGTP, and either dTTP or BiodUTP, all at 3 uM, for 15 minutes at 37^{0}C. The amount of repair synthesis, measured as the ^{32}P to ^{14}C ratio in Cl_3CCO-OH-insoluble material, was then determined. In experiment 2, intact cells were damaged with UV radiation as in experiment 1 and were then incubated with inhibitors of repair synthesis (10 mM hydroxyurea plus 7.7 uM aphidicolin) for 20 minutes at 37^{0}C. These inhibitors caused a rapid accumulation of DNA single-strand breaks (incised and partially polymerized sites; ca.7500 sites per cell). The cells were then permeabilized, washed, and incubated as in experiment 1. In experiment 3, cells were permeabilized, irradiated with 100 J/m^2 UV radiation at 4^{0}C, and incubated at 37^{0}C for 20 minutes in the presence of [^{32}P]dCTP, dATP, dGTP, and either dTTP or BiodUTP, all at 3 uM, and the amount of repair synthesis was determined (Hunting, et al., 1985).

DNA with proteinase K whenever we wish to preserve all the biotin moieties. In the case of DNA to be digested by exonuclease III, to determine the degree of ligation of repair patches, the loss of the biotin group has no effect on the assay since it does not affect the integrity of the DNA.

Furthermore, this assay requires that breakage of the DNA be kept to a minimum and therefore CsTFA gradients are not suited to this problem. Figure 2 shows typical profiles of control and biotin labelled DNA used for electron microscopy. The peak

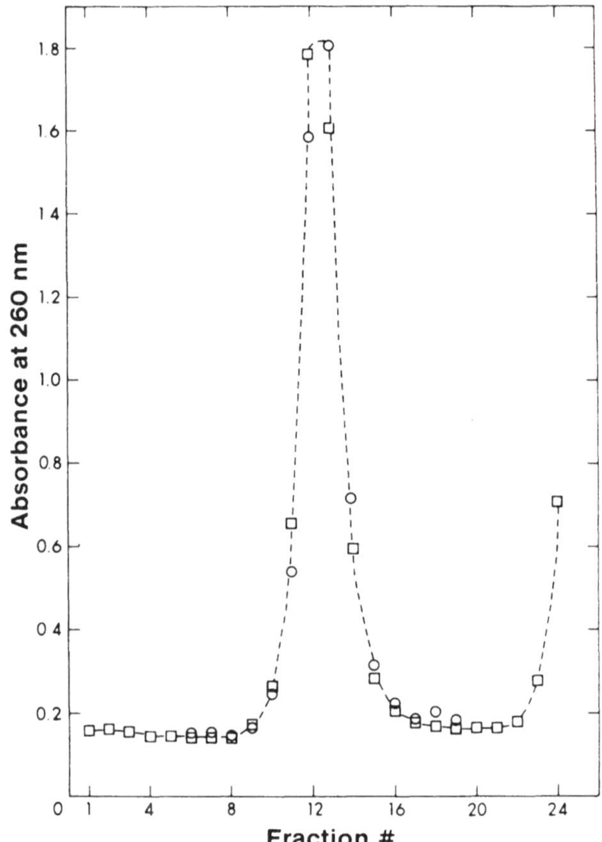

Figure 2. Profile of CsTFA gradients. The washed cell pellet was resuspended in 2 mM Tris (pH 7.4), 1 mM EDTA and 1% sodium N-lauroylsarcosine, incubated 1 hour at 37°C and sheared four times through a 23-gauge needle. The solution was mixed with CsTFA and the density was adjusted to 1.6 g/cc. The solution was centrifuged 16 hours at 50,000 rpm in a Beckman VTi65 rotor. (□), control DNA; (○); biodUMP-labelled DNA.

of repair incorporated radioactivity coincided with the absorbance peak (data not shown).

Ligation of Biotinated Repair Patches

Repair patch ligation is a prerequisite for nucleosome rearrangement. We have previously found that, during UV-induced excision repair, ligation is more rapid than nucleosome rearrangement (Hunting, et al., 1985). As well, inhibitors which indirectly block ligation also inhibit nucleosome rearrangement (Hunting, et al., 1985). Finally, it has been found that only ligated repair patches can be isolated from nucleosomal core DNA (Smerdon, 1986). We, therefore, determined the effect of biodUTP on repair patch ligation, as measured by sensitivity to exonuclease III. The results presented in Figure 3 show that 82% of repair incorporated nucleotides were present in ligated repair patches when repair synthesis took place in the pre- sence of dTTP. The corresponding value for patches synthesized in the presence of biodUTP was 78%, indicating that the presence of biodUMP in the repair patch has little or no effect on ligation.

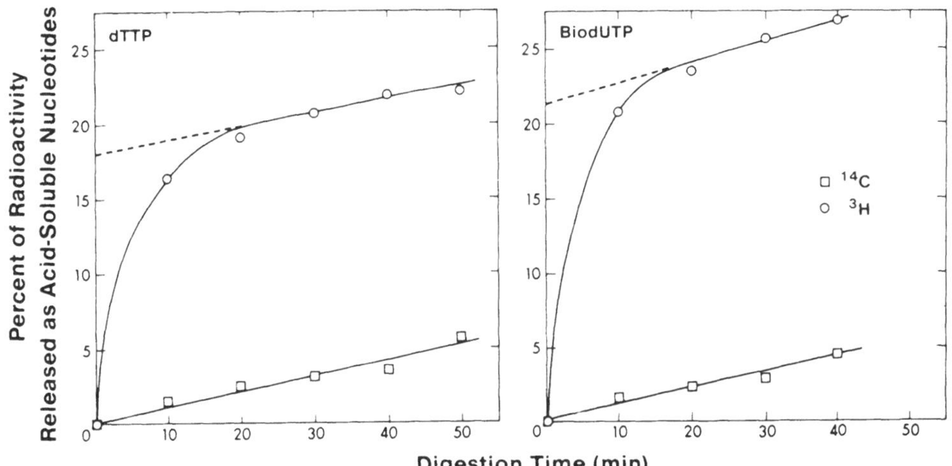

Figure 3. Ligation of repair patches synthesized in the presence of dTTP or biodUTP. DNA purified from permeable cells which had performed DNA repair in the presence of dTTP (left panel) or biodUTP (right panel) was digested with exonuclease III, and the release of acid-soluble nucleotides from repair patches [^{3}H] (○) and bulk DNA [^{14}C] (□) was determined. The final slope of the[^{3}H] digestion curve was extrapolated to zero time to give the fraction of repair-incorporated nucleotides present in unligated patches (Hunting, et al., 1985).

Effect of BiodUTP on the Restoration of Chromatin Structure During Excision Repair

Repair-incorporated nucleotides are initially very sensitive to staphylococcal nuclease, but with time they acquire the same nuclease resistance as nucleotides in bulk chromatin. This process has previously been shown to occur in human AG1518 cells permeabilized by the same protocol used in the present work. As shown in Figure 4, when nuclei were isolated immediately following a 10 minute pulse, the repair-incorporated nucleotides were much more sensitive to staphylo-

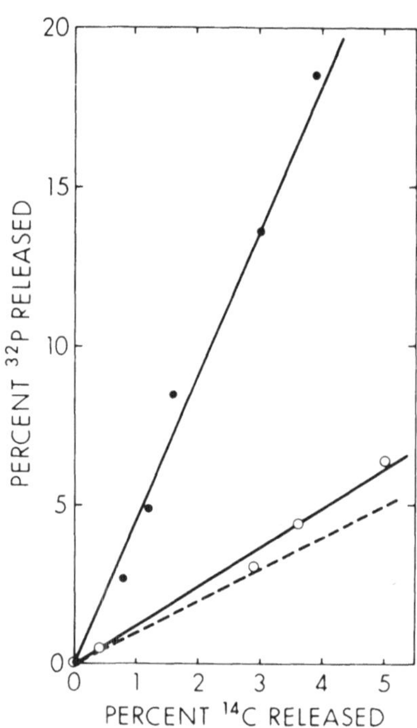

Figure 4. Restoration of chromatin structure during DNA excision repair in permeable cells in the presence of BiodUTP. Permeable cells were damaged with UV radiation, incubated for 10 minutes in a reaction mixture containing 0.1 μM [^{32}P]dCTP, dGTP, dATP, and biodUTP(●), then chased for 50 minutes in a reaction mixture containing 3 μM dCTP, dGTP, dATP, and biodUTP (○). Nuclei were prepared and digested with staphylococcal nuclease (Materials and Methods) and the release of acid-soluble nucleotides from repair patches [^{32}P] and bulk DNA [^{14}C] was measured. The dashed line represents equal sensitivity of repair and bulk nucleotides (i.e., restoration of chromatin structure completed) (Hunting, et al., 1985).

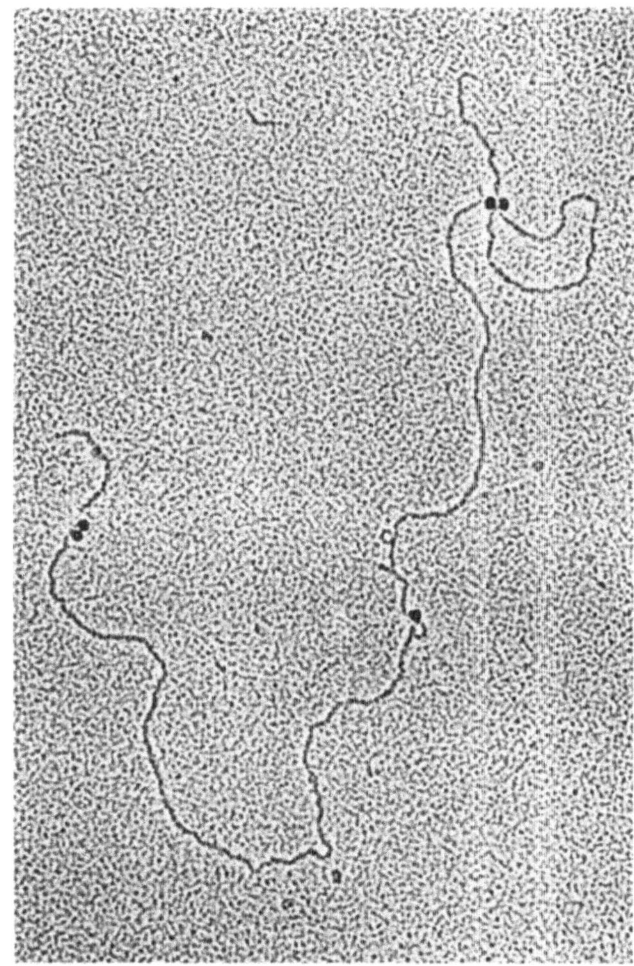

Figure 5. Electron microscopic visualization of biotin-labeled repair patches using streptavidin-ferritin. DNA was purified from permeable cells which had performed UV-induced repair synthesis in the presence of biodUTP. The DNA was incubated with streptavidin-ferritin and the unbound protein was removed by gel filtration (Materials and Methods). The DNA was visualized using the Kleinschmidt cytochrome C technique.

coccal nuclease than bulk DNA. Following a 50 minute chase period, repair-incorporated nucleotides were only slightly more sensitive to staphylococcal nuclease than bulk nucleotides, thus demonstrating that the presence of biodUMP in repair patches does not prevent the restoration of chromatin structure during excision repair (nucleosome rearrangement).

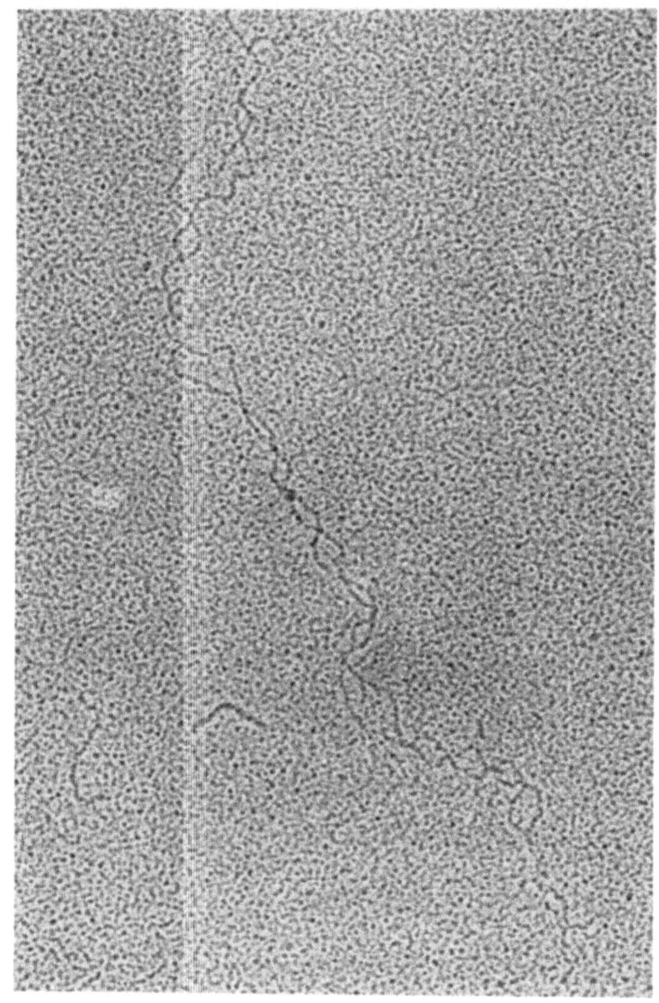

Figure 6. Electron microscopic visualization of a biotin-labelled repair patch located at one of the sites of psoralen crosslinking of DNA in permeable cells. Permeable cells which had performed DNA excision repair in the presence of biodUTP were incubated with 10 µg/ml of trimethylpsoralen and irradiated on ice with 300,000 J/m^2 365 nm radiation (Materials and Methods). The DNA was purified on CsTFA gradients as described (Materials and Methods). The purified DNA was incubated overnight at room temperature with a 10-fold excess of streptavidin-ferritin and the excess protein was removed by gel filtration (Materials and Methods). The DNA was denatured by an overnight incubation at room temperature in a buffer containing glyoxyl (0.4 M) and formamide (72% v/v). The DNA was visualized using the Kleinschmidt method (Materials and Methods).

Electron Microscopic Visualization of Sites of Excision Repair

Streptavidin conjugated to ferritin was used to visualize the location of biotins in the DNA by electron microscopy. As shown in Figure 5, the location of biodUMP-labelled repair patches can be visualized by electron microscopy of native DNA. No streptavidin-ferritin was bound to DNA from UV-damaged permeable cells which underwent repair in the presence of dTTP.

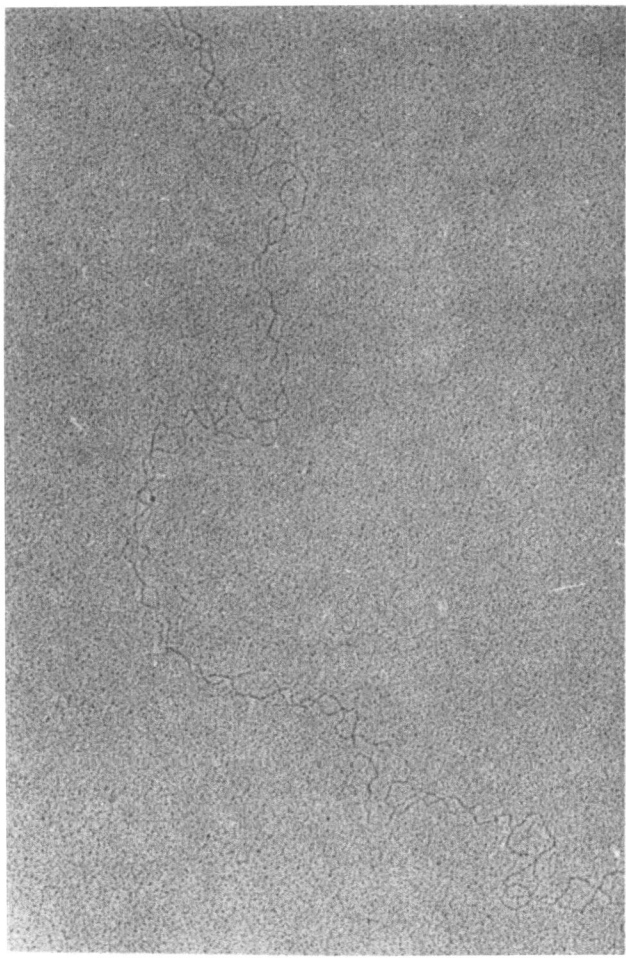

Figure 7. Electron microscopic visualization of a biotin-labelled repair patch located between sites of psoralen cross-linking of DNA in permeable cells. Experimental details are given in the legend of Figure 6.

Electron Microscopic Visualization of Chromatin Structure at Sites of Excision Repair

As discussed above, incubation of cells with psoralen followed by exposure to 365 nm radiation results in cross-linking of linker DNA but not nuclesome core DNA. Thus, denaturation of such DNA followed by electron microscopy allows one to visualize the position of nucleosomes in the cell. We wished to use this technique in combination with the visualization of biotin-labelled repair patches in order to determine the chromatin structure at sites of excision repair. Since the denaturation step involves incubation of the DNA with high concentrations of glyoxyl (0.4 M) and formamide (72% v/v), we considered the possibility that this treatment could disrupt the interaction between streptavidin-ferritin and biotin. However, as shown in Figure 6, the streptavidin-ferritin is still bound to the DNA following denaturation. We compared labelling frequencies for native and denatured forms of the same samples and observed no differences, arguing that the streptavidin-biotin interaction withstands the denaturation conditions. Interestingly, the modification of DNA by glyoxyl is rapidly reversed under the conditions used for spreading the DNA. Within 5 minutes of formation of the hypophase, one could observe formation of double-stranded DNA and a resultant decrease in the size of the bubbles of single- stranded DNA. More than 90% of the repair patches were located in internucleosomal (linker) DNA following a 15 minute incubation of the permeable cells with biodUTP (Figure 6). Less than 10% of the repair patches were located in uncross-linked nucleosomal core DNA (Figure 7).

DISCUSSION

We have shown that DNA excision repair patches can be labelled with biodUTP and that the presence of this modified nucleotide has only a slight inhibitory effect on repair synthesis and little or no effect on subsequent steps, including the restoration of chromatin structure. Further, we have used psoralen, which cross-links linker but not core DNA, to determine the chromatin structure at sites of DNA repair. After a 15 minute pulse, more than 90% of UV-induced repair patches were located in linker DNA and nucleosomal structure was not substantially perturbed at or near sites of repair. For example, if histones were removed at sites of excision repair, the DNA in this region would have been completely cross-linked, as has been observed for actively transcribed ribosomal DNA (Sogo, et al., 1984).

In order to account for the initial, transient sensitivity of repair patches to nucleases, it was proposed several years ago that nucleosome cores either slide or unfold during excision repair (Lieberman, et al., 1979), although recently the unfolding hypothesis has been favored (Smerdon, 1986; Watkins and Smerdon, 1985; Lan and Smerdon, 1985; Nissen, et al., 1986; Sidik and Smerdon, 1987). Lan and Smerdon have shown that, following what they refer to as the nucleosome refolding step, repair patches are preferentially located near the ends of the nucleosomal core DNA (Lan and Smerdon, 1985). With further incubation, the repair patches eventually become randomized within the nucleosomal DNA. They propose that this randomization results from limited sliding of the nucleosomes (Lan and Smerdon, 1985; Nissen, et al., 1986; Arnold, et al., 1987). Our data are not consistant with a substantial unfolding of the nucleosomal core DNA during excision repair. We propose that repair synthesis occurs in linker DNA and that nucleosome core movement then results in the appearance of newly synthesized repair patches in nucleosomal core DNA, thus rendering them resistant to staphylococcal nuclease. Further nucleosomal movement, at a slower rate, would then randomize the location of the repair patches within the nucleosomal core DNA. Our findings raise several questions, including whether nucleosome sliding is constitutive or induced and what mechanism causes sliding.

Attempts have been made to mimic the restoration of chromatin structure during excision repair using purified nuclei, with limited success. Incubation of intact or H1-depleted nuclei in buffers containing KCl up to 0.625 M resulted in nucleosomal displacement but even this high concentration of salt resulted in the appearance of only 50% of repair-incorporated nucleotides into nucleosomal core DNA (Watkins and Smerdon, 1985). It is not known why complete randomization of the nucleosomes could not be achieved but one possibility is that post-transcriptional modifications of the histones, such as attachment of poly(ADP-ribose), are required to facilitate movement of nucleosomes. Since many of these modifications are susceptible to rapid degradation, it is possible that these modifications were lost during the isolation of the nuclei in the above mentioned study.

The work of Zolan, et al., provides further evidence that nucleosome movement can occur in vivo (Zolan, et al., 1982). Cells were damaged with angelicin plus UVA radiation, which causes damage primarily in linker DNA. The subsequent repair patches were initially very sensitive to staphylococcal nuclease, but with time became resistant, consistant with the displacement of nucleosome cores onto what was originally internucleosomal DNA.

It has been proposed, based on inhibitor studies, that topoisomerases are involved in DNA repair (e.g., Mattern and Scudiero, 1981; Collin, et al., 1980) and, until now, there was also a theoretical reason for suspecting their involvement. We have previously shown that the rate of repair patch ligation is substantially faster than the rate of restoration of nucleosomal structure and proposed that ligation always preceded the restoration of nucleosomal structure (Hunting, et al., 1985). This conclusion was confirmed by Smerdon who showed that all repair patches present in nucleosomal DNA are ligated (Smerdon, 1986). If nucleosomes were removed during excision repair, incision would result in relaxation of the DNA supercoils which are normally restrained by the histone octamer. Ligation of the repair patch prior to restoration of the nucleosomal structure would then require a topoisomerase. Our finding that the nucleosomal structure is not disrupted during repair suggests that topoisomerases should not be required for the restoration of chromatin structure during repair.

It is interesting that repair patches synthesized in response to bulky adducts, such as thymine dimers and psoralens have an average length similar to that of linker DNA. The mechanism which determines repair patch length is not known, but our finding that repair synthesis occurs in linker DNA suggests a possible mechanism for regulating the maximum size of repair patches, given that the adjacent nucleosome core should serve as a barrier to repair synthesis. Thus, the distance between the site of incision and the nucleosome core in the 5' direction would determine the maximum length of a given repair patch. This of course does not rule out the operation of other patch termination mechanisms, such as the limited processivity of the DNA repair polymerases.

REFERENCES

Arnold, G.E., Dunker, A.K., and Smerdon, M.J., 1987, Limited nucleosome migration can completely randomize DNA repair patches in intact human cells, J. Mol. Biol., 196:433.

Beard, P., 1978, Mobility of histones on the chromosome of simian virus 40, Cell, 15:955.

Bodell W.J., and Cleaver, J.E., 1981, Transient conformation changes in chromatin during excision repair of ultraviolet damage to DNA, Nucleic Acids Res., 9:304.

Bodell, W.J., Kaufmann, W.K., and Cleaver, J.E., 1982, Enzyme digestion of intermediates of excision repair in human cells irradiated with ultraviolet light, Biochemistry, 21:6767.

Collin, A.R.S., Downes, C.S., and Johnson, R.T., 1980, Cell cycle related variations in UV damage and repair capacity in Chinese hamster (CHO-Kl) cells, J. Cell. Physiol., 103:179.

Davis, R.W., Simon, M., and Davidson, N., 1971, Electron microscope heteroduplex methods for mapping regions of base sequence homology in nucleic acids, Methods Enzymol., 210:413.

Dresler, S.L., Roberts, J.D., and Lieberman, M.W., 1982, Characterization of deoxyribonucleic acid repair synthesis in permeable human fibroblasts, Biochemistry, 21: 2557.

Hunting, D.J., Dresler, S.L., and de Murcia, G., 1985, Incorporation of biotin-labeled deoxyuridine triphosphate into DNA during excision repair and electron microscopic visualization of repair patches, Biochemistry, 24:5729.

Hunting, D.J., Dresler, S.L., and Lieberman, M.W., 1985, Multiple conformational states of repair patches in chromatin during DNA excision repair, Biochemistry, 24:3219.

Hunting, D.J., and Gowans, B.J., 1987, Inhibition of repair patch ligation by an inhibitor of Poly(ADP-ribose) synthesis in normal human fibroblasts damaged with ultraviolet radiation, Molec. Pharmacology, 33:358.

Kohn, K.W., Erickson, L.C., Ewig, R.A.G., and Friedman, C. A., 1976, Fractionation of DNA from mammalian cells by alkaline elution, Biochemistry, 15:4629.

Lan, S.Y., and Smerdon, M.J., 1985, A nonuniform distribution of excision repair synthesis in nucleosome core, Biochemistry, 24: 7771.

Langer, P.R., Waldrop, A.A., and Ward, D.C., 1981, Enzymatic synthesis of biotin-labeled polynucleotides: novel nucleic acid affinity probes, Proc. Natl. Acad. Sci., 78:6633.

Lieberman, M.W., Smerdon, M.J., Tlsty, T.D., and Oleson, F. B., 1979, The role of chromatin structure in DNA repair in human cells damaged with chemical carcinogens and ultraviolet radiation, in: "Environmental Carcinogenesis", P. Emmelot and E. Kriek, eds, Elsevier, Amsterdam.

Mattern, M.R., and Scudiero, D.A., 1981, Dependence of mammalian DNA synthesis on DNA supercoiling. III. Characteristics of the inhibition of replicative and repair type DNA synthesis by novobiocin and nalidixic acid, Biochim. Biophys. Acta, 653:248.

Menissier, J., Hunting, D.J., and de Murcia, G., 1985, Electron microscopic mapping of single-stranded discontinuities in cauliflower virus DNA by means of the biotin-aviding technique, Anal. Biochem., 148:339.

Nissen, K.A., Lan, S.Y., and Smerdon, M.J., 1986, Stability of nucleosome placement in newly repaired regions of DNA, J. Biol. Chem., 261:8585.

Oleson, F.B., Mitchell, B.L., Dipple, A., and Lieberman, M. W., 1979, Distribution of DNA damage in chromatin and its relation to repair in human cells treated with 7-bromomethylbenz(a) anthracene, Nucleic Acids Res., 7:1343.

Ryoji, M., and Worcel, A., 1985, Structure of two distinct types of minichromosomes that are assembled on DNA injected in Xenopus oocytes, Cell, 40:923.

Sidik, K., and Smerdon, M.J., 1984, Nuclease sensitivity of repair-incorporated nucleotides in chromatin and nucleosome rearrangement in human cells damaged by methyl methanesulfonate and methylnitrosourea, Carcinogenesis, 5:245.

Sidik, K., and Smerdon, M.J., 1987, Rearrangement of nucleosome structure during excision repair in Xeroderma pigmentosum (group A) human fibroblasts, Carcinogenesis, 8:733.

Smerdon, M.J., 1983, Rearrangements of chromatin structure in newly repaired regions of deoxyribonucleic acid in human cells with sodium butyrate or hydroxyurea, Biochemistry, 22:3516.

Smerdon M.J., 1986, Completion of excision repair in human cells. Relationship between ligation and nucleosome formation, J. Biol. Chem., 261:244.

Smerdon, M.J., Kastan, M.B., and Lieberman, M.W., 1979, Distribution of repair-incorporated nucleotides and nucleosome rearrangement in the chromatin of normal and Xeroderma Pigmentosum human fibroblasts, Biochemistry, 18:3732.

Smerdon, M.J., and Lieberman, M.W., 1978, Nucleosome rearrangement in human chromatin during UV-induced DNA repair synthesis, Biochemistry, 75:4238.

Smerdon, M.J., and Lieberman, M.W., 1980, Comparison of the distribution within chromatin of DNA repair synthesis occurring at different times after UV radiation, Biochemistry, 19:2992.

Smerdon, M.J., Tlsty, T.D., and Lieberman, M.W., 1978, Distribution of ultraviolet-induced DNA repair synthesis in nuclease sensitive and resistant regions of human chromatin, Biochemistry, 17:2377.

Smerdon, M.J., Watkins, J.F., and Lieberman, M.W., 1982, Effect of histone H1 removal on the distribution of ultraviolet-induced deoxyribonucleic acid repair synthesis within chromatin, Biochemistry, 21:3879.

Sogo, J.M., Ness, P.J., Widmer, R.M., Parish, R.W., and

Koller, Th., 1984, Psoralen-crosslinking of DNA as a probe for the structure of active nucleolar chromatin, J. Mol. Biol., 178:897.
Tlsty, T.D., and Lieberman, M.W., 1978, The distribution of DNA repair synthesis in chromatin and its rearrangement following damage with N-acetoxy-2-acetylaminofluorene, Nucleic Acids Res., 5:3261.
Watkins, J.F., and Smerdon, M.J., 1985, Nucleosome rearrangement in vitro. 1. Two phases of salt-induced nucleosome migration in nuclei, Biochemistry, 24:7279.
Watkins, J.F., and Smerdon, M.J., 1985, Nucleosome rearrangement in vitro. 2. Formation of nucleosomes in newly repaired regions of DNA, Biochemistry, 24:7288.
Zolan, M.E., Smith, C.A., Calvin, N.M., and Hanawalt, P.C., 1982, Rearrangement of mammalian chromatin structure following excision repair, Nature, 299:462.

POLY(ADP-RIBOSYL)ATION REACTIONS AND MODULATION OF CHROMATIN STRUCTURE

Gilbert de Murcia[1], Gérard Gradwohl[1], Alice Mazen[1], Josiane Ménissier-de Murcia[1], Ann Huletsky[2] and Guy Poirier[2]

[1]IBMC du CNRS, Laboratoire de Biochimie 2, 15 rue Descartes, 67084 Strasbourg Cedex, France

[2]Centre de Recherche de l 'Hôtel-Dieu de Québec, 11, Côte du Palais, Québec, Canada, G9R 2J6

INTRODUCTION

Posttranslational modifications of histones are generally considered as potential modulators of chromatin structure during DNA transcription and replication. One of them, the addition of poly(ADP-ribose) is ubiquitous to eukaryotes and is mediated at the expense of the NAD pool by the chromatin bound enzyme, poly ADP-ribose polymerase. This highly conserved enzyme is strictly DNA-dependent and is inactive unless stimulated by DNA strand breaks (Ueda and Hayaishi, 1985; Gaal and Pearson, 1985; Althaus and Richter, 1987).

Poly(ADP-ribosyl)ation reations have been related to the regulation of several chromatin functions especially those including intermediate nicking and resealing of DNA strands; in fact a number of results have been accumulated associating this modification to the recovery of mammalian cells from DNA damage. However, these evidences are almost exclusively based on experimental approaches involving enzyme inhibitors (Shall, 1984).

To overcome this problem, we have set up in vivo reconstitution conditions under which the synthesis and degradation of poly(ADP-ribose), catalyzed, respectively, by purified calf thymus poly(ADP-ribose)polymerase and bull testis poly(ADP-ribose)glycohydrolase, were assessed using calf thymus polynucleosome chains as polymer receptors. Changes of the nucleosomal structure were examined by electron microscopy and ultracentrifugation. In order to understand the molecular basis of the complex mechanism leading to the enzyme activation and modification of an acceptor, we have recently developed new experimnntal approaches that will be discussed in the second part of this article.

DNA Repair Mechanisms and Their Biological Implications in Mammalian Cells
Edited by M.W. Lambert and J. Laval
Plenum Press, New York

Poly(ADP-ribosyl)ation of nuclear proteins

Conformational changes of chromatin structure induced by poly(ADP- ribosyl)ation of histone H1. Purified calf thymus poly(ADP-ribose)polymerase, reassociated with freshly prepared calf thymus polynucleosomes, utilizes in vitro the nicks created in the nucleosomal DNA by the purification procedure (de Murcia, et al.,1986) to synthesize poly(ADP-ribose) bound to acceptors. A typical time course of incorporation of ^{32}P-NAD is shown in Figure 1. After 60 minutes, the ADP-ribosylation reaction was stopped with 10 mM nicotinamide, then poly(ADP-ribose)glycohydrolase was added. As indicated by the rapid decrease of acid insoluble radioactivity, the polymer produced both by the automodification reaction or by the modification of nucleosomal acceptors was hydrolyzed up to 95% within 60 minutes under our incubation conditions (Figure 1).

In parallel experiments, non radioactive NAD (200 µM) was used in order to visualize the conformational changes of chromatin induced by the poly(ADP-ribosyl)ation reaction. After 0 minutes (Figure 2a), 60 minutes (Figure 2b) and 120 minutes (Figure 2c), an aliquot of the incubation medium was fixed and spread on positively charged carbon coated grids as

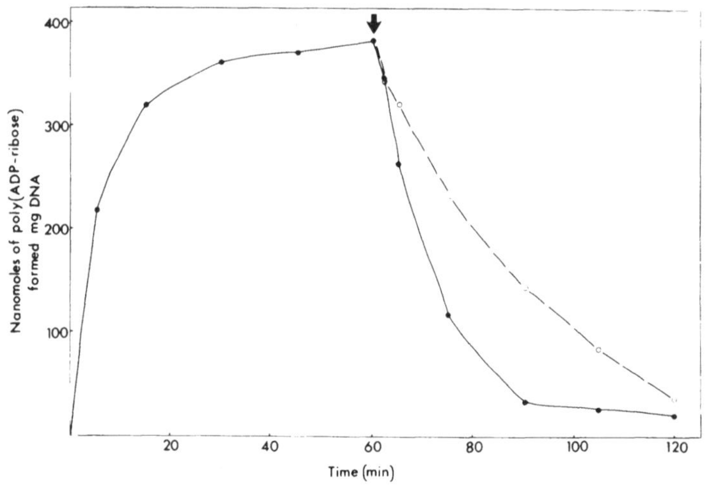

Figure 1. Time course of poly(ADP-ribose) synthesis and degradation. Calf thymus nucleosomes were incubated at 25^{0}C in the presence of purified poly(ADP-ribose)-polymerase(2 µg/OD unit) and ^{32}P-NAD (200µM final). After 60 minutes (indicated by the arrow), the synthesis reaction was stopped with nicotinamide (10 mM final) and poly(ADP-ribose)glycohydrolase was added (solid line 2 µg/OD unit, broken line 1 µg/OD unit). Then the reaction mixture was incubated for 60 minutes to follow the degradation of poly(ADP-ribose). At different time intervals samples were precipitated and transferred on Whatman GF/C filters.

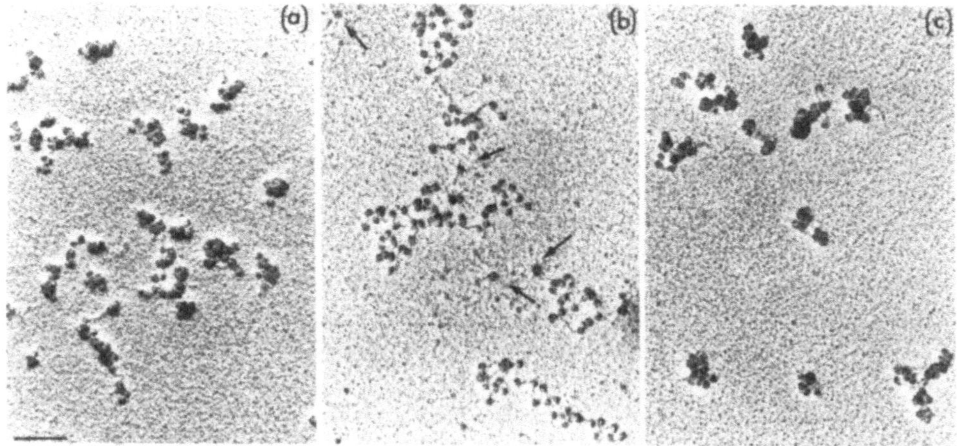

Figure 2. Modulation of chromatin synthesis and degradation of poly(ADP-ribose). (a) 0 minutes incubation time; (b) relaxation of chromatin superstructure after 60 minutes of poly(ADP-ribosyl)ation reaction; (c) recondensation of superstructure after 60 minutes incubation of poly(ADP-ribosyl)ated chromatin with glycohydrolase. The arrows in b point to automodified poly(ADP-ribose)polymerase molecules. The spreading was performed at 40 mM NaCl, 10 mM Tris buffer, pH 7.4, and 0.2 mM EDTA. The bar indicates 100 Å.

described previously (De Murcia and Koller,1981). Following 60 minutes of incubation, poly(ADP-ribosyl)ated nucleosomes (Figure 2b) exhibit a fully opened structure as compared to the control (Figure 2a). Highly automodified poly(ADP-ribose)polymerase molecules are also visible (Figure 2b, arrows). They are detached from DNA and surrounded by long chains of poly(ADP-ribose). When modified nucleosomes were incubated with poly(ADP-ribose)glycohydrolase (Figure 2c), the chromatin structure was found recondensed in a manner similar to the control and most of the polymerase molecules were again bound to DNA (Figure 2c).

The sedimentation coefficients of the three chromatin samples, corresponding to Figure 2a-c, are displayed in Table 1. One can see that the relaxation of chromatin superstructure is correlated with a decrease in the S value. At the end of the polymer hydrolysis reaction (120 minutes), the sedimentation coefficient is again very close to the control, thus demonstrating that the compact structure seen by electron microscopy also exists in solution.

Calf thymus nucleosomes were incubated as indicated in the legend to Figure 1. After 0, 60, and 120 minutes of incubation, histones were extracted, subjected to acid urea gel electrophoresis and autoradiographed as described previously (Aubin et al.,1982). After 60 minutes of (ADP-ribosyl)-ation (Figure 3, slot c), the hyper ADP-ribosylated form of

Table 1. Sedimentation Coefficient of Poly(ADP-ribosyl)ated Chromatin Samples

Reaction Conditions	Time	S_{20},W
Chromatin + poly(ADP-ribose)polymerase (control)	0 min	61.5 ± 0.3 S
Poly(ADP-ribosyl)ated chromatin	60 min	53.3 ± 0.5 S
Poly(ADP-ribosyl)ated chromatin treated with glycohydrolase	120 min	59.1 ± 1.2 S

histone H1 was predominant; 60 minutes later (slot d) most of the highly modified histone H1 species have been converted to faster migrating intermediates due to the hydrolysis of the polymer chains by glycohydrolase.

From these results it clearly appears that there is a positive correlation between the synthesis and the degradation of histone H1 bound poly(ADP-ribose) and the reversible conformational changes of chromatin superstructure detected by EM and ultracentrifugation.

Auto-poly(ADP-ribosyl)ation of the poly(ADP-ribose)polymerase

In the absence of any acceptor, the enzyme catalyzes its

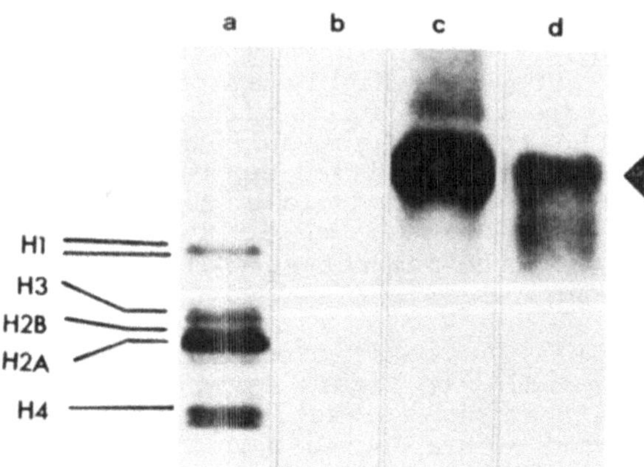

Figure 3. Acid urea polyacrylamide gel of ^{32}P-(ADP-ribosyl)-ated histones. Nucleosomes were incubated as described in Figure 1. At different reaction times, histones were extracted and electrophoresis was performed in a 1 M acetic acid, 6 M urea, 15% polyacrylamide gel. The gel was dried and autoradiographed. (a) stained gel (control); b-d autoradiograms corresponding to incubation times of 0, 60 and 120 minutes. The arrow points to the hyper(ADP-ribosyl)ated form of histone H1.

own poly(ADP-ribosyl)ation in the presence of DNA as cofactor and NAD as substrate (Okazaki, et al.,1980; Nishikimi, et al., 1982; Ferro and Olivera, 1982; de Murcia, et al., 1983).

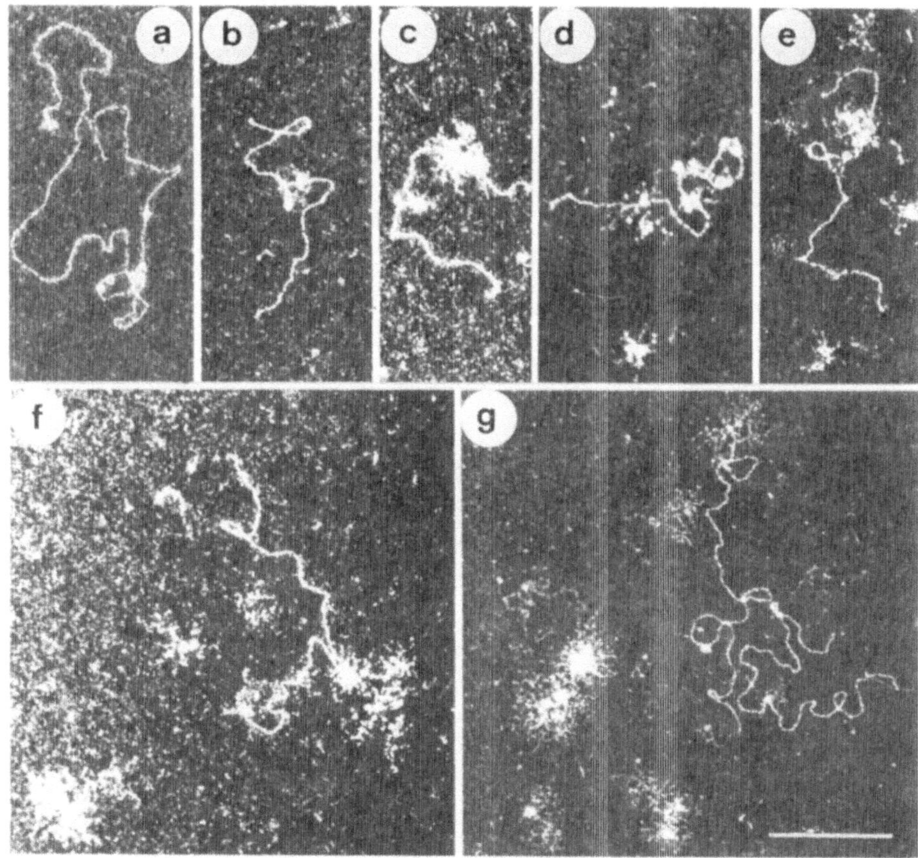

Figure 4. Kinetics of the automodification reaction visualized by dark field electron microscopy. Samples were incubated in 25 mM Tris-HCl buffer, pH 8, 10 mM $MgCl_2$, 0.4 mM dithiothreitol, and 100 μM of cold NAD during: (a) 0 minutes, (b) 5 minutes, (c) 10 minutes, (d) 15 minutes, (e) 30 minutes, (f) 45 minutes, (g) 60 minutes. The bar indicates 1000 Å.

As observed by dark field electron microscopy, the enzyme, which appears as a globular dot bound to DNA (Figure 4a), becomes progressively more dense, in terms of staining, after 15 minutes of ADP-ribosylation (Figure 4b) as compared to the control (Figure 4a). Polymerase molecules are surrounded by numerous points of uranyl acetate which also stains the products of the reaction (see arrows, Figure 4c). After 45 minutes of reaction, the electron density of enzyme molecules is strongly increased (Figure 4d), some material is detached from DNA and adsorbed onto the carbon film. With increasing time the proportion and size of this material increase, whereas, the quantity of DNA bound enzyme molecules decreases.

To visualize the product of the automodification reaction, after 60 minutes of ADP-ribosylation, DNase I and proteinase K resistant material was spread by the cytochrome C technique (Figure 5). This polymer was identified as poly(ADP-ribose), since no material was observed in control experiments in which NAD was omitted. The polymer presents a branched structure; these branches are up to 300 ADP-ribose residues for the largest molecules observed. This result fully confirms the proposed branched structure of poly(ADP-ribose), as reported by Miwa, et al.,1979.

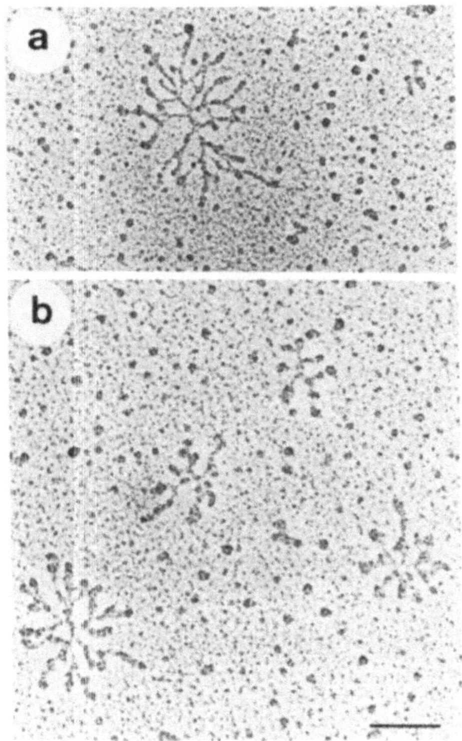

Figure 5. Visualization of poly(ADP-ribose) purified from automodified enzyme. The bar represents 1000 Å.

ADP-ribosylation of topoisomerase I. A DNA topoisomerase I activity was found to copurify with calf thymus poly(ADP-ribose)polymerase. As is the case for other nuclear enzymes, we demonstrated that the relaxing activity of either the contaminating or the purified topoisomerase I could be inhibited following poly(ADP-ribosyl)ation (Figure 6) (Mandel, et al.,1983). It was further shown that calf thymus type II topoisomerase was also inhibited in vitro by the same post-translational modification. The basis of this inhibition was found to be a disruption of the strand cleavage reaction (Darby, et al.,1985).

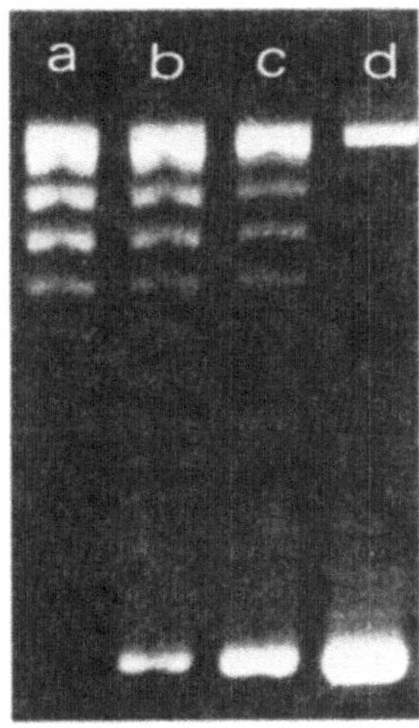

Figure 6. Activity of topoisomerase I following its ADP-ribosylation with increasing concentration of NAD. (a) 0 µM NAD, (b) 1 µM NAD, (c) 10 µM NAD, (d) 100 µM NAD.

The three acceptors of poly(ADP-ribose) described above have been formerly identified *in vivo*, after recovery from DNA damage, as modified nuclear proteins (Adamietz and Rudolph, 1984; ThiMan and Shall, 1982; Thraves and Smulson,1982). It is likely that the same mechanism decreasing the affinity of the modified acceptor for DNA is involved in the relaxation of chromatin structure, as well as in the inhibition of the polymerase and the DNA topoisomerases. It is remarkable that both are related to the nicking and resealing of DNA strand interruptions.

Poly(ADP-ribose)polymerase: a novel finger protein

Localization of the zinc-binding sites in the DNA-binding domain of the bovine poly ADP-ribose polymerase. Like other classes of proteins involved in nucleic acid binding, poly-(ADP-ribose)polymerase is a zinc-metalloenzyme (Zahradka and Ebiduzaki, 1984), and it was suggested that a metal-binding site is involved as part of its interaction with DNA.

We have attempted to localize the zinc binding sites with regard to the DNA-binding domain of the enzyme. For this purpose we have developed a simple and rapid technique for detection of zinc-binding proteins and peptides using radioactive zinc (^{65}Zn) on electroblots (Mazen, *et al*., 1988). The selectivity and the sensitivity of the method were tested using zinc-metalloenzymes (carbonic anhydrase, thermolysin, and alcohol dehydrogenase), as well as nucleic acid binding poteins containing potential metal-binding sequences (TFIIIA, methionyl-tRNA synthetase, T_4 gene 32 protein). Finally, we applied this technique to the bovine poly(ADP-ribose)polymerase and to the different proteolytic fragments. Figure 7 represents the zinc binding pattern of the holoenzyme (116 KD) and of the chymotrypsin and papain digestion fragments transferred to nitrocellulose. The binding of ^{65}Zn was detected in the 66, 46 and 29 KD fragments, obtained by mild protease digestion, which have been reported to contain DNA-binding sites. In order to correlate the ability of some domains to bind both zinc and DNA, blots of the enzyme and digestion fragments were incubated at room temperature in a sealed plastic bag containing 0.2 µg of ^{32}P-labeled nicktranslated DNA ($5x10^7$ cpm) in 1 ml of binding buffer. After washing out the unbound probe, the blots were air dried and autoradiographed.

The results displayed in Figure 8 demonstrate that the same bands (66, 46, and 29 KD as well as the whole enzyme) were labeled with ^{65}Zn, with ^{32}P nick translated DNA, and were also immunostained by the monoclonal antibody C_{19} which was found to inhibit the enzyme activity (Lamarre, et al.,1988). These results, taken together, prove that the zinc-binding sites are localized in the 29 KD N-terminal fragment, which is a part of the 46 KD DNA binding domain recognized by the $C^1{}_9$ antibody.

Using Energy Dispersive X-Ray Fluorescence (EDXRF), we found that purified calf thymus poly(ADP-ribose)polymerase contained 2 zinc atoms per enzyme molecule (Mazen,et al.1988). The hypothesis that the poly(ADP-ribose)polymerase contains two zinc fingers in its DNA binding domain was fully confirmed by the recent publication of the cDNA sequence (Uchida, et al., 1987). Figure 9 shows the two homologous sequences containing the repeated motif $C-X_2-C-X_{28,30}-H-X_2-C$; they are located in the DNA binding domain (amino acid position 2-97, and 106-207).

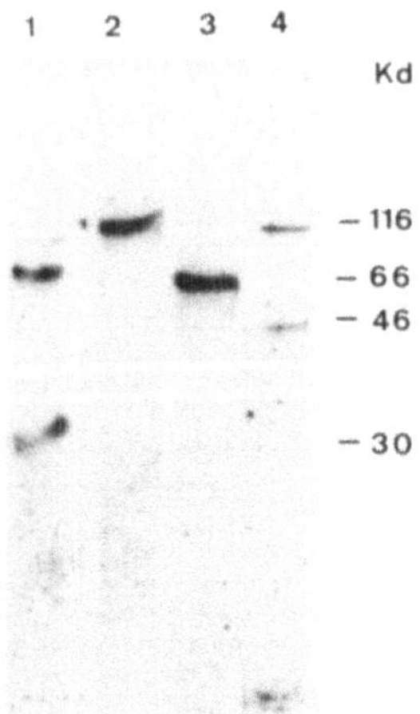

Figure 7. Selective binding of ^{65}Zn to different proteolytic fragments of poly ADP-ribose polymerase. Lane 1, molecular weight markers (bovine serum albumin, 67 KD; carbonic anhydrase, 30 KD); Lane 2, calf thymus poly ADP-ribose polymerase; lane 3, chymotrypsin digestion of poly(ADP-ribose)polymerase; lane 4, papain digestion of poly(ADP-ribose)-polymerase.

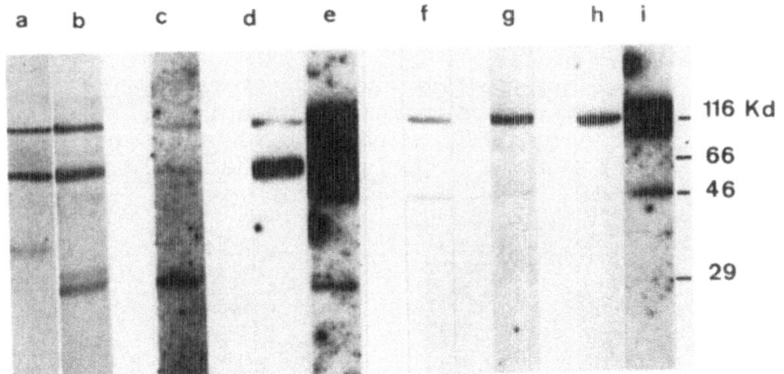

Figure 8. Western blots of trypsin (a-e) and papain fragments (f-i) of calf thymus poly ADP-ribose polymerase. Immunoblot incubated with monoclonal antibody C^2_{10} (a) or C^1_9 (b,f); autoradiogram after incubation with ^{65}Zn (c,g) or with nick translated ^{32}P DNA (d,h 2 hours of exposure; e,i, 16 hours of exposure).

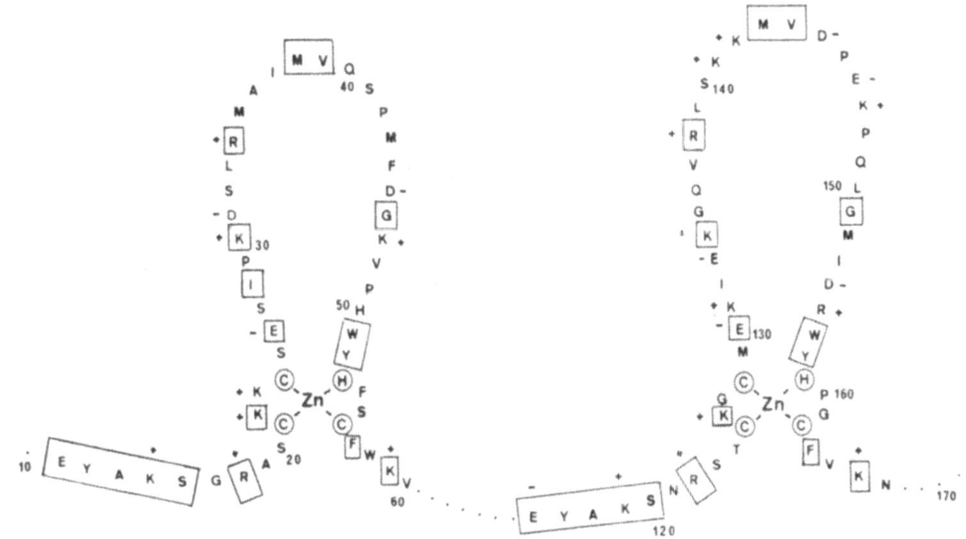

Figure 9. Schematic representation of the two zinc fingers located in the N-terminal part of the DNA binding domain.

Since its discovery in Xenopus transcription factor IIIA (TFIIIA), the "zinc finger" structural motif has been involved in numerous regulatory proteins (see for reviews: Berg,1986; Klug and Rhodes, 1987; Evans and Hollenberg,1988). Our results suggest strongly that poly(ADP-ribose)polymerase could well be a new member of the "zinc finger" protein family.

<u>Poly ADP-ribose polymerase interacts with a nick.</u> There is abundant literature concerning the activation of poly(ADP-ribose)polymerase by DNA strand breaks (see Althaus and Richter, 1987, for recent review). It has been suggested that

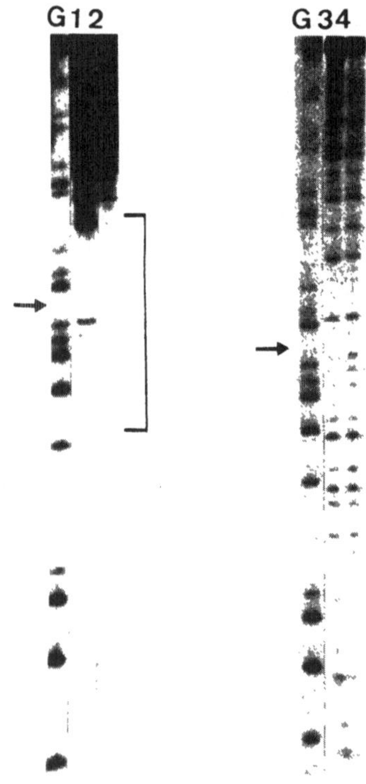

Figure 10."Blot and footprint" experiment. Purified poly(ADP-ribose)polymerase (1 μg/cm) was electrophoresed, transferred to nitrocellulose and incubated with labeled DNA containing (lane 2) or not (lane 4) a nick. The filter-bound protein DNA complex was then digested with DNAse I (20 ng/ml). Lanes 1 and 3: controls of DNase I degradation product of protein free-DNA (with or without nick, respectively). Lane G: sequencing reaction with dimethylsulfate.

the enzyme is stimulated by a nick or by a double strand break on DNA via an allosteric mechanism affecting its K_m for NAD. In order to get some insight into the molecular mechanism of the enzyme activation, we have set up footprint experiments using the south western technique described above. The transferred enzyme was incubated with a 66 mer deoxyoligonucleotide, containing a single-strand break created by the annealing of two 33 mer complementary to the long one. After washing out the unbound probe, the filter was autoradiographed wet during 30 minutes at 4^0C. The 116 KD band was cut out and used directly in the DNase I footprint experiment, as described by Huet and Sentenac (1987). As a control experiment, the same DNA without a nick was used. The results displayed in Figure 10 demonstrate that the immobilized enzyme binds around the DNA break protecting 7 nucleotides upstream of the nick and 9 nucleotides downstream. No visible footprint was obtained in the absence of nick.

Conclusion

In higher eukaryotes, in response to a DNA-damaging agent, DNA is nicked during the excision-repair process. This strongly stimulates poly(ADP-ribose)polymerase which in turn synthesizes poly ADP-ribose by consumption of the NAD pool. This post-translational protein modification is probably the most drastic in terms of affecting the charge and size of nuclear protein acceptors during the cell life. The target proteins are exclusively DNA binding proteins involved either in chromatin architecture or in DNA metabolism. In both cases, the consequence of this covalent addition of poly(ADP-ribose)-chains to nuclear proteins is to modulate DNA-protein interactions. The resulting relaxation of chromatin structure, triggered by poly(ADP-ribosyl)ation of histone H1, could be one of the early events in the excision-repair pathway, thus, providing increased accessibility of the damaged DNA to repair enzymes. Indeed, it has been reported that repair of DNA damage in eukaryotic cells involves local disruption of the chromatin structure and that repair incorporated nucleotides are transiently sensitive to nuclease digestion (Zolan, et al.,1982). This process, termed nucleosomal rearangement, could well be related to the modulation of chromatin structure induced by poly(ADP-ribosyl)ation described in this article. Simultaneously, the inhibition of the topoisomerase activities would decrease or block all chromatin functions involving DNA topology; this in term would impair cell cycle progression until DNA continuity was restored.

In trying to evaluate the metabolic role of poly(ADP-ribosyl)ation, it is perhaps useful to consider that the poly(ADP-ribose)polymerase, which is present in large amounts in the nucleus (Yamanaka, et al, 1988; Ludwig, et al., 1988), could also have a still unknown structural role. It is tempting to speculate that the poly(ADP-ribose)polymerase gene has evolved through the fusion of several functional domains from different proteins; the resulting polypeptide appears to be extremely conserved during evolution pointing to an essential role of the enzyme for the survival of eukaryotes.

ACKNOWLEDGEMENTS

This work was carried out in Prof. G. Dirheimer's laboratory and was supported by the Ligue Nationale Francaise contre le Cancer (Délégation Départementale du Haut-Rhin).

REFERENCES

Adamietz, P., and Rudolph, A., 1984, ADP-ribosylation of

nuclear proteins in vivo, J.Biol. Chem., 259:6841-6848.

Althaus, F.R., and Richter, C.R., 1987, ADP-ribosylation of proteins, Mol. Biol. Biochem. Biophys., 37:1-126.

Aubin, R.J., Dam, V.T., Miclette, J., Brousseau, and Y., Poirier, G.G., 1982, Chromosomal protein poly(ADP-ribosyl)ation in pancreatic nucleosomes. Can. J. Biochem., 60:295-305.

Berg, J., 1986, Potential metal-binding domains in nucleic acid binding proteins, Science, 232:485-487.

Darby, M.K., Schmitt, B., Jongstra-Bilen, I., and Vosberg, H.P., 1985, Inhibition of calf thymus type II DNA topoisomerase by poly(ADP-ribosylation). EMBO J., 4:2129-2134.

De Murcia, G., Jongstra-Bilen, J., ittel, M., Mandel, P., and Delain, E., 1983, Poly(ADP-ribose)polymerase automodification and interaction with DNA : electron microscopic visualization. EMBO J., 2:543-548.

De Murcia, G., and Koller, Th., 1981, The electron mlcroscopic appearance of soluble rat liver chromatin mounted on different supports, Biol. Cell, 40:165-174.

De Murcia, G., Huletsky, A., Lamarre, D., Gaudreau, A., Pouyet, J., Daune, M., and Poirier, G.G., 1986, Modulation of chromatin superstructure induced by poly(ADP-ribose)synthesis and degradation, J. Biol. Chem., 261: 7011- 7018.

Evans, R.M., and Hollenberg, S.M., 1988, Zinc fingers: Gilt by association, Cell, 52:1-3.

Ferro, A.M., and Olivera, B.M., 1982, Poly(ADP-ribosylation) in vitro. Reaction parameters and enzyme mechanism, J. Biol. Chem., 257:7808-7813.

Gaal, J.C., and Pearson, C.K., 1985, Eukaryotic nuclear ADP-ribosylation reactions. Biochem. J., 230:1-18.

Huet, J., and Sentenac, A., 1987, TUF, the yeast DNA-binding factor specific for UASrpg upstream activating sequences :Identification of the protein and its DNA-binding domain Proc. Natl. Acad. Sci. USA., 84:3648-3652.

Kameshita, I., Matsuda, Z., Taniguchi, T., and Shizuta, Y., 1984, Poly(ADP-ribose)synthetase, separation and identification of three proteolytic fragments as the substrate-binding domain, the DNA binding domain and the automodification domain, J. Biol. Chem., 259:4770-4776.

Klug, A., and Rhodes, D., 1987, "Zinc fingers" : a novel protein motif for nucleic acid recognition, TIBS, 12:464-469.

Lamarre, D., Talbot, B., De Murcia, G., Laplante, C., Leduc, Y., Mazen, A., and Poirier, G., 1988, Structural and functional analysis of Poly(ADP-ribose)polymerase: an immunological study, Biochim. Biophys. Acta, 950:147-160.

Ludwig, A., Behnke, B., Hotlund, J., and Hilz, H., 1988, Immunoquantitation and size determination of intrinsic poly(ADP-ribose)polymerase from acid precipitates, J. Biol. Chem., 263:6993-6999.

Mandel, P., Jongtra-Bilen, J., Ittel, M.E., De Murcia, G., Delain, E., Niedergang, C., and Vosberg, H.P., 1983, Some electron microscopic aspects of poly(ADPR)polymerase DNA interactions and of auto-poly(ADP-ribosyl)ation reaction, Proceedings of the Princess Takamatsu imperial cancer research fund, Symposium on ADP-ribosylation and cancer, edited by M. Miwa, Tokyo, pp 77.

Mazen, A., Gradwohl, G., De Murcia, G., 1988, Zinc binding proteins detected by protein blotting, Anal. Biochem., 172:39-42.

Mazen et al., in preparation.
Miwa, M., Saikawa, N., Yamaizumi, Z., Nishimura, S., and Sugimura, T., 1979, Structure of poly(adenosine diphosphate ribose) : identification of 2'-(1''-ribosyl-2''- or 3''-)(1'''-ribosyl))adenosine-5',5'',5''-tris(phosphate) as a banch linkage, Proc. Natl. Acad. Sci. USA, 76:595- 599.
Nishikimi, M., Ogasawara, K., Kameshita, I., Taniguchi, T., and Shizuka, Y., 1982) Poly(ADP-ribose synthetase. The DNA binding domain and the automodification domain, J. Biol. Chem., 257:6102-6105.
Okazaki, H., Niedergang, C., and Mandel, P., 1980, Adenosine diphosphate ribosylation of histone Hl by purified calf thymus polyadenosine diphosphate ribose polymerase, Biochimie, 62:147-157.
Shall, S., 1984, ADP-ribose in DNA repair: a new component of DNA excision repair, Adv. Radiat. Biol., 11 :1-69.
Thi Man, N., and Shall, S., 1982, The alkylating agent, dimethylsulfate, stimulates ADP ribosylation of histone H and other proteins in permeabilized mouse lymphoma (L1210) cells, Eur. J. Biochem.,126:83-88.
Thraves, P.J., and Smulson, M.E., 1982, Acceptors for the poly ADP-ribosylation modification of chromatin structure are altered by carcinogen induced DNA damage, Carcinogenesis, 3:1143-1148.
Uchida, K., Morita, T., Sato, T., Ogura, T., Yamashita, R., Noguchi, S., Suzuki, H., Nyunoya, H., Miwa, M., and Sugimura, T., 1987, Nucleotide sequence of a full-length cDNA for human fibroblast poly(ADP-ribose)polymerase, Biochem. Biophys. Res. Commun., 148:617-622.
Ueda, K., and Hayaishi, O., 1985, ADP-ribosylation, Annu. Rev. Biochem., 54:73-100.
Yamanaka, H., Penning, C.A., Willis, E.H., Wasson, P., and Carson, S.A., 1988, Charaterization of human poly(ADP-ribose)polymerase with autoantibodies. J. Biol. Chem., 263:3879-3883.
Zahradka, P., and Ebisuzaki, K., 1984, Poly(ADP-ribose)polymerase is a zinc metalloenzyme, Eur. J. Biochem., 142:503-509.
Zolan, M.E., Smith, C.A., Calvin, N.M., and Hanawalt, P.C., 1982, Rearrangement of mammalian chromatin structure following excision repair, Nature, 299:462-464.

VARIATION OF NUCLEAR ADP-RIBOSYL TRANSFERASE IN RAT LIVER CARCINOGENESIS AND IN SYNCHRONIZED HELA CELLS

U. Bertazzoni[1], C. F. Cesarone[2], R. Izzo[1], L. Scarabelli[2], I. Colombo[1], M. Orunesu[2] and A. I. Scovassi[1]

[1]Istituto di Genetica Biochimica Evoluzionistica del CNR, Via Abbiategrasso 207, 27100 Pavia, Italy

[2]Istituto di Fisiologia Generale, Facoltá di Scienze, Universitá di Genova, C.so Europa 26, 16132 Genova, Italy

INTRODUCTION

The chromatin-bound enzyme ADP-ribosyl transferase (ADPRT) is responsible for post-translational modification of several nuclear acceptor proteins (including the enzyme itself and histones H1 and H2B) and appears to play a central role in DNA repair (Shall, 1984). By using activity gel and immunoblot techniques we have recently shown that a depletion of ADPRT activity occurs in rat liver during exposure of animals to 2-acetylaminofluorene (2-AAF) (Cesarone, et al., 1988) according to the experimental in vivo system of Teebor and Becker (1971). This model is based on a discontinuous feeding regimen with the carcinogen 2-AAF given to rats during four consecutive cycles, each one composed of 3-weeks of treatment followed by 1 week of recovery.

In more recent experiments, we followed rat liver ADPRT activity daily during the first week of 2-AAF treatment and after 2 or 3 weeks of recovery, following 3 weeks of treatment with 2-AAF. In addition, we report here the results obtained by analyzing the activity of ADPRT in two other experimental systems for in vivo hepatocarcinogenesis, namely those introduced by Solt and Farber (Solt, et al., 1977) and Druckrey (1967).

A useful model for studying functional changes in enzyme activity is provided by regenerating rat liver, a system that closely reflects the physiological induction of DNA synthesis and cell proliferation. In order to measure the rate of ADP-ribosylation reactions catalyzed by ADPRT, ^{32}P-NAD, which is the specific substrate for ADPRT, was incorporated into nuclei prepared from liver samples obtained at different hours after

DNA Repair Mechanisms and Their Biological Implications in Mammalian Cells
Edited by M.W. Lambert and J. Laval
Plenum Press, New York

hepatectomy. ADPRT activity was also measured in extracts prepared at various regeneration times.

The possible variation in ADPRT activity during the progression of the cell cycle has not yet been clearly defined (Scovassi, et al., 1987). To this end we have synchronized HeLa cells in S phase by a double block with aphidicolin, and we have followed ^{3}H-thymidine incorporation and ADPRT activity. To possibly obtain further information about the in vivo acceptor proteins, we have incorporated ^{3}H-adenosine into HeLa cells at various phases of the cell cycle and analyzed cellular extracts by SDS-PAGE after hydrolysis of the bulk of DNA and RNA.

MATERIALS AND METHODS

Experimental design for in vivo carcinogenesis. The three models used for in vivo carcinogenesis (Teebor and Becker, 1971; Solt, et al., 1977; Druckrey, 1967) are illustrated in Figure 1. In the Teebor and Becker system, rats were placed on a 4-cycle discontinuous feeding regimen, each cycle consisting of a 3-week exposure to a diet containing 0.05% 2-AAF followed by a week of recovery with the standard diet. Animals were sacrificed each day during the first week of

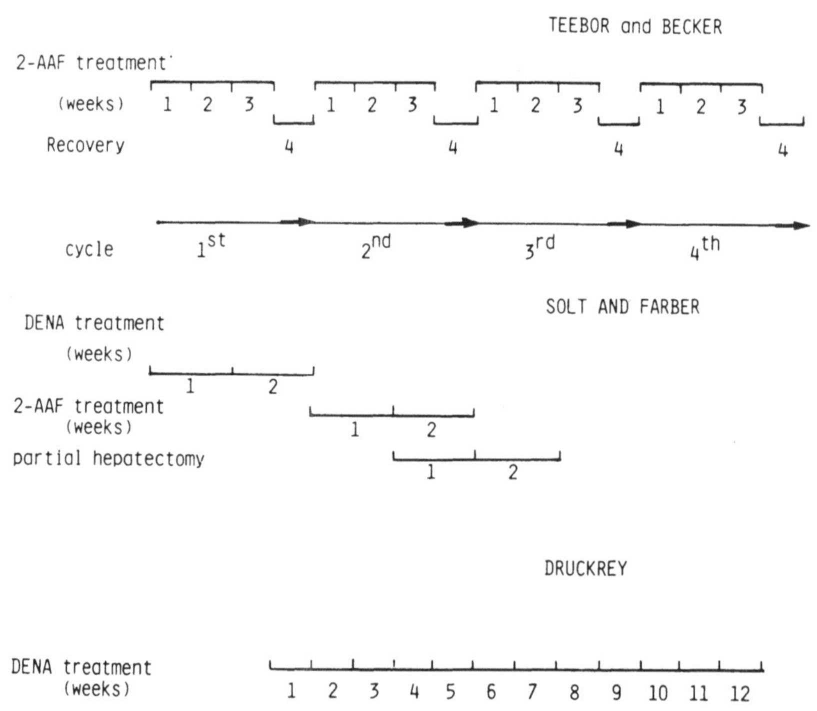

Figure 1. Experimental models for rat liver carcinogenesis. Other details for each system are described in Materials and Methods.

treatment and at the end of the 3rd and of the 4th week of each cycle. Recovery from the first cycle of treatment was obtained by treating the rats with a standard diet for 2-3 weeks. Livers were rapidly removed, washed in cold PBS, frozen in liquid nitrogen and stored at -80^{0}C.

The Solt and Farber model consists of a single injection of diethylnitrosamine (DENA, 200 mg/kg b.w.), followed by 2 weeks of a standard diet, 2 weeks of a diet with 0.035% 2-AAF, and a final week with a standard diet. After one week of exposure to 2-AAF, animals were submitted to partial hepatectomy. Livers were obtained from animals sacrificed 5 hours after DENA injection, before partial hepatectomy and 3 days after.

The Druckrey model is characterized by a continuous 12 week treatment with DENA diluted in drinking water (10 mg/kg/day). DENA was administered to rats weighing 120 g at the beginning of the experiment. This particular exposure gives rise to hepatocellular carcinomas with a mean induction time of 14 weeks. Animals were sacrificed at 2, 4, 6 and 12 weeks from the beginning of exposure. Male albino Wistar rats were used for all experiments.

Preparation of extracts and ADPRT assay. Aliquots of 0.2 g of frozen livers, obtained at various times, were minced in 10 vol of extraction buffer (1.5 M NaCl, 50 mM Tris HCl, pH 7.5, 0.5 mM DTT, 1 mM EDTA, 1 mM PMSF, 1 µM pepstatin and 10 mM Na bisulfite), sonicated and centrifuged as described (Cesarone, et al., 1988). The supernatant fraction was immediately used for an enzymatic assay and an activity gel. Enzyme activity was measured by a biochemical assay specific for ADPRT as previously described (Scovassi, et al., 1986). Catalytic peptides of ADPRT were detected by means of the activity gel technique following our standard procedure (Scovassi, et al., 1987).

Partial hepatectomy. Partial hepatectomies (65-70%) were performed by the technique of Higgins and Anderson (1931) on adult rats (180-200 g) under slight anaesthesia. Animals were sacrificed at 1, 2, 4, 8, 12 and 18 hours after operation.

^{32}P-NAD incorporation in rat liver nuclei. Crude preparations of rat liver nuclei were obtained as previously described (Cesarone, et al., 1979). Nuclei were resuspended in saline EDTA (75 mM NaCl, 24 mM EDTA, pH 7.5), at a final concentration of about 1 x 10^8/ml. DNA content was measured according to Cesarone, et al.(1979) and the level of DNA damage was determined by DNA elution in alkaline buffer (Bolognesi, et al., 1981). For ^{32}P-NAD incorporation, 25 µl of nuclear suspension were added to 100 µl of reaction mixture containing 100 mM Tris HCl, pH 8.0, 10 mM $MgCl_2$, 100 µM adenine-2,8-^{32}P-NAD (specific activity 1500 cpm/pmol), and 1 mM DTT. Incubation was carried out at 37^{0}C; aliquots of 25 µl were withdrawn at 1.5, 3 and 10 minutes, spotted on glass-fiber discs, and processed according to the method of Bollum (1966).

HeLa cells synchronization. Synchronization of HeLa cells was obtained by a double block with aphidicolin as previously reported (Pedrali-Noy, et al., 1980), and is illustrated in Figure 2. At various times after aphidicolin release (2, 4,

8, 12 and 18 hours) aliquots of cells were used for ^{3}H-thymidine and ^{3}H-adenosine incorporation. For activity gel analysis, about $5x10^6$ cells were resuspended with 100 µl of extraction buffer, sonicated and centrifuged as reported (Scovassi, et al., 1987).

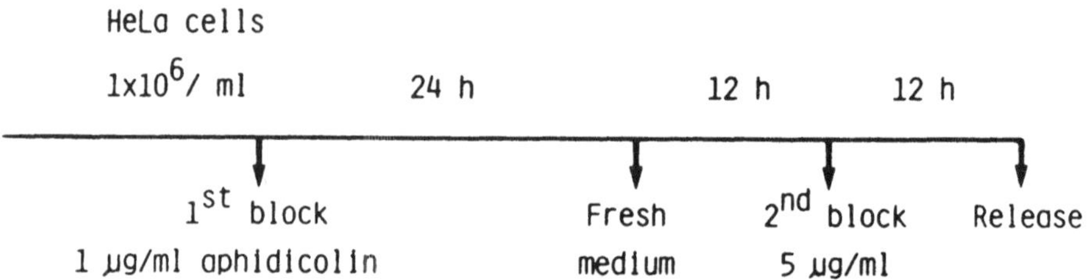

Figure 2. HeLa cells synchronization by a double block of aphidicolin (see Materials and Methods).

^{3}H-adenosine incorporation. To $2x10^6$ cells, resuspended in 300 µl of medium, 50 µl of ^{3}H-adenosine (51 Ci/mmol, 1 mCi/ml) were added. Incubation was carried out for 2 hours at 37^{0}C and after centrifugation, the cellular pellet was washed and resuspended with 100 µl of 0.2 M Tris HCl, pH 7.5, 10 mM $MgCl_2$, 10 mM EtSH, 1 mM EDTA, 10 mM Na bisulfite, 1 mM PMSF, 1 µM pepstatin, and 20% glycerol. Cells were disrupted by sonication; DNA and RNA were digested by DNase I and RNase A (40 µg/ml of each) at 25^{0}C for 15 minutes. Samples were then loaded on a 7.5% polyacrylamide gel; at the end of the run, the gel was washed, incubated with Enlightning, dried and autoradiographed.

RESULTS

Variation of ADP-ribosyl transferase in rat liver carcinogenesis

We have recently reported that, when exposing rats during four discontinuous cycles to a diet containing 0.05% 2-AAF, the active band of ADPRT, clearly evident in untreated livers, is no longer detectable after the first cycle and that a subsequent increase of activity occurred during following cycles (Cesarone, et al., 1988). These results were now confirmed by analyzing the same enzyme by means of a biochemical assay. In Table 1, we report the values of enzyme activity in extracts prepared from the liver of animals submitted to 4 cycles of treatment. It is evident that a drastic drop in activity occurs after the first cycle, followed by a tendency to increase in later cycles.

In order to verify whether the effect induced by 2-AAF is also evident at early times in the first cycle, ADPRT activity was followed during the first week of treatment. No significant variations in ADPRT activity bands were evident during the first six days of exposure to 2-AAF, suggesting that a longer time of treatment is needed to deplete the enzyme (results not shown). As shown in Figure 3, it appears that the drop in ADPRT activity can be reversed. In fact, when the carcinogen is no longer supplied with the diet, the activity of ADPRT tended to return, after 2 weeks of recovery, to the level of the control.

TABLE 1. ADPRT ACTIVITY IN THE LIVER OF RATS TREATED WITH 2-AAF

Treatment	Specific Activity (units x 10^{-4}/mg proteins)
Standard diet	270
2-AAF: I cycle	60
II cycle	64
III cycle	80
IV cycle	100

The ADPRT assay was performed in total extracts from rat livers obtained at the end of each 2-AAF cycle. The preparation of extracts is described in Methods; the values reported represent the average of three independent experiments. One unit of enzymatic activity corresponds to 1 nmole of NAD incorporated into acid-insoluble material in 1 minute at 25^{0}C.

Beside the Teebor and Becker system, we have also experimented with other models for rat hepatocarcinogenesis, these are the Solt and Farber and the Druckrey systems, which are described in the Methods and in Figure 1.

Using these two experimental systems, we analyzed the activity of ADPRT in livers of rats sacrificed at different times. As seen in Figure 4 (panel A), in the Solt and Farber model a single injection of diethylnitrosamine (DENA) resulted in a significant decrease of activity (lane 2) and the subsequent administration of 2-AAF induced a further drop in activity (lane 3). Partial hepatectomy had no influence on this effect (lane 4). Using the Druckrey model, no significant variation in enzyme band intensity was noted in samples obtained after 2, 4, 6 and 12 weeks of oral administration of DENA, as shown in panel B of Figure 4.

To possibly understand further the role of ADPRT during the regeneration process which follows partial hepatectomy, we analyzed the catalytic polypeptides of ADPRT and the extent of DNA damage by the alkaline elution technique. It appeared that during the first hours after partial hepatectomy, DNA was not significantly damaged and the activity tended to be constant until 8 hours (Figure 5, lanes 1-5) and to decrease sharply at 12 and 18 hours (lanes 6 and 7).

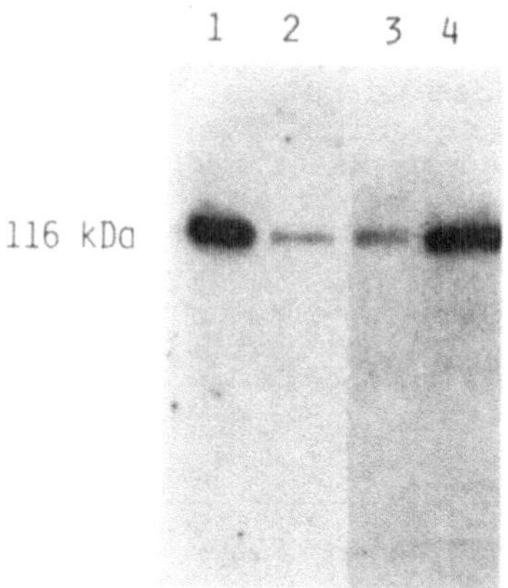

Figure 3. Activity gel analysis of ADPRT in liver extracts from rats exposed to 2-AAF. Lane 1: control; lane 2: after one cycle of diet containing 2-AAF; lanes 3 and 4: experiment as in lane 2, followed by one and two weeks of a standard diet, respectively.

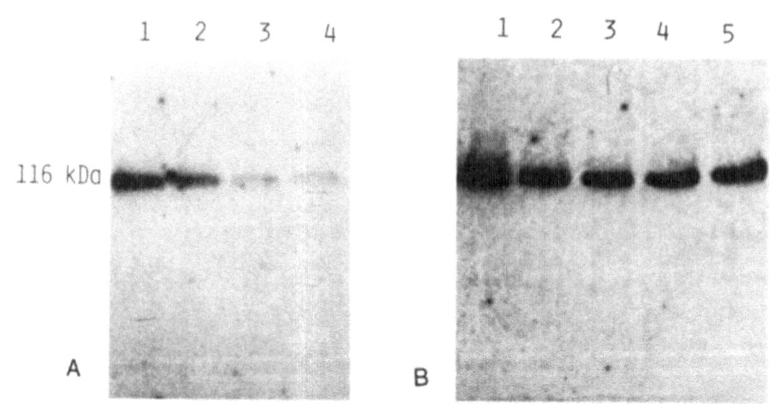

Figure 4. Activity gel analysis of ADPRT in liver extracts from rats treated according to Solt and Farber (panel A) and Druckrey (panel B) models. A) lane 1: control; lane 2: 5 hours after DENA injection; lane 3: 1 week of 2-AAF treatment after DENA injection; lane 4: 3 days after partial hepatectomy performed during 2-AAF treatment; B) lane 1: control; lanes 2-5: 2,4,6 and 12 weeks of continuous treatment with DENA, respectively. For experimental procedures see Materials and Methods.

The reaction of ADP-ribosylation was also followed by measuring the incorporation of ^{32}P-NAD in nuclei isolated from livers at 6 and 12 hours of regeneration. Preliminary results, shown in Figure 6, suggest that at 6 hours after partial hepatectomy the incorporation is higher than in controls whereas it becomes significantly lower after 12 hours, thus confirming the data obtained with the activity gel.

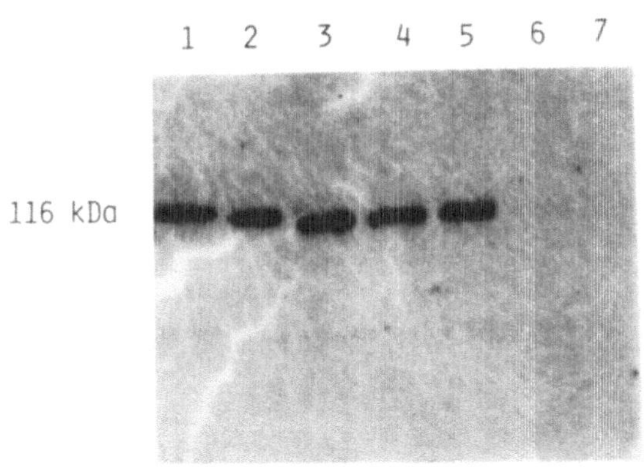

Figure 5. Activity gel analysis of ADPRT in extracts of regenerating rat liver. Lane 1: control; lanes 2-7: 1, 2, 4, 8, 12 and 18 hours after partial hepatectomy, respectively.

Variation of ADP-ribosyl transferase activity in synchronized HeLa cells

We have studied the functional role of ADPRT during the cell cycle in HeLa cells synchronized by a double block with aphidicolin. The rate of replicative DNA synthesis was measured following ^{3}H-thymidine incorporation (Figure 7, panel A). This reached a maximum at 4 hours after aphidicolin release and returned to the initial level after 8 hours, which correlated with the end of S phase. The level of ADPRT, visualized by activity gel analysis, is shown in panel B. The intensity of the 116 kDa band tended to remain constant during the whole S period, G_2 and M phases, as previously observed (Scovassi, et.al., 1987). This indicates that no significant variation in enzyme activity occurs during the cell cycle.

To obtain information about the protein acceptors which are ADP-ribosylated within the cell, we have labeled ADP-ribose by incubating dividing HeLa cells with ^{3}H-adenosine. This precursor is incorporated into ATP and NAD and, consequently, into ADP-ribose. In order to measure selectively the ADP-ribosylated proteins, cellular extracts were treated with RNase and DNase to eliminate radioactive adenosine incorporated into nucleic acids and radioactive proteins were analysed on SDS-PAGE. The results of a typical experiment are shown in Figure 8. Using a 7.5% polyacrylamide gel, protein bands of 200, 175, 120, 86 and 28 kDa were visualized. The 120 kDa band represented the autoribosylated enzyme, the 86 kDa could possibly arise from the ADP-ribosylation of a heat-shock protein and the lower band possibly represented histone H2B. The 175 kDa band could possibly correspond to the ADP-ribosylation of topoisomerase II.

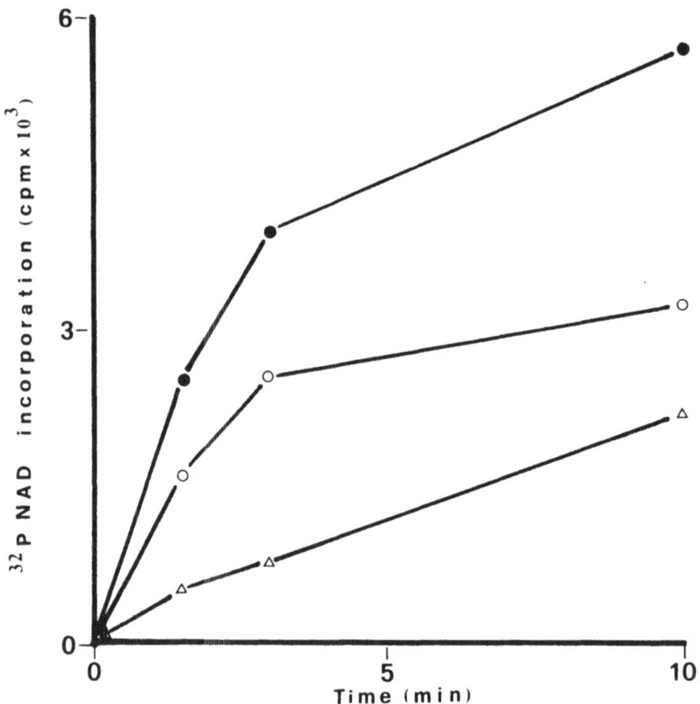

Figure 6. ^{32}P-NAD incorporation in nuclei of regenerating rat liver. O—O, control; , ●—●, 6 hours; Δ—Δ 12 hours after partial hepatectomy.

DISCUSSION

Experimental evidence indicates that a decreased capacity in DNA repair activity may represent the basic event in the initiation of carcinogenesis. For this reason, many enzymes involved in DNA repair have been analyzed during the early steps of the carcinogenic process (Cesarone, et al., 1988).

It has been demonstrated that the enzyme ADP-ribosyl transferase plays a central role in the modulation of cellular response to DNA damage. We have analyzed the activity of this enzyme in different sytems of rat hepatocarcinogenesis. For this purpose, we have followed the experimental models of

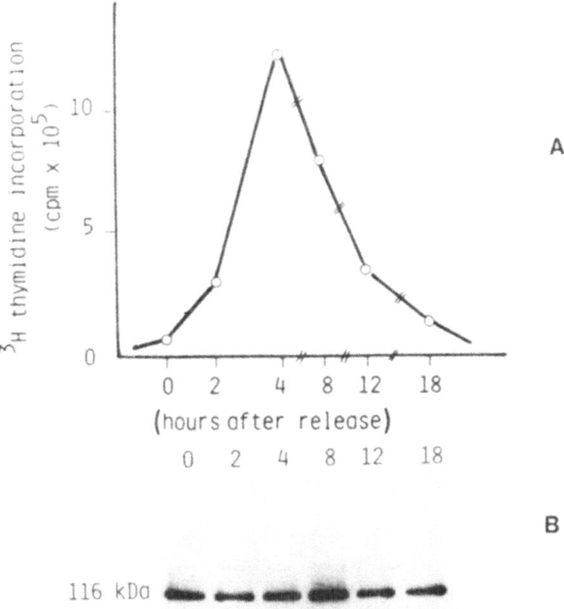

Figure 7. Synchronization of HeLa cells in S phase by a double block of aphidicolin. Panel A: ^{3}H-thymidine incorporation at 0, 2, 4, 8, 12 and 18 hours after aphidicolin release. Panel B: activity gel analysis of ADPRT performed on extracts from cells obtained at the same times after aphidicolin release. Experimental details are reported in Materials and Methods.

Teebor and Becker (1971), Solt and Farber (1977), and Druckrey (1967) and studied the properties of the enzyme in samples of livers obtained at different times. The results previously obtained using the Teebor and Becker model showed a loss of ADPRT activity after the first cycle of treatment of rats with the carcinogen 2-acetylaminofluorene (2-AAF) (Cesarone, et al., 1988). We have now demonstrated that the effect of a 3-week treatment with 2-AAF can be reversed by feeding the same animals for 2 weeks with a standard diet.

To study the effect of diethylnitrosamine (DENA), a carcinogenic initiator, and of the subsequent administration of the promoter 2-AAF on ADPRT activity, the experimental model of Solt and Farber was followed. The results obtained indicate that ADPRT activity was strongly decreased after 2-AAF treatment thus confirming the effect already observed using the Teebor and Becker model (Cesarone, et al., 1988). In fact, when DENA alone was continuously given to rats according to the Druckrey model, the level of the enzyme remained constant, suggesting that an initiator alone is not effective in regulating enzyme activity.

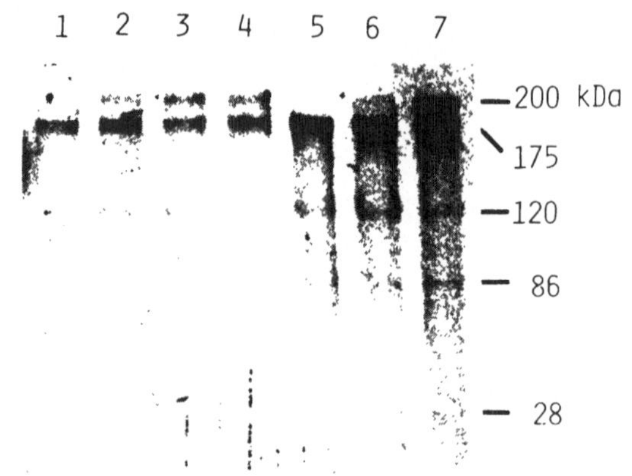

Figure 8. Analysis by SDS-PAGE of acceptor proteins after ^{3}H-adenosine incorporation in synchronized HeLa cells. Lanes 1-7: 0, 4, 8, 12, 20, 24 and 48 hours after aphidicolin release, respectively.

To possibly understand further the role of ADPRT in DNA synthesis and cell proliferation, we have measured the extent of ^{32}P-NAD incorporation in nuclei isolated from regenerating rat liver. Preliminary results suggest that a rapid increase in NAD incorporation occurred 6 hours after partial hepatectomy, followed by a significant decrease after 12 hours.

Concerning the possible role of ADPRT during the progression of the cell cycle, no conclusive evidence has been obtained so far. We have analyzed the enzyme in synchronized HeLa cells and have shown that no significant variations in ADPRT activity occurred during the different phases of the cell cycle (Scovassi, _et al._, 1987).

To obtain further information on changes in the pattern of ADP-ribosylation of nuclear proteins, we have also analyzed the proteins ADP-ribosylated _in vivo_ during the cell cycle. Our data indicate that several proteins were modified by ADPRT and among these, of particular interest, is a peptide of 175 kDa, which could correspond to topoisomerase II.

ACKNOWLEDGMENTS

The authors thank Dr. P. Giannoni for his excellent assistance. This work was supported by Italian National Research Council (C.N.R.) Special Project "Oncology" and by contract B16-158 of the Radiation Programme of the Commission of the European Communities, contribution n. 2483. R.I. is the recipient of a fellowship from the Italian Association for Cancer Research (AIRC-Milan).

REFERENCES

Bollum, F.J., 1966, in:"Proc. in Nucleic Acid Research" (G.L. Cantoni and D.R. Davies, eds), Harper & Row, New York p. 296.

Bolognesi, C., C.F. Cesarone and L. Santi, 1981, Evaluation of DNA damage by alkaline elution technique after in vivo treatment with aromatic amines, _Carcinogenesis_, 2:265.

Cesarone, C.F., C. Bolognesi and L. Santi, 1979, Improved microfluorometric DNA determination in biological material using 33258 Hoechst, Anal. Biochem., 100:188.

Cesarone, C.F., A.I. Scovassi, L. Scarabelli, R. Izzo, M. Orunesu and U. Bertazzoni, 1988, Depletion of Adenosine Diphosphate-Ribosyl Transferase activity in rat liver during exposure to N-2-acetylaminofluorene: effect of thiols, Cancer Res., 48:3851.

Druckrey, H., 1967, Quantitative aspects in chemical carcinogenesis, IUCC Monogr. Series, 7:60.

Higgins, G.M. and R.M. Anderson, 1931, Experimental pathology of the liver, Arch. Pathol., 12:187.

Pedrali-Noy, G., S. Spadari, A. Miller-Faures, A.O.A. Miller, J. Kruppa and G. Koch, 1980, Synchronization of HeLa cell cultures by inhibition of DNA polymerase α with aphidicolin, Nucleic Acids Res., 8:377.

Scovassi, A.I., R. Izzo, E. Franchi and U. Bertazzoni, 1986, Structural analysis of poly(ADP-ribose)polymerase in higher and lower eukaryotes, Eur. J. Biochem., 159:77.

Scovassi, A.I., M. Stefanini, P. Lagomarsini, R. Izzo and U. Bertazzoni, 1987, Response of mammalian ADP-ribosyl tranferase to lymphocyte stimulation, mutagen treatment and cell cycling, Carcinogenesis, 8:1295.

Shall, S., 1984, ADP-ribose in DNA repair: a new component of DNA excision repair, Adv. Radiat. Biol., 11:1.

Solt, D.B., A. Medline and E. Farber, 1977, Rapid emergence of carcinogen-induced hyperplastic lesions in a new model for the sequential analysis of liver carcinogenesis, Am. J. Pathol., 88:595.

Teebor, G.W. and F.F. Becker, 1971, Regression and persistence of hyperplastic nodules induced by N-2-fluorenylacetamide and their relationship to hepatocarcinogenesis, Cancer Res., 31:1.

DNA REPAIR PROCESSES IN MUTANT CELLS DEFICIENT IN POLY(ADP-RIBOSE) SYNTHESIS

Nathan A. Berger, Satadal Chatterjee, Ming-Feng Cheng, Denise Meckler, Shirley J. Petzold and Sosamma J. Berger.

Departments of Medicine and Biochemistry, Ireland Cancer Center, University Hospitals of Cleveland Case Western Reserve University, Cleveland, Ohio 44106, U.S.A.

Poly(ADP-ribose) polymerase is a eukaryotic enzyme that uses NAD as substrate to synthesize homopolymers of ADP-ribose which are covalently linked to protein acceptors such as the enzyme itself (Carter and Berger, 1982). The enzyme is predominantly located in the nucleus in an inactive form; it is activated by DNA strand breaks that result from a variety of treatments that cause DNA damage (Berger, et al., 1979; Cohen and Berger, 1981). Treatment of cells with agents such as 3 aminobenzamide (3AB) to inhibit poly(ADP-ribose) polymerase interferes with the repair of DNA strand breaks and suggests that poly(ADP-ribose) synthesis is required for DNA repair (Durkacz, et al., 1980). Other studies suggest that poly(ADP-ribose) synthesis may also be involved in DNA replication and cell differentiation (Hayaishi and Ueda, 1977). Since many of these studies are based on the use of enzyme inhibitors which also affect other cellular process, the function of poly(ADP-ribose) polymerase has not yet been determined with certainty.

One approach to determine the function of an enzyme is to study mutant cell lines defective in activity of that enzyme. In this report, we describe the development and characteristics of several mutant cell lines deficient in their ability to synthesize poly(ADP-ribose). The first approach to develop poly(ADP-ribose) polymerase deficient mutants was based on our demonstration that high levels of DNA damage could induce cells to commit suicide by sufficiently activating poly(ADP-ribose) polymerase to consume cellular NAD pools resulting in cessation of glycolysis, failure to generate ATP, rundown of energy metabolites and subsequent cell death due to loss of energy dependent functions (Berger, 1985; Sims, et al., 1983; Das and Berger, 1986). We reasoned that cells with low or absent levels of poly(ADP-ribose) polymerase would not deplete their NAD and ATP levels in response to treatment with high levels of DNA damage and might

DNA Repair Mechanisms and Their Biological Implications in Mammalian Cells
Edited by M.W. Lambert and J. Laval
Plenum Press, New York

therefore survive such a treatment. Thus, we treated V-79 cells with high levels of MNNG and ultimately cloned several cell lines with markedly reduced activities of poly(ADP-ribose) polymerase (Chatterjee, et al., 1987). Table 1 shows two mutant cell lines derived by this approach, ADPRT 54 and ADPRT 351. These cells have less than 14% of the enzyme activity measured in parental cell lines. They also have a tetraploid DNA content and an approximate 2-fold content of protein compared to parental V-79 cells. Whether the tetraploid DNA content is invariably or coincidentally associated with poly(ADP-ribose) polymerase deficiency remains to be determined.

Table 1. Characteristics of V79 and Derivative Cell Lines

Cell line	Poly(ADPR) polymerase (ADPR incorporated pmol/µg DNA/min)	NAD Level (pmol per 10^6 cells)	DNA Content (µg/10^6 cells)	Protein Content (µg/10^6 cells)
V79	90.4	690	6.6	118
N3(-Nam)	71.5	16	6.5	177
ADPRT 54	12.3	795	13.6	217
ADPRT 351	4.6	838	15.3	298

We also took another approach to develop cell lines with reduced ability to synthesize poly(ADP-ribose) by cloning mutant cells capable of growing in the absence of added nicotinamide. Under normal conditions, V-79 cells require exogenously supplied nicotinamide to synthesize and maintain cellular NAD pools. N3 cells derived from V-79 cells are capable of continuous growth in the absence of nicotinamide with extremely low cellular levels of NAD. As shown in Table 1, the NAD level in N3 cells is less than 3% of that in the parental cells. These cells have apparently normal levels of poly(ADP-ribose) polymerase, however, enzyme activity is severely restricted due to lack of substrate. The DNA and protein contents of N3 cells are similar to the parental V-79 cells. When nicotinamide content is restored in the growth medium, the N3 cells rapidly synthesize and restore normal cellular levels of NAD. By growing the cells in the absence or presence of nicotinamide we can easily establish low or normal intracellular NAD levels and study processes in the same cell line in the presence of restricted or normal poly(ADP-ribose) synthesis.

When cells are permeabilized and supplied with radioactive NAD, they incorporate radioactive ADP ribose into poly-(ADP-ribose)(Berger, et al., 1978a; Berger, et al., 1978b; Berger, et al., 1979). Total cellular activity of poly(ADP-ribose) polymerase can be determined in this system by treating the permeable cells with an excess of DNase to make sufficient strand breaks to maximally activate the enzyme (Berger, et al., 1978a). The product of this reaction can be solubilized, run on SDS polyacrylamide gels and examined by autoradiography to provide a quantitative assessment of the ADP-ribosylated proteins (Carter and Berger, 1982; Surowy and Berger, 1983). When V-79 cells are examined by this approach, there is a heavy band of radioactivity at a molecular weight of 116,000 corresponding to the ADP-ribosylated enzyme (Chatterjee, et al., 1987). There is also a heavy smear of radioactivity extending from 116,000 to the top of the gel (Surowy and Berger, 1983; Chatterjee, et al., 1987). The latter represents enzyme with attached poly(ADP-ribose) chains of increasing lengths. There are also radioactively labelled bands in the low molecular weight portions of the gel corresponding to ADP-ribosylated histones and other chromosomal proteins (Surowy and Berger, 1983). When N3 cells are evaluated by the same procedure and supplemented with adequate amounts of [^{32}P]-NAD, they show the normal pattern of ADP-ribosylated enzyme at 116,000 extending to the top of the gel as well as the ADP-ribosylation of the low molecular weight bands. These studies are as expected and support the notion that the N3 cells have normal poly(ADP-ribose) polymerase but synthesize low levels of polymer in the intact cells due to restriction of substrate.

When ADPRT 54 and 351 are examined by permeabilization, treatment with DNAse, incubation with [^{32}P]-NAD and autoradiography, they show several abnormalities. First, the ADP-ribosylated band of enzyme at 116,000 is very faint. This indicates that the enzyme in the mutant cells is of normal size but has very low activity. Second, they fail to show the normal pattern of radioactivity moving to the top of the gel showing that the enzyme is defective in elongating poly(ADP-ribose) chains. Third, there is a decrease in radioactive labelling of bands in the low molecular weight region of the gel suggesting that in the mutant cells, the enzyme is defective in ADP-ribosylation of other chromosomal proteins.

The studies outlined above show that poly(ADP-ribose) polymerase is quantitatively deficient and qualitatively abnormal in the ADPRT 54 and 351 cell lines. The enzyme appears more normal in the N3 cells but can not synthesize normal amounts of poly(ADP-ribose) when the cells are grown in nicotinamide deficient media.

Table 2 shows some of the phenotypic characteristics of the poly(ADP-ribose) synthesis deficient mutants. All three cell lines have prolonged doubling times compared to the parental V-79 cell line. This suggests a requirement for poly(ADP-ribose) synthesis in the cell cycle and the normal process of cell replication. When the N3 cells are supplemented with nicotinamide to restore their NAD content they are then capable of shortening their generation time. Previous studies have shown an increase in sister chromatid exchange occurring when cells are exposed to inhibitors of poly(ADP-ribose) polymerase (Oikawa, et al., 1980). In the present study we found that all three cell lines have a high spontaneous rate of sister chromatid exchange. This is almost an order of magnitude greater than found in the V-79 cells and is similar to the increase in SCE produced by treating V-79 cells with 3AB. Treatment of the mutant cells with 3AB further

TABLE 2. Phenotypic Characteristics of V79 and Derivative Cell Lines

Cell Lines	Generation Time (Hrs)	SCE per Chromosome
V79	10	0.29 ± 0.12
V79 + 3AB		2.94 ± 0.34
N3 (-Nam)	24	2.69 ± 0.59
N3 (+Nam)	17	0.35 ± 0.12
ADPRT 54	68	1.78 ± 0.19
ADPRT 351	50	3.21 ± 0.51

increases their level of SCE. Restoring NAD levels in the N3 cells alters the baseline SCE to return to normal. Thus, three methods of interfering with poly(ADP-ribose) synthesis - i.e., inhibition by 3AB, mutant cells with defective enzyme activity or mutant cells with relative substrate deficiency all result in increased SCE.

Figure 1 shows the results of testing the mutant cells for sensitivity to MNNG. The left panel shows that 3AB sensitizes V-79 cells to the cytotoxic effects of MNNG. ADPRT-351 cells are hypersensitive to MNNG and appear to be similar to the V-79 cells treated with 3AB. However, the 351 cells can be even further sensitized to MNNG by treatment with 3AB. This suggests that 3AB is a more effective inhibitor of poly(ADP-ribose) polymerase in the mutant cells where the enzyme is considerably reduced before the inhibitor is added. The right panel in Figure 1 shows that the N3 cells are similar to the 351 cells in their response to MNNG. Thus, the N3 cells are hypersensitive to MNNG compared to V-79 and they can be sensitized even further by treatment with 3AB.

In Figure 2, cells were treated for 1 hour with 1 µg/ml MNNG and 3 mM 3AB. At the end of the treatment period, the cells were washed free of drug and then resuspended in either medium alone or medium containing 3 mM 3AB. At the time of drug removal and at successive time intervals, cells were examined for DNA strand breaks using an alkaline unwinding technique and hydroxyapatite chromatography (Ahnstrom and Erixon, 1981). The ordinate labelled single strand DNA is an operational expression of DNA strand breaks. Figure 2 shows that all cells treated by this protocol have a high level of DNA strand breaks at the end of the drug treatment period. However, ADPRT-351 cells had higher initial strand breaks than did V-79 and N3 had higher levels of breaks than did 351. The V-79 cells showed an initial rapid rate of strand break repair followed by a somewhat slower second repair phase. The V-79 cells treated with 3AB showed some prolongation of the rapid rate of repair with minimal effects on the slow rate.

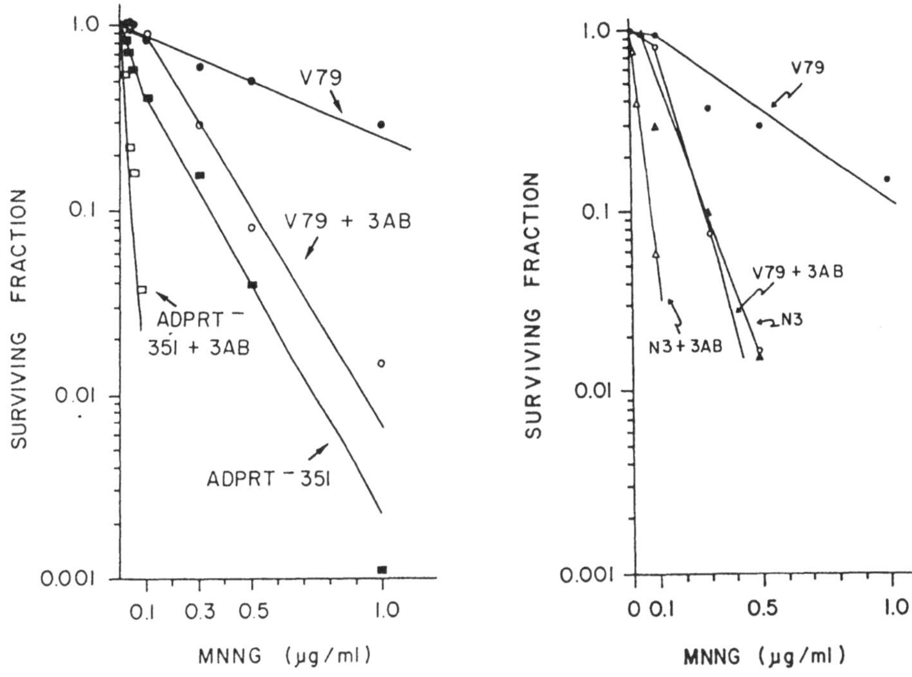

Figure 1. Survival curves for V79 cells and poly(ADP-ribose) synthesis deficient mutant cell lines exposed to MNNG in the presence or absence of 3AB. Left, V79 and 351 cells. 351 cells are deficient in poly(ADP-ribose) polymerase. Right, V79 and N3 cells. N3 cells grown in absence of nicotinamide.

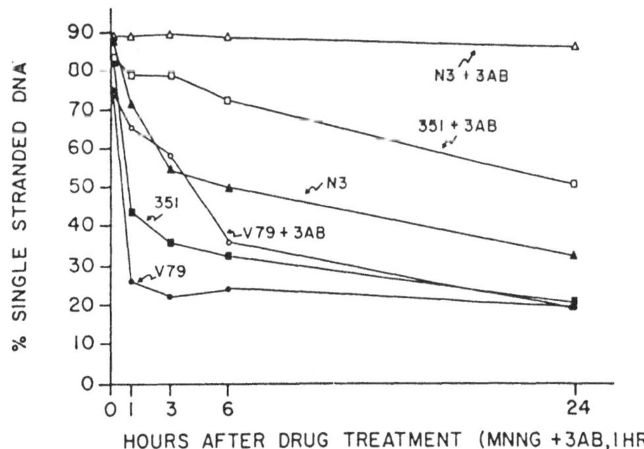

Figure 2. DNA strand break repair in V79 and mutant cells exposed to 1 μg/ml MNNG and 3 mM 3AB for 1 hour. Experimental details as in text.

351 cells showed a slight decrease in the amount of DNA repair carried out by the rapid process and normal activity of the slower repair process. In the presence of 3AB, 351 cells showed a marked deficiency in the rapid repair phase of DNA strand breaks without much alteration in the slow phase. The N3 cells showed greater reduction of the rapid phase relative to the 351. When the N3 cells were incubated in 3AB the rapid phase of DNA strand break repair was completely inhibited. Under these conditions, there was also complete inhibition of the second or slow phase of DNA repair.

These observations suggest that V-79 cells respond to MNNG treatment with at least two phases of DNA strand break repair. Interference with poly(ADP-ribose) synthesis activity in the mutant cells interferes with the rapid phase of DNA strand break repair. In the cells with maximal inhibition of poly(ADP-ribose) synthesis , i.e., N3 cells treated with 3AB, strand break repair appeared to be completely inhibited.

In summary, we have developed mutant V-79 cell lines deficient in their ability to synthesize poly(ADP-ribose). The phenotypic characteristics of these cell lines include prolonged generation times and increased levels of spontaneous sister chromatid exchanges. Each cell line is more sensitive than the parental V-79 cells to the cytotoxic effects of MNNG. Each cell line demonstrated delayed repair of DNA strand breaks. These studies indicate that poly(ADP-ribose) synthesis is required for normal DNA strand break repair and its deficiency results in delayed repair of DNA strand breaks, increased sensitivity to MNNG, increased SCE and prolonged generation times. These mutant cell lines should provide an important tool for mechanistic studies of poly(ADP-ribose) in the DNA replication and repair processes.

ACKNOWLEDGMENT

These studies were supported in part by NIH grants CA-35983, GM32647 and CA43703. MFC was supported by an NIH Training Grant T32HL07147 and DM by a Silber Student Fellowship from the American Cancer Society, Cuyahoga County Chapter.

REFERENCES

Ahnstrom, G., and Erixon, K, 1981, Measurement of strand breaks by alkaline renaturation and hydroxyapatite chromatography, in: "DNA Repair, A Laboratory Manual of Research Procedures," E.C. Friedberg and P.E. Hanawalt, eds., Marcel Dekker Inc., New York.

Berger, N.A., 1985, Poly(ADP-ribose) in the cellular response to DNA damage,, Rad. Res., 101:4-15.

Berger, N.A., Adams, J.W., Sikorski, G.W., Petzold, S.J., and Shearer, W.T., 1978a, Synthesis of DNA and poly(adenosine diphosphate ribose) in normal and chronic lymphocytic leukemia lymphocytes, J. Clin. Invest., 62:111-118.

Berger, N.A., Weber, G., and Kaichi, A.S., 1978b, Characterization and comparison of poly(adenosine diphosphoribose) synthesis and DNA synthesis in nucleotide permeable cells, Biochim. Biophys. Acta, 519:87-104.

Berger, N.A., Sikorski, G.W,, Petzold, S.J., and Kurohara, K.K., 1979, Association of poly(adenosine diphosphoribose) synthesis with DNA damage and repair in normal human lymphocytes, J. Clin. Invest., 63:1164-1171.

Carter, S.G., and Berger, N.A., 1982, Purification and characterization of human lymphoid poly(adenosine diphosphate ribose) polymerase, Biochemistry, 21:5475-5481.

Chatterjee, S., Petzold, S.J., Berger, S.J., and Berger, N.A., 1987, Strategy for selection of cell variants deficient in poly(ADP-ribose) polymerase, Exp. Cell. Res., 172: 245-257.

Cohen, J.J., and Berger, N.A., 1981, Activation of poly(adenosine diphosphate ribose) polymerase with UV irradiated and UV endonuclease treated SV40 minichromosome, Biochem. & Biophys. Res. Communs., 98:268-274.

Das, S.K., and Berger, N.A., 1986, Alterations in deoxynucleoside triphosphate metabolism in DNA damaged cells: Identification and consequences of poly(ADP-ribose) polymerase dependent and independent processes, Biochem. Biophys. Res. Communs., 137:1153-1158.

Durakacz, B.W., Omidiji, O., Gray, D.A., and Shall, S., 1980, (ADP-ribose)n participates in DNA excision repair, Nature, 283:593-593.

Hayaishi, O., and Ueda, K., 1977, Poly(ADP-ribose) and ADP ribosylation of proteins, Ann. Rev. Biochem., 46:95-116.

Oikawa, A., Tohda, H., Kanai, M., Miwa, M., and Sugimura, T., 1980, Inhibitors of poly(adenosine diphosphate ribose) polymerase induce sister chromatid exchanges, Biochem. Biophys. Res. Communs., 97:1311-1316.

Sims, J.L., Berger, S.J., and Berger, N.A., 1983, Poly(ADP-ribose) polymerase inhibitors preserve nicotinamide adenine dinucleotide and adenosine 5'-triphosphate pools in DNA-damaged cells: Mechanism of stimulation of unscheduled DNA synthesis, Biochemistry, 22:5188-5194.

Surowy, C.S., and Berger, N.A., 1983, Unique acceptors of poly(ADP-ribose) in resting, proliferating and DNA-damaged human lymphocytes, Biochim. Biophys. Acta, 740:8-18.

CO-RECESSIVE INHERITANCE: A MODEL FOR DISEASES ASSOCIATED WITH DEFECTIVE DNA REPAIR

W. Clark Lambert and Muriel W. Lambert
Department of Pathology
UMDNJ-New Jersey Medical School
Newark, NJ 07103

SUMMARY

We have proposed a genetic model, "co-recessive inheritance" (C-RI), in which an individual must be homozygous or hemizygous for defective alleles at more than one of a specific set of genetic loci in order to express a given trait, particularly a disease associated with one or more defective DNA repair processes. One or more dominant alleles at other loci could conceivably also be required to produce the trait. The model accounts for a number of paradoxical aspects of the sun-sensitive, cancer-prone inherited disease, xeroderma pigmentosum (XP), including the large number of complementation groups despite a biochemically limited associated DNA repair defect, the existence of some individuals who appear to have the DNA repair defect but not clinical XP, the apparent existence of the DNA repair defect in association with other inherited diseases, the co-existence of XP and Cockayne's syndrome in two different complementation groups of XP, siblings with markedly different degrees of severity of XP in one family and transmission of the disease in an X-linked manner in another, and the seeming paradox of a disease associated with a marked defect in a DNA repair mechanism but not associated with an obvious increase in incidence of internal cancer. Compared with autosomal recessive inheritance, whether with one gene or with several genes, the model predicts lower frequencies of affected siblings and fewer first cousin matings between parents of affected individuals. Since it was first put forward, further evidence supporting the model in XP has emerged, including defective repair of UV irradiation induced damage in cellular DNA of cells from additional individuals who do not have XP and a deficiency in killer T-lymphocytes as well as in a catalase in some XP patients. The model predicts that a large proportion of the general population is a carrier of one or more of these defective genes, many of which encode for products active in DNA repair mechanisms. This prediction is valid even if the model holds true for only one or more complementation groups of one or a few very rare disorders. Such genes may be important in the etiology of a number of common conditions involving DNA repair processes, such as neurodegenerative diseases, cancer and aging.

DNA Repair Mechanisms and Their Biological Implications in Mammalian Cells
Edited by M.W. Lambert and J. Laval
Plenum Press, New York

INTRODUCTION

Although the following discussion focuses largely on the inherited, sun-sensitive, cancer-prone disease, xeroderma pigmentosum (XP), the C-RI model may also apply to other diseases, both those associated with defective DNA repair and other disorders.

XP is characterized in most cases by more or less extreme hypersensitivity to ultraviolet radiation (UVR), UVR associated skin and eye changes, and development of numerous cancers in light-exposed tissues, usually at a very abnormally early age. In some patients neurological defects also develop (Andrews, 1983; Cleaver and Bootsma, 1975; Cleaver, 1985; Jung, 1986; Kraemer, 1980, 1981; Kraemer, et al., 1987; Lambert, et al., 1985; Lambert and Lambert, 1987; Robbins, 1987; Robbins, et al., 1974). In most cases of XP, there is a defect in the initial, incision step in the nucleotide excision-repair pathway which acts on pyrimidine cyclobutane dimers, pyrimidine-pyrimidine 6-4 photoproducts and probably other adducts induced in DNA by exposure to UVR. This pathway is initiated by a DNA endonuclease, which is probably also influenced by other chromatin proteins, which incises the DNA (i.e., cleaves the single affected strand) at or near the site of damage, followed by a series of steps about which there is some controversy (Andrews, 1983; Cleaver, 1985; Friedberg, et al., 1981; German, et al., 1980; Kraemer, 1980, 1981; Lambert, et al., 1985; Lambert and Lambert, 1987; Lehmann and Karran, 1981; Lindahl, 1982; Regan and Setlow, 1974). Although, in various organisms, several pathways exist, each of them terminates in a polymerase-mediated step, in which new DNA is synthesized to fill a gap left by these other processes, and a DNA ligase-mediated step in which the strand is resealed. If the first step is defective, subsequent repair-associated DNA synthesis is necessarily diminished. Cells in culture from such excision-deficient (E-D) XP patients therefore show diminished non-S phase, repair-associated synthesis of DNA following exposure to UVR and to a number of other agents, determined, among other methods, by diminished uptake of labeled thymidine visualized by autoradiography (Andrews, 1983; Cleaver and Bootsma, 1975; Cleaver, 1985; German, et al., 1980; Kraemer, 1980, 1981; Lambert, et al., 1985; Lambert and Lambert, 1987; Robbins, 1987; Robbins, et al., 1974). This autoradiographic technique, known as the unscheduled DNA synthesis (UDS) assay, has become the most widely accepted method for establishing a diagnosis of E-DXP.

The UDS assay has also become the basis for a now standard method for detecting genetic heterogeneity and assigning patients to genetic subgroups in E-DXP. If heterokaryons derived from two different E-DXP patients are exposed in culture to UVR, they often show normal or near normal levels of UDS. This cell fusion - UDS system has formed the basis for classification of E-DXP patients into complementation groups, defined as follows: Pairs of patients whose cells complement each other in the UDS assay following UVL exposure in culture are said to be in different complementation groups and those whose cells fail to complement each other in culture are classified in the same complementation group. In this way 9 complementation groups of E-DXP, labeled A through I, have now been identified despite the biochemically limited nature of the DNA repair defect (Cleaver, 1985; Cleaver and Bootsma, 1975; DeWeerd-Kastelein, et al., 1972; Fischer, et al., 1985; Jung, 1986; Lambert, et al., 1985; Lambert and Lambert, 1987; Robbins, et al., 1974).

XP has been thought to be transmitted by autosomal recessive inheritance, and, indeed, the pedigrees of most patients are consistent with that mode although one more consistent with autosomal dominant inheritance (Anderson and Begg, 1950) and one with sex-linked inheritance (Stefanini, et al., 1980) have been reported. In our proposed model (C-RI; Lambert and Lambert, 1985) for the inheritance of XP, more than one genetic locus is involved in each patient for whom the model applies. Although a number of other human oligogenic models have been proposed, none has been for a disease of similar rarity to that of XP (Hodge, 1981; Hodge and Spence, 1981; Hodge, et al., 1981; Witkamp, 1981; Falk, et al., 1983). Another type of oligogenic model proposes that two different defective alleles occupying the same locus in a heterozygous manner may be associated with expression of a trait, such as a disease state. Such combinations of alleles producing a trait/disease are known as compounds (Tolleshang, et al., 1982). We have been unable to construct a model, however, based on compounds, which would explain the available information on XP. The model we have proposed, which we have termed "co-recessive inheritance" (Lambert and Lambert, 1985), not only accounts for the large number of complementation groups of E-DXP identified to date but also is consistent with a number of other observations which are difficult to reconcile with an autosomal recessive mode of inheritance for E-DXP.

DEFINITION: CO-RECESSIVE INHERITANCE

The co-recessive inheritance model hypothesizes that there are certain genetically transmitted traits, particularly diseases, in which the abnormal phenotype is expressed if and only if the individual is homozygous or hemizygous for defective alleles at a number, greater than one, of a distinct set of genetic loci, not necessarily linked to each other. The number of loci may vary depending on which loci are involved but it is always greater than one.

DISCUSSION

Complementation Groups

The large number of complementation groups that have been found in E-DXP to date appear to us to be improbable if the disorder is transmitted by autosomal recessive inheritance, given the highly specific nature of the DNA repair defect. Although it is true that a similar or slightly larger number of excision-deficient complementation groups in yeast have been identified (Friedberg, 1988; Haynes and Kunz, 1981; Lawrence, 1982), and six complementation groups have been identified in Chinese Hamster Ovary (CHO) cells (Thompson, et al., 1987), the number of complementation groups of E-DXP that have been identified would be expected to be a low estimate of the number of genotypes that may cause this biochemical defect. This is because some such genotypes may produce a lethal phenotype, others may be excision-deficient but not show clinical XP, others may fail to complement each other in fused cells, and still others may, because of the relative rarity of XP and the relatively small number of cases that have been subjected to complementation analysis, simply not yet have been discovered. Evidence that the number of complementation groups of E-DXP identified to date is indeed a low estimate exists in the form of clinical heterogeneity among patients in group A (Sato, et al., 1987), group C (Kraemer, 1980, 1981; Kraemer, et al., 1987; Lambert and Lambert, 1987) and group D (Ichihashi, et al., 1988) and differences in cloning efficiency following UVL exposure in cells from patients in Group A (Andrews, et al., 1978; Moshell, et al., 1981).

The C-RI model offers a straightforward explanation for these large numbers of complementation groups, as well as for additional defects reported in XP patients, as follows:

Let us define the following terms: N, total number of genetic loci in a co-recessive inheritance system; n, total number of these co-recessive gene loci that must be homozygous or hemizygous for defective alleles in order for the trait to be expressed; $C_{n,N}$, theoretical maximum number of complementation groups in a co-recessive inheritance system with n and N defined as above, assuming that there are no individuals homozygous or hemizygous for defective alleles at >n loci in the system.

$C_{n,N}$, is therefore the number of ways that an individual can be homozygous or hemizygous for defective alleles at n loci out of a defined set of N loci so as to both have clinical XP and have cells that show depressed UDS following UVR exposure in culture. This is given by combinatorial mathematics (Feller, 1960) as:

$$C_{n,N} = \frac{N!}{n!(N-n)!} \qquad (1)$$

Eqn. 1 explains how a rather small number of co-recessive loci can account for a large number of complementation groups. For example, if N = 5 and n = 2, $C_{n,N}$ = 10; if N = 6 and n = 3, $C_{n,N}$ = 20, and if N = 8 and n = 4, $C_{n,N}$ = 70. This is further shown in Table 1A.

If one considers the possibility that there may exist some individuals homozygous or hemizygous for defective alleles at >n co-recessive genetic loci, the number of possible complementation groups becomes considerably larger. For example, let us define a new term, $C_{MAX,N}$ as follows:

$C_{MAX,N}$, theoretical maximum number of complementation groups in a co-recessive inheritance system with $n \geq 2$, n and N defined as above. It follows quite simply from combinatoric mathematics and the binomial theorem that:

$$C_{MAX,N} = 2^N - N - 1 \quad (2)$$

In the above examples, if N = 5, $C_{MAX,N}$ = 26, IF N = 6, $C_{MAX,N}$ = 57, and if N = 8, $C_{MAX,N}$ = 247. This is further shown in Table 1B. Many individuals homozygous or hemizygous for defective alleles at >n loci would probably not be viable, however, so that $C_{MAX,N}$ may not be a realistic estimate of the possible number of complementation groups in the system. On the other hand, $C_{n,N}$ may be a low estimate.

Table 1. Co-recessive Inheritance Gives Many More Possible Complementation Groups Than There Are Genetic Loci in the System

(A) Generation of 15 possible complementation groups of XP from six loci (i.e., N = 6) in a co-recessive system with n = 2. N and n defined as in the text.

Co-recessive Loci: 1,2,3,4,5,6
Complementation Groups with Involved Loci:

A:1,2	F:2,3	K:3,5
B:1,3	G:2,4	L:3,6
C:1,4	H:2,5	M:4,5
D:1,5	I:2,6	N:4,6
E:1,6	J:3,4	O:5,6

(B) Generation of 42 additional complementation groups assuming n >2.

P:1,2,3	Z:2,3,4	AK:1,2,3,5	AV:2,3,5,6
Q:1,2,4	AA:2,3,5	AL:1,2,3,6	AW:2,4,5,6
R:1,2,5	AB:2,3,6	AM:1,2,4,5	AX:3,4,5,6
S:1,2,6	AC:2,4,5	AN:1,2,4,6	AY:1,2,3,4,5
T:1,3,4	AD:2,4,6	AO:1,2,5,6	AZ:1,2,3,4,6
U:1,3,5	AE:2,5,6	AP:1,3,4,5	BA:1,2,3,5,6
V:1,3,6	AF:3,4,5	AQ:1,3,4,6	BB:1,2,4,5,6
W:1,4,5	AG:3,4,6	AR:1,3,5,6	BC:1,3,4,5,6
X:1,4,6	AH:3,5,6	AS:1,4,5,6	BD:2,3,4,5,6
Y:1,5,6	AI:4,5,6	AT:2,3,4,5	BE:1,2,3,4,5,6
	AJ:1,2,3,4	AU:2,3,4,6	

There may, of course, be yet other loci at which one must be homozygous or hemizygous to show the trait but which do not contribute to abnormal results in assays used for studies of complementation. Recent evidence (reviewed below) suggests that this may indeed be true for some patients with E-DXP.

Dominant genes may also conceivably participate in a co-recessive system.

Carrier Frequencies of Co-recessive Genes

Perhaps the most interesting and relevant prediction of the co-recessive inheritance model is that frequencies of defective alleles at co-recessive loci and of individuals who are carriers of one or more such defective alleles are much higher than for autosomal recessive inheritance and, in fact, may include a substantial proportion of the general population, even for extremely rare diseases such as E-DXP. Although it is widely accepted among geneticists that oligogenic systems for genetically transmitted diseases are associated with higher carrier frequencies than are ordinary mendelian systems, we are not aware of an analysis of this type for a disease of rarity similar to that of XP. This analysis of XP indicates that the discrepancy between carrier frequencies and frequencies of the various complementation groups of the disease is likely to be vastly greater, in some cases orders of magnitude greater, than for other oligogenic systems studied to date. An essential difference from many other systems is that they hypothesize homozygosity for defective alleles at one _or_ at one or more other loci whereas the co-recessive inheritance model hypothesizes homozygosity or hemizygosity for defective alleles at one _and_ at one or more other loci. The predicted carrier frequencies are much higher for the latter case than for the former.

In the following computations, we assume that individuals within the same complementation group are homozygous or hemizygous for defective alleles at the same loci and that no individual is affected at >n such loci. Let us define the following terms: q_i, prevalence of defective alleles at locus i in the general population; α_i, q_1/q_i; Z, observed frequency of the trait (i.e., clinical XP) of a given complementation group in the general population; f_i, frequency of individuals carrying defective alleles at locus i (both homozygotes and heterozygotes).

If we assume all co-recesive genes in the system are themselves autosomal recessive, then, from the Hardy-Weinberg relationship and mathematical probability and combinatoric analysis (Mathematical Appendix I):

$$q_i = \frac{Z^{1/2n}}{\alpha_i} \prod_{k=1}^{n} \alpha_k^{1/n} \tag{3}$$

and, since, from the Hardy-Weinberg relationship we have $f_i = 2q_i\ (1 - 1/2\ q_i)$, Eqn. (3) may be expressed as (Mathematical Appendix I):

$$f_i = \frac{2Z^{1/2n}}{\alpha_i} \prod_{k=1}^{n} \alpha_k^{1/n} \left\{ 1 - \frac{Z^{1/2n}}{2\alpha_i} \prod_{k=1}^{n} \alpha_k^{1/n} \right\} \tag{4}$$

F, the frequency of individuals in the general population carrying a defective allele at any of the loci involved in this complementation group, may then be readily obtained from the f_is by the following combinatorial equation (Mathematical Appendices I and II):

$$F = 1 - \prod_{i=1}^{n} (1 - f_i) \tag{5}$$

to give:

$$F = 1 - \prod_{i=1}^{n} 1 - \frac{2Z^{1/2n}}{\alpha_i} \prod_{k=1}^{n} \alpha_k^{1/n} \left\{ 1 - \frac{Z^{1/2n}}{2\alpha_i} \prod_{k=1}^{n} \alpha_k^{1/n} \right\} \tag{6}$$

It is apparent that, even for systems in which n = 2 and Z is very small, F constitutes a significant proportion of the general population. For example, if $n = 2$, $\alpha_1 = \alpha_2 = 1$, and $Z = 10^{-7}$, $q_i = 1.78\%$, $f_i = 3.5\%$ and $F = 6.9\%$. Thus 6.9% of the population would be carriers of defective alleles at either or both of the co-recessive gene loci. Furthermore, if $n>2$, q_i, f_i and F all become much larger; in the above system, for example, if $n = 3$ and $\alpha_1 = \alpha_2 = \alpha_3 = 1$, then $q_i = 6.81\%$, $f_i = 13.3\%$ and $F = 34.8\%$. Thus, considering only this one complementation group, over one-third of the population would be carriers of defective alleles at one or more of the co-recessive gene loci even though the incidence of the trait (i.e., disease within this complementation group) is only 10^{-7}. Carrier frequencies corresponding to 2-5 loci and varying disease frequencies are shown in Figure 1. This is perhaps made even clearer by direct comparison of F for co-recessive versus oligogenic autosomal recessive inheritance (for $\alpha_i = 1$; q_i, standard oligogenic; $\underline{q}_i$, co-recessive; Mathematical Appendix II):

$$q_i = \left\{ 1 - \left(1 - \underline{q}_i^{\,2n} \right)^{1/n} \right\}^{1/2} ; \quad \underline{q}_i = \left\{ 1 \quad \left(1 - q_i^{2} \right)^{n} \right\}^{1/2n} \tag{7}$$

repeated substitution into the above functions of which gives (for $\alpha_i = 1$, F, standard oligogenic, $\underline{F}$, co-recessive; Mathematical Appendix II):

$$\underline{F} = 1 - \left[1-2 \left[1 - \left\{ 1 - \left[1 - (1-F)^{1/2n} \right]^{2} \right\}^{n} \right]^{1/2n} \left\{ 1 - 1/2 \left[1 - \left\{ 1 - \left[1 - (1-F)^{1/2n} \right]^{2} \right\}^{n} \right]^{1/2n} \right\} \right]^{n} \tag{8}$$

Figure 2 shows values of $\underline{F}$ compared with values of F. Points shown refer to $Z = 10^{-8}$, 10^{-7}, 10^{-5}, 10^{-4}, 10^{-3}, and 10^{-2}, left to right on each function, respectively. It is readily apparent that, even for n = 2 and Z small, $\underline{F}$ is a significant proportion of the general population, whereas F is not, with values of $\underline{F}$ escalating rapidly for higher n. Although these values of $\underline{F}$ are smaller if $\alpha_i \neq 1$, we have separately determined that unless variation in α_i is quite large, $\underline{F}$ remains extremely high (Lambert, W.C., et al., manuscript in preparation). These formulations are, of course, subject to such other considerations as genetic drift and founder effects, which may modify the results obtained from them in any specific instance.

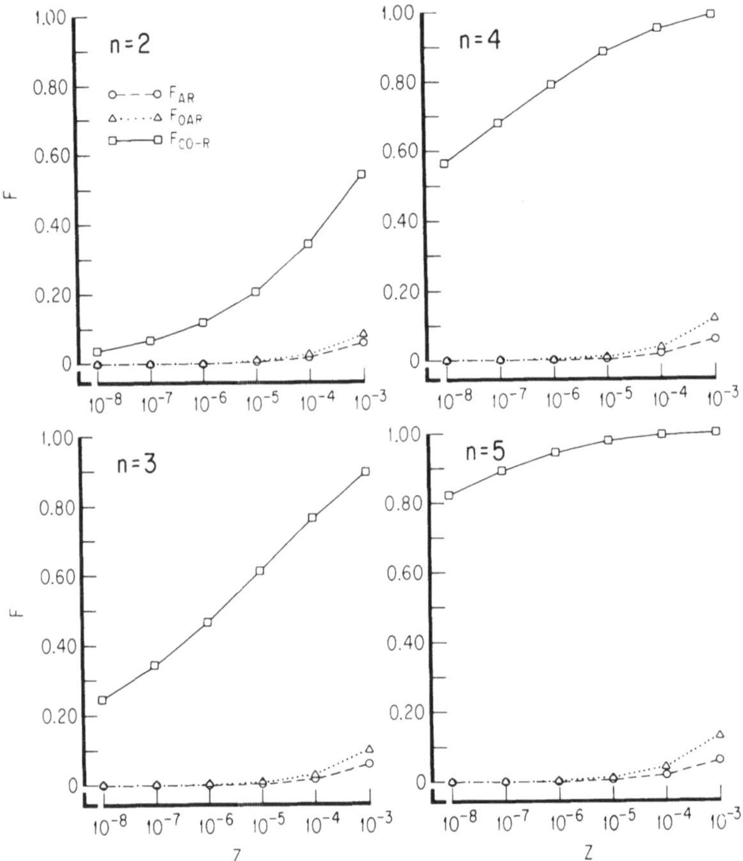

Figure 1. Carrier frequencies for defective alleles in clinically normal individuals for traits of incidence, Z, with carrier frequencies, F_{AR}, autosomal recessive inheritance; F_{OAR}, oligogenic autosomal recessive inheritance; and F_{C-RI}, co-recessive inheritance. All oligogenic autosomal recessive and co-recessive carrier frequencies computed assuming all involved genes equally prevalent. n, number of loci involved. 1.0 = 100% of the normal population.

The above formulations, and many of those that follow, apply to co-recessive genes which are themselves inherited in an autosomal recessive manner. Slight modifications would be necessary for co-recessive genes transmitted by sex-linked recessive inheritance. The above computations also assume a somewhat idealized and therefore simple system. In nature, the system is likely to be more complex, with, for example, n varying depending upon which defective alleles are present at which co-recessive loci.

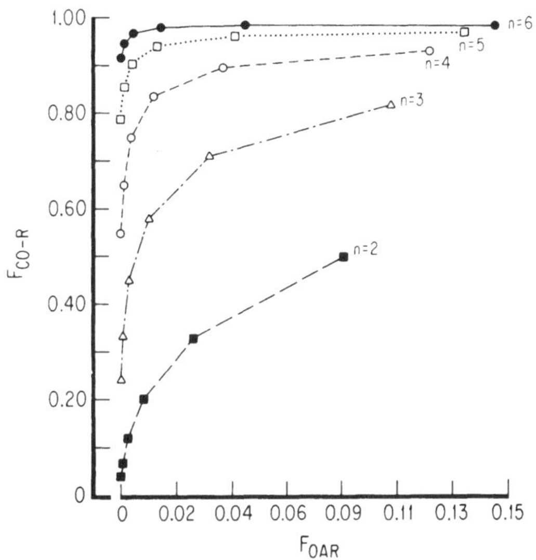

Figure 2. Direct comparison of the carrier frequencies associated with F_{OAR}, oligogenic autosomal inheritance versus F_{C-RI}, co-recessive inheritance. Note difference in scale. 1.0 = 100% of the population. Points on each line refer to trait expression frequencies of 10^{-8}, 10^{-7}, 10^{-6}, 10^{-5}, 10^{-4}, 10^{-3}, and 10^{-2}, respectively. Values of n, defined as for Figure 1, are as shown. All computations assume equally prevalent genes.

Although the incidence of XP is not precisely known at present, it has been estimated to be approximately 2×10^{-6}, whereas the incidence of ataxia-telangiectasia, another inherited disease with defective DNA repair and an increased incidence of cancer, has been estimated to be approximately 25×10^{-6} (Kraemer, et al., 1980; 1987; Swift, et al., 1987). Localization of an ataxia telangiectasia gene on human chromosome 11 has recently been reported (Gatti, et al., 1989). However, those authors did not consider the possibility that more than one gene may have been involved in the patients they studied. The co-recessive inheritance model predicts that carriers for these diseases might easily constitute a substantial proportion of the general population, even if only one or more complementation groups of one or a few rare disorders are transmitted by this mechanism.

The Co-recessive Inheritance Model and Xeroderma Pigmentosum

The co-recessive inheritance model offers explanations for a number of old and recent observations in XP that are difficult, although not impossible, to reconcile with autosomal recessive inheritance: All of the members of two different complementation Groups, B (of which there is only one member) and H, of XP have also had unequivocal phenotypic expression of a quite distinct, light-sensitive, DNA-repair deficient, recessively transmitted disorder, Cockayne's syndrome (Kraemer, 1980, 1981; Kraemer, et al., 1987; Lehmann, 1987; Lambert, et al., 1985; Lambert and Lambert, 1987; Moshell, et al., 1983; Robbins, et al., 1983; Schweiger, et al., 1987). If the Cockayne's syndrome locus were a co-recessive locus for these complementation groups of XP or if the two diseases were both co-recessively transmitted but had one or more common co-recessive locus, this puzzling finding would be readily accounted for. An alternative explanation is that the two diseases may be due to defects in pathways that share enzymes (Lehmann, 1987).

The well-documented X-linked inheritance of XP in one Italian family (Stefanini, et al., 1980) would be readily explained by the model if it were postulated that at least one co-recessive locus were on the X chromosome. Marked differences in severity as well as different manifestations of XP among affected siblings could be explained by postulating that one or more of their affected loci were different. The very high carrier frequencies postulated by the model would provide a source for affected alleles at different loci in this instance. These high carrier frequencies might also provide an explanation for the apparent dominant transmission of XP in one family (Anderson and Begg, 1950), because they would markedly increase the chance that successive generations of a family would inherit defective alleles at at least n loci.

Depressed UDS following UV irradiation in vitro has been reported in cells from individuals who did not show phenotypic XP: a first degree relative of an XP patient (Kraemer, Slor and Andrews, 1983), a patient with an entirely different sun-sensitive disorder, hidroa vacciniforme (Andrews, Halasz and Poh-Fitzpatrick, 1983) and a number of patients with yet another disease, which is sometimes sun-sensitive, trichothiodystrophy (Lehmann, et al., 1988; Nuzzo, et al., 1988; Stefanini, et al., 1987). The co-recessive model, with its high carrier frequencies, would readily account for these findings if it were also postulated that n required for clinical XP were larger than n required for abnormal UDS. Abnormal UDS would, then, be particularly likely to occur in close relatives of XP patients, and, because of the high carrier frequencies, could also occur in a small but significant proportion of the remainder of the population, who might show some manifestation of light sensitivity other than XP.

Finally, the fact that the XP "gene" has proven extremely difficult to transvect from normal human cells into XP cells in culture to restore normal UDS following UV irradiation (Keijzer, et al., 1987; Lambert, et al., 1988) despite one earlier claim of success (Takano, et al., 1982), could be explained if transfer of two or more genes were necessary. A mouse gene that restores the XP defect in complementation group A has recently been successfully transvected into the genome of XP cells in culture (Tanaka, et al., 1989), but only partial restoration was achieved. In transvection experiments, only a very small proportion of the recipient cell population is successfully transvected. If two or more genes have to be transvected, the proportion of such successful multiple transvections would be likely to be too low to be detectable. There are other explanations, however, for the difficulties that have been encountered, for example, the normal human gene may simply be too large to transvect easily.

Recent Evidence for Additional Defects in XP

Recently there have been well documented reports of additional abnormalities associated with XP which do not appear to involve defective DNA repair mechanisms. Norris, et al. (1988) have shown that some XP patients have a functional defect in their killer T cells, and a defective cellular catalase activity has been reported in cells from an XP patient in complementation group A (Crawford, et al., 1988), although they found no deficiency in cells from patients in groups C and D or the variant group. Since catalase is an enzyme active in phagocytosis and cell to cell killing mechanisms, this enzyme defect may be the underlying mechanism for the killer T cell deficiency. Alternatively, these may constitute separate deficiencies.

These non-DNA repair related defects in XP potentially expand the scope of the C-RI model. They offer an additional possible explanation for the differences among XP patients within complementation groups, discussed above. Beyond this, however, they do not contribute to the anticipated number of complementation groups in XP or in other DNA repair-deficient diseases as determined by DNA repair assays. On the other hand, the above formulations regarding carrier frequencies, and those given below regarding sibling frequencies and consanguinity rates, are as affected by these non-DNA repair defects as are deficiencies in DNA repair genes. Thus, the existence of these additional defects in XP implies that the carrier frequencies, not only for those genes but also for affected DNA repair genes, are much higher than would be expected if these non-DNA repair deficiencies were not present and presumably required to produce clinical E-DXP.

Baysian Analysis of Frequencies of Affected Siblings in Disorders Transmitted by Co-Recessive Inheritance

A disease or trait inherited in a co-recessive manner

would be expected to be associated with a low frequency of affected siblings, since a sibling of an affected individual, to be affected, must inherit 2n defective alleles, rather than 2 such alleles. For autosomal recessive inheritance, with both parents heterozygous, the chance of each parent passing a defective allele is 1/2, so that the chance of a sibling having the trait is $(1/2)^2 = 1/4$. Were the chance of a co-recessive gene being passed also 1/2, the chance of a sibling having the disease, considering just the n loci at which the proband was homozygous for defective alleles, would then be $(1/2)^{2n}$.

Three factors tend to increase the sibling frequency, defined here as the chance of a sibling of an affected individual having the disease, however, above $(1/2)^{2n}$. First, either parent may be homozygous for defective alleles at one or more of the n co-recessive loci at which the proband is homozygous for defective alleles. Second, if the sibling were not homozygous (or hemizygous) for defective alleles at one or more the n loci at which the proband is homozygous (or hemizygous), he or she may nevertheless be homozygous or hemizygous for defective alleles at one or more of the remaining loci, of which there are N-n possibilities. Third, the father may be hemizygous, or the mother homozygous, for a relevant defective allele at an X-linked locus. All of these factors are made much more significant by the prediction that the defective alleles in co-recessive inheritance systems are extremely common.

Applying Baysian analysis to the former of these factors, if the frequency of a co-recessive allele, i, in the population is q_i, then, from the Hardy-Weinberg relationship (Mathematical Appendix I), the proportion of individuals homozygous for this allele is q_i^2 and that of heterozygous carriers is $2q_i(1-q_i)$. The parents of the proband must be one or the other of these, barring a new mutation. Also since the probability of a homozygous and heterozygous parent passing a defective allele to the proband is 1 and 1/2, respectively, the chance that each parent of an XP patient is homozygous for each of the autosomally transmitted co-recessive alleles is given by:

$$f_{hom} = \frac{q_i^2}{q_i^2 + 1/2\ 2q_i\ (1-q_i)} = q_i \qquad (9)$$

The chance that the parent would transmit a defective allele at this locus to a sibling is 1X chance of being homozygous + 1/2X chance of being heterozygous, which is given by:

$$f_{sib} = 1\ (q_i) + 1/2\ (1-q_i) = \frac{1 + q_i}{2} \qquad (10)$$

Thus, considering only the n loci at which the proband is homozygous for defective alleles, the expected frequency of the trait in a sibling is given by:

$$F_{sib} = \left\{\frac{1}{2}\right\}^{2n} \prod_{i=1}^{n} (1+q_i)^2 \qquad (11)$$

The chance of acquiring an additional autosomally transmitted co-recessive gene, j, in a homozygous configuration at a single locus not involved in the proband is q_j^2. The overall chance of acquiring such an additional co-recessive gene is thus:

$$F_{add} = 1 - \prod_{j=1}^{N-n} (1-q_j^2) \qquad (12)$$

This factor raises F_{sib} slightly above the value given above.

For a sex-linked co-recessive gene on the X chromosome, the probability that the mother is homozygous is q_{sl}, the prevalence of this defective allele at this locus. The chance of her transmitting this gene to another son is therefore:

$$f_{sib,sl,m} = 1\,(q_{sl}) + 1/2\,(1 - q_{sl}) = \frac{1 + q_{sl}}{2} \qquad (13)$$

The probability of both parents transmitting this defective allele to a daughter is this probability times q_{sl}, the chance that the father is a carrier; this is:

$$f_{sib,sl,f} = q_{sl}\left\{\frac{1 + q_{sl}}{2}\right\} = \frac{q_{sl} + q_{sl}^2}{2} \qquad (14)$$

These probabilities are also valid for co-recessive sex-linked loci at which the proband is not hemizygous or homozygous, but at which the mother is a carrier for a defective allele.

These probabilities would vary widely for different families and complementation groups of various diseases. They are also susceptible to introduction of bias by other factors. Because of these constraints examination of the sibling frequency of a disease is of limited value in assessing whether it is transmitted by co-recessive inheritance, even though the expected sibling frequency is much lower than for autosomal or sex-linked recessive inheritance.

<u>Baysian Analysis of Consanguinity in Diseases Transmitted by Co-recessive Inheritance</u>

One method of testing the validity of the co-recessive inheritance model is to examine pedigrees of individuals to determine the frequency and degree of consanguinity. For disorders transmitted as a co-recessive trait, this is expected to be much lower than for autosomal recessive disorders of the same frequency occurring in the same or a similar population, as the following theoretical considerations make clear:

In performing our pedigree analysis, we will limit our considerations of inbreeding to first cousin marriages. This is done for several reasons:

1. Pedigree data, as obtained by physicians reporting to the Xeroderma Pigmentosum Registry or compiling case reports, is usually fairly accurate for first cousin, or closer, consanguineous matings in parents of individuals with xeroderma pigmentosum, but accuracy and reliability fall off dramatically for more distant relationships. This is likely also to be true for other inherited diseases.

2. Matings consanguineously closer than first cousin are rare in most cultures, including all American, European and Japanese population groups under study.

3. Matings consanguiniously more distant than first cousin produce lower inbreeding coefficients and are individually less significant, although in aggregate they may be important.

4. We can relate our counts of first cousin matings with the same statistic in the relevant background population in each case.

5. First cousin relationships are much easier to deal with theoretically than are total consanguinity relationships.

For an autosomal recessive disorder of prevalence, Z, associated with a recessive allele, a, of frequency $q = Z^{1/2}$, occurring in a population in which the proportion of first cousin matings is C, the expected (Baysian) proportion of first cousin relationships in the parents of affected individuals, K, is given by the Weinberg-Dahlberg equation (Mathematical Appendix III) as:

$$K = \frac{C\ (15\ q + 1)}{C + q\ (16-C)} \qquad (15)$$

which may also be expressed as:

$$K = \frac{C\ (15\ Z^{1/2} + 1)}{C + Z^{1/2}\ (16-C)} \qquad (16)$$

For a co-recessively transmitted condition, however, in which n loci are involved (Mathematical Appendix III):

$$K = \frac{C \prod_{i=1}^{n} (15q_i + 1)}{(2)^{4n} (1-C) \prod_{i=1}^{n} q_i + C \prod_{i=1}^{n} (15q_i + 1)} \tag{17}$$

For the special case in which all q_i are equal, this becomes:

$$K = \frac{C (15q_i + 1)^n}{(2)^{4n} (1-C) q_i^n + C (15q_i + 1)^n} \tag{18}$$

which, since under these conditions $q_i^{2n} = Z$, may also be expressed as:

$$K = \frac{C (15 Z^{1/2n} + 1)^n}{(2)^{4n} (1-C) Z^{1/2} + C (15 Z^{1/2n} + 1)^n} \tag{19}$$

These equations (15-19) apply to a single complementation group. Where more than one complementation group is under consideration, the contributions of each must be summed together using the above functions for each complementation group separately.

These relationships indicate that the expected frequency of first cousin matings in parents of patients with xeroderma pigmentosum is far lower for complementation groups in which inheritance is by the co-recessive mechanism compared with autosomal recessive inheritance. For the special case in which all $\alpha_i = 1$, we have, for $Z = 2X10^{-6}$, the current best estimate for the overall frequency of xeroderma pigmentosum (from Eqn. 19):

C	K, AUTOSOMAL RECESSIVE	K, CO-RECESSIVE n = 2	n = 3
.01	.35	.06	.03
.02	.51	.12	.06
.03	.60	.17	.09
.04	.68	.22	.12
.05	.72	.26	.15

with 1.0 = 100% of the population.

The overall incidence of first cousin marriages in the U.S. is probably approximately C = 0.01 (Levitan and Montagu, 1971). Thus, even though background levels of C for each population in which xeroderma pigmentosum patients occurs is a factor, the difference in K for co-recessive versus autosomal recessive inheritance is sufficiently great to allow epidemiologic analysis to show a difference. In the above computations we accepted $Z = 2X10^{-6}$ for the complementation group under consideration, it should actually be lower since Z holds for all xeroderma pigmentosum. On the other hand, Z itself is an estimate. A lower value of Z enhances the above differences.

The C-RI model also predicts that, since homozygosity for alleles in parents of patients with XP is a predisposing factor, consanguinity in their parents becomes a factor, whereas this is never the case in autosomal recessive inheritance. We are aware of at least one family in which this has, in fact, been found. This type of pedigree analysis is a promising approach for testing the co-recessive inheritance hypothesis. Studies underway by the Xeroderma Pigmentosum Registry, and by other registries, should help to approach this problem.

Testing the Model in the Laboratory

Definitive testing of the co-recessive inheritance model in the laboratory is likely to prove a difficult task, although eventually the delineation of human repair processes at the molecular level should provide the answer. One must first of all choose the correct individuals or cell lines for study. The model may not apply to all complementation groups of E-DXP, and some cases of other inherited disorders, particularly those associated with defective DNA repair and/or replication, such as Cockayne's disease, ataxia-telangiectasia, Blooms' syndrome, and Fanconi's anemia, may also be transmitted in this manner. Findings supporting the co-recessive model would include detection of multiple biochemical defects possibly related to DNA-repair mechanisms in cells within a single complementation group of E-DXP, ability to restore normal DNA repair capability in E-DXP cells of one complementation group by transfer of any of more than one gene or gene products, and requirement of transfer of more than one gene or gene product to an E-DXP cell to completely restore normal DNA-repair capability in that cell. Failure of one or more E-DXP cell lines to complement cells in more than one complementation group of E-DXP, thus overlapping complementation groups, would provide convincing evidence for the validity of the C-RI model, and is specifically predicted by it, whereas it is inherent in autosomal recessive inheritance models that no such overlapping complementation group exists.

We and others have found abnormalities in activities, or their cofactors, of more than one endonuclease in XPA cells (Bickley and Lambert, 1988; Kuhnlein, et al., 1976; Lambert, et al., 1983, 1988; Lambert and Parrish, this volume; Witte and Thielmann, 1979), and abnormalities in XPA proteins have been reported (Amari, et al., 1983; Bickley and Lambert, 1988; Kuhnlein, et al., 1983). There is also some evidence that the mechanism for repair of pyrimidine dimers is not precisely the same as that for repair of 6-4 photoproducts induced by UV radiation in cellular DNA; thus failure to repair these adducts may represent two different defects present in E-DXP (Cleaver, et al., 1987; Mitchell, et al., 1985). The recently discovered additional killer T cell and catalase deficiencies in XP have already been referred to. However, all of these defects could conceivably be due to defective alleles at a single locus, with numerous alterations resulting form this single deficiency. In individual cases, there could also be deletion mutations involving several, or even multiple, loci.

The standard practice for complementation group analysis in most laboratories at present is to discontinue further testing once a cell line fails to complement indicator cells from a defined group, so that cell lines that overlap complementation groups may exist even among those that have already been assigned. The partial or lack of success, despite attempts by numerous very competent laboratories, to transvect human genes defective in E-DXP cells has already been discussed. The partial complementation reported by a number of laboratories could be due to one of several genes or gene products being replaced, consistent with the C-RI model. However, this may be due simply to technical factors. On the other hand, they may have obtained spuriously high levels of correction due to introducing extra amounts of their factor(s). We have recently found greater than 100% correction in XPA cells by introducing the enzymes we have isolated from normal human chromatin by electroporation (Tsongalis, et al., submitted). This may be due to the introduction of enzyme into XP cells above the amount present in normal cells. Finally, the validity of the model may be tested by comparison of pedigrees of E-DXP, obtained form the literature and from such sources of data as the Xeroderma Pigmentosum Registry, which is maintained at the office of one of us (WCL), with predictions based on Baysian probabilities as discussed above.

The Co-recessive Inheritance Model and Common Disorders

Although the high carrier frequencies of defective alleles predicted for co-recessive E-DXP loci are consistent with predictions of other oligogenic models of human disease, we are unaware of another model in which the discrepancy between carrier frequency and disease prevalence is nearly as great as for the co-recessive model. Moreover, these very high carrier frequencies have significant implications not only for the pathophysiology of E-DXP but also for a number of broader questions, such as DNA repair mechanisms, mutagenesis, and mutation-induced carcinogenesis, teratogenesis and possibly such other processes as aging, that relate to it.

Cairns (1981) has argued, noting that E-DXP patients have not been found to have an immediately evident large increase in morbidity or mortality from the types of internal cancer that commonly affect the general population, that the "commonly fatal cancers are not caused by the kind of lesions of DNA that XP patients are specifically unable to repair". Against this, however, is the fact that XP cells have been found to be defective in their response not only to UVR but also to a number of other agents which damage DNA (Amacher and Lieberman, 1977; Bredberg, et al., 1982; Cleaver, 1985; Lambert and Parrish, this volume; Leadon, et al., 1981). Moreover, because of the very high carrier frequencies for defective alleles predicted by the co-recessive model, development of internal cancers in carriers might be related both to the carrier state and to chance environmental exposure to certain carcinogens, and E-DXP patients would not be at so markedly an increased risk to develop such cancers as might otherwise be anticipated. Indeed, some normal individuals might well carry as many or more defective alleles at co-recessive loci related to E-DXP than some E-DXP patients. Thus, although E-DXP patients might be expected to be more susceptible to internal cancer than normal individuals, the difference might be expected to be detectable only by detailed epidemiologic analysis, which has not yet been done. Such a relatively subtle increase has been reported, however, by Kraemer, et al. (1984; 1987), using the limited data on E-DXP available at present, although it has not been found in all studies (English and Swerdlow, 1987; Pippard, et al., 1988).

Close relatives of patients with the repair-deficient disease, ataxia telangiectasia, have been reported to be at increased risk for cancer and possibly other diseases as well in some but not all studies (Pippard, et al., 1988; Swift and Chase, 1983; Swift, et al., 1987). Moreover, considerable evidence for defective DNA repair in otherwise healthy, but cancer-prone, individuals has now been found: Abo-Darub, et al. (1978) reported decreased DNA repair levels in lymphocytes cultured from otherwise healthy patients with the sun-induced pre-cancerous cutaneous lesion, actinic (solar) keratosis. Hypersensitivity to G2 chromatid radiation damage has been reported in the common, dominantly inherited, melanoma-prone condition, familial dysplastic nevus syndrome (Sanford, et al., 1987). Kovacs and Langemann (1988) reported defective DNA repair in a large kindred in which there was an increased occurrence of cancer. A number of authors have reported markedly heterogeneous levels of susceptibility to mutagens and/or carcinogens in the normal human population, and have noted that these are associated with a predisposition for the development of various types of cancer (Gantt, et al., 1986; Glickman, 1980; Hsu, et al., 1985; 1986; 1989; Parshad, et al., 1983; Pero, et al., 1983). If, as has been proposed by some authors (Cairns, 1981), carcinogenesis is a multi-step process, with a series of heritable changes occurring in cells destined to become cancerous, defective DNA repair may play a role at more than one of them. This would have the effect of compounding the influence of the defect on the carcinogenic process. This may also be true for other disease etiologies possibly related to deficiencies in DNA repair systems. On the other hand, as Lehmann (1987) has pointed out, at least one disease, Cockayne's syndrome, is associated with defective DNA repair without a documented increase in cancer. DNA repair and mutagenesis is also likely to be involved in the pathophysiology of aging (Gensler and Bernstein, 1981; Kirkwood, 1989; Park and Ames, 1988). Moreover, these mechanisms may play a role in the etiopathogenesis of diseases not thought in the past to be related to DNA repair and/or mutagenesis. A possible association between mutagenesis and atherosclerosis has also been suggested (Bridges, 1987).

The co-recessive inheritance model may, in addition to its possible relationship to DNA repair mechanisms, also be applicable for any biochemical or cellular process for which partially or totally overlapping mechanisms may exist, and thus may relate to a number of disease processes in addition to those we have discussed. Teliologically, such overlapping mechanisms would be most likely to exist for the most critical functions of the organism, so that carriers of defective alleles at these loci would be likely to develop other problems. Because of the high carrier frequencies associated with even very rare disorders, with only two or a few genes involved, inherited by corecessive inheritance (Figures 2 and 3), the model has important implications for human disease even if it applies for just one or more complementation groups of only one or a very few rare diseases.

ACKNOWLEDGEMENT

The valuable assistance of D.D. Parrish, W.S. Tanz, and G.J. Tsongalis in the preparation of the figures and in numerous discussions of the C-RI model is gratefully acknowledged. Supported by N.I.H. grant AM35148.

REFERENCES

Abo-Darub, J.M., R. Mackie and J. Pitts, 1978. DNA repair deficiency in lymphocytes from patients with actinic keratoses, Bull. Cancer (Paris), 65:357-362.

Amacher, D.E. and M.W. Lieberman, 1977. Removal of acetylaminofluorene from the DNA of control and repair-deficient human fibroblasts, Biochem. Biophys. Res. Commun., 74:285-290.

Amari, N.M.B., W.C. Lambert and M.W. Lambert, 1983. Differences in nuclear non-histone proteins in xeroderma pigmentosum and normal human lymphoblastoid cells, Fed. Proc., 42:1296.

Anderson, T.E. and M. Begg, 1950. Xeroderma pigmentosum of mild type, Br. J. Dermatol., 62:402-407.

Andrews, A.D., 1983. Xeroderma pigmentosum, in: J.L. German III (Ed.), Chromosome Breakage and Repair, Alan R. Liss, New York, pp. 63-83.

Andrews, A.D., S.F. Barnett and J.H. Robbins, 1978. Xeroderma pigmentosum neurological abnormalities correlate with colony-forming ability after ultraviolet irradiation, Proc. Natl. Acad. Sci. USA, 75:1984-1988.

Andrews, A.D., C.L.G. Halasz and M.B. Poh-Fitzpatrick, 1983. Abnormally low UV-induced unscheduled DNA synthesis in cells from a patient with hydroa vacciniforme, J. Cell. Biochem., Suppl. 7B:209.

Bickley, L.K. and M.W. Lambert, 1988. DNA binding of an apurinic/apyrimidinic DNA endonuclease activity from xeroderma pigmentosum cells, Cell Biol. Int. Rep., 12:231-237.

Bohr, V.A., D.H. Phillips and P.C. Hanawalt, 1987. Heterogeneous DNA damage and repair in the mammalian genome, Cancer Res., 47:6426-6436.

Bredberg, A., B. Lambert and S. Soderhall, 1982. Induction and repair of psoralen cross-links in DNA of normal human and xeroderma pigmentosum fibroblasts, Mutation Res., 93: 221-234.

Bridges, B.A., 1987. International Commission for Protection against Environmental Mutagens and Carcinogens. Topic No. 1. Are somatic mutations involved in atherosclerosis? Mutat. Res., 182:301-302.

Cairns, J., 1981. The origin of human cancers, Nature (London), 289:353-357.

Cleaver, J.E., 1985. DNA repair deficiences, in: R. Fleischmajer (Ed.), Progress in Diseases of the Skin, Vol. 2, Grune and Stratton, New York.

Cleaver, J.E. and D. Bootsma, 1975. Xeroderma pigmentosum: biochemical and genetic characteristics, Annu. Rev. Genet., 9:19-38.

Cleaver, J.E., F. Cortes, L.H. Lutze, W.F. Morgan, A.N. Player and D.L. Mitchell, 1987. Unique DNA repair properties of a xeroderma pigmentosum revertant, Molec. Cell Biol., 7:3353-3357.

Crawford, D., I. Zbinder, R. Moret and P. Cerutti, 1988. Antioxidant enzymes in xeroderma pigmentosum fibroblasts, Cancer Res., 48:2132-2134.

De Weerd-Kastelein, E.A., W. Keijzer and D. Bootsma, 1972. Genetic heterogeneity of xeroderma pigmentosum demonstrated by somatic cell hybridization, Nature (London), 238: 80-83.

English, J.S. and A.J. Swerdlow, 1987. The risk of malignant melanoma, internal malignancy and mortality in xeroderma pigmentosum patients, Br. J. Dermatol., 117:457-461.

Falk, C.T., N.R. Mendell and P. Rubinstein, 1983. Effect of population associations and reduced penetrance on observed and expected geneotype frequencies in a simple genetic model: Application to HLA and insulin dependent diabetes mellitus, Ann. Hum. Genet., 47:161-165.

Feller, G.W., 1960. An introduction to Mathematical Probability and its Applications, Vol. 1, Wiley, New York, 67-88.

Fischer, E., W. Keijzer, H.W. Thielman, O. Popanda, E. Bohnert, L. Elder, E.G. Jung and D. Bootsma, 1985. A ninth complementation group in xeroderma pigmentosum, XPI, Mutation Res. (DNA Repair Reports), 145:217-225.

Friedberg, E.C., 1988. Deoxyribonucleic acid repair in the yeast Saccharomyces cerevisiae, Microbiol. Rev., 52:70-102.

Friedberg, E.C., C.T.M. Anderson, T. Bonura, R. Cone, E.H. Rodany and R.J. Reynolds, 1981. Recent developments in the enzymology of excision repair of DNA, Prog. Nucl. Acid Res. Mol. Biol., 26:197-215.

Gantt, R., R. Parshad, M.P. Floyd and K.K. Sanford, 1986. Biochemical evidence for deficient DNA repair leading to enhanced G_2 chromatid radiosensitivity and susceptibility to cancer, Rad. Res., 108:117-126.

Gatti, R.A., I. Berkel, E. Boden, G. Braedt, P. Charmley, P. Concannum, F. Ersey, et al., 1989. Localization of an ataxia telangiectasia gene to chromosome 11q 22-23, Nature, 336:577-580.

Gensler, H.L. and H. Bernstein, 1981. DNA damage as the primary cause of aging. Quart. Rev. Biol., 56:279-303.

German J.L., III, W.M. Generoso, M.D. Shelby and F.J. de Serres (Eds.), 1980. DNA Repair and Mutagenesis in Eukaryotes, Liss, New York, 429-439.

Glickman, B.N., 1980. DNA repair and its relationship to the origins of human cancer, In: F.J. Cleton and J.W.I.M. Simmons (eds.), Genetic Origin of Tumor Cells, M. Nijhoff, The Hague, 25-51.

Haynes, R.H. and B.A. Kunz, 1981. DNA repair and mutagenesis in yeast, in: J. Strathern, E. Jones and J. Broach (Eds.), The Molecular Biology of the Yeast Saccharomyces, Cold Spring Harbor Laboratory, Cold Spring Harbor, NY, 371-414.

Hodge, S.E., 1981. Some epistatic two-locus models of disease, I. Relative risks and identity-by-descent distributions in affected sib pairs, Am. J. Hum. Genet., 33:381-395.

Hodge, S.E. and M.A. Spence, 1981. Some epistatic two-locus models of disease, II. The confounding of linkage and association, Am. J. Hum. Genet., 33: 396-406.

Hsu, T.C., D.A. Johnston, L.M. Cheng, D. Ramkissom, S.P. Schwartz, J.M. Jessup, R.J. Winn, L. Shirley and C. Furlong, 1989. Sensitivity to genotoxic effects of bleomycin in humans: Possible relationship to environmental carcinogenesis, Int. J. Cancer, 43:403-409.

Hsu, T.C., D. Ramkissoon and C. Furlong, 1986. Differential susceptibility to a mutagen among human individuals: synergistic effect on chromosome damage between bleomycin and aphidicolin, Anticancer Res., 6:1171-1176.

Hsu, T.C., L.M. Cherry and N.A. Samaan, 1985. Differential mutagen susceptibility in cultured lymphocytes of normal individuals and cancer patients, Cancer Genet. Cytogenet., 17: 307-313.

Ichihashi, M, K. Yamamura, T. Hiramoto and Y. Fujiwara, 1988. No apparent neurologic defect in a patient with xeroderma pigmentosum complementation group D, Arch. Dermatol., 124:256-260.

Jung, E.G., 1986. Xeroderma pigmentosum, Int. J. Dermatol., 25:629-633.

Kano, Y. and Y. Fujiwara, 1983. Defective thymine dimer excision and its characteristic catalysis by cell-free extracts, Carcinogenesis, 4:1419-1424.

Keijzer, W., M. Stefanini, D. Bootsma, A. Verkerk, A.H. Geurts van Kessel, J.F. Jongkind and A. Westerveld, 1987. Localization of a gene involved in complementation of the defect in xeroderma pigmentosum group A cells on human chromosome I, Exp. Cell Res., 167:490-501.

Kirkwood, T.B.L., 1989. DNA, mutations and aging, Mutation Res., 219:1-7.

Kovacs, E. and H. Langemann, 1988. Defective DNA repair in a large family having a high occurrence of cancer, Oncology, 45:444-447.

Kraemer, K.H., 1980. Xeroderma pigmentosum: a prototype disease of environmental-genetic interaction, Arch. Dermatol., 116:541-542.

Kraemer, K.H., 1981. Xeroderma pigmentosum, in: D.J. Demis, R.L. Dobson and J. McGuire (Eds.), Clinical Dermatology, Harper and Row, Hagerstorm MD, Unit 19-7, 1-33.

Kraemer, K.H., H. Slor and A. Andrews, 1983. A new form of xeroderma pigmentosum: Reduced DNA repair without neoplasia, J. Invest. Dermatol., 80:331.

Kraemer, K.H., M.M. Lee and J. Scotto, 1984. DNA repair protects against cutaneous and internal neoplasia: Evidence from xeroderma pigmentosum, Carcinogenesis, 5: 511-514.

Kraemer, K.H., M.M. Lee and J. Scotto, 1987. Xeroderma pigmentosum. Cutaneous, ocular, and neurologic abnormalities in 830 published cases. Arch. Dermatol., 123:241-250.

Kuhnlein, U., S.S. Tsang, O. Lokken, S. Tong and D. Twa, 1983. Cell lines fron xeroderma pigmentosum complementation group A lack a single-stranded-DNA-binding activity, Bioscience Reports, 3:667-674.

Kuhnlein, U., E.E. Penhoet and S. Linn, 1976. An altered apurinic DNA endonuclease activity in group A and group D xeroderma pigmentosum fibroblasts, Proc. Natl. Acad. Sci. USA, 73:1169-1173.

Lambert, C., L.B. Couto, W.A. Weiss, R.A. Schultz, L.H. Thompson and E.C Friedberg, 1988. A yeast DNA repair gene partially complements defective excision repair in mammalian cells, EMBO J., 7:3245-3253.

Lambert, M.W., W.C. Lambert and D.E. Lee, 1983a. Defective DNA endonuclease activities on DNA damaged by intercalating drugs in xeroderma pigmentosum cells, J. Cell. Biochem., Suppl. 7B:201.

Lambert, M.W., W.C. Lambert and A.O. Okorodudu, 1983b. Nuclear DNA endonuclease activities on partially apurinic/-apyrimidine DNA in normal human and xeroderma pigmentosum lymphoblastoid and mouse melanoma cells, Chem. Biol. Interact., 46:109-120.

Lambert, W.C., A.D. Andrews, J. German and K.H. Kraemer, 1985. The etiology and pathogenesis of xeroderma pigmentosum, in: B. Theirs and R. Dobson (Eds.), The Etiology of Skin Diseases, Churchill-Livingstone, New York.

Lambert, W.C. and M.W. Lambert, 1985. Co-recessive inheritance: A model for DNA repair, genetic disease and carcinogenesis, Mutat. Res. (DNA Repair Reports), 145: 227-234.

Lambert, W.C. and M.W. Lambert, 1987. DNA repair deficiency and cancer in xeroderma pigmentosum, Cancer Rev., 7:56-81.

Lawrence, C.W., 1982. Mutagenesis in Saccharomyces cerevisiae, Adv. Genet., 21:173-254.

Leadon, S.A., R.M. Tyrrell and P.A. Cerutti, 1981. Excision repair of aflatoxin B1-DNA adducts in human fibroblasts, Cancer Res., 41:5125-5129.

Lehmann, A.R., 1987. Cockayne's syndrome: Defective DNA repair without cancer, Cancer Rev., 7:82-101.

Lehmann, A.R. and P. Karran, 1981. DNA repair, Int. Rev. Cytol., 72:101-146.

Lehmann, A.R., C.F. Arlett, B.C. Broughton, S.A. Harcourt, H. Steingrimadottir, M. Stefanini, A. Malcolm, R. Taylor, T. Natarajan, S. Green, et al., 1988. Trichothiodystrophy, a human DNA repair disorder with heterogeneity in the cellular response to ultraviolet light, Cancer Res., 48:6090-6096.

Levitan M. and A. Montagu, 1971. Textbook of Human Genetics. Oxford University Press, New York, 513.

Lindahl, T., 1982. DNA repair enzymes, Annu. Rev. Biochem., 51:61-87.

Mitchell, D.L., C.A. Haipek and J.M. Clarkson, 1985. (6-4) Photoproducts are removed from the DNA of UV-irradiated mammalian cells more efficiently than cyclobutane pyrimidine dimers, Mutat. Res., 143:109-112.

Mortelmans, K., E.C. Friedberg, H. Slor, G. Thomas and J.E. Cleaver, 1976. Defective thymine dimer excision by cell-free extracts of xeroderma pigmentosum cells, Proc. Natl. Acad. Sci. (U.S.A.), 73:2757-2761.

Moshell, A.N., R.E. Tarone, S.A. Newfield, A.D. Andrews and J.H. Robbins, 1981. A simple and rapid method for evaluating the survival of xeroderma pigmentosum lymphoid lines after irradiation with ultraviolet light, In Vitro, 17:299-307.

Moshell, A.N., M.B. Ganges, M.A. Lutzner, H.G. Coon, S.F. Barrett, J.M. Dupuy and J.H. Robbins, 1983. A new patient with both xeroderma pigmentosum and Cockayne syndrome comprises the new xeroderma pigmentosum complementation group H, J. Cell Biochem., Suppl. 7B:202.

Norris, P.G., G.A. Limb, A.S. Hamblin and J.L.M. Hawk, 1988. Impairment of natural-killer-cell activities in xeroderma pigmentosum, N. Engl. J. Med., 319:1668-1669.

Nuzzo, R., M. Stefanini, M. Rocchi, A. Casati, R. Colognola, P Lagomarsini, S. Marinoni and R. Scozzari, 1988. Chromosome and blood marker studies in families of patients affected by xeroderma pigmentosum and trichothiodystrophy, Mutat. Res., 208-159-161.

Okorodudu, A.O., W.C. Lambert and M.W. Lambert, 1982. Nuclear deoxyribonuclease activities in normal and xeroderma pigmentosum lymphoblastoid cells, Biochem. Biophys. Res. Commun., 108:576-584.

Parshad, R., K.K. Sanford and G.M. Jones, 1983. Chromatin damage after G_2 phase X-irradiation of cells from cancer prone individuals implicates deficiency in DNA repair, Proc. Natl. Acad. Sci. (USA), 80:5612-5616.

Pero, R.W., D.G. Miller, M. Lipkin, M. Markowitz, S. Gupta, S.J. Winawer, W. Enker and R. Good, 1978. Reduced capacity for DNA repair synthesis in patients with or genetically predisposed to colorectal cancer, J. Nat. Cancer Inst., 70:867-875.

Pippard, E.C., A.J. Hall, D.J. Barker and B.A. Bridges, 1988. Cancer in homozygotes and heterozygotes of ataxia-telangiectasia and xeroderma pigmentosum in Britain, Cancer Res., 48: 2929-2932.

Regan, J.D. and R.B. Setlow, 1974. Two forms of repair in the DNA of human cells damaged by chemical carcinogens and mutagens, Cancer Res., 34:3318-3325.

√Robbins, J.H., 1987. Xeroderma pigmentosum. Defective DNA repair causes skin cancer and neurodegeneration (Clinical Conference), J.A.M.A., 260:384-388.

Robbins, J.H., K.H. Kraemer, M.A. Lutzner, B.W. Festoff and H.G. Coon, 1974. Xeroderma pigmentosum: An inherited disease with sun sensitivity, multiple cutaneous neoplasms and abnormal DNA repair, Ann. Int. Med., 80: 221-248.

Robbins, J.H., A.N. Moshell, M.A. Lutzner, M.B. Ganges and J.M. Dupuy, 1983. A new patient with both xeroderma pigmentosum and Cockayne syndrome is in a new xeroderma pigmentosum complementation group, J. Invest. Dermatol., 80:331.

Sanford, K.K., R.E. Tarone, R. Parshad, M.A. Tucker, M.H. Greene and G.M. Jones, 1987. Hypersensitivity to G2 chromatid radiation damage in familial dysplastic naevus syndrome, Lancet, 2(8568):1111-1116.

Sato, K., M. Watatani, M. Ikenaga, T. Kozuba, Y. Kitano, K. Yoshikawa, T. Mimak, J. Abe and T. Sugita, 1987. Sensitivity to UV radiation of fibroblasts from a Japanese group A xeroderma pigmentosum patient with mild neurological abnormalities, Br. J. Dermatol., 116:101-108.

Schweiger, M., B. Auer, H.J. Burtscher, M. Hirsch-Kauffmann, H. Klocker and R. Schneider, 1987. The Fritz-Lippman lecture: DNA repair in human cells. Biochemistry of the hereditary diseases Fanconi's anemia and Cockayne syndrome, Eur. J. Biochem., 165:235-242.

Stefanini M., W. Keijzer, L. Dalpra, R. Elli, M.N. Porro, B. Nicoletti and F. Nuzzo, 1980. Differences in the levels of UV repair and in clinical symptoms in two sibs affected by xeroderma pigmentosum, Hum. Genet., 54: 177-182.

Stefanini, M., P. Lagomarsini, R. Giorgi and F. Nuzzo, 1987. Complementation studies in cells from patients affected by trichothiodystrophy with normal or enhanced UV photosensitivity, Mutat. Res., 191:117-119.

Swift, M. and C. Chase, 1983. Cancer and cardiac deaths in obligatory ataxia-telangiectasia heterozygotes [letter], Lancet, 1 (8332):1049-1050.

Swift, M., P.J. Reitnauer, D. Morrell and C.L. Chase, 1987. Breast and other cancers in families with ataxia-telangiectasia, N. Engl. J. Med., 316:1289-1294.

Takano, T., M. Noda and T.-A. Tamura, 1982. Transvection of cells from a xeroderma pigmentosum patient with normal human DNA confers UV resistance, Nature (London), 296: 269-270.

Tanaka, K., I. Satokata, Z. Ogita, T. Uchida and Y. Okada, 1989. Molecular cloning of a mouse DNA repair gene that complements the defect of group-A xeroderma pigmentosum, Proc. Natl. Acad. Sci. USA, 86:5512-5516.

Thielman, H.W. and I. Witte, 1980. Correlation of the colony-forming abilities of xeroderma pigmentosum fibroblasts with repair-specific DNA incision reactions catalyzed by cell free extracts, Arch. Toxicol., 44:197-207.

Thompson, L.H., E.P. Salazar, K.W. Brookman, C.C. Collins, S.A. Stewart, D.B. Busch and C.A. Weber, 1987. Recent progress with the DNA repair mutants of Chinese hamster ovary cells, J. Cell Sci., 6(Suppl.):97-110.

Tolleshang, H., J.L. Goldstein, W.J. Schneider and M.S. Brown, 1982. Posttranslational processing of the LDL receptor and its genetic disruption in familial hypercholesterolemia, Cell, 30:715-724.

Witkamp, L.R., 1981. HLA and disease: Predictions for HLA haplotype sharing in families, Am. J. Hum. Genet., 33: 776-784.

Witte, I. and H.W. Thielman, 1979. Extracts of xeroderma pigmentosum group A fibroblasts introduce less nicks into methyl methanesulfonate treated DNA than extracts of normal fibroblasts, Cancer Letters, 6:129-136.

MATHEMATICAL APPENDIX I: CARRIER FREQUENCIES IN CO-RECESSIVE INHERITANCE

From the Hardy-Weinberg expression:

$$q_i^2 + 2q_i(1 - q_i) + (1 - q_i)^2 = 1 \qquad (20)$$

with: q_i^2, frequency of individuals in the population homozygous for an allele at locus i; $2q_i(1 - q_i)$, frequency of individuals in the population that are carriers of this allele; $2q_i(1 - 1/2q_i) = f_i$, frequency of individuals in the population either homozygous or heterozygous for this allele at locus i.

Since homozygosity for defective alleles at n loci is required for the trait to be observed:

$$z = \prod_{i=1}^{n} q_i^2 \qquad (21)$$

By definition:

$$\alpha_i = \frac{q_1}{q_i} \; ; \; q_m = \frac{\alpha_r q_r}{\alpha_m} \tag{22}$$

Therefore (. ̇ .):

$$Z = \prod_{k=1}^{n} \left\{ \frac{\alpha_i \, q_i}{\alpha_k} \right\}^2 = \alpha_i^{2n} q_i^{2n} \prod_{k=1}^{n} \alpha_k^{-2} \tag{23}$$

Therefore (. ̇ .):

$$q_i = \frac{Z^{1/2n}}{\alpha_i} \prod_{k=1}^{n} \alpha_k^{1/n} \tag{3}$$

And, since, from the Hardy-Weinberg relationship we have $f_i = 2q_i\ (1 - 1/2\ q_i)$, Eqn. (3) may be expressed as

$$f_i = \frac{2Z^{1/2n}}{\alpha_i} \prod_{k=1}^{n} \alpha_k^{1/n} \left\{ 1 - \frac{Z^{1/2n}}{2\alpha_i} \prod_{k-1}^{n} \alpha_k^{1/n} \right\} \tag{4}$$

Eqn. (5) is derived in a manner analogous to that for Eqn. (24), below (Mathematical Appendix II).

MATHEMATICAL APPENDIX II: DIRECT COMPARISON OF GENE AND CARRIER FREQUENCIES

The incidence, Z, of a trait transmitted by oligogenic autosomal inheritance, which is identical to the risk of an individual having the trait, is related to that of each of its genetic subgroups, Z_i, each associated with homozygosity for a different autosomal recessive gene of frequency q_i, at its own locus, i, by the expression:

$$Z = 1 - \prod_{i=1}^{n} (1 - Z_i) \tag{24}$$

which is readily explained as follows: The probability of not having a trait of a given subgroup, i, is 1 - the probability of having it, since the two must add up to 1. This is $(1 - Z_i)$. The probability of not having the trait in any subgroup is the product of all of these probabilities. This is

$$\prod_{i=1}^{n} (1 - Z_i) \tag{25}$$

The probability of having the trait in any one (or more) genetic subgroup, Z, is 1 - the probability of not having the trait in any genetic subgroup, since the two must add up to unity. This is

$$Z = 1 - \prod_{i=1}^{n} (1 - Z_i) \tag{24}$$

In autosomal recessive inheritance, $Z = q^2$, and $q = Z^{1/2}$ (Mathematical Appendix I). This also holds for each q_i and Z_i in oligogenic autosomal recessive inheritance, since each gene acts independently of the others to produce the trait in homozygous individuals. Thus, $Z_i = q_i^2$; $q_i = Z_i^{1/2}$. Substituting this value of Z_i into Eqn. (24) gives

$$Z = 1 - \prod_{i=1}^{n} (1 - q_i^2) \tag{26}$$

For co-recessive inheritance we have, from Mathematical Appendix I:

$$q_i = \frac{Z^{1/2n}}{\alpha_i} \prod_{k=1}^{n} \alpha_k^{1/n} \tag{3}$$

If all $\alpha_i = 1$, this becomes

$$q_i = Z^{1/2n} \ ; \ Z = q_i^{2n} \tag{27}$$

Substituting this value of Z into Eqn (26) and designating q_i for co-recessive inheritance as $\underline{q}_i$, we have

$$\underline{q}_i^{2n} = 1 - \prod_{i=1}^{n} (1 - q_i^2) \tag{28}$$

which, for all $\alpha_i = 1$; (i.e., for all q_i equal) gives:

$$\underline{g}_i^{2n} = 1 - (1 - q_i^{2})^{n} \qquad (29)$$

algebraic rearrangement of which gives:

$$q_i = \left\{ 1 - \left(1 - \underline{g}_i^{2n} \right)^{1/n} \right\}^{1/2} ; \; \underline{g}_i = \left\{ 1 - \left(1 - q_i^{2} \right)^{n} \right\}^{1/2n} \qquad (7)$$

For any autosomal recessively transmitted gene, i, the carrier frequency of that allele, f_i, is given by (Mathematical Appendix I):

$$f_i = 2q_i \,(1 - 1/2\, q_i) \qquad (30)$$

algebraic rearrangement of which gives:

$$q_i^{2} - 2q_i + f_i = 0 \qquad (31)$$

This is an equation of the general type, $ax^2 + bx + c = 0$, where $a = 1$, $b = -2$, and $c = f_i$. The general quadratic function

$$x = \frac{-b \pm (b^2 - 4ac)^{1/2}}{2a} \qquad (32)$$

gives, for Eqn (31):

$$q_i = \frac{-(-2) \pm [(-2)^2 - 4\,(1)(f_i)]^{1/2}}{2} \qquad (33)$$

$$= \quad 1 \pm (1 - f_i)^{1/2}$$

Since neither f_i nor q_i can exceed unity, our final expression is:

$$q_i = 1 - (1 - f_i)^{1/2} \qquad (34)$$

Equations (30), (31), and (34) apply for both oligogenic autosomal recessive and co-recessive inheritance.

Equation (8) is obtained by simple algebraic substitution in the above functions, as follows:

1. Substitute the value for q_i obtained from Eqn (34) into the expression for $\underline{q}_i$ given in Eqn. (7).

2. Substitute the value of $\underline{q}_i$ obtained in step 1 into Eqn. (30).

3. Eqn. 5, for all f_i equal, may be re-written as

$$F = 1 - (1 - f_i)^n \tag{35}$$

algebraic rearrangement of which gives

$$f_i = 1 - (1 - F)^{1/n} \tag{36}$$

This holds for both types of inheritance. Substitute this value of f_i into the function for $\underline{f}_i$ obtained in step 2.

4. Substitute the function for $\underline{f}_i$ obtained in step 3 into the function for $\underline{F}$ given in step 3.

MATHEMATICAL APPENDIX III: THE WEINBERG-DAHLBERG EXPRESSION

This expression, which was derived independently by Weinberg in 1920 and Dahlberg in 1929 (Levitan and Montagu, 1971), is sometimes called Dahlberg's formula. It is easily derived from the laws of Mendelian inheritance, as follows:

By definition, an individual who is the product of a first-cousin mating has one set of double great-grandparents (i.e., great-grandparents from whom the patient is descendent via two pathways). At any locus on an autosomal chromosome there are two alleles; the chance of each of these double great-grandparents of the individual passing any given allele to any one particular offspring (i.e., to one of the individual's grandparents) is therefore 1/2. Similarly, the chance of this same allele being passed to each of the individual's parents (i.e., the chance of a grandparent passing it to one of their offspring) is $(1/2)^2 = 1/4$ and the chance of the individual receiving for this allele from each parent is $(1/2)^3 = 1/8$. The probability of receiving this particular allele from both parents is thus $(1/8)^2 = 1/64$. The chance that the individual is homozygous for any of the four alleles present at this locus in his or her double great grandparents (i.e., two alleles per double great grandparent x 2) is therefore 4 (1/64) = 1/16, and the probability that the individual is not homozygous for any of these four alleles is 1 - 1/16 = 15/16. Since the probability that each of these four alleles is defective is, by definition, q, the gene frequency, the probability of the individual receiving this defective allele as a result of the first-cousin mating is (1/16)xq. The probability that the individual is homozygous for the defective allele at this locus as a result of random mating, and not as a result of the first-cousin mating, is (15/16)x(chance of receiving this defective allele from each parent)^2 = (15/16)xq^2. Thus the total probability of an individual who is the product of a first-cousin mating being homozygous for a defective allele is:

$$(1/16)q + (15/16)q^{2} \tag{37}$$

If C is the proportion of matings in the population under study that are between first cousins, then C is also the proportion of children in the population that are the product of such matings. Thus, for any given allele, the proportion of individuals in the population homozygous for the allele and who are the product of first-cousin matings is:

$$C\left\{(1/16)\, q + (15/16)\, q^{2}\right\} = \frac{Cq + 15Cq^{2}}{16} \tag{38}$$

The proportion of individuals in the population who are <u>not</u> the product of first-cousin matings is (1-C), and in these individuals the probability of being homozygous for this allele is q^2. Thus the proportion of individuals in the population who are not the product of a first-cousin mating but who are homozygous for the allele is:

$$(1-C)q^{2} \tag{39}$$

The sum of these two expressions [(38) + (39)] is the total proportion of the population which is homozygous for this allele, regardless of whether their parents were first cousins. This is given by:

$$(1 - C)\, q^{2} + \frac{Cq + 15Cq^{2}}{16} = \frac{16q^{2} - Cq^{2} + Cq}{16} \tag{40}$$

K, defined as the proportion of individuals with the trait (i.e., who are homozygous for this allele) who are the product of first-cousin matings is, therefore, [from expression (38) and (40)]:

$$K = \frac{\dfrac{Cq + 15Cq^{2}}{16}}{\dfrac{16q^{2} - Cq^{2} + Cq}{16}} = \frac{Cq + 15Cq^{2}}{16q^{2} - Cq^{2} + Cq} \tag{41}$$

algebraic rearrangement of which gives:

$$K = \frac{C\ (15q + 1)}{C + q\ (16-C)} \tag{15}$$

which is the Weinberg-Dahlberg expression.

For co-recessive inheritance, the probability of an individual who is the product of a first-cousin mating showing the trait is the combined probability of being homozygous for all of the n genes in the system. This is:

$$\prod_{i=1}^{n} \left\{ (1/16)\ q_i + (15/16)\ q_i^2 \right\} \tag{42}$$

whereas the probability of an individual who is not the product of a first-cousin mating showing the trait is:

$$\prod_{i=1}^{n} q_i^2 \tag{43}$$

Since there are, proportionally, C individual in the first group and 1 - C in the second, we can obtain K in a manner entirely analogous to the above derivation, as follows:

$$K = \frac{C \prod_{i=1}^{n} q_i \left\{ (1/16) + (15/16)\ q_i \right\}}{(1 - C) \prod_{i=1}^{n} q_i^2 + C \prod_{i=1}^{n} q_i \left\{ (1/16) + (15/16)\ q_i \right\}} \tag{44}$$

algebraic rearrangement of which gives:

$$K = \frac{C \prod_{i=1}^{n} (15q_i + 1)}{(2)^{4n}\ (1-C) \prod_{i=1}^{n} q_i + C \prod_{i=1}^{n} (15q_i + 1)} \tag{17}$$

MAMMALIAN DNA LIGASES AND THE MOLECULAR DEFECT IN BLOOM'S SYNDROME

Tomas Lindahl, Anne E. Willis[1], Dana D. Lasko[2], and Alan Tomkinson

Imperial Cancer Research Fund
Clare Hall Laboratories, South Mimms, Herts
EN6 3LD, U.K.

INTRODUCTION

DNA ligases catalyse the formation of phosphodiester bonds at strand breaks in DNA with adjoined 3'OH and 5'P termini. Studies on conditional lethal mutants of yeast, *E. coli*, and phage T4 have established the involvement of DNA ligase in the replication, repair, and recombination of DNA (Engler and Richardson, 1982; Barker, *et al.*, 1985).

Eukaryotic cells invariably seem to employ ATP to generate the ligase-AMP covalent complex that interacts with nicked DNA, whereas *E. coli* and *B. subtilis* require NAD as ligase cofactor. However, the basic mechanism by which DNA ligase repairs a DNA strand break remains the same in mammalian cells as in bacteria, with transfer of the AMP moiety from ligase-AMP to the 5' terminus of the chain break to generate a covalent DNA-AMP reaction intermediate prior to closure of the strand interruption (Söderhall and Lindahl, 1973a, 1976; Teraoka, *et al.*, 1986).

MAMMALIAN DNA LIGASE I

The major DNA ligase activity in mammalian cells is due to a high-molecular weight enzyme we have designated DNA ligase I. This enzyme is present at a much higher level in proliferating than in non-proliferating cells, and increases 10-15 fold during rat liver regeneration (Söderhall and Lindahl, 1975; Söderhall, 1976). The enzyme has a half-life

---

[1]Present address: Department of Biochemistry, University of Cambridge, U.K.

[2]Fellow of The Jane Coffin Childs Memorial Fund for Medical Research.This investigation has been aided by a grant from The Jane Coffin Childs Memorial Fund for Medical Research.

reported to be only 0.5 hours <u>in vivo</u> (Teraoka and Tsukada, 1985), and is very susceptible to proteolysis <u>in vitro</u>. The latter property has made the purification of intact DNA ligase I a difficult task, and the generation of enzymatically active fragments during preparative work has caused considerable confusion in the field. The major 130 kDa form of the enzyme has been claimed to be derived from an extremely labile 200 kDa precursor (Teraoka and Tsukada, 1985, 1986). However, we have not detected this 200 kDa form in our experiments, in spite of using extraction buffers containing several different types of protease inhibitors (Fig. 1). The 130 kDa form is cleaved by proteolysis to an 85-90 kDa active fragment (Fig. 1), and also to other minor fragments.

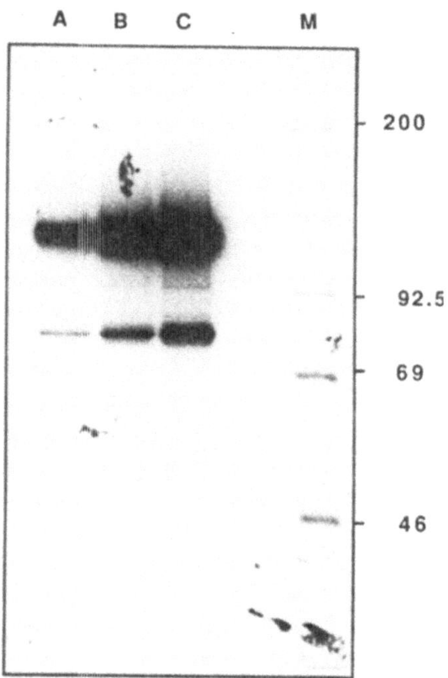

Figure 1. Sizes of active fragments of DNA ligase I. The enzyme was purified from calf thymus by methods to be described elsewhere. The protein (1.5 µg, 1.4 x 10^{-5} Weiss units) was incubated at 22^{0}C for 40 minutes in a total volume of 50 µl of 50 mM Tris HCl, pH 7.8, 10 mM $MgCl_2$, 5 mM dithiothreitol, and 20 µM ATP containing 10 µCi [α-^{32}P]ATP. The sample was dialyzed as a drop on top of a Millipore VSWP filter for 20 minutes against 50 mM Tris HCl, pH 7.5, 50 mM NaCl, 0.5 mM dithiothreitol, 1 mM EDTA. Aliquots (A: 5 µl; B: 10 µl; C: 15 µl) were subjected to electrophoresis through a 7.5% SDS-polyacrylamide gel, and AMP-adducted polypeptides were visualized by autoradiography (3 hour exposure to Kodak XAR film). The molecular weight markers were myosin, phosphorylase b, bovine serum albumin, and ovalbumin.

DNA ligase I has a markedly asymmetric shape (Söderhall and Lindahl, 1973b; Teraoka and Tsukada, 1982) so size determinations of the enzyme activity by hydrodynamic methods are not very accurate, and molecular weight measurements based only on gel filtration data (Chan, et al., 1987) are gross overestimates.

The 90 kDa fragment of DNA ligase I was the first ligase activity described in mammalian cells (Lindahl and Edelman, 1968). The high molecular weight of DNA ligase I was established several years later (Beard, 1972; Söderhall and Lindahl, 1973b; Teraoka and Tsukada, 1986). It is interesting that the 90 kDa fragment retains the low K_m for the ATP cofactor (5 x 10^{-7}M) and the efficient DNA ligation capacity of the high-molecular weight form of the enzyme. Moreover, DNA ligases from bacteria, yeast, and Drosophila (Rabin, et al., 1986) have molecular weights of 70,000-90,000. It seems likely

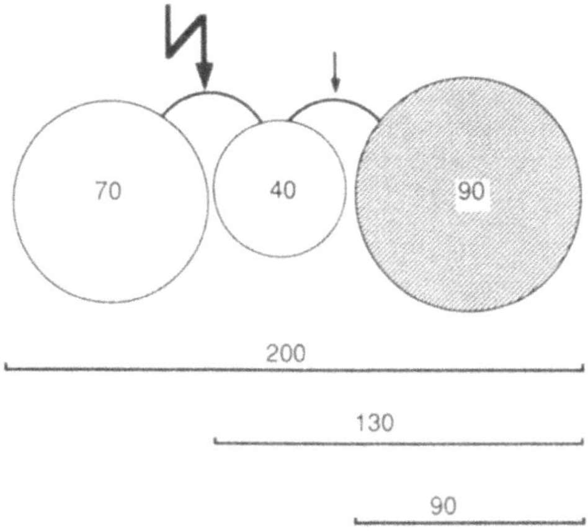

Figure 2. Possible domain structure of mammalian DNA ligase I. It is controversial whether the 130 kDa form, or a very labile 200 kDa form, is the primary translation product. The hatched area indicates the catalytic domain. The sizes of the arrows reflect the relative susceptibilities of the putative hinge regions to proteolysis in crude cell extracts.

that mammalian DNA ligase I contains distinct domains united by protease-sensitive hinge region(s). A model of DNA ligase I based on these observations is shown in Figure 2. Removal of long sequences on the N-terminal side of the 90 kDa fragment allows the catalytic domain to function in an essentially unaltered fashion as a DNA ligase. The other part of the polypeptide may be involved in regulatory events such as protein-protein interactions in cell nuclei, or possess a different, unknown enzymatic activity. In a comparison between the DNA ligases of the two very distantly related yeasts, *Saccharomyces cerevisiae* and *Schizosaccharomyces pombe*, it was observed that the cell cycle regulation of ligase expression is different, and that the two protein sequences show extensive homology in their C-terminal region, but little or no homology in their N-terminal parts (Barker, et al., 1987). Mammalian DNA ligase I may also have a C-terminal region similar to those of the yeast enzymes but a variable N-terminal region, which in this case would have greatly expanded in size into a large separate domain or would represent the product of a gene fusion event.

MAMMALIAN DNA LIGASE II

DNA ligase II was discovered during purification of DNA ligase I from calf thymus by hydroxyapatite chromatography (Söderhall and Lindahl, 1973b). DNA ligase II adsorbs strongly to hydroxyapatite and appears as a separate late peak of enzyme activity on gradient elution. Subsequent experiments showed that DNA ligase II is not induced upon cell proliferation (Söderhall and Lindahl, 1975; Söderhall, 1976). It has a lower molecular weight, and a 50 times higher K_m for ATP than ligase I. Polyclonal antibodies which neutralize DNA ligase I activity neither inhibit nor bind to DNA ligase II (Söderhall and Lindahl, 1975). DNA ligase II is able to join interruptions in a DNA strand hydrogen-bonded to an RNA template strand, an activity not present in ligase I (Arrand, et al., 1985). These data firmly establish the existence of DNA ligase II as an enzyme distinct from DNA ligase I.

DNA ligase II is a labile enzyme, and a number of groups have encountered difficulties working with it, and separating it from proteolytic fragments of DNA ligase I. Sarasin and co-workers have investigated mammalian DNA ligases by the "activity gel" assay, involving SDS-polyacrylamide gel electrophoresis of crude protein fractions followed by renaturation attempts according to an empirical procedure and enzyme assays in situ. Several active fragments of DNA ligase I activity could be detected in this fashion, whereas ligase II activity could not be clearly identified (Mezzina, et al., 1987). It would appear that DNA ligase II does not renature effectively after denaturation with SDS. Similarly, Teraoka and Tsukada (1982) failed to find DNA ligase II in impure preparations of calf thymus ligase I and suggested that "DNA ligase II is a product of partial proteolysis of DNA ligase localized in the nucleus as a single molecular species". However, in subsequent work the same authors were able to detect DNA ligase II, and achieved the feat of purifying this 68 kDa protein to apparent homogeneity. Polyclonal antibodies raised against homogenous DNA ligase II did not recognize DNA ligase I (Teraoka, et al., 1986). In their earlier studies, a selective loss of DNA ligase II apparently occurred during an initial batch purification of crude cell extracts with calcium phosphate gel; ligase II adsorbs strongly to this gel, as it does to hydroxyapatite, and had not been eluted together with ligase I.

The physiological role of DNA ligase II is unknown. One possibility might be that the enzyme is involved in certain forms of short-patch excision-repair, perhaps together with DNA polymerase β and poly ADP-ribose polymerase, since these three enzymes which interact with DNA termini appear to lack counterparts in lower eukaryotes (Lindahl, 1982). There is conflicting evidence whether DNA ligase II is induced after MMS treatment of cells (Creissen and Shall, 1982; Teraoka, et al., 1986). Another possibility is that DNA ligase II is required for repair of accidental DNA strand breaks in the DNA-RNA hybrids occurring during active transcription.

DNA LIGASES IN ACUTE LYMPHOBLASTIC LEUKEMIA

Recently, Rusquet, et al. (1988), reported that extracts of leukemic T cells from patients with acute lymphoblastic leukemia (ALL) show low or undetectable DNA ligase activity, possibly indicating a ligase deficiency in such cells. Since the major ligase activity, DNA ligase I, has a short half-life in vivo, these data may perhaps be explained by enzyme decay after collection and storage of the leukemic cells from the patients, before the cells could be homogenized and assayed for enzyme activity. To avoid this technical problem, we have employed a representative human T-cell line of ALL origin which can be propagated in the laboratory. (Several such T-cell lines are available from the NIGMS Human Genetic Mutant Cell Repository.) We have investigated line J-6, a cloned subline of Jurkat, which expresses the surface CD3 antigen as a characteristic T-cell marker. In a parallel experiment with the control B-cell line Raji, which has been studied previously (Willis and Lindahl, 1987), size-fractionated extracts of J-6 and Raji were indistinguishable in that they both contained 1.3 microunits of DNA ligase I activity per mg of extracted protein and also similar amounts of DNA ligase II activity. We conclude that the reported DNA ligase deficiency in ALL is not observed with a T-cell line representative of the disorder.

NORMAL DNA LIGASE ACTIVITIES IN SEVERAL HUMAN INHERITED SYNDROMES

Lymphoblastoid cell lines representative of a variety of inherited disorders with known or suspected DNA repair defects were analyzed for DNA ligase activity. The experimental details have been published elsewhere (Willis and Lindahl, 1987). In brief, the cells were disrupted and extracted with a buffer of physiological ionic strength. Both DNA ligases I and II are quantitatively recovered in this fashion, and no further ligase activity can be extracted with buffers of higher ionic strength up to 4 M NaCl (Söderhall and Lindahl, 1975). Whole cell extracts are preferred to nuclear extracts, because DNA ligase I leaches out of cell nuclei in isotonic sucrose. The crude cell extracts were treated with Polymin-P to remove DNA and reduce viscosity prior to size fractionation. No significant amounts of DNA ligase activity were precipitated with Polymin-P at the ionic strength (0.25) and protein concentration employed. (When the Polymin-P step was excluded, the results in the subsequent size fractionation were the same, except that the peaks of DNA ligase activity were less well resolved from each other due to the presence of nucleic acids in the extract). Size fractionation by FPLC on a Superose-12 column was employed prior to the enzyme assays, to separate DNA ligases I and II from each other and from nucleases and other low-molecular weight factors that might cause interference in quantitative assays. Elution profiles have been published (Willis and Lindahl, 1987; Willis, et al., 1987; Lehmann, et al., 1988). Control cells (HeLa, Raji, and lymphoblastoid cell lines established from normal individuals) exhibited a peak of DNA ligase I containing 65% of the total ligase activity, followed by a smaller distinct peak of DNA ligase II activity. Similar results were obtained with cell lines representative of ataxia-telangiectasia, Fanconi's anemia, Cockayne's syndrome, Werner's syndrome, xeroderma pigmentosum (including the variant complementation group) and Friedreich's ataxia (Figure 3). Thus, there is no apparent DNA ligase defect in any of these disorders. During this survey, no cell line defective in DNA ligase II was ever identified.

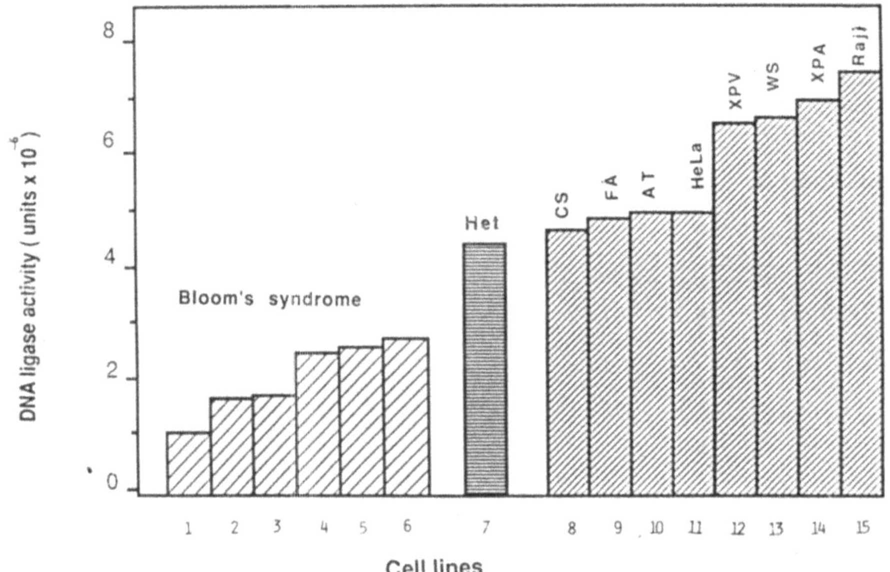

Figure 3. DNA ligase I activity in extracts of human cell lines representative of various inherited syndromes. The experimental data are from Willis and Lindahl, 1987, and Willis, et al., 1987. The total DNA ligase I activity in a crude cell extract of 10^8 cells (4 mg protein) is shown. The BS lines employed were: 1, GM8505; 2, GM3403; 3, W67-4; 4, 1004 (Mennonite); 5, AA87-5-1; 6, D86-1-2. Line 7 is AA87-4 from a BS heterozygote.

REDUCED DNA LIGASE I ACTIVITY IN BLOOM'S SYNDROME

The clinical and cytogenetical aspects of Bloom's syndrome (BS) have been reviewed (for a recent summary, see German and Takebe, 1989). Patients with this recessively inherited disorder exhibit high cancer frequency and stunted growth, and their cells show increased frequencies of chromosome breakage and sister chromatid exchange. All investigated cases fall within a single genetic complementation group (Weksberg, et al., 1988). BS cells show a decreased rate of joining of large DNA replication intermediates (Giannelli, et al., 1977; Ockey and Saffhill, 1986) and are hypersensitive to methylating and ethylating agents (Kurihara, et al.,1987; Lehmann, et al., 1988). When cell lines representative of the syndrome were investigated by exactly the same experimental procedure as that employed with lines from other disorders, a reduced amount of DNA ligase I activity was consistently observed in extracts. The results for five BS cell lines of Ashkenazi Jewish origin, and one line of Mennonite origin, are shown in Figure 3. A sixth independent strain of Ashkenazi origin, GM3498, was shown recently to have a similar anomalously low level of DNA ligase I (Willis, et al., 1989). A single BS heterozygote line could not be distinguished clearly from the control group, although it exhibited lower DNA ligase I activity than any of the controls (Figure 3).

In these experiments, the presence of normal levels of DNA ligase II activity in the size-fractionated extracts served as an internal control. Several other enzyme activities, such as DNA polymerases and (Parker and Lieberman, 1977) and uracil-DNA glycosylase (Vilpo and Vilpo, 1988), are also present at a normal levels in BS cell extracts.

Two BS lines of Anglo-Saxon and Japanese origin showed a different type of DNA ligase I anomaly. In these cases, the enzyme occurred in a dimeric form not detected in the control cells (Willis, et al., 1987; Lehmann, et al., 1988). Another line, 46BR, has related but unusual properties (Lehmann, et al., 1988). The data indicate that different mutations may have occurred in the Ashkenazi cases vs most of the non-Jewish isolates. In addition to our results, Chan, et al.(1987), have reported on a reduced level of DNA ligase I activity in BS, as well as on the presence of ligase I dimers in control cells and an apparent 20% reduction in molecular weight of the BS enzyme. The two latter observations are not in agreement with our data.

STRUCTURAL DEFICIENCY IN DNA LIGASE I FROM BS CELLS

The decreased amount of DNA ligase I activity in BS cell extracts could reflect either a structural alteration in the enzyme, or reduced synthesis (or increased degradation) due to a regulatory mutation. Direct evidence for the former concept was provided by our finding of anomalous heat-sensitivity of the DNA ligase I from BS cells. This abnormal lability is an intrinsic property of the enzyme which persists during protein purification (Willis and Lindahl, 1987; Willis, et al., 1987).

Further evidence in this direction has been obtained by immunoblotting experiments. Two groups have prepared rabbit antisera against calf thymus DNA ligase I (Söderhall and Lindahl, 1975; Teraoka and Tsukada, 1982). Both antisera also recognize ligase I from other mammals, including man, albeit at lower efficiency. Our antiserum, which was prepared against highly purified but non-homogenous DNA ligase I, seems more active in the latter regard.

Cell extracts from the BS cell line, GM3403 and the normal lymphoblastoid line, GM1953, were fractionated by SDS gel electrophoresis and compared by immunoblotting (Figure 4). These extracts have a 4-fold difference in DNA ligase I activity (cf. Figure 3). In contrast, they contained the same amount of DNA ligase I protein, present in the 130 kDa and 90 kDa forms (Figure 4, cf. Figure 1). Similar data have also been obtained by J. Chan and F. Becker (pers. commun.), who used an aliquot of the same DNA ligase I antiserum. These data show that the anomalously low amount of ligase I activity

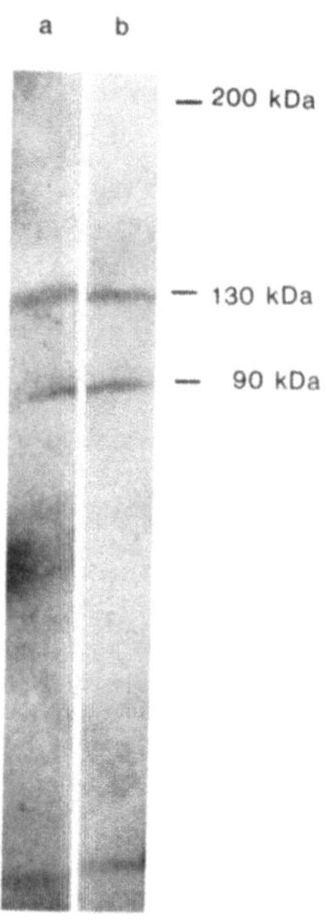

Figure 4. Immunoblot of DNA ligase I in cell extracts from the BS line GM3403 (a) and the control lymphoblastoid line GM1953 (b). Cell extracts were made and treated with Polymin-P as described (Willis and Lindahl, 1987). Fifty µg protein of each such extract were fractionated by SDS polyacrylamide gel electrophoresis on a 10% gel. After transfer of the protein to nitrocellulose, the filter was blocked with 10% horse serum and then incubated for 10 hours at 10^0C with a rabbit antiserum raised against calf thymus DNA ligase I, followed by washing of the filter with a buffer containing 10% horse serum. The secondary antibody employed was goat anti-rabbit IgG conjugated to horseradish peroxidase (Nordic Immunologicals). Antigen-antibody complexes were visualized by reaction of the peroxidase with 4-chloro-1-naphtol and H_2O_2 (Simanis and Lane, 1985).

in BS cell extracts is due to a reduced turnover number of the enzyme itself, indicating that it is faulty in comparison with the normal protein. The results are in agreement with the abnormal heat-sensitivity of the BS enzyme. These data indicate that BS cells have a point mutation in the structural gene for DNA ligase I. Furthermore, a BS revertant cell line with normal sister chromatid exchange, GM4408, had also regained a normal DNA ligase I (Willis, et al., 1989). Final confirmation of a DNA ligase I deficiency associated with BS should come from the cloning and sequencing of the gene for DNA ligase I, and this project is now being pursued in several laboratories.

REFERENCES

Arrand, J.E., Willis, A.E., Goldsmith, I., and Lindahl, T., 1986, Different substrate specificities of the two DNA ligases of mammalian cells, J. Biol. Chem., 261:9079.

Barker, D.G., Johnson, A.L., and Johnston, L.H., 1985, An improved assay for DNA ligase reveals temperature-sensitive activity in cdc9 mutants of Saccharomyces cerevisiae, Mol. Gen. Genet., 200:458.

Barker, D.G., White, J.H.M., and Johnston, L.H., 1987, Molecular characterisation of the DNA ligase gene, CDC17, from the fission yeast Schizosaccharomyces pombe, Eur. J. Biochem., 162:659.

Beard, P., 1972, Polynucleotide ligase in mouse cells infected by polyoma virus, Biochim. Biophys. Acta, 269:385.

Chan, J.Y.H., Becker, F.F., German, J., and Ray, J.H., 1987, Altered DNA ligase I activity in Bloom's syndrome cells, Nature, 325:357.

Creissen, D., and Shall, S., 1982, Regulation of DNA repair ligase activity by poly(ADP-ribose), Nature, 296:271.

Engler, M.J., and Richardson, C.C., 1982, DNA ligases, in "The Enzymes, 3rd ed., Vol. XV B", P.D. Boyer, ed. , p.3, Academic Press, New York.

German, J., and Takebe, H., 1989, Bloom's syndrome. The disorder in Japan, Clin. Genetics, in press.

Giannelli, F., Benson, P.E., Pawsey, S.A., and Polani, P.E., 1977, Ultraviolet light sensitivity and delayed DNA-chain maturation in Bloom's syndrome fibroblasts, Nature, 265:466.

Kurihara, T., Inoue, M., and Tatsumi, K., 1987, Hypersensitivity of Bloom's syndrome fibroblasts to N-ethyl-N-nitrosourea, Mutat. Res., 184:147.

Lehmann, A.R., Willis, A.E., Broughton, B.C., James, M.R., Steingrimsdottir, H., Harcourt, S.A., Arlett, C.F., and Lindahl, T., 1988, Relation between the human fibroblast strain 46BR and cell lines representative of Bloom's syndrome, Cancer Res., 48:6343.

Lindahl, T., 1982, Worthy of a detour, Nature, 298:424.

Lindahl, T., and Edelman, G.M., 1968, Polynucleotide ligase from myeloid and lymphoid tissues, Proc. Natl. Acad. Sci. USA., 61:680.

Mezzina, M., Rossignol, J.M., Philippe, M., Izzo, R., Bertazzoni, U., and Sarasin, A., 1987, Mammalian DNA ligase. Structure and function in rat-liver tissues, Eur. J. Biochem., 162:325.

Ockey, C.H., and Saffhill, R., 1986, Delayed DNA maturation, a possible cause of the elevated sister-chromatid

exchange in Bloom's syndrome, *Carcinogenesis*, 7:53.
Parker, V.P., and Lieberman, M.W., 1977, Levels of DNA polymerases , , and in control and repair-deficient human diploid fibroblasts, *Nucleic Acids Res.*, 4:2029.
Rabin, B.A., Hawley, R.A., and Chase, J.W., 1986, DNA ligase from Drosophila melanogaster embryos, *J. Biol. Chem.*, 261:10637.
Rusquet, R.M., Feon, S.A., and David, J.C., 1988, Association of a possible DNA ligase deficiency with T-cell acute leukemia, *Cancer Res.*, 48:4038.
Simanis, V., and Lane, D.P., 1985, An immunoaffinity purification procedure for SV40 large T antigen, *Virology*, 144:88.
Söderhall, S., 1976, DNA ligases during rat liver regeneration, *Nature*, 260:640.
Söderhall, S., and Lindahl, T., 1973a, Mammalian deoxyribonucleic acid ligase. Isolation of an active enzyme-adenylate complex, *J. Biol. Chem.*, 248:672.
Söderhall, S., and Lindahl, T., 1973b, Two DNA ligase activities from calf thymus, *Biochem. Biophys. Res. Commun.*, 53:910.
Söderhall, S., and Lindahl, T., 1975, Mammalian DNA ligases. Serological evidence for two separate enzymes. *J. Biol. Chem.*, 250:8438.
Söderhall, S., and Lindahl, T., 1976, DNA ligases of eukaryotes, *FEBS Lett.*, 67:1.
Teraoka, H., and Tsukada, K., 1982, Eukaryotic DNA ligase. Purification and properties of the enzyme from bovine thymus, and immunochemical studies of the enzyme from animal tissues, *J. Biol. Chem.*, 257:4758.
Teraoka, H., and Tsukada, K., 1985, Biosynthesis of mammalian DNA ligase, *J. Biol. Chem.*, 260:2937.
Teraoka, H., Sumikawa, T., and Tsukada, K., 1986, Purification of DNA ligase II from calf thymus and preparation of rabbit antibody against calf thymus DNA ligase II, *J. Biol. Chem.* , 261:6888.
Teraoka, H., and Tsukada, K., 1986, Immunochemical analysis of molecular forms of mammalian DNA ligases I and II, *Biochim. Biophys. Acta*, 873:297.
Vilpo, J.A., and Vilpo, L.M., 1989, Normal uracil-DNA glycosylase activity in Bloom's syndrome cells, *Mutation Res.*, 210:59.
Weksberg, R., Smith, C., Anson-Cartwright, L., and Maloney, K., 1988, Bloom syndrome: A single complementation group defines patients of diverse ethnic origin, *Am. J. Hum. Genet.*, 42:816.
Willis, A.E., and Lindahl, T., 1987, DNA ligase I deficiency in Bloom's syndrome, *Nature*, 325:355.
Willis, A.E., Spurr, N.K., and Lindahl, T., 1989, Concomitant reversion of the characteristic phenotypic properties of a cell line of Bloom's syndrome origin, *Carcinogenesis*, 10:217.
Willis, A.E., Weksberg, R., Tomlinson, S., and Lindahl, T., 1987, Structural alterations of DNA ligase I in Bloom syndrome, *Proc. Natl. Acad. Sci. USA.*, 84:8016.

DETECTION OF UNIQUE ANTIGENIC LESIONS IN THE URACIL DNA GLYCOSYLASE FROM BLOOM'S SYNDROME

Michael A. Sirover, Gita Seal, Thomas M. Vollberg, Barbara L. Cool, Kilian Brech and Seth J. Karp

Fels Institute for Cancer Research and Molecular Biology and the Department of Pharmacology Temple University School of Medicine Philadelphia, PA 19140

INTRODUCTION

Human cells contain singular DNA repair pathways to excise critical lesions from DNA (Lindahl, 1982; Teebor and Frenkel, 1983; Friedberg, 1985, Strauss, 1985). Two major excision repair pathways have been identified. In nucleotide excision repair, the DNA adduct is released within an oligonucleotide after sequential action of an undetermined number of enzymes. In contrast, in base excision repair, the DNA adduct is released in the initial enzymatic step of the pathway. This reaction is catalyzed by a DNA glycosylase which cleaves the base-sugar glycosyl bond. Human cells may contain a family of DNA glycosylases each of which may be responsible for the removal of a specific modified base. Thus, the uracil DNA glycoslase removes uracil from DNA. Uracil may arise in DNA by the deamination of cytidine residues (Hayatsu, 1977) or by the incorporation of 5'dUMP during DNA synthesis (Bessman, et al., 1958). The former, by definition, would be a mutagenic event if left unrepaired. Uracil excision results in the formation of an apyrimidinic site in DNA. That site is incised by the apurinic/apyrimidinic acid (AP) endonuclease to form a single strand break. Subsequent catalysis by a DNA polymerase and DNA ligase would be minimal requirements for the completion of base excision repair.

Bloom's syndrome is an autosomal recessive human genetic disease characterized by low body weight at birth, stunted growth, cutaneous rash and by immunological deficiency (Bloom, 1966, German, 1969). Individuals with Bloom's syndrome are predisposed to infection and are cancer-prone (German, et al., 1977, 1984). Bloom's syndrome cells are characterized by high rates of chromosomal aberrations especially sister chromatid exchanges (Chaganti, et al., 1974; Bryant, et al., 1979). Thus, this disorder is considered as one of the "chromosomal breakage" human genetic syndromes. Recently, several laboratories reported that Bloom's syndrome cells were hypermutable

DNA Repair Mechanisms and Their Biological Implications in Mammalian Cells
Edited by M.W. Lambert and J. Laval
Plenum Press, New York

with spontaneous mutation frequencies 5-10 fold higher than those observed in normal human cells (Gupta and Goldstein, 1980; Warren, et al., 1981; Vijayalaxmi, et al., 1983). Further, Doniger, et al. (1983) reported that Bloom's syndrome cells were singularly susceptible to transformation by DNA transfection. However, no difference was observed in the excision of DNA adducts or in the level of DNA repair enzymes in confluent Bloom's syndrome cells. In particular, the excision of N-acetoxyacetylaminofluorene adducts was comparable to that observed in confluent normal human cells (Remson, 1980). Further, the level of the apurinic acid endonuclease and that of the O^6-methylguanine methyltransferase were identical in confluent Bloom's syndrome or normal human cells (Inoue, et al., 1977; Kim, et al., 1986).

Recent studies demonstrated that the capacity for DNA repair was dependent on the proliferative state of the cell (Sirover, 1989). Several experimental approaches demonstrated an increase in repair capacity as non-cycling cells were recruited to proliferate by a variety of mitogenic stimuli. These determinations included proliferative-dependent increases in: the activities of individual DNA repair enzymes as measured by in vitro biochemical assay using synthetic polynucleotide substrates, DNA repair synthesis after exposure to DNA damaging agents as measured by the incorporation of deoxyribonucleoside triphosphate precursors during the DNA polymerase step of the repair pathway and, lastly, the direct removal of defined DNA lesions as measured by HPLC analysis or sequence analysis of DNA isolated from damaged cells.

Using synchronized cells, the temporal sequence with which DNA repair pathways were enhanced relative to the induction of DNA replication was examined. In particular, we demonstrated that human cells actively regulated DNA repair pathways during the defined pattern of gene expression observed during the cell cycle (Gupta and Sirover, 1980, 1981a,b, 1984a,b; Dehazya and Sirover, 1986, Kim, et al., 1986). Thus, nucleotide excision repair after uv irradiation and base excision repair after MMS exposure were enhanced prior to the induction of DNA synthesis. Each pathway was decreased at maximal levels of DNA replication. The increase in base excision repair corresponded to the induction of the uracil and the hypoxanthine DNA glycosylases as measured by in vitro biochemical assay. These increases in enzyme activity were observed in the absence of any exogeneous DNA damaging agent. Thus, they would reflect the regulation of these base excision repair enzymes as a normal event within the cell cycle. It was considered that this program of gene expression was designed to function as a proof-reading mechanism. This increase in DNA repair capacity would be used to pre-screen DNA before DNA replication in the cell cycle to remove potentially miscoding lesions to prevent mutagenesis or carcinogenesis (Sirover and Gupta, 1983).

In hypermutable cells from cancer-prone Bloom's syndrome patients, the proliferative-dependent regulation of DNA repair pathways was altered. Analysis of the cell cycle expression of DNA repair in Bloom's syndrome cells demonstrated that they were unable to enhance either the nucleotide excision repair or the base excision repair pathway prior to the induction of DNA replication (Gupta and Sirover, 1984a). Instead, although the fold increase in each repair pathway was comparable to that observed in normal human cells, each pathway was increased coordinate with DNA replication. This aberration included the temporal alteration in the induction kinetics of both the uracil DNA glycosylase or the hypoxanthine DNA glycosylase (Gupta and Sirover, 1984a; Dehazya and Sirover, 1986). These results were observed during the cell cycle without exposure to DNA damaging agents. Thus, this alteration would reflect an alteration in gene expression intrinsic to Bloom's syndrome. It was considered that this defect in gene function would prevent the prescreening of DNA prior to its replication. This would result in the copying of miscoding lesions before they could be repaired. In the course of these investigations, using a panel of anti-human placental uracil DNA glycosylase monoclonal antibodies, we observed that the Bloom's syndrome uracil DNA glycosylase was not recognized by one of the monoclonal antibodies. This immunological alteration appeared to be characteristic of Bloom's syndrome cells.

CHARACTERIZATION OF THE NORMAL HUMAN URACIL DNA GLYCOSYLASE

To begin to examine the regulation of DNA repair enzymes at the molecular level, a series of monoclonal antibodies was prepared against partially purified human placental uracil DNA glycosylase (Arenaz and Sirover, 1983). The enzyme was purified by DEAE-cellulose, phosphocellulose and hydroxylapatite column chromatography. The glycosylase fraction from the hydroxylapatite column step was used as an antigen. Splenocytes from immunized animals were isolated for cell fusion. Hybridomas producing antibodies against the enzyme preparation were identified by Enzyme Linked Immunosorbent Assay (ELISA). Each positive hybridoma was cloned twice by limited dilution and tested for anti-glycosylase activity in an enzyme immunoprecipitation assay. Three of the hybridomas were selected for further study based on the ability of the monoclonal antibodies to inhibit the partially purified human placental enzyme (phosphocellulose fraction). The characteristics of each antibody are described in Table 1. Monoclonal antibody 1.05 was derived from a spontaneous hybridoma from a cell fusion using splenocytes from a non-immunized animal. Each monoclonal antibody was purified from hybridoma supernatants by ammonium sulfate fractionation and DEAE-cellulose column chromatography.

TABLE 1. Characteristics Of Monoclonal Antibodies To The Human Placental Uracil DNA Glycosylase

Hybridoma Line	Myeloma Parental Strain	ELISA	Enzyme Immunoprecipitation (% of 1.05 control)
1.05	SP2/O	-	100
37.04.12	SP2/O	+	44
40.10.09	P3X63 Ag8.653	+	13
42.08.07	SP2/O	+	41

A series of experiments was performed to verify that the monoclonal antibodies prepared against the partially purified uracil DNA glycosylase recognized determinants on the glycosylase molecule itself. First, the human placental uracil DNA glycosylase was purified 3700 fold to apparent homogeneity by a six step chromatographic protocol (Seal, et al., 1987). This purification protocol included antibody affinity columns using the glycosylase antibodies 40.10.09 and 42.08.07. As defined by SDS-gel electrophoresis, the glycosylase was a single polypeptide of molecular weight 37,000 daltons. A similar molecular weight was obtained by Sephadex G-100 gel filtration analysis.

The immunological reactivity of the purified glycosylase was then examined. Each of the three monoclonal antibodies (37.04.12, 40.10.09 and 42.08.07) recognized the homogeneous glycosylase by ELISA. Over the range of the concentration curve (5-50 ng), there was an equivalent recognition using antibodies 37.04.12 and 40.10.09. This suggested that both had equal affinity for the glycosylase protein. In contrast, antibody 42.08.07 had a 2-fold greater immunoreactivity. Thus, it would have a greater affinity for the glycosylase protein and would recognize a different antigenic determinant on the glycosylase molecule.

To directly demonstrate that each of the three monoclonal antibodies inhibited glycosylase activity each antibody was incubated with the highly purified enzyme. Equal concentrations of each glycosylase antibody or the negative control antibody 1.05 were used. The enzyme-antibody complex was then sedimented through a 10-35% glycerol gradient as previously described (Arenaz and Sirover, 1983; Seal and Sirover, 1986). Fractions were collected from the top of the gradient and glycosylase activity was determined. As shown in Figure 1A, after incubation with negative control antibody 1.05, uracil DNA glycosylase sedimented as a single peak near the top of the gradient. This sedimentation pattern was identical to that observed when the partially purified enzyme was preincubated with mouse IgG or with BSA (Arenaz and Sirover, 1983). However, incubation with each of the three anti-uracil DNA glycosylase monoclonal antibodies resulted in a reduced level of glycosylase activity. The greatest inhibition was observed with monoclonal antibodies 37.04.12 (Figure 1A) and 40.10.09 (Figure 1B). A smaller degree of inhibition was observed with antibody 42.08.07 (Figure 1C).

The mechanisms through which human cells synthesized the 37,000 dalton glycosylase enzyme were then examined. Initial immunoblot analysis was performed using each of the three monoclonal antibodies with the purified glycosylase species. With antibody 40.10.09, a single immunoreactive protein of 37,000 daltons was detected. In contrast, immunoblot analysis with either monoclonal antibody 37.04.12 or 42.08.07 was unsuccessful. The immunoblot results obtained using monoclonal antibody 40.10.09 corresponded directly with the molecular weight of the glycosylase as detected by both SDS-gel electrophoresis and by gel filtration. Thus, immunoblot analysis was performed with the human placental uracil DNA glycosylase from the initial purification steps, using monoclonal antibody 40.10.09, to determine whether glycosylase precursor species could be detected. As shown in Figure 2, these results demon-

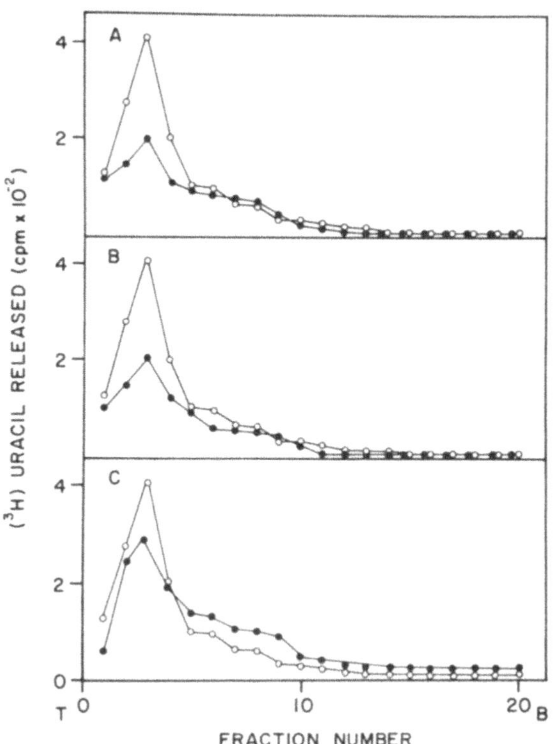

Figure 1. Glycerol gradient analysis of anti-uracil DNA glycosylase activity of monoclonal antibodies 37.04.12, 40.10.09 and 42.08.07. Highly purified placental uracil DNA glycosylase was incubated for 120 minutes at 4^{0}C with equal concentrations of each monclonal antibody or with negative control antibody 1.05. The enzyme-antibody mixture was then centrifuged through a 5 ml 10-35% glycerol gradient as previously described (Arenaz and Sirover, 1983, Seal and Sirover, 1986). Fractions were collected from the top of the gradient. Uracil DNA glycosylase activity was determined as previously described (Seal, et al., 1987). (- O -), preincubation with control antibody 1.05 (A-C); (- ● -), preincubation with antibodies 37.04.12 (A), 40.10.09 (B) or 42.08.07 (C).

strated that, even in crude enzyme samples, only one immunoreactive glycosylase species of 37,000 daltons could be detected. Thus, human placenta did not contain isoenzymes or immunoreactive glycosylase polypeptides which were removed during the purification procedure. To determine the uniqueness of this observation, a similar analysis was performed using crude extracts from several different human cell sources.

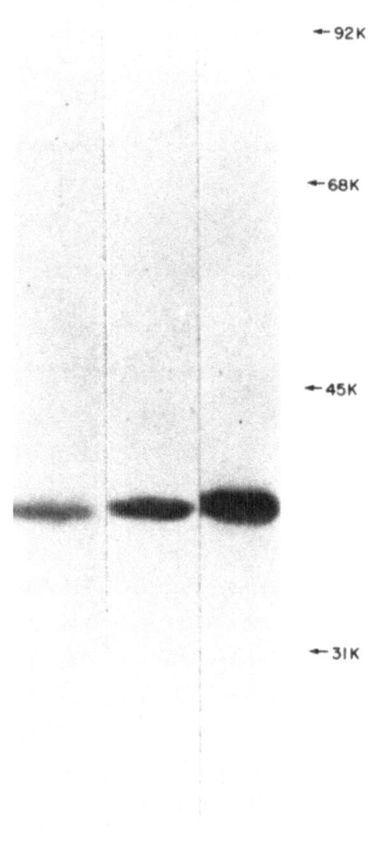

Figure 2. Immunoblot analysis of crude human placental uracil DNA glycosylase. Fractions II-IV (DEAE-,Phospho- and DNA-cellulose fractions, respectively) from the glycosylase purification were analysed. SDS-gel electrophoresis and immunoblot analysis were performed as previously described (Seal, et al., 1987). Lanes 1-3 correspond to fractions II-IV, respectively (50 μg fractions II, III; 40 μg fraction IV). Reprinted from Seal, et al., 1987, by permission of the publisher.

Identical results were observed using crude cell extracts from a variety of human cell strains including normal human cell fibroblasts (CRL-1222, WI-38), SV-40 transformed WI-38 cells and from two human progeroid cell strains (AG-3199, AG-6917). Thus, these results suggest that human cells do not contain precursor glycosylase polypeptides.

The biosynthesis of the normal human uracil DNA glycosylase was then examined to determine whether transitory precursor glycosylase polypeptides might exist. Total and poly (A) containing RNA was isolated from human placenta and analyzed by in vitro translation (Vollberg, et al., 1987a). [^{35}S]-methionine radiolabeled protein was immunoprecipitated with either the negative control monoclonal antibody 1.05 or with glycosylase monoclonal antibody 40.10.09. After SDS electrophoresis no immunoprecipitated proteins were observed using the control antibody 1.05. In contrast, using the glycosylase monoclonal antibody 40.10.09, two protein bands of molecular weight 37,000 and 24,000 daltons were observed. The Poly (A) containing RNA was then sedimented through a 10-40% linear neutral sucrose gradient. Fractions were collected from the bottom of the gradient, in vitro translation was performed on each fraction followed by immunoprecipitation by antibody 40.10.09 and subsequent SDS gel electrophoresis. This analysis documented that the 37,000 dalton polypeptide was encoded by a 16 S Poly (A) RNA while the 24,000 dalton polypeptide was synthesized from an 11 S Poly (A) RNA. However, immunoprecipitation of protein from crude cell extracts of [^{35}S]-methionine labeled cells demonstrated that only the 37,000 dalton protein could be observed in vivo. These results demonstrated that the normal human uracil DNA glycosylase was not synthesized as a precursor polypeptide which required a post-translational proteolysis to produce the catalytically active 37,000 dalton glycosylase molecule. The results also suggest that the 24,000 dalton immunoreactive polypeptide observed during in vitro translation was due to RNA degradation and did not reflect a smaller precursor polypeptide which might be synthesized in vivo.

ALTERED IMMUNOREACTIVITY OF THE BLOOM"S SYNDROME URACIL DNA GLYCOSYLASE

As previously described, Bloom's syndrome cells can be characterized by a series of alterations in the regulation of DNA repair during the cell cycle (Gupta and Sirover, 1984a; Dehazya and Sirover, 1986; Kim, et al., 1986). In the course of those experiments, we examined the induction of the uracil DNA glycosylase through two successive cell cycles (Figure 3, Vollberg, et al., 1987b). Bloom's syndrome cells (GM-2548) were synchronized at the G1/S border by incubation with complete media containing 0.5 mM hydroxyurea for 18 hours. Proliferation was initiated by removal of the drug and addition of fresh media. The Bloom's syndrome cells immediately entered the S phase of the first cell cycle with maximum levels of DNA synthesis 6 hours after removal of the drug. DNA synthesis then declined and was reinitiated in a second cell cycle with a maximum rate at 24 hours after proliferation was initiated. This temporal progression of DNA replication in two successive cell cycles in Bloom's syndrome cells was indistinguishable from that observed for normal human cells

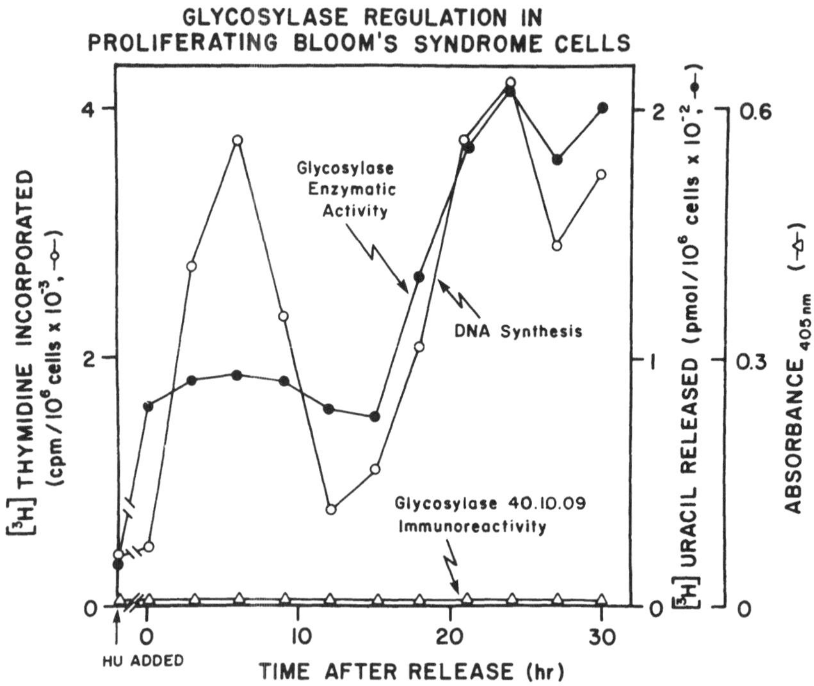

Figure 3. Immunochemical characterization of Bloom's syndrome uracil DNA glycosylase during synchronous growth. Bloom's syndrome cells (GM-2548; 1 x 10^6 cells/100mm culture dish) were incubated with 0.5 mM hydroxyurea for 18 hours. Cell growth was initiated by media exchange. DNA synthesis (-o-) was determined by [^{3}H]thymidine incorporation. Uracil DNA glycosylase activity (-●-) was quantitated by the release of ethanol soluble radioactivity from a polynucleotide substrate containing [^{3}H]uracil. Immunoreactivity of the crude cell extracts (-Δ-) was measured by ELISA. Reprinted from Vollberg, et al., 1987b, by permission of the publisher.

after hydroxyurea synchronization (Dehazya and Sirover, 1986).

The induction of the Bloom's syndrome uracil DNA glycosylase was then examined. Cells were collected at the indicated intervals. Crude extracts were prepared by sonication at 60 Watts for 20 seconds at 4^{0}C using a Braunosonic needle probe. Cell debris was pelleted by centrifugation at 2300 x g for 10 minutes at 4^{0}C. The supernatant fraction was removed and used to quantitate glycosylase biochemical and immunochemical activity. Glycosylase regulation was measured by two criteria. First, the activity of the enzyme was measured by in vitro bioassay. For this analysis, a polynucleotide substrate containing [^{3}H]uracil was prepared in a DNA polymerase reaction as previously described (Sirover, 1979, Gupta and Sirover, 1984a,b). Glycosylase activity was detected by the release of [^{3}H]uracil into the ethanol soluble supernatant fraction.

Basal levels of glycosylase activity could easily be detected at the interval at which hydroxyurea was added. This activity increased in the first cell cycle coordinate with the induction of the first round of DNA replication. As the cells entered the second round of DNA replication, the glycosylase was increased further. This second increase in glycosylase activity paralleled the second round of DNA replication in the second cell cycle after release from the hydroxyurea block. A similar cyclic variation was observed in normal human cells. Further, in both cell types, the hypoxanthine DNA glycosylase was induced in both cell cycles in a periodic fashion (Dehazya and Sirover, 1986).

Second, the immunoreactivity of the glycosylase in the crude cell extracts was examined by ELISA using the identical samples in which glycosylase activity was quantitated. Although basal levels of uracil DNA glycosylase activity could be detected by biochemical assay when the cells were initially exposed to 0.5 mM hydroxyurea, no ELISA reactivity could be observed using the anti-human glycosylase antibody 40.10.09. Similarly, no ELISA reactivity with that antibody could be observed during the induction of the glycosylase in the first cell cycle after release from the hydroxyurea block. Lastly, no immunoreactivity was observed after the sequential induction of the glycosylase in the second cell cycle. Thus, surprisingly, using the 40.10.09 monoclonal antibody, no immunoreactivity was observed at any of the intervals examined even though dramatic increases in glycosylase enzyme activity could be observed.

Two control studies were performed. First, using the identical cell-free extracts, ELISA reactivity of the uracil DNA glycosylase was examined with glycosylase monoclonal antibody 42.08.07. In this instance, normal immunoreactivity was observed. This included recognition of basal glycosylase levels at the time of hydroxyurea addition; increased ELISA reactivity during the first induction of the uracil DNA glycosylase; and a further increase in ELISA immunoreactivity during the second increase in glycosylase activity. Second, mixing experiments were performed with equal amounts of the normal human glycosylase preparation and a comparable preparation from Bloom's syndrome cells. No inhibition of the ELISA reactivity of the normal human enzyme was observed. This result verified that the altered immunogenicity of the Bloom's syndrome enzyme was not due to the presence of some inhibitor of the 40.10.09 antibody which may have been present in the enzyme preparation.

To determine whether this antigenic defect may be characteristic of Bloom's syndrome, the enzyme was partially purified from five Bloom's syndrome cell strains derived from five separate individuals who presented clinically with this human genetic disorder. Each enzyme was examined by ELISA in comparison with the normal human fibroblast enzyme. Using antibodies 37.04.12 and 42.08.07, identical immunoreactivity was observed with each of the six human uracil DNA glycosylases over the entire concentration range which was examined (Figure 4A, 4B, respectively). In contrast, a different result was observed using antibody 40.10.09 as the first antibody in the ELISA (Figure 4C). This antibody recognized the enzyme from normal human fibroblasts. However, the monoclonal antibody was unable to recognize any of the Bloom's syndrome

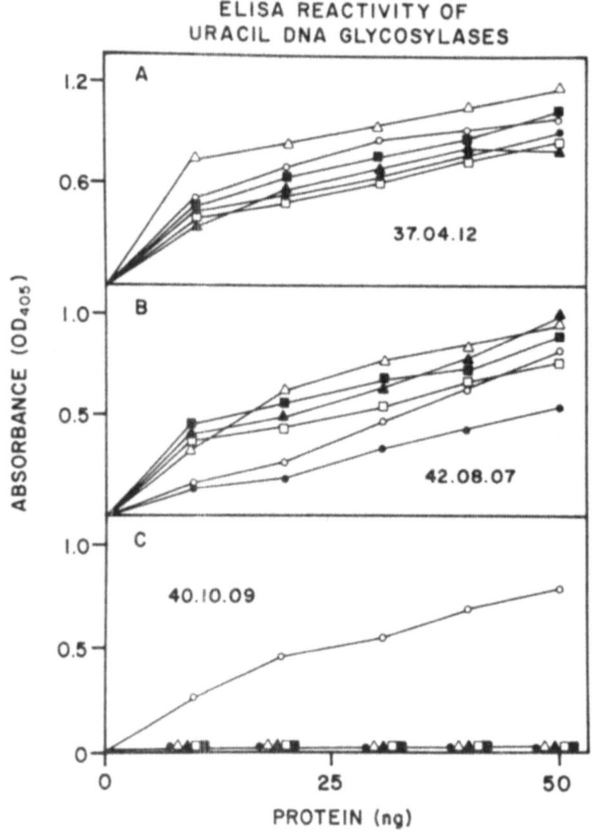

Figure 4. Altered immunoreactivity of Bloom's syndrome uracil DNA glycosylases. Human uracil DNA glycosylases were purified by DEAE and Phosphocellulose chromatography. ELISA was performed in triplicate with each of the indicated monoclonal antibodies over the protein concentration curve. ELISA with antibody 37.04.12 (A); antibody 42.08.07 (B); antibody 40.10.09 (C). Normal human glycosylase CRL-1222 (-O-); Bloom's syndrome glycosylase GM-1492 (-△-); GM-2548 (-●-); GM-3402 (-▲-); GM-3498 (-□-); GM-3510 (-■-). Published in Seal, et al., Proc. Natl. Acad. Sci., 1988.

uracil DNA glycosylases. No immunoreactivity was observed over the entire concentration range nor detected even when a 100 fold excess of the Bloom's syndrome enzyme was used (Results not shown).

To determine whether this result was unique to Bloom's syndrome cells, the ELISA was performed using the uracil DNA glycosylase from a variety of normal human cell sources as well as cells from a number of other human genetic disorders. Using crude extracts, the ELISA was performed over a range of protein concentrations (10-100 ng) and compared to that of a normal human cell strain (CRL-1222) at each of those concentrations. The absorbance ratios were calculated at each interval in the concentration curve then summed for statistical analysis. As shown in Table 2, each of the three anti-human placental uracil DNA glycosylase monoclonal antibodies quantitatively recognized the human enzyme from a number of normal human sources. These included the enzyme from three fibroblast

TABLE 2. Immunoreactivities of Human Uracil DNA Glycosylases*

Source	Immunoreactivities of monoclonal antibodies (relative to normal human skin fibroblasts, CRL 1222)		
	37.04.12	42.08.07	40.10.09
Normal human cells			
Skin fibroblasts (CRL 1222)	1.00	1.00	1.00
Human placenta	0.92 ± 0.09	0.92 ± 0.19	0.90 ± 0.10
Lung fibroblasts (CCL 75)	1.02 ± 0.10	0.89 ± 0.16	0.94 ± 0.12
Skin fibroblasts (GM 5879)	0.98 ± 0.04	1.28 ± 0.07	0.91 ± 0.11
Lymphocytes	1.32 ± 0.23	1.04 ± 0.08	1.19 ± 0.26
Transformed cells			
SV-40 transformed lung fibroblasts (CCL 75.1)	0.92 ± 0.14	1.12 ± 0.12	1.14 ± 0.21
Bloom's syndrome fibroblasts			
GM 1492	1.02 ± 0.12	1.15 ± 0.07	0.00
GM 2548	1.08 ± 0.09	1.11 ± 0.08	0.00
GM 3402	1.01 ± 0.03	1.13 ± 0.03	0.00
GM 3498	0.99 ± 0.00	0.94 ± 0.10	0.00
GM 3510	0.81 ± 0.05	1.23 ± 0.22	0.00
Ataxia telangiectasia fibroblasts			
GM 0367	1.17 ± 0.23	0.91 ± 0.24	0.94 ± 0.18
GM 2052	0.85 ± 0.22	0.96 ± 0.08	1.03 ± 0.19

Table 2 Continued

Xeroderma pigmentosum fibroblasts			
CRL 158 (XP-C)	0.98 ± 0.07	1.08 ± 0.20	0.81 ± 0.09
CRL 1258 (XP-D)	1.11 ± 0.25	0.93 ± 0.05	0.93 ± 0.04
GM 3614 (XP-variant)	0.82 ± 0.04	0.84 ± 0.04	0.99 ± 0.04
Tay-Sachs flbroblasts			
GM 2968	1.20 ± 0.25	1.00 ± 0.06	0.98 ± 0.24
GM 4863	0.80 ± 0.08	1.07 ± 0.08	0.99 ± 0.18
Familial hypercholesterolemia fibroblasts			
GM 1355	1.07 ± 0.10	1.01 ± 0.08	1.04 ± 0.30
GM 2408	0.65 ± 0.13	1.09 ± 0.06	0.75 ± 0.13
Galactosemia fibroblasts			
GM 1209	0.89 ± 0.13	0.89 ± 0.19	1.16 ± 0.08
GM 1908	1.09 ± 0.13	1.05 ± 0.07	1.10 ± 0.26
Progeroid fibroblasts			
AG 3911	0.99 ± 0.09	0.98 ± 0.09	1.08 ± 0.17
AG 6917	1.03 ± 0.12	1.14 ± 0.07	1.01 ± 0.06

*The concentration dependent recognition of the uracil DNA glycosylase in the ELISA by the respective anti-human uracil DNA glycosylase monoclonal antibody, over a protein concentration range of 10-50 ng, was determined. Ratios were calculated by dividing the A_{405} of the cell strain at each concentration by the A_{405} of the normal human fibroblast cell strain (CRL 1222) at that concentration. Ratios were determined at a minimum of 4 separate protein concentrations within the concentration curve. The lack of recognition of the uracil DNA glycosylase in the ELISA by the respective anti-human uracil DNA glycosylase monoclonal antibody was examined over the protein concentration range of 10-50 ng as well as at 10 µg of protein (100-fold excess). Ratios were calculated as described above. The numbers of each human cell strain or line refers to their designations by either the American Type Culture Collection or the Human Genetic Cell Repository in Camden, New Jersey. Published in Seal, et al., Proc. Natl. Acad. Sci., 1988.

cell strains, from human placenta as well as from human lymphocytes. In addition, the immunoreactivity of an SV-40 transformed derivative of one of the fibroblast cell strains was examined. Within the range of statistical error, there appeared to be an equal recognition of the enzyme from each normal human cell source in the ELISA.

To examine further the specificity of the defect in immunoreactivity of the Bloom's syndrome uracil DNA glycosylase, ELISA was performed using crude cell extracts from a variety of fibroblasts from individuals with defined genetic disorders. In particular, individuals with xeroderma pigmentosum (XP) or ataxia telengectasia (AT) are cancer prone and display defined hypersensitivities to specific DNA damaging agents. Similarly, in common with Bloom's syndrome, they display a high level of chromosomal aberrations. However, each was reported to normally regulate the temporal induction of the uracil DNA glycosylase during the cell cycle (Gupta and Sirover, 1984b). As shown in Table 2, the uracil DNA glycosylase from two different XP complementation groups as well as that from an XP-variant cell strain were normally recognized by each of the three anti-uracil DNA glycosylase monoclonal antibodies. Similar results were observed with the glycosylase from two different AT cell strains.

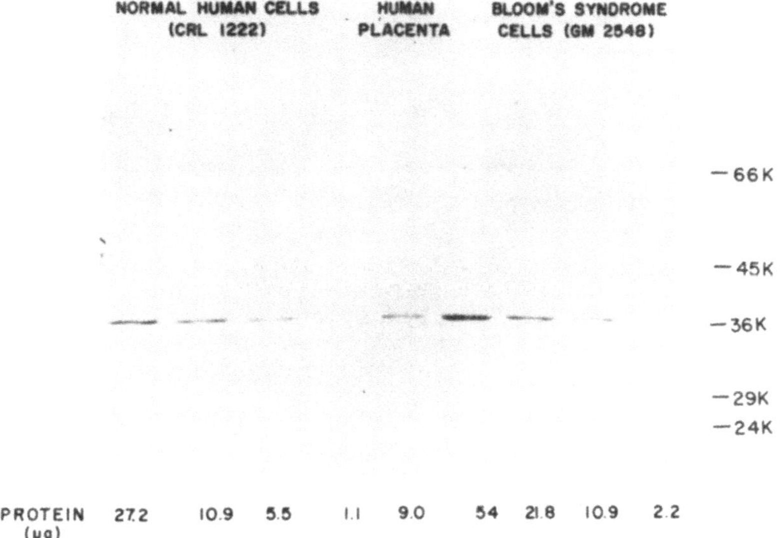

Figure 5. Immunoblot analysis of human uracil DNA glycosylases. Crude cell extracts were prepared from normal human (CRL-1222) and Bloom's syndrome (GM-2548) cells. SDS-gel electrophoresis, electroblotting and immunoblot analysis were performed as previously described. Reprinted from Vollberg, et al., 1987b, by permission of the publisher.

To further define the uniqueness of the defect in the Bloom's syndrome uracil DNA glycosylase, the ELISA was performed using the glycosylase from fibroblast cell strains isolated from individuals with diverse genetic disorders. These included cell strains from individuals with progeria,

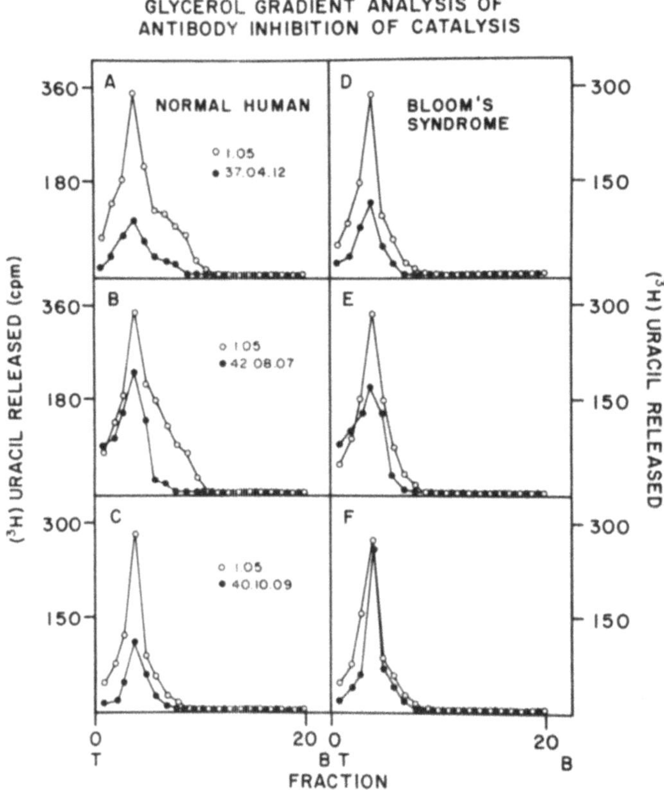

Figure 6. Inhibition of enzyme catalysis by anti-human placental uracil DNA glycosylase monoclonal antibodies. Equivalent concentrations of each glycosylase monoclonal antibody or the negative control antibody 1.05 were used (normal human fibroblasts, CRL 1222; Bloom's syndrome, GM 2548). Published in Seal, et al., Proc. Natl. Acad. Sci., 1988.

a disease characterized by premature aging, as well as Tay-Sachs disease and two other syndromes which contain metabolic disorders. Normal ELISA reactivity was observed with each of the eight different cell strains with each of the three glycosylase monoclonal antibodies. These results further demonstrate the specificity of the antigenic defect in the uracil DNA glycosylase from the five Bloom's syndrome cell strains.

Immunoblot analysis was performed to determine whether the Bloom's syndrome enzyme represented a different glycosylase species and whether that glycosylase contained the antigenic determinant recognized by the 40.10.09 antibody (Figure 5). Using the GM-2548 cell strain, a 37,000 Dalton

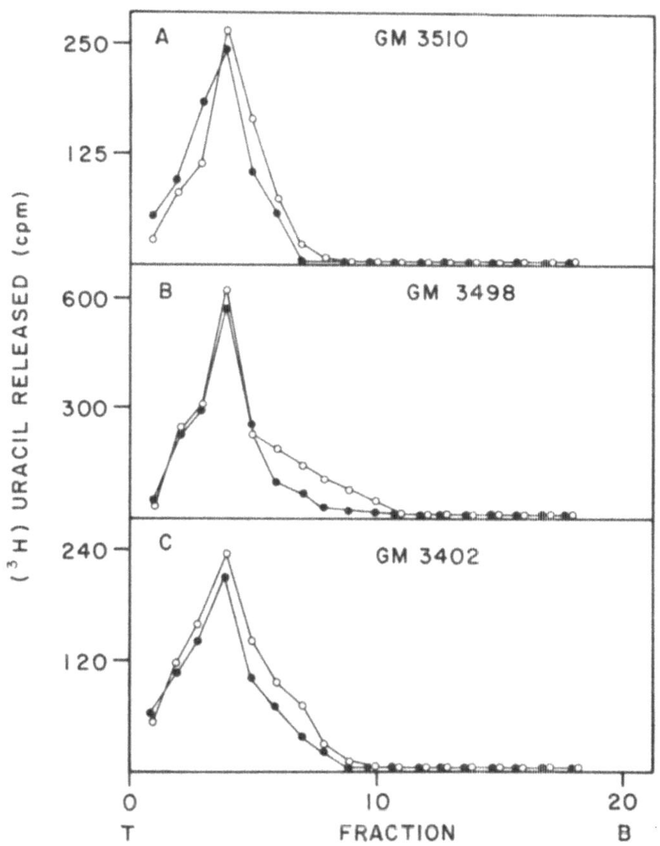

Figure 7. Lack of inhibition of Bloom's syndrome uracil DNA glycosylase with monoclonal antibody 40.10.09. Open circles - incubation with control antibody 1.05, closed circles - incubation with glycosylase monoclonal antibody 40.10.09.

polypeptide was detected identical to that observed in normal human fibroblasts and in human placenta. It also appeared that the 40.10.09 antibody had an equal affinity for each of the human enzymes. Similarly, a single 37,000 Dalton polypeptide band was observed in the immunoblot analysis when the cell free extracts from the four other Bloom's syndrome cell strains were used (Seal, et al., 1988). These results suggest that each of the Bloom's syndrome enzymes contains the 40.10.09 antigenic determinant which could be recognized by that antibody when the Bloom's syndrome glycosylase was denatured.

To determine whether the altered immunoreactivity of the Bloom's syndrome enzymes could be detected by a different criteria, inhibition of glycosylase enzyme activity by each antibody was examined. The glycosylase from either normal human skin fibroblasts (CRL-1222) and Bloom's syndrome cells (GM-2548) were used. Each enzyme was preincubated with each of the three antibodies. The antigen-antibody preparation was then centrifuged through a 10-35% glycerol gradient. Each gradient was fractionated and glycosylase activity determined. As a negative control, each enzyme was preincubated with monoclonal antibody 1.05. Using monoclonal antibody 37.04.12, comparable enzyme inhibition was observed when either the normal human or the Bloom's syndrome glycosylase were used (Figure 6A, 6D, respectively). Similar results were obtained when each glycosylase was preincubated with antibody 42.08.07 (Figure 6B, 6E, respectively).

In contrast, a difference was observed when either glycosylase was preincubated with antibody 40.10.09. As expected, preincubation of the normal human enzyme with that antibody reduced enzyme activity compared to that observed when the enzyme was preincubated with the negative control antibody 1.05 (Figure 6C). However, there was no comparable decrease in enzyme activity when the Bloom's syndrome glycosylase was preincubated with the 40.10.09 antibody (Figure 6F).

In this experiment, the level of activity was identical to that observed for the Bloom's syndrome enzyme when it was preincubated with the negative control antibody 1.05. Similar results were observed when each of the other four Bloom's syndrome uracil DNA glycosylases were preincubated with the 40.10.09 antibody (Figure 7; in addition see Vollberg, _et al._, 1987b). Thus, the altered immunoreactivity of the uracil DNA glycosylase from five separate Bloom's syndrome cell strains was observed using a second criteria independent of the ELISA protocols initially used to document this alteration.

SUMMARY AND PERSPECTIVES

These studies suggest that Bloom's syndrome may be characterized by a defined structural mutation in the uracil DNA glycosylase protein. This defect is defined by the altered immunological response by the Bloom's syndrome glycosylase molecule. In particular, an anti-human placental uracil DNA glycosylase monoclonal antibody, 40.10.09, failed to immunologically react with the native glycosylase purified from five Bloom's syndrome cell strains. In contrast, normal reactivity was observed with two other anti-glycosylase monoclonal antibodies. Each cell strain was derived from separate individuals who clinically presented with this human genetic disorder. Four of the individuals were Ashkenazi Jewish, the fifth was an American Black. Thus, this aberrant immunoreactivity does not appear to be restricted to a polymorphism characteristic of a particular ethnic group. Instead this abnormality would be a distinguishing feature of this human genetic disorder. The lack of immunoreactivity of the Bloom's syndrome glycosylase was not related to the proliferative state of the cell and, therefore, would not appear to be the result of the absence of a proliferative-dependent glycosylase isoenzyme. Immunoblot analysis demonstrated that the altered immunoreactivity of the Bloom's syndrome enzyme resulted from an alteration within the glycosylase molecule which in the native conformation masked the antigenic sequence recognized by the 40.10.09 antibody. Lastly, the specificity of the aberrant immunoreactivity of the Bloom's syndrome uracil DNA glycosylase was confirmed by the normal reaction of the glycosylase from five normal human cell types as well as thirteen other abnormal human cell strains.

The results summarized in this report demonstrate that the Bloom's syndrome uracil DNA glycosylase may be characterized by a unique antigenic defect. However, it is equally clear that Bloom's syndrome cells do not display a reduced amount of this base excision repair enzyme. In particular, determinations of the specific activity of the Bloom's syndrome enzyme from the GM-1492 and GM-2548 cell strains demonstrate an equal or greater amount of the glycosylase as compared to normal human cells or to cells from individuals with other genetic disorders. Thus, it is reasonable to suggest that the structural mutation in the Bloom's syndrome enzyme does not affect catalysis as measured by _in vitro_ assay. However, this amino acid alteration could affect the

TABLE 3. DNA Repair Defects In Bloom's Syndrome Cells

Category	Repair Deficiency	Reference
Temporal alteration in Cell Cycle Regulation	Nucleotide Excision Repair	Gupta and Sirover, 1984
	Base Excision Repair	Dehazya and Sirover, 1986
	Uracil DNA Glycosylase	Kim, _et al._, 1986
	Hypoxanthine DNA Glycosylase	
No Cell Cycle Regulation	O^6-Methylguanine Methyltransferase	Kim, _et al._, 1986
Reduced Levels of Enzyme Activity (Cell Cycle Related?)	DNA Ligase	Willis and Lindahl; Chan, _et al._, 1987; Willis, _et al._, 1987
Mutant Enzyme Protein	Uracil DNA glycosylase	Vollberg, _et al._, 1987; Seal, _et al._, 1988

binding of the Bloom's syndrome enzyme to damage within chromatin or its potential physical association with other DNA repair enzymes (Seal and Sirover, 1986). Alternatively, one can not exclude the possibility that the Bloom's syndrome enzyme is not characterized by an amino acid substitution within the glycosylase. Instead, there could be a defect in a post-translational modification. At present, it is unknown whether such modifications of the glycosylase occur in normal human cells.

Recent evidence suggests that Bloom's syndrome cells can be characterized by a series of cellular DNA repair deficiencies. As shown in Table 3, four different types of alterations have been identified. However, it is currently unclear how each of these deficiencies in Bloom's syndrome cells relate to one another. Each would appear to be independent of the others and it is uncertain at the present time how each, by itself, could account for the multiplicity of defects characteristics of this human genetic disorder. In addition, there is an apparent inconsistency in the identification of four separate and independent defects in Bloom's syndrome as it has been identified as an autosomal, recessive human genetic disease. That observation would presumably locate the defect in Bloom's syndrome within a single gene. However, although no direct evidence has as yet been presented, one can not exclude the possibility of the existence of complementation groups within this disorder.

Thus, it is tempting to speculate that none of the DNA repair defects described in Table 3 represent the actual molecular mechanism which underlies this human genetic disease. Instead, one can postulate that Bloom's syndrome may arise due to a molecular defect in a regulatory gene which is required for the cell cycle expression of DNA repair. Each of the four categories described in Table 3 share the common characteristic of an aberrant regulation of DNA repair gene expression during cell proliferation. This would include not only the temporal alterations in the regulation of nucleotide excision repair and of base excision repair, but also, the failure to enhance O^6-alkylguanine alkyltransferase activity during the cell cycle. This common failure of repair gene expression would also include the observations of decreased DNA ligase activity in Epstein-Barr Virus transformed Bloom's syndrome human lymphoblastoid cell lines. As a transformed cell population, they would contain a significantly higher proportion of proliferating cells.

Each of these hypotheses may be verified experimentally. In particular, the use of molecular technology is extremely attractive to isolate and characterize these DNA repair genes in both normal human cells and in Bloom's syndrome cells. Through these investigations one may rigourously define the existence of individual DNA repair enzyme species; the nucleotide sequence of the normal human uracil DNA glycosylase gene; the existence of specific alterations in this nucleotide sequence in the Bloom's syndrome uracil DNA glycosylase gene; the determination of common regulatory sequences which may be required for the cell cycle regulation of DNA repair genes; and whether such common regulatory sequences may be altered in Bloom's syndrome.

ACKNOWLEDGEMENTS

This research was funded by grants to MAS from the National Institutes of Health (CA-29414), from the National Science Foundation (DCB-8416295) and by the W.W. Smith Charitable Trust and by grants to the Fels Institute for Cancer Research and Molecular Biology from the National Institutes of Health (CA-12227) and from the American Cancer Society (SIG-6).

REFERENCES

Arenaz, P. and Sirover, M.A., 1983, Isolation and characterization of monoclonal antibodies to the uracil DNA glycosylase from human placenta, Proc. Natl. Acad. Sci., USA, 80:5822.

Bessman, M.J., Lehman, I.R., Adler, J., Zimmerman, S.B., Simns, E.S., and Kornberg, A., 1958, Enzymatic synthesis of DNA. III. The incorporation of purine and pyrimidine analogues into DNA. Proc. Natl. Acad. Sci., USA, 44:633.

Bloom, D., 1966, The syndrome of congenital telangiectatic erythema and stunted growth, J. Pediatrics, 68:103.

Bryant, E.M., Hoehn, H., and Martin, G.M., 1979, Normalization of sister chromatid exchange frequencies in Bloom's syndrome by euploid cell hybridization. Nature, 279:795.

Chaganti, R.S.K., Schonberg, S. and German, J., 1979, A many fold increase in sister chromatid exchanges in Bloom's syndrome lymphocytes. Proc. Natl. Acad. Sci., USA, 71:4508.

Chan, J.Y.H., Becker, F.F., German, J., and Ray, J.H., 1987, Altered DNA ligase I activity in Bloom's syndrome cells. Nature, 325:357.

Dehazya, P. and Sirover, M.A., 1986, Regulation of hypoxanthine DNA glycosylase in normal human and Bloom's syndrome fibroblasts. Cancer Res., 46:3756.

Doniger, D., DiPaolo, J.A., and Popescu, N.C., 1983, Transformation of Bloom's Syndrome Fibroblasts by DNA Transfection. Science, 222:1144.

Friedberg, E.C., 1985, DNA Repair, New York, W.H. Freeman and Co.

German, J., 1969, Bloom's syndrome. I. Genetic and clinical observations in the first 27 patients. Am. J. Hum. Genet., 21:196.

German, J., Bloom, D. and Passarge, E., 1977, Bloom's syndrome, V. Surveillence for cancer in affected families. Clin. Gen., 12:162.

German, J., Bloom, D. and Passarge, E., 1984, Bloom's syndrome. XI. Progress report for 1983, Clin. Genet., 254:166.

Gupta, R.S. and Goldstein, S., 1980, Diphtheria toxin resistance in human fibroblast cell strains from normal and cancer-prone individuals, Mutat. Res., 73:331.

Gupta, P.K. and Sirover, M.A., 1980, Sequential stimulation of DNA repair and DNA replication in normal human cells, Mutation Res., 72:273.

Gupta, P.K. and Sirover, M.A., 1981a, Cell cycle regulation of DNA repair in normal and repair deficient human cells, Chem. Biol. Int., 36:19.

Gupta, P.K. and Sirover, M.A., 1981b, Stimulation of the

nuclear uracil DNA glycosylase in proliferating human fibroblasts, Cancer Res., 41:3133.
Gupta, P.K. and Sirover, M.A., 1984a, Altered temporal expression of DNA repair in hypermutable Bloom's syndrome cells, Proc. Natl. Acad. Sci. USA, 81:757.
Gupta, P.K. and Sirover, M.A., 1984b, Regulation of DNA repair in serum stimulated xeroderma pigmentosum cells, J. Cell Biol., 99:1275.
Hayatsu, H., 1977, Co-operative mutagenic actions of bisulfite and nitrogen nucleophiles, J. Mol. Biol., 115:19.
Inoue, T., Hirano, K., Yokoiyama, A., Kata, T., and Kato, H., 1977, DNA repair enzymes in ataxia telangiectasia and Bloom's syndrome fibroblasts, Biochim. Biophys. Acta, 479:497.
Kim, S., Vollberg, T.M., Ro, J.Y., Kim, M. and Sirover, M.A., 1986, O^6-methylguanine methyltransferase increases before S-phase in normal human cells but does not increase in hypermutable Bloom's syndrome cells, Mutat. Res., 173:141.
Lindahl, T., 1982, DNA Repair Enzymes, Annu. Rev. Biochem., 51:61.
Remsen, J.F., 1980, Repair of damage by N-acetoxy-2-acetyl aminofluorene in Bloom's syndrome, Mutat. Res., 72:151.
Seal, G. and Sirover, M.A., 1986, Physical association of the human base excision repair enzyme uracil DNA glycosylase with the 70,000 dalton catalytic subunit of DNA polymerase alpha, Proc. Natl. Acad. Sci. USA, 83:7608.
Seal, G., Arenaz, P. and Sirover, M.A., 1987, Purification and properties of the human placental uracil DNA glycosylase, Biochim. Biophys. Acta., 925:226.
Seal, G., Brech, K., Karp, S.J., Cool, B.J. and Sirover, M.A., 1988, Immunological lesions in human uracil DNA glycosylase: Association with Bloom's Syndrome, Proc. Natl. Acad. Sci. USA, 85:2339.
Sirover, M.A., 1979, Induction of the DNA repair enzyme uracil DNA glycosylase in stimulated human lymphocytes, Cancer Res., 39:2090.
Sirover, M.A., 1989 "Cell cycle regulation of DNA repair enzymes and pathways" in: Transformation of Human Fibroblasts (G.P. Milo and B.C. Casto, Eds.) CRC Press, Boca Raton, Florida. In Press.
Sirover, M.A. and Gupta, P.K., 1983, "Regulation of DNA Repair in Human Cells" in: Human Carcinogenesis (C.C. Harris and H.N. Autrup, Eds.) Academic Press, New York, 255.
Strauss, B.S., 1985, Cellular Aspects of DNA Repair, Adv. Can. Res., 45:45.
Teebor, G.W. and Frenkel, K., 1983, The Initiation of DNA-Excision Repair, Adv. Cancer Res., 38:23.
Vijayalaxmi, Evans, H.J., Ray, J.H. and German, J., 1983, Bloom's syndrome: evidence for an increased mutation frequency in vivo, Science, 221:851.
Vollberg, T.M., Cool, B.L. and Sirover, M.A., 1987a, Biosynthesis of the human base excision repair enzyme uracil DNA glycosylase, Cancer Res., 47:123.
Vollberg, T.M., Seal, G. and Sirover, M.A., 1987b, Monoclonal antibodies detect conformational abnormality of uracil DNA glycosylase in Bloom's syndrome cells, 8:1725.
Warren, S.T., Schultz, R.A., Chang, C.C., Wade, M.H. and Trosko, J.E., 1981, Elevated spontaneous mutation rate in Bloom's syndrome fibroblasts, Proc. Natl. Acad. Sci. USA, 78:3133.

Willis, A.E. and Lindahl, T., 1987, DNA ligase I deficiency in Bloom's syndrome, Nature, 325:355.
Willis, A.E., Weksberg, R., Tomlinson, S. and Lindahl, T. 1987. Structural alterations of DNA ligase I in Bloom's syndrome, Proc. Natl. Acad. Sci. USA, 84:8016.

DNA LIGASE ACTIVITY IN HUMAN CELLS FROM NORMAL DONORS AND FROM PATIENTS WITH BLOOM'S SYNDROME AND FANCONI'S ANEMIA

Mauro Mezzina[1], Silvano Nocentini[3], Jeannette Nardelli[2], A. Ivana Scovassi[4], Umberto Bertazzoni[4] and Alain Sarasin[1]

[1]Laboratoire de Génétique Moléculaire, [2]Laboratoire de Cancer et Differentiation, Institut de Recherches Scientifiques sur le Cancer, b.p. n[0] 8, 94802 Villejuif, France; [3]Institut Curie, Section de Biologie, 26, Rue D'Ulm 75231 Paris, France; [4]Istituto di Genetica del CNR, Via Abbiategrasso n[0] 207, 27100 Pavia, Italy

ABSTRACT

DNA ligase was analyzed in human cells obtained from normal individuals and from Bloom's syndrome (BS) and Fanconi's anemia (FA) patients. The level of enzyme activity in untransformed and transformed FA cell lines was identical to that of control cells while it was significantly higher in BS cell lines. These results were confirmed after purification of the BS or control cell enzymes on fast protein liquid chromatography from crude extracts and from ammonium sulfate pellets. After precipitation with polymin-P a lower level of ligase activity was found in BS cells. The structural property of the enzyme, investigated by activity gel analysis, showed that active polypeptides of identical sizes are obtained from control or BS cell lysates or extracts and that ligase activity could be recovered from the polymin-P pellet of BS cell extracts. These results indicate that FA and BS do not appear to be deficient in DNA ligase activity.

INTRODUCTION

Some inherited human diseases characterized by a high incidence of cancer have been reported to present a deficiency in DNA ligase. In particular, Fanconi's anemia (FA) (Fanconi, 1976) and Bloom's syndrome (BS) (Bloom, 1954; German, 1969) cells exhibit physiological characteristics which could be explained by a DNA ligase defect. FA cells present, among other abnormalities (Sasaki and Tonomura, 1973; Latt, et al., 1975; Fujiwara, et al., 1977; Joenjie and Oostra, 1983; Poll, 1984; Moustacchi, and Diatloff-Zito, 1985), a decreased growth rate, an increased generation time (Sasaki, 1975; Weksberg, et al.,

1979; Dutrillaux, et al., 1982) and a delay in the rejoining of repaired DNA strands following UV-irradiation (Hirsch-Kauffmann, et al., 1978). In agreement with these observations, a decreased activity of DNA ligase in both fibroblasts and lymphoblasts from FA patients has been reported (Hirsch-Kauffmann, et al., 1978), but this result has not been confirmed by others (Willis, and Lindhal, 1987).

BS cells have also been described as having cellular abnormalities suggestive of a DNA ligase deficiency such as: longer generation time (Giannelli, et al., 1977), increased chromosomal instability (German, et al., 1965), high level of sister chromatid exchanges (SCE) (Chaganti, et al., 1974), defects in semi-conservative DNA synthesis, i.e., slow progression of DNA replication forks, and delayed rates of conversion of nascent DNA chains into mature forms (Hand and German, 1975; Kapp, 1982; Ockey and Saffhill, 1986; Friedberg, et al., 1979), and abnormal responses to DNA damage (Friedberg, et al., 1979; Hirschi, et al., 1981; Selcky, et al., 1979; Smith and Paterson, 1982). It has recently been shown that in several Epstein-Barr-virus (EBV)-transformed lymphoblastoid cells or Simian-Virus-40 (SV40)-transformed fibroblasts, derived from BS, the first chromatographic form of DNA ligase eluting by gel filtration was reduced, compared to DNA ligase activity observed in control cells (Willis and Lindhal, 1987; Chan, et al., 1987; Willis, et al., 1987). This form of enzyme, called DNA ligase I, has been described as involved in DNA replication in several mammalian systems, including human cells; it is the major, large sized 180-480 kDa polypeptide after gel filtration (Söderhall and Lindhal, 1976; Spadari, 1976; Chan, and Becker, 1985), or a 130-200 kDa polypeptide after polyacrylamide gel electrophoresis (Teraoka, and Tsukada, 1982, 1985; Mezzina, et al., 1987). It has thus been suggested that a mutation in the DNA ligase gene producing an altered enzyme could be responsible for the genetic defect of BS (Willis, et al., 1987).

In this work we extended DNA ligase analysis to several other untransformed and transformed fibroblast and lymphoblast cell lines obtained from control, FA and BS donors. DNA ligase measurements of total enzyme activity were carried out in the crude cell extracts and in the partially-purified enzyme after fast protein liquid chromatography (FPLC). The identification of catalytic polypeptides was performed using the activity gel assay (Mezzina, et al., 1984). Our results indicate that no significant difference in the specific activity of the enzyme was found in FA compared to control cells. In BS cells, DNA ligase does not seem to be reduced, but in contrast, it is significantly higher than in control cells.

RESULTS

In a first set of experiments, total ligase activity from control, BS and FA cells was measured in crude extracts. The enzymatic assay is based on the conversion of poly(dA) $\cdot$ ^{32}P-oligo(dT) substrate into an alkaline phosphatase-resistant form (Mezzina and Nocentini, 1978). In our standard conditions, the enzyme activity was measured from first order kinetics curves. Results showed that: (1) the specific activity of DNA ligase in untransformed fibroblasts is significantly higher ($P < 0.02$) in BS cells as compared to control cells; (2) in transformed cell lines, both fibroblasts and lymphoblasts, the levels of DNA ligase in BS cells are significantly higher ($P \ll 0.001$) than in control cells; and (3) in FA cells no significant difference ($P < 0.5$ or 0.9) in the enzyme activity was observed either in untransformed fibroblasts or in transformed lymphoblasts with regard to control cells (Table 1).

Table 1. DNA ligase specific activity (SA) in human cells from control (C), Bloom's Syndrome (BS) and Fanconi's anemia (FA) donors.

UNTRANSFORMED CELL LINES		TRANSFORMED CELL LINES	
fibroblasts	DNA ligase SA	lymphoblasts	DNA ligase SA
1BR/3(C)	6.4 ± 0.6(6)	GM3299(C)	14.7 ± 5.4(20)
HEL(C)	7.1 ± 1.2(5)	GM3403C(BS)	31.1 ± 8.4(14)
YBL6(BS)	9.4 ± 1.9(4)		P << 0.001
	P < 0.02	AHH1(C)	12.1(2)
GM1492(BS)	10.4 ± 1.6(3)	HSC62(FA)	10.9(2)
	P < 0.01	HSC99(FA)	11.8
F145(FA)	6.01 ± 2.1(8)	fibroblasts	
	P < 0.5	HGOV5(C)	10.7 ± 2.0(7)
F150(FA)	5.66 ± 1.9(10)	GM8505(BS)	17.0 ± 2.8(7)
	P < 0.5		P < 0.001
F71(FA)	6.04 ± 1.0(5)		
	P < 0.9		

Enzyme extraction was carried out with buffer containing 0.5 M NaCl and 0.53 Triton X-100. The data refer to pmoles$\cdot$min^{-1}. mg^{-1} of proteins of oligo(dT) converted into alkaline phosphatase-resistant form (enzyme units), ± standard deviation of the mean. Number of experiments is indicated in parenthesis. Values of P were derived from the Student t-test.

A partial enzyme purification from control and BS lymphoblasts was carried out by FPLC of crude extracts, ammonium sulfate (AmS) pellets and polymin-P supernatant fractions as described (Willis and Lindahl, 1987). Results, presented in Figure 1, show that: (1) when cell extracts were immediately processed for FPLC, the enzyme activity was eluted in a major peak of about 180 kDa for both control and BS cells (compare panels A and D, closed symbols). When total units were calculated in active fractions for the same amount of protein loaded, BS cells showed a higher extent of enzyme activity than control cells (18 and 4.3 units recovered, respectively, for 2.3 mg of proteins loaded) (Figure 2A and D), confirming thus, the results on total activity detected in crude extracts (Table 1). About 10% of DNA ligase activity was eluted close to the void volume (Vo) of the column. When the same extracts were submitted to AmS precipitation and then analyzed by FPLC, DNA ligase was eluted in multiple forms: a 180 kDa form, a 90 kDa form, and about 50% of the activity which eluted close to Vo in both control and BS materials. Thus, we observed again a similar pattern of elution between control and BS cells and a higher total activity was recovered in BS material than in control (16.5 and 5.7 units, respectively, for 3.3 mg of pro-

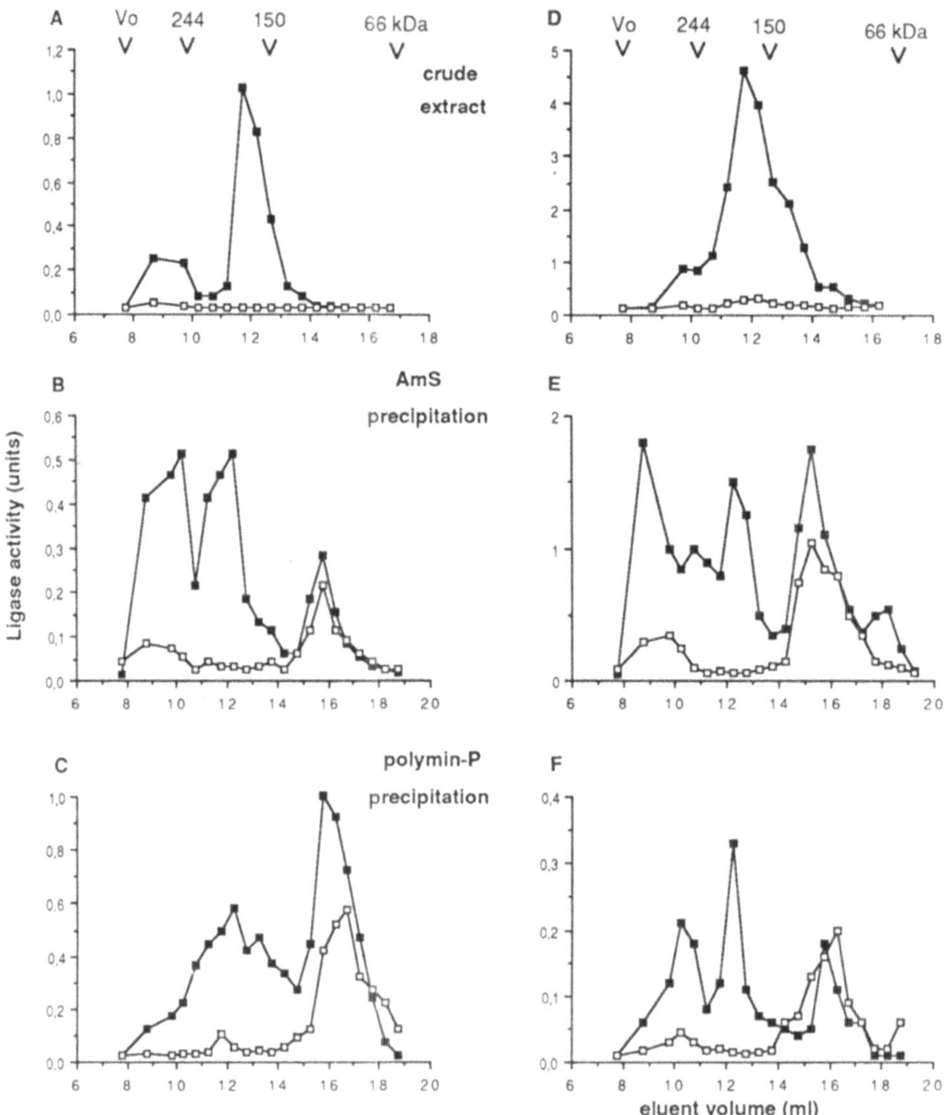

Figure 1. Size-fractionation of DNA ligase from control and BS cells through FPLC. Enzyme fractions from GM3299 control (panels A-C) and and GM3403C BS (panels D-F) cells were: crude extracts (A, D), pellet from ammonium sulfate (AmS) precipitation (B, E) and supernatant fraction from polymin-P precipitation (C, F). The conversion of poly(dA)·oligo(dT) (■) or poly-(rA)·oligo(dT) (□) substrates into alkaline phosphatase-resistance were expressed as units in 50 µl of each fraction as in the legend of Table 1.

teins) (Figure 2B and E). The precipitation of crude extract from control cells by polymin-P did not reduce the enzyme recovery (6.5 units for 2.6 mg of protein) (compare panels C to A and B). In contrast, the activity recovered after the same precipitation of BS extract was strongly reduced (1.8 units for

2.6 mg of protein) with respect to that measured in crude extract and after AmS precipitation (compare panel F to D and E). This residual activity appeared in three peaks of 240, 180 and 90 kDa. When column fractions were assayed using a poly-ly(rA)·^{32}P-oligo(dT) substrate, only the 90 kDa peaks were able to convert such a substrate into alkaline phosphatase-resistant form (Figure 1 A-F, open symbols). According to previous reports (Willis and Lindhal, 1987; Willis, et al., 1987), this 90 kDa peak would correspond to the ligase II form. Since DNA ligase I has been described as a high molecular weight form and it does not function on a poly(rA)·^{32}P-oligo (dT) substrate (Arrand, et al., 1986), we conclude that the major form of activity (the 180 kDa peak), identified after size-fractionation of cell extracts in these experiments, corresponds to the DNA ligase I. Thus, it appears that not only the total DNA ligase activity (Table 1) but also the high molecular weight form fractionated through FPLC, corresponding to DNA ligase I, is significantly higher in BS than in control cells. The precipitation of cell extracts by polymin-P seems to be a crucial step leading to the loss of DNA ligase activity in BS material (Figure 1F). This could be explained either by a considerable degradation of the BS enzyme occurring during this step or by its precipitation with nucleic acids in the pellet. To verify this point, it was necessary to measure ligase activity both in the supernatant fraction and in the pellet fractions in this step. This assay was not possible, because of the presence of polymin-P which precipitates the ligase substrate. However, we were able to verify this point by the activity gel technique, which eliminates the polymin-P.

Activity gels allow us to identify the DNA ligase active polypeptides without enzyme purifications (Mezzina, et al., 1984, 1985) from normal human and BS cells. Results showed that almost all enzyme activity seems to be correlated to a major 130 kDa polypeptide for both control and BS cells (Figure 2). The lower 43 kDa polypeptide did not contain DNA ligase activity, as revealed when the size of ^{32}P-oligo(dT) in the gel was analyzed on a DNA sequencing gel (not shown), as already described (Mezzina, et al., 1984, 1985, 1987). When cells were directly lysed inside the gel, a higher 170 kDa band was also visible (lanes 1, 2), which disappeared in crude extracts and during further purifications (lanes 3-10), probably due to the instability of this higher species. Lower molecular mass (90, 80 and 60 kDa) active polypeptides were also visible in these enzyme fractions. The intensity of all these polypeptides in BS materials (lanes 2, 4 and 6) was equal or higher than that in controls (lanes 1, 3 and 5). The ammonium sulfate precipitation of crude extracts did not affect the enzyme polypeptide pattern (lanes 5 and 6), whereas, in the supernatant fraction after polymin-P precipitation, the active 130 kDa species, still present in the supernatant fraction from control cells (lane 7), disappeared almost completely in BS materials (lane 8).

The reverse situation was obtained when the protein fraction, which co-precipitated with nucleic acids in the pellet, was extracted with 0.5 M NaCl buffer and then analyzed: the BS pellet contained the 130 kDa polypeptide to a much higher extent (Figure 2, lane 10) than did the control pellet (lane

9). This indicates, therefore, that the loss of DNA ligase activity observed in the polymin-P supernatant fraction of BS cell extracts after FPLC analysis (Figure 1F) appears to be due to the precipitation of the enzyme with the nucleic acids and not to its degradation.

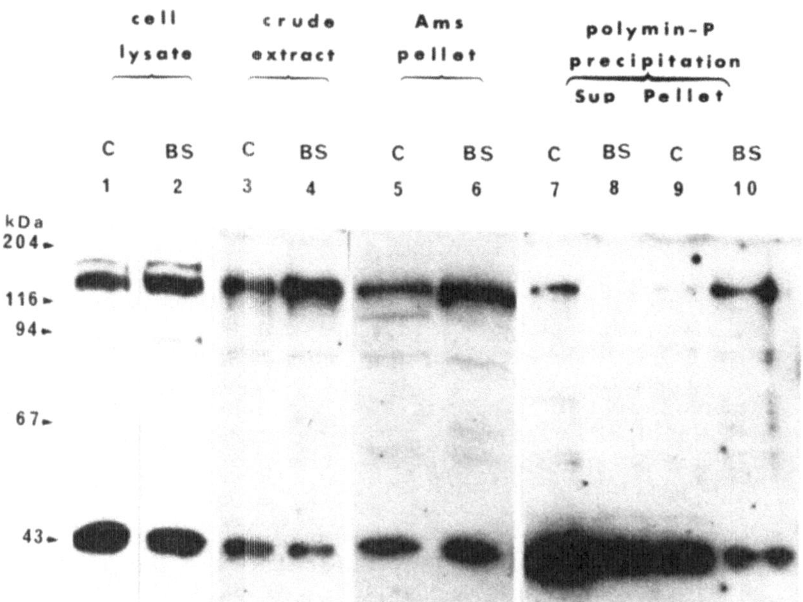

Figure 2. Activity gel analysis of catalytic polypeptides of different purifications of human DNA ligase in control and BS cell lysates, corresponding to 5 x 10^5 cells (lanes 1, 2). 30 µg of protein from crude extracts (3, 4) or pellets from AmS precipitation (5, 6), and 10 µg of protein from the supernatant fraction (7, 8) or the pellet extracted from polymin-p precipitation (9, 10) were used from control (1, 3, 5, 7 and 9) or BS (2, 4, 6, 8, 10) cells.

DISCUSSION

We have analyzed DNA ligase activity in human cell lines derived from the inherited diseases FA and BS, previously described to exhibit reduced levels of DNA ligase. Our results indicate that the level of enzyme activity in cell extracts from these patients is not reduced, and in the case of BS, is even significantly higher than in control cells. In our experiments we used DNA ligase substrates, size-fractionation procedures and, in part, cell lines similar or identical to those used by others in previous studies (Willis and Lindhal, 1987; Willis, et al., 1987). However, our experimental methods employed for preparing cell extracts or enzyme fractions for further analysis, differed. It is possible, therefore, that the apparent discrepancy between our and previous results with FA and BS cells could be explained by the different procedures used. In fact, the more or less drastic physical methods employed for obtaining crude extracts, involving buffers with high or low salt concentrations, with or without detergent, or further precipitation with polymin-P, could lead to the solubilization of different extents of DNA ligase activity. Moreover, it is possible that the composition of proteins/ nucleic acids associated with DNA ligase could differ in the extracts from normal or pathological cell lines. Previous results showing reduced DNA ligase activity in FA cell extracts could be explained by the fact that in these experiments a freezing-thawing procedure was used to prepare cell extracts (Hirsch-Kauffmann, et al., 1978). In our experiments, when cells were lysed without sonication, we sometimes found reduced levels (50%) of enzyme activity in some FA cell lines, whereas, by using our standard sonication method, we never found such variations in FA cells (Table 1), thus indicating that in the latter case more enzyme could be recovered in FA cell extracts.

For BS cells, the DNA ligase activity appears, in our experiments, significantly increased over that in control cells. This increase was observed regardless of which type of analysis was carried out on crude extracts. Results from these experiments revealed the integrity of DNA ligase in BS cells, which appears as a major 180 kDa form on FPLC (Figure 1A, D), or a major 130 kDa polypeptide, when analyzed by activity gels (Figure 2). The extent of this increase (1.5 to 4 fold) is comparable to that previously observed in monkey or human cells after treatment with DNA-damaging agents (Mezzina and Nocentini, 1978; Mezzina, et al., 1982a, 1982b). During repair of lesions, single-stranded DNA breaks occur and more DNA ligase seems to be necessary for joining DNA strands. In BS cells a similar situation may be constitutive, since the genetic material is submitted to spontaneous or free radical-induced breakage (German, et al., 1965; Emerit and Cerutti, 1981) and rejoining events. It is, therefore, likely that the high level of DNA ligase activity in BS cells could be a consequence of the abnormal chromosomal instability. Alpha and beta DNA polymerase levels in BS cells are not significantly different from those in control cells (data not shown), indicating thus that the increased DNA ligase level does not correspond to a general enhancement of DNA metabolizing enzymes.

In our experiments, we found a particular behaviour of the BS DNA ligase during polymin-P precipitation since this enzyme almost completely co-precipitated with nucleic acids. This effect was not found with the enzyme from control cells. We do not know the exact reason for this difference, which could be due either to a qualitative modification of the BS ligase or to an increase in the number of breaks in the cellular BS DNA which, being the natural substrate of the enzyme, brings down the ligase during the polymin-P precipitation. This interaction between DNA ligase I and nucleic acids has been reported for human placenta extracts (Bath and Grossman, 1986). This interpretation allows us to understand the apparent discrepancy between our results and those previously published by others (Willis and Lindhal, 1987; Willis, et al., 1987).

REFERENCES

Arrand, J.E., Willis, A.E., Goldsmith, I. and Lindhal, T., 1986. J. Biol. Chem., 261:9079-9082.

Bath, R. and Grossman, L., 1986. Archives of Biochem.Biophys., 244:801-812.

Bloom, D., 1954. Am. J. Dis. Child., 88:754-759.

Chaganti, R.S.K., Schomberg, S. and German, J., 1974. Proc. Natl. Acad. Sci. USA, 71:4508-4512.

Chan, J.Y.H. and Becker, F.F., 1985. Carcinogenesis, 6:1275-1277.

Chan, J.Y.H., Becker, F.F., German, J. and Ray, J.H., 1987. Nature, 325:357-359.

Dutrillaux, B., Aurias, A., Dutrillaux, A.M., Buriot, D., Prieur, M., 1982. Human Genet., 62:327-332.

Emerit, I., and Cerruti, P., 1981, Proc. Natl. Acad. Sci. USA, 78:1868-1872.

Fanconi, G., 1976. Semin. Hematol., 4:233-240.

Friedberg, E.C., Ehmann, U.K. and Williams, J.I., 1979. Adv. Rad. Biol., 8:85-174.

Fujiwara, Y., Tatsumi, M., Sasaki, M.S., 1977. J. Mol. Biol., 113:63S-649.

German, I., Archibald, R.and Bloom, D., 1965. Science, 148:506-507.

German, Y., 1969. Birth Defects, 5:117-119.

Giannelli, F., Benson, P.F., Pansey, S.A. and Polani, P.E., 1977. Nature, 265:466-469.

Hand, R. and German, J., 1975. Proc. Natl. Acad. Sci. USA, 72:758-762.

Hirschi, M., Netrawali, M.S., Remsen, J.F. and Cerutti, P.A., 1981. Cancer Res., 41:2003-2007.

Hirsch-Kauffmann, M., Schweiger, M., Wagner, E.F., and Sperling, K., 1978. Human Genet., 45:25-32.

Joenje, H. and Oostra, A.B., 1983. Human Genet., 65:99-101.

Kapp, L.N., 1982. Biochem. Biophys. Acta, 696:226-227.

Latt, S.A., Stetler, E., Jengens, L.A., Buchanan, A.R., Gerald, P.S., 1975. Proc. Natl. Acad. Sci. USA, 72:40664070.

Mezzina, M. and Nocentini, S., 1978. Nucleic Acids Res., 5: 4337-4328.

Mezzina,M., Suarez, H.G., Cassingena, R. and Sarasin, A., 1982a. Nucleic Acids Res., 10:573-584.

Mezzina, M., Nocentini, S. and Sarasin, A., 1982b. Biochimie, 64:743-748.

Mezzina, M., Rossignol,J.M., Philippe, M., Izzo, R., Bertazzoni, U. and Sarasin, A., 1987. Eur. J. Biochem., 162:325-332.

Mezzina, M., Sarasin, A., Politi, N. and Bertazzoni, U., 1984. Nucleic Acids Res., 12:5109-5122.

Mezzina, M., Franchi, E., Izzo, R., Bertazzoni, U., Rossignol, J.M. and Sarasin, A., 1985. Biochem. Biophys. Res. Comm., 132:857-863.

Moustacchi, E. and Diatloff-Zito, C., 1985. Human Genet., 70:236-242.

Ockey, C.H. and Saffhill, R., 1986. Carcinogenesis, 7:53-57.

Poll, E.H.A., 1984. Ph.D. Vrije Universiteit, Amsterdam.

Sasaki, M.S., Tonomura, A., 1973. Cancer Res., 33:1829-1836.

Sasaki, M.S., 1975. Nature, 257:501-503.

Selsky, C.A., Henson, P., Weichselbaum, R.R. and Little, J.B., 1979. Cancer Res., 39:3392-3399.

Schroeder, T.M., 1982. Cytogenet. Cell Genet., 13:119-132.
Smith, P.Y. and Paterson, M.C., 1982. Photochem. Photobiol., 36:333-343.
Söderhall, S. and Lindhal, T., 1976. FEBS Lett, 67:1-7.
Spadari, S., 1976. Nucleic Acids Res., 3:2155-2166.
Teraoka, H. and Tsukada, K., 1982. J. Biol. Chem., 257:4758-4763.
Teraoka, H. and Tsukada, K., 1985. J. Biol. Chem., 260:2937-2940.
Weksberg, R., Buchwald, M., Sargent, P., Thompson, M.W., and Siminovitch, L., 1977. J. Cell Physiol., 101:311-324.
Willis, A.E. and Lindahl, T., 1987. Nature, 325:355-357.
Willis, A.E., Weksberg, R., Tomlinson, S., and Lindahl, T., 1987. Proc. Natl. Acad. Sci. USA, 84:8016-8020.

PROCESSING OF PHOTOINDUCED CROSS-LINKS AND MONOADDUCTS IN HUMAN CELL DNA: GENETIC AND MOLECULAR FEATURES

E. Moustacchi, D. Papadopoulo, D. Averbeck, D. Fraser and C. Diatloff-Zito

Institut Curie - Biologie, URA 1292 CNRS
26 rue d'Ulm
75231 Paris cedex 05, France

INTRODUCTION

The interest in cellular processing of DNA interstrand cross-links and monoadducts stems from a number of observations including their induction by a variety of agents present in the human environment. Indeed, several antitumoral drugs (mitomycin C, cis-platinum, nitrogen mustards), environmental pollutants (alkylating agents) and products such as furocoumarins, used in photochemotherapy of certain skin diseases and in cosmetology, have been shown to produce interstrand cross-links (CL) between either purines or pyrimidines in DNA. In all cases described so far, the production of CL is accompanied by the induction of monoadducts (MA) in various proportions according to the agent used and the treatment conditions. Also, in some cases oxidative type DNA damage is concomitantly induced by cross-linking agents (mitomycin C, for instance) but this type of damage will not be considered further in this report. The presence of such agents in a variety of practical situations provides compelling reasons for predicting and understanding their genotoxic/carcinogenic potentials. Moreover, they constitute attractive model systems to study basic DNA repair mechanisms. In this respect, furocoumarins are of particular interest. These tricyclic aromatic compounds contain two reactive sites, the 3-4 pyrone and the 4'-5' furan double bonds. They intercalate into DNA and form covalent adducts only with pyrimidine bases after exposure to near ultraviolet light (UVA); when activated by UVA either of the reactive sites of the furocoumarins can react specifically with the 5,6-double bond of pyrimidine bases forming cyclobutane-type monoadducts on the furan or pyrone-side (MA_f or MA_p) between a psoralen molecule and a pyrimidine base. Upon absorption of a second photon of 365 nm, a fraction of the 4',5'-furan side MA (MA_f) can react with a pyrimidine base on the opposite strand by engaging its 3,4-pyrone double bond to form an interstrand CL (for review, see Cimino, et al., 1985).

Psoralens photoinduced MA and CL are stable to more bio-

DNA Repair Mechanisms and Their Biological Implications in Mammalian Cells
Edited by M.W. Lambert and J. Laval
Plenum Press, New York

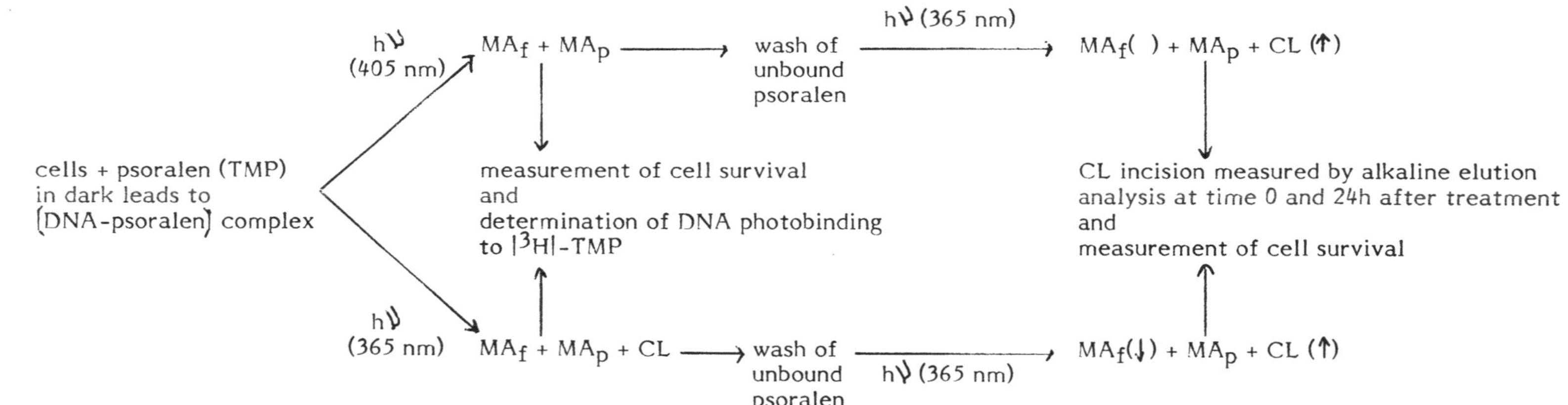

Fig. 1 Reirradiation protocol. Exponentially growing human fibroblasts (normal 1BR/3) and Fanconi's anemia FA 150) are incubated with TMP 5 μM for 10 min in the dark, then irradiated with a first dose of 365 nm or 405 nm monochromatic radiation. Unbound TMP is washed out and cells are exposed to a second dose of 365 nm radiation. For the same total amount of lesions, a fraction of MA_f is reduced (↓) by the second irradiation by their conversion into CL (↑).

chemical treatments and are not chemically lost from the DNA *in vivo*. Their induction can be quantitatively controlled by the UVA fluence, and they do not require metabolic activation to react with DNA. The ratio of CL to MA, at a constant total number of adducts, can be controlled by a double irradiation protocol after elimination of unbound psoralen molecules. Moreover, in certain irradiation conditions (405 nm) only MA are formed by otherwise bifunctional psoralens. Since these advantageous features are not shared by other cross-linking agents, most of our studies on MA and CL processing by normal and defective human cells deal with psoralen photoaddition.

Mutants characterized by a more or less specific hypersensitivity to cross-linking agents have been isolated from yeast (Henriques and Moustacchi, 1980; Cassier and Moustacchi, 1981), hamster (Robson and Hickson, 1986; Zdzienicka and Simons, 1987) and mouse cells (Hama-Inaba, *et al.*, 1983, 1988). This hypersensitivity is shared by cells derived from patients affected by Fanconi's anemia (FA), an inherited autosomal recessive disorder in humans. Comparative studies on the processing of psoralen photoinduced lesions in normal and FA cells are reported here as well as attempts to correct this defect by normal DNA sequences.

I) Influence of the presence of MA on the repair of CL in normal cells, comparison with FA

Psoralen photoaddition to cellular DNA provides an interesting situation in which the influence of the presence of the two types of MA (i.e. MA_f and MA_p), in different proportions, on the incision repair step of DNA interstrand CL can be quantitatively studied.

Using the highly photoreactive furocoumarin, 4,5',8-trimethylpsoralen (TMP), in combination with monochromatic 405 nm (only MA induced) or 365 nm (MA + CL induced) radiation, as well as the reirradiation protocol depicted in Figure 1, we show that:

1) In normal human fibroblasts, the presence of large amounts of MA inhibits CL incision. This is true for single doses of 365 nm radiation leading to a total number of adducts per 10^8 bp varying from 440 to 22,000 and $K_c (x\ 10^{-2})$ values (measuring CL induction) from 4.6 to 153.0. For the three doses of 365 nm tested in combination with TMP treatment (5×10^{-6} M), no incision of CL was detected as measured by alkaline elution.

2) When a fraction of the MA, i.e., the MA_f, is reduced by their conversion into CL (reirradiation protocol), incision of CL is observed. One must note that, although the total number of adducts rests the same, following either the first or the second 365 nm dose, after the double irradiation procedure, the proportion of CL is increased. For instance, when 440 adducts per 10^8 bp are induced, a single dose of 365 nm radiation leads to a K_c of 4.6×10^{-2} and no incision of CL is detected. In contrast, for the same total of adducts, a reirradiation with 365 nm leads to a K_c of 73×10^{-2} and 44% of the CL are incised after 24 hours of post-treatment incubation (Table 1) (for details, see Papadopoulo, *et al.*, 1988).

Table 1 Incision of TMP plus light induced CL and crosslinkable MA (MA_f, furan side) in normal and FA fibroblasts after 24h of post treatment incubation and at a constant number of total adducts induced per 10^8 base pairs, i.e. 440.

First dose (kJm^{-2})	Second dose of 365 nm (kJm^{-2})	CL induction ($K_c \times 10^{-2}$)	% CL incision		% MA_f incision	
			Normal	FA 150	Normal	FA 150
0.08 (365 nm)	0.0	4.6	0	-	29.5	-
	0.8	56.9	32.1	-	-	-
	1.2	73.1	44.4	-	-	-
4.0 (405 nm)	0.0	CL undetect.	-	-	40	8
	1.2	38.7	46	25	-	-

3) For the same number of total adducts, the amount of MA_f is smaller after 405 nm irradiation in comparison to 365 nm irradiation. This is demonstrated by the differential rate of conversion of MA_f into CL after a second irradiation at 365 nm. In this case, the incision of the CL induced by the second irradiation at 365 nm is not prevented.

Also, in contrast to TMP, after a single treatment with 8-methoxypsoralen (8-MOP) plus 365 nm, 50 to 90% of the CL are removed in the first 24 hours after treatment (Kaye, et al., 1980; Gruenert and Cleaver, 1985; Nocentini, 1986; Papadopoulo, et al., 1987). Hearst, et al. (1984) reported that TMP plus 365 nm induces considerably higher levels of MA_f in comparison to 8-MOP. Consequently, it is likely that the better repairability of 8-MOP plus 365 nm induced CL compared to TMP is related to the difference in the amount of MA_f induced.

4) The incision of CL induced after 8-MOP plus 365 nm radiation is present in FA cells belonging to the two genetic complementation groups A and B (Duckworth-Rysiecki, et al., 1985; Moustacchi, et al., 1987). However, the process is slower and the final amount of CL incised is lower in FA cells than in normal cells (Papadopoulo, et al., 1987). Using the 405 nm-365 nm reirradiation protocol with TMP, we demonstrate that for a constant number of total adducts (20 or 440 per 10^8 bp) and different ratios of CL over MA (K_c equals 16 or 39 x 10^{-2}), the incision of CL is systematically hampered in FA compared to normal cells (Table 1 and Averbeck, et al., 1988). In other words, in normal cells, the incision of CL is progressively diminished by increasing amounts of MA_f. We show that this inhibition is even more pronounced in FA cells.

It can be asked, why, despite the fact that the total number of 365 nm photoinduced TMP adducts per 10^8 base pairs is low, the repair of CL can be inhibited by the presence of MA and even more specifically by the furan-side MA. The first explanation which comes in mind is that MA_f may trap the incision-excision repair complex which may then not be available to efficiently incise CL. In contradiction with this hypothesis is the fact that when the number of total adducts is enhanced by a factor of 100, the efficiency of incision changes very little (Papadopoulo, et al., 1987). Alternatively, the stabilization of the helical structure of DNA due to MA_f (Shi and Hearst, 1986) may lead to conformational alterations which in turn may change the way in which repair enzymes recognize or have access to DNA CL. Such stabilized regions may well be located at the specific sequences defined as "strong sites" which are the preferential targets for psoralen derivatives (Boyer et al., 1988). For 254 nm UV-induced pyrimidine dimers, cluster formation has been observed (Lam and Reynolds, 1987). Similarly, the production of MA_f at given DNA sites is likely to occur in clusters which in turn may favor the induction of a CL at the closest vicinity compatible with a favorable sequence. These possibilities are not mutually exclusive and merit investigation using DNA sequencing methodology combined with specific enzymatic digestion.

II) The sensitivity of normal and FA cells to MA and repair of cross-linkable furan-side MA

It is generally accepted that FA cells are more or less

specifically hypersensitive to DNA cross-linking agents. However, as mentioned above, MA are always induced concomitantly to CL. In order to precisely define the contribution of MA and CL to the particular sensitivity of FA cells, we determined the lethal effects of different types of adducts formed, using the same bifunctional furocoumarin TMP in combination with either 405 nm or 365 nm radiations. We have also investigated the extent to which the sensitivity of FA cells is related to their capacity to repair both CL and cross-linkable MA_f.

1. As mentioned above, in the range of doses used in biological experiments, TMP in combination with 405 nm radiation induces only MA. We show that FA cell lines from both complementation groups are more sensitive than normal cells to TMP photoinduced MA. FA group B cells are about 5 fold more sensitive whereas group A cells are only about 3 fold more sensitive than normal human cells (see for details Averbeck, _et al._, 1988). It should be recalled that after treatments inducing a mixture of CL and MA, i.e., TMP or 8-MOP and 365 nm radiation, for instance, FA group A cells are more sensitive than FA group B cells (Papadopoulo, _et al._, 1987; Moustacchi, _et al._, 1988). FA cells are only marginally susceptible to cytotoxicity due to DNA damage induced by UV (Klocker, _et al._, 1985), ionizing radiation (Dritschillo, _et al._, 1984; Duckworth-Rysiecki and Taylor, 1985) and monofunctional compounds such as decarbamoyl-mitomycin C, methylmethane sulfonate or 4-nitroquinoline oxide (Weksberg, _et al._, 1979; Sasaki, 1978). The hypersensitivity of FA cells to N-methyl-N'-nitro-N-nitrosoguanidine (MNNG) (Ishida and Buchwald, 1982) and to TMP photoinduced MA appear to be exceptions to this general rule.

2. As seen above, it is established that only MA_f are susceptible to conversion into CL by additional exposure to 365 nm light. After incision, these MA_f lose their convertibility. This property allows us to follow, as a function of time of incubation after treatment with 405 nm radiation, the proportion of MA_f which are incised; at different time intervals after the first dose, cells are treated with a second dose of 365 nm radiation. Subsequent alkaline elution measurement of the number of CL's allows us to calculate the incision rate of the initial MA's. Table 1 shows that 24 hours after treatment, FA cells incise approximately 4 times less efficiently than normal cells cross-linkable MA_f (8% against 40% for normal cells). The result is likely to be related to the higher sensitivity of FA cells to TMP plus 405 nm light-induced MA.

It can be asked, why, in contrast to other monofunctional agents, FA cells turn out to be hypersensitive to TMP photoinduced MA. The excision repair pathway is assumed to operate on bulky DNA adducts that cause major helical modifications. The bacterial uvr ABC excinuclease participates in the repair of DNA lesions induced by MNNG (Van Houten and Sancar, 1987). In other words, besides the lesions repaired by the methyl transferase and glycosylase, MNNG produces lesions which are recognized and processed by the excision complex. The MA induced by TMP plus 405 nm light are likely to produce perturbations in the stability of the double helix. It can be proposed that minor perturbations produced by monofunctional agents and radiations are processed in FA cells as in normal ones. It would be only for high levels of helical perturbations that the incision complex in FA cells would manifest a defect. In other words, with respect to the relative sensitivity of FA cells to genotoxic agents, it is not the mono- or bifunctional nature of the DNA adducts which matters but the degree of helical modification induced.

III) Towards the cloning of FA correcting DNA sequences

Correction of the abnormal response to DNA cross-linking agents of FA fibroblasts by transfection with high molecular weight normal human DNA has been achieved (Diatloff-Zito, et al., 1986). Since in our conditions exogenous correcting DNA could not be distinguished from the host DNA, the correction of sensitivity to mitomycin C (MMC) of FA cells with DNA from mouse cells has been attempted. A similar strategy using inter-species complementation by DNA transfection has been effective for identification and isolation of several human genes concerned with excision repair (Westerveld, et al., 1984; Rubin, et al., 1983). This was performed with UV-sensitive hamster mutants used as recipients of the human DNA.

1) The protocol which allowed correction of FA fibroblasts by transfection with normal human DNA was re-adopted when using mouse DNA. We show that the degree of correction of MMC sensitivity of FA group B cells (FA 145) with DNA from mouse lymphoma L5178Y cells is close to 100%. The relative D_{37} (lethal dose leaving 37% survivors) of transfected FA 145 cells compared to untransfected FA 145 equals 2.75, whereas the relative D_{37} of FA 145 cells over that of normal 1BR/3 cells equals 3.9. Subsequently, DNA from transfected cells was hybridized with ^{32}P labelled C_0t1 mouse probe, which detects only highly repetitive interspersed mouse sequences, and it was shown that FA 145 human transfectants contain a noticeable amount of foreign mouse DNA sequences. This is true even two months after transfection (for details see Diatloff-Zito and Moustachi, 1988).

Secondary transfectants were also generated by transfecting FA 145 cells with high molecular weight DNA extracted from primary FA 145 transfectants. Survival curves to MMC were established, and it was shown that FA 145 secondary transfectants were only 1.5 times more resistant than the untransfected FA 145 cell line. Also, the hybridization signal of the secondary transfectant's DNA with the C_0t1 mouse probe is below the level of detection by Southern blot analysis. Secondary transformants obtained by transfer of human DNA from primary transformants into the X-ray sensitive Chinese hamster ovary cell line EM9 also displayed intermediate resistance to EMS and X-rays and weak hybridization to human Alu sequences (Spiro, et al., 1986). Such partial resistance may be due to the fact that expression of more than one gene may be required, following transfection, for acquisition of full resistance. Alternatively, it is possible that the gene responsible for CL and MA processing was damaged during transfection and selection and that this gene is now only partially active. High variability (increase in mutation rates and in rearrangements of DNA fragments) has indeed been reported for transfected DNA (Calos, et al., 1983).

After the presence of mouse DNA in FA primary transfectants was demonstrated by Southern blotting, a genomic library in phage lambda was constructed and screened for the presence of mouse DNA as well as for correction of FA cells.

2) Mouse DNA from L5178Y cells was found to be ineffective in correcting MMC sensitivity of FA cell lines belonging to complementation group A (FA 150 and E.K.) in spite of several trials. However, when mouse DNA extracted from total mouse embryo was used for transfection of FA group A cells, a partial correction of the MMC sensitivity was observed. It is of interest to note that, in comparison to FA group B cells, FA cells belonging to group A are: a) more sensitive in terms of clonogenic cell survival, b) appear to be more affected with respect to removal of CL (Papadopoulo, et al., 1987), and c) demonstrate a lack of recovery of a normal rate of DNA semi-conservative synthesis (Moustacchi, et al., 1987) following a DNA cross-linking treatment.

It seems that in FA group A cells, in which the defect is more pronounced, it is more difficult to supply, by transfection with mouse DNA, the proper sequences suitably expressed. Moreover, DNA from adult (lymphoma) and embryonic mouse cells are not in the same state of competence for expression of gene(s) necessary for correction of FA group A cells.

Correction of FA fibroblasts hypersensitivity to the effects of diepoxybutane, another DNA cross-linking agent, was achieved by transfection of Chinese hamster lung cell DNA (Shaham, et al., 1987). Taken together with our results, these findings indicate that DNA sequences correcting the FA defect are present not only in human but also in rodent DNA.

3) Mitomycin C-sensitive mutants derived from mouse lymphoma L5178Y cells (Hama-Inaba, et al., 1983) share with FA several features: a) they are highly sensitive to the lethal effect of a number of DNA cross-linking agents whereas they are clearly less sensitive to their monofunctional counterparts (Hama-Inaba, et al., 1988); b) the mouse mutants have been classified by somatic hybridization into two genetic complementation groups I and II (Hama-Inaba, et al., 1983) which, as in FA, demonstrate either a normal capacity of recovery of the rate of DNA semi-conservative synthesis after a cross-linking treatment (group II of mouse mutants equivalent to group B in FA) or an absence of such a recovery (group I of mouse mutant equivalent to group A in FA) (for details, see Moustacchi, et al., 1988); c) cytogenetic analysis demonstrates high rates of chromosomal aberrations in the mouse mutants compared to the original L5178Y cell line which is clearly enhanced by treatment with MMC (Rosselli and Moustacchi, 1988). This is comparable to the well established spontaneous and induced chromosomal instability which is one of the major characteristics of FA. Such mouse mutants can serve as recipients for transfection with human DNA with a selectable marker.

Optimum conditions for transfection of these mouse mutants which grow in suspension were established. Assays by classical transfection by the Ca-phosphate procedure having failed, a range of electroporation conditions were tried using several parameters: cellular viability and uptake of a fluorescent dye, challenge for MMC resistance of surviving cells and expression of the neo[R] gene held by the pCD vector containing the cDNA inserts of a human DNA library (Chen and Okayama, 1987). A transfection frequency in the order of 5 x 10^{-5} for the neo gene was found using the optimum conditions for the MCN 151 mouse mutant (30 nF at 5 kV and 3 pulses with 8 seconds between pulses using an "Apelex" Electropulsing Unit). Some clones are now being examined for MMC resistance.

The same experiments were carried out on FA lymphoblastoid cell lines and optimum conditions differ from those found for the mouse lymphoblastoid cells. The cDNA library derived from normal mouse cells could also be used in the near future since the subsequent rescue would be similar to that for the human library.

The strategies described here are complementary of each other and should hopefully lead to the cloning of sequences correcting the hypersensitivity of FA cells to DNA cross-linking agents.

CONCLUSION

It has been established that FA cells are hypersensitive to DNA cross-linking agents, including psoralen derivatives in combination with UVA. Although incision of CL following post-treatment incubation takes place in FA cells, the processing of these lesions is hampered compared to normal cells; the incision kinetics is slower and the final amounts of CL incised are lower in FA relative to normal cells. FA complementation group A cells are more affected than group B cells both in terms of cell survival and in terms of CL repair. To our surprise in the absence of CL, FA cells showed a higher sensitivity than normal cells to the cytotoxicity of MA induced by TMP and 405 nm, in this case, complementation group B cells being more sensitive than group A cells. Both cell lines have in fact a reduced incision capacity of MA relative to normal cells, this is especially true for furan-side MA.

Thus, FA cells are altered in the processing of both CL and MA of a specific structural type. In the presence of high amounts of MA_f, the incision of CL induced by TMP photoaddition (and not 8-MOP) is blocked and it is only when the proportion of MA_f declines that incision of CL can take place. This phenomenon is even more pronounced in FA cells.

By attempting to clone the normal sequences correcting the FA sensitivity to cross-linking agents, we hope to unravel the nature of the defect.

ACKNOWLEDGMENTS

This work was supported by grants from CNRS, INSERM (Contract 852017), CEA (Saclay) France, CEE Grant BIO-151 F and the Ligue National contre le Cancer.

REFERENCES

Averbeck, D., Papadopoulo, D. and Moustacchi, E., 1988. Repair of 4,5',8-trimethylpsoralen plus light-induced DNA damage in normal and Fanconi's anemia cells. Cancer Res., 48:2015.

Boyer, V., Moustacchi, E. and Sage, E., 1988. Sequence specificity in photoreaction of various psoralen derivatives with DNA: role in biological activity, Biochemistry, 27: 3011.

Calos, M.P., Lebkowski, J.S. and Botchan, M.R., 1983. High mutation frequency in DNA transfected into mammalian cells, Proc. Natl. Acad. Sci. USA, 80:3015.

Cassier, C., and Moustacchi, E., 1981. Mutagenesis by mono- and bifunctional alkalylating agents in yeast mutants sensitive to photoaddition of furocoumarins (pso), Mutation Res., 84:37-47.

Chen, C., and Okayama, H., 1987. High-efficiency transformation of mammalian cells by plasmid DNA. Mol. Cell. Biol., 7:2745.

Cimino, G.D., Gamper, H.B., Isaacs, S.T. and Hearst, J.E., 1985. Psoralens as photoreactive probes of nucleic acid structure and function: organic chemistry, photochemistry. Ann. Rev. Biochem., 54-1151.

Diatloff-Zito, C., Papadopoulo, D., Averbeck, D. and Moustacchi, E., 1986. Abnormal response to DNA cross-linking agents of Fanconi anemia fibroblasts can be corrected by transfection with normal human DNA, Proc. Natl. Acad. Sci. USA, 83:7034.

Diatloff-Zito, C., and Moustacchi, E., 1989. Complementation of mitomycin C hypersensitivity in Fanconi's anemia fibroblasts transfected with high molecular weight mouse DNA. Submitted.

Dritschillo, A., Brennan, T., Weichselbaum, E.R., and Mossman, K.L., 1984. Response of human fibroblasts to low dose rate irradiation. Radiation Res., 100:387.

Duckworth-Rysiecki, G. and Taylor, A.M., 1985. Effects of ionizing radiation on cells from Fanconi's anemia patients. Cancer Res., 45:416.

Duckworth-Rysiecki, G., Cornish, K., Clarke, C.A. and Buchwald, M., 1985. Identification of two complementation groups in Fanconi's anemia. Somatic Cell and Mol. Genet., 11:35.

Gruenert, D.C., and Cleaver, J.E., 1985. Repair of psoralen-induced cross-links and monoadducts in normal and repair deficient human fibroblasts. Cancer Res., 45:5399.

Hama-Inaba, H., Hieda-Shiomi, N., Shiomi, T. and Sato, K., 1983. Isolation and characterization of mitomycin-C-sensitive mouse lymphoma cell mutants. Mutation Res., 108:405.

Hama-Inaba, H., Sato, K. and Moustacchi, E., 1988. Survival and mutagenic responses of mitomycin C-sensitive mouse lymphoma cell mutants to other DNA cross-linking agents. Mutation Res., 194:121.

Hearst, J.E., Isaacs, S.T., Kanne, D., Rapoport, H. and Straub, K., 1984. The reaction of the psoralens with deoxyribonucleic acid. Q. Rev. Biophys., 17:1.

Henriques, J.A.P. and Moustacchi, E., 1980. Isolation and characterization of pso mutants sensitive to photoaddition of psoralen derivatives in Saccharomyces cerevisiae. Genetics, 95:273.

Kaye, J., Smith, C.A. and Hanawalt, P.C., 1980. DNA repair in human cells containing photoadducts of 8-methoxypsoralen or angelicin. Cancer Res., 40:696.

Klocker, H., Burtscher, H.J., Auer, B., Hirsch-Kauffmann, M.

and Schweiger, M., 1985. Fibroblasts from patients with Fanconi's anemia are not deficient in excision of thymine dimer. Eur. J. Cell. Biol., 37:240.

Lam, L.H. and Reynolds, R.J., 1987. DNA sequence dependence of closely opposed cyclobutyl pyrimidine dimers induced by UV radiation. Mutation Res., 178:167.

Moustacchi, E., Papadopoulo, D., Diatloff-Zito, C. and Buchwald, M., 1987. Two complementation groups of Fanconi's anemia differ in their phenotypic response to a DNA-crosslinking treatment. Human Genet., 75:45.

Moustacchi, E., Averbeck, D., Diatloff-Zito, C. and Papadopoulo, D., 1988. Phenotypic and genetic heterogeneity in Fanconi's anemia, fate of cross-links, correction of the defect by DNA transfection, In: "Clinical and Experimental Aspects of Fanconi anemia", Eds.: T.M. Schroeder-Kurth and G. Obe. Springer-Verlag, Berlin (in press).

Nocentini, S., 1986. DNA photobinding of 7-methylpyrido/3,4-cl psoralen and 8-methoxypsoralen. Effects on macromolecular synthesis, repair and survival in cultured human cells. Mutation Res., 161:181.

Papadopoulo, D., Averbeck, D. and Moustacchi, E., 1987. The fate of 8-methoxypsoralen-photoinduced DNA interstrand cross-links in Fanconi's anemia cells of defined genetic complementation groups. Mutation Res., 184:271.

Papadopoulo, D., Averbeck, D. and Moustacchi, E., 1988. High level of 4,5',8-trimethylpsoralen photoinduced furan-side monoadducts can block cell-link removal in normal human cells. Photochem. Photobiol., 47:321.

Robson, C. and Hickson, I.D., 1986. Genetic analysis of mitomycin C-sensitive mutants of a Chinese hamster ovary cell line. Mutation Res., 163:201.

Rosselli, F. and Moustacchi, E., 1989. Chromosomal hypersensitivity in mutant MCN-151 mouse cells exposed to mitomycin C, Mutation Res., 225:115.

Rubin, J.S., Joyner, A.L., Bernstein, A. and Whitmore, G.F., 1983. Molecular identification of human DNA repair gene following DNA mediated gene transfer. Nature, 306:206.

Sasaki, M.S., 1978. Fanconi's anemia. A condition possibly associated with a defective DNA repair, In: "DNA Repair Mechanisms", ICN-UCLA Symposia on Molecular and Cellular Biology, Eds.: P.C. Hanawalt, E.C. Friedberg and C.F. Fox. Academic Press, 9:675.

Shaham, M, Adler, B., Ganguly, S. and Chaganti, R.S.K., 1987. Transfection of normal human and Chinese hamster DNA corrects diepoxybutane induced chromosomal hypersensitivity of Fanconi anemia fibroblasts. Proc. Natl. Acad. Sci. USA, 84:5853.

Shi, Y.B. and Hearst, J.E., 1986. Thermostability of double-stranded deoxyribonucleic acids: Effects of covalent additions of a psoralen, Biochemistry, 25:5895.

Spiro, I.J., Barrows, L.R., Kennedy, K.A. and Ling, C.C., 1986. Transfection of a human gene for the repair of X-ray and EMS-induced DNA damage. Radiation Res., 108:146.

Van Houten, B. and Sancar, A., 1987. Repair of N-methyl-N-1-nitro-N-nitroso-guanidine induced damage by ABC excinuclease, J. Bacteriol., 169:540.

Weksberg, R., Buchwald, M., Sargent, P. and Siminovitch, L., 1979. Specific cellular defects in patients with Fanconi's anemia, J. Cell Physiol., 101:311.

Westerfeld, A., Hoeijmakers, J.H.J., Van Duin, M., De Witt, J., Odijk, H., Pastink, A., Wood, R.D. and Bootsma, D.,

1984. Molecular cloning of a human DNA repair gene. Nature, 310:425.
Zdzienicka, M.Z. and Simons, J.W.I.M., 1987. Mutagen-sensitive cell lines are obtained with a high frequency in V-79 Chinese hamster cells. Mutation Res., 178:235.

PARALLELS BETWEEN NUCLEOTIDE EXCISION REPAIR IN HUMAN CELLS AND E. COLI

Richard D. Wood

Imperial Cancer Research Fund
Clare Hall Laboratories
South Mimms, Herts EN6 3LD, U.K.

INTRODUCTION

A good deal is known about the mechanism of nucleotide excision repair in the bacterium Escherichia coli. This information has been used as a conceptual framework for the design and interpretation of many studies of repair in human and other mammalian cells. To what extent is this approach justified? Does the information now available encourage further emphasis on analogies between repair in bacteria and humans? These questions have often been posed and have been analyzed in depth in several important reviews (for example, Hanawalt, et al., 1979; Hall and Mount, 1981; Friedberg, 1985; Sedgwick, 1986). In this article I briefly summarize some recent information which suggests that an understanding of excision repair in E. coli will continue to be instructive as the details of excision repair in human cells are elucidated.

A REPAIR SYSTEM FOR A BROAD RANGE OF SUBSTRATES

Nucleotide excision repair in E. coli is performed by a multisubunit enzyme consisting of the UvrA,B, and C proteins, and also requires UvrD protein, a DNA polymerase, and DNA ligase (Husain, et al., 1985; Caron, et al., 1985). A variety of types of damage are substrates for this complex, as shown by genetic and biochemical studies. For example, mutants in uvr A,B, or C have defects in the removal of at least two ultraviolet light (UV) induced photoproducts: cyclobutane pyrimidine dimers (Boyce and Howard-Flanders, 1964), and (6-4) pyrimidine dimers (Franklin, et al., 1984). Using the purified proteins in a reconstituted system, both types of pyrimidine dimers have been found to be substrates for the UvrABC nuclease (Myles, et al., 1987).

Genetic analysis has shown that many other types of DNA damage are substrates for the Uvr repair system. In addition, direct in vitro incision by UvrABC has been demonstrated for many lesions, including psoralen monoadducts and cross-links

DNA Repair Mechanisms and Their Biological Implications in Mammalian Cells
Edited by M.W. Lambert and J. Laval
Plenum Press, New York

(Seeberg, 1981), cis-diamminedichloroplatinum(II) adducts (Sancar and Rupp, 1983), and lesions formed by N-acetoxyacetyl-aminofluorene (Fuchs and Seeberg, 1984). These mutagens generate lesions that distort the double helix structure by causing a sharp bend or partial unwinding of the helix (Pearlman, et al., 1985; Rao and Kollman, 1985; Tomic, et al., 1987; Husain, et al., 1988; Haran and Crothers, 1988).

In humans, mutants in DNA excision repair occur naturally in the form of at least one inherited disease, xeroderma pigmentosum (XP) (Cleaver, 1983). A total of 9 XP complementation groups have been identified to date, termed XP-A through XP-H, and XP-V (de Weerd-Kastelein, et al., 1972; Kraemer, et al., 1987). Studies of the kinetics of complementation suggest that the different genetic complementation groups are deficient in different proteins (Gianelli, et al., 1982).

Like uvr mutants of E. coli, XP cells show defective removal of UV-induced cyclobutane pyrimidine dimers (Zelle and Lohman, 1979) and pyrimidine (6-4) pyrimidone dimers (Mitchell, et al., 1985). XP cells also exhibit deficiencies in removal of chemical adducts generated by mutagens such as acetylaminofluorene (Amacher and Lieberman, 1977), photoactivated psoralens (Kaye, et al., 1980; Bredberg, et al., 1981), benzpyrene (Yang, et al., 1980), and aflatoxin B1 (Leadon, et al., 1981). Such results imply that the XP gene products, like the Uvr gene products, form part of a repair system with broad substrate specificity.

CONSEQUENCES OF GENETIC HETEROGENEITY

The presence of multiple complementation groups in XP suggests the existence of a multisubunit repair complex analogous to the E. coli UvrABCD system. In a large complex, some components may be essential, while others may be partially dispensible. For example, uvrD mutants of E. coli are less UV sensitive than uvrA or B mutants (Kuemmerle and Masker, 1980). Similarly, cells from different XP groups show a wide range of residual DNA repair and sensitivity to DNA damaging agents (Andrews, et al., 1978; Cleaver and Gruenert, 1984; Edwards, et al., 1987).

INCISION AT DAMAGE REQUIRES ATP

The multisubunit UvrABC nuclease cuts a damaged DNA strand on each side of the lesion in an ATP-dependent reaction (Sancar and Rupp, 1983; Yeung, et al., 1987). In addition, the UvrD gene product (DNA helicase II) is a DNA-dependent ATPase.

It has been known for some time from studies with permeabilized cell systems that repair replication in human cells is stimulated by ATP (Ciarocchi and Linn, 1978; Smith and Hanawalt, 1978). More recent studies with permeabilized human cells indicate that the damage-specific incision step of DNA after exposure to UV light requires ATP (Dresler and Lieberman 1983; Kaufmann and Briley, 1987), as in E. coli.

SIMILARITIES IN PATCH SIZE AND THE STRUCTURE OF EXCISED FRAGMENTS

The Uvr system in E. coli excises DNA damage as a short oligonucleotide (Sancar and Rupp, 1983; Caron, et al., 1985; Kumura, et al., 1985). The shortest repair patches that are formed in vivo after UV damage in E. coli have been found to be between 13-16 nucleotides long in cells with normal DNA polymerase I (Ben-Ishai and Sharon, 1978). In human cells, DNA damage also seems to be excised within short oligonucleotides (LaBelle and Linn, 1982; Weinfeld, et al., 1986). Careful measurements of the DNA repair patches that are formed after UV irradiation of human cells (under conditions where exonucleolytic enlargement of the excision gap is minimized) indicate a similar patch size of 10-20 nucleotides(Th'ng and Walker, 1985).

HUMAN CELLS AND E. COLI ARE KNOWN TO HAVE SIMILAR REPAIR ENZYMES

It is worth remembering that human cells and E. coli have already been found to have similar DNA repair enzymes (other than those involved in nucleotide excision repair). A number of DNA repair enzymes from human cells have been identified which serve to correct DNA bases altered by deamination, alkylation, and ring saturation or ring fragmentation. All of these activities closely resemble enzymes found previously in E. coli (Lindahl, 1982).

BIOCHEMICAL STUDIES

Biochemical studies of nucleotide excision repair in human cells are still at an early stage, but several promising systems have emerged based on complementation of XP cells (Yamaizumi, et al., 1985; Vermeulen, et al., 1986; Nishida, et al., 1988; Wood, et al., 1988).

In the system described by Wood, et al.(1988), soluble extracts from normal human cell lines can promote repair replication of plasmid DNA damaged by ultraviolet light and photoactivated psoralen. More recent work indicates that diamminedichloroplatinum damaged DNA is also a substrate for such repair (J. Hansson and R. Wood, unpublished). Repair replication after ultraviolet irradiation is strictly dependent on the presence of ATP in the reaction mixture, and leads to repair patches with sizes less than 100 nucleotides. Extracts from XP cells generally show deficient repair. However, mixing cell extracts from two different selected XP complementation groups leads to increased repair activity, by in vitro complementation. Using this and other systems, it should soon be possible to compare human and bacterial excision repair at the protein level.

REFERENCES

Amacher, D.E., and Lieberman, M.W., 1977, Removal of acetylaminofluorene from the DNA of control and repair-deficient human fibroblasts, Biochem. Biophys. Res. Comm., 74:285.

Andrews, A.D., Barrett, S.F., and Robbins, J.H. , 1978, Xeroderma pigmentosum neurological abnormalities correlate with colony-forming ability after ultraviolet radiation, Proc. Natl. Acad. Sci. USA, 75:1984.

Ben-Ishai, R. and Sharon, R., 1978, Patch size and base composition of ultraviolet light-induced repair synthesis in toluenized Escherichia coli, J. Mol. Biol., 120:423.

Boyce, R. and Howard-Flanders, P., 1964, Release of ultraviolet light-induced thymine dimers from DNA in E. coli K-12, Proc. Natl. Acad. Sci. USA, 51:293.

Bredberg, A., Lambert, B., and Soderhall, S., 1981, Induction and repair of psoralen cross-links in DNA of normal human and xeroderma pigmentosum fibroblasts, Mutat. Res., 93:221.

Caron, P.R., Kushner, S.R., and Grossman, L., 1985, Involvement of helicase II (uvrD gene product) and DNA polymerase I in excision mediated by the uvrABC protein complex, Proc. Natl. Acad. Sci. USA, 82:4925.

Ciarrocchi, G. and Linn, S., 1978, A cell-free assay measuring repair DNA synthesis in human fibroblasts, Proc. Natl. Acad. Sci. USA, 75:1891.

Cleaver, J.E., 1983, Xeroderma pigmentosum., in "The Metabolic Basis of Inherited Disease", 5th ed., J.B. Stanbury, J.B. Wyngaarden, D.S. Frederickson, J.L. Goldstein and M.S. Brown, eds., McGraw-Hill, New York.

Cleaver, J.E. and Gruenert, D.C., 1984, Repair of psoralen adducts in human DNA: differences among xeroderma pigmentosum complementation groups, J. Invest. Dermatol., 82:311.

de Weerd-Kastelein, E.A., Keijzer, W., and Bootsma, D., 1972, Genetic heterogeneity of xeroderma pigmentosum demonstrated by somatic cell hybridization, Nature New Biol., 238:80.

Dresler, S.L. and Lieberman, M.W., 1983, Requirement of ATP for specific incision of ultraviolet-damaged DNA during excision repair in permeable human fibroblasts, J. Biol. Chem., 258:12269.

Edwards, S., Fielding, S., and Waters, R., 1987, The response to DNA damage induced by 4-nitroquinoline-1-oxide or its 3-methyl derivative in xeroderma pigmentosum fibroblasts belonging to different complementation groups: evidence for different epistasis groups involved in the repair of large adducts in human DNA, Carcinogenesis, 8:1071.

Franklin, W.A., and Haseltine, W.A., 1984, Removal of UV light-induced pyrimidine-pyrimidone (6-4) products from Escherichia coli DNA requires the uvrA, uvrB, and uvrC gene products, Proc. Natl. Acad. Sci. USA, 81:3821.

Friedberg, E., 1985, "DNA repair", W.H. Freeman and Co., New York.

Fuchs, R.P.P. and Seeberg, E., 1984, pBR322 plasmid DNA modified with 2-acetylaminofluorene derivatives: transforming activity and in vitro strand cleavage by the Escherichia coli uvrABC endonuclease, EMBO J. , 3:757.

Giannelli, F., Pawsey, S.A., and Avery, J.A., 1982, Differences in patterns of complementation of the more common groups of xeroderma pigmentosum: possible implications, Cell, 29:451.

Hall, J.D., and Mount, D.W., 1981, Mechanisms of DNA replication and mutagenesis in ultraviolet-irradiated bacteria and mammalian cells, Prog. Nucleic Acid Res. Mol. Biol., 25:53.

Hanawalt, P.C., Cooper, P.K., Ganesan, A.K., and Smith, C.A., 1979, DNA repair in bacteria and mammalian cells, Ann. Rev. Biochem., 48:783.

Haran, T.E., and Crothers, D.M., 1988, Phased psoralen cross-links do not bend the DNA double helix, Biochemistry, 27:6967.

Husain, I., Van Houten, B., Thomas, D.C., Abdel-Monem, M., and Sancar, A., 1985, Effect of DNA polymerase I and DNA helicase II on the turnover rate of Uvr ABC excison nuclease, Proc. Natl. Acad. Sci. USA, 82:6774.

Husain, I., Griffith, J., and Sancar, A., 1988, Thymine dimers bend DNA, Proc. Natl. Acad. Sci. USA, 85:2558.

Kaufmann, W.K. and Briley, L.P., 1987, Reparative strand incision in saponin-permeabilized human fibroblasts, Mutation Res., 184:237.

Kaye, J., Smith, C.A., and Hanawalt, P.C., 1980, DNA repair in human cells containing photoadducts of 8-methoxypsoralen or angelicin, Cancer Res., 40:696.

Kraemer, K.H., Lee, M.M., and Scotto, J., 1987, Xeroderma pigmentosum: cutaneous, ocular, and neurologic abnormalities in 830 published cases, Arch. Dermatol., 123:241.

Kuemmerle, N.B., and Masker, W.E., 1980, Effect of the uvrD mutation on excision repair, J. Bacteriol., 142:535.

Kumura, K., Sekiguchi, M., Steinum, A.L., and Seeberg, E., 1985, Stimulation of the UvrABC enzyme-catalyzed repair reactions by the UvrD protein (DNA helicase II), Nucl. Acids Res., 13:1483.

La Belle, M. and Linn, S., 1982, In vivo excision of pyrimidine dimers is mediated by a DNA N-glycosylase in Micrococcus luteus but not in human fibroblasts, Photochem. Photobiol., 36:319.

Leadon, S.A., Tyrrell, R.M., and Cerutti, P.A., 1981, Excision repair of aflatoxin B1-DNA adducts in human fibroblasts, Cancer Res., 41:5125.

Lindahl, T., 1982, DNA repair enzymes, Annu. Rev. Biochem., 51:61.

Mitchell, D.L., Haipek, C.A., and Clarkson, J.M., 1985, (6-4) photoproducts are removed from the DNA of UV-irradiated mammalian cells more efficiently than cyclobutane pyrimidine dimers, Mutation Res., 143:109.

Myles, G.M., Van Houten, B. and Sancar, A., 1987, Utilization of DNA photolyase, pyrimidine dimer endonucleases, and alkali hydrolysis in the analysis of aberrant ABC excinuclease incisions adjacent to UV-induced DNA photoproducts, Nucl. Acids Res., 15:1227.

Nishida, C., Reinhard, P., and Linn, S., 1988, DNA repair synthesis in human fibroblasts requires DNA polymerase delta, J. Biol. Chem., 263:501.

Pearlman, D.A., Holbrook, S.R., Pirkle, D.H., Kim, S.H., 1985, Molecular models for DNA damaged by photoreaction, Science, 227:1304.

Rao, S.N., and Kollman, P.A., 1985, Conformations of deoxydodecanucleotides with pyrimidine (6-4)-pyrimidone photoadducts, Photochem. Photobiol., 42:465.

Sancar, A. and Rupp, W.D., 1983, A novel repair enzyme: UVRABC excision nuclease of Escherichia coli cuts a DNA strand on both sides of the damaged region, Cell, 33:249.

Sedgwick, S.G., 1986, Stability and change through DNA repair, in "Accuracy in Molecular Processes", D.J. Galas, T.B.L. Kirkwood and R.F. Rosenberger, ed. Chapman and Hall, London.

Seeberg, E., 1981, Strand cleavage at psoralen adducts and pyrimidine dimers in DNA caused by interaction between semi-purified uvr+ gene products from Escherichia coli, Mutat. Res., 82:11.

Smith, C. A. and Hanawalt, P.C., 1978, Phage T4 endonuclease V stimulates DNA repair replication in isolated nuclei from ultraviolet-irradiated human cells, including xeroderma pigmentosum fibroblasts, Proc. Natl. Acad. Sci. USA, 75:2598.

Th'ng, J.P.H. and Walker, I.G., 1985, Excision repair of DNA in the presence of aphidicolin, Mutat. Res., 165:139.

Tomic, M.T., Wemmer, D.E., and Kim, S.H., 1987, Structure of a psoralen cross-linked DNA in solution by nuclear magnetic resonance, Science, 238:1722.

Vermeulen, W., Osseweijer, P., de Jonge, A.J.R., and Hoeijmakers, J.H.J., 1986, Transient correction of excision repair defects in fibroblasts of 9 xeroderma pigmentosum complementation groups by microinjection of crude human cell extracts, Mutat. Res., 165:199.

Weinfeld, M., Gentner, N.E., Johnson, L.D., and Paterson, M.C., 1986, Photoreversal-dependent release of thymidine and thymidine monophosphate from pyrimidine dimer-containing DNA excision fragments isolated from ultraviolet-damaged human fibroblasts, Biochemistry, 25:2656.

Wood, R.D., Robins, P., and Lindahl, T., 1988, Complementation of the xeroderma pigmentosum DNA repair defect in cell-free extracts, Cell, 53:97.

Yamaizumi, M., Sugano, T., Asahina, H., Okada, Y., and Uchida, T., 1986, Microinjection of partially purified protein factor restores DNA damage specifically in group A of xeroderma pigmentosum cells, Proc. Natl. Acad. Sci. USA, 83:1476.

Yang, L.L., Maher, V.M., and McCormick, J.J., 1980, Error-free excision of the cytotoxic and mutagenic N2-deoxyguanosine DNA adduct formed in human fibroblasts by (±)7 ,8 -dihydroxy, 9 ,10 -epoxy-7,8,9,10-tetrahydrobenzo[a]pyrene, Proc. Natl. Acad. Sci. USA, 77:5933.

Yeung, A.T, Mattes, W.B., Oh, E.Y., Yoakum, G.H. and Grossman, L., 1987, The purification of the Escherichia coli UvrABC incision system, Nucl. Acids Res., 14:8535.

Zelle, B. and Lohman, P.H.M., 1979, Repair of UV-endonuclease-susceptible sites in the seven complementation groups of xeroderma pigmentosum A through G, Mutation Res., 62:363.

MECHANISMS OF RESISTANCE TO IONISING RADIATIONS: GENETIC AND MOLECULAR STUDIES ON ATAXIA-TELANGIECTASIA AND RELATED RADIATION-SENSITIVE MUTANTS

John Thacker, Reg Wilkinson, Anil Ganesh and Phillip North

Cell & Molecular Biology Division, MRC Radiobiology Unit, Chilton, Didcot, Oxon, OX11 ORD England

INTRODUCTION

In mammalian cells the mechanisms of resistance to damage caused by ionising radiations (IR) are largely unknown. Knowledge of the types of damage induced and of the enzymes, identified mostly in lower organisms, which act upon that damage may suggest that certain mechanisms are likely to prevail. However, in this respect, the diversity of damage caused by IR in the genetic material (and presumably in some other cellular molecules) leads to some uncertainty; a large number of different types of altered chemical products have been identified in irradiated DNA (Hutchinson, 1985; Teoule, 1987). This suggests that a large number of enzymes may be involved in recovery from IR damage, although some will be more important than others when considering a given cellular response. IR-induced cell killing, for example, has been linked to the production of DNA strand breakage, especially double-strand breaks (dsb). In E. coli there are a number of genes affecting in the repair of dsb; mutations in these genes lead to a reduction in the efficiency of dsb rejoining to varying extents and to enhanced sensitivity to IR (e.g., Sargentini and Smith 1986). Probably the best example to illustrate this link is the use of an X-ray-sensitive yeast mutant rad54-3 which is also temperature-sensitive: it can repair dsb at 23^{0}C but not at 36^{0}C and shows a quantitative recovery of radiation resistance as the temperature is shifted to allow dsb repair (Frankenberg-Schwager, et al., 1988). Additionally, in yeast it has been found that the induction of approximately one dsb by IR is equivalent to one lethal event in the absence of repair (Ho, 1975; Resnick and Martin, 1976; Frankenberg, et al., 1981). These analyses show that mutants sensitive to IR are potentially useful in defining mechanisms, although such mutants have to be carefully characterized before they can be of value.

DNA Repair Mechanisms and Their Biological Implications in Mammalian Cells
Edited by M.W. Lambert and J. Laval
Plenum Press, New York

Table 1. Mammalian cell mutants sensitive to ionising radiations

Species	*Parent cells*	*Mutant designation*	*(reference)*
Human	-	Ataxia-telangiectasia	(Taylor, et al., 1975)
Mouse	L5178Y	M10	(Sato and Hieda, 1979)
Hamster	CHO	EM7, EM9	(Thompson, et al., 1980)
	"	XR-1	(Stamato, et al., 1983)
	"	xrs-1-xrs-7	(Jeggo and Kemp, 1983)
	"	BLM2	(Robson, et al., 1985)
	V79	irs1, irs2, irs3	(Jones, et al., 1987)
	"	V-C4, V-E5, V-G8	(Zdzienicka and Simons,1987)
	CHO	irs1SF	(Fuller and Painter, 1988)

IR sensitivity occurs in certain human disorders, such as ataxia-telangiectasia, or has been selected in mutants of cultured mammalian cells. A list of some of these mutants is given in Table 1. These mutants have been used to establish the number of genetic groups determining radiation resistance and the involvement of specific types of damage (especially dsb) in radiation responses.

ATAXIA-TELANGIECTASIA (A-T)

In addition to sensitivity to IR, this disorder gives progressive neuronal degeneration, variable immunodeficiency and a predisposition to cancer. Cellular markers of irradiated A-T cells, compared to normal cells, include enhanced cell killing, an elevated level of chromosome damage, and resistance to DNA synthesis delay (McKinnon, 1987).

The cellular markers may be used to establish the genetic similarity of cell cultures derived from different A-T patients. Thus, by fusing single cells from different patients and estimating either the radioresistance of DNA synthesis or the level of chromosome damage it has been found that at least 4 genetic (complementation) groups exist

(Jaspers, et al., 1985). Recently, we have attempted to use X-ray-induced cell killing in A-T x A-T fusion hybrid populations as a means of estimating genetic identities (Thacker and Debenham, 1988). To do this, transformed lines of A-T and normal human cells were used because of their indefinite lifespan (Day, et al., 1980; Huschtscha and Holliday, 1983; Mayne, et al., 1986). Analysis of the fusion hybrids, by chromosome counts and by DNA fingerprinting, suggested that all were true hybrids, although the transformed lines used are chromosomally abnormal. X-ray cell killing curves showed that fusion hybrids of transformed AT4BI x AT5BI cells were as sensitive as the individual parent cells, suggesting that they are in the same genetic group (Thacker and Debenham 1988). However, it was previously found, using the radioresistance of DNA synthesis on single-cell hybrids, that these two A-T patients are in different genetic groups (Jaspers, et al., 1985).

We have now checked the radioresistance of DNA synthesis in our fusion hybrids to see if this cellular marker has dissociated from X-ray killing, as would be predicted from previous data. However, we found that it has not; that is, the AT4BI x AT5BI hybrids tested to date show A-T-like radioresistant DNA synthesis after irradiation, while hybrids made between A-T cells and wild-type (MRC5V1) cells show wild-type DNA synthesis inhibition. Thus, either the two methods of measuring DNA synthesis (by autoradiography in single-cell hybrids or by TCA-precipitable counts in hybrid cell populations) are not reporting the same genetic entity or there are artefacts in the methods. A possible artefact in the use of transformed cell lines and the selection of clonal hybrids is the loss of genes determining the characteristic A-T responses. We have examined the karyotypes of the A-T lines in more detail but do not find that there are losses of specific chromosomes in all cells (it should be remembered that if the two A-T cell lines do complement, there should be two copies of each complementing gene [chromosome] in each line so that the hybrid would have to lose 4 genes [chromosomes] to become non-complementing). We cannot at present comment further on these conflicting data, except to say that, contrary to our previous suggestion (Thacker and Debenham, 1988) and some other observations (Lehmann, et al., 1986; Painter, 1986), our data seem to link genetically the cellular markers of enhanced cell killing and radioresistant DNA synthesis in A-T cells.

In a number of laboratories it has been shown that A-T cells have no apparent deficiency in rejoining IR-induced DNA dsb when measured by gradient or elution techniques (review: McKinnon, 1987). However, such techniques do not have the resolution to measure a small fraction of unrepaired breaks (Lehmann, 1982) and do not measure the fidelity of the rejoining process. Indeed it has been suggested from cytogenetic observations (Taylor, 1978; Cornforth and Bedford, 1985) that an increased level of unrepaired chromosome breaks occurs in A-T cells compared to normal cells after irradiation. We have investigated the ability of A-T cells to repair correctly "model" dsb in recombinant DNA molecules, in an attempt to identify a defect in the fidelity of the rejoining process (Cox, et al., 1984; 1986; Debenham, et al., 1988a). After breakage with restriction endonucleases at specific sites within an encoded gene, the molecules are transferred into cell cultures and the activity of the gene is monitored. These experiments showed that A-T cells had a significantly reduced ability to rejoin these dsb with fidelity, but to date have only been performed with one A-T cell line (transformed AT5BI) compared to a small number of transformed normal lines (review: Thacker 1989). We have not been able to overcome technical difficulties in attempting to extend the use of our rejoin assay to two other available transformed A-T lines.

To bypass these difficulties, and to facilitate molecular analysis of the rejoin process, we have recently undertaken a study of this process in vitro. Recombinant molecules which can be assessed rapidly in bacterial cells are treated with cell extracts from A-T and normal cultures. After some experimentation with different cell extraction procedures and assay conditions we now have observed the rejoining of

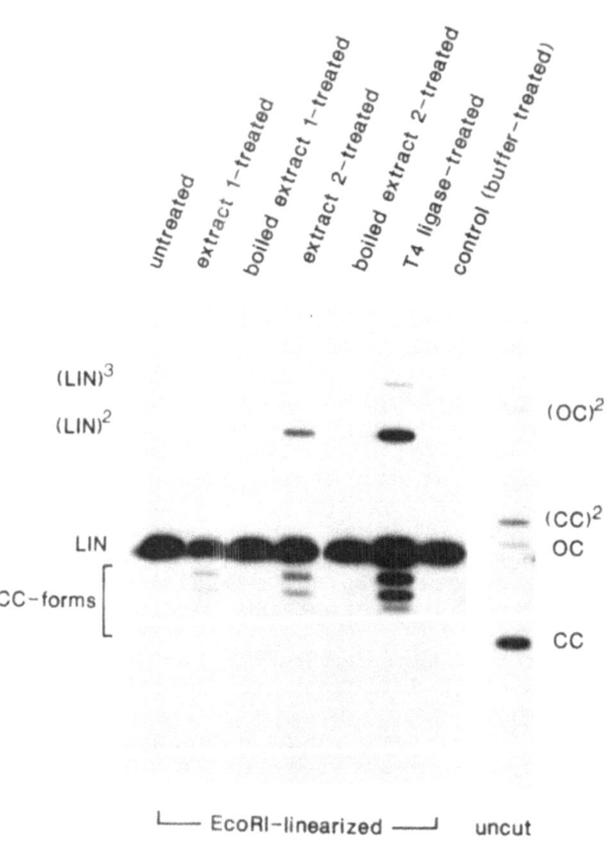

Figure 1. The rejoining of endonuclease-induced double-strand breaks in plasmid DNA by human cell-free extracts. Southern blot of extract-treated DNA (extract 1 has a lower protein concentration than extract 2), ligase-treated DNA, and controls (boiled extract or buffer alone) hybridized to radioactively-labelled plasmid. CC = closed circular, OC = open circular, LIN = linear molecules.

cohesive-end dsb by human cell extracts. Gel electrophoresis and Southern blotting of treated molecules shows the appearance of recircularized monomers as well as linear dimers and trimers, etc. (Figure 1). These rejoined molecules give an approximately 5-fold increase in bacterial transformation frequency over either untreated or boiled extract controls. We are extending these studies to look at the rejoining of different types of termini at the dsb, and the rejoining abilities of A-T cell extracts relative to those of normal cells.

HAMSTER irs MUTANTS

We have recently isolated 3 new IR-sensitive mutants of V79 hamster cells and have completed their preliminary characterization (Jones, et al., 1987, 1988a; unpublished data). These mutants are in separate genetic groups from each other and from a number of other IR-sensitive mutants isolated in different laboratories (with one exception - see below). The irs mutants are all about 3-fold more sensitive than V79 to X-rays (cf. A-T cells compared to normal human cells) but have different spectra of sensitivities to agents other than IR.

We have examined two other responses of relevance to radiation-induced repair processes: the rejoining of radiation-induced DNA strand breaks and the radioresistance of DNA synthesis. DNA breakage and rejoining was assessed by alkaline and neutral sucrose gradient sedimentation to study single- and double-strand breaks, respectively. Rejoining was assessed over a 30 minute period for single-strand breaks and for up to 8 hours after irradiation for dsb. In summary of a number of such experiments, no significant difference was found between the V79 parent cells and any of the mutants irs1, irs2, and irs3 in the induction or rejoining of single or dsb. However, the mutant xrs-1, found previously by elution techniques to have a reduced ability to repair dsb (Kemp, et al., 1984), showed the expected reduction in dsb rejoining using the neutral sedimentation method. An example of these data is shown in Figure 2.

It should be noted that, while irs1 shows no evidence of reduced dsb rejoining efficiency in these experiments, it has been found to have a 3-fold reduction in fidelity of rejoining endonuclease-induced dsb in the molecular assay described above (Debenham, et al., 1988b).

The radioresistance of DNA synthesis was measured in V79 cells and the irs mutants by dual labelling of cells (with ^{14}C-thymidine prior to irradiation, to give a measure of overall synthetic activity, and with ^{3}H-thymidine after irradiation). Expressing the post-irradiation DNA synthesis activity relative to that of unirradiated cells, we find that irs2 cells show less depression of synthesis than V79 cells, although the difference is less marked than for A-T cells relative to normal human cells. The irs3 mutant shows a similar depression of synthesis to V79 cells, but irs1 cells appear to have a response inbetween V79 and irs2. We are at present investigating these differences in time-course experiments.

To assess the effect of the irs mutant genes on another cell response, we have measured the frequency of radiation-induced mutation compared to that in V79 cells. Interestingly, the irs2 and irs3 mutants do not differ significantly from V79 in their dose-responses for HPRT-deficient mutant induction. A parallel study with irs1 has presented some difficulties due to low cloning efficiencies under the conditions of the experiment, but preliminary data suggest that this mutant has an elevated frequency of radiation-induced mutation per unit dose, relative to V79.

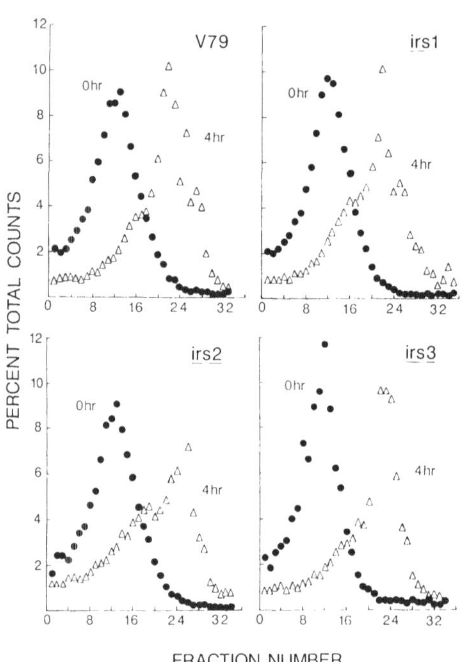

Figure 2. Examples of neutral sucrose gradient profiles of hamster cell DNA after irradiation (200 Gy) and either zero or 4 hours incubation at 37^0C. Experimental conditions were based on those of Bloecher (1982).

DISCUSSION

Comparison of the data for A-T cells with that for the irs mutants presents some interesting parallels but also some puzzles. Thus, none of these mutants appears to have an obvious defect in DNA strand-break rejoining, measured in biochemical experiments, despite the evidence that this is an important lesion in cell killing (see Introduction). We could simply suggest that alternative types of lesions are important to the survival of these mutants, perhaps because of the loss of repair enzymes specific to those lesions. However, we have previously shown, using a molecular assay for the rejoining of endonuclease-induced dsb, that if the _fidelity_ of this process is assessed both A-T cells and _irs1_ are defective (review: Thacker 1989). This defect in correct rejoining would not be revealed in biochemical assays based on sedimentation or elution.

A further link is now seen in the measurements of the radioresistance of DNA synthesis. A-T cells show a resistant phenotype compared to normal human cells, and this is also found for _irs2_ and perhaps for _irs1_ cells compared to the parental V79 cells. The V79 hamster mutants isolated by Zdzienicka and Simons (see Table 1) also show radioresistant DNA synthesis (Zdzienicka, _et al._, 1988) and we have shown that these mutants all fall into the same complementation group as _irs2_ (J. Thacker and R.E. Wilkinson, unpublished). Our A-T fusion hybrid analyses now seem to reinforce the genetic link between enhanced radiation-induced cell killing and radioresistant DNA synthesis.

Again, the mutation responses of the irs cells are not uniform: _irs2_ and _irs3_ show a "normal" mutational response while preliminary data for _irs1_ shows a hypermutable response. Evidently, the _irs2_ and _irs3_ data suggest that the defect in these mutants affects lesions leading to cell killing but not those leading to mutation. The relatively sparse data on the mutational response of A-T cells (Arlett and Harcourt, 1978) could also be interpreted similarly. However, A-T cells appear to be like _irs1_ cells in having increased frequencies of chromosomal damage (Jones et al.,1988b).

In attempting to draw some conclusions from these data, two further points may be noted:

(1) a link between IR-sensitivity and radioresistant DNA synthesis has been found in some other organisms; e.g. _Neurospora_: 2 of 8 IR-sensitive mutants show this dual phenotype (Koga and Schroeder, 1987).

(2) two different observations suggest, however, that failure to reduce DNA synthesis after irradiation does not account for the enhanced sensitivity of these types of mutants. Thus, normal human cells held in no-growth conditions after irradiation will recover from IR damage while A-T cells held under the same conditions will not (Cox 1982); i.e. A-T cells remain radiosensitive under conditions of little or no DNA synthesis. Secondly, "revertants" of A-T cell lines have recently been derived which show normal radiosensitivity but retain radioresistant DNA synthesis (Lehmann, _et al._, 1986; Painter, 1986).

Therefore, it seems that there is a genetic, but not a functional, link between cell killing sensitivity and the radioresistance of DNA synthesis. It is conceivable that a single gene could supply or regulate both responses (Lehmann, _et al._, 1986), but it is difficult to allow for more than one complementation group on this basis (i.e., each covering the same dual function). An alternative possibility is that the genetic link between these two responses is at the level of a common precursor of factors synthesised in a branched biochemical pathway. If the products of one (or more) A-T gene(s) catalyze the early stages of a biochemical pathway for these factors it (they) would potentially affect both responses. However, if the biochemical pathway for these factors branches after the action of the A-T gene product(s), mutations in genes affecting later steps of the pathway could affect only one response.

However, it will be seen that this type of explanation only accounts for some of the observed mutant phenotypes; these suggest that there is more than one pathway for the production of factors involved in IR-damage repair (and/or that species-specific differences exist), as well as different genes coding for enzymes directly involved in repair. In addition to this diversity of phenotype, a number of human disorders have been described recently which resemble A-T but which have either a partial phenotype or additional features. For example, there are A-T-like patients with intermediate radiosensitivity and "normal" inhibition of DNA synthesis (e.g., Fiorilli, et al., 1985); A-T-like patients with intermediate radiosensitivity and radioresistant DNA synthesis (e.g., Taylor, et al., 1987); a chromosome-breakage disorder not classifiable as A-T but which has similar levels of radiosensitivity and radioresistant DNA synthesis (the Nijmegen breakage syndrome; e.g., Taalman, et al., 1983); and other non-A-T disorders showing radioresistant DNA synthesis and various degrees of radiosensitivity (Barenfeld, et al., 1986). While some of these disorders may represent "leaky" variants of A-T, taken together with the observation that the majority of the rodent mutants listed in Table 1 are in different complementation groups (Jones, et al., 1988a; J. Thacker and R.E. Wilkinson, unpublished), it seems likely that a complex mechanism of genetic control of resistance to radiation damage in mammalian cells exists.

ACKNOWLEDGEMENTS

This work was supported in part by CEC Contract B16-E-144-UK. We are very grateful to all those who supplied mutant and transformed cell lines for this work.

REFERENCES

Arlett, C.F., and Harcourt, S.A., 1978, Cell killing and mutagenesis in repair-defective human cells, in: "DNA Repair Mechanisms", P.C. Hanawalt, E.C. Friedberg and C.F. Fox, eds., Academic Press, New York, p. 633.

Barenfeld, L.S., Pleskach, N.M., Bildin, V.N., Prokofjeva, V.V., and Mikhelson, V.M., 1986, Radioresistant DNA synthesis in cells of patients showing increased chromosomal sensitivity to ionising radiation, Mutation Res., 165:159.

Bloecher, D., 1982, DNA double strand breaks in Ehrlich ascites tumour cells at low doses of X-rays. I. Determination of induced breaks by centrifugation at low speed, Int. J. Radiat. Biol., 42:317.

Cornforth, M.N., and Bedford, J.S., 1985, On the nature of a defect in cells from individuals with ataxia-telangiectasia, Science, 227:1589.

Cox, R., 1982, A cellular description of the repair defect in ataxia-telangiectasia, in: "Ataxia-telangiectasia", B.A. Bridges and D.G. Harnden,eds., Wiley, Chichester, p. 141.

Cox, R., Masson, W.K., Debenham, P.G., and Webb, M.B.T., 1984, The use of recombinant DNA plasmids for the determination of DNA repair and recombination in cultured mammalian cells, Br. J. Cancer, 49, Supp.6:67.

Cox, R., Debenham, P.G., Masson, W.K., and Webb, M.B.T., 1986, Ataxia-telangiectasia: a human mutation giving high-frequency misrepair of DNA double-stranded scissions, Mol. Biol. Med., 3:229.

Day, R.S., Ziolkowski, C.H.J., Scudiero, D.A., Meyer, S.A., Lubiniecki, A.S., Giradi, A.J., Galloway, S.M., and Bynum, G.D., 1980, Defective repair of alkylated DNA by human tumour and SV40-transformed human cell strains, Nature, 288:724.

Debenham, P.G., Webb, M.B.T., Stretch, A., and Thacker, J., 1988a, Examination of vectors with two dominant selectable genes for DNA repair and mutation studies in mammalian cells, Mutation Res., 199:145.

Debenham, P.G., Jones, N.J., and Webb, M.B.T., 1988b, Vector-mediated DNA double-strand break repair in normal and radiation-sensitive Chinese hamster V79 cells, Mutation Res., 199:1.

Fiorilli, M., Antonelli, A., Russo, G., Crescenzi, M., Carbonari, M., and Petrinelli, P., 1985, Variant of ataxia-telangiectasia with low level radiosensitivity, Hum. Genet., 70:274.

Frankenberg, D., Frankenberg-Schwager, M., Bloecher, D., and Harbich, R., 1981, Evidence for DNA double-strand breaks as the critical lesions in yeast cells irradiated with sparsely or densely ionising radiation under oxic or anoxic conditions, Radiat. Res., 88:524.

Frankenberg-Schwager, M., Frankenberg, D., and Harbich, R., 1988, Exponential or shouldered survival curves result from repair of DNA double-strand breaks depending on post-irradiation conditions, Radiat. Res., 114:54.

Fuller, L.B., and Painter, R.B., 1988, A Chinese hamster ovary cell line hypersensitive to ionising radiation and deficient in repair replication, Mutation Res., 193:109.

Ho, K.S.Y., 1975, Induction of DNA double-strand breaks by X-rays in a radiosensitive strain of the yeast Saccharomyces cerevisiae, Mutation Res., 30:327.

Hutchinson, F., 1985, Chemical changes induced in DNA by ionising radiation, Prog. Nucleic Acid Res., 32:115.

Huschtscha, L.I., and Holliday, R., 1983, Limited and unlimited growth of SV40-transformed cells from human diploid MRC-5 fibroblasts, J. Cell Sci., 63:77.

Jaspers, N.G.J., Painter, R.B., Paterson, M.C., Kidson, C, and Inoue, T., 1985, Complementation analysis of ataxia telangiectasia, in: "Ataxia-telangiectasia: Genetics, Neuropathology, and Immunology of a Degenerative Disease of Childhood", R.A. Gatti and M. Swift, eds., Liss, New York, p. 147.

Jeggo, P.A., and Kemp, L.M., 1983, X-ray sensitive mutants of Chinese hamster ovary cell line, isolation and cross-sensitivity to other DNA-damaging agents, Mutation Res., 112:313.

Jones, N.J., Cox, R., and Thacker, J, 1987, Isolation and cross-sensitivity of X-ray sensitive mutants of V79-4 hamster cells, Mutation Res., 183:279.

Jones, N.J., Cox, R., and Thacker, J., 1988a, Six complementation groups for ionising-radiation sensitivity in Chinese hamster cells, Mutation Res., 193:139.

Jones, N.J., Thompson, L.H., Stewart, S.A., Tucker, J.D., Minkler, J.L., and Carrano, A.V., 1988b. Characterization of the ionising radiation sensitive mutants irs1 and irs2, J. Cell. Biochem., 12A:315.

Kemp, L.M., Sedgwick, S.G., and Jeggo, P.A., 1984, X-ray sensitive mutants of CHO cells defective in double-strand break rejoining, Mutation Res., 132: 189.

Koga, S.J., and Schroeder, A.L., 1987, Gamma-ray sensitive mutants of Neurospora crassa with characteristics analogous to ataxia-telangiectasia cell lines, Mutation Res., 183:139.

Lehmann, A.R., 1982, The cellular and molecular responses of ataxia-telangiectasia cells to DNA damage, in: "Ataxia telangiectasia", B.A. Bridges and D.G. Harnden,eds., Wiley, Chichester, p. 83.

Lehmann, A.R., Arlett, C.F., Burke, J.F., Green, M.H.L., James, M.R., and Lowe, J.E., 1986, A derivative of an ataxia-telangiectasia cell line with normal radiosensitivity but A-T like inhibition of DNA synthesis, Int. J. Radiat. Biol., 49:639.

Mayne, L.V., Priestley, A., James, M.R., and Burke, J.F., 1986, Efficient immortalization and morphological transformation of human fibroblasts by transfection of SV40 DNA linked to a dominant marker, Exp. Cell Res., 162:530.

McKinnon, P.J., Ataxia-telangiectasia: an inherited disorder of ionising radiation sensitivity in man, Hum. Genet., 75:197

Painter, R.B., 1986. Inhibition of mammalian cell DNA synthesis by ionising radiation, Int. J. Radiat. Biol., 49:771.

Resnick, M.A, and Martin, P., 1976, Repair of double-strand breaks in the nuclear DNA of Saccharomyces cerevisiae and its genetic control, Mol. Gen. Genet., 143:119.

Robson, C.N., Harris, A.L., and Hickson, I.D., 1985, Isolation and characterization of Chinese hamster ovary cell lines sensitive to mitomycin C and bleomycin, Cancer Res., 45:5304.

Sargentini, N.J., and Smith, K.C., 1986, Quantitation of the involvement of the recA, recB, recC, recF, recJ, recN, lexA, radA, radB, uvrD, and umuC genes in the repair of X-ray induced DNA double strand breaks in Escherichia coli, Radiat. Res., 107:58.

Sato, K., and Hieda,N., 1979, Isolation and characterization of a mutant mouse lymphoma cell sensitive to methyl methanesulphonate and X-rays, Mutation Res., 78:167.

Stamato, T.D., Weinstein, R., Giaccia, A., Mackenzie, L., 1983, Isolation of cell cycle-dependent gamma ray sensitive Chinese hamster ovary cell, Somatic Cell Genet., 9:165.

Taalman, R.D.F.M., Jaspers, N.G.J., Scheres, J.M.J.C., de Wit, J., and Hustinx, T.W.J., 1983. Hypersensitivity to ionising radiation, in vitro, in a new chromosomal breakage disorder, the Nijmegan Breakage Syndrome, Mutation Res., 112:23.

Taylor, A.M.R., 1978, Unrepaired DNA strand breaks in irradiated ataxia telangiectasia lymphocytes suggested from cytogenetic observations, Mutation Res., 50:407.
Taylor, A.M.R., Flude, E., Laher, B., Stacey, M., McKay, E., Watt, J., Green, S.H., and Harding, A.E., 1987. Variant forms of ataxia telangiectasia, J. Med. Genet., 24:669.
Taylor, A.M.R., Harnden, D.G., Arlett, C.F., Harcourt, S.A., Lehmann, A.R., Stevens, S., and Bridges, B.A., 1975, Ataxia telangiectasia: a human mutation with abnormal radiation sensitivity, Nature, 258:427.
Teoule, R., 1987, Radiation-induced DNA damage and its repair, Int. J. Radiat. Biol., 51:573.
Thacker, J., 1989, The use of integrating DNA vectors to analyse the molecular defects in ionising radiation-sensitive mutants of mammalian cells including ataxia-telangiectasia, Mutation Res., 220:187.
Thacker, J., and Debenham, P.G., 1988, The molecular basis of radiosensitivity in the human disorder ataxia-telangiectasia, in: "Mechanisms and Consequences of DNA Damage Processing", E. Friedberg and P. Hanawalt, eds., Liss, New York, p. 361.
Thompson, L.H., Rubin, J.S., Cleaver, J.E., Whitmore, G.F., and Brookman, K., 1980, A screening method for isolating DNA repair-deficient mutants of CHO cells, Somatic Cell Genet., 6:391.
Zdzienicka, M.Z., Jaspers, N.G.J., and Simons, J.W.I.M., 1988, New complementation groups in repair-deficient Chinese hamster cells and their characteristics, J. Cell. Biochem., 12A:330.
Zdzienicka, M.Z., and Simons, J.W.I.M., 1987, Mutagen-sensitive cell lines are obtained with a high frequency in V79 Chinese hamster cells, Mutation Res., 178:235.

DEFECTIVE DNA SYNTHESIS AND/OR REPAIR IN AMYOTROPHIC LATERAL SCLEROSIS

W. Clark Lambert, Warren S. Tanz, Gregory J. Tsongalis and Anthony O. Okorodudu

University of Medicine and Dentistry of New Jersey-New Jersey Medical School and Graduate School of Biomedical Sciences
Newark, New Jersey 07103

INTRODUCTION

Amyotrophic lateral sclerosis (ALS) is a progressive neurodegenerative disease with onset in adulthood characterized by loss of motor neurons with gradual paralysis of multiple muscle groups, eventuating in death (Bradley and Krasin, 1982a; 1982b; Eisen and Hudson, 1987; Mitsumoto, et al., 1988). There is considerable clinical heterogeneity. For example, some cases have onset in the fourth decade, whereas most cases begin much later, in the sixth or seventh decades. The cause of the disease is unknown, although multiple etiologies, including a viral infection, an autoimmune phenomenon, and a toxic process have been proposed (Bradley and Krasin, 1982a; 1982b; Eisen and Hudson, 1987; Mitsumoto, et al., 1988; Robison and Bradley, 1984). Several authors have hypothesized that a defective DNA repair mechanism may be present in certain neurodegenerative diseases, resulting in gradual loss of cells due to accumulation of chemical alterations in cellular DNA. Cell loss may occur only in certain regions or may affect tissues throughout the body, with loss of cells from the central nervous system, which are both irreplaceable and individually critical in function, producing clinical disease before significant signs or symptoms result from loss of cells from other organ systems (Bradley and Krasin, 1982a; 1982b; Robbins, 1983; 1987c; Robison and Bradley, 1984; Tandan, et al., 1987). Assays thought to be related to defective DNA repair mechanisms have, however, revealed either no alterations or relatively minor or unconvincing alterations, often with extensive overlap with normal cells, in cells derived from patients with ALS (Lambert, et al., 1987; Robbins, et al., 1989; Scudiero, et al., 1983; Tandan, et al., 1985, 1987; Vijaylaxmi, et al., 1985).

Another progressive neurodegenerative disease, ataxia-telangiectasia (A-T) or the Louis-Barr syndrome, is known to be associated with marked hypersensitivity, both clinically

DNA Repair Mechanisms and Their Biological Implications in Mammalian Cells
Edited by M.W. Lambert and J. Laval
Plenum Press, New York

and in cells obtained from affected individuals in culture, to gamma/X-irradiation and to radio-mimetic drugs such as bleomycin (McKinnon, 1987; Painter, 1985; Paterson and Smith, 1979; Rudolph, et al., 1989; Rudolph and Latt, 1989; Thacker, 1989). Moreover, cells from A-T patients continue to undergo DNA synthesis at near-normal rates following gamma/X-irradiation in culture, whereas similarly treated cells from normal individuals show marked inhibition of DNA synthesis (Painter, 1985; Young and Painter, 1989). This phenomenon has been termed "radiation-resistant DNA synthesis" (Lavin, et al., 1989; Lavin and Schroeder, 1988; Painter, 1985; Young and Painter, 1989). It is seen in some but not all cell types hypersensitive to gamma/X irradiation (Lavin, et al., 1989; Young and Painter, 1989). A similar phenomenon, occurring in cells from such non-human eukaryotes as Neurospora and Drosophila following exposure to a variety of DNA damaging agents to which they are also hypersensitive, has been reported by a number of laboratories; it has been termed "damage resistant DNA synthesis" (DRDS) (Koga and Schroeder, 1987; Lavin and Schroeder, 1988). It has also been reported that DRDS can be induced following exposure of HeLa cells to UV irradiation by prior treatment with fluorodeoxyuridine (Brozmanova, 1984; Brozmanova, et al., 1985; Lavin and Schroeder, 1988).

To investigate whether ALS is associated with a defect in a cellular DNA repair mechanism, we chose to study cultured, Epstein-Barr transformed lymphoblastoid cells, since in other diseases examined to date associated with a DNA repair deficiency the defect has been found in multiple cell types (Lambert and Lambert, 1987), and non-transformed neural ALS cells are both difficult to culture and almost impossible to obtain. Moreover, since lymphocytes in ALS patients are not known to be altered in any clinically apparent way, the chance that an abnormality identified in them is an epiphenomenon is quite small, in contrast to neural cells, which are undergoing marked degenerative changes associated with numerous secondary and tertiary events. Because ALS is not known to be etiopathogenically associated with exposure to a known DNA damaging agent or mutagen, we selected agents which produce lesions which occur spontaneously in cellular DNA. We therefore chose to study the response of ALS lymphoblastoid cells in culture to exposure to the DNA alkylating agent, methyl methanesulfonate (MMS), which produces large numbers of apurinic/apyrimidinic (AP) sites in DNA (Lawley and Shad, 1972). AP sites are one of the most commonly occuring spontaneous alterations in mammalian cellular DNA (Kunkel, et al., 1981; Lindahl and Nyberg, 1972; Lindahl, 1977). Moreover, this rate of occurrance has been postulated to be several times higher in cells from aged individuals compared to young subjects, due to accumulation of unrepaired alkylations that tend to undergo depurination (Park and Ames, 1988). ALS occurs primarily in older persons.

We found that cells from patients with ALS with onset in the sixth or seventh decades show marked DRDS following exposure in culture to the DNA alkylating agent, methyl methanesulfonate (MMS), to which they also show slight to moderate, but reproducible, hypersenitivity (Tanz, et al., 1988). Neither cells from normal subjects nor one cell line from an early onset case of ALS showed this phenomenon.

In attempting to separate the two phenotypes we had observed in ALS cells, hypersensitivity and DRDS following MMS exposure, we found that artificial inhibition of cellular replicative DNA synthesis during and just prior to the brief (one hour) period of exposure to MMS produced a marked protective effect in cells from patients with late onset ALS, but provided no protection to cells from normal subjects. The protective effect lasts for several days, with kinetics resembling persistence of cell viability during prolonged inhibition of DNA synthesis, without introduction of a DNA damaging agent, as previously reported by one of us (WCL) (Lambert and Studzinski, 1967). This phenomenon, which we have termed "paradoxical rescue by inhibition of DNA synthesis" (PRIDS), thus constitutes a second major alteration in DNA synthesis and/or repair in ALS. How it relates to the DRDS we have found in ALS cells exposed to MMS is unclear at present but is under active investigation. These findings provide a basis for assay systems for further investigation of the etiopathogenesis of ALS, and suggest that defective DNA synthesis and/or repair, or cellular processes related to them, play an important role in the mechanisms responsible for development of this disease.

MATERIALS AND METHODS

Cell lines selected for study and culture conditions

(a) Cell lines. Lymphoblastoid cell lines (transformed with Epstein-Barr virus) from normal individuals and from patients with ALS have been selected for study. In all diseases examined to date associated with a DNA repair deficiency, the defect has been present in such cultured lymphoblastoid cells (Lambert and Lambert, 1987), and it is thus anticipated that where a defect of this type is present in ALS, it is detectable in cultured lymphoblastoid cells. Our studies to date corroborate this concept.

Both normal and ALS cell lines were originally obtained from the Institute for Medical Research (IMR), Camden, New Jersey, or courtesy of Dr. J.H. Robbins, NIH, Bethesda, MD, and are being routinely propagated by us. These lymphoblastoid cells grow rapidly, at very similar rates, and thus also readily provide the cell bulk necessary for enzyme isolation and analysis.

(b) Conditions for cell culture. The lymphoblastoid cell lines were grown in suspension culture at 37^{0}C in sealed flasks in RPMI 1640 medium buffered to physiological pH with Hepes buffer (Grand Island Biological Co. (GIBCO), in 15% horse serum plus GMS (Growth Media Supplement)(GIBCO). The lymphoblastoid cells grow without attaching to any surface and without being constantly stirred or agitated in any way. Careful monitoring of the effect of periodic agitation on growth kinetics of four different cell lines showed that agitation was necessary at least each 24 hours to ensure maximal cell proliferation. This is, therefore, routinely done in all experiments. The effect of various feeding and splitting regimens prior to seeding showed that cultures had optimal growth if previously split 1:4 or more every three to four days. The growth kinetics of all lymphoblastoid cell cultures handled in this way showed a maximal proliferation rate at 24 hours after 1:4 split-feeding; therefore, control cell cultures are routinely harvested and experiments performed at 24 hours. The normal and ALS lymphoblastoid cell lines selected for study were selected to grow at similar rates under these conditions. Care was taken that the serum concentration is constantly maintained at at least 13%, because we have found XP lymphoblastoid cells to be markedly sensitive to small decreases in serum concentrations below 12% (Lambert and Lambert, 1983).

All cultures were routinely tested for mycoplasma using an autoradiographic technique. Also, since it has been reported that fluorescent lighting within laboratories and brief exposure to ultraviolet light routinely used in many laboratories for germidical purposes may produce chromosomal abnormalities and DNA damage and may affect DNA repair mechanisms of cultured cells (Bradley, et al., 1978; Parshad, et al., 1978), steps were taken to ensure that the cells were not exposed to UV light and that other light exposure was minimal.

Drugs

Hydroxyurea (HU) and thymidine (TdR) were obtained from Sigma (St. Louis, Mo.); methyl methanesulfonate (MMS) was obtained from Aldrich (Milwaukee, WI). MMS was made up fresh from stock for each experiment, and sterilized by filtration through a 0.22 µm filter. HU was prepared from a 100 mM stock solution, which was made by dissolving HU into sterile H_2O. This stock solution was then filter sterilized and kept at 4^0C. This has a shelf life of 6 months. TdR was made up fresh from stock for every experiment.

Treatment of cells with MMS

Cells in logarithmic growth phase were centrifuged for 10 minutes at 800xG and seeded to 2.0 - 2.5 x 10^5 cells/ml. Cells were then treated with MMS for one hour and washed three times (with centrifugations 10 minutes at 800xG) with Hank's balanced salt solution (HBSS). Cells were then reseeded to 2.0 x 10^5 cells/ml in medium and cultured at 37^0C. Cell counts were performed at 1.0, 20, and 45 hours post-treatment, and at other time points for various procedures as indicated in the results section. Cell growth and viabilities were recorded by gently removing aliquots of cells, diluting them 1:1 in 0.2% trypan blue dye in phosphate buffered saline, and counting on a hemocytometer after incubating at room temperature for 5 to 15 minutes.

PRIDS Assay

Cell cultures were treated as above with MMS with the following modifications: Prior to MMS treatment, cells were resuspended into medium containing either 10 mM HU for one hour or 2 mM TdR for four hours. After these treatment times, cells were exposed to MMS as described above, but with HU or TdR added as well. After one hour of exposure to MMS, all agents were removed as described above.

DNA Synthesis Determination

^{3}H-Tdr uptake was measured by placing 1.0 x 10^6 cells into 2 ml of medium and incubating at 37^0C for 2 hours with

10 μCi/ml ^{3}H-methylthymidine (S.A. 61 Ci/mmole; ICN, Irvine, CA). Cells were then washed once in ice-cold PBS, and centrifuged 10 minutes at 800xG. The cells were resuspended into 2 ml 5% trichloroacetic acid (TCA) for at least 30 minutes on ice to precipitate cellular DNA. The supernatant fluid was discarded, and the (DNA) precipitate washed onto pre-wetted filters (Millipore HA-type 0.45 μm, Bedford, MA). The tubes holding the precipitates were washed three times with 95% ethanol, which was passed through the filter. The filters were washed in the following order (all ice-cold): 2 ml saline, 4 ml 5% TCA, 2 ml 95% ethanol, and 2 ml 100% ethanol. The filters were put into scintillation vials and allowed to air dry. Scintillation fluid (Ecoscint, National Diagnostics) was added, and filters were counted 2-5 minutes in a Beckman liquid scintillation beta counter (Fullerton, CA).

Autoradiography

The ability of both normal and ALS cells to undergo replicative DNA synthesis after treatment with MMS and/or HU was measured in the following manner: After treatment, $3x10^6$ cells were resuspended into 2 ml medium containing 10 μCi/ml ^{3}H-methylthymidine (S.A. 61 Ci/mmole; ICN, Irvine, CA) and incubated for 2 hours at 37^0C. Cells were then washed with PBS, and the pellets were gently smeared onto glass slides. The smears were fixed in methanol: acetic acid (3:1)), dehydrated (95% ethanol) and hydrated (distilled water) several times, and dipped into Kodak NTB3 nuclear emulsion. Slides were exposed for 7 days at 4^0C and developed using Kodak D19 developer. These slides were then stained with Giemsa and grains/nuclei were counted in random fields. Mean numbers of grains per nucleus were determined in 500 nuclei per slide.

RESULTS

Methyl methane-sulfonate (MMS) was toxic to all cell lines with progressive loss of viability, determined by ability to exclude 0.1% trypan blue dye, observed at 20 and 45 hours after a one hour exposure to doses of the agent ranging from 0.5 to 2.0 mM. Loss of viability was more marked, however, in cell lines derived from patients who developed ALS in the sixth and seventh decades of life than in lines derived from normal subjects. By studying cell lines derived from age-matched normal subjects, we have determined that this greater loss of viability in ALS cells was not simply due to the age of the subject.

Figure 1 shows cell viabilities, expressed as percentage of viable cells in similarly treated cultures of normal cells, 45 hours after exposure to varying doses of MMS, as shown. Cell viabilities in normal cell cultures, against which these values have been normalized, were 77%, 60%, 25% and 8%, for 0.5, 1.0, 1.5 and 2.0 mM MMS, respectively. Two of the three ALS cell lines tested were hypersensitive to killing by MMS, the third, derived from a patient who developed ALS in the fourth decade of life, showed no difference from normal cells

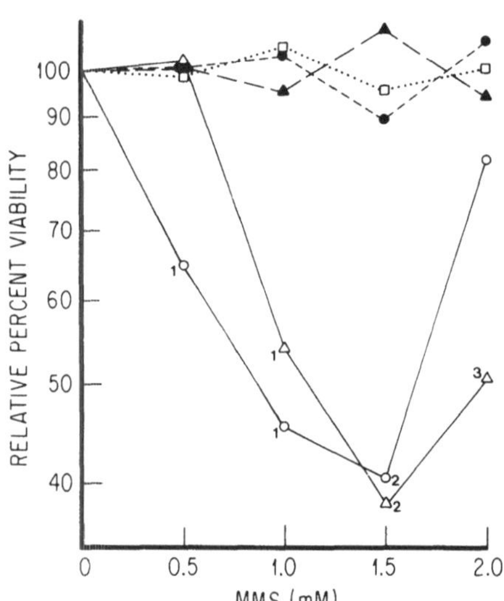

Figure 1. Viabilities, as determined by exclusion of trypan blue dye, of normal and ALS cell lines 45 hours following a one hour exposure to MMS in the concentrations shown. All values have been normalized to the mean viabilities of similarly treated normal cells (=100). Cell lines as follows: normals: closed circles (GM01989), closed triangles (GM03299); ALS cell lines: open circles (RB04073), open triangles (RB04075), open squares (RB04532)(atypical ALS cell line, see text). Points shown represent the mean of four separate experiments with 2000 cells counted per point. Numbers refer to significant reductions of viability from normal in a two-tailed Student's t test, as follows: 1, $p<0.001$; 2, $p<0.025$; 3, $p<0.05$.

in this or any other assay. The two hypersensitive ALS cell lines, derived from patients who developed ALS in the sixth and seventh decades of life, showed different threshold doses of sensitivity, with one line unaffected by 0.5 mM MMS and the other severely affected. Both showed maximum sensitivity, relative to normals, at 1.5 mM MMS. At 2.0 mM MMS and at higher doses, not shown, all cell types showed marked decreases in viability, so that differences between them are difficult to determine.

Figure 2 shows total cell counts, normalized as for Figure 1, in separate experiments on cultures of normal and ALS cell lines treated with MMS at varying doses. Although total numbers of cells (i.e., both those excluding and those not excluding 0.1% trypan blue dye on hemocytometer counts) were decreasing in all cultures, there were relatively more cells in the cultures of ALS cells hypersensitive to killing by the agent than in similarly treated cultures of normal cells. Thus the hypersensitive ALS cells were both dying more rapidly and, parodoxically, replicating more rapidly than normal cells following exposure to MMS, indicating that damage-resistant DNA synthesis (DRDS) was taking place in the ALS cells, resulting in an increased rate of cell division compared with normal cells.

Figure 3 shows ^{3}H-thymidine uptake in ALS and normal cells following exposure to 1.5 mM MMS monitored by scintillation counts (Chart A) and by grain counts on autoradiograms (Charts B and C). Results were normalized to those obtained in untreated cells of the same types; results in untreated cells varied no more than 15%. An unequivocal increase in rate of DNA synthesis in both ALS cell types following MMS exposure compared to similarly treated normal cells was observed at all times using both methods ($p < 0.001$). This increase, on autoradiography, was similar regardless of whether the proportion of cells with more than 100 grains per cell (Chart B) or mean number of grains per nucleus (Chart C) was determined.

Rates of thymidine uptake in ALS cells were higher than those of control cells 12 hours after treatment, as determined by scintillation counts. We attribute this at present to partial synchronization of the cells by the MMS treatment, but this is undergoing further study.

We then attempted to separate the two phenotypes we had observed in ALS cells following MMS exposure, more rapid cell killing and more rapid cell replication, by monitoring the former while inhibiting the latter. Figures 4 and 5 show results obtained when cell replication was inhibited, during exposure to 2.0 mM MMS, either by hydroxyurea (Figure 4) or by a high concentration of thymidine (Figure 5). Doses of hydroxyurea and thymidine selected were those just sufficient to produce stationary cultures, as determined by cell counts, over a 24 hour period of continuous treatment. Hydroxyurea was applied one hour prior to and during the one hour exposure to MMS; thymidine was applied four hours prior to and during the one hour exposure to MMS. This was to ensure that inhibition of DNA synthesis was in effect during exposure to MMS, since we have separately determined that thymidine treatment requires two to three hours to bring about inhibition of DNA synthesis in these cells. Following the one hour exposure to MMS, all inhibitors were removed. Thus DNA synthesis was inhibited for only a few hours in these cells. Charts C and

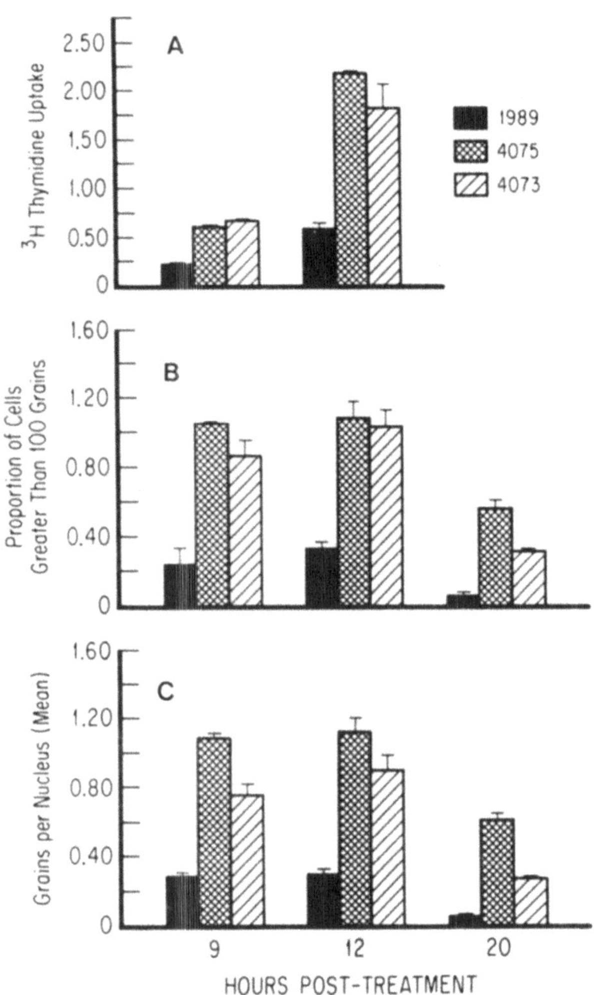

Figure 3. ^{3}H thymidine uptake during a 2 hour pulse at varying times after a one hour exposure to 1 mM MMS, as determined by scintillation counts (Chart A) and autoradiography (Charts B and C). All values have been normalizedd to levels in untreated cells of the same type (=1.0). Results shown represent the means of five experiments. Vertical bars refer to standard error of the mean.

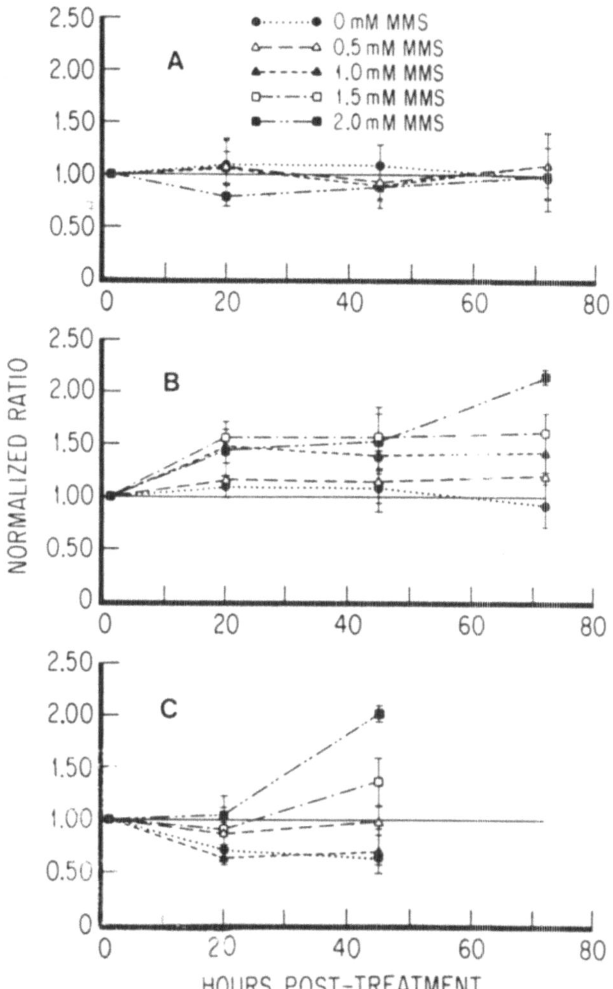

Figure 2. Total numbers of cells, both excluding and not excluding trypan blue dye, in cultures of lymphoblastoid cells following a one hour exposure to MMS at the concentrations shown, at various times after treatment. All values have been normalized to total cell numbers in cultures of similarly treated normal cells. The mean of five separate experiments is shown, vertical bars represent standard error of the mean for each point. Panel A, normal aged cell line GM05600; Panel B, ALS cell line RB4075; Panel C, ALS cell line 4073.

D of both figures clearly show that there was a marked protective effect, with increases in both cell number (both figures, Chart C) and proportion of viable cells (both figures, Chart D), in both ALS cell lines exposed to MMS in the presence of an inhibitor of DNA synthesis, compared to similarly treated normal cells and normal and ALS cells

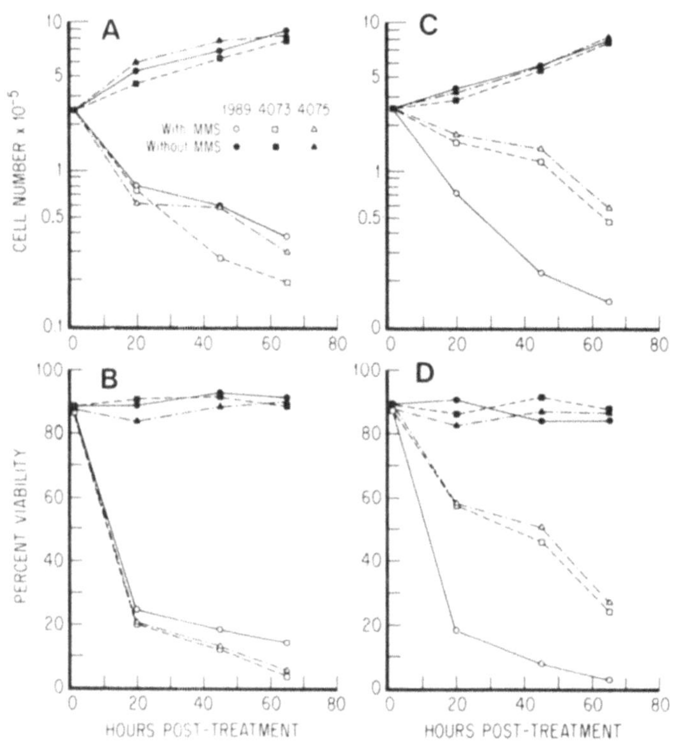

Figure 4. Numbers of cells (Charts A and C) and proportion of cells (Charts B and D) excluding 0.1% trypan blue dye, as determined by hemocytometer counts in cell cultures treated with MMS for one hour with (Charts C and D) 2 mM thymidine and without (Charts A and B) 10 mM hydroxyurea applied one hour before and during the one hour period of MMS treatment. Twenty experiments were performed with similar results. Results from a typical experiment are shown.

treated with MMS without DNA synthesis inhibition (both figures, Charts A and B). This effect was significant in all experiments ($p < 0.001$) and lasted for at least 45 hours, after which there was a gradual loss of viability in ALS cells treated with both types of agents. No such effect was observed in any normal cell line. We have termed this phenomenon "paradoxical rescue by inhibition of DNA synthesis" (PRIDS).

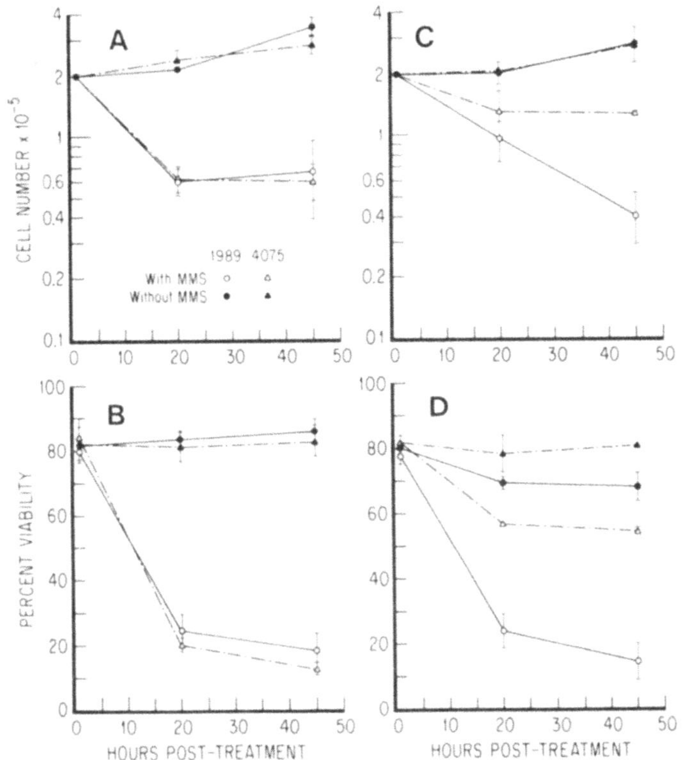

Figure 5. Numbers of cells (Charts A and C) and proportion of cells (Charts B and D) excluding 0.1% trypan blue dye, as determined by hemocytometer counts, in cell cultures treated with MMS for one hour with (Charts C and D) and without (Charts A and B) 2 mM thymidine applied for four hours before and during the one hour period of MMS treatment. Points represent results from four experiments. Vertical bars represent standard error of the mean for each point.

These experiments have been repeated using a total of four normal lymphoblastoid cell lines, two of which were from age-matched subjects, and four cell lines from patients with late-onset ALS, with the same results (Table 1). In addition, three xeroderma pigmentosum (XP) lymphoblastoid cell lines, one each from complementation groups A, C and D, have been used as controls. We have separately shown that two of these three XP cell lines show both DRDS and PRIDS following exposure to ultraviolet (UV) irradiation, to which they are extremely hypersensitive, provided that the dose of UV radiation is carefully selected to avoid masking the DRDS and PRIDS effects due to the toxicity of the UV radiation. The third XP cell line shows DRDS and is being tested for PRIDS following lower doses of UVC irradiation. XP cells do not show hypersensitivity to MMS in our experiments, however. XP cells also failed to show either DRDS or PRIDS following exposure to MMS over a range of doses. These results are summarized in Table 1.

Table 1. DRDS[1] and PRIDS[2] in normal, ALS, and XP lymphoblastoid cells treated with MMS[3] or UVR[4].

Cell Line	Disease	Age[5]	Sex	MMS		UVR	
				DRDS	PRIDS	DRDS	PRIDS
GM01989	Normal	33	M	-	-	-	-
GM03299	Normal	8	F	-	-	-	-
GM05600A	Normal	76	M	-	NT[6]	-	NT
GM03657	Normal	68	M	-	NT	NT	NT
RB04073	ALS	68	M	+	+	-	-
RB04075	ALS	72	M	+	+	-	-
GM08883A	ALS	61	F	+	+	-	-
GM08886A	ALS	64	F	+	+	-	-
RB04532	ALS (Variant)	40	F	-	NT	NT	NT
GM02345	XP-A[7]	8	F	-	-	+	+
GM02249	XP-C	16	M	-	-	+	-[8]
GM02485	XP-D	14	F	-	-	+	+

[1]DRDS = Damage resistant DNA synthesis.
[2]PRIDS = Paradoxical rescue by inhibition of DNA synthesis.
[3]MMS = methyl methanesulfonate.
[4]UVR = Ultraviolet radiation (UVC, 254 nm).
[5]Age refers to age of individual at time cells for culture were obtained.
[6]NT = Not tested.
[7]XP-A = Xeroderma pigmentosum, complementation group A.
XP-C = Xeroderma pigmentosum, complementation group C.
XP-D = Xeroderma pigmentosum, complementation group D.
[8]XP-C cells treated failed to show DRDS or PRIDS at higher doses of UVC irradiation but showed DRDS at lower doses. PRIDS at lower doses is currently being tested.

DISCUSSION

The alterations detected in the DRDS and PRIDS assays in cells from patients with sporadically occurring, late onset ALS contrast strikingly with results using other assays for defective DNA repair mechanisms in ALS reported both by ourselves and by others. Several years ago Scudiero, et al. (1983), reported that cultured human cell lines derived from patients with a number of neurological disorders, including ALS, did not show hypersensitivity to the DNA alkylating agent, N-methyl-N'-nitro-N-nitrosoguanidine (MNNG), and Vijaylaxmi, et al. (1985) found no abnormalities in sister chromatid exchanges in ALS cells treated with mitomycin C or with the alkylating agent, ethyl methanesulfonate (EMS). However, it has been found that O^6-alkylguanine bases in cellular DNA are repaired by a specific mechanism requiring degradation of an entire protein for each adduct repaired (Day, et al., this volume). Thus the relatively large proportion of O^6-alkyl-guanine sites produced by MNNG and EMS in mammalian cells would be likely to account for much of their toxicity. We thought it more likely that repair of apurinic/-apyrimidinic sites (APS) in cellular DNA is defective in ALS, since this, along with single-strand breaks, is the predominant spontaneously occuring adduct in human cellular DNA in vivo (Kunkel, et al., 1981; Lindahl, 1977; Lindahl and Nyberg, 1972), and thus choose MMS, which produces a high proportion of APS but few O^6-alkylguanine adducts in cultured human cells (Lawley and Shah, 1972). Moreover, it has been hypothesized that the rate of occurrance of APS is markedly increased in cells of elderly individuals due to depurination of unrepaired alkylations, especially at the N^7 of guanine, that accumulate during the lifetime of the subject (Park and Ames, 1988). Thus defective repair of APS is a good candidate for a deficiency in DNA repair, if it exists, in ALS.

Our results show that there was, indeed, a greater loss of viability in ALS lymphoblastoid cells in culture following a one hour exposure to MMS than in normal lymphoblastoid cells grown under identical, carefully controlled conditions (Lambert, et al., 1986). However, the difference is relatively small and requires these special experimental conditions (nearly identical growth rates; large numbers of cells counted) to be detected. Tandan, et al. (1985; 1987) detected abnormal UDS, increased numbers of DNA strand breaks detected by the alkaline elution assay, and decreased cell survival of cultured ALS cells continuously exposed to MMS. However, their study showed extensive overlap between results obtained in normal versus ALS cells. More recently, Robbins, et al. (1989) reported that they were unable to detect abnormalities in UDS in cultured ALS fibroblasts following exposure to MMS. The most straightforward interpretation of all of these results is that the most sensitive assays to detect abnormalities in ALS were simply not used. It has now been documented in several laboratories, for example, that results using the UDS assay may not correlate with cell survival and may not reflect DNA repair events in a proportional manner (Andrews, et al., 1978; Cleaver, et al., 1987; Mitchell, et al., 1985). Also, the anomalies we have found may be due to one or more abnormalities in ALS cells that primarily effect other systems, such as DNA replication, and only secondarily affect DNA repair systems.

The list of neurodegenerative diseases in which abnormal responses to DNA damage have been suggested includes, besides amyotrophic lateral sclerosis and ataxia-telangiectasia, certain patients with xeroderma pigmentosum (Andrews, et al., 1978; Robbins, et al., 1987c), Cockayne's disease (Lehmann, 1987), Alzheimer's disease (Li and Kaminskas, 1985; Robbins, et al., 1985a; Robison and Bradley, 1985; Robison, et al.,

1985, 1987; Scudiero, et al., 1986), Parkinson's disease (Nove, et al., 1987; Robbins, 1987b; Robbins, et al., 1983; 1985a), Huntington's disease (McGovern and Webb, 1982; Moshell, et al., 1980; Nove, et al., 1987; Scudiero, et al., 1981), Friedreich's ataxia (Chamberlain, et al., 1982), Usher's syndrome (Nove, et al., 1987; Robbins, et al., 1984), Duchenne muscular dystrophy (Nove, et al., 1987; Robbins, et al., 1984), and familial dysautonomia (Riley-Day syndrome; Scudiero, et al., 1981) as well as an animal model for neurodegenerative disease, the wasted mouse (Lavin and Schroeder, 1988; Woloschak, et al., 1987), and Down's syndrome (Otsuka, et al., 1985). For many of these disorders, however, results are more or less inconclusive, with extensive overlap between normal and diseased cells in conventional assays. Alterations in DNA repair mechanisms due to aging may also play a role in the etiology of these diseases (Roth, et al., 1989), as may an increased rate of spontaneous occurrance of APS in the elderly at sites of unrepaired alkylated purines accumulated during the aging process (Park and Ames, 1988). It is possible that the DRDS and PRIDS assays, using the appropriate DNA damaging agents, will be useful in examining these as well as other neurodegenerative disorders.

Pathological examination of diseased tissues in ALS and the wobbler mouse, which is a putative model for it, has shown a number of changes, including decreased cellular content and rate of synthesis of RNA and protein, which are consistent with an underlying defect in DNA metabolism as the etiology of the disease (Bradley, et al., 1983; Davidson and Hartmann, 1981; Davidson, et al., 1981; Mann and Yates, 1974; Murakami, et al., 1980; 1981). These changes are relatively non-specific, however, and it is therefore possible to propose many hypotheses to account for them, so that they offer weak support, at best, for a proposed alteration in DNA metabolism as the etiology of ALS. ALS occurs in certain isolated populations, possibly as a result of exposure to dietary toxins (Eisen and Hudson, 1987; Mitsumoto, et al., 1988). These may be related to an alteration in DNA metabolism or may be toxic to motor neurons by a entirely different mechanism. Indeed, our own results suggest that not all cases of ALS are associated with altered DNA metabolism.

If, as the present study suggests, there is a defect in DNA metabolism in ALS cells, this would be expected to be a heritable defect. Most cases of ALS are not thought to be inherited, however, although familial cases have been reported (Bradley and Krasin, 1982a; 1982b; Eisen and Hudson, 1987; Mitsumoto, et al., 1988). Robbins, et al. (1985b) proposed a somatic mutation to account for this discrepancy in non-familial cases of neurodegenerative diseases associated with defective DNA repair mechanisms. An alternate possibility is suggested by the co-recessive inheritance model (Lambert and Lambert, this volume). That model proposes that defects in DNA repair mechanisms are quite common in the general population. Thus one, or, in combination, more than one such defect might easily occur in different individuals in an apparently spontaneous manner, and also occur in others in a familial pattern with no clear cut mode of inheritance analyzed by classical mendelian genetics. Whatever the cause, it has now been well documented that there is marked heterogeneity in the normal human population in sensitivity to agents which damage cellular DNA. This implies marked heterogeneity in DNA repair systems in apparently normal individuals (Hsu, et al., 1989; reviewed in Lambert and Lambert, this volume).

The underlying mechanisms responsible for damage-resistant DNA synthesis are unknown, although it has been possible to subdivide the phenomenon into two components, replicon initiation and chain elongation. Both processes are abnormal in ataxia-telangiectasia cells (Lavin, et al., 1987; Young and Painter, 1989). It is possible that the underlying mechanism for DRDS is in some aspect of DNA enzymology or chromatin structure that, although it causes DRDS and possibly also PRIDS, is not primarily involved with either DNA synthesis or repair. Indeed, since motor neurons are non-dividing cells, it is difficult, without invoking such a mechanism, to understand how a defect in a replicative DNA synthesis mechanism could adversely affect them, unless some aspect of replicative DNA synthesis were inappropriately activated in these cells. An abnormality in DNA topoisomerase II has been suggested in ataxia-telangiectasia, but reports have been inconsistent (Davies, et al., 1989; Singh, et al., 1988; Smith and Malkinson, 1989). How such a defect would relate to DRDS in those cells is unclear. A gene for ataxia-telangiectasia has recently been mapped to chromosome 11q22-23, but the function of this gene remains unknown (Gatti, et al., 1989).

The kinetics of the PRIDS phenomenon are quite similar to those of the "unbalanced growth, or "thymineless death" phenomenon in mammalian cells, in which loss of viability occurs following persistent selective inhibition of cellular DNA synthesis (Lambert and Studzinski, 1967). In bacteria, this is associated with an interval of approximately one generation time of the organism in which no loss of viability occurs followed by abrupt, rapid loss of viability if DNA synthesis continues to be inhibited beyond this interval (Lambert, 1969). Although a similar course was once thought to occur in mammalian cells, one of us (WCL) showed many years ago that viability in these cells is lost only after several generation times of selective inhibition of DNA synthesis in cells in culture (Lambert and Studzinski, 1967). In the PRIDS system, DNA synthesis is inhibited for only a few hours, during which a one hour exposure to MMS occurs. The MMS is itself toxic, to both normal and ALS cells, and remains as or more toxic in normal cells following simultaneous brief inhibition of DNA synthesis. In the abnormally responding ALS cells, however, viability is largely preserved, but only for a period of several generation times for these cells (approximately 28 hours per generation time), after which it is lost fairly rapidly over a 48 hour interval. We do not know what relationship, if any, exists between the unbalanced growth and PRIDS phenomena, nor do we know whether this abnormal response of ALS cells is due to a defect in cellular DNA replication versus a DNA repair system, or both. The fact that the PRIDS phenomenon is produced equally well by hydroxyurea and by high doses of thymidine, both of which inhibit DNA synthesis by unbalancing precursor pools by very different mechanisms (Collins and Oates, 1987; Lambert, 1969; Lambert and Studzinski, 1967; Menth, 1989), indicates that inhibition of DNA synthesis is, indeed, the underlying requirement. DNA polymerase alpha, which is required for replicative DNA synthesis, is selectively inhibited by deoxynucleotide triphosphate pool imbalances, whereas other DNA polymerases, such as delta, are required for repair synthesis (Collins and Oates, 1987; Dresler, et al., 1988; Nishida, et al., 1988; Orlando, et al., 1988). Further investigations regarding this and other aspects of the PRIDS phenomenon, such as the effect of using different inhibitors of DNA synthesis and/or of different DNA polymerases and its relationship to cell cycle events, are a central focus of activity in our laboratory, as is investigating whether additional cell lines, particularly those derived from patients with other neurodegenerative disorders, may show the effect.

The DRDS and PRIDS phenomenon we have found in XP cells following exposure to UV irradiation is no better understood, at present, than these same phenomena observed in ALS cells following exposure to MMS. It is unclear whether they have any relationship to the neurodegenerative component of both disorders. The normal response of ALS cells to UV irradiation which we observed is coroberated by the results of Tandan, et al. (1987) using different assays; whereas the normal response of XPA cells to MMS exposure which we observed in all assays is similar to results obtained, using yet other assays, by Mirzayans, et al. (1988).

The protection of ALS cells by hydroxyurea, a drug currently used for chemotherapy of neoplasms of the central nervous system, suggests the possibility of a therapeutic application in ALS, either using hydroxyurea or some other agent based on in vitro research on the PRIDS phenomenon. Moreover, the DRDS and PRIDS phenomena may form the basis for a diagnostic test for ALS or other neurodegenerative diseases in persons predisposed to develop these diseases. Finally, by introducing normal genes or gene products into ALS cells in culture and assaying for abolition of the DRDS and PRIDS phenomena after MMS exposure, it may be possible to identify and begin to isolate and study the genes and their products responsible for development of ALS. This is now underway in our laboratory using the isolated chromatin non-histone protein system we have developed (Lambert and Parrish, this volume) and an electroporation system which we have recently shown to be able to restore UDS in XP cells to above normal levels following treatment with psoralen plus long wavelength ultraviolet (UVA) irradiation (Tsongalis, et al., submitted).

ACKNOWLEDGEMENT

We thank Mr. Robert Lockwood for his technical assistance. Portions of this research were supported by a grant from The Amyotrophic Lateral Sclerosis Association.

REFERENCES

Andrews, A.D., S.F. Barrett and J.H. Robbins, 1978. Xeroderma pigmentosum neurological abnormalities correlate with colony-forming ability after ultraviolet radiation, Proc. Natl. Acad. Sci. USA, 75:1984-1988.

Bradley, M.O., L.C. Erickson and K.W. Kohn, 1978. Non-enzymatic DNA strand breaks induced in mammalian cells by fluorescent light, Biochem. Biophys. Acta, 520:11-20.

Bradley W.G., P. Good, C.G. Rasool and L.S. Adelman, 1983. Morphometric and biochemical studies of peripheral nerves in amyotrophic lateral sclerosis, Ann. Neurol., 14:267-277.

Bradley, W.G. and F. Krasin, 1982a. A new hypothesis of the etiology of amyotrophic lateral sclerosis: The DNA hypothesis, Arch. Neurol., 39:677-680.

Bradley, W.G. and F. Krasin, 1982b. DNA hypothesis of amyotrophic lateral sclerosis, Adv. Neurol., 36:493-500.

Brozmanova, J., 1984. Stimulation of postirradiation DNA synthesis in ultraviolet radiated HeLa cells by fluorodeoxyuridine, Neoplasma, 31:169-173.

Brozmanova, J., B.I. Synzynys, F. Masek and A.S. Saenko, 1985. Effect of 5-fluorodeoxyuridine on DNA replication in ultraviolet irradiated HeLa cells, Stud. Biophys., 108:97-108.

Chamberlain, S. and P.D. Lewis, 1982. Studies of cellular hypersensitivity to ionizing radiation in Friedreich's ataxia, J. Neurol. Neurosurg. Psychiatry, 45:1136-1138.

Cleaver, J.E., F. Cortes, L.H. Lutze, W.F. Morgan, A.N. Player and D.L. Mitchell, 1987. Unique DNA repair properties of a xeroderma pigmentosum revertant, Molec. Cell Biol., 7:3353-3357.

Collins, A. and D.J. Oates, 1987. Hydroxyurea: Effects on deoxyribonucleotide pool sizes correlated with effects on DNA repair in mammalian cells, Eur. J. Biochem., 169:299-305.

Davidson, T.J. and H.A. Hartmann, 1981. RNA content and volume of motor neurons in amyotrophic lateral sclerosis: II. The lumbar intumescence and nucleus dorsalis, J. Neuropathol. Exp. Neurol., 40:187-192.

Davidson, T.J., H.A. Hartmann and P.C. Johnson, 1981. RNA content and volume of motor neurons in amyotrophic lateral sclerosis: I. The cervical swelling, J. Neuropathol. Exp. Neurol., 40:32-36.

Davies, S.M., A.L. Harris and I.D. Hickson, 1989. Overproduction of topoisomerase II in an ataxia telangiectasia fibroblast cell line: Comparison with a topoisomerase II-overproducing hamster cell mutant, Nucl. Acids Res., 17:1337-1351.

Dresler, S.L., B.J. Gowans, R.M. Robinson-Hill and D.J. Hunting, 1988. Involvement of DNA polymerase delta in DNA repair synthesis in human fibroblasts at late times after ultraviolet irradiation, Biochemistry, 27:6379-6383.

Eisen, A.A. and A.J. Hudson, 1987. Amyotrophic lateral sclerosis: Concepts in pathogenesis and etiology, Can. J. Neurol. Sci., 14:649-652.

Ganges, M.B., R.E. Tarone, H.X. Jiang, C. Hauser and J.H. Robbins, 1988. Radiosensitive Down syndrome lymphoblastoid lines have normal ionizing-radiation-induced inhibition of DNA synthesis, Mutat. Res., 194:251-256.

Gatti, R.A., I. Berkel, E. Boder, G. Braedt, P. Charmley, P. Concannon, F. Ersoy, T. Foroud, N.G. Jaspers, K. Lange, et al., 1988. Localization of an ataxia-telangiectasia gene to chromosome 11 q 22-23, Nature, 336:577-580.

Hsu, T.C., D.A. Johnston, L.M. Cherry, D. Rambisson, S.P. Schantz, J.M. Jessup, R.J. Winn, L. Shirley and C. Furlong, 1989. Sensitivity to genotoxic effects of bleomycin in humans: Possible relationship to environ-

mental carcinogenesis, Int. J. Cancer, 43:403-409.
Koga, S.J. and A.L. Schroeder, 1987. Gamma-ray-sensitive mutants of Neurospora crassa with characteristics analogous to ataxia telangiectasia cell lines, Mutat. Res., 183:139-148.
Kunkel, T.A., C.W. Shearman and L.A. Loeb, 1981. Mutagenesis in vitro by depurination of OX174 DNA, Nature (London), 291:349-351.
Lambert, W.C., 1969. Synchrony and Reversible Unbalanced Growth Induced in HeLa Cells by Excess Thymidine, Doctoral Dissertation, Thomas Jefferson University, Philadelphia, PA.
Lambert, W.C. and M.W. Lambert, 1983. Increased sensitivity of a line of xeroderma pigmentosum lymphoblastoid cells to serum deprivation in vitro, In Vitro, 19:621-624.
Lambert, W.C. and M.W. Lambert, 1987. DNA repair deficiency and cancer in xeroderma pigmentosum, Cancer Rev., 7:56-81.
Lambert, W.C., A.O. Okorodudu and M.W. Lambert, 1986. Hypersensitivity of ALS lymphoblastoid cells in culture to the mutagen, methyl methanesulfonate, Neurology, 36:136.
Lambert, W.C. and G.P. Studzinski, 1967. Recovery from prolonged unbalanced growth induced in HeLa cells by high concentrations of thymidine, Cancer Res., 27:2364-2369.
Lavin, M.F., P. Bates, P. LePoidevin and P. Chen, 1989. Normal inhibition of DNA synthesis following gamma-irradiation of radiosensitive cell lines from patients with Down's syndrome and Alzheimer's disease, Mutat. Res. (DNA Repair), 218:41-47.
Lavin, M.F. and A.L. Schroeder, 1988. Damage-resistant DNA synthesis in eukaryotes, Mutat. Res., 193:193-206.
Lawley, P.D. and S.A. Shah, 1972. Reaction of alkylating mutagens and carcinogens with nucleic acids: detection and estimation of a small extent of methylation at O^6 of guanine in DNA by methyl methanesulphonate in vitro, Chem-Biol. Interact., 5:286-288.
Lehmann, A.R., 1987. Cockayne's disease. Defective DNA repair without cancer, Cancer Rev., 7:82-101.
Li, J.C. and L.E. Kaminskas, 1985. Deficient repair of DNA lesions in Alzheimer's disease fibroblasts, Biochem. Biophys. Res. Commun., 129:733-738.
Lindahl, T., 1977. DNA repair enzymes acting on spontaneous lesions in DNA. In W.W. Nichols and W.G. Murphy, eds., Cellular Senescence and Somatic Cell Genetics, Symposia Specialists, Miami, 225-234.
Lindahl, T. and B. Nyberg, 1972. Rate of depurination of native DNA, Biochemistry, 11:3610-3618.
Mann, D.M.A. and P.O. Yates, 1974. Motor neuron disease: The nature of the pathogenic mechanism, J. Neurol. Neurosurg. Psych., 37:1036-1046.
McGovern, D. and T. Webb, 1982. Sensitivity to ionising radiation of lymphocytes from Huntington's chorea patients compared to controls, J. Med. Genet., 19:168-174.
McKinnon, P.J., 1987. Ataxia-telangiectasia: An inherited disorder of ionizing-radiation sensitivity in man. Progress in the elucidation of the underlying biochemical defect, Hum. Genet., 75:197-208.
Menth, M., 1989. The molecular basis of mutations induced by deoxynucleoside pool imbalances in mammalian cells, Exp. Cell Res., 181:305-316.

Mitchell, D.L., C.A. Haipek and J.M. Clarkson, 1985. (6-4) Photoproducts are removed from the DNA of UV-irradiated mammalian cells more efficiently than cyclobutane pyrimidine dimers, Mutat. Res., 143:109-112.

Mitsumoto, H., M.R. Hanson and D.A. Chad, 1988. Amyotrophic lateral sclerosis. Recent advances in pathogenesis and therapeutic trials, Arch. Neurol., 45:187-202.

Mirzayans, R., M. Liuzzi and M. Paterson, 1988. Methyl methanesulfonate-induced DNA damage and its repair in cultured human fibroblasts: Normal rates of induction and removal of alkalai-labile sites in xeroderma pigmentosum (group A) cells, Carcinogenesis, 9:2257-2263.

Moshell, A.N., O. Barrett, R.E. Tarone and J.H. Robbins, 1980. Radiosensitivity in Huntington's disease: Implications for pathogenesis and presymptomatic diagnosis, Lancet, 1:9-11.

Murakami, T., F.L. Mastaglia and W.G. Bradley, 1980. Reduced protein synthesis in spinal anterior horn neurons in the wobbler mouse mutant, Exp. Neurol., 67:423-432.

Murakami, T., F.L. Mastaglia, D.M.A. Mann and W.G. Bradley, 1981. Abnormal RNA metabolism in spinal motor neurons in the wobbler mouse, Muscle Nerve, 4:407-412.

Nishida, C., P. Reinhard and S. Linn, 1988. DNA repair synthesis in human fibroblasts requires DNA polymerase delta, J. Biol. Chem., 263:501-510.

Nove, J., R.E. Tarone, J.B. Little and J.H. Robbins, 1987. Radiation sensitivity of fibroblast strains from patients with Usher's syndrome, Duchenne muscular dystrophy and Huntington's disease, Mutat. Res., 184:29-38.

Orlando, P., R. Gersmia, C. Frusciante, B. Tedeschi and P. Grippo, 1988. DNA repair synthesis in mouse spermatogenesis involves DNA polymerase beta activity, Cell Differ., 23:221-230.

Otsuka, F., R.E. Tarone, L.R. Seguin and J.H. Robbins, 1985. Hypersensitivity to ionizing radiation in cultured cells from Down syndrome patients, J. Neurol. Sci., 69:103-112.

Painter, R.B., 1985. Altered DNA synthesis in irradiated and unirradiated ataxia-telangiectasia cells. In: Ataxia-telangiectasia: Genetics, Neuropathology, and Immunology of a Degenerative Disease of Childhood, Alan R. Liss, New York, 89-100.

Painter, R.B., 1980. A replication model for sister-chromatid exchange, Mutat. Res., 70:337-341.

Park, J.W. and B.N. Ames, 1989. 7-methylguanine adducts in DNA are normally present at high levels and increase on aging: Analysis by HPLC with electrochemical detection, Proc. Natl. Acad. Sci. USA, 85:7467-7470,Correction,9508.

Parshad, R., K.K. Sanford, G.M. Jones and R. Tarone, 1978. Fluorescent light-induced chromosome damage and its prevention in mouse cells in culture, Proc. Natl. Acad. Sci. USA, 75:1830-1833.

Paterson, M.C. and P.J. Smith, 1979. Ataxia telangiectasia: An inherited human disorder involving hypersensitivity to ionizing radiation and related DNA-damaging chemicals, Ann. Rev. Genet., 13:291-318.

Robbins, J.H., 1987a. Incorrect priority claim for the DNA-damage hypothesis (letter), Arch. Neurol., 44:579-583.

Robbins, J.H., 1987b. Parkinson's disease, twins and the DNA-damage hypothesis (letter), Ann. Neurol., 21:412.

Robbins, J.H. 1987c. Xeroderma pigmentosum. Defective DNA repair causes skin cancer and neurodegeneration (Clinical conference), J.A.M.A., 260:384-388.

Robbins, J.H., 1983. Hypersensitivity to DNA-damaging agents in primary degenerations of excitable tissue. In: E. Friedberg and B. Bridges, eds., Cellular Responses to DNA Damage, Alan R. Liss, New York, 671-700.

Robbins, J.H., M.B. Ganges, A.N. Moshell, R.E. Tarone and R.A. Brumback, 1989. Amyotrophic lateral sclerosis fibroblasts have normal rates of unscheduled DNA synthesis induced by the DNA-damaging alkylating chemical, methyl methane-sulfonate, Clin. Res., 37:139A.

Robbins, J.H., F. Otsuka, R.E. Tarone, R.J. Polinsky, R.A. Brumback and L.E. Nee, 1985a. Parkinson's disease and Alzheimer's disease: hypersensitivity to x-rays in cultured cell lines, J. Neurol. Neurosurg. Psych., 48:916-923.

Robbins, J.H., F. Otsuka, R.E. Tarone, R.J. Polinsky, R.A. Brumback, A.N. Moshell, L.E. Nee, M.B. Ganges and S.J. Cayeux, 1983a. Radiosensitivity in Alzheimer disease and Parkinson disease, Lancet, 1:468-469.

Robbins, J.H., R.N. Polinsky and A.N. Moshell, 1983b. Evidence that lack of deoxyribonucleic acid repair causes death of neurons in xeroderma pigmentosum, Ann. Neurol., 13: 682-684.

Robbins, J.H., D.A. Scudiero, F. Otsuka, R.E. Tarone, R.A. Brumback, J.D. Wirtschafter, R.J. Polinsky, S.F. Barrett, A.N. Moshell, R.G. Scarpinato, M.B. Ganges, S.A. Mayer and R.E.W. Clatterbuck, 1984. Hypersensitivity to DNA damaging agents in cultured cells from patients with Usher's syndrome and Duchenne muscular dystrophy, J. Neurol. Neurosurg. Psych., 47:391-398.

Robbins, J.H., R.E. Tarone, F. Otsuka, A. Moshell, R.J. Polinsky and L.E. Nee, 1985b. Somatic mutations in DNA-repair systems: The cause of Parkinson's disease, Alzheimer's disease, and other nonfamilial primary neuronal degenerations, Neurology, 35 (Suppl 1):191.

Robison, S.H. and W.G. Bradley, 1985. Impaired DNA repair replication in Alzheimer's disease cells. In: Hutton JH and Kenny AD (eds.), Senile Dementia of the Alzheimer Type, Alan R. Liss, New York, 205-218.

Robison, S.H. and W.G. Bradley, 1984. DNA damage and chronic neuronal degenerations, J. Neurol. Sci., 64:11-20.

Robison, S.H., J.S. Munzer, R. Tandan and W.G. Bradley, 1987. Alzheimer's disease cells exhibit defective repair of alkylating agent-induced DNA damage, Ann. Neurol., 21:250-258.

Robison, S.H., J.S. Munzer, R. Tandan, R. Bradley and W.G. Bradley, 1985. Repair of alkylated DNA is impaired in Alzheimer's disease cells, Neurology, 35 (Suppl 1):217-218.

Roth, M., L.R. Emmons, M. Haner, H.J. Muller and J.M. Boyle, 1989. Age-related decrease in an early step of DNA-repair of normal human lymphocytes exposed to ultraviolet-irrad-iation, Exp. Cell. Res., 180:171-177.

Rudolph, N.S., H. Nagasawa, J.B. Little and S.A. Latt, 1989. Identification of ataxia telangiectasia heterozygotes by flow cytometric analysis of x-ray damage, Mutat. Res., 211:19-29.

Rudolph, N.S. and S.A. Latt, 1989. Flow cytometric analysis

of x-ray sensitivity in ataxia telangiectasia, Mutat. Res., 211:31-41.

Scudiero, D.A., R.A. Brumback, R.E. Tarone, B.F. Clatterbuck and J.H. Robbins, 1983. Amyotrophic lateral sclerosis and spinal muscular atrophy fibroblasts are not hypersensitive to killing by a DNA damaging agent, Clin. Res., 31:292A.

Scudiero, D.A., S.A. Meyer, B.E. Clatterbuck, R.E. Tarone and J.H. Robbins, 1981. Hypersensitivity to N-methyl-N'-nitro-N-nitrosoguanidine in fibroblasts from patients with Huntington disease, familial dysautonomia and other primary neuronal degenerations, Proc. Natl. Acad. Sci. USA, 78:6451-6455.

Scudiero, D.A., R.J. Polinsky, R.A. Brumback, R.E. Tarone, L.E. Nee and J.H. Robbins, 1986. Alzheimer disease fibroblasts are hypersensitive to the lethal effects of a DNA-damaging chemical, Mutat. Res., 159:125-131.

Singh, S.P., R. Mohamed, C. Salmond and M.F. Lavin, 1988. Reduced DNA topoisomerase II activity in ataxia-telangiectasia cells, Nucl. Acids Res., 16:3919-3929.

Smith, P.J. and T.A. Malkinson, 1989. Cellular consequences of overproduction of DNA topoisomerase II in an ataxia-telangiectasia cell line, Cancer Res., 49:1118-1124.

Tandan, R. and W.G. Bradley, 1985. Amyotrophic lateral sclerosis: Part 2. Etiopathogenesis, Ann. Neurol., 18:419-431.

Tandan, R., S.H. Robison, J.S. Munzer and W.G. Bradley, 1987. Deficient DNA repair in amyotrophic lateral sclerosis cells, J. Neurol. Sci., 79:189-203.

Tandan, R., S.H. Robison, J.S. Munzer and W.G. Bradley, 1985. Deficient DNA repair in amyotrophic lateral sclerosis cells. Neurology, 35 (Suppl 1):73.

Tanz, W., G.J. Tsongalis and W.C. Lambert, 1988. Paradoxical impairment of down regulation of replicative DNA synthesis in ALS cells by methyl methanesulfonate, Clin. Res., 36:814A.

Thacker, J., 1989. The use of integrating DNA vectors to analyze the molecular defects in ionizing radiation-sensitive mutants of mammalian cells including ataxia telangiectasia, Mutat. Res., 220:187-204.

Vijaylaxmi, B. Pentland, M.S. Newton, J.D. Mitchell and H.J. Evans, 1985. Spontaneous and mutagen-induced sister chromatid exchange in motor neuron disease, Mutat. Res., 150:355-358.

Woloschak, G.E., M. Rodriguez and C.J. Krco, 1987. Characterization of immunologic and neuropathologic abnormalities in wasted mice, J. Immunol., 138:2493-2497.

Young, B.R. and R.B. Painter, 1989. Radioresistant DNA synthesis and human genetic diseases, Hum. Genet., 82:113-117.

GENETIC ANALYSIS IN TRICHOTHIODYSTROPHY REPAIR DEFICIENT CELLS CONFIRMS THE OCCURRENCE OF XERODERMA PIGMENTOSUM GROUP D MUTATION IN UNRELATED PATIENTS

Miria Stefanini, Paola Lagomarsini and Fiorella Nuzzo

Istituto di Genetica Biochimica ed Evoluzionistica, CNR
Via Abbiategrasso, 207-27100 Pavia, Italy

Xeroderma Pigmentosum (XP) represents the best studied human DNA-repair mutant in which photosensitivity and high incidence of sunlight-induced skin cancers comply with cellular impairement in the capacity to repair UV-induced DNA damage. Genetic heterogeneity in XP has been demonstrated by complementation analysis. Ten complementation groups have been identified to date, termed XP-A through XP-I and XP Variant (XP-V). The most common groups are A, C, D and V; groups E, F and G are thus far represented by few families, whereas, from groups B and H only one patient is known (reviewed by Jung, 1986).

A peculiar phenotype is present in both the sole members of groups B and H; in fact, these patients have been reported as having Cockayne's syndrome in addition to XP. XP has also been found sporadically associated with other hereditary disorders; in these patients the concurrence of the two pathological conditions can be considered fortuitous because the XP-associated anomaly is not infrequent in the population, and because they represent single cases (Nishigori, et al., 1986; reports quoted by Kraemer, et al.,1987).

The association of XP with trichothiodystrophy (TTD) appears to be completely different. In fact, although TTD is extremely rare, in all patients so far analysed in which the clinical features typical of TTD are accompanied by photosensitivity, a DNA repair defect has been demonstrated, and the presence of the same mutation present in XP group D cells has been ascertained by complementation analysis.

TTD is a rare autosomal recessive disorder; major diagnostic criteria are brittle hair with reduced sulfur content, mental and physical retardation, and unusual facies (Nuzzo and Stefanini, 1988). A relevant feature in TTD is the presence of photosensitivity in about 20% of the reported patients (Table 1).

DNA Repair Mechanisms and Their Biological Implications in Mammalian Cells
Edited by M.W. Lambert and J. Laval
Plenum Press, New York

The presence of photosensitivity in four Italian TTD patients led us to investigate the efficiency of their DNA repair mechanisms. Cellular hypersensitivity to UV light was observed in the patients' lymphocytes; the inhibition of the

Table 1. TTD cases without (A) and with (B) photosensitivity

Reference	Sex	Age (years)	Place of origin
A			
Salfeld and Lindley, 1963	M	10	Germany*
Pollit, _et al_., 1968	F,M	5,3	GB*
Tay, 1971	2M,F	0.2,11,9	China
Jackson, _et al_., 1974	14M,11F	0-62	Switzerland
Leupold, 1979	F	13	Germany*
Jorizzo, _et al_., 1980	F	20	USA*
Price, _et al_., 1980	M	5	USA*
Braun-Falco, _et al_., 1981	F,M	10,4	Germany*
Howell, _et al_., 1981	6M,6F	0-39	Mexico
Jorizzo, _et al_., 1982	M	8	USA*
Venencie, _et al_., 1982	-	-	France*
Happle, _et al_., 1984	M	5	Germany*
Morel, _et al_., 1984	-	-	France*
Przedborski, _et al_., 1985	M,2F	17,5,2	Morocco
King, _et al_., 1986	M	4	Scotland
Larregue, _et al_., 1986	-	-	France*
Larregue, _et al_., 1986	-	-	France*
Prost, _et al_., 1986	F	1	France*
Meynadier, _et al_., 1987	M	23	France*
Fois, _et al_., 1988	F	8	Italy
Lehmann, _et al_., 1988	M	6	GB*
Van Neste, _et al_., 1988	F	3	Italy
B			
Calderon and Gonzales-Cantu, 1979	3F	6,14,19	Mexico*
Price, _et al_., 1980	M	8,5	USA*
Diaz-Perez and Vasquez, 1983	F,M	18,14	Spain*
Van Neste, _et al_., 1985	F	5	Poland
Lucky, _et al_., 1984	M	16	Italy
King, _et al_., 1984**	F	2.5	Scotland
Yong, _et al_., 1984	M	10	Italy
Stefanini, _et al_., 1986	4F	4,7,8.5,18	Italy
Van Neste, _et al_., 1988	M	-	France*
Kleijer, 1988***	F	3	Netherlands*
Sarasin, 1988***	M	2	France*
Sarasin, 1988***	F	5	France*

*Place where the patient was studied.
**Photosensitivity not confirmed at last examination (Lehmann, _et al_., 1988).
***Personal communication.

rate of DNA synthesis after mitogen stimulation was significantly higher than in cells from normal donors, whereas, it was in the normal range following treatment with mono- and bi-functional alkylating agents (Figure 1).

Accordingly, the analysis of survival and of RNA synthesis recovery in fibroblasts after UV light indicated a greater UV sensitivity in TTD cells compared to normal and TTD heterozygous cells (Figure 2). The abnormal response to UV light was associated with a remarkable decrease in the capacity to perform UV-induced DNA synthesis, which was reduced to 10% of normal level. Genetic analysis of the repair defect was performed by complementation studies in heterokaryons obtained by fusion of TTD cells and XP group A, C, and D fibroblasts. The results of these investigations clearly demonstrated that the same genetic defect was present

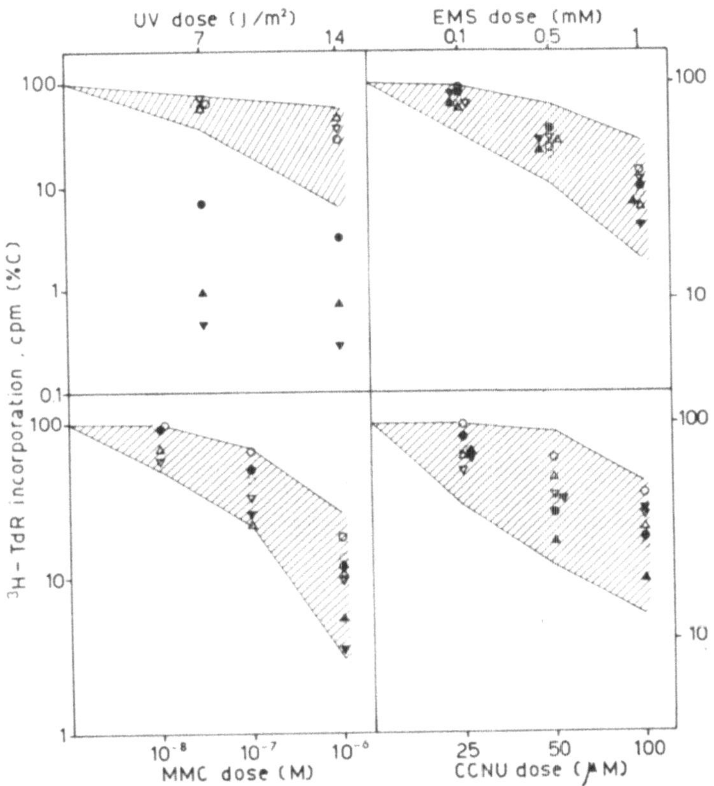

Figure 1. Rate of DNA synthesis in stimulated lymphocytes after exposure of G_0 cells to mutagens [UV light; ethylmethanesulfonate (EMS); mitomycin C (MMC); 1-2-chloroethyl-3-cyclohexyl-1-nitrosourea(CCNU)]. The incorporation values of ^{3}H-thymidine (^{3}H-TdR) in treated samples are expressed as a percentage of those in untreated samples (C). The shadowed area indicates the range of variability (mean value ± 95% confidence limits) of the values in 11 healthy subjects. TTD patients: black symbols; TTD patients' parents: white symbols.

in all four TTD patients and that the repair alteration was due to the presence of the XP-D mutation (Stefanini, et al., 1986).

To clarify whether this cellular and genetic situation was constantly associated with the TTD phenotype, we investigated DNA repair proficiency in cells from a TTD patient without signs of photosensitivity. Cells from this patient showed normal capacity to perform unscheduled DNA synthesis (UDS) after UV and to complement the defect in TTD repair-deficient cells, clearly demonstrating that the mutation determining TTD is independent of the set of XP mutations (Stefanini, et al., 1987). Thus, in the TTD patients with clinical and cellular UV sensitivity, two different mutations, one determining TTD and the other XP-D, are present.

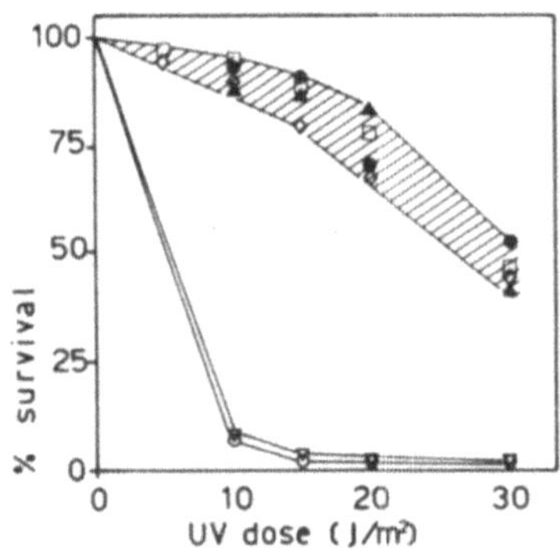

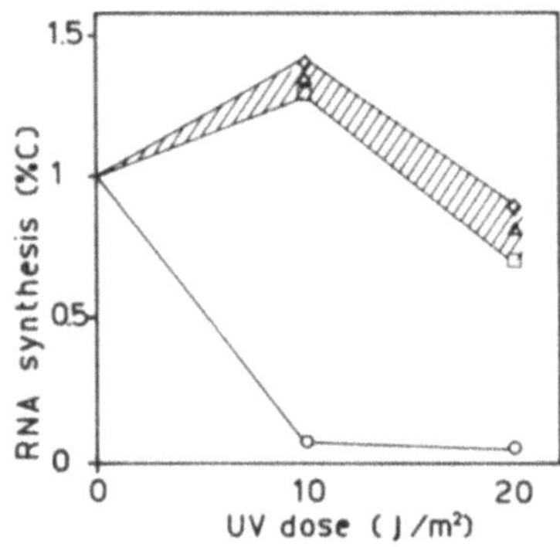

Figure 2. UV-sensitivity in fibroblasts from the TTD4PV patient (O), from its parents (●,▲), from the XP group D patient XP3NE (▽), and from normal donors (□,△,◇). Left panel: survival after UV in stationary cells. Right panel: recovery of RNA synthesis after UV in stationary cells. ^{3}H-Uridine incorporation values in irradiated cultures are expressed as percentage of those in unirradiated cultures (C).

To verify whether the association of TTD and XP-D was due to a sporadic event, we extended DNA repair investigations to other TTD patients. In collaboration with Drs. C.F. Arlett and A. Lehmann (MRC Cell Mutation Unit, University of Sussex, Brighton, U.K.), we had the opportunity to perform genetic analysis on cells from three English TTD patients (Lehmann, et al., 1988). As shown in Figure 3, one cell strain (TTD1GL) is repair-proficient and appears to be able to complement the UDS defect in the repair-deficient TTD cells from the Italian patients.

The other two cell strains (TTD2GL and TTD1BI) are repair deficient; TTD2GL fibroblasts show UDS levels reduced to 10% of normal. Complementation is not observed after fusion with repair-deficient TTD PV cells, whereas, complementation occurs after fusion with the two repair-proficient TTD cell strains (TTD1GL, TTD5PV). The cell strain TTD1BI has a lower degree of UV sensitivity, corresponding to a 50% reduction of UDS. In heterokaryons obtained by fusion of TTD1BI cells with TTD PV cells, in which repair is more strongly affected, there is not full complementation, but the UDS values approach those typical of the less sensitive TTD1BI parental cells. As expected, normal values of UDS are observed after fusion of TTD2GL or TTD1BI cells with XP-A and XP-C cells.

These results indicate that the genetic defect in repair-deficient TTD cells from English patients is the same as that present in the other TTD UV-sensitive cells, i.e., the XP-D mutation.

More recently we studied three other TTD patients; one case was kindly referred to us by Dr. W. Kleijer (Erasmus University, Rotterdam, The Netherlands) and the others by Dr. A. Sarasin (IRSC, Paris, France). All of these three patients show diminished cellular ability to perform UV-induced DNA repair synthesis, with UDS levels ranging between 20% and 50% of normal. Fusion of these three cell strains with repair-deficient TTD cells does not restore normal UDS levels in heterokaryons.

In conclusion, our DNA repair investigations, extended to six patients from different countries, confirmed that TTD

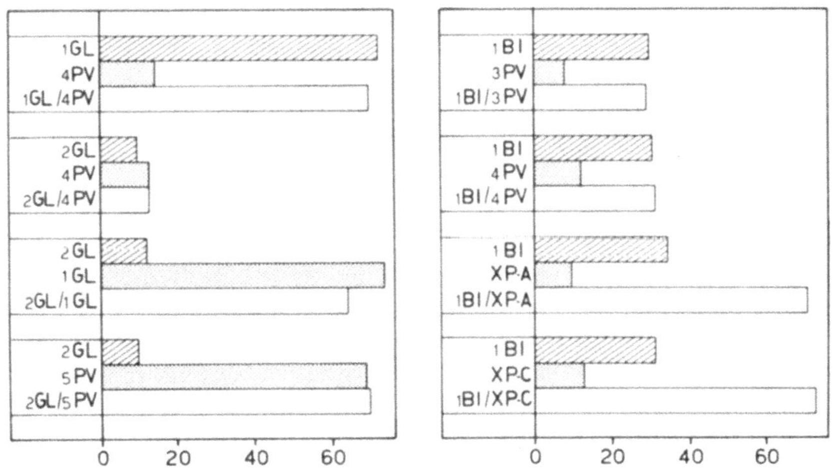

Figure 3. Complementation analysis in heterokaryons obtained by fusion of repair-deficient TTD cells (TTD3PV, TTD4PV, TTD2GL, TTD1BI) with repair proficient TTD cells (TTD5PV, TTD1GL) and XP cells from groups A and C (XP-A: XP25RO; XP-C: XP9PV). The prefix TTD was omitted. The mean number of grains over the nuclei in 50 mononuclear parental cells and in 25 heterokaryons are reported.

cells may have normal or enhanced UV sensitivity. Among the patients showing the DNA repair defect, different degrees of alteration in UDS levels are present. However, genetic analysis indicates that in all the repair-deficient TTD cells, the defect is always due to the presence of the XP-D mutation.

Also, cells from different XP-D patients, such as UV sensitive TTD cells, are characterized by different residual levels of UDS. From a review of the literature, it appears clearly that XP patients belonging to group D show a great heterogeneity both at clinical and cellular levels. Remarkable differences among patients concern the neurological abnormalities, the age at onset of skin symptoms, the presence of skin tumors and the age at onset of first tumors, and finally the degree of the impairment in the capacity to repair the UV-induced damage (Table 2).

Table 2. XP-D cases

Reference	Patients' number*	Fibroblast analysis UDS**	CFA**	Clinical data neur. anom.	skin tumors	age*** (years)
Andrews, <u>et al</u>., 1978	7(5)	25-50+	10	+	+	9,9,18,?
Pawsey, <u>et al</u>., 1979	3	10-20		-	-	5,7,11
	1	30		-	-	18
Kato, <u>et al</u>., 1985	1	25		-	-	29
	2(1)	40		+	+	13,20
	2	40		-	+	9,20
Fisher, <u>et al</u>., 1982	7	15-23	10-23	-	+	2,3,4,5 10,15,17
Jung, <u>et al</u>., 1986	3	28-51	23-44	-	+	12,22,27
	1	29	15	-	?	?
	2	46-50	20-28	-	-	11,38
Ichihashi, <u>et al</u>., 1988	2	40		-	-	6,8
	2	25-45	10	-	+	31,41

*The number of unrelated patients in parenthesis.
**Unscheduled DNA synthesis (UDS) and colony-forming ability (CFA) are reported as % of normal.
***Age at onset of first cancer for patients showing skin tumors (+); age at last examination for patients with no skin tumors (-).
+UDS levels lower than those reported by Andrews, <u>et al</u>., 1978, have been observed in the XP2NE, XP3NE cells (10-20% of normal) and in XP5BE, XP6BE cells (2-10% of normal) (Bootsma, <u>et al</u>., 1977, our data).

There is one peculiar aspect in the association of XP-D with TTD: in all the TTD patients showing clinical and cellular hypersensitivity to UV light and even in those where the genetic defect appears to be identical to that of XP-D, tumors and severe skin alterations are absent. This fact could be related to the young age of the patients (the maximum age is 21 years), or it may be hypothesized that the association with TTD gives rise to a different phenotypic expression of the XP-D mutation. The response to UV light in repair-deficient TTD cells has been recently analyzed by Lehmann, et al., (1988) by measuring the efficiency of repair processes using different assays. The results of this detailed study point out that the heterogeneous expression of the XP-D mutation in TTD/XP-D patients is also present at biochemical and cytological levels. Further clinical and cellular investigations are needed to clarify these intriguing aspects of the TTD/XP-D association.

As to the incidence of the TTD/XP-D concurrence (assuming that all the TTD cases where clinical photosensitivity has not been reported are repair-proficient), TTD appears to be associated with XP-D in about 40% of the families with TTD affected members (Table 3). This fact indicates that the association cannot be considered fortuitous and allows some speculations on the genetic basis of the TTD/XP-D phenotype.

Table 3. TTD cases

	No. of patients	No. of families	No. of unrelated families
without photosensitivity	62	36	21
with photosensitivity*	19(10)	15(9)	13(7)
total	81	51	34

In parenthesis the number of cases where XP-D mutation has been ascertained by complementation analysis.
*The patient TTD1BI which despite the normal clinical photosensitivity shows cellular UV-sensitivity due to the presence of XP-D mutation has been included in this group (Lehmann, et al.,1988).

The simplest explanation of the complex phenotype is the occurrence of two independent mutations (XP-D and TTD), each in homozygous conditions. However, it is highly improbable that this association occurred, by chance, in several unrelated patients. In the three Italian families with four affected members (TTD1-4PV patients reported in this paper and in Stefanini et al., 1986), we performed an accurate search for consanguinity based on the reconstruction of geneological trees, the study of blood genetic markers, and the analysis of surnames. These studies led us to conclude that it is highly probable that the patients are carriers of a genetic defect, identical by descent, as a consequence of multiple remote inbreeding (Nuzzo, et al., 1989). Even for a single original mutated genotype, it is required that the two mutations (maintained together through several generations and now present in at least six heterozygous individuals) are very close to each other on the same chromosome or affect the same gene. Obviously in this case a chromosome alteration (such as a translocation or a deletion), involving both the TTD and XP-D loci, could be responsible for the TTD/XP-D phenotype. To verify this possibility, we performed cytogenetic studies using high resolution banding techniques in lymphoblastoid cell lines from three patients (TTD1-3PV) and from six heterozygotes. A normal karyotype was found in all the individuals, indicating that a cytologically detectable chromosome rearrangement does not account for the complex phenotype of the patients (Nuzzo, et al., 1988).

At present, the identification of other TTD/XP-D patients with different degrees of repair-deficiency suggests as improbable a chromosomal alteration. However, it cannot be excluded that microdeletions of different lengths are present in different patients.

Several other explanations may account for the TTD/XP-D concurrence; for example, the heterozygosity for two different defective alleles occupying the same locus (i.e.,compounds) in the trans configuration or a more complex genetic situation according to the co-recessive model of inheritance proposed by Lambert and Lambert (1985).

Further cellular and genetic analysis of the TTD/XP-D phenotype may offer new insight in understanding the genetic basis of the association. So far, the typing of 25 blood markers with known chromosomal locations in three patients, in six heterozygous individuals and in twenty-two relatives from the Italian families, allowed us to conclude that the pathologic condition is not linked to any of the informative markers analyzed (Nuzzo, et al., 1988).

ACKNOWLEDGEMENTS

This work was supported by grants from the National Research Council of Italy (C.N.R. Target Project "Biotechnology") and from the Associazione Italiana per la Ricerca sul Cancro (A.I.R.C.). P. Lagomarsini is the recipient of a fellowship from the A.I.R.C. The authors thank Mrs D. Tavarnè for typing the manuscript.

REFERENCES

Andrews, A.D., Barrett, S.F., and Robbins, J.H., 1978, Xeroderma pigmentosum neurological abnormalities correlate with colony-forming ability after ultraviolet radiation. Proc. Natl. Acad. Sci. USA, 75:1984.

Bootsma, D., 1977, Defective DNA repair and cancer, in: "Research in Photobiology" Castellani A. ed. Plenum Press New York, 455.

Braun-Falco, O., Ring, J., Butenandt, O., Selzle, D., and Landthaler, M., 1981, Ichthyosis vulgaris, Minderwuchs, Haardysplasie, Zahnanomalien, Immundefekte, psychomotorische Retardation und Resorptionsstörungen. Hautarzt, 32:67.

Calderon, R., and Gonzalez-Cantu, N., 1979, Kinky hair, photosensitivity, broken eyebrows and eyelashes, and non-progressive mental retardation. J. Pediatr., 95: 1007.

Diaz-Perez, J.L., and Vasquez, J.A., 1983, Flattened hair syndrome: a new disease. Arch. Dermatol., 119:854.

Fisher,E., Thielmann, H.W., Neundörfer, B., Rentsch, F.J. Edler, L. and Jung, E.G., 1982, Xeroderma pigmentosum patients from Germany: clinical symptoms and DNA repair characteristics. Arch. Dermatol. Res., 274:229.

Fois, A., Balestri, P., Calvieri, S., Zampetti, M., Giustini, S., Stefanini, M., and Lagomarsini, P., 1988, Trichothiodystrophy without photosensitivity: biochemical, ultrastructural and DNA repair studies. Eur. J. Pediatr., 147:439.

Happle, R., Traupe, H., Gröbe, H., and Bonsmann, G., 1984, The Tay syndrome (congenital ichthyosis with trichothiodystrophy). Eur. J. Pediatr., 141:147.

Howell, R.R., Arbisser, A.I., Parsons, D.S., Scott, C.I., Frausdadt, U., Collie, W.R., Marshall, R.N., and Ibarra, O.C., 1981, The Sabinas syndrome. Am. J. Hum. Genet 33:957.

Ichihashi, M., Yamamura, K., Hiramoto, T., and Fujiwara, Y., 1988, No apparent neurologic defect in a patient with xeroderma pigmentosum complementation group D. Arch. Dermatol., 124:256.

Jackson, C.E., Weiss, L., and Watson, J.H.L., 1974, "Brittle" hair with short stature, intellectual impairment and decreased fertility: an autosomal recessive syndrome in an Amish kindred. Pediatrics, 54:201.

Jorizzo, J.L., Crounse, R.G., and Wheeler, C.E., 1980, Lamellar ichthyosis, dwarfism, mental retardation and hair shaft abnormalities. J. Am. Acad. Dermatol., 2: 309.

Jorizzo, J.L., Atherton, D.J., Crounse, R.G., and Wells, R.S., 1982, Ichthyosis, brittle hair, impaired intelligence, decreased fertility and short stature (IBDS syndrome). Br. J. Dermatol., 106:705.

Jung, E.G., 1986, Xeroderma pigmentosum. Int. J. Dermatol., 25:629.

Jung, E.G., Bohnert, E. and Fischer, E., 1986, Heterogeneity of xeroderma pigmentosum (XP); variability and stability within and between the complementation groups C, D, E, I and variants. Photodermatology, 3:125.

Kato, T., Akiba, H., Seiji, M., Tohda, H. and Oikawa, 1985, Clinical and biological studies of 26 cases of xeroderma pigmentosum in northeast district of Japan. Arch. Dermatol. Res., 277:1.

King, M.D., Gummer, C.L., and Stephenson, J.B.P., 1984, Trichothiodystrophy-neurotrichocutaneous syndrome of Pollit: a report of two unrelated cases. J. Med. Genet. 21:286.

King, M.D., Arlett, C.F., Lehmann, A.R., Hayne, L.V., and Stephenson, J.B.P., 1986, Clinical and cellular studies in trichothiodystrophy. 24th Annual Symposium SSIEM, Amersfoort, p. 158 (Abstr).

Kraemer, K.H., Lee, M.M. and Scotto, J., 1987, Xeroderma pigmentosum. Cutaneous, ocular, and neurologic abnormalities in 830 published cases. Arch. Dermatol., 123:241.
Lambert, W.C. and Lambert M.W., 1985, Co-recessive inheritance: A model for DNA repair, genetic disease and carcinogenesis, Mutat. Res., 145:227.
Larrègue, M., Ottavy, N., Bressieux, J.M., and Lorette, J., 1986. Bebé collodion. Trente-deux nouvelles observations. Ann. Dermatol. Venereol., 113:773.
Lehmann, A.R., Arlett, C.F., Broughton, B.C., Harcourt, S.A., Steingrimsdottir, H., Stefanini, M., Taylor, A.M.R., Natarajan, A.T., Green, S., King, M.D., McKie, R.M., Stephenson, J.B.P., and Tolmie, J.L., 1988, Trichothiodystrophy: a human DNA-repair disorder with heterogeneity in the cellular response to ultraviolet light. Cancer Res., 48:6090.
Leupold, D., 1979, Ichthyosis congenita, Katarakt, Schwachsinn, Ataxie, Osteosklerose und Abwehrdefekt - ein eigenständiges Syndrom? Monatsschr Kinderheilkd, 127: 307.
Lucky, P.A., Kirsch, N., Lucky, A.W., and Carter, D.M., 1984, Low-sulfur hair syndrome associated with UVB photosensitivity and testicular failure. J. Am. Acad. Dermatol., 11:340.
Meynadier, J., Guillot, B., Barnéon, G., Djian, B., and Lévy, A., 1987. Trichothiodystrophie. Ann. Dermatol. Venereol., 114:1529.
Morel, P., Cahuzac, P., and Vanleberghe, O., 1984, Trichothiodystrophie. Cas n^0 80, Soc. Fr. Dermatol. Journées Parisiennes, mars 1984.
Nishigori, C., Miyachi, Y., Takebe, H., and Imamura S., 1986, A case of Xeroderma pigmentosum with clinical appearance of dyschromatosis symmetrica hereditaria. Pediatric Dermatology, 3:410.
Nuzzo, F., and Stefanini, M., 1989, The association of xeroderma pigmentosum with trichothiodystrophy: a clue to a better understanding of XP-D?, in: "DNA Damage and Repair", Castellani A. ed. Plenum Press, New York, p. 61.
Nuzzo, F., Stefanini, M., Rocchi, M., Casati, A., Colognola, R., Marinoni, S., and Scozzari, R., 1988, Chromosome and blood marker studies in families of patients affected by xeroderma pigmentosum and trichothiodystrophy. Mutat. Res., 208:159.
Nuzzo, F., Zei, G., Stefanini, M., Colognola, R., Santachiara, A.S., Lagomarsini, P., Marinoni, S., and Salvaneschi, L., 1989. Search for consanguinity within and among families of patients with trichothiodystrophy associated with xeroderma pigmentosum, J. Med. Genet., in press.
Pawsey, S.A., Magnus, I.A., Ramsay, C.A., Benson, P.F., and Giannelli, F., 1979, Clinical, genetic and DNA repair studies on a consecutive series of patients with xeroderma pigmentosum. Quart. J. Med., 190:179.
Pollit, R.J., Jenner, F.A., and Davies, M., 1986, Sibs with mental and physical retardation and trichorrhexis nodosa with abnormal amino acid composition of the hair. Arch. Dis. Child., 43:211.
Price, V.H., Odom, R.B., Ward, W.H., and Jones, F.T., 1980, Trichothiodystrophy. Sulfur-deficient brittle hair as a marker for a neuroectodermal symptom complex. Arch. Dermatol., 116:1375.
Prost, de Y., Lemaistre, R., and Dupré, A., 1986, Tricho-

thiodystrophie associée a une ichtyose et a un retard statural et psychomoteur. Ann. Dermatol. Venereol., 113:1016.

Przedborski, S., Ferster, A., Song, M., Tonnesen, T., Ketelbant, P., and Vamos, E., 1985, Brittle hair, intellect ual impairement, decreased fertility and short stature (BIDS) syndrome in three sibs. J. Neurol., 232:127.

Salfeld, K. , and Lindley, M.J. , 1963, Zur Frage der Merkmalskombination bei Ichthyosis vulgaris mit Bambushaarbildung und ektodermaler Dysplasie. Dermatol. Wochenschr. 147:118.

Stefanini, M., Lagomarsini, P., Arlett, C.F., Marinoni, S., Borrone, C., Crovato, F., Trevisan, G., Cordone, G., and Nuzzo, F., 1986, Xeroderma pigmentosum (complementation group D) mutation is present in patients affected by trichothiodystrophy with photosensitivity. Hum. Genet., 74: 107.

Stefanini, M., Lagomarsini, P., Giorgi, R., and Nuzzo, F., 1987, Complementation studies in cells from patients affected by trichothiodystrophy with normal or enhanced UV-photosensitivity. Mutat. Res., 191:117.

Tay, C.H., 1971, Ichthyosiform erythroderma, hair shaft abnormalities, and mental and growth retardation. Arch. Dermatol., 104:4.

Van Neste, D., and Boré, P., 1983, Trichothiodystrophye: une étude morphologique et biochimique. Ann. Dermatol. Venereol., 110:409.

Van Neste, D., Caulier, B., Thomas, P., and Vasseur, F., 1985, PIBIDS: Tay's syndrome and xeroderma pigmentosum. J. Am. Acad. Dermatol., 12:372.

Van Neste, D., Degreef, H., Van Haute, N., Van Hee, J., Vandermaesen, J. J., Taieb, A., Maleville, A., Fontan, D., and Bakry, N., 1988, TTD variante: une variante clinique de trichothiodystrophie:aspects cliniques à propos de deux cas non publiés. Nouv. Dermatol., 7:49.

Venencie, P.Y., Dupré, A., Gouttières, F., and Saurat, J.H., 1982, Trichotiodystrophie. Cas N^{0} 78, Soc. Fr. Dermatol., Journées Parisiennes, Paris, mars 1982.

Yong,S.L., Cleaver, J.E., Tullis, G.D., and Johnston, M.M., 1984, Is trichothiodystrophy part of the xeroderma pigmentosum spectrum? Am. J. Hum. Genet., 36:82S.

CHINESE HAMSTER CELL LINES DEFECTIVE IN DNA REPAIR

Magorzata Z. Zdzienicka, J.W.I.M. Simons and P.H.M. Lohman

Department of Radiation Genetics and Chemical Mutagenesis, State University of Leiden, Sylvius Laboratories, Wassenaarseweg 72, 2333 AL Leiden, The Netherlands

SUMMARY

Four different general categories of mutants defective in cellular response to UV, X-rays, cross-linking and alkylating agents were isolated from Chinese hamster cell lines. Novel types of mutants are described. A new 7th complementation group in the class of UV-sensitive mutants was found. Amongst two complementation groups of X-ray-sensitive mutants, one group resembles Ataxia-telangiectasia cells. Three complementation groups were identified amongst mutants sensitive to cross-linking agents; one mutant (V-H4) is homologous to Fanconi's anemia.

INTRODUCTION

DNA damage can be converted to mutations, chromosomal aberrations and/or lead to cell death. In higher organisms it may contribute to malignancy and possibly aging. DNA repair processes serve to protect organisms from deleterious effects of DNA damaging agents. The genetic and biochemical complexity of DNA repair processes is reflected by the multiple complementation groups identified in cancer-prone genetic disorders, such as xeroderma pigmentosum (XP), Fanconi's anaemia (FA) and ataxia telangiectasia (AT) (Fischer, et al., 1985; Duckworth-Rysiecki, et al., 1985; Murnane and Painter, 1982; Jaspers and Bootsma, 1982).

The study of the mechanisms of DNA repair depends heavily on the availability of mutants defective in DNA repair. Due to the existence of many DNA-repair-deficient mutants in micro-organisms, mechanisms of DNA repair have been clarified. Many genes have been cloned and their biochemical function established (for review see Friedberg, 1985, 1987).

Recently it became evident that repair mutants of rodent cells provide an important tool for isolating human genes involved in DNA repair pathways (Westerveld, et al., 1984;

DNA Repair Mechanisms and Their Biological Implications in Mammalian Cells
Edited by M.W. Lambert and J. Laval
Plenum Press, New York

Weber, et al., 1988). It has been shown that both ERCC1 and ERCC2 human genes largely correct all impaired functions in the defective Chinese hamster cells, suggesting a homology of human DNA repair genes with respect to function of the defective genes in the DNA-repair deficient Chinese hamster mutants (Westerveld, et al., 1984; Zdzienicka, et al., 1987; Weber, et al., 1988). During the last few years many mutants of rodent cells hypersensitive to DNA-damaging agents have been isolated and characterized (for review see Collins and Johnson, 1987; Hickson and Harris, 1988). But so far, a mutant equivalent to a known genetic human disorder such as XP, FA, AT, etc. has not yet been reported amongst the isolated rodent mutants (Stefanini, et al., 1985; Thompson, et al., 1985).

The relative ease of isolation of the presumably recessive repair-deficient mutations results probably from functional hemizygosity of the hamster genome (Siciliano, et al., 1983; Siminovitch, 1976); thus, each established cell line may have a different part of the genome in a functionally haploid state. In most cases a mutant isolation frequency of about 1 per 10^3 for repair-deficient mutants has been found in rodent cells.

In this report the characteristics of mutants of Chinese hamster ovary (CHO) and V79 cells hypersensitive to DNA-damaging agents, isolated in our laboratory, is presented. Four different general categories of mutants sensitive to: UV, X-ray-irradiation, cross-linking and alkylating agents have been investigated, and their characteristics described.

RESULTS AND DISCUSSION

Isolation of mutants

The detailed procedure of mutant isolation has been described earlier (Zdzienicka and Simons, 1987; Zdzienicka, et al., 1988c). Briefly, the Chinese hamster cells were mutagenized with ethyl nitrosourea (ENU), giving a surviving fraction of ~1%, and after allowing expression of the induced mutations, a replica plating technique (96 wells plates) was used. In order to isolate different categories of mutants, the cells were screened at the same time for hypersensitivity to different DNA damaging agents. One V79 cell line appeared to be an exceptionally rich source of mutants in comparison to other Chinese hamster cell lines. The frequency of the isolated mutagen-sensitive clones was about 30 times higher than that observed usually for CHO cells. This remarkably high frequency suggests that there are large areas of the genome in this V79 cell line which are functionally hemizygous.

UV-sensitive mutants

Genetic diversity of UV-sensitive mutants of CHO cells has been shown earlier and already 6 genetic complementation groups have been identified (Thompson, et al., 1981; Thompson and Carrano, 1983; Thompson, et al., 1985, 1987). The three UV-sensitive mutants isolated in our laboratory are listed in Table 1. The genetic complementation analysis, by testing the UV-sensitivity of proliferating hybrids of these UV-sensitive mutants fused with representative mutants of the six different complementation groups, revealed that one mutant (V-B11) represents a new 7th complementation group (Zdzienicka, et al., 1988b).

It has been reported that the first five complementation groups are defective in performing the incision step of the nucleotide excision repair pathway after exposure to UV (Thompson, et al., 1982). The V-B11 mutant has also been shown to be defective in incision; only ~30% of the incision activity of that observed in wild-type V79 cells was found (Zdzienicka, et al., 1988b).

Table 1. UV-sensitive mutants.

Mutant	Parental line	Hyper-sensitivities[a]	Process(es) defective	Complementation group
V-H1	V79	UV, 4NQO[a]	No dimer removal and partially defective (6-4) photoproducts removal[b], ~70% of UDS, ~50% of repair replication and 50% incision of the wild-type cells.	2[c]
C-A6	CHO9	UV, 4NQO	--	2[c]
V-B11	V79	UV	~30% incision of the wild-type cells.	7(the only representant).

[a]4-nitroquinoline-1-oxide (4NQO).
[b]Mitchell et al., to be published
[c]The numbering of groups 1 and 2 was recently changed to correspond to the numbers of the complementing human genes, ERCC1 and ERCC2.

Two other mutants: V-H1 of V79 and C-A6 of CHO cells belong to the second complementation group (Zdzienicka, et al., 1988a), which has been already well characterized. However, V-H1 has the extraordinary and interesting properties of extreme sensitivity to UV killing (the most sensitive mutant amongst UV-sensitive mutants) combined with a high level of nucleotide excision repair (Zdzienicka, et al., 1988a). The other mutants of this complementation group show a complete lack of incision (Thompson, et al., 1982). This heterogeneity within one complementation group suggests that the repair gene may have more than one functionally important domain or that the gene is not involved in the incision per se, but is involved in, e.g., preferential repair of active genes. As the recently isolated human ERCC2 gene (Weber, et al., 1988) is able to complement the defect of all the mutants of the 2nd complementation group, the molecular basis of this phenotypic heterogeneity is open for analysis.

The diversity of UV-sensitive mutants manifest in complementation groups in Chinese hamster cells (Thompson, et al., 1987; Zdzienicka, et al., 1988) and the observed heterogeneity amongst one complementation group (Zdzienicka, et al., 1988a) indicate a great complexity of the nucleotide excision repair pathways in mammalian cells. The existence of 9 different complementation groups in XP cell lines (Fischer, et al., 1985) suggests a probability for isolation of additional complementation groups in Chinese hamster cells; as in humans, only mutations which allow nearly normal development of the fetus can be detected.

X-ray-sensitive mutants

The four X-ray-sensitive mutants isolated in our laboratory fall into two complementation groups. Their general characteristics are presented in Table 2. The XR-V15B mutant belongs to the same complementation group as the X-ray-sensitive (xrs) mutants of CHO cells (Jeggo and Kemp, 1983; Jeggo, 1985). Biochemical analysis confirms this finding as XR-V15B like xrs mutants show a decreased ability to rejoin double strand breaks induced by X-rays. One interesting difference between the XR-V15B mutant and xrs mutants is that heritable variations in gene activity in xrs mutants were caused by epigenetic alterations by methylation, whereas, methylation seems to be not involved in shutting off the gene in XR-V15B (Jeggo and Holliday, 1986; Zdzienicka, et al., 1988c).

Three X-ray-sensitive mutants V-C4, V-E5 and V-G8 belong to one complementation group despite some differences in the phenotype (e.g., V-G8 does not show increased sensitivity to UV, but V-C4 is more than 2-fold more sensitive to UV than the parental line) (Zdzienicka, et al., submitted). Our results indicate that these mutants represent the first rodent cell mutants which show unique phenotypic characteristics strongly resembling cells derived from AT patients, as they show a lack of DNA synthesis inhibition after X-ray irradiation, together with normal repair of DNA single- and double-strand breaks, and also chromosomal instability, which are the main characteristics of AT cells (Zdzienicka, et al., 1989). But a final conclusion on the question whether these mutants are the counterparts of AT cells can only be reached from complementation studies between these mutants and AT cells.

To characterize the mutants better and to find the best selective condition for cells in transfection experiments, we checked the sensitivity of these mutants to additional DNA-damaging agents and we have found that all these mutants are sensitive to bleomycin (Zdzienicka, et al., 1989) and to adriamycin (data not shown). Amongst them V-E5 is the most hypersensitive to adriamycin (6-fold increased sensitivity in comparison to wild-type V79 cells). This high hypersensitivity to adriamycin will allow proper selective conditions for the transfection experiments. As adriamycin generates free

Table 2. X-ray-sensitive mutants.

Mutant[a]	Hypersensitivities[b]	Other features
XR-V15B	X-ray, BLM, 4NQO, MMS	defective double-strand break repair; complementation group as xrs-1-6
V-C4	X-ray, ADR, BLM, UV, 4NQO, MMC, MMS	radioresistant DNA synthesis, chromosomal instability and normal double- and single-strand break repair
V-E5	X-ray, ADR, BLM, MMC, MMS	
V-G8	X-ray, ADR, BLM, MMS	

[a]V79 is the parental line of these mutants.
[b]Bleomycin (BLM), adriamycin (ADR), mitomycin (MMC), methyl methane-sulfonate (MMS).

radicals (Sinha, et al., 1984) and also is thought to intercalate into DNA causing DNA strand breaks (Tewey, et al., 1984), the hypersensitivity of the mutants to adriamycin and only a slightly increased sensitivity to MMC indicates that these mutants are mainly impaired in repair of damages induced by free radicals.

Cross-linking agent-sensitive mutants

Four Chinese hamster V79 mutants (V-C8, V-H4, V-H11 and V-B7), sensitive to mitomycin C (MMC), were isolated (Zdzienicka and Simons, 1987). Results of complementation analysis indicate that they belong to at least three different complementation groups. To determine whether one of these mutants represents a novel complementation group, complementation analysis was performed with different mutants exhibiting increased sensitivity to MMC.

To determine the genetic complementation group of the MMC-sensitive mutants, UV- and X-ray-sensitive mutants, which exhibit increased sensitivity to MMC, were also used. Two Chinese hamster X-ray-sensitive mutants isolated on the basis of a moderately increased sensitivity to X-ray, irs1 isolated by Jones, et al. (1987) and irs1SF isolated by Fuller and Painter (1988), are dramatically (more than 50-fold) hypersensitive to MMC (Jones, et al., 1987; Thompson, personal communication). So far, the results presented in Table 3 show that all tested mutants were complementing each other, indicating the recessive nature of mutations and the involvement of many different genetic loci in the repair pathway of MMC-induced lesions. Mutant V-H4 complements all known Chinese hamster mutants sensitive to MMC and, therefore, represents a new complementation group. Our results indicate that this V-H4 mutant is an equivalent of the FA gene (Arwert and Zdzienicka, in preparation).

Table 3. Mutants sensitive to cross-linking agents.

Mutant[a]	Hypersensitivities[b]	Complementation analysis
V-C8	MMC, UV, 4NQO, X-ray	Different from V-H4, V-H11, UV20, UV41, irs1SF[c]
V-H4	MMC, cis-DDP, 4NQO	Different from V-C8, V-H11, MMC-1,3,4, irs1[c], irs1SF, UV20,UV41
V-H11	MMC, 4NQO	Different from V-C8, MMC-1,3,4, UV20, UV41
V-B7	MMC, UV	--

[a]V79 is the parental line of these mutants.
[b]Mitomycin C (MMC), 4-nitroquinoline-1-oxide (4NQO), cis-diamminedichloroplatinum II (cis-DDP).
[c]irs1TOR was kindly provided by Dr. J. Thacker, and irs1SF by Dr. L. Thompson)

Alkylating agent-sensitive mutants

Two mutants of CHO and V79 cell lines (C-G11 and V-24B), sensitive to monofunctional alkylating agents, were recently isolated in our laboratory (Table 4). Based on the D_{10} value (the dose resulting in 10% of cell survival), C-G11 and V-24B are 7- and 3-fold more sensitive to MMS, respectively, than their parental cell lines (data not shown). As the C-G11 mutant is not more sensitive to UV-radiation than the wild-type cells it may indicate that only the base excision repair pathway is defective in this mutant. However, increased sensitivity to monofunctional alkylating agents may not necessarily be related to a DNA repair deficiency. A biochemical pathway distinct from DNA repair may be defective. Meuth (1983) has shown that deoxycytidine kinase-deficient mutants are hypersensitive to DNA alkylating agents, supporting the idea that biochemical pathways distinct from DNA-repair pathways can play an important role in the response of cells to DNA damaging agents. The phenotype of C-G11 is similar to that found in the MMS-sensitive mutants of CHO cells isolated recently by Robson and Hickson (1987). Complementation analysis will soon reveal whether they belong to the same complementation group or not.

Table 4. Monofunctional alkylating agents-sensitive mutants.

Mutant	Parental line	Hypersensitivities[a]
V-24B	V79	EMS, MMS, UV
C-G11	CHO9	MMS, EMS (not sensitive to UV)

[a]Methyl methanesulfonate (MMS); ethyl methanesulfonate (EMS).

GENERAL REMARKS

DNA-repair deficient mutants provide an important tool for investigating the biochemistry of DNA repair. It has been shown that, in attempts to isolate human DNA repair genes, the rodent DNA-repair mutants may also be exploited as recipients for transfection with DNA (Rubin, et al., 1983, 1985; Westerveld, et al., 1984; Weber, et al., 1988). However, although the cell fusion data are far from complete, none of the so far tested Chinese hamster mutants belongs to a known complementation group of a repair deficient human syndrome (Stefanini, et al., 1985; Thompson, et al., 1985), except V-H4 which did not complement FA. Therefore, it is to be expected that many more different mutants can be isolated including those complementing human hereditary disorders. To get a whole set of DNA-repair deficient mutants, it is essential to isolate also conditional mutants (temperature-sensitive) as some gene-product may be essential for the viability of the cell.

For studies of DNA-repair pathways, it is important to establish genetic complementation groups for the different categories of DNA-repair deficient mutants. As cross-sensitivity to mutagens often occurs, it should be considered that an already identified complementation group for one agent may be identical with a complementation group for another agent. Chinese hamster mutants, previously isolated on the basis of increased sensitivity to MMC by Robson, et al.(1985) fall into 4 complementation groups (Robson and Hickson, 1986). However, one of these mutants (MMC-2) appeared also to belong to the third complementation group of UV-sensitive mutants (Thompson, et al., 1987). Amongst seven complementation groups of UV-sensitive Chinese hamster mutants, only two groups (1 and 4) show a dramatically increased sensitivity to MMC (Hoy, et al., 1985: Zdzienicka, et al., 1988b). Therefore, it is of great importance to establish the degree of cross-sensitivity of isolated mutants to many different mutagenic agents, such as UV-, X-ray irradiation, cross-linking and alkylating agents, and then perform additional genetic complementation analysis to avoid misclassification.

Also, it should be considered that some of the mutants may be double mutants, e.g., the 27-1 mutant of the third complementation group of UV-sensitive mutants. Nucleotide excision repair of UV-induced DNA lesions and repair of DNA alkylation products operate by two distinct pathways (Friedberg, 1985). Surprisingly, it has been shown that the UV-sensitive mutant 27-1 of CHO cells, isolated by Wood and Burki (1982), is also hypersensitive to monofunctional alkylating agents such as MMS, EMS and ENU (Zdzienicka and Simons, 1986). Recently Kaina (1987) has reported that the hypersensitivity of 27-1 cells to methylating mutagens can be corrected by transfection with human DNA, without restoring the hypersensitivity of the mutant to UV. This result suggested that 27-1 is a double mutant having defects in the repair pathway of UV-induced lesions and also in the removal of DNA alkylation lesions. As 27-1 belongs to the third complementation group of UV-sensitive mutants of Chinese hamster cells (Thompson, et al., 1987), we have tested whether another UV-sensitive mutant (UV24) of the same complementation group exhibits also cross-sensitivity to alkylating agents, assuming that the possibility for a double mutation in UV24 is almost impossible. It turned out that UV24-1 is not more sensitive to MMS than the parental AA8 cell line (data not shown), whereas 27-1 is 5-6 times more sensitive to MMS than wild-type CHO-9 cells (Zdzienicka and Simons, 1986). This result strongly supports the suggestion that 27-1 is a double mutant which is in agreement with many data showing that UV-induced DNA lesions and DNA methylation products are repaired by different pathways (Friedberg, 1984).

Another complication is that mutants of the same complementation group can be heterogenous, therefore a well defined phenotype, without the complementation analysis, cannot be the basis of a judgement whether mutants belong to different complementation groups. Phenotypic heterogeneity has been observed amongst members of the same complementation group of X-ray-sensitive or UV-sensitive mutants of Chinese hamster cells (Jeggo and Kemp, 1983; Zdzienicka, et al., 1988a; Zdzienicka, et al., 1989).

As the complete set of mutagent-sensitive cell lines would allow reconstitution of repair pathways in mammalian cells and also would help recognition of the most important DNA lesions responsible for the different biological endpoints, there is a great need for new well characterized mutants. Hopefully, rodent mutants will also provide the tool for the isolation of human genes complementing defects of such hereditary disorders as XP, AT and FA.

ACNOWLEDGEMENTS

We thank I. Neuteboom for his skillful technical assistance. This work was supported by the Foundation of Basic Medical Research (MEDIGON), contract No. 13-23-91, EURATOM, contract No. ENV-534-NL and the J.A. Cohen Institute, Interuniversity Research Institute for Radiopathology and Radiation Protection, The Netherlands.

REFERENCES

Collins, A. and R.T Johnson, 1987, DNA repair mutants in higher eukaryotes, J. Cell Sci. Suppl., 6:61.

Duckworth-Rysiecki, G., Cornish, K., Clarke, C.A., and Buchwald, M., 1985, Identification of two complementation groups in Fanconi anemia, Somatic Cell and Molec. Genet., 11:35.

Fischer, E., Keijzer, W., Thielmann, H.W., Popanda, O., Bohnert, E., Edler, L., Jung, E.G., and Bootsma, D., 1985, A ninth complementation group in Xeroderma pigmentosum, Mutation Res., 145:217.

Friedberg, E.C., 1985, DNA repair, Freeman and Co., New York.

Friedberg, E.C., 1987, The molecular biology of nucleotide excision repair of DNA: Recent progress, J. Cell Sci. Suppl., 6:1.

Fuller, L.F., and Painter, R.B., 1988, A Chinese hamster ovary cell line hypersensitive to ionizing radiation and deficient in repair replication, Mutation Res., 193:109.

Hickson, I.D., and Harris, L., 1988, Mammalian DNA repair - use of mutants hypersensitive to cytotoxic agents, Trends in Genetics, 4:101.

Hoy, C.A., Thompson, L.H., Mooney, C.L., and Salazar, E.P., 1985, Defective DNA cross-link removal in Chinese hamster cell mutants hypersensitive to bifunctional alkylating agents, Cancer Res., 45:1737.

Jaspers, N.G.J., and Bootsma, D., 1982, Genetic heterogeneity in AT studied by cell fusion, Proc. Natl. Acad. Sci. USA, 79:2641.

Jeggo, P.A., and Kemp, L.M., 1983, X-ray-sensitive mutants of Chinese hamster ovary cell line. Isolation and cross-sensitivity to other DNA-damaging agents, Mutation Res., 112:313.

Jeggo, P.A., 1985, Genetic analysis of X-ray-sensitive mutants of the CHO cell line, Mutation Res., 146:265.

Jeggo, P.A., and Holliday, R., 1986, Azacytidine-induced reactivation of a DNA repair gene in Chinese hamster ovary cells, Mol. and Cellular Biol., 6:2944.

Jones, N.J., Cox, R., and Thacker, J., 1987, Isolation and cross-sensitivity of X-ray-sensitive mutants of V79-4 hamster cells, Mutation Res., 183:279.

Kaina, B., 1987, Correction of alkylation hypersensitivity of CHO-W27-1 cells by transfection with human DNA, Carcinogenesis, 8:1935.

Meuth, M., 1983, Deoxycytidine kinase-deficient mutant of Chinese hamster ovary cells are hypersensitive to DNA alkylating agents, Mutation Res., 110:383.

Murnane, J.P., and Painter, R.B., 1982, Complementation of the defects in DNA synthesis in irradiated and unirradiated AT cells, Proc. Natl. Acad. Sci. USA, 79:1960.

Robson, C.N., and Hickson, I.D., 1986, Genetic analysis of mitomycin C-sensitive mutant of a Chinese hamster ovary cell line, Mutation Res., 163:201.

Robson, C.N., and Hickson, I.D., 1987, Isolation of alkylating agents-sensitive Chinese hamster ovary cell, Carcinogenesis, 8:601.

Rubin, J.S., Joyner, A.L., Bernstein, A., and Whitmore, G.F., 1983, Molecular identification of a human DNA repair gene following DNA-mediated gene transfer, Nature, 306:206.

Rubin, J.S., Prideaux, V.R., Willard, H.F., Dulhanty, A.M., Whitmore, G.F., and Bernstein, A., 1985, Molecular cloning and chromosomal localization of DNA sequences associated with human DNA repair gene, Molec. Cell. Biol., 5:398.

Siciliano, M.J., Stallings, R.L., Adair, G.M., Humphrey, R.M., and Siciliano, J., 1983, Provisional assignment of TPI, GPI, and PEPD to Chinese hamster autosomes 8 and 9: a cytogenetic basis for functional haploidy of an autosomal linkage group in CHO cells, Cytogenet. Cell Genet., 35:15.

Siminovitch, L., 1976, On the nature of hereditable variations in culture somatic cells, Cell, 7:1.

Sinha, B.K., Trush, M.A., Kennedy, K.A., and Mimnaugh, E.G., 1984, Enzymatic activation and binding of adriamycin to nuclear DNA, Cancer Res., 44:2892.

Stefanini, M., Keijzer, W., Westerveld, A., and Bootsma, D., 1985, Interspecies complementation analysis of xeroderma pigmentosum and UV-sensitive Chinese hamster cells, Exp. Cell Res., 161:373.

Tewey, K.M., Chen, G.L., Nelson, E.M., and Liu, L.F., 1984, Intercalative antitumor drugs interfere with the breakage-reunion reaction of mammalian DNA with topoisomerase II, J. Biol. Chem., 259:9182.

Thompson, L.H., Busch, D.B., Brookman, K., Mooney, C.L., and Glaser, D.A., 1981, Genetic diversity of UV-sensitive DNA repair mutants of Chinese hamster ovary cells, Proc. Nat. Acad. Sci., U.S.A., 78:3734.

Thompson, L.H., and Carrano, A.V., 1983, Analysis of mammalian cell mutagenesis and DNA repair using in vitro selected CHO cell mutants. In: Cellular responses to DNA damage, UCLA Symposia on Molecular and Cellular Biology, New Series, Vol. 11 (ed. E.C. Friedberg and B.A. Bridges), pp 125-143, New York, Alan R. Liss.

Thompson, L.H., Mooney, C.L., and Brookman, K.W., 1985, Genetic complementation between UV-sensitive CHO mutants and xeroderma pigmentosum fibroblasts, Mutation Res., 150:423.

Thompson, L.H., Salazar, E.P., Brookman, K.W., Collins, C.C., Stewart, S.A., Busch, D.B., and Weber, C.A., 1987, Recent progress with the DNA repair mutants of Chinese hamster ovary cells. In: The Molecular Biology of DNA Repair, J. Cell Sci. Suppl., 6:97.

Weber, C.A., Salazar, E.P., Stewart, S.A., and Thompson, L.H., 1988, Molecular cloning and biological characterization of a human gene, ERCC2, that corrects the nucleotide excision repair defect in CHO UV5 cells, Mol. Cell. Biol., 8:1137.

Westerveld, A., Hoeijmakers, J.H.J., Duin, M. van, Wit, J. de, Odijk, H., Pastink, A., Wood, R.D., and Bootsma, D., 1984, Molecular cloning of a human DNA repair gene, Nature, 310:425.

Wood, R.D., and Burki, H.J., 1982, Repair capability and the cellular age response for killing and mutation induction after UV, Mutation Res., 95:505.

Zdzienicka, M.Z., and Simons, J.W.I.M., 1986, Analysis of repair processes by the determination of the induction of cell killing and mutations in two repair-deficient Chinese hamster ovary cell lines, Mutation Res., 166:59.

Zdzienicka, M.Z., and Simons, J.W.I.M., 1987, Mutagen-sensitive lines are obtained with a high frequency in V79 Chinese hamster cells, Mutation Res., 178:235.

Zdzienicka, M.Z., Roza, L., Westerveld, A., Bootsma, D., and Simons, J.W.I.M., 1987, Biological and biochemical consequences of the human ERCC1 repair gene after transfection into a repair-deficient CHO cell line, Mutation Res., 183:69.

Zdzienicka, M.Z., Schans, G.P. van der, Westerveld, A., Zeeland, A.A. van, and Simons, J.W.I.M., 1988a, Phenotypic heterogeneity within the first complementation group of UV-sensitive mutants of Chinese hamster cell lines, Mutation Res., 193:31.

Zdzienicka, M.Z., Schans, G.P. van der, and Simons, J.W.I.M., 1988b, Identification of a new seventh complementation group of UV-sensitive mutants in Chinese hamster cells, Mutation Res., 194:165.

Zdzienicka, M.Z., Tran, Q., Schans, G.P. van der, and Simons, J.W.I.M., 1988c, Characterization of an X-ray sensitive mutant of V79 Chinese hamster cells, Mutation Res., 194:239.

Zdzienicka, M.Z., Jaspers, N.G.J., Schans, G.P. van der, Natarajan, A.T., and Simons, J.W.I.M., 1989. Ataxia-telangiectasia-like Chinese hamster V79 cell mutants with radioresistant DNA synthesis, chromosomal instability, and normal DNA strand break repair, Cancer Res., 49:1481-1485.

HUMAN DNA REPAIR AND RECOMBINATION GENES

Larry H. Thompson, Christine A. Weber, and Nigel J. Jones

Biomedical Sciences Division
Lawrence Livermore National Laboratory
P.O. Box 5507, Livermore, California 94550

ABSTRACT

Several genes involved in mammalian DNA repair pathways were identified by complementation analysis and chromosomal mapping based on hybrid cells. Eight complementation groups of rodent mutants defective in the repair of UV radiation damage are now identified. At least seven of these genes are probably essential for repair and at least six of them control the incision step. The many genes required for repair of DNA cross-linking damage show overlap with those involved in the repair of UV damage, but some of these genes appear to be unique for cross-link repair. Two genes residing on human chromosome 19 were cloned from genomic transformants using a cosmid vector, and near full-length cDNA clones of each gene were isolated and sequenced. Gene ERCC2 efficiently corrects the defect in CHO UV5, a nucleotide excision repair mutant. Gene XRCC1 normalizes repair of strand breaks and the excessive sister chromatid exchange in CHO mutant EM9. ERCC2 shows a remarkable 52% overall homology at both the amino acid and nucleotide levels with the yeast RAD3 gene. Evidence based on mutation induction frequencies suggests that ERCC2, like RAD3, might also be an essential gene for viability.

INTRODUCTION

Unrepaired damage to the DNA molecules in somatic cells will likely produce mutations if DNA replication preceeds repair of the damage. Mutations are generally assumed to be the starting point for cellular changes that can lead to malignancy. The critical relationship between repair and carcinogenesis is borne out by the studies of human genetic disorders such as xeroderma pigmentosum (XP), ataxia telangiectasia, and Bloom's syndrome, which have various defects in the repair or related metabolism of DNA (Hanawalt and Sarasin, 1986; Friedberg, 1985). Individuals with these disorders have a substantially increased predisposition to cancer. Thus, DNA repair pathways must play an essential role

DNA Repair Mechanisms and Their Biological Implications in Mammalian Cells
Edited by M.W. Lambert and J. Laval
Plenum Press, New York

in maintaining genetic integrity and the growth controls that ensure cellular differentiation.

In order to understand how cells respond to DNA damage, it is necessary to determine the genetic and biochemical details of particular repair pathways. A major pathway is nucleotide excision repair, which acts on bulky chemical adducts and photoproducts from ultraviolet (UV) radiation, i.e., cyclobutane dimers and pyrimidine(6-4)pyrimidone[(6-4)] photoproducts. This pathway has been the focus of much study in different organisms. In E. coli, the uvrA, uvrB, and uvrC gene products are essential for the initial damage recognition and incision step (Friedberg, 1985; 1987). In the small eukaryote S. cerevisiae, five genes are absolutely required for incision, and at least five other genes appear to be involved in the pathway but are not essential for incision (Friedberg, 1985; 1987).

For mammalian cells, we do not yet have a good estimate of the number of genes participating in the incision step of nucleotide excision repair, but the number appears to be larger than in yeast. In XP cell lines, nine complementation groups have been repored in which there are varying degrees of repair deficiency in this pathway (Fischer, et al., 1985). A recent study presents evidence that the groups designated D and H may, in fact, be allelic as direct measurement of repair incision failed to show complementation in hybrid cells derived from these groups (Johnson, et al., 1989).

To facilitate studying various repair pathways, many laboratories have isolated mutant lines on the basis of hypersensitivity to DNA damaging agents. The first reports of mutants having pronounced sensitivity (2-10 fold) were isolated in CHO (Chinese hamster ovary) cells, a line widely used for genetic studies (Gottesman, 1985). These mutants showed enhanced killing with UV radiation, ethyl methanesulfonate, or mitomycin C (Thompson, et al., 1980; Busch, et al., 1980). Since then, a great variety of mutant lines that show hypersensitivity to DNA-damaging agents have been obtained in the near-diploid CHO and V79 hamster lines (see review by Hickson and Harris, 1988), as well as in mouse lymphoma and other cell lines (reviewed by Collins and Johnson, 1987). One recent report, based on screening simultaneously for a variety of mutagen-sensitive phenotypes, described a very high induced mutant frequency of 2×10^{-2} (Zdzienicka, et al., 1987b), which may be due to de partial hemizygosity of the V79 cells (Thacker, 1981).

The isolation and classification of mutants into complementation groups has now greatly exceeded the rate at which mutants are being characterized at the molecular level. For a few mutants, some identification of the nature of the biochemical defects has been achieved, i.e., defects in single-strand or double-strand break rejoining, alteration in polymerase alpha, defective incision at bulky adducts, or altered topoisomerase II (see Table 1 in Hickson and Harris, 1988). In this article, we discuss recent developments toward cloning and characterizing mammalian DNA repair genes, as well as recent information about the biochemical defects in several mutant cell lines that are currently being investigated.

RESULTS AND DISCUSSION

Complementation Groups of Mutations Affecting Nucleotide Repair

One of our goals has been the identification of new complementation groups of mammalian cell mutants that are likely to be defective in nucleotide excision repair. A summary of these complementation studies is given in Table 1. Initially our laboratory described four groups based on a high degree of UV sensitivity, with emphasis on mutant lines UV20, UV5, UV24, UV41, and later a fifth group (UV135), all of which were isolated in CHO cells. The representative mutants from each of these groups had little or no repair activity in response to UV radiation and all showed marked sensitivity in terms of cell killing and mutation induction (Busch, et al., 1980; Thompson, et al., 1980; 1981; 1982a). The similar degree of hypersensitivity and the observed complementation suggested that each mutation affects a diferent gene in the same pathway; each gene appears necessary for incision to occur (Thompson, et al., 1982a). Presumably each mutation inactivated a protein essential for repair function. Subsequently, a sixth complementation group for UV sensitivity was identified in CHO (Thompson, et al., 1987b), but this line (UV61) was appreciably less sensitive than the representative members of groups 1 through 5.

Mutant UV61 was recently studied in more detail and some insight was obtained concerning its intermediate UV sensitivity (Thompson, et al., 1989). Using a radioimmunoassay for (6-4) photoproducts, repair of these lesions appears to be normal in UV61. However, using a chromatographic assay, no removal of cyclobutane dimers from bulk DNA could be detected during 24 hour postirradiation, indicating that this defect is likely responsible for the 2.5-fold hypersensitivity of this mutant (Thompson, et al., 1989). The rate of strand incision during the first 90 minutes after UV irradiation was normal in UV61, suggesting that the repair seen in normal cells at early times can be attributed predomminantly to (6-4) photoproducts. It is presently unclear whether UV61 has any repair activity for cyclobutane dimers in actively transcribed genes. If not, then this mutant provides a particularly useful system for evaluating the relative contributions of these two major classes of UV photoproducts to cell killing, mutation, and other biological endpoints.

Knowing the reason for the partial repair activity seen in UV61 is of much importance. In general, a mutant phenotype could arise from a cell having both a normal and a mutant allele. However, since mutations affecting repair are usually recessive in cell hybrids, we shall assume that residual repair capacity reflects the expression of only a mutant allele. One possibility with UV61 is that the hamster ERCC6 gene product is normally needed for the repair of cyclobutane dimers but not for (6-4) photoproducts. In this case, the mutation may have fully inactivated the repair protein. A second possibility is that the mutation is "leaky" and occurs in a protein that is normally required for repair of both classes of UV photoproducts. A partially defective protein may have lost its affinity for cyclobutane dimers, which are normally repaired poorly in bulk DNA of hamster cells, while

Table 1. Complementation Groups of UV-Sensitive Rodent Cell Lines

Group	Representatives	Degrees of UV-sen.[a]	Degrees of MMC-sen.[a]	Ref.(resp.)[b]
1[c]	UV20,UV4,UVL10,43-3B	6	90	1,2,3,4
2[c]	UV5,UV57,UVL1,V-H1	6	3.5	1,2,3,5
3	UV24,27-1,MMC-2	6	3.5	1,6,6
4	UV41,UV47	6	90	1,2
5	UV135;Q31	6;4	3.5;1.0	7,8
6	UV61;US46	2.5;4	?;2.7	6,9
7	V-B11	2	2	10
8	US31	4	4	9

[a]In most instances the hypersensitvity refers to the first mutant listed. Numbers refer to ratios of D_0's or, in some cases with mitomycin C (MMC), to the differential cytotoxicity ratio as defined by Hoy, et al., 1985. For groups 5 and 6, values are given for both hamster and mouse mutants.

[b]References indicate the studies in which the complementation group assignments were made:
1. Thompson, et al., 1981
2. Hoy, et al., 1985
3. G. Adair, personal comm.
4. Wood and Burki, 1982
5. Zdzienicka, et al., 1988a
6. Thompson, et al.,1987b
7. Thompson and Carrano,1983
8. Thompson, et al., 1987a
9. Thompson, et al., 1988d
10. Zdzienicka, et al., 1988b

[c]The numbering of groups 1 and 2 was recently interchanged to correspond to the numbers of the complementing human genes, ERCC1 and ERCC2: See announcement accompanying Thompson, et al., 1988d.

maintaining its affinity for (6-4) photoproducts. In support of this idea is the isolation of a revertant of UV61 in which the mutant gene appears to have been amplified ~10-fold (J. Hoeijmakers, personal communication). Presumably by overproducing an altered protein product the revertant has restored repair activity sufficiently to produce normal UV resistance.

Another observation that is pertinent to understanding the biochemical defect in UV61 is its sensitivity to bulky chemical mutagens. With 7-bromomethylbenz[a]anthracene, the degree of hypersensitivity to killing is almost as great as in the mutant UV5 (Thompson, et al., 1989), which appears to be fully deficient in the removal of DNA adducts produced by this compound (Thompson, et al., 1984). These results suggest that the normal protein encoded by the gene that is mutated in UV61 is an essential one for the nucleotide excision repair pathway. The mouse mutant, US46, which also belongs to complementation group 6, appears to be fully repair-deficient, based on its degree of hypersensitivity to killing by UV radiation (Shiomi, et al., 1982). From these observations, we conclude that the ERCC6 gene appears to be essential for the nucleotide excision repair pathway.

Complementation groups 7 and 8 in Table 1 are each

represented by single members isolated from V79 hamster cells and mouse lymphoma cells, respectively. The V-B11 mutant shows only about 2-fold hypersensitivity and an intermediate level of incision afer UV irradiation (Zdzienicka, et al., 1988b), but there is no evidence as to whether this phenotype involves a leaky mutation in an essential gene or loss of function in a gene that is only partially required for repair. The mouse mutant line US31, while not characterized biochemically in terms of its repair defect, has a degree of UV sensitivity similar to the fully repair deficient (incision deficient) mutant Q31 (Sato and Setlow, 1981; Shiomi, et al., 1982). These results suggest that the gene involved in complementation group 8 is also essential for the UV repair pathway. Thus, altogether, the preceeding results suggest that in rodent cells at least seven genes (ERCC1-ERCC6 and ERCC8) are required for UV damage repair, which is two more than appear to be necessary in yeast.

Chromosomal Mapping of Repair Genes

The use of interspecific hybrid cells provides a valuable way to determine how well human genes can complement rodent mutations and, at the same time, to localize the complementing genes on specific chromosomes and regions of chromosomes. We have obtained primary hybrids afer cell fusion between rodent mutants and human lymphocytes by selecting for complementation of the repair defect. Selection against mutants defective in nucleotide excision repair has been performed using repeated exposure to UV radiation, or continual exposure to mitomycin C (MMC) for mutants that show extreme hypersensitivity to this agent. Chromosomal analysis of a set of resistant hybrid clones (usually 20-30) will establish which human chromosome correlates with the repair proficient phenotype. Segregation of the complementing chromosome from resistant hybrids, by growth in normal medium, allows confirmation of the assignment by checking for the acquisition of sensitivity in such subclones. Breakage of the complementing chromosome occurs frequently in this system. However, this property can be used advantageously to localize a repair gene to a particular chromosomal region by analyzing hybrids that retain a small portion of the chromosome using previously mapped DNA probes.

A summary of chromosomal assignments for human repair genes made to date using rodent mutants is given in Table 2. Complementation groups 1 through 5 map to four different human chromosomes. Groups 1 and 2 are complemented by the genes ERCC1 and ERCC2, respectively, both on chromosome 19. Recent results based on pulsed field gel electrophoresis indicate that these two genes are < 280 kb apart and contained on a single Not I restriction fragment (H. Mohrenweiser, personal comunication). Whether this tight linkage of two genes involved in the same repair pathway has functional significance is unclear. In Table 2, ERCC gene numbers have been assigned to each of the human genes that complement, or which are expected to complement (ERCC7 and ERCC8), each of the UV sensitve rodent mutations. By comparison, the chromosomal assignment of the gene for only one XP group has been determined so far. XP-F cells were partially complemented to UV resistance by chromosome 15 in hybrids produced by microcell-mediated chromosome transfer (Schultz, et al., 1988). In addition to the two nucleotide excision repair mutants discussed above, the EM9 mutant is also corrected by a gene (XRCC1) on human chromosome 19 (Siciliano, et al., 1986). This finding of three repair genes on chromosome 19 is at least partly a reflection of the fact that the homologous loci lie on a hemizygous chromosome in CHO cells (Thompson, et al., 1988a). Hemizygosity greatly favors the isolation of recessive mutations.

Cloning and Characterization of ERCC Repair Genes

Several of the human nucleotide excision repair genes (designated ERCC = Excision Repair Cross Complementing) listed in Table 2 have been cloned and characterized in some detail. The status of these cloning efforts is gven in Table 3. The first human repair gene to be cloned was ERCC1, obtained by screening a cosmid library made from a secondary transformant and probing for a closely linked plasmid marker sequence (Westerveld, et al., 1984). A reconstructed, functional cDNA was obtained (van Duin, et al., 1986), and the intron-exon junctions were determined (van Duin, et al., 1987). ERCC2 was isolated in our laboratory from the cosmid library of a

Table 2. Identifying and Mapping Human DNA Repair Genes Using Rodent Mutants

Mutant	UV Group	Chromosome	Reference[a]	Gene name[b]
UV20,43-3B	1	19	1,2	ERCC1
UV5	2	19	3	ERCC2
UV24	3	2	4	ERCC3
UV41	4	16	3	ERCC4
UV135,Q31	5	13	4,6	ERCC5
UV61	6	ND[c]	-	ERCC6
VB11	7	ND	-	ERCC7?[d]
US31	8	ND	-	ERCC8?[e]
EM9	-	19	5	XRCC1

[a]References for chromosomal assignments:
1. Thompson, et al., 1985 2. van Duin, et al., 1986
3. Siciliano, et al., 1987 4. Thompson, et al., 1987a
5. Siciliano, et al., 1986 6. Hori, et al., 1983

[b]XRCC = X-ray Repair Cross Complementing; ERCC = Excision Repair Cross Complementing

[c]ND = not determined

[d]Question marks signify that correction of the rodent mutaton by a human gene has not yet been demonstrated.

[e]The mouse mutant US31 has not yet been evaluated for a presumptive defect in nucleotde excision repair.

secondary transformant by screening for human Alu-family repetitive sequences (Weber, et al., 1988a; Thompson, 1988). Gene XRCC1 was also obtained this way (see below). However, this approach has proven not to be generally applicable since the gene ERCC4 seems to lack repetitive elements of this type

(Dulhanty, et al., 1988). In the case of ERCC3, a portion of the gene has been difficult to isolate (Weeda, et al., 1988), but recently a functional cDNA clone of 2.9 kb was obtained from the Okayama pcD2 expression library (Chen and Okayama, 1987) (J. Hoeijmakers, personal communication). Efforts to isolate the full-length cDNA of ERCC6 are ongoing (J. Hoeijmakers, personal communication).

A cDNA clone of ERCC2, isolated from the pcD2 expression library, conferred partial UV resistance to UV5 cells 24 hours afer transfection, but did not confer stable resistance (Weber, et al., 1988b). Analysis of the 5' end of the cosmid-borne genomic sequence indicates that this cDNA contains part of an intron at its 5' end and lacks the first five base pairs of protein coding sequence as well as the 5'-untranslated region (Weber, et al., 1988c). The deduced amino acid sequence of ERCC2 was found to have a striking 52% identity with the yeast RAD3 encoded protein, and the two encoded proteins are quite similar in length (760 a.a. vs. 778 a.a.). Several regions of ERCC2 have at least 70% nucleotide homology with RAD3. Since RAD3 is an essential protein for viability (Naumovski and Friedberg, 1983, 1986; Higgins, et al., 1983) and has both ATPase activity (Sung, et al., 1987b) and helicase activity (Sung, et al., 1987a), the question arises as to whether ERCC2 is also an essential gene. A comparison of mutation induction frequencies for point mutagens versus a frame-shift agent ICR170 in CHO cells has provided highly suggestive evidence that this mammalian gene is also essential (Busch, et al., 1988). Point mutagens produced the expected relative recovery of mutants in complementation groups 1 and 2. However, when ICR170 was used, the frequency of mutants

Table 3. Status of Cloning and Analyzing Human Repair Genes

Gene name	Cloned?	References[a]	Homologous gene in S. cerevisiae	Gene size	Size of putative protein
ERCC1	yes	1,2,3	RAD10	15 kb	297 a.a.
ERCC2	yes	4,5	RAD3	19 kb	760 a.a.
ERCC3	yes	6	?		
ERCC4	no	7			
ERCC5	no	8			
ERCC6	partly	9	?	~100 kb	
ERCC7	no				
ERCC8	no				
XRCC1	yes	10	?	~33 kb	

[a]References pertaining to gene cloning are as follows:

1. Westerveld, et al., 1984
2. van Duin, et al., 1986
3. van Duin, et al., 1987
4. Weber, et al., 1988a
5. Weber, et al., 1988c
6. Weeda, et al., 1988
7. Dulhanty, et al., 1988
8. MacInnes, et al., 1988
9. J.H.J. Hoeijmakers, pers.comm.
10. Thompson, et al., 1988b

recovered in complementation group 2 was 40-fold lower than expected, relative to the frequency of mutants induced in group 1. ICR170 would be expected to disrupt the hamster ERCC2 protein by producing premature termination codons, and thus lethal mutations, if a regon of ERCC2 has an essential function.

The ERCC1 gene shows homology with the yeast gene RAD10. However, the encoded proteins differ considerably in size, and the homology is limited to certain regions (van Duin, et al., 1986). Nevertheless, evidence has been presented that RAD10 can provide some restoration of the repair defect in the CHO mutants of (newly defined) complementation group 1 (Lambert, et al., 1988). Thus, it is of considerable interest, in view of their similarity, to test whether RAD3 and ERCC2 show any interchangeability of function. Looking ahead, the fact that the first two cloned human repair genes to be analyzed show homology with known yeast genes suggests that many other human genes will probably have counterparts in the yeast system.

Thus, these results for ERCC1 and ERCC2 point to a strong interspecies similarity of repair proteins. At the functional level, ERCC2 restored UV survival, mutation frequencies, and strand incision to levels that were indistinguishable from those of normal CHO cells (Weber, et al., 1988a). ERCC1 gave less efficient, but very substantial, correction for cell survival with either UV radiation or MMC or for dimer removal (Westerveld, et al., 1984; Zdzienicka, et al., 1987a). In addition, ERCC1 and ERCC2 were tested for specificity of complementation, and each gene corrected only CHO mutants in the complementation group that was used for isolating the gene (van Duin, et al., 1988; Weber, et al., 1988a).

Cloning and Characterization of a Gene That Complements Mutant EM9

Our laboratory also isolated the human gene identified as XRCC1 (see Table 2), which corrects the CHO mutant EM9. EM9 cells are noted for their very high level of baseline sister chromatid exchange (SCE) and defective strand-break rejoining (Thompson, et al., 1982b). All complementation studies have been performed using chlorodeoxyuridine as the selective agent. XRCC1 was cloned from a cosmid library of a tertiary transformant (Thompson, et al., 1987b; 1988b), and a functional cDNA clone was obtained from the pcD2 library (Chen and Okayama, 1987). Based on SCE frequency as a sensitve measure of complementation, two cosmid clones give 100% corection, but the cDNA clone consistently gives 80% correction (Thompson, et al., 1988d). This correction is stable, unlike the transient correction seen with the incomplete cDNA of ERCC2. The XRCC1 genomic clones also efficiently correct the hypersensitivity of EM9 to ionizing radiation and completely restore the normal rate of strand-break rejoining after irradiation (Thompson, et al., 1988b). Analysis of the nucleotide sequence of XRCC1 cDNA with respect to open reading frames and candidate translational start codons suggests that a portion of the protein coding region is missing at the 5' end. This interpretation is consistent with the incomplete correction obtained in transformants. The efficient correction seen with the gene itself suggests a repair protein having a high degree of conservation among mammals.

Table 4. Hamster Mutants Having Pronounced Sensitivity to the Cross-Linking Agent MMC But Little or No UV Sensitivity

Mutant	Cell line	Degree of UV-sen.[a]	Degree of MMC-sen.[a]	Refs.[a]
irs1SF	CHO	2	100	1
irs1	V79	2-3	50	2
irs3	V79	1	7	2
V-C8	V79	2	110	3
V-H4	V79	1	33	3
V-H11	V79	1	8	3
UV-1	CHO	2	10	4,5
MMC-1	CHO	1	5	6,7
MC5	CHO	1	8	8

[a]References are as follows:
1. Fuller and Painter, 1988
2. Jones, et al., 1987
3. Zdzienicka and Simons, 1987
4. Stamato and Waldren, 1977
5. Waldren, et al., 1983
6. Robson, et al., 1985
7. Robson and Hickson, 1986
8. Thompson, et al., 1980

Possible Role of ERCC2 in Excision Repair

A highly unusual feature of the mutants in UV complementation group 2 was recently recognized. In Table 1, four mutants are listed in this group, and they all have similar sensitivity to killing by UV radiation. However, mutants V-H1 and UVL1 show intermediate levels of repair, while mutants UV5 and UV57 have no detectable repair (Mitchell, et al., 1988; Zdzienicka, et al., 1988a). These properties suggest that mutants V-H1 and UVL1 are performing repair in a way that it is not biologically effective for restoring survival. Recent studies have shown that the survival of CHO cells correlates with repair of cyclobutane dimers in active genes (Bohr, et al., 1985; 1986; 1987). Very little repair seems to occur in bulk DNA in rodent cells. Thus, repair of nontranscribed regions of DNA may have little impact on cell survival. One hypothesis to explain the phenotypic heterogeneity of mutants in group 2 is that the (hamster) ERCC2 protein has at least two functions. One function, perhaps the ability to interact with the repair complex, may be essential for all repair throughout the genome. Mutants such as UV5 and UV57 would lack this function. Another function may be one that helps determine active gene repair. The other mutants, V-H1 and UVL1, could lack this function while retaining part of the generalized repair function. In these mutants, the remaining repair would no longer be channeled to active genes. A third possible function associated with the ERCC2 repair protein was suggested above, one required for cell growth. The possibility of additional functions is suggested by the finding that mutations in RAD3 can increase spontaneous mitotic recombination and mutation without affecting UV sensitivity (Montelone, et al., 1988).

Other Hamster Mutants That Are Sensitive to Cross-linking Agents

The most extreme sensitivities found among mammalian repair mutants pertain to DNA cross-linking agents. This finding suggests that cross-links are very toxic if not repaired and that cells normally have highly efficient mechanisms for coping with these lesions. In view of the 10-100 fold increases in mutant sensitivity, it appears that most all cross-links are normally rendered innocuous, either by unhooking or actual excision. The UV sensitive mutants presented in Table 1, all show varying degrees of hypersensitivity to the cross-linking agent MMC. These results point to overlap of the genetic pathways for repairing cross-links and UV damage. However, cross-link repair appears to require some genes that play only a minor role, or none at all, in UV dimer repair.

Table 4 lists several reported mutants (not a comprehensive list) that have the property of substantial hypersensitivity to the cross-linking agent MMC while showing little or no change in UV sensitivity. Some of these mutants show extreme degrees of MMC sensitivity, similar to the nucleotide excision repair mutants in UV complementation groups 1 and 4 of Table 1. For example, irs1SF, irs1, and V-C8 have 50-100 fold hypersensitivity. Interestingly, irs1SF and irs1 were both isolated on the basis of their increased sensitivity to ionizing radiation, with each mutant being 2-3 fold more sensitive than its parental line. (These findings point out that the most pronounced aspect of a mutant's phenotype may be distantly related to its mode of isolation). It is clear that MMC-sensitive mutants belong to many different complementation groups. For example, irs1, irs1SF, UV20, and UV41 are in different groups (L. Thompson and N. Jones, unpublished data). Further analysis indicates additional groups (Robson and Hickson, 1986; M. Zdzienicka, personal communication).

ACKNOWLEDGEMENTS

The work was performed under the auspices of the U.S. Department of Energy by the Lawrence Livermore National Laboratory under contract No. W-7405-ENG-48.

REFERENCES

Bohr, V.A., Smith, C.A., Okumoto, D.S., and Hanawalt, P.C., 1985, DNA repair in an active gene: removal of pyrimidine dimers from the DHFR gene of CHO cells is much more efficient than in the genome overall, Cell, 40:359-369.

Bohr, V.A., Okumoto, D.S., and Hanawalt, P.C., 1986, Survival of UV-irradiated mammalian cells correlates with efficient DNA repair in an essential gene, Proc. Natl. Acad. Sci. U.S.A., 83:3830-3833.

Bohr, V.A., Phillips, D.H., and Hanawalt, P.C., 1987, Heterogeneous DNA damage and repair in the mammalian genome, Cancer Res., 47:6426-6436.

Busch, D.B., Cleaver, J.E., and Glaser, D.A., 1980, Large scale isolation of UV-sensitive clones of CHO cells, Somat. Cell Genet., 6:407-418.

Busch, D.B., Greiner, C., Lewis, K., Ford, R., Adair, G., and Thompson, L., 1989, Summary of complementation groups of UV-sensitive CHO cell mutants isolated by large-scale screening, Mutagenesis, 4:349-354.

Chen, C., and Okayama, H., 1987, High-efficiency transformation of mammalian cells by plasmid DNA, Mol. Cell. Biol., 7:2745-2752.

Collins, A., and Johnson, R.T., 1987, DNA repair mutants of higher eukaryotes, J. Cell Sci., Suppl., 6, 61-82.

Dulhanty, A.M., Rubin, J.S., and Whitmore, G.F., 1988, Complementation of the DNA repair defect in a CHO mutant by human DNA that lacks highly abundant repetitive sequences, Mutat. Res., 194:207-217.

Fischer, E., Keijzer, W., Thielmann, H.W., Popanda, O., Bohnert, E., Edler, L., Jung, E.G., and Bootsma, D., 1985, A ninth coplementation group in xeroderma pigmentosum, XP I, Mutat. Res., 145:217-225.

Friedberg, E.C., 1985, "DNA Repair," W.H. Freeman, New York.

Friedberg, E.C., 1987, The molecular biology of nucleotide excision repair of DNA: recent progress, J. Cell Sci., Suppl. 6:1-23.

Fuller, L.F., and Painter, R.B. 1988, A Chinese hamster ovary cell line hypersensitive to ionizing radiation and deficient in repair replication, Mut. Res., 193:109-121.

Gottesman, M.M., 1985, "Molecular Cell Genetics," John Wiley and Sons, New York.

Hanawalt, P.C., and Sarasin, A., 1986, Cancer-prone hereditary diseases with DNA processing abnormalities, Trends Genet., 2:124-129.

Hickson, I.D., and Harris, A.L., 1988, Mammalian DNA repair: use of mutants hypersensitive to cytotoxic agents, Trends Genet., 4:101-106.

Higgins, D.R., Prakash, S., Reynolds, P., Polakowska, R., Weber, S., Prakash, L., 1983, Isolation and characterization of the RAD3 gene of Saccharomyces cerevisiae and inviability of RAD3 deletion mutants, Proc. Natl. Acad. Sci. U.S.A., 80:5680-5684.

Hori, T.A., Shiomi, T., and Sato, K., 1983, Human chromosome 13 compensates a DNA repair defect in UV-sensitive mouse cells by a mouse-human cell hybridizatiion, Proc. Natl. Acad. Sci. U.S.A., 80:5655-5659.

Hoy, C.A., Thompson, L.H., Mooney, C.L., and Salazar, E.P., 1985, Defective cross-link removal in Chinese hamster cell mutants hypersensitive to bifunctional alkylating agents, Cancer Res., 45:1737-1743.

Johnson, R.T., Elliott, G.C., Squires, S., and Joysey, V.C., 1989, Lack of complementation between xeroderma pigmentosum complementation groups D and H, Human Genetics, 81:203-210.

Jones, N.J., Cox, R., and Thacker, J., 1987, Isolation and cross-sensitivity of X-ray-sensitive mutants of V79-4 hamster cells, Mutat. Res., 183:279-286.

Lambert, C., Couto, L.B., Weiss, WN.A., Schultz, R.A., Thompson, L.H., and Friedberg,E.C, 1988, A yeast DNA repair gene partially complements defective excision repair in mammalian cells, EMBO J., 7:3245-3253.

Mitchell, D.L., Humphrey, R.M., Adair, G.A., Thompson, L.H., and Clarkson, J.M., 1988, Repair of (6-4)photoproducts correlates with split-dose recovery in UV-irradiated normal and hypersensitive rodent cells, Mut. Res., 193: 53-63.

Montelone, B.A., Hoekstra, M.F., and Malone, R.E., 1988, Spontaneous mitotic recombination in yeast: The hyper-recombinational rem1 mutations are alleles of the RAD3 gene, Genetics, 119:289-301.

Naumovski,L., and Friedberg, E.C., 1983, A DNA repair gene required for the incision of damaged DNA is essential for viability in yeast., Proc. Natl. Acad. Sci. U.S.A., 80:4818-4821.

Naumovski, L., and Friedberg, E.C., 1986, Analysis of the essential and excision repair functions of the RAD3 gene in Saccharomyces cerevisiae by mutagenesis, Mol. Cell. Biol., 6:1218-1227.

MacInnes, M.A., Okinaka, R.T., Chen, D.J., Nickols, J.W., Tesmer, J.G., McCoy, L.S., and Strniste, G.F., 1988, Molecular-genetic evidence for identification of the ERCC-5 human DNA excision repair gene, J. Cell. Biochem., Suppl. 12A:319.

Robson, C.N., Harris, A.L., and Hickson, I.D., 1985, Isolation and characterization of Chinese hamster ovary cell lines sensitive to mitomycin C and bleomycin, Cancer Res., 45:5304-5309.

Robson, C.N., and Hickson, I.D., 1986, Genetic analysis of mitomcyin C-sensitive mutants of a Chinese hamster ovary cell lines, Mut. Res., 163:201-208.

Sato, K. and Setlow, R.B., 1981, DNA repair in a UV-sensitive mutant of a mouse cell line, Mutat. Res., 84:443-455.

Schultz, R.A., Saxon, P.J., Glover, T.W., Stanbridge, E.J., and Friedberg, E.C., 1988, Phenotypic complementation of xeroderma pigmentosum cells by transfer of single human chromosomes. In: Friedberg, E. and Hanawalt, P., eds, "Mechanisms and Consequences of DNA Dammage Processing," UCLA Symposia on Mol. Biol., New Series, Vol. 83, Alan R. Liss, New York, pp. 343-348

Shiomi, T., Hieda-Shiomi, N., and Sato, K., 1982, Isolation of UV-sensitive mutants of mouse L5178Y cells by a cell suspension spotting method, Somat. Cell Genet., 8:329-345.

Siciliano, M.J., Bachinski, L., Dolf, G., Carrano, A.V., and Thompson, L.H., 1987, Chromosomal assignments of human DNA repair genes that complement Chinese hamster ovary (CHO) cell mutants, Cytogenet. Cell Genet., 46:691-692.

Siciliano, M.J., Carrano, A.V., and Thompson, L.H., 1986, Assignment of a human DNA-repair gene associated with sister-chromatid exchange to chromosome 19, Mutat. Res., 174:303-308.

Stamato, T.D., and Waldren, C.A., 1977, Isolation of UV-sensitive variants of CHO-K1 by nylon cloth replica plating, Somat. Cell Genet., 3:431-440.

Sung, P., Prakash, L., Matson, S.W., and Prakash, L., 1987a, RAD3 protein of Saccharomyces cerevisiae is a DNA helicase, Proc. Natl. Acad. Sci. U.S.A. , 84:8951-8955.

Sung, P., Prakash, L., Weber, S., and Prakash, L., 1987b, The RAD3 gene of Saccharomyces cereyvisiae encodes a DNA-dependent ATPase, Proc. Natl. Acad. Sci. U.S.A., 84: 6045-6049.

Thacker, J., 1981, The chromosomes of a V79 Chinese hamster line and a mutant subline lacking HPRT activity, Cytogenet. Cell Genet., 29:1625.

Thompson, L.H., 1988, Use of Chinese hamster ovary cell mutants to study human DNA repair genes, in: "DNA Repair, Vol.3," E. Friedberg and P. Hanawalt, eds., Marcel Dekker, New York, pp. 115-132.

Thompson, L.H., Bachinski, L., Weber, C.A., Stallings, R., and Siciliano, M.J., 1988a, Complementation of repair gene mutations on the hemizygous chromosome 9 in CHO cells: a third DNA repair gene on human chromosome 19, Genomics, submitted.

Thompson, L.H., Brookman, K.W., Dillehay, L.E., Mooney, C.L., and Carrano, A.V., 1982a, Hypersensitivity to mutation and sister-chromatid-exchange induction in CHO cell mutants defective in incising DNA containing UV lesions, Somat. Cell Genet., 8:759-773.

Thompson, L.H., Brookman, K.W., Dillehay, L.E., Carrano, A.V., Mazrimas, J.A., Mooney, C.L., Minkler, J.L., 1982b, A CHO-cell strain having hypersensitivity to mutagens, a defect in DNA strand-break repair, and an extraordinary baseline frequency of sister chromatid exchange, Mutat. Res., 95:427-440.

Thompson, L.H., Brookman, K.W., Jones, N.J., Collins, C.C., and Carrano, A.V., 1988b, Molecular cloning of a human gene, XRCC1, involved in strand-break repair and sister chromatid exchange, in preparation.

Thompson, L.H., Brookman, K.W., and Mooney, C.L. 1984, Repair of DNA adducts in asynchronous CHO cells and the role of repair in cell killing and mutation induction in synchronous cells treated with 7-bromomethylbenz[a]anthracene. Somat. Cell Mol. Genet., 10:183-194.

Thompson, L.H., Busch, D.B., Brookman, K., Mooney, C.L., and Glaser, D.A. 1981. Genetic diversity of UV-sensitive DNA repair mutants of Chinese hamster ovary cells, Proc. Natl. Acad. Sci. U.S.A., 78:3734-3737.

Thompson, L.H. and Carrano, A.V. 1983, Analysis of mammalian cell mutagenesis and DNA repair using in vitro selected CHO mutants, in: Cellular Responses to DNA Damage, UCLA Symposia on Molecular and Cellular Biology," New Series, Vol. 11, E.C. Friedberg and B.A. Bridges, eds., Alan R. Liss, New York, 125-143.

Thompson, L.H., Carrano, A.V., Sato, K., Salazar, E.P., White, B.F., Stewart, S.A., Minkler, J.L., and Siciliano, M.J. 1987a. Identification of nucleotide excision repair genes on human chromosome 2 and 13 by functional complementation in hamster hybrids, Somat. Cell Mol. Genet., 13:539-551.

Thompson, L.,H., Mitchell, D.L., Regan, J.D., Bouffler, S.D., Stewart, S.A., Carrier, W.L., Nairn, R.S. and Johnson, R.T., 1989. CHO mutant UV61 removes (6-4) photoproducts but not cyclobutane dimers, Mutagenesis, 4:140-146.

Thompson, L.H., Mooney, C.L., Burkhart-Schultz, K., Carrano, A.V., and Siciliano, M.J., 1985, Correction of a nucleotide-excision-repair mutation by human chromosome 19 in hamster human hybrid cells, Somat. Cell Mol. Genet., 11: 87-92.

Thompson, L.H., Rubin, J.S., Cleaver, J.E., Whitmore, G.F., and Brookman, K., 1980, A screening method for isolating DNA repair-deficient mutants of CHO cells, Somat. Cell Genet., 6:391-405.

Thompson, L.H., Salazar, E.P., Brookman, K.W., Collins, C.C., Stewart, S.A., Busch, D.B., and Weber C.A., 1987b, Recent progress with the DNA repair mutants of Chinese hamster

ovary cells, J. Cell Sci., Suppl., 6: 97-110.
Thompson, L.H., Shiomi, T., Salazar, E.P., and Stewart, S.A., 1988c, An eighth complementation group of rodent cells hypersensitive to ultraviolet radiation, Somat. Cell Mol. Genet., 14:605-612.
Thompson, L.H., Weber, C.A., and Carrano, A.V., 1988d, Human DNA repair genes, in: "Mechanisms and Consequences of DNA Damage Processing," UCLA Symposia on Mol. Biol., New Series, Vol. 83, E. Friedberg and P. Hanawalt, eds., Alan R. Liss, New York, pp. 289-293.
Van Duin, M., de Wit, J., Odijk, H., Westerveld, A., Yasui, A., Koken, M.H.M., Hoeijmakers, J.H.J., and Bootsma, D., 1986, Molecular characterization of the human excision repair gene homology with the yeast DNA repair gene RAD10, Cell, 44:913-923.
Van Duin, M., Koken, M.H.M., van Den Tol, J., Ten Dijke, P., Odijk, H., Westerveld, A., Bootsma, D., and Hoeijmakers, J.H.J., 1987, Genomic characterization of the human DNA excision repair gene ERCC1, Nuc. Acids Res., 15:9195-9213.
Van Duin, M. , Janssen, J.H., de Wit, J., Hoeijmakers, J.H.J., Thompson, L.H., Bootsma, D., and Westerveld, A., 1988, Transfection of the cloned human excision repair gene ERCC-1 to UV-sensitive CHO mutants only corects the repair defect in complementation group-2 mutants, Mutat. Res., 193:123-130.
Waldren, C., Snead, D., and Stamato, T., 1983, Restoration of normal resistance to killing and of post-replication recovery (PRR) in CHO-UV-1 cells by transformation with hamster or human DNA, in: "Cellular Responses to DNA Damage, UCLA Symposia on Molecular and Cellular Biology," New Series, Vol.11, E.C. Friedberg and B.A. Bridges, eds., Alan R. Liss, New York, 637-646.
Weber, C.A., Salazar, E.P., Stewart, S.A., and Thompson, L.H., 1988a, Molecular cloning and biological characterization of a human gene, ERCC2, that corrects the nucleotide excision repair defect in CHO UV5 cells, Mol. Cell Biol., 8:1137-1146.
Weber, C.A., Salazar, E.P., Stewart, S.A., and Thompson, L.H., 1988b, Cloning of a human gene and its cDNA that correct the nucleotide excision repair defect in CHO complementation group 1, J. Cell. Biochem., Suppl., 12A:329.
Weber, C.A., Salazar, E.P., Stewart, S.A., and Thompson, L.H., 1988c, ERCC2: cDNA cloning and molecular characterization of a human nucleotide excision repair gene with homology to the yeast RAD3 gene, EMBO J., submitted.
Weeda, G., van Ham, R.C.A., Masurel, R., Hoeijmakers, J.H.J., Westerveld, A., Bootsma, D., and van der Eb, A.J., 1988, Molecular cloning of part of the human excision repair gene ERCC-3, J. Cell. Biochem., Suppl., 12A:303.
Westerveld, A., Hoeijmakers, J.H.J., van Duin, M., de Wit, J., Odijk, H., Pastink, A., Wood, R.D., and Bootsma, D., 1984, Molecular cloning of a human DNA repair gene, Nature, 310:425-429.
Wood, R.D., and Burki, H.J., 1982, Repair capability and the cellular age response for killing and mutation induction after UV, Mutat. Res., 95:505-514.
Zdzienicka, M.Z., Roza, L., Westerveld, A., Bootsma, D., and Simons, J.W.I.M., 1987a, Biological and biochemical consequences of the human ERCC-1 repair gene after transfection into a repair deficient CHO cell line, Mutat. Res., 183:69-74.

Zdzienicka, M.Z. , and Simons, J.W.I.M., 1987b, Mutagen-sensitive cell lines are obtained with a high frequency in V79 Chinese hamster cells, Mutat. Res., 178:235-244.

Zdzienicka, M.Z., van der Shans, G.P., Westerveld, A., van Zeeland, A.A., Simons,J.W.I.M., 1988a, Phenotypic heterogeneity within the first complementation group of UV-sensitive mutants of Chinese hamster cells, Mutat. Res., 193:31-41.

Zdzienicka, M.Z., van der Shans, G.P., and Simons, J.W.I.M., 1988b, Identification of a new seventh complementation group of UV-sensitive mutants in Chinese hamster cells, Mutat. Res., 194:165-170.

MOLECULAR GENETIC DISSECTION OF MAMMALIAN EXCISION REPAIR

Jan H.J. Hoeijmakers [1], Geert Weeda[2], Christine Troelstra[1], Marcel van Duin[1], Andries Westerveld[1], Alex van der Eb[2] and Dirk Bootsma[1]

[1]Dept. of Cell Biology & Genetics, Erasmus University, PO Box 1738, 3000 DR Rotterdam The Netherlands

[2]Dept. of Tumor Virology, State University Leiden Wassenaarseweg 72, 2333 AL Leiden The Netherlands

ABSTRACT

To elucidate the mechanism of excision repair in mammalian cells we have focussed on the cloning of human genes involved in this process. The strategy followed consists of transfection of human genomic DNA to excision deficient CHO mutants from different complementation groups (c.g.), followed by selection of repair proficient transformants and 'rescue' of the transferred human sequences responsible for correction of the repair defect. Applying this strategy to mutants 43-3B (c.g.1), 27-1 (c.g.3) and UV-61 (c.g.6) we have cloned the complementing human genes ERCC-1, -3 and -6, respectively.

The predicted ERCC-1 gene product (297 amino acids) has a mozaic structure: its first 214 amino acids show overall sequence homology to the yeast excision repair protein RAD10. In addition, its C-terminus shares domainal homology with part of the *E.coli* uvrA protein and with the C-terminus of uvrC. ERCC-1 and RAD10 also share an exceptional type of gene configuration: both genes harbour overlapping antisense transcription units in their 3' regions.

The ERCC-3 gene specifies a predicted protein of 782 amino acids. It has no yeast homolog known thusfar, however, Zoo-blot analysis indicates that this gene is also strongly conserved in evolution.

The recently isolated ERCC-6 gene appeared to have a size of approximately 100 kb. Its analysis is in progress.

INTRODUCTION

The occurrence of damage in DNA induced by chemical or

DNA Repair Mechanisms and Their Biological Implications in Mammalian Cells
Edited by M.W. Lambert and J. Laval
Plenum Press, New York

physical agents can have dramatic consequences. DNA-lesions interfere with vital, cellular processes such as transcription and replication. Furthermore, damaged sites can be converted into parmenent mutations, which in turn can inactivate or alter crucial genes, with cell death, inherited disorders or onset of carcinogenesis as the most severe, possible outcome. To counteract the deleterious effects of DNA injury, all living organisms have equipped themselves with intracellular defense mechanisms, consisting of an intricate network of DNA repair pathways (see for an extensive review on DNA repair, Friedberg 1985). Considering the enormous size of the genome, the wide spectrum of different types of lesions and the complex chromatin structure, DNA repair in higher organisms must be a sophisticated and elaborative process. Several, recent observations shed some light onto the intricacies associated with DNA repair. The discovery that mammalian cells preferentially repair (the transcribed strand of) expressed genes as described by Hanawalt and coworkers for the removal of UV-induced pyrimidine dimers (Bohr, et al., 1985, Mellon, et al., 1986) provides an elegant solution to the problems posed by DNA lesions hampering transcription. The recent finding that the yeast repair gene RAD6 encodes a ubiquitin conjugating enzyme specific for histones 2A and 2B (Jentsch, et al., 1987) and thought to be implicated in chromatin remodeling suggests the existence of tight interactions between chromatin structure and dynamics and repair events.

One of the best characterized repair systems in the cell is the excision repair pathway, which focusses on the category of lesions causing a relatively strong deformation of the regular DNA structure, such as pyrimidine dimers and (6-4) photoproducts (both induced by UV) and bulky DNA adducts. In E.coli this process involves the cooperation of at least 6 gene products: the uvrA and B proteins which form a complex scanning the DNA for abnormalities in its structure, the uvrC endonuclease, cleaving the damaged strand on both sites around the lesion, the uvrD helicase catalyzing release of the damage containing oligonucleotide leaving a 12-13 b. single-stranded gap, and finally, DNA polymerase I and ligase to fill in the gap and to ligate the newly synthesized DNA to the existing strand (for recent reviews see Sancar and Sancar, 1988, Grossman, et al., 1988). The mechanism of this system in eukaryotes is poorly understood. However, considerable, biochemical complexity can be predicted on the basis of the large number of complementation groups within excision deficient eukaryotic mutants. In yeast at least 10 different mutants constitute the excision deficient RAD3 epistasis group (for a review on UV-sensitive yeast mutants see Haynes and Kunz 1981). In mammals two classes of repair mutants are known: patients suffering from the autosomal recessive human disorder xeroderma pigmentosum (XP)(reviewed by Kraemer, et al., 1987) and laboratory induced, repair deficient mutant cell lines of rodent origin (reviewed by Collins and Johnson 1987). In each of these classes 9 resp. 8 complementation groups have been identified (Fischer, et al., 1985, Thompson, et al., 1988). The relationship between these two categories of mutants is largely unknown. Since the absence of overlap between the mutants in both species found thusfar (Thompson, et al., 1985; Stefanini, et al., 1987) holds for all groups, this might imply that in total 17 or more genetic loci are implicated in early steps of the excision pathway. One way to get entrance into the mechanism of excision repair is by molecular cloning of the genes involved. Here we summarize our recent progress with respect to isolation and characterization of 3 human DNA repair genes complementing excision deficient Chinese hamster ovary (CHO) cell mutants.

STRATEGY FOR CLONING MAMMALIAN REPAIR GENES

One of the most straightforward strategies for the isolation of mammalian repair genes involves genomic DNA transfer to repair deficient mutants followed by selection for transferrants resistant against DNA damaging agents and recombinant DNA techniques to retrieve the transfected, correcting gene (Hoeijmakers, et al., 1987, 1988). Often cloned markers encoding a dominantly selectable trait are included in the transfection to permit preselection of the small proportion of cells competent for DNA uptake and stable expression. As primary transformants, in general, have integrated much more exogenous DNA than the gene of interest a second round of transfection is frequently carried out to get rid of the irrelevant, co-integrated sequences in primary transformants. We have applied this approach to the repair deficient CHO mutants 43-3B and 27-1 (complementation groups 1 and 3 respectively, generously obtained from Dr. R.D. Wood, ICRF, London)(Wood and Burki 1982) and UV61 (complementation group 6, generously provided by Dr. L.H. Thompson, Livermore)(Thompson, et al., 1987), and in this way isolated the correcting human genes ERCC-1, -3 and -6.

RESULTS AND DISCUSSION

Isolation of the ERCC-6 gene

CHO A-A8 mutant UV61 constitutes together with mouse lymphoma mutant US46 rodent complementation group 6 (see Thompson, et al., elsewhere this volume). In contrast to members of CHO complementation groups 1-5, UV61 is only moderately sensitive to UV, as shown by its repair phenotype summarized in Table I. A remarkable property of the mutant concerns the finding of Thompson, Mitchell and coworkers that the rate of removal of UV-induced (6-4) photoproducts is normal but that excision of pyrimidine dimers and bulky DNA adducts is severely impaired (L. Thompson, personal communication). UV61, therefore, appears to discriminate between different types of lesions that are substrate for the same (excision repair) pathway. One possible explanation is that the gene product affected in UV61 normally functions in the removal or recognition of only one subset of DNA lesions repaired by nucleotide excision (i.e. pyrimidine dimers and bulky adducts) but not of (6-4) photoproducts. Alternatively, it is possible that a protein, involved in excision of both categories of DNA injury, is mutated in UV61 in such a way that only the affinity for one type of lesion (pyrimidine dimers, bulky adducts) is disturbed but not that for the other (6-4 lesions).

We have followed the scheme of genomic DNA transfection, outlined above, for the cloning of the ERCC-6 gene. The moderate UV-sensitivity of the mutant (see Table I) made selection, on the basis of regained UV-resistance, more difficult. Nevertheless, with a notably low frequency (1 in approximately 10^5 transformants, i.e., 30-50x lower as with ERCC-1) UV-resistant, primary and secondary transformants were obtained. No linked transfer between the putative, correcting human repair gene in the DNA of a primary transformant with a closely situated dominant marker could be achieved in secondary transfection. Furthermore, some of the transformants did not contain detectable human sequences, suggesting that they were revertants. However, one secondary transformant (ST-1) clearly retained human DNA. From a lambda library constructed from the DNA of ST-1, the entire human insert

TABLE 1. Repair Phenotype of Excision Repair Deficient CHO Mutants 43-3B, 27-1 and UV61.

Mutant	43-3B	27-1	UV61	Ref.[a]
Complement.group[b]	1	3	6	1,2
Degree of sensitivity				
to:[c] UV	6-7x	6-7x	2.7x	3,4
7BrMe BA	(~4x)	(~4x)	~3x	5,4
MM-C	100x	2x	(2.7x)	3,6
ENU	1.5-2x	5-6x[d]	?	3
MMS	1.5x	7-8x[d]	?	3
X-rays, bleomycin	wt.	wt.	?	3
UV-induced mutagenesis	30x	5-12x	4-6x	3,4
Rate of incision	(<10%)	<10%	wt.	4
Dimer removal (24 hrs)	n.d.	(n.d.)	n.d.	7,4
Removal of (6-4) photoproducts (6 hrs)		(n.d.)	wt.	
UDS (% of wt)	5-20%	5-20%	70%	8,9

[a]References: 1. Wood and Burki (1982); 2. Thompson, et al., (1987); 3. Zdzienicka and Simons (1986); 4. Thompson, et al., pers. comm.; 5. Thompson, et al., (1981); 6. Thompson, et al., (this volume); 7. Zdzienicka, et al., (1987); 8. Westerveld, et al. (1984); 9. our unpublished results.

[b]Recently, the numbering of complementation groups 1 and 2 was interchanged to match with the numbers of the human complementing genes ERCC-1 and -2. (Thompson and Bootsma, 1988).

[c]Degree of sensitivity is expressed as ratio of D10 mutant and D10 wild type (D10 is doses required to give 10% survival). Abbreviations used: UV: ultraviolet light; 7 BrMeBA: 7 bromomethylbenz(a)anthracene; MM-C: mitomycin-C, ENU: ethylnitrosourea; MMS: methylmethanesulphonate.

[d]Not a characteristic feature of complementation group 3 mutants.
nd = not detectable.
between brackets: values determined for other members of the same complementation group.

(approximately 125 kb) was cloned by hybridization with human cot-1 DNA and by chromosome walking procedures. The insert was subsequently mapped. It appeared to be notably poor in human, repeat containing sequences. Unique probes derived from different regions spread over the entire, cloned human insert were isolated and used to hybridize to DNA of independent, other transformants. When the same sequences are 'coinherited' by other transformants, this must mean that they are part of the human ERCC-6 gene itself or closely flanking the gene. From the results of this 'coinheritance' analysis using DNA of 6 independent, transformants we deduce that the cloned ERCC-6 gene probably has a size of approximately 100 kb. The very large size of the gene might explain the low transfection frequency. As far as we know no other genes of this size have hithereto been cloned using the strategy of genomic DNA transfection. In addition, evidence was found for several UV61 revertants, one of which appeared to have amplified the endogenous, mutated CHO-ERCC-6 gene, suggesting that the mutation in UV61 is leaky. This might provide support in favour of the idea, put forward above, that a partially defective repair protein is responsible for the peculiar feature of UV61 in that it removes (6-4) photoproducts but not pyrimidine dimers.

Isolation and characterization of the ERCC-3 gene

Cotransfection of human, genomic DNA together with dominant selectable marker genes to the excision repair deficient mutant 27-1 of CHO complementation group 3, was used to clone the ERCC-3 gene. (For the phenotype of mutant 27-1 see Table I). Transfection of DNA of independent primary transformants with or without addition of extra copies of the dominant marker yielded several, UV resistant secondary transformants, one of which was generated by linked transfer of at least one copy of the dominant marker and the correcting, human gene. Cosmid clones containing sequences directly flanking the dominant marker from this secondary transformant did not harbor the ERCC-3 gene. Apparently, the gene was situated at a greater distance from the marker copy. Therefore, lambda and cosmid clones were isolated which retained a human, repeat containing fragment that on Southern blots appeared to be common between all primary and secondary transformants. Unique human probes from the surrounding region were used to hybridize to DNA of the same panel of transformants to more precisely define the segment 'coinherited' by all transformants. In this way the borders of the ERCC-3 gene were determined, however, one end (that later turned out to be the 3' terminus) appeared to be refractory to cloning. Subsequently, unique, human segments were identified which cross-hybridized to Chinese hamster genomic DNA. Since such conserved sequences usually are indicative of a coding function, these probes were employed to screen cDNA libraries. This resulted in the isolation of cDNA clones, which ,upon transfection, corrected the UV sensitivity and UDS of 27-1 to wild type levels with a very high efficiency. However, the sensitivity of 27-1 to alkylating agents (which is a unique property of this mutant, but not a general feature of the complementation group, see Table I) is not corrected after introduction of the functional ERCC-3 cDNA. This suggests that the sensitivity to alkylating agents is caused by a second mutation.

Characterization and functional analysis of the gene and cDNA revealed the following features:

1. The gene is approximately 35 kb and assigned to chromosome 2. It encodes a mRNA of 3 kb, which is not inducible by UV irradiation in HeLa cells.
2. Correction by transfection of the cDNA is limited to mutants of CHO-complementation group 3.
3. Based on Southern and Northern analysis and transfection or microinjection of ERCC-3 cDNA in XP cells, we conclude that ERCC-3 is probably not involved in XP-A, C, D, E, G, H and I (XP-B and -F not tested).
4. The cDNA encodes a predicted protein of 782 amino acids, with putative nucleotide binding, helix-turn-helix DNA binding, nuclear location signal and histone binding domains.
5. The ERCC-3 amino acid sequence bears no significant homology to known repair genes of yeast and *E.coli.* However, Zoo-blot analysis indicates that the gene is strongly conserved.

Characterization of the ERCC-1 gene

ERCC-1 corrects CHO mutants of complementation group 1. As shown by Table I mutants of this group are very sensitive to UV-light, carcinogens causing bulky adducts and even more to cross-linking agents such as mitomycin-C. Transfection of the ERCC-1 gene fully compensates for the wide spectrum of impaired repair functions of the mutant (Westerveld, *et al.*, 1984; Zdzienicka, *et al.*, 1987; Bohr, *et al.*, 1988, Darroudi, *et al.*, 1989), indicating that ERCC-1 represents the human counterpart of the mutated gene in 43-3B. This is further supported by the finding that the ERCC-1 gene only corrects mutants of complementation group 1 and not of the other groups (van Duin, *et al.*, 1988a). We have investigated whether ERCC-1 is coincidently also involved in the human repair syndromes XP, Cockayne's Syndrome (CS) and Fanconi's anemia. From these experiments (which include Southern and Northern blot analysis of the ERCC-1 gene in representative cell lines of each complementation group, as well as transfection or microinjection of the cloned ERCC-1 gene or cDNA into the same cells), we conclude that ERCC-1 is very likely not the mutated gene in XP (all complementation groups) nor in CS A or B (van Duin, *et al.*, 1989b).

The molecular architecture of the ERCC-1 gene and its expression have been elucidated in great detail and is summarized in Figure 1 (van Duin, *et al.*, 1986, 1987, 1988b; Hoeijmakers, *et al.*, 1986). As reported previously, the gene spans a region of 15-17 kb on chromosome 19q13.2 and is composed of 10 exons. The 72 bp, coding exon 8 (hatched box in Figure 1) is subject to alternative splicing yielding two transcripts of 1.1 and 1.0 kb (pcDE and pcDE-72, respectively, see Figure 1), only the larger of which is necessary and sufficient to correct the CHO 43-3B mutation. The function of the 1.0 kb mRNA is as yet unknown. In addition to alternative splicing, the ERCC-1 gene also displays alternative polyadenylation site selection, yielding a transcript with a longer 3' untranslated region. ERCC-1 transcripts are found at a low, basal level in all mouse tissues and stages of embryogenesis analysed, and does not seem to be induced upon UV-irradiation in HeLa cells.

The 1.1 kb ERCC-1 mRNA encodes a protein of 297 amino acids. Comparison with consensus sequences of functional protein domains has pointed to the presence of a potential nuclear location signal (NLS) and a 'helix-turn-helix' DNA binding motive. Antibodies raised against an oligonucleotide with the amino acid sequence of the putative 'ERCC-1-NLS' specifically react with nuclear proteins in immunofluorescence preparations of fixed human fibroblasts, consistent with the idea that this region specifies an NLS. However, definite proof for the existence of the postulated domains awaits experimental verification at the protein level.

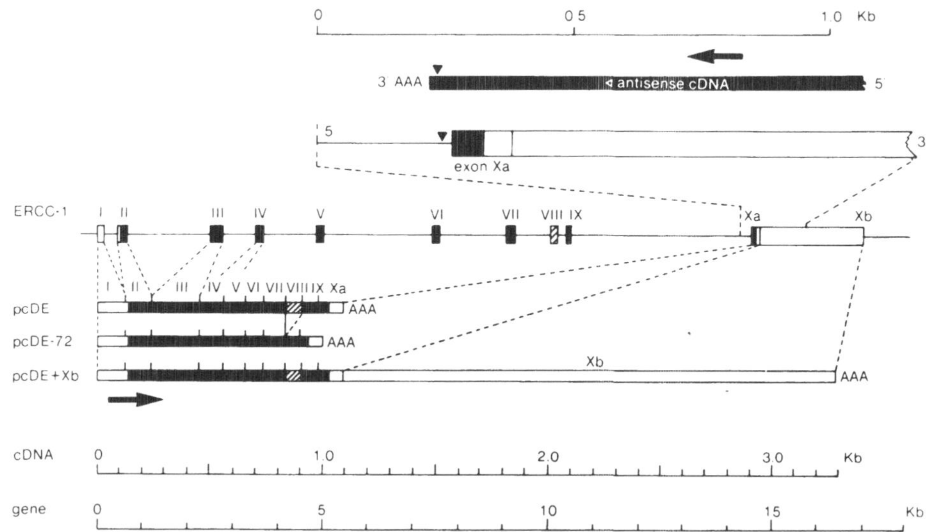

Figure 1. Architecture of the ERCC-1 locus.
Lower part: ERCC-1 transcripts pcDE, pcDE-72 (lacking exon VIII) and pcDE+Xb (alternatively polyadenylated). Upper part: enlargement of the ERCC-1 exon X region and position of the partial ASE-1 antisence cDNA. Filled boxes: coding (parts of) exons; open boxes: non-coding (part of) exons; hatched box: the alternatively spliced exon VIII. Filled triangle: polyadenylation signal of the ASE-1 antisense RNAs. Arrows indicate the 5' to 3'direction. Note the difference in scale for the gene and transcripts.

Computer comparison of the ERCC-1 amino acid sequence with known repair proteins of lower organisms revealed striking homology with the predicted amino acid sequence of the yeast excision repair protein RAD10 (van Duin, et al., 1985b). The central portion of ERCC-1 containing the tentative DNA binding domain and the C-terminal half of RAD10 display a high level of similarity. The N-termini of both proteins harbor only scattered homology, consistent with the finding that also most differences between the human and mouse ERCC-1 genes are concentrated in the N-terminal part (van Duin, et al., 1988b). Taken together, these findings suggest that ERCC-1 and RAD10 are descendants of the same ancestral gene and, hence, have analogous functions. The only, major difference between the two proteins is the fact that ERCC-1 is 83 amino acids longer than RAD10. At the position where the homology with RAD10 stops a stretch of approximately 40 amino acids begins, with significant similarity with part of the E.coli excision repair protein uvrA (Hoeijmakers, et al., 1986). Intriguingly, at the point where this homology terminates again another region of identity turns up: this time between the carboxy terminus of ERCC-1 and that of uvrC (Doolittle, et al., 1986). Therefore, it appears that ERCC-1 is a mozaic gene composed of domains present in different repair proteins of lower organisms. The functions exerted by the parts shared between ERCC-1 and the microbial repair polypeptides are not established. However, mutation studies have shown that the C-terminal 'extension' of ERCC-1, not present in RAD10, is essential for its function. Deletions, frameshift or point mutations in this region abolish the ability of ERCC-1 to compensate for the 43-3B defect (van Duin, et al., 1988b, unpublished results).

From detailed analysis of the 3' region of the ERCC-1 gene it became apparent that this part of ERCC-1 is also used by another gene. Sequence analysis of partial cDNA clones and hybridization with strand specific probes showed that this gene is transcribed from the opposite strand and that the 3' terminus of its 2.6 kb mRNA overlaps with exon 10 and terminates within intron 9 of ERCC-1 (van Duin, et al., 1989a, see Figure 1). Also the mouse ERCC-1 locus appears to harbour an antisense gene. Independently, Prakash and coworkers (Rochester) discovered that the yeast RAD10 gene region also contains a 3' overlapping, antisense gene (van Duin, et al., 1989a). The occurrence of this type of gene configuration is rare. In rodents, Drosophila, and yeast only a few examples have been reported (Adelman, et al., 1987; Williams and Fried, 1986; Chen, et al., 1987; Henikoff, et al., 1986; Spencer, et al., 1986; Barker, et al., 1985; Hahn, et al., 1988). To our knowledge ERCC-1/ASE-1 represents the first case in the human genome. The evolutionary conservation of this exceptional type of gene configuration between the human ERCC-1 and its yeast homolog RAD10 strongly suggests that it has an important biological function and extends the homology between the two genes to their gene organization. At present the function of the antisense genes is unknown as well as whether they are homologous to each other.

CONCLUDING REMARKS

The advent of recombinant DNA technology has opened new perspectives to the study of the mechanism of mammalian DNA repair. The cloning of many repair genes is in progress and the characterization of the genes isolated thusfar has revealed a striking, evolutionary conservation of the human gene products (ERCC-1, -2 and probably -3) with repair proteins of yeast and in the case of ERCC-1 even with parts of the uvrA and C polypeptides of E.coli. This implies that functional aspects of the microbial repair proteins are also relevant for the homologous mammalian counterparts. It remains to be seen to what extend the nucleotide excision pathway as a whole is conserved. The functions (known or presumed on the basis of the predicted amino acid sequence) of the human excision repair genes cloned until now, as well as their yeast counterparts are summarized in Table II. The table includes also the repair protein specifically correcting the XP-A defect as determined in our microneedle injection assay and of which the mouse gene is cloned recently by Tanaka and coworkers (Kyoto, personal communication).

TABLE II. Summary Properties of Cloned, Human Excision Repair Genes

Gene	Size[a] Gene	Size[a] Protein	S.cerevisiae Homolog	Function	Ref.
ERCC-1	15-17 kb	297 aa	RAD10	DNA binding?	1
ERCC-2	19 kb	760 aa	RAD3	DNA helicase? vital function?	2,3
ERCC-3	35 kb	782 aa	unknown	DNA, nucleotide + chrom.binding?	4
ERCC-6	100 kb	?	?	?	
XPAC	20-40 kb	40-45 kD	?	DNA binding	5,6

[a]Protein size of the <u>ERCC</u>-genes is predicted on the basis of the nucleotide sequence (aa: amino acids; kb, kilo base pairs). References: 1. Van Duin, <u>et al.</u> (1985), 2: Thompson, <u>et al.</u> (1989, this volume), 3. Sung, <u>et al.</u> (1988), 4. This paper, 5. cloning of the XP-A correcting mouse gene: H. Tanaka (Kyoto) personal communication, 6. Size and DNA binding properties of XP-A correcting protein: A.P.M. Eker (Rotterdam) personal communication.

Undoubtedly, this list of genes will extend in the coming years. We employ the evolutionary conservation to try to isolate repair genes of higher species using sequence homology with cloned repair genes of <u>S. cerevisiae</u>. The cloning of repair genes permits overproduction and purification of the encoded gene products as a prelude to the investigation of their functional properties. Eventually, the intricate excision repair process has to be dissected into its biochemical components in order to understand its mechanism, specificity and fidelity and its role in the (prevention of) carcinogenesis.

ACKNOWLEDGEMENTS

We gratefully acknowledge the excellent technical assistance of Mrs. H. Odijk and Mr. J. de Wit and R.C.A. van Ham. We thank Dr. R.D. Wood (London) and Dr. L.H. Thompson

(Livermore) for generously providing the repair deficient CHO mutants and Mrs. R. Boucke for typing the manuscript. This work was financially supported by MEDIGON, Foundation of Medical and Health Research in the Netherlands, contract no. 900.501.091 and EURATOM, contract no. BJ6-141-NL.

REFERENCES

Adelman, J.P., Bond, C.T., Douglass, J., and Herbert, E., 1987, Two mammalian genes transcribed from opposite strands of the same locus, Science, 235:1514-1517.

Baker, D.G., White, J.H.M., and Jonhston, L.H., 1985, The nucleotide sequence of the DNA ligase gene (CDC9) from Saccharomyces cerevisiae: a gene which is cell-cycle regulated and induced in response to DNA damage, Nucl. Acids Res., 13:8323-8337

Bohr, V.A., Smith, C.A., Okumoto, D.S., and Hanawalt, P.C., 1985, DNA repair in an activegene: removal of pyrimidine dimers from the DHFR gene of CHO cells is much more efficient than in the genome overall, Cell, 40:359-369.

Bohr, V.A., Chu, E.H.Y., van Duin, M., Hanawalt, P.C., and Okumoto, D.S., 1988, Human repair gene restores normal pattern of preferential DNA repair in repair defective CHO cells, Nucl. Acids Res., 16:7397-7403.

Chen, C., Malone, T., Beckendorf, S.K., and Davis, R.L., 1987, At least two genes reside within a large intron of the dunce gene of Drosophila, Nature, 329:721-724.

Collins, A., and Johnson, R.T., 1987, DNA repair mutants in higher eukaryotes, J. Cell Sci. Suppl., 6:61-82.

Darroudi, F., Westerveld, A., and Natarajan, A.T., 1989, Cytogenetical characterization of Chinese hamster 43-3B transferrants with the amplified or non-amplified human repair gene ERCC-1, Mutat. Res., (in the press).

Doolittle, R.F., Johnson, M.S., Husain, I., van Houten, B., Thomas D.C., and Sancar, A., 1986, Domainal evolution of a prokaryotic DNA repair protein and its relationship to active transport proteins, Nature, 323:451-453.

Fischer, E., Keijzer, W., Thielmann, H.W., Popanda, O., Bohnert, E., Edler, L., Jung, E.G., and Bootsma, D. 1985, A ninth complementation group in xeroderma pigmentosum, XP-I, Mutat. Res., 145:217-225.

Friedberg, E.C., 1985, DNA repair, Freeman and Company, San Francisco.

Grossman, L., Caron, P.R., Mazur, S.J., and Oh, E.Y., 1988, Repair of DNA-containing pyrimidine dimers, FASEB J., 2:2996-2701.

Hahn, S., Pinkham, J., Wei, R., Miller, R., and Guarente, L., 1988, The HAP3 regulatory locus of Saccharomyces cerevisiae encodes divergent overlapping transcripts, Mol. Cell. Biol., 8:655-663.

Haynes, R.H., and Kunz, B.A., 1981, DNA repair and mutagenesis in yeast, in: Strathern, J.N., Jones, E.W. and Broach, J.R. (Eds.)., The Molecular Biology of yeast Saccharomyces, Cold Spring Harbor Laboratory Publications, New York, pp. 371-414.

Henikoff, S., Keene, M., Fechtel, K., Fristrom, J., 1986, Gene within a gene: nested Drosophila genes encode unrelated proteins on opposite strands, Cell, 44:33-42.

Hoeijmakers, J.H.J., van Duin, M., Westerveld, A., Yasui, A., and Bootsma, D., 1986, Identification of DNA repair genes in the human genome, Cold Spring Harbor Symp.Quant.Biol. vol. LI, 91-101.

Hoeijmakers, J.H.J., Odijk, H., and Westerveld, A., 1987, Differences between rodent and human cell lines in the amount of integrated DNA after transfection, Exp. Cell. Res., 169:111-119.

Hoeijmakers, J.H.J., Westerveld, A., and Bootsma, D., 1988, Methods and strategies for molecular cloning of mammalian genes by DNA mediated genetransfer, In: Friedberg, E.C. and Hanawalt, P.C. (eds.), DNA repair, a laboratory manual for research procedures, Marcel Dekker Inc., New York, Vol. 3:181-201.

Jentsch, S., Mc Grath, J.P., and Varshavsky, A., 1987, The yeast DNA repair gene RAD6 encodes a ubiquitin-conjugating enzyme, Nature, 329:131-134.

Kraemer, K.H., Lee, M.M., and Scotto, J., 1987, Xeroderma pigmentosum, Arch. Dermatol., 123:241-250.

Mellon, I., Spivak, G., and Hanawalt, P.C., 1987, Selective removal of transcription-blocking DNA damage from the transcribed strand of the mammalian DHFR gene, Cell, 51:241-249.

Sancar, A., and Sancar, G.B., 1988, DNA repair enzymes, Ann. Rev. Biochem., 57:29-67.

Spencer, C., Gietz, D., and Hodgetts, R., 1986, Overlapping transcription units in the dopa decarboxylase region of Drosophila, Nature, 322:279-281.

Stefanini, M., Keijzer, W., Westerveld, A., and Bootsma, D., 1985, Interspecies complementation analysis of xeroderma pigmentosum and UV-sensitive Chinese hamster cells, Exp. Cell Res., 161:373-380.

Sung, P., Prakash, L., Weber, S., and Prakash, L., 1987, The RAD3 gene of Saccharomyces cerevisiae encodes a DNA-dependent ATPase, Proc. Natl. Acad. Sci. U.S.A., 84:6045-6049.

Thompson, L.H., Busch, D.B., Brookman, K., Mooney,C.L., and Glaser, D.A., 1981, Genetic diversity of UV-sensitive DNA repair mutants of Chinese hamster ovary cells, Proc. Natl. Acad. Sci. U.S.A., 78:3734-3737.

Thompson, L.H., Mooney, C.L., and Brookman, K.W., 1985, Genetic complementation between UV-sensitive CHO mutants and xeroderma pigmentosum fibroblasts, Mutat. Res., 150:423-429.

Thompson, L.H., Salazar, E.P., Brookman, K.W., Collins, C.C., Stewart, S.A., Busch, D.B., and Weber, C.A., 1987, Recent progress with the DNA repair mutants of Chinese hamster ovary cells, J. Cell Sci. Suppl., 6:97-110.

Thompson, L.H., and Bootsma, D., 1988, Designation of mammalian complementation groups and repair genes, In: Friedberg, E.C., Hanawalt, P.C. (eds.), Mechanisms and Consequences of DNA Damage Processing. Alan R. Liss Inc., New York, UCLA Symposia on Molecular and Cellular Biology New Series, 83:279.

Thompson, L.H., Shiomi, T., Salazar, E.P., and Stewart, S.A., 1988, An eighth complementation group of rodent cells hypersensitive to ultraviolet radiation, Somat. Cell Mol. Genet., 14:605-612.

Van Duin, M., de Wit, J., Odijk, H., Westerveld, A., Yasui, A., Koken, M.H.M., Hoeijmakers, J.H.J., and Bootsma, D., 1986, Molecular characterization of the human excision repair gene ERCC-1: cDNA cloning and amino acid homology with the yeast DNA repair gene RAD10, Cell, 44:913-923.

Van Duin, M., Koken, M.H.M., van den Tol., J., ten Dijke, P., Westerveld, A., Bootsma, D., and Hoeijmakers, J.H.J., 1987, Genomic characterization of the human excision repair gene ERCC-1, Nucl. Acids. Res., 15:9195-9213.

Van Duin, M., Janssen, J.H., de Wit, J., Hoeijmakers, J.H.J., Thompson, L.H., Bootsma, D., and Westerveld, A., 1988a, Transfection of the cloned human excision repair gene ERCC-1 to UV-sensitive CHO mutants only corrects the repair defect in complementation group 2 mutants, Mutat. Res., 193:123-130.

Van Duin, M., van den Tol, J., Warmerdam, P., Odijk, H., Meijer, D., Westerveld, A., Bootsma, D., and Hoeijmakers, J.H.J., 1988b, Evolution and mutagenesis of the mammalian excision repair gene ERCC-1. Nucl. Acids Res., 16:5305-5322.

Van Duin, M., van den Tol., J., Hoeijmakers, J.H.J., Bootsma, D., Rupp, I.P., Reynolds, P., Prakash, L., and Prakash, S., 1989a, Conserved pattern of antisense overlapping transcription in the homologous human ERCC-1 and yeast RAD10 DNA repair gene regions, Molec. Cell Biol., 9 (in press).

Van Duin, M., Vredeveldt, G., Mayne, L.V., Odijk, H., Vermeulen, W., Weeda, G., Klein, B., Hoeijmakers, J.H.J., Bootsma, D., and Westerveld, A., 1989b, The cloned human DNA excision repair gene ERCC-1 fails to correct xeroderma pigmentosum complementation groups A through I. Mutat. Res., (in press).

Westerveld, A., Hoeijmakers, J.H.J., van Duin, M., de Wit., J., Odijk, H., Pastink, A., Wood, R., and Bootsma, D., 1984, Molecular cloning of a human DNA repair gene, Nature, 310:425-429.

Williams, T., and Fried, M., 1986, A mouse locus at which transcription from both DNA strands produces mRNAs complementary at their 3' ends, Nature, 322:275-277.

Wood, R.D., and Burki, H.J., 1982, Repair capability and the cellular age response for killing and mutation induction after UV, Mutat. Res., 95:505-514.

Zdzienicka, M.Z., and Simons, J.W.I.M., 1986, Analysis of repair processes by the determination of the induction of cell killing and mutations in two repair-deficient Chinese hamster ovary cell lines, Mutat. Res., 166:59-69.

Zdzienicka, M.Z., Roza, L., Westerveld, A., Bootsma, D., and Simons, J.W.I.M., 1987, Biological and biochemical consequences of the human ERCC-1 repair gene after transfection into a repair deficient CHO cell line, Mutat. Res., 183:69-74.

THE CLONING AND CHARACTERIZATION OF A CANDIDATE GENE FOR THE CORRECTION OF THE XERODERMA D DEFECT

Janet E. Arrand*, Neil M. Bone, and Robert T. Johnson

CRC Mammalian Cell DNA Repair Research Group
Department of Zoology, University of Cambridge
Downing Street, Cambridge CB2 3EJ, UK

ABSTRACT

DNA from a repair competent hamster cell line, ligated to a dominant selectable marker, was transfected into immortalised xeroderma D cells. Two rounds of transfection resulted in the isolation of a cell line with UV-induced excision repair capability similar to that of a primary transfectant but with few integrated copies of both marker and hamster DNA sequences. Cosmid rescue of the dominant marker from secondary transfectant DNA gave rise to several clones, two of which contain hamster DNA sequences and confer increased survival to UV on transfection into two, independently-immortalised xeroderma D cell lines, but not on transfection into a xeroderma A line. Restriction enzyme analysis revealed a 13 kilobase overlap between these 2 cosmids; no similarities to the previously isolated UV excision repair genes ERCC1 and 2 were evident. A non-repetitive portion of the overlap region, when used to probe total cellular RNA, showed that the gene is expressed to varying degrees in all cell lines tested.

INTRODUCTION

The genetic disease xeroderma pigmentosum (XP) is autosomal and recessive. Its classic symptoms are multiple skin lesions, (including cancers) at sites exposed to ultraviolet (UV) light, and neurological disorders (Kraemer, et al., 1987). So far, 10 complementation groups have been identified, each group potentially representing a defect in a different protein of a repair complex (Giannelli, et al., 1982; Wood, et al., 1982). However, all groups share, to different extents, the common inability to perform the first, incision step of UV-induced excision repair (Fornace, et al.,

*Present address: Gray Laboratory, Mount Vernon Hospital, Northwood, Middlesex, HA62JR, UK.

DNA Repair Mechanisms and Their Biological Implications in Mammalian Cells
Edited by M.W. Lambert and J. Laval
Plenum Press, New York

1976). Representative cells of several of the complementation groups have been immortalised either by viral transformation or by cell fusion (Collins and Johnson, 1987). In addition to the 10 XP complementation groups, a further 6 non-overlapping groups of hamster excision repair mutants have been characterised (Thompson, et al., 1985). This battery of cell lines should facilitate the isolation of many genes involved in UV-induced excision repair. Two such genes have recently been obtained via the transfer of normal human DNA into defective hamster cells (Westerveld, et al., 1984; Weber, et al., 1988). Many other attempts have failed due to the low quantities of DNA integrated into SV40 immortalised human XP cells (Hoeijmakers, et al., 1987; Mayne, et al., 1988) coupled with high reversion frequencies (Royer-Pokora and Haseltine, 1984).

Here we report the partial, but stable correction of the excision repair defect in a UV-sensitive, non-reverting HeLa/-XPD hybrid cell line (Johnson, et al., 1985) by neo gene-tagged normal hamster DNA which is taken up efficiently by these cells and stably integrated at low copy number (Arrand, et al., 1987; Arrand, et al., 1988). Further, we show that cosmid rescue of the neo gene from a secondary transfectant has resulted in the isolation of the correcting hamster DNA sequences, enabling us to assess their function in independently immortalised XP cells.

METHODS

Cells, growth conditions, transfection, repair assays and DNA and RNA hybridisations have been described previously (Arrand, et al., 1987). For primary transfection, 10^6 A31-27 cells were exposed for 16 hours to 20 µg pro-CHO DNA partially digested with Sau3AI to an average size of 50 kilobase pairs (kb) and ligated to a 2-fold excess of Bam HI digested p272 poly (Brady, et al., 1985). Secondary transfections were performed using high molecular weight primary transfectant DNA sheared by passage through a 23 g needle to an average size of 100 kb.Selection with 500 µg/ml G418 was commenced 48 hours after removal of calcium phosphate precipitates and was maintained throughout. UV selection at 99% killing doses for A31-27 was carried out at 96 hours and 120 hours.

Cosmid rescue of integrated neo genes was effected using cosmid arms from pHSG250 (Km^s), Gigapack gold packaging mixes and ED8767 $recA^-$ host bacteria as described by Brady, et al. (1985), with modifications and precautions suggested by Little (1987). Cosmid rescuants were maintained in ED8767 either as actively growing cells or as frozen glycerol stocks. Cosmid DNA was prepared by scaling up a mini-prep method (Little, 1987). This yielded up to 100 µg per 250 ml culture.

RESULTS

Primary and secondary transfection, using neo gene-tagged genomic DNA, produced $G418^R$ colonies at a frequency of 10^{-3}. Of these, 2 in 10^5 survived UV selection in the primary transfection, yielding 5 independent $G418^R$ UV^R cell lines; one secondary transfectant was obtained at a frequency of 2.5 x 10^{-5}.

Figure 1, Panel A shows the sensitivity to UV killing of A31-27 (the XPD/HeLa hybrids used as recipients of CHO DNA in the primary and secondary transfections) compared to that of the DNA-donor CHO cells and the primary and secondary transfectants which survived UV selection. The primary and secondary transfectants show very similar UV survival characteristics, each having a 2-3 fold increase in resistance to doses between 5 and 8 Jm^{-2} compared to A31-27. This resistance does not approach the level shown by the CHO DNA donor, but the increase is stably maintained when cells are passaged continuously in medium containing G418.

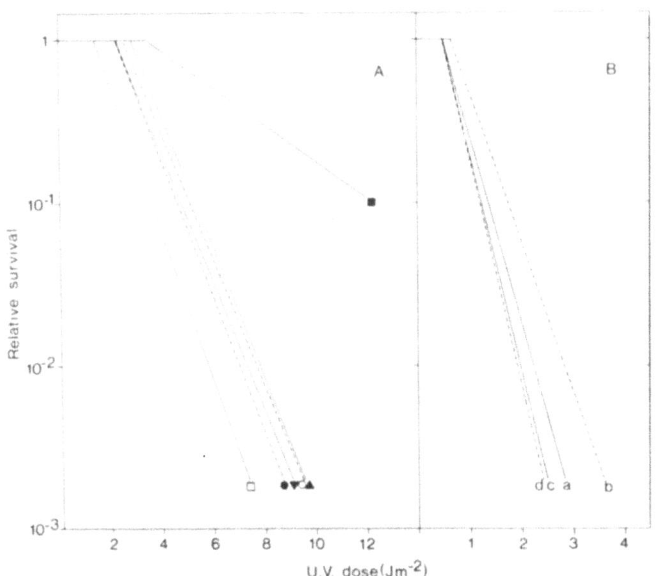

Figure 1. Colony survival of UV irradiated cells. Curves fitted by least squares analysis; 5 doses, 2 determinations per dose. Panel A: □, A31-27 XPD/HeLa hybrid); ■, pro⁻CHO; ▲, primary pro⁻xA31-27 transfectant; ▼, secondary transfectant; O, ●, pX1 transfectants of A31-27. Panel B: a. XP6Be (XPD, SV40 transformed); b, pX1 transfected XP6Be; c, XP2OS SV (XPA, SV40 transformed); d, pX1 transfected XP2OS SV. (Transfectant survivals are shown as dashed lines.)

This increase in resistance to killing by UV is a result of the enhanced ability of the transfectants to incise irradiated cell DNA at damaged sites and subsequently carry out repair DNA synthesis as shown in Figure 2. Panel A shows the accumulation of single-strand breaks, produced in the presence of inhibitors of DNA synthesis, following exposure to a UV dose of 6 Jm^{-2} for recipient A31-27 XPD/HeLa hybrids, transfectants and CHO cells. It is clear that the primary and secondary transfectants once again show very similar proper-

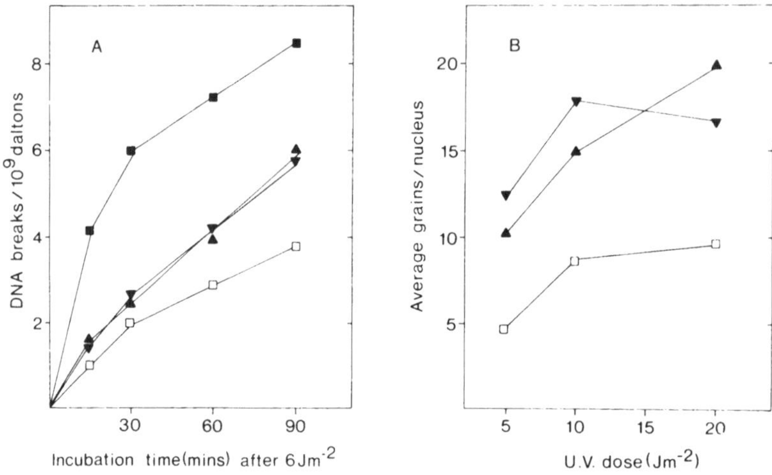

Figure 2. Analysis of repair functions in the XPD/HeLa hybrid and its transfectants. Panel A: Repair related break accumulation measured by alkaline unwinding (mean of two experiments) Panel B: UV dose response of unscheduled DNA synthesis. Cells were incubated with [Me-^{3}H] thymidine for 60 minutes prior to autoradiography; 100 nuclei scored at each dose. All symbols as in Figure 1.

ties in accumulating 2-fold more breaks than the recipient cell over the time course of this experiment, but do not express the full incision capability of the donor hamster cell. In the absence of inhibitors of DNA synthesis, no breaks are seen in the DNA of any of these cell types (data not shown). Panel B shows the 2-3 fold increase in unscheduled repair DNA synthesis (UDS) which is conferred on A31-27 cells by the incoming CHO cell DNA. Once again, though wild-type levels are not achieved, the increased activity observed is consistent with the increased survival of the transfectants to UV in this dose range. Cells transfected with a dominant marker plasmid alone show no increase in survival to UV or in the repair functions measured here (Arrand, _et al._, 1987).

Southern blot analysis of the XPD/HeLa hybrid and its transfectants, shown in Figure 3, reveals that a very small amount of CHO and dominant marker DNA is integrated into the genomic DNA of the primary transfectantBy comparison with the copy number reconstructions, it can be seen that 5-10 copies of _neo_ have been acquired along with 2-5 copies of the hamster _Alu_-equivalent sequence, which is visible in bands of approximately 7 kb, 4.2 kb and 3.8 kb in EcoRI-digested DNA (track 1). The 4.2 kb and 3.8 kb bands are also faintly visible in the EcoRl-digested DNA of the secondary transfectant (track 3). This secondary transfectant, however, has fewer copies of the Alu repeat (1-2) and an associated reduction in _neo_ gene copy number. It should be noted that there is some cross-hybridization between the hamster _Alu_ repeat probe and human DNA, resulting in a detectable back ground in track 7 (A31-27 DNA alone) and in all transfectant tracks. This is insignificant when compared to the hybridization of the probe to hamster cell DNA (tracks 5 and 6).

Very high molecular weight DNA, prepared from the secondary transfectant (Little, 1987), was partially digested with Sau3AI to an average size of 40 kb, fractionated by sucrose gradient centrifugation and ligated to an equimolar mixture of cosmid arms prepared from pHSG250 (Brady, et al., 1985). The total product of a 2 µg ligation was packaged in phage λ heads and plated on E.coli ED8767 in 4 150 mm petri dishes. Selection with kanamycin (15 µg/ml) gave rise to 7 cosmid neo gene rescuants, 2 of which (pX1 and pX3) were shown by hybridization to contain hamster Alu-equivalent sequences.

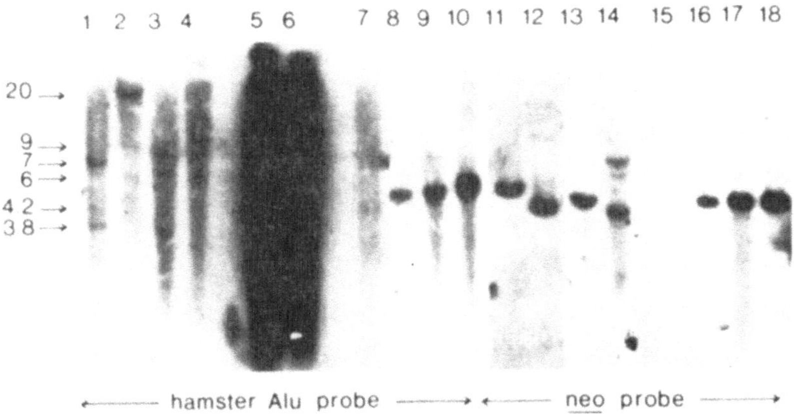

Figure 3. Southern blot analysis of cellular DNA of the XPD/HeLa hybrid and pro⁻CHO genomic DNA-transfected derivatives. 20 ug digests were electrophoresed through 0.8% agarose gels, blotted and hybridised to ^{32}P-labelled probes as indicated. For copy number reconstructions, plasmids pNB137 (Bone and Arrand, in press) (hamster Alu) or p272 poly15 (neo) were added to 20 ug hybrid cell DNA before digestion. Tracks 1, 11, primary transfectant, Eco RI; 1, 12, primary, Bam HI; 3, 13, secondary, Eco RI; 4, 14, secondary, Bam HI; 5, 6, 5 ug pro⁻CHO, Eco RI and Bam HI; 7-10, hybrid plus 0, 1, 2, 5 copies pNB137, Eco RI; 15-18, as 7-10 but with p272 poly, Bam HI. Fragment sizes in kb are indicated on the left.

Restriction endonuclease analysis of pX1 and pX3 (Figure 4) shows that pX1 is a deleted molecule of 21.1 kb. This deletion must have occured immediately after phage infection of ED8767 since re-analysis of immediate glycerol freezings of pX1 are similarly deleted. pX3 is 47.5 kb in size and is unstable, since deleted and rearranged forms have been found. Parallel restriction digests of the 2 cosmids reveal several common bands which detailed mapping has localised to a 13 kb

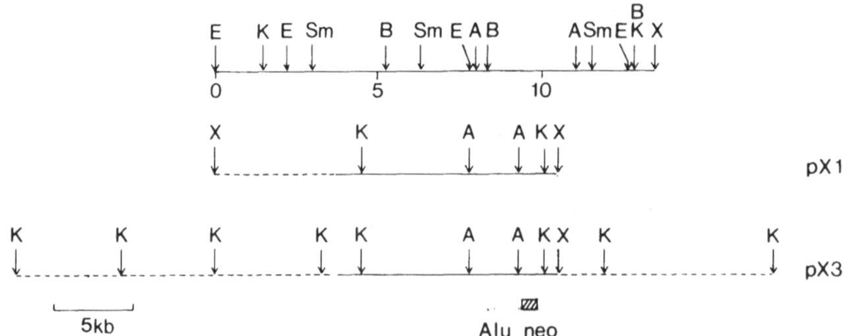

Figure 4. Restriction endonuclease analysis of cosmid rescuants pXl and pX3. Restriction maps were generated using single and double digests and were checked by cross-hybridization. Hamster Alu and neo sequences were located by Southern blot hybridisation. The overlap region is shown as a detailed enlargement at the top of the figure. Abbreviations: A, ApaI; B, BamHI; E, EcoRI; K, KpnI; Sm, SmaI; X, XhoI.

overlap shown in the upper section of Figure 4.

DNAs from cosmids pXl and pX3 have been transfected into the XPD/HeLa hybrid, a second independently-immortalised XPD cell line XP6Be (Protic-Sablijic, et al., 1986) and an SV40-transformed XPA fibroblast XP20S-SV (Takebe, et al., 1974). 13 of 18 G418^R, UV-selected pXl transfectants of A31-27 show a stable increase in survival to UV (see Figure 1, panel A). pX3 transfectants are unstable and lose UV resistance on passaging. 6 out of 12 XP6Be pXl transfectants, selected for

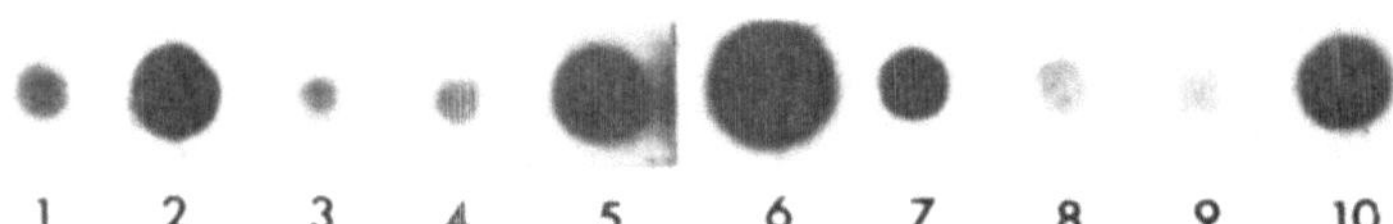

Figure 5. Dot blot analysis of total cellular RNA. Approximately 4 μg dots hybridised to pXl/pX3 overlap. Dots and Cerenkov count ratios (normalised by comparison to human aprt-hybridised dots): 1, pro$^-$CHO (3.64); 2, XPD/HeLa hybrid (24.15); 3, primary transfectant (2.54); 4, secondary (2.42); 5, pXl transfectant (10.76); 6, EJ30 human bladder carcinoma cell (1204); 7, Human embryo lung fibroblasts (10.5); 8, XP102LO XPD parent of hybrid (1.5); 9, GM8207 (1); 10, XP20S SV (46).

G418 resistance only, show a stable 2-3 fold increase in survival to a 99% killing UV dose, similar to that conferred on A31-27 (Figure 2, panel B). 8 out of 12 XP6Be pX3 transfectants show a transient increase in UV resistance. 20 G418[R] pX1 and pX3 XP2OS-SV transfectants were tested for increased survival to UV. No increase (transient or stable) was detected (Figure 2, panel B).

A portion of the 13 kb overlap between pX1 and pX3, which contains few or no repetitive elements (data not shown), was used to probe dot blots of total cellular RNA from normal human and XP cells (both fibroblast and SV40- or hybrid-immortalised). The results are shown in Figure 5. Low level, but detectable RNA transcripts are seen in the CHO donor cells, the primary and secondary transfectants, the XPD fibroblast parent of the XPD/HeLa hybrid, and XP6Be, the SV40 immortalised XPD fibroblast. Intermediate levels are detected in the XPD/HeLa hybrid, a human embryo fibroblast, the XPA cell line and the pX1-transfected A31-37; very high levels appear in the human bladder carcinoma line.

Variations in ploidy and inaccuracies in measuring the small RNA concentrations were corrected by hybridising identical dot blots to a human _aprt_ gene probe (Murray, _et al._, 1984).

DISCUSSION

The data presented here indicate a partial, but stable, correction of the excision repair defect in xeroderma D, but not in xeroderma A cells. This correction is evident at early and late stages of repair and is reflected in a greater resistance to UV irradiation. It differs, therefore, from the effects seen on introduction of the yeast _RAD1_ gene into the XPD/HeLa hybrid, when an increase in incision is not reflected in higher survival (our unpublished observations). The partial nature of the correction may reflect low levels of an incision enzyme in hamster cells where sites of DNA damage are mostly tolerated by the cells' replication machinery (Mitchell, _et al._, 1988) and only removed in actively transcribed regions (Bohr, _et al._, 1985). However, more DNA breaks are seen to accumulate in CHO cells than in the primary and secondary transfectants, so the correcting DNA sequences are not expressing the full hamster potential. This could be due to: a) loss or rearrangement of sequences during transfection, b) inefficient use of a hamster promoter in human cells, c) gene dosage effects causing an imbalance in the proteins which assemble into a repair complex (Wood, _et al._, 1988), or d) insufficient homology between the hamster and human proteins to allow correct functioning of the repair mechanism.

The small amount of DNA integrated during transfection of the XPD/HeLa hybrid cell line (approximately 50 kb) has facilitated the isolation of the correcting hamster sequences after only two rounds of transfection. Furthermore, _Alu_-containing EcoR1 fragments of approximately 3.8 kb and 4.2 kb, which are common to the primary and secondary transfectants, are recovered in at least one cosmid rescuant, and the total number of _neo_ genes recovered by cosmid rescue reflects their copy number in the secondary transfectant.

The cosmids recovered deleted sequences and rearranged frequently, perhaps reflecting the nature of the gene which they carry. Two of the cosmids, which cross-hybridise to hamster Alu sequences, show a 13 kb overlap of which 2 kb is derived from the original p272 poly tag. Therefore, the minimum size of the gene responsible for the correction of the XPD defect is 11 kb, since both pX1 and pX3 confer at least a transient increase in resistance to UV irradiation on recipient cells. pX3 is substantially rearranged outside the overlap region when compared to pX1; we speculate that a terminal portion of the gene may be missing in this construct and this may account for its transient effectiveness.

In both pX1 and pX3, the neo gene is in very close proximity to sequences which cross-hybridise strongly to the hamster Alu probe. This explains their linked transfer during transfection and the efficient cosmid rescue. The overlap region, though of similar size to the previously isolated repair genes ERCC1 and 2, shows no obvious similarity in restriction pattern. The small size of the gene probably accounts for the relative ease of its transfer by transfection.

A non-repetitive probe derived from the overlap region between cosmids pX1 and pX3 hybridises to total RNA in all the cell types tested here. The low signal seen in the CHO cells may again indicate the favouring of damage tolerance over excision by these cells. Human cells show large variations in cross-hybridising RNA levels; repair proficient cells, XPD/HeLa hybrids and XPA cells having higher levels than the transfectants which, in turn, show stronger signals than the XPD fibroblasts and SV40-transformed XPD cells. This preliminary analysis gives no information on accuracy of splicing, transcript length or translation of active protein.

We conclude that although the hamster sequences isolated here may not represent an entire repair gene, they do contain regions which are expressed, with unknown accuracy, in all cell types tested and can therefore be used to isolate the homologous human gene from existing cDNA and genomic DNA libraries.

ACKNOWLEDGEMENTS

We thank Jacquie Northfield, Nigel Mann, and Peggy Pawley for technical assistance and Shoshana Squires and Colin Sharpe for advice on incision experiments and RNA analysis. This research was supported by the Cancer Research Campaign of which R.T.J. is a Research Fellow.

REFERENCES

Arrand, J.E., N.M. Bone, S. Squires and R.T. Johnson, 1988. Isolation of hamster DNA sequences which partially restore excision repair in xeroderma D cells, In: "Mechanisms and consequences of DNA damage processing". UCLA Symposia on molecular and cellular biology, New series, Vol. 83. E. Friedberg and P. Hanawalt eds. Alan R. Liss Inc., New York.

Arrand, J.E., S. Squires, N.M. Bone and R.T. Johnson, 1987. Restoration of UV-induced excision repair in xeroderma D cells transfected with the denV gene of bacteriophage T4. EMBO J., 6:3125.

Bohr, V.A., C.A. Smith, D.S. Okumoto and P.C. Hanawalt, 1985. DNA repair in an active gene: removal of pyrimidine dimers from the DHFR gene of CHO cells is much more efficient than in the genome overall. Cell, 40:359.

Bone N.M. and J.E. Arrand, A cloned probe for hamster Alu-equivalent repetitive sequences. Gene (in press).

Brady, G., A. Funk, J. Mattern, G. Schutz and R. Brown, 1985. Use of gene transfer and a novel cosmid rescue strategy to isolate transforming sequences. EMBO J., 4:2583.

Collins, A. and R.T. Johnson, 1987. DNA repair mutants in higher eukaryotes. J. Cell Sci. Suppl., 6:61.

Fornace, A.J., K.W. Kohn and H.E. Kann, 1976. DNA single-strand breaks during repair of UV damage in human fibroblasts and abnormalities of repair in xeroderma pigmentosum. Proc. Natl. Acad. Sci. USA, 73:39.

Giannelli,F., S.A. Pawsey and J.A. Avery, 1982. Differences in patterns of complementation of the more common groups of xeroderma pigmentosum: possible implications. Cell, 29:451.

Hoeijmakers, J.H.J., H. Odijk and A. Westerveld, 1987. Differences between rodent and human cells in the amount of integrated DNA after transfection. Exp. Cell Res., 169:111.

Johnson, R.T., S. Squires, G.C. Elliott, G.L.E. Koch and A. J. Rainbow, 1985. Xerodemna pigmentosum D-HeLa hybrids with low and high ultraviolet sensitivity associated with normal and diminished DNA repair ability, respectively. J. Cell Sci., 74:115.

Kraemer, K.H., M.M. Lee and J. Scotto, 1987. Xeroderma pigmentosum: cutaneous, ocular, and neurologic abnormalities in 830 published cases. Arch. Dermatol., 123:241.

Little, P.F.R., 1987. Choice and use of cosmid vectors, In: "DNA cloning volume III: A practical approach". D.M. Glover ed. IRL Press Oxford.

Mayne, L.V., T. Jones, S.W. Dean, S.A. Harcourt, J.E. Lowe, A. Priestley, H. Steingrimsdottir, H. Sykes, M.H.L. Green and A.R. Lehmann, 1988. SV40-transfomned normal and DNA-repair-deficient human fibroblasts can be transfected with high frequency but retain only limited amounts of integrated DNA. Gene, 66:65.

Mitchell, D.L., R.M. Humphrey, G.A. Adair, L.H. Thompson and J.M. Clarkson, 1988. Repair of (6-4) photoproducts correlates with split-dose recovery in UV-irradiated normal and hypersensitive rodent cells. Mutat. Res., 193:53.

Murray, A.M., E. Drobetsky and J.E. Arrand, 1984. Cloning the complete human adenine phosphoribosyl transferase gene. Gene, 31:233.

Protic-Sablijic, M., S. Seetharam, M.M. Seidman and K.H. Kraemer, 1986. An SV40-transformed xeroderma pigmentosum group D cell line: establishment, ultra-violet sensitivity, transfection efficiency and plasmid mutation induction. Mutat. Res., 166:287.

Royer-Pokora, B. and W.A. Haseltine, 1984. Isolation of UV-resistant revertants from a xeroderma pigmentosum group A cell line. Nature, 311:390.

Takebe, H., S. Nii, M. Ishii and H. Utsumi, 1974. Comparative studies of host cell reactivation, colony forming ability and excision repair after UV irradiation of xeroderma pigmentosum, normal human, and some other mammalian cells. Mutat. Res., 25:383.

Thompson, L.H., C.L. Mooney and K.W. Brookman, 1985. Genetic complementation between UV-sensitive CHO mutants and xeroderma pigmentosum fibroblasts. Mutat. Res., 150:423.

van Duin, M., M.H.M. Koken, J. van den Tol, P.ten Dijke, H. Odijk, A. Westerveld, D. Bootsma and J.H.J. Hoeijmakers, 1987. Genomic characterisation of the human DNA excision repair gene ERCC1. Nucleic Acids Res., 15:9195.

Weber, C.A., E.P. Salazar, S.A. Stewart and L.H. Thompson, 1988. Molecular cloning and biological characterisation of a human gene, ERCC2, which corrects the nucleotide excision repair defect in CHO UV5 cells. Mol. Cell. Biol., 8:1137.

Westerveld, A., J.H.J. Hoeijmakers, M. van Duin, J. de Wit, H. Odijk, A. Pastink, R.D. Wood and D. Bootsma, 1984. Molecular cloning of a human DNA repair gene. Nature, 310:425.

Wood, R.D., P. Robbins and T. Lindahl, 1988. Complementation of the xeroderma pigmentosum DNA repair defect in cell-free extracts. Cell, 53:97.

CHARACTERISATION OF CHINESE HAMSTER OVARY CELL MUTANTS HYPERSENSITIVE TO DNA DAMAGING AGENTS

C.N. Robson*, H.D. Lohrer, P.R. Hoban, C. Johnston, T. Robson, C. Charlton, J. Reid, A.L. Harris*, and I.D. Hickson*

Cancer Research Unit, Medical School, University of Newcastle upon Tyne, NE2 4HH, U.K.

ABSTRACT

We have isolated 13 mutants of Chinese hamster ovary (CHO) cells exhibiting hypersensitivity to DNA damaging agents. These mutants represent at least 9 different genetic complementation groups. Here, we review genetic and biochemical data available for these mutants, and describe our progress towards the cloning of the human genes which correct the phenotype of three of them (MMS-1, MMS-2 and BLM-2). We also consider the advantages of using expression cDNA libraries for gene transfer.

INTRODUCTION

The work of our laboratory has principally been directed towards elucidating the mechanisms of DNA repair in mammalian cells. Our approach, like that of several other laboratories, has centered on a study of mutants of cultured hamster cells exhibiting hypersensitivity to DNA damaging agents. We have concentrated on the use of cytotoxic drugs as selective agents, in contrast to most groups who have used UV or ionizing radiation. However, it has subsequently become clear that there is considerable overlap between the pathways controlling cellular resistance to radiation and to DNA damaging drugs. This is not altogether surprising, since the likely lethal lesions induced by both classes of DNA damaging agents are structurally similar.

Analysis of mutants is desirable for a number of reasons. For example, a clue to the *in vivo* role of a repair enzyme is often provided through such analysis. A related usage is the

*Present address: ICRF, Institute of Molecular Medicine, John Radcliffe Hospital, Oxford, OX3 904, U.K.

DNA Repair Mechanisms and Their Biological Implications in Mammalian Cells
Edited by M.W. Lambert and J. Laval
Plenum Press, New York

purification of repair proteins via the *in vitro* complementation of mutant extracts. An equally important role for mutants is to provide drug/radiation-sensitive host strains for the cloning of repair genes by DNA-mediated gene transfer. The potential for using mammalian cell mutants for this purpose is demonstrated by the dramatic progress in the last ten years that has taken place in our understanding of repair pathways in *E. coli*. Although purification of bacterial repair enzymes had been reported prior to the advent of gene cloning, progress was greatly facilitated by the availability of strains overproducing the desired gene product following the cloning of the appropriate gene. It is now routine to prepare milligram quantities of homogeneous protein, and this provides the means to carry out a wide range of biochemical and physical studies.

Although a number of human cell lines have been generated from patients suffering genetic disorders which lead to a defect in DNA metabolism (e.g., xeroderma pigmentosum and ataxia telangiectasia), these cells are not always the ideal subjects for study. In particular, these cell lines are restricted to a limited range of mutations that are compatible with human development. Human cell lines also present difficulties for gene transfer experiments, as they generally incorporate and stably maintain small quantities of transfected DNA.

To circumvent these problems, we and others have used Chinese hamster ovary (CHO) cells as the starting point for the isolation of mutants. These cells have historically been a good source of a wide range of stable mutants and are adequate recipients of transfected DNA.

We have previously identified 13 CHO cell mutants isolated on the basis of hypersensitivity to the monofunctional alkylating agent methylmethanesulfonate (designated MMS-1 to -6), the bifunctional alkylating agent mitomycin C (MMC-1 to -5), or the radiomimetic agent bleomycin (BLM-1 and -2). This paper reviews progress towards identifying and characterising the defects in these mutants.

MATERIALS AND METHODS

Cell culture

All cells were grown in Ham's F10 medium supplemented with 5% foetal calf serum, 5% newborn calf serum, glutamine (3 mM) and antibiotics (penicillin 100 U/ml, streptomycin 100 μg/ml and nystatin 50 U/ml).

Carcinogen treatment

Mitomycin C (Kyowa Hakko, Japan) and bleomycin (Lundbeck) were dissolved in water and stored at 4^0C. MMS and MNNG(Sigma) were dissolved in ethanol immediately prior to use.

Inhibition of DNA synthesis

The rate of semi-conservative DNA replication was determined by modification of the method described by Jeggo

(1985). Briefly, 2 x 10^5 cells were plated on 35 mm petri dishes and prelabeled with [^{14}C]-thymidine (16 hours, 0.01 µCi/ml). Cells were exposed to bleomycin for 30 minutes, incubated for 60 minutes in fresh growth medium, and then grown in medium containing [^{3}H]-thymidine (5 µCi/ml). After 30 minutes, cells were harvested by scraping and transferred to 10% trichloroacetic acid, 1% sodium pyrophosphate. Precipitated DNA was collected on Whatman GF/C filters, washed with trichloroacetic acid, rinsed with ethanol and air dried, before transferring to liquid scintillant. The ratio of ^{3}H to ^{14}C dpm was taken as a measure of the overall rate of DNA synthesis. Within a single experiment, each point was a mean of at least 2 separate dishes and was expressed as a percentage of parallel treated control values.

The rate of recovery of DNA synthesis was measured by following the recovery of DNA synthesis to control levels for cells treated for 30 minutes with 100 µg/ml bleomycin for recovery periods up to 7 hours.

DNA ligase assay

The method was essentially as Willis and Lindahl (1987). Briefly, approximately 5 x 10^7 cells were lysed in the presence of protease inhibitors, and their nucleic acid removed by PEI precipitation. The supernatant was then applied to an FPLC Superose-12 column (Pharmacia). The column was eluted at 0.4 ml/minute and 200 µl fractions were collected. Ligase assays were performed using 40 µl aliquots in a final reaction volume of 50 µl. A poly(dA)•oligo(dT) substrate was used to measure both DNA ligase I and II activities and a poly(rA)•oligo(dT) substrate was used to specifically measure DNA ligase II activity.

Southern hybridisation

Genomic DNA was isolated by standard methods. DNAs were digested with restriction enzymes, electrophoresed and transferred to filters by the method of Southern (1975).

DNA transfection

Foreign DNA was introduced into CHO cells using a modification of the $CaPO_4$ precipitation method of Graham and Van der Eb (1973) or the polybrene method of Kawai and Nishazawa (1984).

In the $CaPO_4$ method, cells were exposed to the precipitate for 16 hours under 3% CO_2. Following this, the cells were treated for 2-4 minutes with 27.5% DMSO, before being returned to growth medium.

In the polybrene method, cells were exposed to DNA in growth medium containing 22.5 µg/ml polybrene (Aldrich) for 6 hours. The cells were then DMSO shocked as before.

Selective medium for the bacterial gpt or neo genes contained 10 µg/ml mycophenolic acid, 25 µg/ml adenine, 250 µg/ml xanthine, 10 µg/ml thymidine (MAXT medium) or 1 mg/ml G418, respectively.

RESULTS AND DISCUSSION

Review of genetic data on CHO mutants

We have previously reported that mutants MMC-1 to -5 and BLM-1 and -2 represent 6 different genetic complementation groups (Robson and Hickson, 1986; Robson, et al., 1988), MMC-1 and MMC-5 being genetically identical. MMC-2 is our only mutant hypersensitive to UV light (Robson, et al., 1985) and is representative of UV complementation group 3, as defined by Thompson, et al.(1981). BLM-2 is our only X-ray sensitive isolate and differs genetically from the xrs series (Robson, et al., 1988), from irs-1 to -4 (J. Thacker, personal communication) and from EM9 and XR-1 (P. Jeggo, personal communication).

We have recently carried out a genetic analysis of mutants MMS-1 to - 6. This has shown that MMS-1, -2, -3 and -6 are genetically distinct (unpublished data). Mutant MMS-2, which is sensitive to DNA cross-linking agents such as mitomycin C and cis-platinum, is also distinct from MMC-1, which is unique among the MMC series in being UV-resistant, but sensitive to mitomycin C and cis-platinum. We are currently testing whether C-G11-3, an MMS sensitive CHO-9 derivative isolated by M. Zdzienicka, represents the same group as any of our mutants.

Characterisation of BLM-2

BLM-2 is greater than 10-fold hypersensitive to bleomycin and is cross-sensitive to some degree to a wide variety of agents, including UV light, X-rays, and mono- and bi-functional alkylating agents. BLM-2 shows delayed rejoining of both single- and double-strand breaks induced by bleomycin or neocarzinostatin (NCS) (Robson, et al., Mutat. Res., 1989, in press). However, far from being sensitive to NCS, this mutant is around 3-fold more resistant than the parental line. This appears to be due to a very much reduced accumulation of DNA strand breaks in BLM-2 after NCS treatment. Whether this complex phenotype is the result of a single genetic alteration is not certain.

Both bleomycin and NCS generate apurinic/apyrimidinic (AP) sites by oxidation of the sugar moiety of DNA. It has recently been shown (Povirk and Houlgrave, 1988) that virtually all NCS induced AP sites, and a proportion of those induced by bleomycin, are unusual in being associated with a closely opposed break in the complementary strand. Whether the generation or subsequent processing of these complex lesions is altered in BLM-2 cells in such a way as to lead to enhanced killing by bleomycin but protection against NCS toxicity is currently being investigated.

The X-ray and bleomycin sensitive phenotype of BLM-2 is similar to that exhibited by cell lines derived from ataxia telangiectasia (AT) patients. The similarity is somewhat stronger when it is considered that both BLM-2 and AT cells are also sensitive to VP16 (etoposide), adriamycin and H_2O_2. However, AT cells do not generally show the delayed strand break rejoining evident in BLM-2 cells. A characteristic of AT cells is their failure to inhibit DNA synthesis after radiation or chemical damage. For this reason, we studied DNA synthesis inhibition by bleomycin in BLM-2 cells. Figure 1 shows that there is a dose-dependent inhibition of DNA synthesis in parental CHO-K1 cells. Instead of being resistant to DNA synthesis inhibition by bleomycin, BLM-2 cells show a consistently greater degree of inhibition than CHO-K1 cells at any given drug dose. It should be noted that very high concentrations of bleomycin are needed to depress DNA synthesis.

The rate of recovery of DNA synthesis to control levels following removal of drug is not markedly different for these 2 cell lines (data not shown).

This drug-sensitive response of DNA synthesis in BLM-2 is consistent with the repair defective phenotype of this mutant. The xrs mutants show a similar radiosensitive response to DNA synthesis inhibition (Jeggo,1985) and are double-strand break repair defective (Kemp, et al., 1984).

Assay of DNA ligases

Because of the failure of BLM-2 cells to efficiently rejoin DNA strand breaks induced by bleomycin, we compared the activity of DNA ligase I and II in BLM-2 and CHO-K1 cells. The assays were performed by the method of Willis and

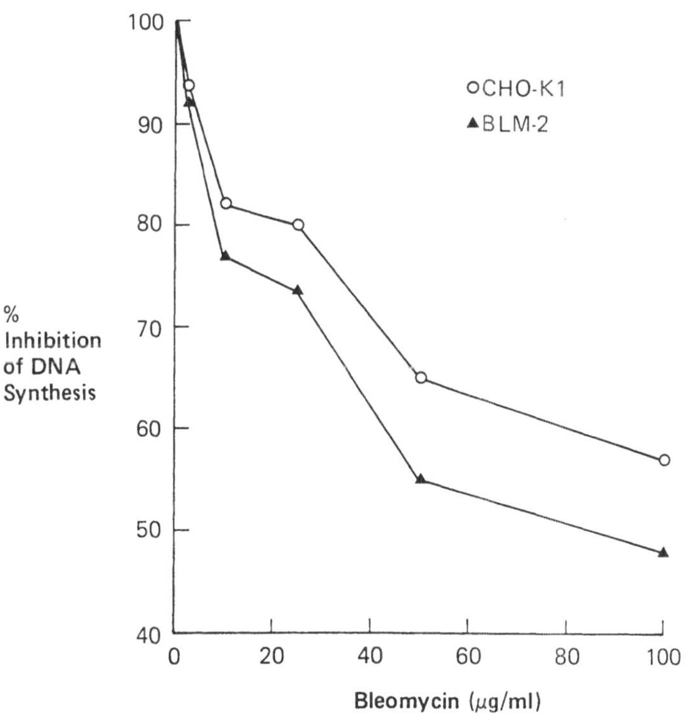

Figure 1. Inhibition of DNA synthesis in CHO-K1 (O) and BLM-2 (▲) cells as a function of the dose of bleomycin. Points represent the mean of at least 2 independent experiments.

Lindahl (1987), using both the poly(dA)·oligo(dT) substrate, which detects both ligase activities, and the poly(rA)·oligo-(dT) substrate, which is specific for DNA ligase II.

A single peak of ligase II activity was reproducibly observed around fraction 20 of the Superose 12 FPLC run at the position in the OD_{280} profile where Willis and Lindahl observed ligase II from human cell lines. This represents an M_r of around 80,000. The level of ligase II activity in BLM-2 and CHO-K1 cells was always very similar in repeated runs (data not shown), indicating that an abnormality in ligase II is unlikely to exist in BLM-2 cells.

The detection of ligase I activity was somewhat less reproducible in our hands, and we observed some variation between experiments. Generally, 2 peaks of activity were seen which failed to show activity with the poly(rA)·oligo(dT) substrate and were thus taken as representing ligase I. Willis and Lindahl found only a single peak of ligase I activity in several human lines. Both of the peaks that we observed represented higher molecular weight species than ligase II. Whether proteolytic degradation of ligase I is responsible for the appearance of 2 activity peaks requires further study. No obvious differences between CHO-K1 and BLM-2 cells were found, although further work will be required to confirm that DNA ligase I is not abnormal in BLM-2 cells.

Several mutants of CHO cells with a defect in DNA strand break repair have been assayed for DNA ligases, but to date no evidence for abnormal DNA ligase I or II activity has been presented (Chan, et al., 1984; Stamato and Hu, 1987). Cells derived from Blooms syndrome patients show an abnormality in DNA ligase I (Willis and Lindahl, 1987), although the cells are only mildly, if at all, sensitive to DNA damaging agents.

Progress towards cloning the human gene correcting the defect in BLM-2, MMS-1 and MMS-2

The main purpose of our isolating mutants was for cloning human genes controlling cellular resistance to chemical damage. Several approaches are under way, using both genomic DNA and an expression cDNA library.

(a) BLM-2

Most progress has been achieved with this mutant. Our initial approach was to transfect BL4-2 cells with a human genomic DNA library constructed in the cosmid pNNL (Grosveld, et al., 1982). This cosmid carries the *gpt* gene for use as a selectable marker (conferring resistance to mycophenolic acid) in CHO cells. The rationale behind using a cosmid library was that a complementing gene would necessarily be of a size suitable for subsequent rescue by cosmid cloning. Also, we considered the possibility that the *cos* sites of the vector may allow rescue by direct *in vitro* packaging of DNA from transfected cells. This possibility has not in practice proved to be viable in our hands, possibly due to the prevalence of DNA rearrangements in transfected DNA.

Transfection of BLM-2 with the cosmid library DNA generally yielded colonies resistant to mycophenolic acid ($MAXT^r$) at a frequency of 3×10^{-4} (somewhat lower than that achievable with CHO-K1 cells). Following MAXT selection, colonies were trypsinized and given three cycles of treatment with bleomycin (1.5 μg/ml for 24 hours, with 24 hours between treatments). Four colonies, out of over 50 that survived this treatment, were chosen and their drug sensitivity tested. All had near wild-type resistance to bleomycin. Southern analysis showed that these transformants contained integrated gpt and human sequences (using BLUR11 "Alu" probe, Deininger, et al., 1981). Figure 2 shows DNA from 2 of these transformants probed with BLUR11, together with a control of CHO-K1 DNA. Hybridisation specific to the Alu probe is evident in tracks 3-6 which are absent from the controls.

The DNA from one of these primary transformants was used to transfect BLM-2 in a second round of selection. Two colonies surviving the MAXT and bleomycin selection protocol were taken. Southern analysis showed that these isolates contained human DNA sequences (not shown).

Figure 3 shows that one of these isolates, designated BLM-2/2^0, has a level of resistance to bleomycin comparable with that of wild-type cells, although it is consistently slightly more sensitive than the control cells.

DNA from the secondary transfectant was used for a third round of transfection into BLM-2. This time, a single colony survived the bleomycin selection (continuous exposure to 0.25 μg/ml bleomycin, no MAXT). This transformant, designated BLM-2/3^0, was found to have a level of stable bleomycin resistance intermediate between that of BLM-2 and CHO-K1 cells, being 4-fold more resistant than BLM-2 but around

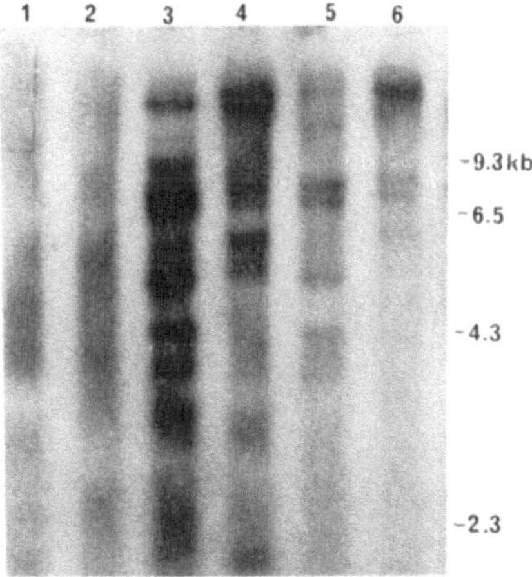

Figure 2. Southern blot analysis of DNA from CHO-K1 cells (tracks 1 and 2) or from 2 bleomycin resistant transfectants (tracks 3-6) probed with the human 'Alu' sequence, BLUR 11. DNA in tracks 1,3 and 5 was digested with EcoRI and that in tracks 2,4 and 6 with BamHI.

3-fold more sensitive than CHO-K1 (Figure 3). This transfectant does not express resistance to MAXT selection and thus co-transfer of gpt and human sequences has obviously not been achieved.

DNA isolated from primary, secondary and tertiary transformants is currently being screened for the presence of common human DNA sequences. DNA from the secondary or the tertiary transformant will be used to create a gene library in a cosmid vector. This library will be screened for cosmids containing human DNA sequences. Such cosmids will be transfected into BLM-2 cells to test for the ability to complement the repair defect in this cell line.

(b) MMS-2

Our strategy for isolating the functional analogue of the gene which is defective in MMS-2 cells has been to complement the mutant by transfer of human genomic DNA and selecting for essentially wild-type resistance to mitomycin C. High molecular weight human DNA, together with the dominant marker plasmid pSV2neo, was co-transfected into the mutant, followed by two sequential selection steps. The neo marker was used

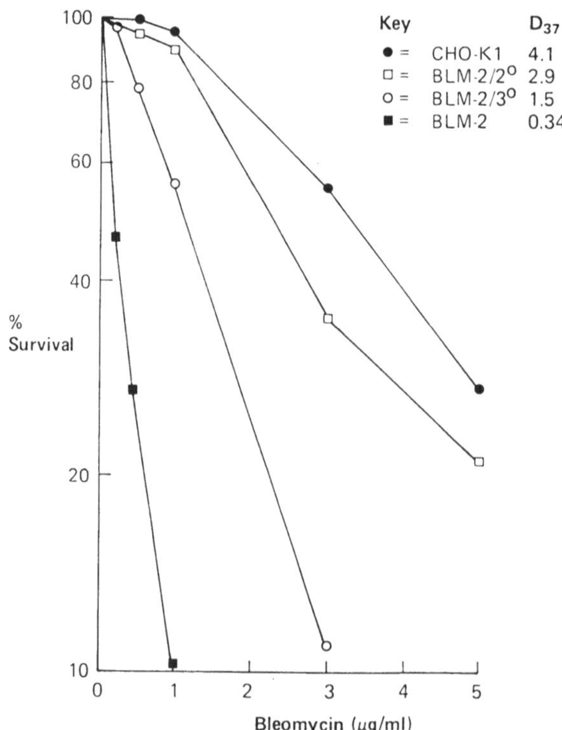

Figure 3. Survival of parental CHO-K1, mutant BLM-2 cells, and the transfectants BLM-2/2^0 and BLM-2/3^0 following a 24 hour exposure to bleomycin. The drug dose reducing survival to 37% of control levels (D_{37}) is given for each cell line.

to select for cells which had taken up "foreign" DNA sequences, followed by repeated treatment with mitomycin C.

Critical parameters of the procedure are: (i) the survival during mitomycin C selection of transfectants whose resistance to the drug has been reverted to wild-type, (ii) the spontaneous reversion rate of MMS-2 cells, and (iii) the efficiency of uptake of "foreign" DNA sequences by the mutant. By all of these criteria, MMS-2 was considered to be a suitable host strain.

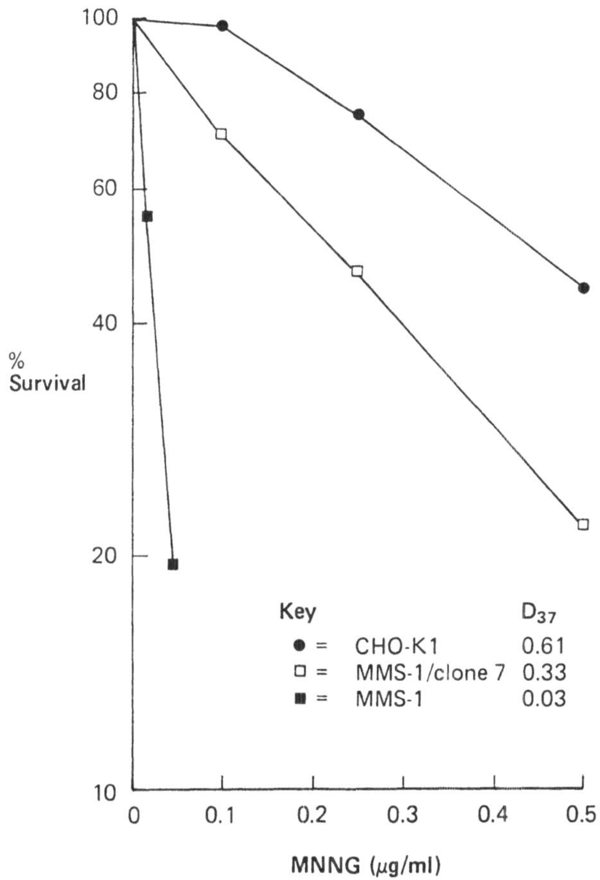

Figure 4. Survival of parental CHO-Kl cells, mutant MMS-1 cells, and the G418-resistant transfectant MMS-l/ clone 7, following a 30 minute exposure to MNNG.

DNA-mediated gene transfer into the mutant MMS-2 using co-transfection of human placental and pSV2neo DNA yielded more than 350,000 G418 resistant clones. These have been screened for resistance to mitomycin C and 9 resistant clones taken for further study. These clones proved in Southern blot experiments to have inherited human repetitive sequences and plasmid sequences (data not shown). Secondary rounds of transfection selecting for the rescue of the neo gene have generated mitomycin C resistant isolates, but as yet the presence of human sequences has not been shown.

Use of pCD2 neo expression cDNA library for DNA mediated gene transfer

We have used the pCD2 neo human expression cDNA library (from H. Okayama) to transfect selected mutants. Transfections were carried out essentially using the modified $CaPO_4$ protocol of Chen and Okayama (1987).

MMS-1

This mutant is around 3-fold hypersensitive to MMS and 10-fold to MNNG. Following transfection with DNA from the pCD2 neo library, cells were exposed to G418, trypsinised, and exposed either to 3 or 4 cycles of MNNG treatment (0.5 µg/ml) at 24 hour intervals. Individual colonies were picked, expanded into bulk cultures and their survival response to MMS and MNNG determined.

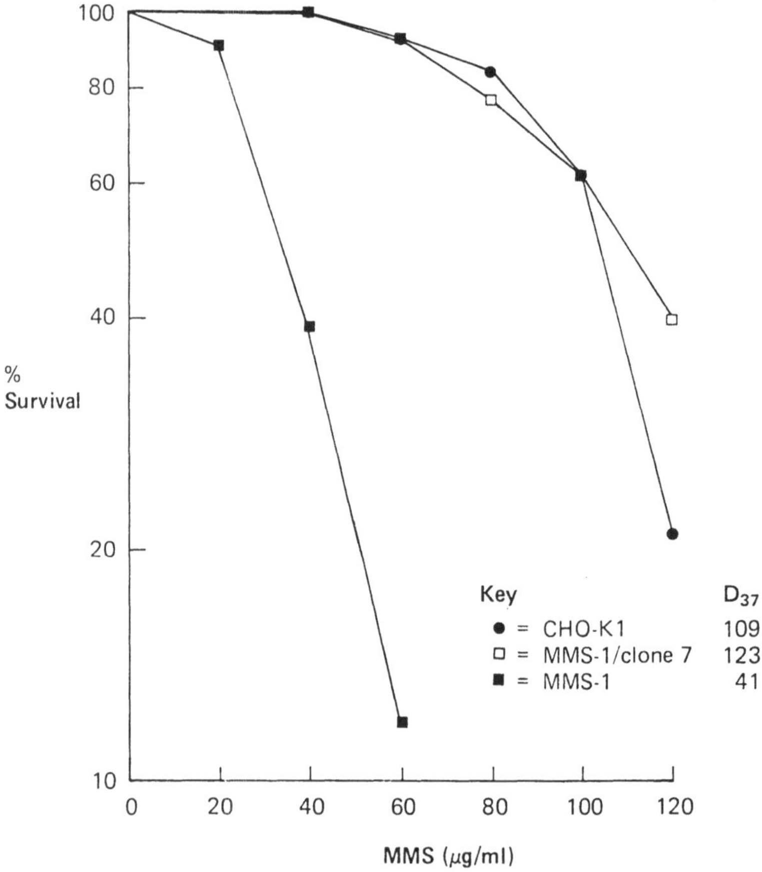

Figure 5. Survival of parental CHO-K1 cells, mutant MMS-1 cells, and the G418-resistant transfectant MMS-1/ clone 7, following a 30 minute exposure to MMS.

From more than 220,000 G418 resistant colonies, 12 clones have been isolated which show resistance to both MMS and MNNG. DNAs from these individual clones have been isolated and are currently being screened for the presence of human and neo sequences.

The survival characteristics of one G418^r transformant (designated MMS-1/clone 7) are shown in Figures 4 and 5. Clone 7 shows near wild-type resistance to both MMS and MNNG and is stably resistant to G418 and MNNG in the absence of selection. We are currently quantifying the approximate copy number of integrated plasmids and estimating the level of stably integrated human DNA among a number of different drug-resistant isolates, including clone 7.

Although exposure of cells to MNNG can induce resistance by a variety of mechanisms, such as elevation in O^6-methylguanine DNA methyl transferase activity, the majority of the MNNG^r isolates reported to date show only low levels of cross-resistance to MMS. We are therefore concentrating our effort on those transfectants in which the MMS-sensitive phenotype has been completely reversed.

The advantages of the cDNA library approach over transfection with genomic DNA are numerous. A very high transfection frequency can be achieved using supercoiled plasmid DNA and the Chen and Okayama $CaPO_4$ protocol. The size of the transferred DNA is small and therefore much more readily rescued from transformants. This allows phage as well as cosmid libraries to be employed. The selectable marker, neo, is physically very close to the cDNA in each plasmid and can therefore be readily co-transferred if secondary transfections are undertaken. This advantage also means that neo would be the marker of choice for probing gene banks, eliminating the necessity of probing for human sequences. It may be possible to rescue plasmids from primary transformants if the number of integration events is sufficiently low to make isolating most or all of the plasmids feasible. Finally, once cloned, an expressing version of the gene of interest, with guaranteed biological activity, is available, broadening the range of studies that can be undertaken.

REFERENCES

Chan, J.Y.H., Thompson, L.H., and Becker, F.F., 1984, DNA-ligase activities appear normal in the CHO mutant EM9. Mutat. Res., 131:209.

Chen, C., and Okayama, H., 1987, High-efficiency transformation of mammalian cells by plasmid DNA. Mol. Cell. Biol., 7:2745.

Deininger, P.L., Jolly, D.J., Rubin, C.M., Friedmann, T., and Schmid, C.W., 1981, Base sequence studies of 300 nucleotide renatured repeated human DNA clones. J. Mol. Biol., 151:17.

Grosveld, F.G., Lund, T., Murray, E.J., Mellor, A.L., Dahl, H.H.M., and Flavell, R.A., 1982, The construction of cosmid libraries which can be used to transform eukaryotic cells. Nucl. Acids Res., 10:6715.

Jeggo, P.A., 1985, X-ray sensitive mutants of Chinese hamster ovary cell line: radio-sensitivity of DNA synthesis. Mutat. Res., 145:171.

Kawai, S. and Nishizawa, M., 1984, New procedure for DNA transfection with polycation and dimethyl sulfoxide. Mol. Cell. Biol., 4:1172.

Kemp, L.M., Sedgwick, S.G., and Jeggo, P.A., 1984, X-ray sensitive mutants of Chinese hamster ovary cells defective in double-strand break rejoining. Mutat. Res., 132:189.

Povirk, L. E., and Houlgrave, C.W., 1988, Effect of apurinic/apyrimidinic endonucleases and polyamines on DNA treated with Bleomycin and Neocarzinostatin: specific formation and cleavage of closely opposed lesions in complementary strands. Biochemistry, 27:3850.

Robson, C.N., Harris, A.L., and Hickson, I.D., 1985, Isolation and characterization of Chinese hamster ovary cell lines sensitive to Mitomycin C and Bleomycin. Cancer Res., 45:5304.

Robson, C.N., and Hickson, I.D., 1986, Genetic analysis of mitomycin C-sensitive mutants of a Chinese hamster ovary cell line. Mutat. Res., 163:201.

Robson, C.N., Hall, A., Harris, A.L., and Hickson, I.D., 1988, Bleomycin and X-ray-hypersensitive Chinese hamster ovary cell mutants: genetic analysis and cross-resistance to neocarzinostatin. Mutat. Res., 193:157.

Southern, E.M., 1975, Detection of specific sequences among DNA fragments separated by gel electrophoresis. J. Mol. Biol., 98:503.

Stamato, T.D., and Hu, J., 1987, Normal DNA ligase activity in a -ray- sensitive Chinese hamster mutant. Mutat. Res., 183:61.

Thompson, L.H., Busch, D.B., Brookman, K., Mooney, C.L. and Glaser, D.A., 1981, Genetic diversity of UV sensitive DNA repair mutants of CHO cells. Proc. Natl. Acad. Sci. USA, 78:3734.

Willis, A.E., and Lindahl, T., 1987, DNA ligase I deficiency in Bloom's syndrome. Nature, 325:355.

ONCOGENE ACTIVATION IN XERODERMA PIGMENTOSUM SKIN TUMORS

Leela Daya-Grosjean[1], Alice de Miranda[1], Horacio Suarez[1], Bertrand Chretien[2], Marie-Françoise Avril[3] and Alain Sarasin[1]

[1]Laboratoire de Génétique Moléculaire, Institut de Recherches Scientifiques sur le Cancer, B.P. no. 8-94802, Villejuif Cedex, France

[2]Clinique Chirurgicale Infantile, Hôpital Necker Enfants Malades, 149, rue de Sèvres, 75015 Paris, France

[3]Service de Dermatologie, Institut Gustave Roussy Rue Camille Desmoulins, 94805, Villejuif Cedex France

There is growing evidence linking oncogene activation to the induction and/or maintenance of carcinogenesis (Knudson, 1986). Data from animal models and cell lines have shown that oncogene activation can occur by four different mechanisms: retrovirus insertion near a proto-oncogene; chromosomal translocations; gene amplification or deletion; and point mutations. In about 20% of the most common forms of human tumors, mutations have been found most often in the proto-oncogenes of the ras family, assayed by the NIH 3T3 transformation system (Barbacid, 1987). Activation is acquired by single point mutations in two domains of the ras protein, p21, (mainly codons 12 and 61) greatly reducing its enzymatic activity in the modulation of signal transduction through transmembrane signalling systems (Barbacid, 1987).

Animal model systems have shown that chemical carcinogens induce tumors at high frequency. The reproducible activation of ras oncogenes in these tumors has made it possible to correlate the activating mutations with the known effects of certain carcinogens. For example, alkylating agents (some chemical carcinogens or anti-tumor drugs) can give rise to O^6-methyl-guanine adducts in the DNA resulting in mutations during replication procedures. In fact, because of the limited fidelity of repair by DNA polymerases, the O^6-Me-G residues can be read as adenine, leading to the frequent generation of G —> A transitions. This was confirmed in a study by Barbacid where it was shown that 100% of the N-nitroso-N-methyl urea (NMU) induced mammary carcinomas of the rat were due to activation of the Ha-ras gene by a point mutation (G —> A

DNA Repair Mechanisms and Their Biological Implications in Mammalian Cells
Edited by M.W. Lambert and J. Laval
Plenum Press, New York

transition) in the second nucleotide of codon 12 of the p21 protein (Barbacid, 1987). Furthermore, the spectra of mutations found in the activated oncogenes in these animal tumor studies correlate well with those found in _in vitro_ studies in bacteria or mammalian cells with the corresponding carcinogens. It would, therefore, be particularly useful to establish a similar causal correlation between proto-oncogene activation and initiation or progression of human cancer.

The repair-deficient human syndrome xeroderma pigmentosum represents one of the best characterized examples of the relationship between exposure to DNA damaging agents and tumorigenesis. In this rare autosomal recessive disorder, an extreme sensitivity of the skin to UV light leads to a high incidence of cutaneous tumors in areas exposed to sunlight (Cleaver, 1968; Hanawalt and Sarasin, 1986). Tumorigenesis is attributed to the complete or partial absence of repair in UV damaged DNA, particularly pyrimidine dimers and pyrimidine (6-4) pyrimidones. The persisting damage in XP DNA acts as a premutagenic lesion which results in point mutations opposite pyrimidines damaged during replication of the irradiated template (Figure 1).

We have analyzed DNA from 8 epithelial tumors (basal or squamous cell carcinomas) obtained from young XP patients using the NIH 3T3 transformation system. DNAs of two independent tumors from the same XP child (BY) were able to induce foci in the 3T3 cells due to the transforming activity of the N-ras gene (Figure 2A) (Suarez, _et al_., 1987). Using the polymerase chain reaction technique for gene amplification and differential hybridization with synthetic oligonucleotide probes (Bos, _et al._, 1986), we were able to determine the presence of a critical point mutation in codon 61 of the N-ras gene (Figure 2B) (Suarez, _et al._, 1988). This results in the

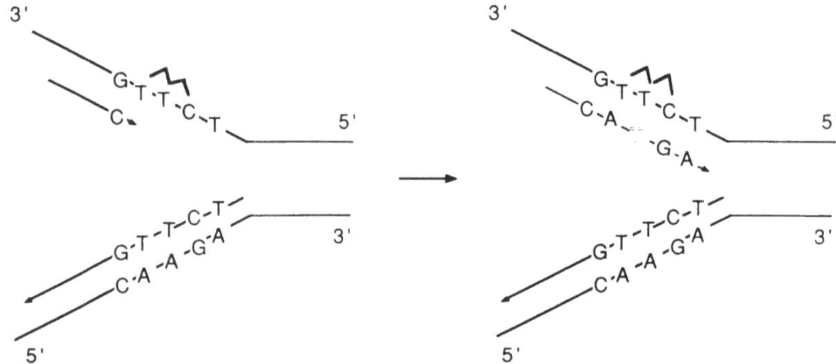

Figure 1. A model for the formation of point mutations during replication. Replication is normally blocked at UV-induced lesions between adjacent pyrimidines: pyrimidine-pyrimidine dimer or pyrimidine (6-4) pyrimidine. When replication eventually proceeds through such lesions, mismatch errors occur resulting in point mutations opposite the lesion. The sequences shown in the model corresponds to those around codon 61 of the N-ras gene (see Figure 3).

substitution of a glutamine by a histidine residue which is sufficient to activate the N-ras gene (Figure 3).

These results strongly suggest that UV-irradiation of the epithelial cells in the exposed skin results in unrepaired lesions at pyrimidines 182-183 or 183-184 of the N-ras gene. Replication errors during the repair process can then result in the incorporation of thymine instead of adenine opposite

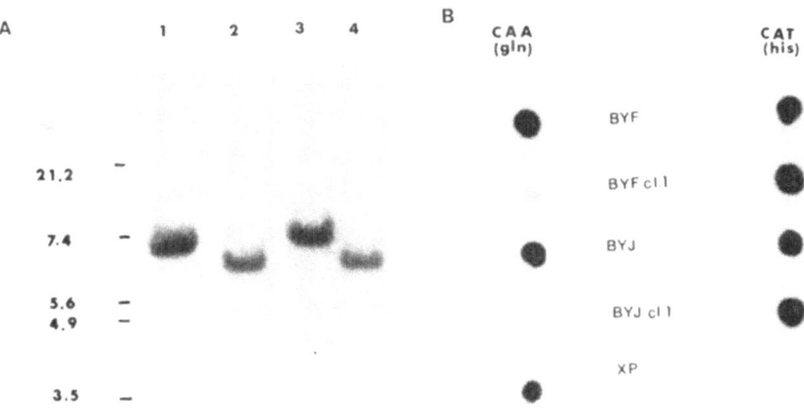

Figure 2. Activated N-ras in XP tumors.
A.) A southern blot analysis of DNA from XP tumors and mouse 3T3 transfectants. 20 μg of high molecular weight DNAs were digested with EcoRI, the fragments separated by electrophoresis on agarose gels and transferred to nitrocellulose filters. A [^{32}P] labelled oncogene probe specific for N-ras was used for hybridization. Lanes 1,3: DNA from XP tumors BYF and BYJ; lanes 2,4: DNA from primary transfectants BYF cl1, and BYJ cl1, corresponding to 3T3 cells transformed by the activated N-ras gene from the XP tumors. The position and molecular weights in Kbp of λ EcoRI fragments are indicated on the left of the figure.

B.) Detection of point mutations by hybridization with specific oligonucleotides. Synthetic oligonucleotides (20 mer) specific for the wild type (CAA) or mutated (CAT) codon 61 of the N-ras gene were used for detecting the mutation responsible for activation in XP tumors. Cellular DNAs were amplified by the polymerase chain reaction (30 cycles), dot-blotted to nitrocellulose filters and hybridized to [^{32}P] labelled probes (Bos, <u>et al.</u>, 1986). The results show that the point mutation revealed in the 3T3 transformants (BYF cl1 and BYJ cl1) is an A —> T transition. The original XP tumors (BYF and BYJ) contain the normal wild type allele (CAA) together with the mutated gene in codon 61 (CAT) and the control untransformed (XP) cells only contain the wild type alleles (CAA).

the thymine involved is the lesion (Figures 1 and 3). This type of mutation has been frequently observed both in vitro and in vivo using probes such as animal viruses (SV40) or shuttle vectors after UV-irradiation (Bourre and Sarasin, 1983; Bredberg, et al., 1986). The activated form of the N-ras gene is not found in the non-transformed fibroblasts derived from the skin of the same patient indicating the predisposition to form tumors is not by a hereditary transmission of the mutated N-ras gene in XP patients.

Human N-ras : codon 61

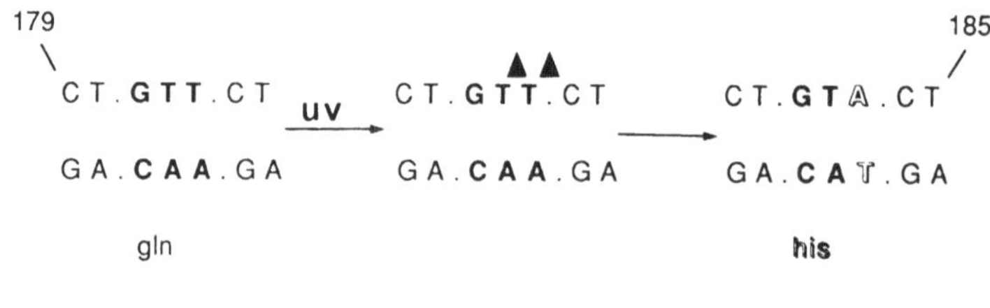

Figure 3. A model for proto-oncogene activation by point mutation. Irradiation by UV light results in the formation of pyrimidine dimers or pyrimidine(6-4)-pyrimidones (TT or TC). In XP cells this type of lesion is not repaired and replication synthesis of the damaged DNA results in a point mutation opposite the lesion. When this occurs in a crucial codon of a proto-oncogene, such as codon 61 of the N-ras gene, the modification of the genetic code results in activation of the gene. Numbers correspond to the nucleotides in the N-ras coding sequence.

Interesting results have also been obtained by Van der Lubbe, et al. (1988), using UV-irradiated cloned cellular N-ras in the NIH 3T3 transformation assay. Analysis of the foci showed that the majority of the transformed cell lines obtained were due to the activation of the N-ras gene in codon 61 at a possible thymine dimer. It was concluded, as in the XP tumors, that lesions induced on the short pyrimidine stretch 182-183-184 resulted in mutation of one of the thymines in codon 61. Specific elimination of the pyrimidine dimers in vitro by UV-endonucleases suggests that this mutagenesis is effectively due to pyrimidine dimers and not the pyrimidine(6-4)pyrimidone photoproduct (Van der Lubbe, et al., 1988).

Hence, it can be summarized that the XP syndrome provides data in a human cancer model comparable to that obtained using carcinogens in animal studies, where the precise mechanism of oncogene activation by a point mutation is correlated with a specific type of adduct on the cellular DNA.

However, it is known that a point mutation activation of a ras gene is not by itself sufficient for tumor development (Barbacid, 1987). For example, some benign skin papillomas induced in animals contain activated ras genes (often Ha-ras) but the majority of these tumors regress and never give rise to cancers (Balmain, et al., 1984). Secondary genetic events therefore seem necessary for the tumor progression after initiation. With this in mind, we have screened the XP epithelial tumors for the existence of other genetic modifications, particularly of other oncogenes. Indeed, the majority of the tumors analyzed exhibited amplification of the Ha-ras and/or myc proto-oncogenes. In particular, the Ha-ras gene was amplified and often rearranged in over 40% of the XP tumors (Figure 4) at a level significantly higher than that seen (1%) in human tumors in general. It would therefore be interesting to determine the relationship between the high level of amplification seen in the XP tumors and the type of lesion responsible for the mutagenesis. It has already been shown that lesions which block DNA replication result in amplification of non-specific DNA sequences, probably due to multiple cycles of reinitiation (Lavi, 1981). Such abortive cycles of replication are frequently induced by pyrimidine dimers (UV-induced lesions) and could therefore constitute the initiating event necessary for human tumor progression. Hence, if a non-repaired DNA lesion results in the random amplification of a gene such as myc or Ha-ras, the overexpression of the corresponding protein could lead to a selective cell proliferation. If the same UV-induced lesions had resulted in a mutated second oncogene (e.g., N-ras), the actively proliferating cell would then be transformed into a tumor cell.

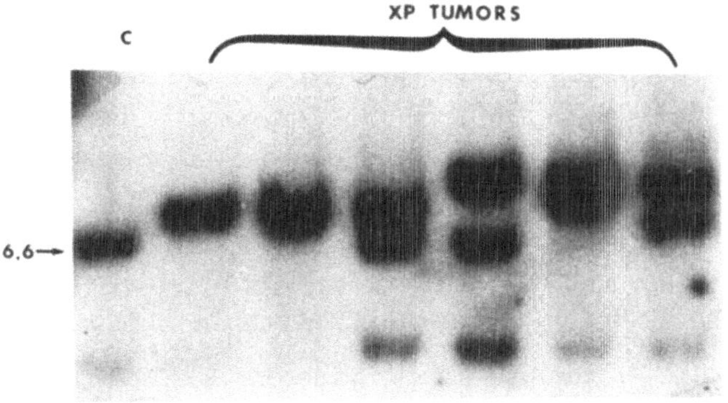

Figure 4. Ha-ras amplification in XP tumors.
Cellular DNA from XP tumors were analyzed after Bam H1 digestion, by the Southern technique using a [^{32}P] labelled Ha-ras probe (pEJ). Amplification and rearrangement are seen in the XP tumor DNAs when compared with the normal fragment (6.6 Kbp) seen in untransformed XP fibroblasts (C).

The data obtained from the XP tumors shows that the hypotheses put forward from the animal studies are applicable to human carcinogenesis. Furthermore, the knowledge of the precise causal agent in tumor development allows a correlation between the primary lesion and the genetic modification involved in neoplasia. Obviously, for most human cancers the primary element generating the deletory genetic modification is not known. One can imagine specific lesions created by physical or chemical carcinogens (ionizing radiations, polycyclic hydrocarbons, nitrosamines, alkylating agents, asbestos or mycotoxins); cryptic lesions arising from spontaneous modifications of bases (deamination of cytosine or Me-cytosine) or errors produced during replication or repair of DNA. As these modifications exhibit a wide spectra of effects it is not surprising, when considering "spontaneous" human tumors, that it is not possible to establish a direct correlation between a specific proto-oncogene modification, the type of cancer and its anatomical localization. A better understanding of the initiating events and mechanisms leading to proto-oncogene activation is therefore essential to improve cancer therapy and for cancer prevention.

REFERENCES

Balmain, A., Ramsden, M., Bowden, G.T. and Smith, J., 1984, Activation of the mouse cellular Harvey-ras gene in chemically induced benign skin papillomas. Nature, 307:658.

Barbacid, M., 1987, Ras genes. Ann. Rev. Biochem., 56:779.

Bos, J.L., Verlaan-de-Vries, M., Marshall, C.J., Veeneman, G.H., Van Boom J.H., and Van der Eb., A., 1986, Human gastric carcinoma contains a single mutated and an amplified normal allele of the Ki-ras oncogene. Nucl. Acids Res., 14:1209.

Bourre, F., and Sarasin, A., 1983, Targeted mutagenesis of SV40 DNA induced by UV light. Nature, 305:68.

Bredberg, A., Kraemer, K.H., and Seidman, M.M., 1986, Restricted ultraviolet mutational spectrum in a shuttle vector propagated in xeroderma pigmentosum cells. Proc. Natl. Acad. Sci. USA, 83:8273.

Cleaver, J.E., 1968, Defective repair replication of DNA in xeroderma pigmentosum. Nature, 218:652.

Hanawalt, P.C., and Sarasin, A., 1986, Cancer prone hereditary diseases with DNA processing abnormalities. Trends in Genetics, 2:124.

Knudson, A.G., 1986, Genetics of human cancer. Ann. Rev. Genetic, 20:231.

Lavi, S., 1981, Carcinogen-mediated amplification of viral DNA sequences in simian virus 40-transformed Chinese hamster embryo cells. Proc. Natl. Acad. Sci. USA, 78:6144.

Suarez, H.G., Nardeux, P.C., Andeol, Y., and Sarasin, A., 1987, Multiple activated oncogenes in human tumors. Oncogene Res., 1:201.

Suarez, H.G., Daya-Grosjean, L., Schlaifer, D., Nardeux, P., Renault, G., Bos, J.L., and Sarasin, A., 1988, Activated oncogenes in skin tumors from a repair deficient syndrome xeroderma pigmentosum. Cancer Res., 49:1223.

Van der Lubbe, J.L.M., Rosdorff, H.J.M., Bos, J.L., and Van der Eb., A.J., 1988, Activation of N-ras by ultraviolet irradiation in vitro. Oncogene Res., 3:9.

HETEROGENEITY IN THE O^6 ALKYLGUANINE DNA ALKYLTRANSFERASE (AGT) ACTIVITY OF HUMAN PERIPHERAL BLOOD LYMPHOCYTES (PBL's)

Bernard Strauss, Daphna Sagher, Jeffrey Schwartz, Theodore Karrison, and Richard Larson

Departments of Molecular Genetics and Cell Biology, Radiation and Cellular Oncology and Medicine, The University of Chicago, Illinois, U.S.A.

The nature of the mechanisms controlling the protein content of mammalian cells and tissues is a central problem of modern biology. Recent studies on the structure of genes and their control elements in eukaryotes indicate the early paradigms based on the _lac_ operon of _E. coli_ (complex as they are) to be much too simple (Watson, _et al._, 1987). It is likely that numerous factors are involved in the determination of the level of activity of any protein, particularly proteins which are developmentally regulated so that their activity differs in different tissues. The repair protein, O^6-alkylguanine DNA alkyltransferase (AGT) is an example of a developmentally regulated protein since it has widely different activities in brain, liver and other tissues (Pegg, 1983). In addition, it has been reported that the peripheral blood lymphocytes (PBL's) from different individuals may vary widely in their AGT activity (Waldstein, _et al._, 1982). In this paper, we report that average AGT activity is characteristic of individuals and of their cells but that the PBL's constitute a heterogeneous population of cells of differing inherent AGT activity. Furthermore, we show that cells may vary in their response to the mutagen N-methyl-N'-nitro-N-nitrosoguanidine by a mechanism which need not involve the AGT protein.

Our interest in this problem is the result of a hypothesis to account for the origin of therapy related acute nonlymphocytic leukemia (t-ANLL). Approximately ten percent of individuals treated with chemotherapy for Hodgkin's disease develop a secondary malignancy. The risk of secondary leukemia is 3.3 percent (Tucker, _et al._, 1988). We asked whether this ten percent is the result of stochastic mutational events or whether there is a subpopulation of patients with inherently greater sensitivity to the chemotherapeutic agents used. We noted that two of the agents, procarbazine and dacarbazine, used in common "cocktails" for the treatment of Hodgkin's disease, are metabolized to a methyldiazonium ion

DNA Repair Mechanisms and Their Biological Implications in Mammalian Cells
Edited by M.W. Lambert and J. Laval
Plenum Press, New York

which reacts with DNA to form O^6-methylguanine (Wiestler, *et al.*, 1984). Since individuals vary in their AGT level as measured in PBL's (Waldstein, *et al.*, 1982), it seemed possible that differences in AGT level could lead to differences in sensitivity to chemotherapy: individuals with inherently low AGT activity would fail to remove O^6-alkylguanine adducts produced during chemotherapy and would be more likely to develop a secondary malignancy. A test of this hypothesis has been published (Sagher, *et al.*, 1988). We did observe a statistically significant lower level of AGT activity in the PBL's of t-ANLL patients as compared with ANLL *de novo* (Table 1) which is what would have been expected. However, there are a number of complicating factors. Untreated individuals with Hodgkin's and non-Hodgkin's lymphoma have lower values of AGT than normal controls. In addition, patients receiving procarbazine do not display an immediate dramatic lowering in their AGT activity as might be expected if the hypothesis were correct. Rather, after several courses of treatment including procarbazine, some, but not all, patients display a slow drift downwards in their AGT level. Furthermore, all these studies were carried out with PBL's as a readily available and representative cell type, but it is in the bone marrow where the critical events occur. We consider the hypothesis as "not proven" and likely to be only one of a number of possible factors in the etiology of secondary leukemia since individuals who have received treatments not including procarbazine may also develop t-ANLL.

The question as to the source of the variability of AGT activity in different individuals and in different cells of the same individual remains open. The data we have support the view that the PBL levels of AGT vary from individual to individual, but that the PBL's for a given individual constitute a population with a heterogenous level of activity which remains fixed under *Epstein Barr virus* (EBV)-induced transformation. In all our experiments we assay AGT activity by measuring the transfer of acid labile radioactivity from DNA to acid stable precipitable radioactivity in protein. The assay is simple and reproducible (Figure 1).

We sampled a group of 5 normal subjects three times over a period of one week to determine short term variability and then again 8 months later to estimate long term variability. Each sample was divided into two aliquots and coded before analysis. Three of these subjects were assayed a fifth time, 13 months after the original determination (Figure 2). The recent determinations confirm our finding that AGT activity as measured in PBL's is a characteristic property of individuals and remains relativily stable over long periods. Estimates of the three components of variability (variability between subjects, between repeated samples from the same subject and between different aliquots from the same sample) were 2.5, 1.0 and 0.4, respectively, i.e., 64 %, 25%, and 12% of the total variation. We have no idea as to whether these levels are familial.

As part of the program of study of t-ANLL at the University of Chicago, we prepare lymphoblastoid lines by Epstein Barr virus (EBV) transformation of PBL's from patients and control individuals. These lines are designed to serve as normal DNA controls for comparison with DNA from leukemia cells. Given the material at hand, we decided to investigate

Table 1. O^6-Alkylguanine DNA Alkyltransferase (fmol/μg DNA) in PBL's.

	GROUP	N	MEAN	± S.E.
(1)	A-ANLL *de novo*	6	7.78	1.72
	B-ANLL in remission	12*	6.90	1.07
(2)	A-t-ANLL	11	4.30	0.58
	B-t-ANLL in remission	2	4.35	2.45
(3)	HD/NHL Untreated	25	4.97	0.42
	HD "	14	5.59	0.53
	NHL "	11	4.19	0.61
(4)	HD/NHL-P	17	3.88	0.44
	HD-P	14	4.03	0.52
	NHL-P	3	3.13	0.58
	HD-R	6	4.40	1.07
	HD/NHL-O	26	5.65	0.72
	HD-O	11	6.09	1.08
	NHL-O	15	5.33	0.99
(5)	HD/NHL in remission	75	5.51	0.33
	HD "	51	5.97	0.40
	NHL "	24	4.53	0.56
(6)	Controls	34	7.05	0.36

1. ANLL *de novo*; Lymphocytes are from patients with less than 10% blasts in the circulating blood; 2. Therapy related ANLL; 3. Previously untreated Hodgkin's patients (HD); previously untreated Non Hodgkin's lymphoma (NHL); 4. HD, NHL patients receiving therapy: P, receiving MOPP and/or ABVD; R, receiving radiotherapy; O, receiving other chemotherapy; 5. HD, NHL patients in remission, > 3 months off therapy; 6. Normal controls; Values for each individual were first averaged over all samples taken. N is the number of individuals whose average value was used to calculate the group mean.

*One outlying value of 29.9 excluded. If this value is included, the mean and SE increase to 8.67 ± 2.02 [from Sagher, *et al.*, 1988].

the relationship between the AGT activity of lymphocytes and of the lymphoblastoid lines derived from such cells. Although the PBL population is mainly T lymphocytes, whereas EBV transformed lines are derived from B cells, we thought this a reasonable comparison because of the observation of Gerson, *et al.* (1985), that the AGT activity of B and T cells is comparable, with the B cells showing slightly lower activity. We also decided to investigate the AGT activity of different lines derived from the same individual since it had previously been reported from this laboratory that high and low AGT lines were obtained from just two samples from the same individual

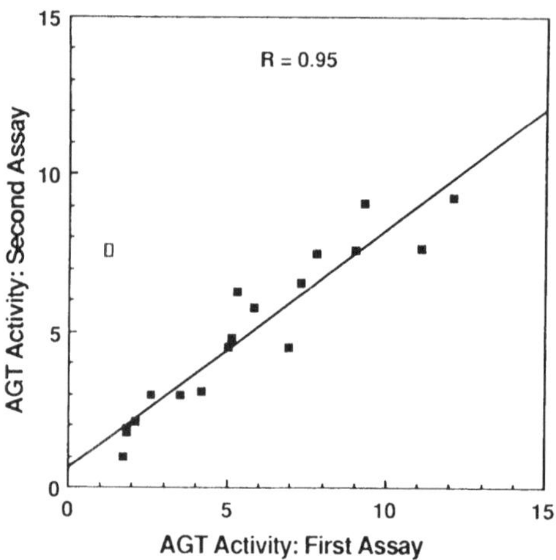

Figure 1. Variability of the alkylguanine transferase assay. The open square represents one outlying assay which was not linear. This value has been excluded from calculation of the correlation coefficient (from Sagher, et al., 1988).

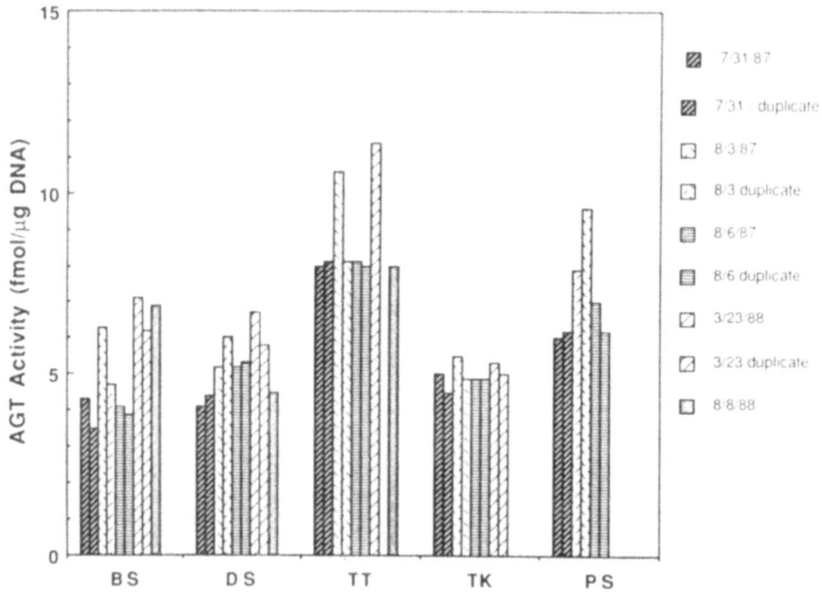

Figure 2. AGT activity in the PBL's of individuals assayed several times over the course of a year. Duplicate, blinded, assays are given the same symbols.

(Sklar & Strauss, 1983). Lawley, et al. (1986), had failed to confirm this observation using a somewhat more extensive sample, and it seemed important to understand the reason for the discrepency. We have presented a preliminary report of these experiments previously (Strauss, et al., in press). We now have data from more individuals.

There is a statistically significant correlation ($R = 0.68$, $p = 0.0003$) between the AGT activity in PBL's and of the lymphoblastoid lines derived from them in the 23 control individuals for whom data are available (Figure 3). However, we found little correlation between lines and the lymphocytes from which they were derived for individuals in remission after treatment for malignancy (mostly Hodgkin's disease and non-Hodgkin's lymphoma; $R = 0.20$, $p > 0.05$ [Not Significant]; Figure 4) or individuals with Hodgkins and non-Hodgkins lymphoma being treated with a regimen including procarbazine ($R = -0.22 = p > 0.05$; Figure 5). The low AGT levels in the PBL's from procarbazine-treated patients reflect the low average values shown in Table 1. In contrast, whether we take data only from individuals in remission (Figure 6) or from all samples (Figure 7), we find there to be correlation between the AGT activity of multiple lines prepared from the same individual ($R = 0.88$; $p < 0.0001$ and $R = 0.59$, $p = 0.006$, respectively). Notwithstanding this correlation there is significant variation in multiple lines derived from different blood samplings of the same individual. Lines derived from blood drawn at different times from the same individual may vary by a factor of nearly four (Table 2).

We followed AGT activity in the lymphocytes of patients undergoing procarbazine therapy. As previously reported (Sagher, et al., 1988) in some, but not all, such patients there is a slow decline over long periods of time to lower AGT

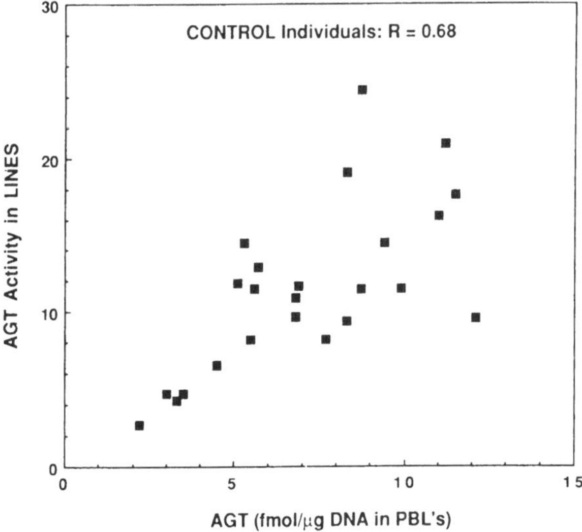

Figure 3. AGT activity in PBL's and in EBV transformed lines from healthy (control) individuals.

Table 2. High and Low AGT Activity from Repeated Sampling.

Patient	Group	Date	PBL AGT	Date	Line AGT*
G	2	7/8/86	5.8	9/24/86	1.8 2.5
	2	8/19/86	3.4	11/10/86	7.0
F	4	7/29/87	4.6	9/15/87	5.2 7.3 5.5
F+	5	9/22/87	4.8	12/10/87 1/12/88	1.7 1.9 3.6

AGT activity is in fmol/µg DNA. The groups are as defined in the legend to Table 1. When multiple lines were prepared and analyzed, each value represents a separate line derived from the PBL sample(*). Patient F is a single individual who moved from treatment (group 4) to remission (F+) (group 5).

PBL values. This low PBL activity is reflected in the average AGT values from these patients (Table 1; Figure 5) and is shown for two individuals in Figure 8. We find that these low

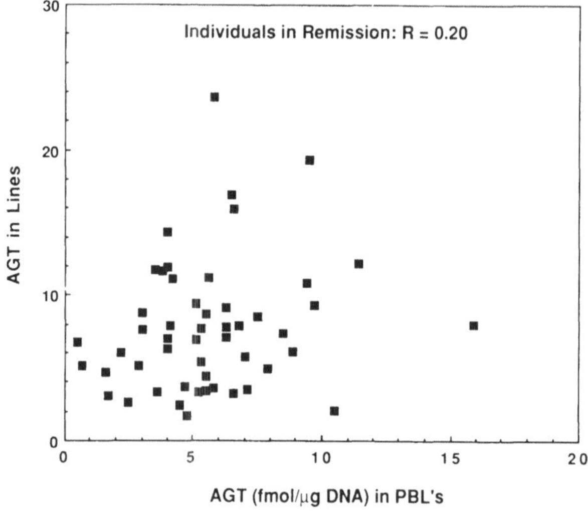

Figure 4. AGT activity in PBL's and in EBV transformed lines from individuals in remission after treatment for malignancy.

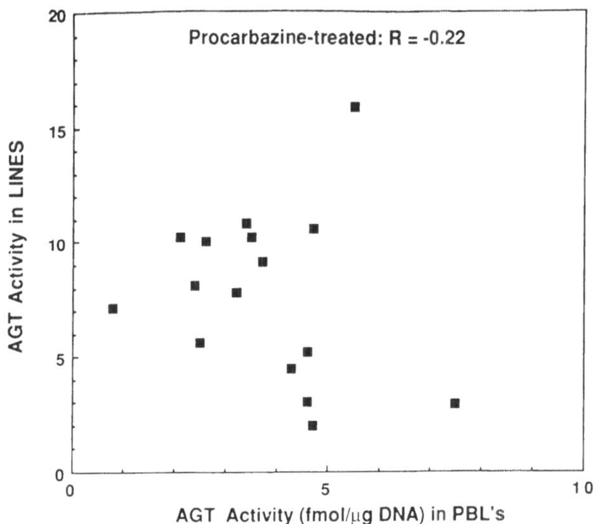

Figure 5. AGT activity in PBL's and in EBV transfomed lines from individuals undergoing treatment for malignancy with a regimen including procarbazine.

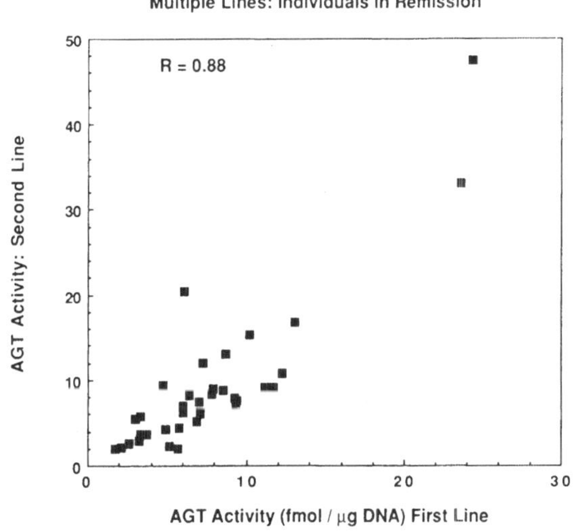

Figure 6. AGT activity in multiple lines derived from the same PBL sample from individuals in remission after treatment for malignancy.

AGT PBL values are not observed in the AGT values of the lymphoblastoid lines derived from these samples. In two of the individuals followed over a long period, the AGT activity in the lines derived from the PBL's did not decline (Figure 8).

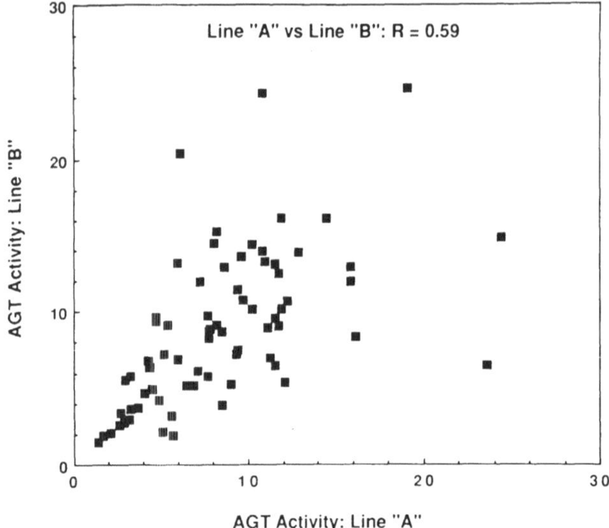

Figure 7. AGT activity in multiple lines derived from the same PBL sample. PBL's from all groups.

In our experiments we expressed our results per μg of DNA as suggested by Gerson, et al. (1986), supposing that the number of AGT molecules per unit of DNA (related to the amount per cell) is the critical factor. It might, therefore, be that our low lines and PBL's represent cells with low protein, or simply smaller cells. In order to control for this factor, we analyzed a set of lymphocytes and lymphoblastoid lines for their activity of uracil-N-glycosylase and compared the activity with AGT activity determined in the same sample. We measured uracil-N-glycosylase because it is an enzyme involved in endogeneous repair and because it has been reported to vary with the cell cycle (Gupta and Sirover, 1984). We found no correlation between the activities of the two enzymes in either lymphocytes or lines ($R = -0.42$, $P > 0.05$ and $R = 0.25$ $p > 0.05$, respectively, Table 3) confirming the result of Myrnes, et al. (1983). Lower AGT activity is not associated with lower glycosylase activity indicating that the activity differences between individuals and lines can not be accounted for by different amounts of total protein. We did observe (Table 3) that uracil-N- glycosylase values in lymphoblastoid lines were about four times higher than in the PBL's, whereas, in this particular group of lines (with L33 excluded) the AGT activities were about equal. We suppose this represents a greater dependence of uracil glycosylase activity on cell cycle.

Our data indicate that individuals have characteristic average AGT values in their PBL's. There is a correlation between PBL and lymphoblastoid line AGT activity for normal individuals but not for individuals who have been treated for malignancy. On the other hand, there is correlation between the AGT activity in multiple lines derived from the same individual sample. How are these findings to be reconciled?

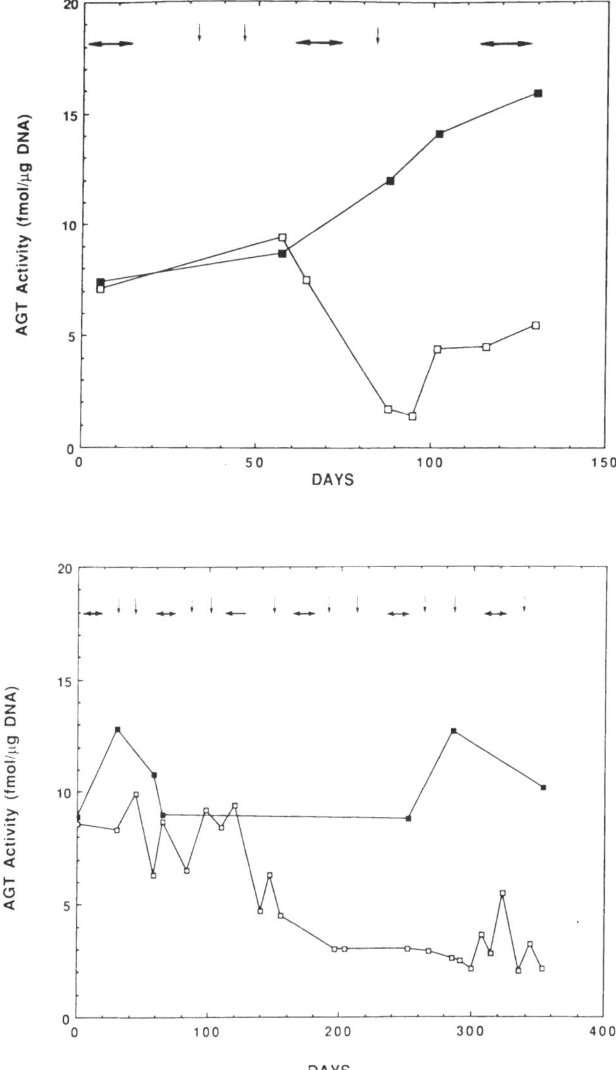

Figure 8, A and B. AGT activity in PBL's and their EBV transformed lines from two patients on MOPP/ABVD therapy including procarbazine. Horizontal lines indicate the duration of procarbazine treatments. Vertical arrows show the times at which dacarbazine was administered. Open symbols, AGT activity in PBL's. Closed symbols AGT activity in lymphoblastoid lines derived from the analyzed PBL samples. Figure 8A. Patient MB; Figure 8B. Patient PM. Data on PBL values from Sagher, et al., 1988.

We suggest that as a result of the treatment, the population of lymphocytes is altered so that only some subclasses are transformable by EBV and that the AGT activity of this sub-

Table 3. AGT and Uracil-N-glycosylase Activity in PBL's and Lymphoblastoid Lines

Designation	Type	AGT Activity	UNGase Activity
727	PBL	5.5	15.5
949	PBL	5.8	16.0
1264	PBL	6.5	20.0
1408	PBL	4.6	>30.0
1413	PBL	4.6	21.0
BS	PBL	6.9	8.1
DS	PBL	4.5	13.3
TT	PBL	8.0	16.5
267C	line	15.0	107.0
496W	line	4.4	34.0
647N	line	4.4	74.0
1167B	line	6.5	64.5
1451B	line	8.3	22.0
LCL	line	1.2	74.0
GM2250A	line	2.8	105.0
L33	line	0.0	15-45
Raji	line	3.0	20-46

AGT activity is given as fmol/µg DNA. Uracil-N-glycosylase activity is in pmol uracil released/µg DNA/30 minutes.

population is different from the average AGT activity in the PBL's. We believe this hypothesis is more plausible than an alternative which postulates that the control mechanisms in the PBL's are altered (or that there are no effective control mechanisms) so that on transformation the lines can take any value; since this second hypothesis does not account for the correlation of multiple lines from the same individual in patients in remission.

We base this explanation on the observed correlation between the AGT activity of PBL's and lymphoblastoid lines in normal subjects indicating that, although the PBL population may be heterogeneous in its AGT activity, AGT activity remains relatively constant for a particular cell even after transformation with EBV. It is known that EBV transformation of cells is multiclonal and that the cultures evolve with time in culture, cells with a growth advantage gradually taking over the population (e.g., Migeon, et al., 1988). If this is so, then the AGT activity of lymphoblastoid cultures should change with time as one or another of the cells becomes dominant. We should also find that eventually some of the lines reach a constant value as the culture becomes clonal. We maintained 24 lines in culture for extended periods of time (Table 4). These cultures were first assayed for AGT activity 30-150 days from their initial infection with EBV when the total number of cells available reached 2-3 x 10^7. They were assayed again as indicated in the table. Some samples were kept frozen in liquid nitrogen during an interval of many months. All cell lines were fed two days before harvest to ensure that they were in a similar, rapidly growing, phase. This precaution

Table 4. Lymphoblastoid AGT Activity after Culture

Line	PBL AGT	Group	Interval days	AGT	Interval days	AGT	Interval days	AGT	Interval days	AGT	Interval days	AGT	Interval days	AGT	Change
105B		1	60	34.8	420	18.6									-
192B	4.6	1	90	7	30+N2+30	11.5	390	19.2	57	37	54	8.6			0
285W	2.5	2	150	4.4	30	9.1	120	8.1							+
133B		4	120	6.6	30+N2+30	1.5	390	0	57	3.1	50	0			-
852A	8.5	4P	90	22.9	14	23.6	2	11.7	12	10.5	4	13	270	11.3	-
852B	8.5	4P	90	22.7	90	18.5	240	11.5	60	21.8	60	9.8			-
128A	14.7	5	150	5.8	50	4.4									0
128B	14.7	5	150	8.1	50	5.7									0
244A	1.4	5	30	0.9	90	1.8	90	2.8	30	2.7					0
244B	1.4	5	60	0.2	60	0.1	270	0.6	240	0.4	210	0.1	90	2.7	0
255A	10.5	5	120	15.4	300	2.6	180	2.1	240	4					-
291A	1.6	5	120	4.7	240	9.4									+
410N		5	90	4.1	210	10.4									+
441N	0.7	5	90	5.1	103	2.2									-
523W	4.2	5	90	9	90	8.6									0
496N	2.2	6	120	2.7	150	3.4	90	4							0
152A		6	90	16.2	210	10.4	210	11.4							0
240C	5.7	6	90	26.2	30+N2+14	3.5	270	7.3							-
237B	7.7	6	90	15.3	300	4									-
272C	11	6	90	16.2	270	8.4									-
526W	5.1	6	150	10.8	30	11.1									0
555N	3	6	150	4.7	120	9.7									+
621W	3.3	6	90	6.8	120	2	14	4.4							0
731F	6.8	6	90	15.7	420	13.6	120	6.3							-

PBL AGT: AGT activity (fmol/μg DNA) in PBL's Interval: Days since last analysis.
N2: frozen in liquid nitrogen. Groups are defined as in the legend to Table 1.

is important because of the reports of variability in AGT activity in cells at different stages in the growth cycle. Gerson (1988) compared AGT levels in a series of PHA stimulated human lymphocyte cultures and found increases after 3 days incubation varying from about 10 to 130 percent in samples from different individuals. It should be pointed out that the lines we followed do not represent a random sample but that many were chosen for study because of their unusually high or low values. Out of the 24 lines, 11 seem to have reached a steady AGT value, 9 have decreased in AGT activity but are not constant while 4 cultures appear still to be increasing. We noticed that changes were still ocurring in some lines after a long period of culture although others remained stable. We conclude that the selective advantage of particular cells in the population may be small so that the evolution of the culture proceeds over a long period. Our inability to clone single lymphoblastoid cells at high efficiency has prevented more direct tests of the hypothesis.

The finding that a number of lines have differing AGT activities that remain stable over relatively long periods permits us to estimate the importance of AGT activity in protecting cells from alkylation damage. We studied the ability of N-methyl-N'-nitro-N-nitrosoguanidine (MNNG), methyl methanesulfonate (MMS) and 1,3 bis(2-chloroethyl)-1-nitrosourea (BCNU) to induce sister chromatid exchange in six lymphoblastoid lines differing in AGT activity (Figure 9). There was a strong negative correlation between AGT activity and induction of SCE's by BCNU ($R = -0.84$; $p < 0.05$) but no correlation was observed for MNNG ($R = 0$; $p > 0.05$). This result is odd in view of the evidence linking O^6-methylguanine to cytotoxic effects (White, et al., 1986; Hall, et al., 1988) but is supported by the results of two groups in Japan (Mitani, et al., 1987; Ikenaga, et al., 1987). Correction for the measured amount of alkylation of DNA produced in different lines by the same concentrations of MNNG does not affect our conclusions. The result with MMS is also surprising since this agent does not produce major amounts of O^6-methylguanine and yet those lines lowest in AGT activity do have greater MMS sensitivity. A difficulty with these experiments, aside from the small sample size, is that we have no idea what other reactions may differ in these lines. Line L33 is the only one in our collection absolutely and continuously devoid of detectable AGT activity both in our hands and in other laboratories (Mitra, personal communication). This line does have detectable and apparently normal 3-methyl adenine glycosylase activity (Sklar and Strauss, 1981), as well as, uracil glycosylase activity comparable to the other lines (Table 3). L33 has been in culture since it was isolated in Littlefield's laboratory in the early 1970's (Sato, et al., 1972) and is derived from a patient with infectious mononucleosis. The results indicate the complexity of the response of these human cells to alkylating agents. They suggest that the response of cells to MNNG depends on more than their AGT content and indicates either that the response of cells to the presence of O^6-methylguanine in the DNA is more complex or that some additional lesion is involved.

The evidence that alkylguanine transferase activity differs in organs, tissues, and even in neighboring cells within the same tissue implies that the activity of this

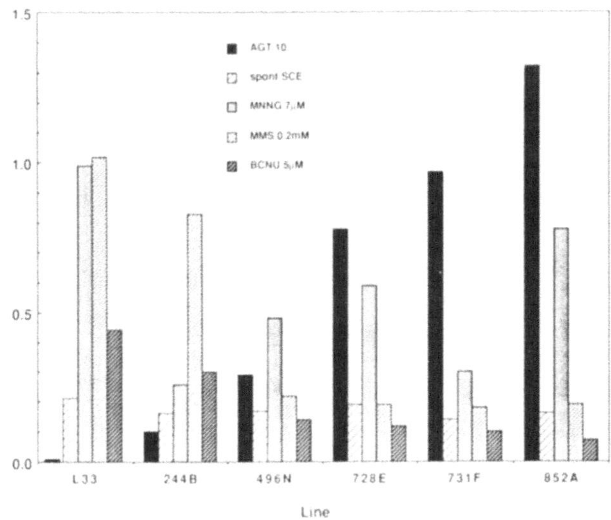

Figure 9. Induction of SCE's in six lymphoblastoid lines of differing AGT activity. Solid black bars indicate the AGT activitv (fmol/μg DNA) divided by ten. Additional bars, SCE's per chromosome induced by treatments as shown.

enzyme is under epigenetic (developmental) control. Such developmentally controlled proteins complicate the analyses as compared to the excision systems in xeroderma pigmentosum, which are genetically controlled and in which all cells derived from an individual display the same characteristics. It is this differentiation control which we suppose accounts for the complexities we encounter. Our data indicate characteristic individual average AGT values along with heterogeneity within the PBL population. We suppose that the distribution is individually (i.e., genetically) determined with a characteristic mean. It should be clear that even though there are characteristic AGT values and there is a correlation between the AGT activity of lymphocytes from normal individuals and their lymphoblastoid lines immediately after transformation, long established lines should not be relied upon as representative of the AGT activity of the original population. We suppose that, for normal individuals, the heterogeneity of AGT activity in lymphoblastoid cultures reflects the heterogeneity of AGT activity in lymphocytes. According to this view, the AGT activity of a particular PBL is a characteristic of that cell and, even after EBV transformation, the AGT activity of the lymphoblast reflects some intrinsic activity of the parent lymphocyte when normal lymphocytes are involved. With regard to the lack of correlation in individuals under therapy, our results can be accounted for by supposing that treatment for malignancy alters the population of B lymphocytes transformable by EBV so that they no longer reflect the AGT activity of the overall population. The observation that cultures may increase or decrease in AGT activity on prolonged culture argues against some adaptive response to the environment or against the level of the AGT protein being the property selected for. We also suggest that the events in cells which result in changes in AGT activity do sometimes, but not always, involve additional systems for the repair of alkylation induced damage. It is such additional changes which result in an apparent correlation between AGT activity and other MNNG-induced effects in some but not in all cases.

Although the factors involved in the regulation and action of AGT activity in cells appear to be overwhelmingly complex, the regulatory features of other pathways in eukaryotic organisms are probably not less complicated. In these cases several tiers of regulatory controls may affect the behavior of particular systems (Watson, et al., 1987). Many investigations in humans have stressed the variability of DNA repair activities (e.g. Freeman, 1988; Setlow, 1988; Oesch, et al., 1987). In no case do we understand the genetics of the variability, except where large effects are involved as in the case of xeroderma pigmentosum. Our observations are best interpreted as indicating that individuals (and their cells) have multiple controls of repair activities with multiple settings of these controls possible. This variation in repair activity within the population might well be the source of an idiosyncratic sensitivity of individuals to DNA damage.

ACKNOWLEDGEMENT

The investigations in this paper were supported by a Program Project Grant (CA40046) from the National Cancer Institute.

REFERENCES

Freeman, S., 1988. Variations in excision repair of UVB-induced pyrimidine dimers in DNA of human skin in situ. J. Invest. Dermatol., 90:814-817.

Gerson, S., 1988. Regeneration of O^6-alkylguanine-DNA alkyltransferase in human lymphocytes after nitrosourea exposure. Cancer Res., 48:5368- 5373.

Gerson, S., Trey, J., Miller, K., and Benjamin, C., 1985. O^6-alkylguanine-DNA alkyltransferase activity in myeloid cells. J. Clin. Invest., 76:2106-2114.

Gerson, S., Trey, J., Miller, K., and Berger, N., 1986. Comparison of O^6-alkylguanine-DNA alkyltransferase activity based on cellular DNA content in human, rat and mouse tissue. Carcinogenesis, 7:745-749.

Gupta,P. and Sirover,M., 1984. Altered temporal expression of DNA repair in hypermutable Bloom's syndrome cells. Proc. Natl. Acad.Sci. USA., 81:757-61.

Hall, J., Kataoka, H., Stephenson, C. and Karran, P., 1988. The contribution of O^6-methylguanine and methylphosphotriesters to the cytotoxicity of alkylating agents in mammalian cells. Carcinogenesis, 9:1587-1593.

Ikenaga, M., Tsujimura, T., Cheng, H., Fujio, C., Zhang, Y., Ishizaki, K., Kataoka, H., and Shima, A., 1987. Comparative analysis of O^6-methylguanine methyltransferase activity and cellular sensitivity to alkylating agents in cell strains derived from a variety of animal species. Mutation Res., 184:161-168.

Lawley, P., Harris, G., Phillips, E., Irving, W., Colaco, C., Lydyard, P., and Roitt, I., 1986. Repair of chemical carcinogen-induced damage in DNA of human lymphocytes and lymphoid cell lines--Studies of the kinetics of removal of O^6-methylguanine and 3-methyladenine. Chem. Biol. Interactions, 57:107-121.

Migeon, B,, Axelman, J. and Stetten, G., 1988. Clonal evolution in human lymphoblast cultures. Am. J. Hum. Genet., 42:742 - 747.

Mitani, H., Ito, K., Fujino, M. and Takebe, H., 1987. Difference in O^6-methylguanine methyltransferase activity among transformed NIH 3T3 cell clones. Mutation Res., 191:201-205.

Myrnes, B., Giercksky, K., and Krokan, H., 1983. Interindividual variation in the activity of O^6-methylguanine-DNA methyltransferase and uracil-DNA glycosylase in human organs. Carcinogenesis, 4:1565-1568.

Oesch, F., Aulmann, W., Platt, K., and Doerjer, G., 1987. Individual differences in DNA repair capacities in man. Arch. Toxicol., 10 (Supp. 1) 172 -179.

Pegg.A., 1984. Properties of the O^6-alkylguanine-DNA repair system of mammalian cells. IARC Sci. Publ., 57:575-80.

Sagher, D., Karrison, T., Schwartz, J., Larson, R., Meier, P. and Strauss, B., 1988. Low O^6-alkylguanine DNA alkyltransferase activity in the peripheral blood lymphocytes of patients with therapy-related acute nonlymphocytic leukemia. Cancer Res., 48:3084-3089.

Sato, K., Slesinski, R. and Littlefield, J., 1972. Chemical mutagenesis at the phosphoribosyltransferase locus in cultured human lymphoblasts. Proc. Natl. Acad. Sci. USA., 69:1244-1248.

Setlow, R., 1988. Relevance of phenotypic variation in risk assessment: the scientific viewpoint. Basic Life Sci., 43:1-5.

Sklar, R., and Strauss, B., 1981. Removal of O^6-methylguanine from DNA of normal and xeroderma pigmentosum-derived lymphoblastoid lines. Nature , 289:417-420.

Sklar, R. and Strauss, B., 1983. O^6-methylguanine removal by competent and incompetent human lymphoblastoid lines from the same male individual. Cancer Res., 43:3316-3320.

Strauss, B.S., Sagher, D., Karrison, T., Larson, R., Meier, P., Schwartz, J., Farber, R., and Weichselbaum, R., 1989. The response of human cells to in vitro methylation damage. In: DNA Damage and Repair (A. Castellani, ed.), Plenum Press, New York, pp. 107-123.

Tucker, M., Coleman, C., Cox, R., Varghese, A., and Rosenberg, S., 1988. Risk of second cancers after treatment for Hodgkin's disease. N.E. J. Med., 318:76-81.

Waldstein, E., Cao, E., Bender, M. and Setlow, R., 1982. Abilities of extracts of human lymphocytes to remove O^6-methylguanine from DNA. Mutat. Res., 95:405-416.

Watson, J., Hopkins, N., Roberts, J., Steitz, J. and Weiner, A., 1987. Molecular Biology of the Gene. Vol. I. General Principles. 4th Edition. Benjamin Cummings. Menlo Park, xxix + 744pp

White, G., Ockey, C., Brennand, J., and Margison, G., 1986. Chinese hamster cells harbouring the Escherichia coli O^6-alkylguanine alkyltransferase gene are less susceptible to sister chromatid exchange induction and chromosome damage by methylating agents. Carcinogenesis, 7:2077 2080.

Wiestler, O., Kleihaus, P., Rice, J. and Ivankovic, S., 1984. DNA methylation in maternal, fetal and neonatal rat tissues following perinatal administration of procarbazine. J. Cancer Res. Clin. Oncol., 198:56-59.

FORMATION AND REPAIR OF ADDUCTS THAT LEAD TO CROSS-LINKS IN DNA TREATED WITH CHLOROETHYLATING AGENTS

Thomas P. Brent, Prescilla E. Gonzaga, Debra G. Smith

Department of Biochemical & Clinical Pharmacology
St. Jude Children's Research Hospital
Memphis, TN 38101

INTRODUCTION

The chloroethylnitrosoureas (CENUs) represent a class of bifunctional alkylating agents that have proved clinically useful for cancer chemotherapy. The group includes carmustine (BCNU), lomustine (CCNU), semustine (MeCCNU) and chlorozotocin (DCNU) (Figure 1). Most recently two novel, non-nitrosourea compounds, mitozolomide and clomesome (Figure 2), have been added tentatively to the CENU class based on the perception that their mechanism of antitumor activity, in common with the CENUs, depends on their chloroethylating function. None of these compounds require metabolic activation because they are unstable and decompose spontaneously in aqueous solution to yield bifunctional alkylating species which can react with nucleophilic sites on cellular macromolecules. It is generally accepted that the formation of cross-links between complimentary strands of duplex DNA represents the crucial mechanism for the cytotoxic, antitumor activity of these drugs.

The exact details of the mechanism for cross-link formation by the chloroethylating drugs are not yet entirely clear. A scheme proposed by Ludlum and Tong (1985) was based on observations by Scudiero et al. (1984) and Erickson et al. (1980) that human cells defective in repair of O^6-methylguanine were hypersensitive to both the cytotoxic and the DNA cross-linking effects of the CENUs. However, the only diadduct identified that might represent the cross-link structure in CENU-treated DNA was 1-(3-deoxycytidyl),2-(1-deoxyguanosinyl)ethane, which does not involve the O^6-position of guanine. This finding led Ludlum to propose that an initial alkylation by a chloroethyl ion at O^6-guanine, is followed by

DNA Repair Mechanisms and Their Biological Implications in Mammalian Cells
Edited by M.W. Lambert and J. Laval
Plenum Press, New York

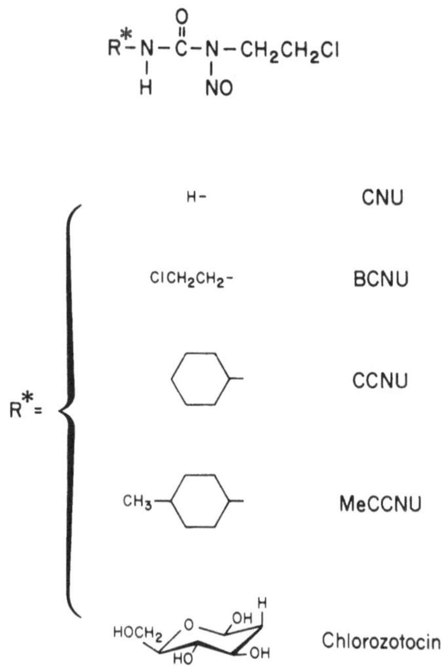

$R^*-NH-C(=O)-N(NO)-CH_2CH_2Cl$

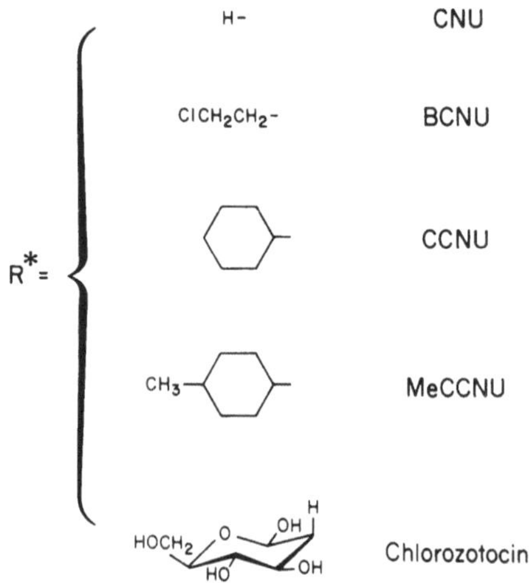

Figure 1. Structures of the common chloroethylnitrosoureas. Abbreviations: CNU, 1-(2-chloroethyl)-1-nitrosourea; BCNU, 1,3-bis(2-chloroethyl)-1-nitrosourea; CCNU,1-(2-chloroethyl)-3-cyclohexyl-1-nitrosourea; MeCCNU, 1-(2-chloroethyl)-3-(trans-4-methylcyclohexyl)-1-nitrosourea; DCNU, chlorozotocin.

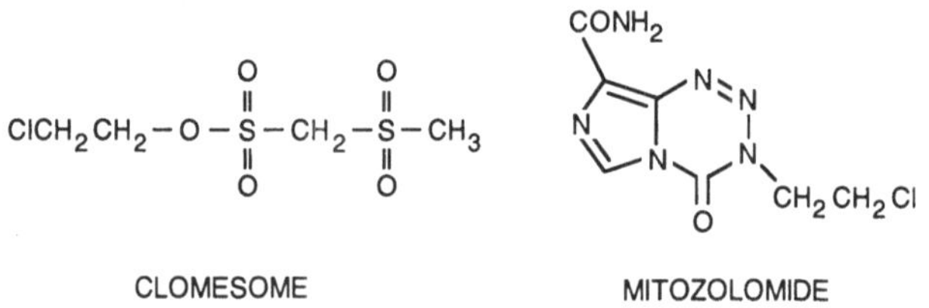

$ClCH_2CH_2-O-S(=O)_2-CH_2-S(=O)_2-CH_3$

CLOMESOME

CONH$_2$ ... N, N, N, N, O, CH_2CH_2Cl

MITOZOLOMIDE

Figure 2. Structures of non-nitrosourea chloroethylating anticancer agents.

intramolecular rearrangement via O^6,N^1-ethanoguanine to yield the observed N^3-cytosine, N^1-guanine ethane cross-link (Figure 3). This scheme is supported by observations that purified O^6-alkylguanine-DNA alkyltransferase (GATase) can suppress formation of BCNU-induced interstrand cross-links (Robins et al., 1983; Brent, 1984). The purified transferase has also been reported to suppress formation of 1-(3-deoxycytidyl,2-(1-deoxyguaninosyl)ethane in isolated CENU-treated DNA (Ludlum et al., 1986). Chloroethyl-O^6-guanine has never been detected

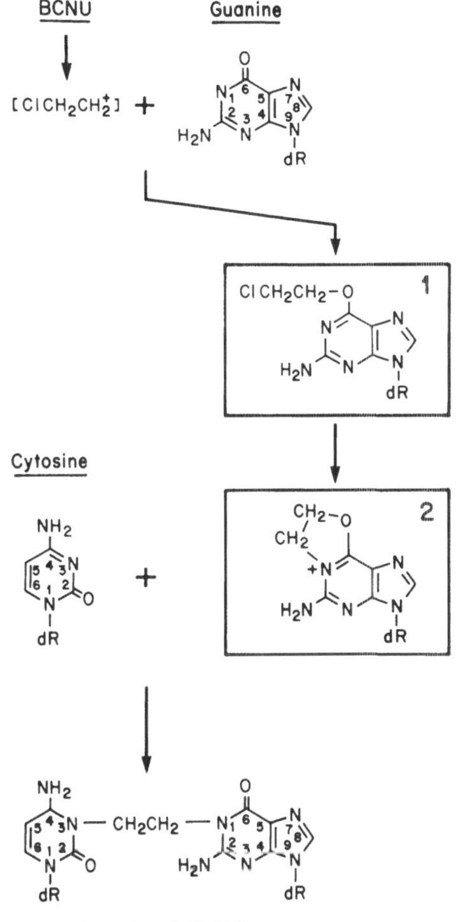

Figure 3. Scheme for formation of a N^3-cytosine; N^1-guanine cross-link product by initial chloroethylation of O^6-guanine and subsequent intramolecular rearrangement from O^6- to N^1-guanine (modified from Tong et al., 1983).

in CENU-treated DNA. However, fluoroethyl-O^6-guanine has been isolated from DNA, and the conversion of this adduct to N^3-cytosine-N^1-guanine ethane has been reported (Tong, et al., 1983). Suppression of this conversion by purified GATase has also been observed.

KINETICS OF DNA INTERSTRAND CROSS-LINK FORMATION

The scheme in Figure 3 has several implications. First, when the initial adduct, O^6-chloroethylguanine is formed, all of the succeeding reactions will proceed in the absence of the drug and will be the same regardless of the drug used. We have tested these predictions by specifically comparing BCNU and clomesome. With BCNU, cross-link formation proceeded relatively slowly at 37°C for at least 8 hours in the continuous presence of the drug (Figure 4). If BCNU was removed after various times of reaction, the DNA continued to become cross-linked in the absence of drug, as predicted, and the number of cross-links so formed provided a measure of precursors in the DNA at the time of drug removal. Such determinations revealed that BCNU induces cross-link precursors relatively rapidly, the maximum level being reached in about 90 minutes at 37°C followed by their slower conversion to cross-links with a $t_{1/2}$ of about 3 hours. The rate for conversion of precursors to cross-links was determined more precisely by treating DNA with BCNU for 90 min, removing the drug and following cross-link formation in the absence of the drug (Figure 5). The reciprocal of the curve for cross-link

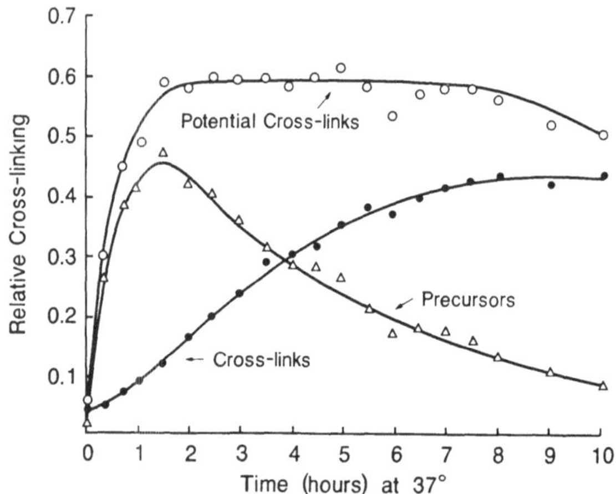

Figure 4. Time course for cross-link formation in DNA treated with 10 mM BCNU. Calf thymus DNA was reisolated at the indicated times by precipitation with ethanol and redissolved in aqueous buffer. Cross-links were measured immediately (● - ●) or after additional incubation at 37°C for 24 hours to produce all potential cross-links (o - o). The difference between initial cross-links and the total potential cross-links for each time point represents cross-link precursors present at that time point (Δ-Δ).

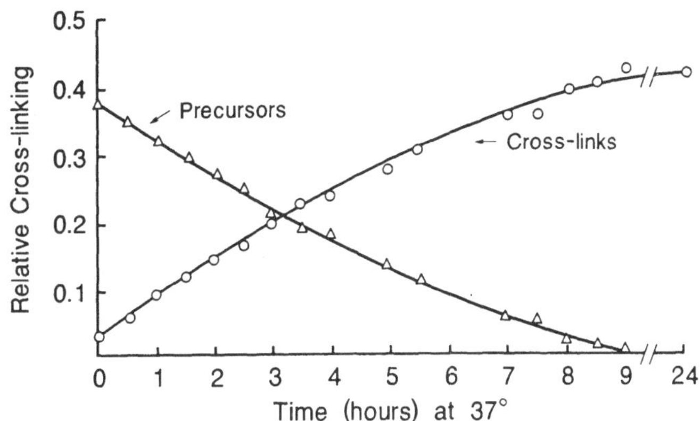

Figure 5. Time course for conversion of BCNU-induced precursors to cross-links. DNA was treated with 10 mM BCNU at 37°C for 90 minutes. After re-isolation, free of drug, the DNA was incubated in aqueous buffer at 37°C for the indicated times, when cross-links were determined (o - o). The difference between cross-links at each time point and the maximum level achieved is taken as a measure of cross-link precursor (Δ-Δ).

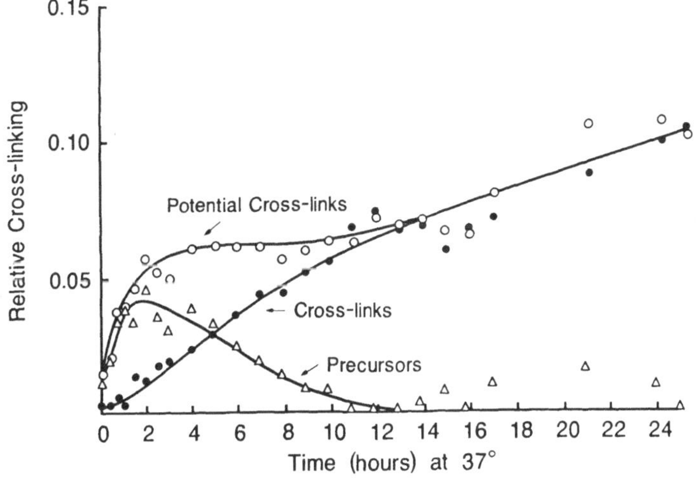

Figure 6. Time course for cross-link formation in calf thymus DNA treated with 20 mM clomesome. The experimental procedure were the same as for Figure 4. The symbols are also the same as Figure 4 (● - ●) overt cross-links, (o - o) total potential cross-links, (Δ-Δ) cross-link precursors.

formation in Figure 5 represents the rate for decay of the precursors, the half-life being approximately 3 hours. The same analysis of cross-links and cross-link precursor formation in DNA treated with clomesome revealed certain similarities. Although clomesome was much less reactive than BCNU, inducing only one-tenth the level of cross-links, the pattern of cross-link precursor formation and decay over a 10-hour period of drug treatment (Figure 6) appeared similar to that with BCNU. The rate of conversion of precursors to cross-links after drug removal (Figure 7) occurred with a $t_{1/2}$ of about 4 hours, which, by comparison with a $t_{1/2}$ of 3 hours for BCNU, suggests that the precursors, and the subsequent reactions for their conversion to cross-links, may well be the same for both these drugs. However, careful calculation of the rate constants for cross-link formation, both in the

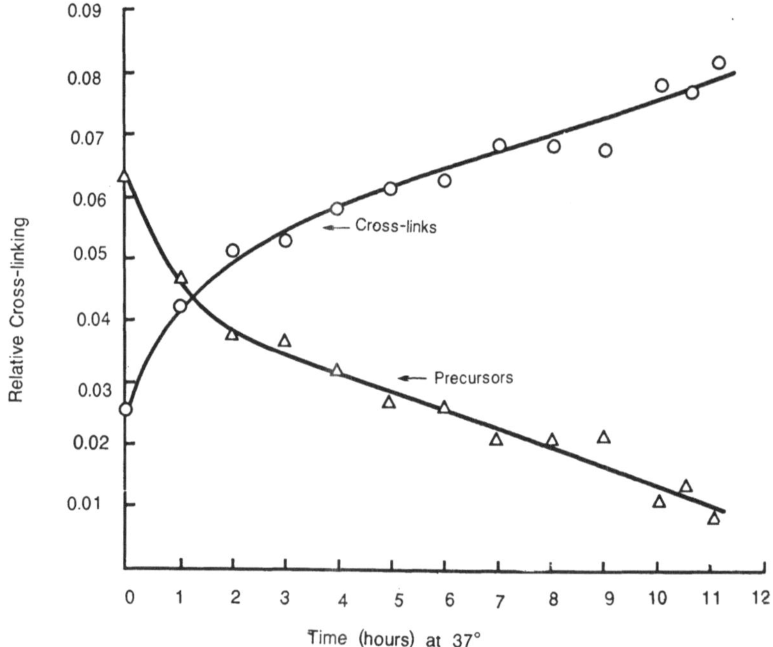

Figure 7. Time course for conversion of precursors induced by 20 mM clomesome to cross-links after drug removal. The experimental procedure and symbols are the same as for Figure 5.

continuous presence of drug, and in DNA after drug removal, indicate significant differences (>3-fold) between BCNU and clomesome (Figure 8). These results raise the possibility that more than one type of initial adduct can lead to the formation of interstrand cross-links. In this regard, Buckley has proposed that whereas clomesome forms interstrand cross-links in DNA by the mechanism shown in Figure 3, CENU-induced cross-link formation occurs by a different mechanism (Figure 9). In this scheme the imidourea form of CENUs forms a tetrahedral intermediate at O^6-dG_1 in a $dG_1dG_2dN_3$ codon. The tetrahedral intermediate, after cyclization, reacts with the adjacent dG_2 to yield (2-(hydroxyazo)ethyl)-O^6-dG, and it is this adduct that ultimately forms the interstrand cross-link.

REPAIR OF CROSS-LINK PRECURSOR ADDUCTS BY ALKYLTRANSFERASE

Another important feature of Ludlum's scheme is that O^6-alkylguanine-DNA alkyltransferase (GATase) may repair precursor adducts, thereby aborting the cross-linking process (Figure 10). When isolated DNA is briefly treated with BCNU and then incubated with the purified human transferase after the drug has been removed, further cross-link formation is suppressed, as predicted (Figure 11). We have observed similar suppression of cross-linking by GATase in DNA treated with all of the other drugs listed in Figures 1 and 2. However, a study comparing BCNU and clomesome revealed what could be a significant difference between the two drugs

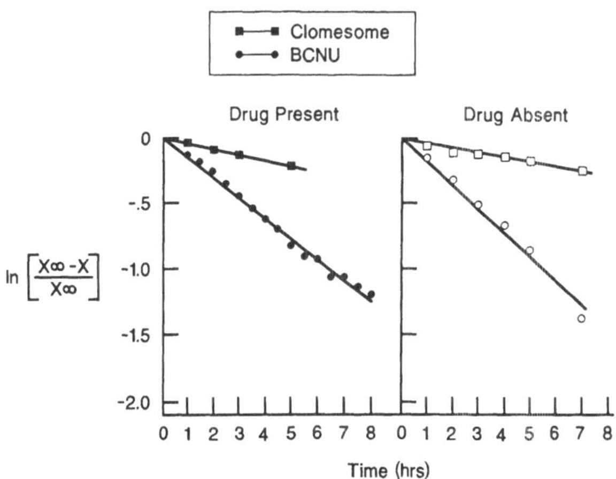

Figure 8. Rate plots for cross-link formation in calf-thymus DNA treated with 10 mM BCNU (●,○) or 20 mM clomesome (■,□). (Modified from Buckley and Brent, 1988).

(Brent, et.al., 1987a). When DNA was treated briefly with each drug to induce approximately equal levels of cross-link precursors, we found that it took four times more GATase to suppress clomesome-induced cross-links compared with those induced by BCNU (Figure 12). One explanation for this discrepancy is that clomesome induces more O^6-guanine adducts that are not cross-link precursors; however, other observations indicate the opposite. Gibson and coworkers have shown that clomesome induces little or no O^6-hydroxyethylguanine in DNA, whereas this is the predominant O^6-guanine adduct in BCNU-treated DNA (Gibson et al., 1986). Since O^6-hydroxyethylguanine is a substrate for GATase (albeit a very poor one),

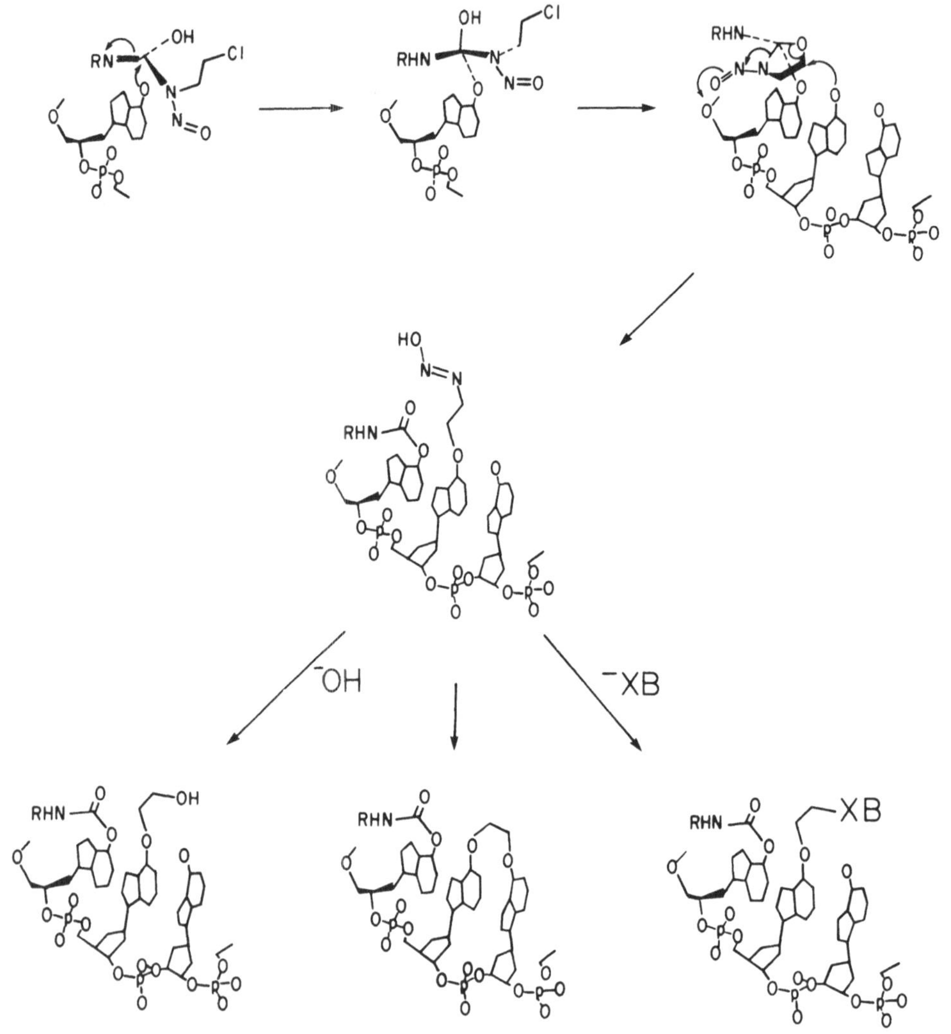

Figure 9. Scheme for cross-link induction by chloroethylnitrosourea via a tetrahedral intermediate. (Modified from Buckley and Brent, 1988).

one might expect the relatively large amounts of this adduct in BCNU-treated DNA would use up more GATase than would clomesome-DNA. Because the reverse situation was found, one must consider the possibility that the cross-link precursor adducts at O^6-guanine, induced by these two drugs, may be different.

FORMATION OF DNA-TRANSFERASE COMPLEX DURING CROSS-LINK PRECURSOR REPAIR

Yet a third implication of the Ludlum model stems from the possibility that GATase may repair the second intermediate in the cross-linking pathway, the exocyclic adduct, O^6,N^1-ethanoguanine. Assuming that the reaction mechanism for this

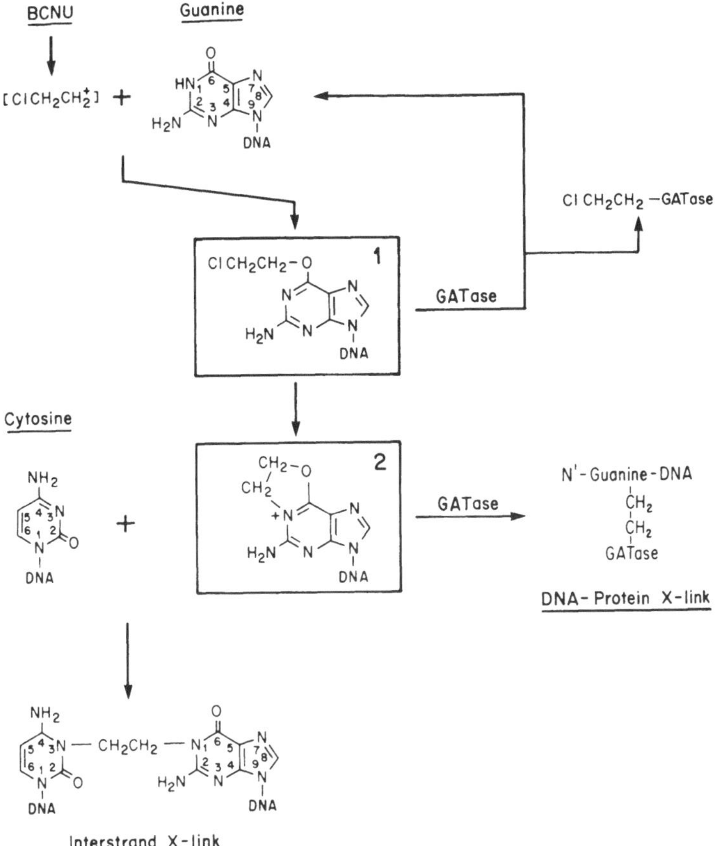

Figure 10. Possible reactions for the alkyltransferase with chloroethylnitrosourea-induced adducts at O^6-guanine in DNA.

adduct with GATase is the same as it is for simple monoadducts at O^6-guanine in DNA, one would expect that the transferase would covalently bond the alkyl moiety after cleaving the carbon-oxygen bond at O^6-guanine. Since the other end of the alkyl group in the case of O^6,N^1-ethanoguanine is still bound to DNA, one would predict a DNA-GATase cross-linked product. We have recently demonstrated the formation of such DNA-GATase complexes using either a protein precipitation assay (Brent et al., 1987b; 1987c) or a gel electrophoresis method (Brent and Remack, 1988).

Selective Precipitation Assay for DNA-GATase Complex

The first approach used selective precipitation of proteins by SDS/KCl under conditions where DNA remained in solution. When untreated radiolabeled DNA was incubated with partially purified preparations of human GATase, none of the radioactivity was precipitated in the assay. However, if the DNA was briefly exposed to BCNU, such that it contained cross-link precursors, then some of the radioactive DNA precipitated with the protein, consistent with the notion of a DNA-GATase

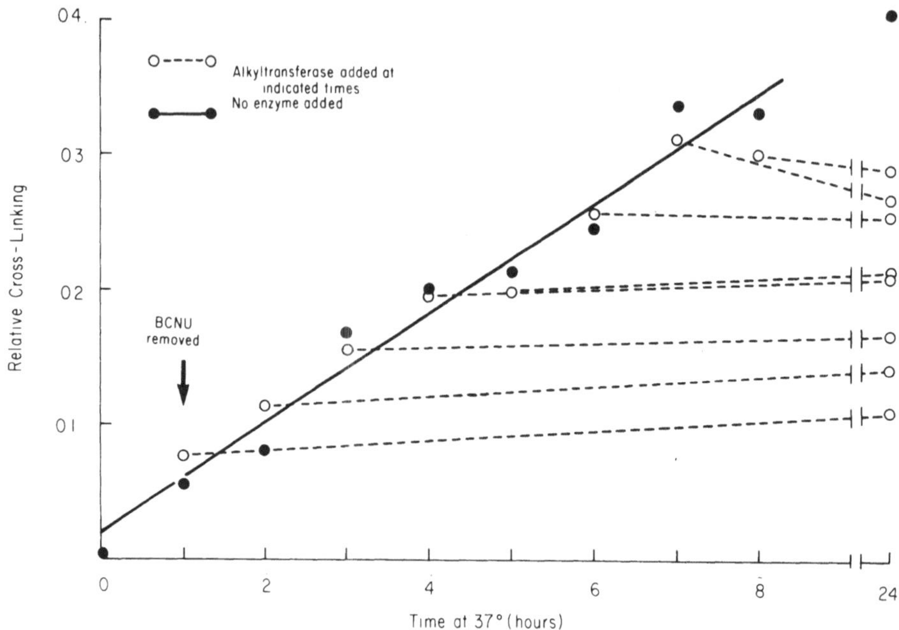

Figure 11. Alkyltransferase-mediated suppression of BCNU-induced cross-links. DNA was pulse-treated for 1 hour with 10 mM BCNU. Relative DNA interstrand cross-linking was determined at the indicated times for up to 23 hours after drug removal (●). Cross-linking was also measured at 24 hours in aliquots to which purified human alkyltransferase was added at the indicated times (O). (From Brent, et al., 1987a).

complex (Table 1). The formation of the complex was absolutely dependent on the presence of transferase activity. When GATase was inactivated either by heating or more specifically by reaction with methylnitrosourea-treated DNA, complex forming activity was lost. Moreover, the properties of the complex-forming activity paralleled those of the alkyltransferase activity in every way tested, i.e., heat inactivation kinetics, inhibitor responses, and chromatographic co-migration (Brent et al., 1987c). Evidence that the substrate for GATase in BCNU-treated DNA is an intermediate in cross-link formation derives from the observation that DNA briefly treated with BCNU (e.g., 1 hr 37°C), when incubated for 15 hours or more after drug removal, can no longer form a complex with GATase (Brent et al., 1987b).

<u>Polyacrylamide Gel Electrophoresis Assay for Oligodeoxynucleotide-GATase Complex</u>

The DNA-protein coprecipitation assay, although strongly

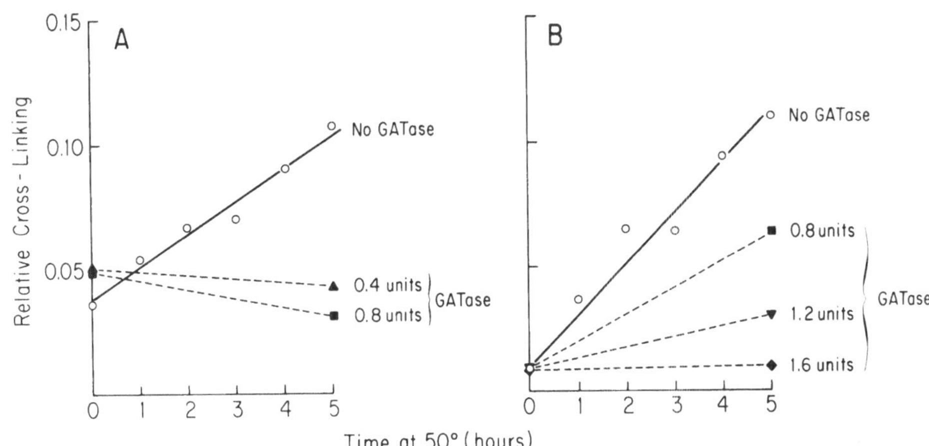

Figure 12. Cross-link suppression by the alkyltransferase (GATase) of DNA pulse-treated with BCNU or Clomesome, so as to contain similar potential for cross-link formation. Calf thymus DNA was treated with 10 mM BCNU for 5 minutes at 37°C (A) or 50 mM Clomesome for 60 minutes at 37°C (B). After drug removal the DNA was incubated at 37°C for 30 minutes with the indicated amounts of purified human alkyltransferase, to allow cross-link precursor repair to occur. The DNA was then heated for up to 5 hours at 50°C, as shown, to convert the remaining precursors to cross-links. (From Brent, et al., 1987a).

implicating the transferase as the crucial protein for formation of the complex, does not directly identify GATase in the complex. Hence we developed another approach to isolate and characterize the complex in which a BCNU-treated synthetic oligodeoxynucleotide of defined length was reacted with GATase and the resulting complex was identified as a distinct band by SDS-polyacrylamide gel electrophoresis (PAGE). Oligonucleotides ranging from 14 to 30 residues long were 5' end-labeled with ^{32}P or ^{35}S, allowed to anneal so as to be double-stranded, and treated briefly with BCNU (e.g., 1 hour at 37°C). After reaction with human GATase, the reaction mixture was analyzed by PAGE and the labeled oligo-GATase complex was identified by autoradiography. A typical gel (Figure 13) reveals a radiolabeled band that is displaced to a higher molecular weight (~30,000 daltons) than that of GATase (25,000 daltons). Similar to the results obtained with the precipitation assay, this band was absolutely dependent on both the presence of active transferase and BCNU treatment of the oligonucleotide in the reaction. Several radioactive bands, corresponding to irrelevant DNA-protein products, can be seen even in those reactions that lacked active transferase or alkylated oligonucleotide. One of the advantages of this

Table 1. Formation of DNA-GATase Complex

Reaction Constituents	$[^3H]$-labeled DNA co-precipitated with protein[a] (dpm)
Untreated DNA, Active GATase	800 ± 163
BCNU-treated DNA, Active GATase	6067 ± 680
BCNU-treated DNA, GATase inactivated by MNU-treated DNA	633 ± 165
BCNU-treated DNA, heat-inactivated GATase	733 ± 249

[a] ^{3}H-thymidine-labeled DNA was treated briefly with 10 mM BCNU before reaction with purified human alkyltransferase. Protein in the reaction mixture was selectively precipitated with 0.25 mM KCl/1% SDS.

assay for the GATase-DNA complex is that it separates out such interfering noise. The band in Figure 13 at 30 kDa is exactly what one would predict for a complex between the 25 kDa GATase and the 4.5 kDa 14-mer oligodeoxynucleotide, and this band is absolutely dependent on the presence of active GATase and BCNU-treatment of the oligonucleotide.

We have used this assay to determine the kinetics of formation and decay of the BCNU-induced adduct that reacts with GATase to form the complex and have compared these kinetics with those for cross-link precursors. Figure 14 shows results from PAGE for the time-course of induction of these adducts that form a complex with GATase, results that closely resemble those for cross-link precursor formation with BCNU (Figure 4). The decay of the complex-forming adduct in BCNU-treated oligonucleotide occurs with a half-life of about 4-5 hours which is of the same order as that for cross-link precursor decay in BCNU-treated calf thymus DNA (Figure 15). Thus the probability that complex forming adducts and cross-link precursors are identical remains strong.

BIOLOGICAL SIGNIFICANCE OF THE DNA-GATase COMPLEX

The biological significance of complex formation by GATase and DNA containing CENU-induced adducts is uncertain. If the adducts involved (i.e. O^6,N^1-ethanoguanine) represent the relatively stable precursors of cross-linking (i.e. $t_{1/2}$ = 3-4 hours), one might expect substantial amounts of DNA-GATase complex to be formed by the transferase. This would not amount to effective repair of the DNA and one might expect further repair of the resultant DNA-protein cross-links to be necessary. Evidence that the *E. coli uvr* genes for bulky adduct excision repair are involved in O^6-alkylguanine repair supports such a notion (Kacinski et al., 1985). It seems most likely, however, that the transferase would repair the initial adduct (O^6-chloroethylguanine) so rapidly <u>in vivo</u> that very little O^6,N^1-ethanoguanine would in fact form. By the same token, all of the transferase would be used up in the suicidal reaction with the O^6-guanine monoadduct, leaving very little to form any complex.

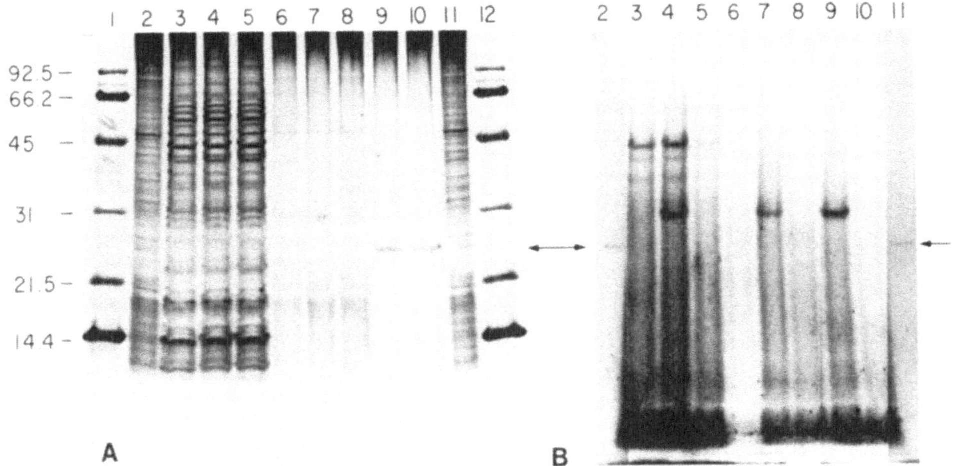

Figure 13. SDS-PAGE of partially purified human GATase after reaction (60 minutes, 37°C) with [^{32}P]-labeled double-stranded 14-mer.
A) Protein banding pattern visualized by silver staining.
B) Autoradiograph of the gel in (A). Lanes 1 and 12 are molecular weight marker proteins; lanes 2 and 11 are [^{3}H-methyl]-labeled GATase, Fraction 4 (6 units); lanes 3, 4, 5 are Fraction 3 (1 unit); lanes 6, 7, 8 are Fraction 4 (1 unit); lanes 9, 10 are Fraction 5 (4 units). Lanes 3 and 6: BCNU-treated 14-mer was reacted with inactivated GATase; lanes 4, 7, and 9: BCNU-treated 14-mer was reacted with active GATase; lanes 5, 8, and 10: untreated 14-mer was reacted with active GATase. (From Brent and Remack, 1988).

A third possibility is that the exocyclic adduct that reacts with the transferase to form a complex represents only a small fraction of the cross-link precursor pool. In preliminary attempts to quantitate the formation of complex, we estimated that only a few percent of synthetic oligodeoxynucleotides become complexed to GATase. Similarly, at best only about one percent of the transferase becomes complexed to oligonucleotide. Thus, we conclude that the adducts

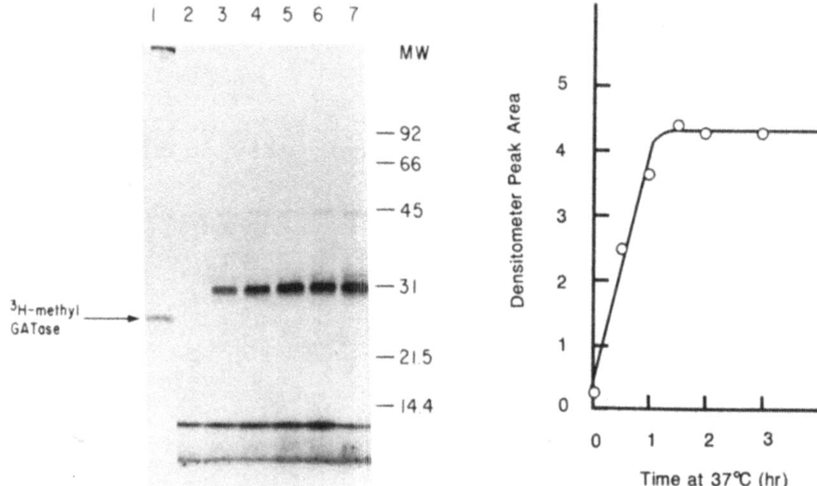

Figure 14. Formation of 30 kDa complex by reaction of GATase (Fraction 4, 1 unit) with 14-mer pretreated with BCNU for different lengths of time. 14-mer was pretreated with BCNU (500 mM) for the following times: 0 minutes, lane 2; 30 minutes, lane 3; 60 minutes, lane 4; 90 minutes, lane 5; 2 hours, lane 6; 3 hours, lane 7; lane 1 was [^{3}H-methyl]-labeled GATase. After incubating the alkylated DNA with GATase, reaction products were analyzed by SDS-PAGE and autoradiography. The radiolabeled 30 kDa bands were quantitated by densitometry and 30 kDa peak areas were plotted against the indicated times of 14-mer exposure to BCNU. (From Brent and Remack, 1988).

forming complex with a GATase represent only a small component of the total pool of O^6-guanine adducts. The identity of the major fraction of stable cross-link precursors remains unclear.

EXPERIMENTAL POTENTIAL OF DNA-GATase COMPLEX

The formation of DNA-GATase complexes has potential beyond that relating to the pharmacological mechanism of DNA interstrand cross-link formation and the biological consequences of this biochemical repair reaction within cells. Complex formation provides us with a novel avenue for isolation and purification of the transferase. In addition, one should be able to derive the nucleoside-amino acid cross-linked diadduct from the isolated complex and determine its precise structure.

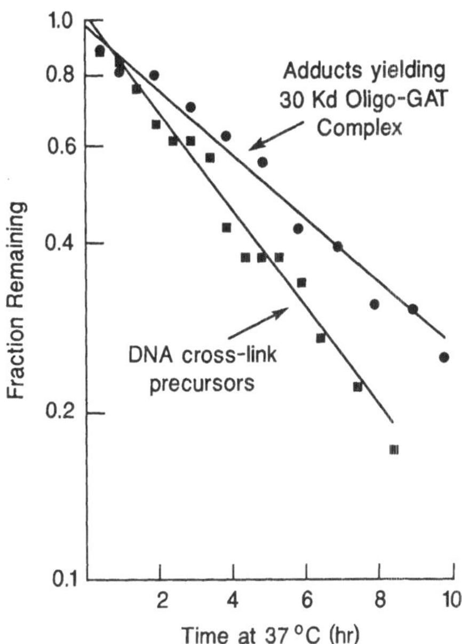

Figure 15. Rate plots for the disappearance from DNA of adducts induced by a brief (90 minutes at 37°C) treatment with BCNU. Adducts, in a synthetic 14-mer, that formed a complex with GATase were determined by the PAGE assay (●-●). DNA cross-link precursors were determined as described in Figure 5 (■-■).

According to the scheme proposed by Ludlum (Figure 3), one expects this diadduct to be N^1guanosine-ethano-cysteine. Confirmation of this structure would provide support for the model, whereas failure to do so would tend to confound it and perhaps provide clues as to the adducts that do in fact constitute cross-link precursors.

Isolation of the complex in our laboratory has entailed the use of oligonucleotides containing a single strand tail made up of about 18 dA residues. After BCNU treatment and reaction with partially purified GATase, noncomplexed oligonucleotides were removed and dA-tailed oligonucleotide complex was bound to an oligo(dT)-cellulose column. Noncomplexed proteins washed through this column under conditions of high salt concentration, so that when the dA-tailed complex was eluted by low salt buffer, it was essentially free of all other protein (Figure 16). We have produced highly purified GATase protein by this affinity method. The purity of the complex obtained by this method will enable us to characterize the GATase protein (e.g. its amino acid sequence) and to analyze the structure of the nucleotide-amino acid linkage.

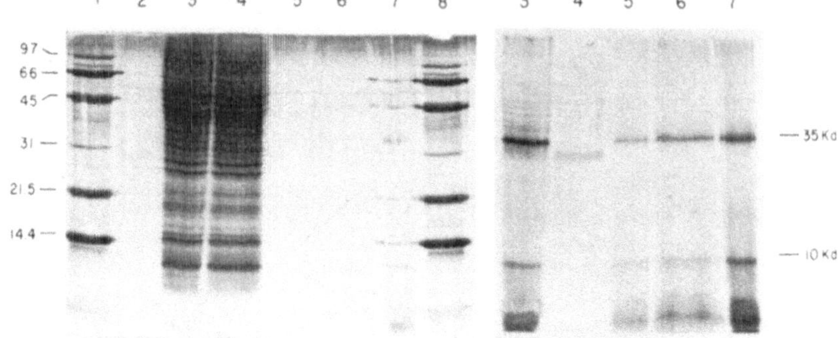

Figure 16. Affinity purification of dA-tailed oligonucleotide-GATase complex. Partially purified (~10 fold) human GATase was reacted with a BCNU-treated oligodeoxynucleotide, 5'dGGGGGGCCCCCC $(A)_{18}$, that was labeled at the 5'end with [δ-^{35}S]ATP. The resulting complex was separated from unreacted oligonucleotide by SDS/KCl precipitation. The redissolved complex was applied to an oligo(dT) cellulose column under conditions of high salt concentration (0.5 M LiCl). SDS-PAGE analysis of the material loaded onto the column is shown in lane 3. Lane 4 is the material that washed through Lanes 5, 6 and 7 are increasing amounts of the material that bound to the column. The bound fraction contained very little protein, while the complex can be seen at 35 kDa in lanes 5, 6, 7 of the autoradiograph and in lane 7 by silver staining. The bulk of the protein and none of the complex washed through the column (lane 4). Free oligonucleotide is apparent at about 10 kDa.

CLINICAL SIGNIFICANCE OF GATase-MEDIATED PRECURSOR REPAIR IN CENU-TREATED HUMAN TUMORS

The Mer$^-$ phenotype, defined as defective host cell reactivation of adenovirus alkylated with MNNG, occurs in only about 20% of human tumor cell lines and is never seen in normal diploid cell lines (Day et al., 1980). Since there appears to be a direct equivalence between the Mer$^-$ phenotype, GATase deficiency and sensitivity to the CENUs, one may speculate that a fraction of human cancers would be Mer$^-$, Gat$^-$ and sensitive to CENUs. We have shown (Brent et al., 1985) that in a series of 6 pediatric rhabdomyosarcoma xenografts in mice, one (Rh28) that was completely deficient in the transferase (Gat$^-$) was curable by a single LD_5 dose of Methyl-CCNU, whereas the two xenograft lines (Rh12 and Rh18) with the highest levels of transferase were totally resistant to the same therapy (Table 2).

Recently, we characterized cultured cell lines that were established from two of these rhabdomyosarcoma xenografts, Rh18 and Rh28, with respect to their alkyltransferase levels and sensitivity to CNU, BCNU and chlorozotocin (Table 3) (Smith and Brent, 1988a). The two lines clearly have maintained their Mer^- phenotype. We also were able to determine, by the alkaline elution method, that few DNA interstrand cross-links were formed in the resistant Rh18 cell line

Table 2. Correlation Between Alkyltransferase Levels and Xenograft Responses to a Therapy with MethylCCNU[a]

Tumor line	Alkyltransferase (fmoles/mg protein)	No. of tumors	Tumor Response[b] CR	PR
Rh 12	658	24	0	4
Rh 18	636	14	0	0
Rh 28	0	25	21	4
Rh 30	186	10	1	9
Rh 35	282	12	2	10
Rh 39	195	10	3	7

[a]MethylCCNU (35 mg/kg) was given as a single i.p. injection.
[b]CR, complete regression of tumor mass. PR, $\geq$50% regression of initial tumor volume at treatment (Modified from Brent et al., 1985).

Table 3. Alkyltransferase Levels and IC_{50} Values for CENU-treated Rhabdomyosarcoma Xenograft-derived Cell Lines[a]

Cell line	Alkyltransferase pmoles/mg protein	IC_{50} (μM) CNU	BCNU	CHLZ[b]
Rh 18	3.8	55	55	60
Rh 28	<0.01	9	9.5	9
Ratio Rh18/Rh28	>380	6.1	5.8	6.7

[a]IC_{50} values were determined from growth inhibition assays.
[b]CHLZ, chlorozotocin

whereas significant cross-linking occurred in the hypersensitive Rh28 line (Figure 17). These two cell lines should prove to be a useful model for studying the Gat phenotype.

The manifestation of Gat⁻ tumors provides a rare opportunity for tumor-selective drug therapy, but first one must establish that Gat⁻ tumors exist in patients analagous to those in cell culture or in xenografts. Second, one must develop assays to distinguish Gat⁺ and Gat⁻ tumor cells in patient material. Alkyltransferase activity has been determined successfully with homogeneous material such as cultured cells and tumor xenografts. Tumor biopsies from patients, however, frequently contain mixed populations, that include host cells as well as necrotic cells, making it impossible to obtain reliable estimates of transferase activity. The approach we have taken to overcome this difficulty has been to develop antibodies as cytological probes for the transferase. This assumes of course that the protein is absent, or antigenically distinct in Mer⁻/Gat⁻ cells. A preliminary report by Yarosh and Ceccoli (1988) suggests that Mer⁻ cells do not react with polyclonal mouse antiserum raised against the partially purified human transferase. However studies with monoclonal antibodies are needed to answer this question

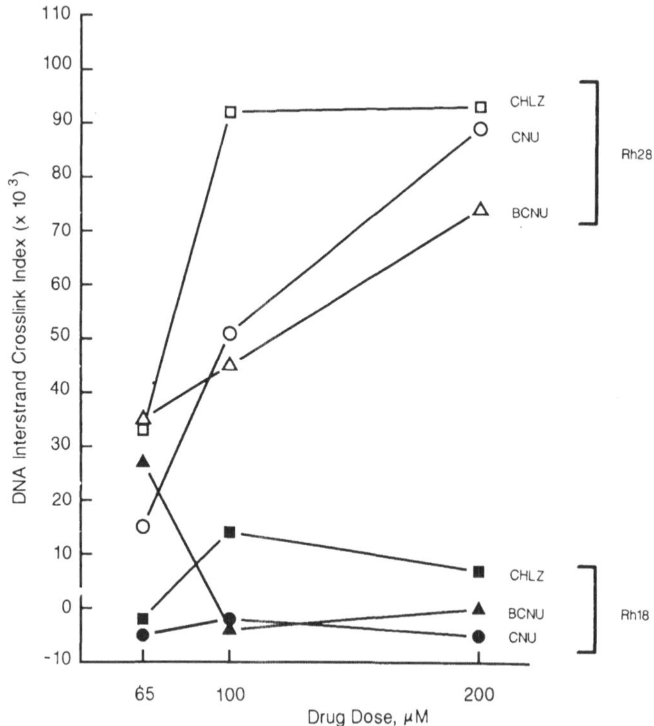

Figure 17. DNA interstrand cross-linking determined by alkaline elution in the Rh18 cell line (closed symbols) compared with the Rh28 cell line (open symbols) 6 hours after a 1 hour treatment with indicated doses of either CNU, BCNU or chlorozotocin (CHLZ) (From Smith and Brent, 1989).

more definitively. Several monoclonal mouse antibodies that are specific for the human transferase have been obtained recently in our laboratory(Brent, et al., 1990).

The highly purified dA-tailed oligonucleotide-GATase complex has been a particularly useful probe for screening and establishing the specificity of such antibodies.

Successful production of antibodies that are highly specific for the human transferase should enable us to develop cytological immunofluorescence assays that can be used on patient biopsy material to identify Mer⁻/Gat⁻ tumors. This information may be used to design chemotherapy with chloroethylating drugs prior to initiation of treatment.

ACKNOWLEDGEMENTS

This work was supported by Grants CA14799, CA21765, CA23099 and CA36888 from the National Cancer Institute and by ALSAC.

REFERENCES

Brent, T.P., 1984. Suppression of cross-link formation in chloroethylnitrosourea-treated DNA by an activity in extracts of human leukemic lymphoblasts. Cancer Res., 44:1887.

Brent, T.P., Houghton, P.J., and Houghton, J.A., 1985. O^6-alkylguanine-DNA alkyltransferase activity correlates with the therapeutic response of human rhabdomyosarcoma xenografts of MeCCNU. Proc. Natl. Acad.Sci. U.S.A., 82:2985.

Brent, T.P., Lestrud, S.O., Smith, D.G., and Remack, J.S., 1987a. Formation of DNA interstrand cross-links by the novel chloro-ethylating agent 2-chloroethyl(methylsulfonyl) methanesulfonate: Suppression by O^6-alkylguanine-DNA alkyltransferase purified from human leukemic lymphoblasts. Cancer Res., 47:3384.

Brent, T.P., Smith, D.G., and Remack, J.S., 1987b. Evidence that O^6-alkylguanine-DNA alkyltransferase becomes covalently bound to DNA containing 1,3-bis(2-chloroethyl)-1-nitrosourea-induced precursors of interstrand cross-links. Biochem. Biophys. Res.Commun., 142:341.

Brent, T.P., Remack, J.S., and Smith, D.G., 1987c. Characterization of a novel reaction by human O^6alkylguanine-DNA alkyltransferase with 1,3-bis(2-chloroethyl)-1-nitrosourea-treated DNA. Cancer Res., 47:6185.

Brent, T.P., and Remack, J.S., 1988. Formation of covalent complexes between O^6-alkylguanine-DNA alkyltransferase and BCNU-treated defined length synthetic oligodeoxynucleotides. Nucleic Acids Res., 16:6779.

Brent, T.P., von Wronski, M., Pegram, C.N., and Bigner, D.D., 1990. Immunoaffinity purification of human O^6-alkylguanine-DNA alkyltransferase using newly developed monoclonal antibodies, Cancer Res. (in press).

Buckley, N., and Brent, T.P., 1988. Structure-activity relations of (2-chloroethyl)nitrosoureas. 2. Kinetic evidence of a novel mechanism for the cytotoxically important DNA cross-linking reactions of (2-chloroethyl)nitrosoureas. J. Am. Chem. Soc., 110:7520.

Day, III, R.S., Ziolkowski, C.H.J., Meyer, S.A., Lubinieki, A.S., Girardi, A.J., Galloway, S.M., and Bynum, G.D., 1980. Defective repair of alkylated DNA by human tumor and SV40-transformed human cell strains. Nature, 288:724.

Erickson, L.C., Laurent, G., Sharkey, N.A., and Kohn, K.W., 1980.DNA cross-linking and monoadduct repair in nitrosourea-treated human tumor cells. Nature, 288:727.

Gibson, N.W., Hartley, J.A., Strong, J.M., and Kohn, K.W., 1986. 2-chloroethyl(methylsulfonyl)methanesulfonate (NSC-338947) a more selective alkylating agent than the chloroethylnitrosoureas. Cancer Res., 46:553.

Kacinski, B.M., Rupp, W.D., and Ludlum, D.B., 1985. Repair of haloethylnitrosourea-induced DNA damage in mutant and adapted bacteria. Cancer Res., 45:6471.

Ludlum, D.B., and Tong, W.P., 1985. DNA modification of the nitrosoureas: Chemical nature and cellular repair, in: Cancer Chemotherapy, Volume 2, F.M. Muggia, ed., Martinus Nijhoff, Boston.

Ludlum, D.B., Mehta, J.R., and Tong, W.P., 1986. Prevention of 1-(3-deoxycytidyl),2-(1-deoxyguanosyl)-ethane cross-link formation in DNA by rat liver O^6-alkylguanine-DNA alkyltransferase. Cancer Res., 46:3353.

Robins, P., Harris, A.L., Goldsmith, T., and Lindahl, T., 1983. Cross-linking of DNA induced by chloroethylnitrosourea is prevented by O^6-methylguanine-DNA methyltransferase. Nucleic Acids Res., 22:7743.

Scudiero, D.A., Meyer, S.A., Clatterbuck, B.E., Mattern, M.R., Ziolkowski, C.H.J., and Day III, R.S., 1984. Sensitivity of human cell strains having different abilities to repair O^6-methylguanine in DNA to inactivation by alkylating agents including chloroethylnitrosoureas. Cancer Res., 44:2467.

Smith, D.G., and Brent, T.P., 1988. Response to treatment with Carmustine (BCNU) of cultured cell lines established from pediatric rhabdomyosarcoma xenografts. Proc. Am. Assoc. Cancer Res., 29:269.

Smith, D.G., and Brent, T.P., 1989. Response of cultured cell lines from rhabdomyosarcoma xenografts to treatment with chloroethylnitrosoureas. Cancer Res., 49:883.

Tong, W.P., Kirk, M.C., and Ludlum, D.B., 1983. Mechanism of action of the nitrosoureas, V: Formation of O^6-(2-fluoroethyl)guanine and its probable role in crosslinking of deoxyribonucleic acid. Biochem. Pharmacol., 32:2011.

Yarosh, D.B., and Ceccoli, J., 1988. Monoclonal antibodies against the human O^6-methylguanine-DNA methyltransferase. Proc. Am. Assoc. Cancer Res., 29:1.

INDUCTION OF FUTILE DNA REPAIR PROCESSES BY BIFUNCTIONAL INTERCALATORS

Bernard Lambert[1], Evelyne Segal-Bendirdjian,[1]
Bernard P. Roques[2] and Jean-Bernard Le Pecq[1]

[1]Institut Gustave Roussy (URA158 CNRS, U140 INSERM) 94805 Villejuif, France

[2]University Paris V (UA498 CNRS, U226 INSERM) 4, av de l'Observatoire 75006 Paris, France

SUMMARY

Ditercalinium is a DNA bifunctional intercalator endowed with antitumor properties. It binds to DNA with high affinity, forming a noncovalent and reversible complex. NMR studies of a ditercalinium oligonucleotide complex have revealed that, in the DNA complex, the linking chain of ditercalinium is located in the large groove of the DNA double helix, and that a bending of the double helix is induced toward the small groove.

It was found that ditercalinium and its antitumor analogues, but not its monomeric derivative, elicit a specific toxicity on polA strains of E. coli. This toxicity is suppressed by the uvrA mutation. In polA strains, ditercalinium induces SOS functions, although no immediate DNA synthesis arrest is observed. In the thermosensitive ligase deficient strain lig7, single-strand breaks accumulate at non-permissive temperature. It is proposed that the structural alteration induced in DNA after ditercalinium binding is recognized by the UVR repair process in E. coli. Because of the noncovalent nature of the DNA-ditercalinium complex, this structural alteration acts as a mock lesion or a lure which triggers a futile and abortive repair process.

In mammalian cells, ditercalinium causes a delayed cytotoxicity. Cells treated with ditercalinium die only 5 to 6 generations after drug exposure. Mitochondrial DNA is rapidly lost, and no damage is observed on nuclear DNA. It is hypothesized that the loss of mitochondrial DNA is related to a malfunctioning of a DNA repair process. DNA structural alterations, caused by agents forming non-covalent and reversible complexes with DNA, appear as a new type of lesion which could, in some conditions, completely fool the DNA repair machinery, and which would then become deleterious for the cells.

DNA Repair Mechanisms and Their Biological Implications in Mammalian Cells
Edited by M.W. Lambert and J. Laval
Plenum Press, New York

INTRODUCTION

Several DNA-intercalating compounds, such as actinomycin D and anthracyclins are widely used as antitumoral agents. It is agreed that DNA binding is a prerequisite to the pharmacological properties of these molecules. In order to search for more active drugs, polyintercalation compounds, characterized by a high DNA binding affinity, have been synthesized (rev. Le Pecq and Roques, 1986 ; Wakelin, 1986). Molecules with two or three intercalating rings, linked by chains of appropriate length and structure to allow the DNA intercalation of each subunit, have been obtained. These molecules bind to DNA with an extremely high binding constant (up to $10^{14}M^{-1}$ for trisintercalating molecules) (Laugâa, et al., 1985). Such molecules could therefore compete with DNA binding proteins, such as RNA polymerase or repressors of which the DNA binding constants are in the range of 10^{10} to $10^{13}M^{-1}$.

Among these molecules, several dimeric molecules derived from 7H-pyridocarbazole, such as ditercalinium (Figure 1), elicited strong antitumor properties on various animal tumor models (Roques, et al., 1979 ; Pelaprat, et al. , 1980).

Ditercalinium bisintercalates into DNA with a binding affinity greater than 10^7M^{-1}. Recently NMR studies, performed on a complex between ditercalinium and a deoxytetranucleotide, allowed us to show that this compound interacted with DNA by the major groove causing some distortion of the double helix, possibly bending towards the small groove (see model of this interaction in Figure 2) (Delbarre, et al., 1987).

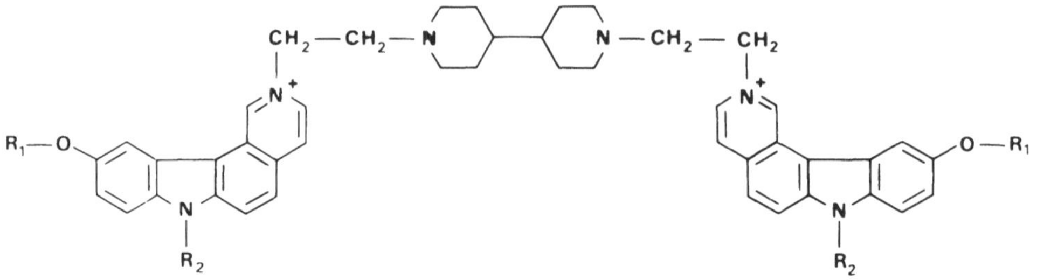

Figure 1. Structure of 7H-pyridocarbazole dimers (ditercalinium series). Ditercalinium: R_1 = OCH3, R_2 = H.

Since ditercalinium elicited strong antitumor properties, it was of interest to study its mechanism of action at the molecular level. This study was performed on two experimental systems : prokaryotic cells and eukaryotic cells. The results which were obtained show that ditercalinium acts according to an amazing mechanism which is completely different from that

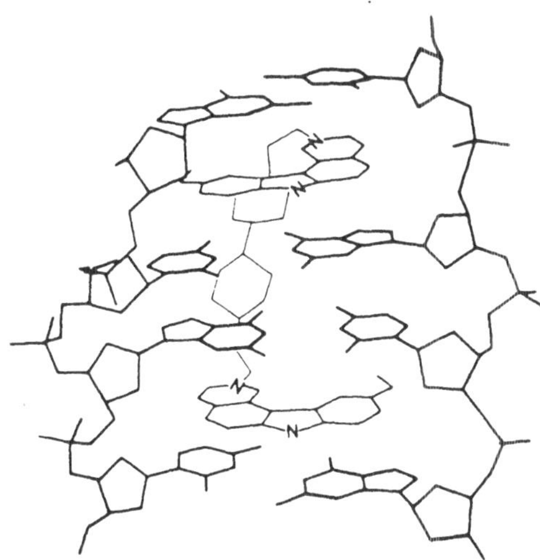

Figure 2. Dreiding model of a d(CpGpCpG)$_2$ minihelix bisintercalated with a ditercalinium molecule and viewed from the minor groove (from Delbarre, et al., 1987).

of antitumor monointercalating compounds and other anticancer agents.

STUDY OF THE MECHANISM OF ACTION OF DITERCALINIUM ON PROKARYOTIC CELLS

In order to have a better understanding of the ditercalinium cytotoxicity, and in order to determine the potential mutagenic properties of this compound, the mechanism of action of ditercalinium was studied on _E.coli_. Since wild-type _E. coli_ cells were found resistant to ditercalinium, an _E. coli_ mutant strain specifically sensitive to ditercalinium was isolated (Lambert and Le Pecq, 1982). Studies of this mutant strain and studies of the cytotoxic action of ditercalinium on this mutant led us to propose that ditercalinium, which forms a reversible and high affinity complex with DNA, interfered with the nucleotide excision repair system. In addition, it was observed that the interference of ditercalinium with this DNA repair system is completely different from that observed in the case of covalent bulky adducts (Lambert, et al., 1988). The different observations which were made, are described below.

polA and lig7 mutations specifically confer sensitivity to ditercalinium

We first isolated, after MNNG mutagenesis, an _E. coli_ ditercalinium sensitive strain. Analysis of this ditercalinium sensitive mutant revealed that it had a very low DNA polymerase I activity. This suggested that a _pol_A mutation could be responsible for the ditercalinium sensitivity. This hypothesis was confirmed by the analysis of the relative ditercalinium sensitivity of genetically defined _pol_A$_1$ and

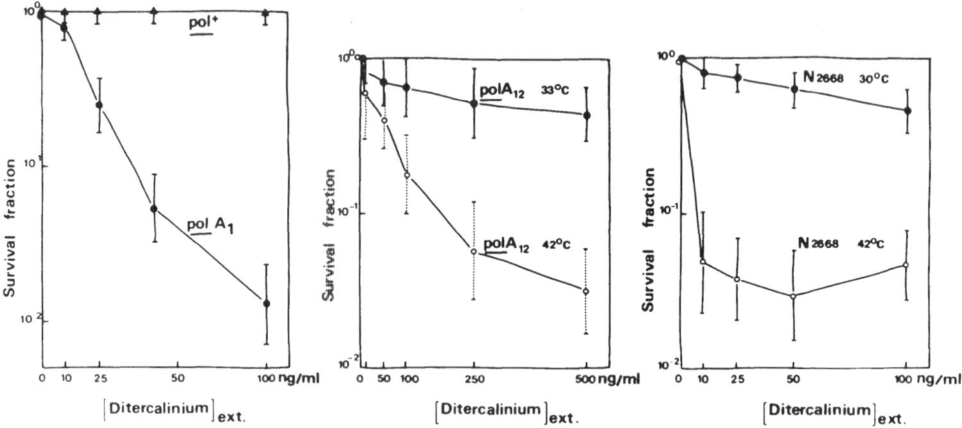

Figure 3. Survival of polA$_1$, PolA$_{12}$, and lig7 mutant strains after ditercalinium treatment. Bacteria were grown in M9 medium containing 50 μg/ml of thymine, at 33°C. At a 0.4 absorbance (λ = 600 nm) bacteria were diluted four fold. 2 ml aliquots of bacterial suspension were treated with various concentrations of ditercalinium and incubated at 33°C or 42°C depending on the thermosensitivity of the polA or lig7 mutations. After 120 minutes of incubation, bacteria were plated on LBT plates. Colonies were counted after an overnight incubation at 33°C (from Lambert, et al., 1988).

polA$_{12}$ mutant strains (see Figure 3). It was observed that ditercalinium was cytotoxic on polA$_1$ cells and on polA$_{12}$ cells, at nonpermissive temperature, when these cells expressed the polA phenotype. In addition, a larger number of polA$^+$ revertant cells were obtained from polA$_1$ cells by reversion of the MMS sensitivity. All these MMS resistant strains were found resistant to ditercalinium.

Furthermore, the lig7 mutant cells were also found ditercalinium sensitive, at nonpermissive temperature, when the ligase activity of this mutant is deficient. However, single mutants deficient in one of the DNA repair functions (uvrA, uvrB, uvrC, uvrD, recA, lexA, ruv, lon, umuC, sfiA, sfiC) were found resistant to ditercalinium.

Ditercalinium sensitivity of polA strain is reverted by uvrA mutation

In order to determine the relationship between ditercalinium sensitivity and polA mutation, ditercalinium resistant cells were isolated by plating polA cells on medium containing a high concentration of ditercalinium. However, analysis of these ditercalinium resistant cells revealed that they had not become pol$^+$ cells but had kept the polA phenotype. In addition, all these ditercalinium resistant polA cells were found more UV sensitive than the original polA cells. This result suggested that the acquired ditercalinium resistance of polA cells resulted from an additional mutation in a gene involved in the nucleotide excision repair system. Construction was performed of double mutants harboring the polA mutation and another mutation in one of the genes involved in the nucleotide excision repair system. Only double mutants harboring the polA mutation and either the uvrA or uvrC mutation were found viable. Analysis of these double mutants revealed that the uvrA mutation and not the uvrC mutation completely suppressed the ditercalinium cytotoxicity in polA cells (Figure 4).

This result was confirmed by a complementation test carried out on the ditercalinium resistant polA cells. These cells, which were found more UV sensitive than the polA cells, were lysogenized either by the λ(UvrA$^+$) or λ(UvrC$^+$) phage. It was observed that only the cells lysogenized with the λ(UvrA$^+$) phage regained their sensitivity to ditercalinium. The polA$_1$ cells sensitivity to ditercalinium can therefore be suppressed by a uvrA mutation and can be restored by the functional expression of the UvrA protein.

Ditercalinium induces only DNA single strand breaks in DNA ligase deficient cells

It has been shown that in polA$_1$ cells, the repair of a DNA covalent adduct, by the nucleotide excision repair system, leads to the accumulation of DNA single-strand breaks (Kanner and Hanawalt, 1970; Monk, *et al.*, 1971). The presence of DNA single-strand breaks was found to be the factor responsible

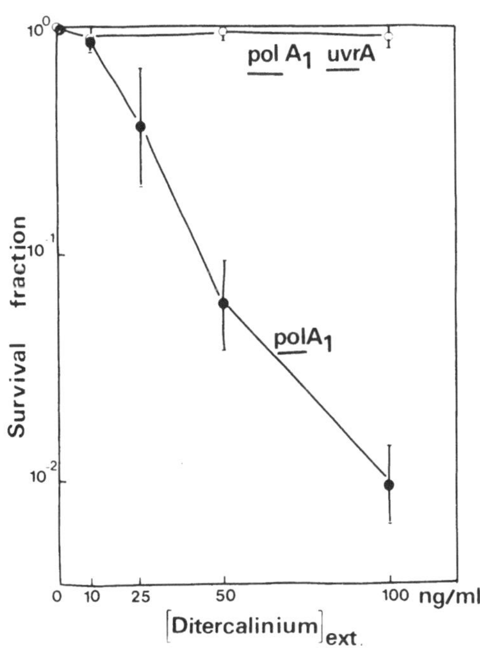

Figure 4. Survival of polA$_1$ uvrA mutant strain after ditercalinium treatment. Bacteria were grown and treated as described in the legend of Figure 3 (from Lambert, *et al.*, 1988).

for the cytotoxic and mutagenic properties of covalent adducts in polA cells. The DNA of $polA_1$ cells and pol^+ cells was analyzed on alkaline sucrose gradients in the absence and in the presence of ditercalinium. No DNA single-strand breaks could be detected in pol^+ and $polA_1$ cells after ditercalinium treatment. In addition, no unscheduled DNA repair synthesis was observed in pol^+ strains after ditercalinium treatment. These observations suggested that the DNA-ditercalinium complex did not lead to a complete repair cycle. However, it was shown that, at nonpermissive temperature, DNA single strand breaks accumulated in the lig7 mutant strain, suggesting that DNA was indeed incised (Lambert, et al., 1988).

Ditercalinium induces an SOS system in $polA_1$ strain

After ditercalinium treatment, filamentation of $polA_1$ cells was observed. This observation suggested that ditercalinium could induce SOS functions in $polA_1$ cells. SOS inducing ability of ditercalinium was measured in $polA_1$ and pol^+ cells lysogenized with λ(sfiA::lacZ) phage. In these strains, the SOS induction can be detected by measuring the β-galactosidase synthesis induction (Quillardet, et al., 1982). The results are shown in Figure 5. Ditercalinium is able to induce β-galactosidase synthesis in $polA_1$ cells and not in pol^+ cells. In addition, the SOS inducing ability of several ditercalinium analogs was studied. It is shown that the length of the linking chain between the two pyridocarbazole rings influences the SOS inducing ability of these analogs. It is interesting to note that the ditercalinium analog (IV), which was found unable to induce β-galactosidase synthesis and which did not elicit cytotoxicity on $polA_1$

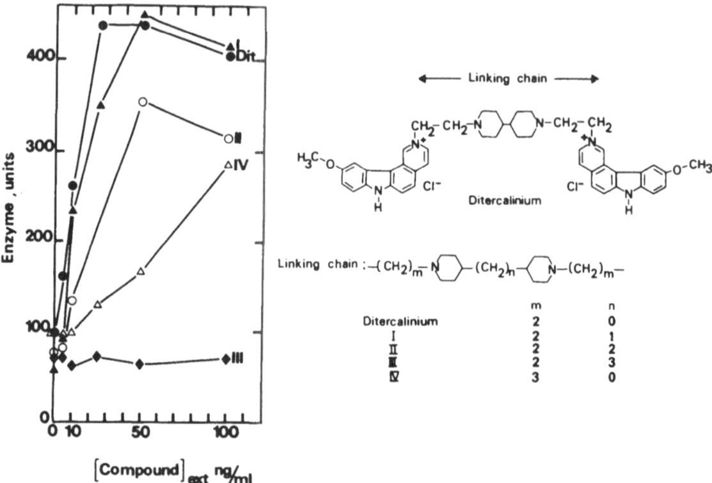

Figure 5. SOS inducing ability of ditercalinium and its analogues on pol^+ [λ(sfiA::lacZ)] and $polA_1$[λ(sfiA::lacZ)] strains. Bacteria were incubated and treated at 37^0C with ditercalinium or its analogues as described (from Lambert, et al., 1988). After 120 minutes of incubation, β-galactosidase synthesis measurement was performed as described by Quillardet, et al., 1982.

cells, is devoided of antitumor activity and cytotoxicity on mammalian cells.

Ditercalinium is therefore able to induce the SOS system in polA$_1$ cells. However, no DNA lesion, known to be able to induce the SOS system, was detected. No DNA single-strand breaks were observed in polA$_1$ cells and, in addition, it was observed that ditercalinium did not inhibit macromolecular synthesis at cytotoxic doses even after 4 hours of treatment. All these different observations led us to propose the following model (Figure 6).

Ditercalinium binds reversibly to DNA and induces a DNA conformational change recognized as a lesion by the UvrAB complex. UvrC protein binds later and DNA is incised (as single-strand breaks are observed in lig7 at 42^{0}C). Ditercalinium analogues with long linking chains, which could bind to DNA without inducing such conformational change, could not be recognized and do not elicit any cytotoxic or SOS inducing effects.

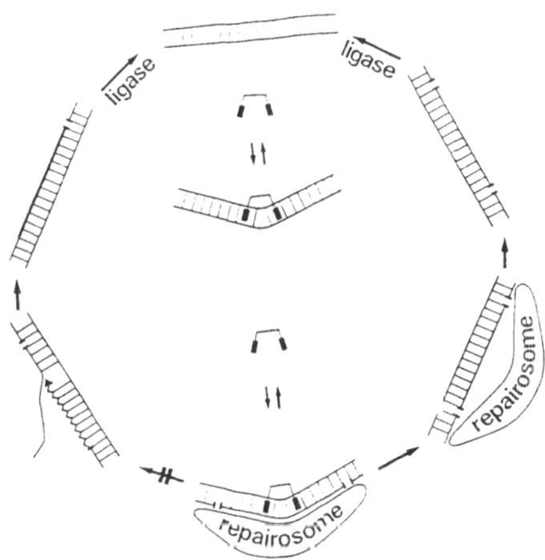

Figure 6. Model of the futile and abortive DNA repair process (from Lambert, et al., 1988).

In wild-type strains, polA and UvrD proteins could not displace the incised DNA fragment because of the large increase of DNA stability caused by the ditercalinium binding (Garbay-Jaureguiberry, et al., 1987). This would account for the lack of unscheduled DNA repair synthesis in wild-type strains and the lack of gaps in DNA after ditercalinium treatment in the polA$_1$ strain and the absence of immediate inhibition of DNA synthesis.

Because ditercalinium is not covalently bound, it will dissociate from the DNA-repairosome complex, leaving the repairosome bound to a DNA of normal structure. In a wild type environment, polA and UvrD gene products increase the turnover of the DNA UvrABC complex (Caron, et al., 1985; Husain, et al., 1985 ; Grossman, et al., 1988 ; Sancar A. and Sancar G., 1988). In the presence of DNA polymerase I, after ditercalinium is released, the repairosome bound to DNA of normal structure could dissociate and single-strand breaks could be rapidly sealed by ligase, accounting for the observed presence of DNA breaks in the lig7 mutant and their absence in the pol^{+} strain. In the same conditions, but in the absence of DNA polymerase I, the UvrABC complex would remain bound to DNA for a much longer time. We propose, therefore, that the presence of a UvrABC complex bound to a normal DNA structure is the factor responsible for cytotoxicity. How such a factor is responsible for cytotoxicity is presently unknown. One could have thought that the uncontrolled SOS induction could lead to cell death. This could be suggested by the fact that ditercalinium cytotoxicity is decreased after chloramphenicol treatment.

However, the strains polA sfiA sfiC, polA lon, polA umuC, which associate polA mutation with a mutation involved in SOS response, appear as sensitive to ditercalinium as polA. Furthermore, in the double mutant polA lexA, SOS functions cannot be induced. However, this strain appears significantly more ditercalinium sensitive than polA, suggesting that some SOS function could contribute to the protection of the cells as observed for covalent adducts (Lambert, et al., 1988).

Another possible explanation would be the deprivation of UvrB protein which is triggered by the binding of the UvrABC complex. $PolA_1$ cells would become phenotypically $polA_1$ UvrB cells. Such a combination of mutations is lethal.

Nevertheless, according to such a model, ditercalinium, by inducing a DNA conformational change similar to the one caused by covalent adducts, would act as a mock lesion for the repair system because of its noncovalent nature. Attempts to repair such a mock lesion would lead to a futile DNA repair process lethal in a polA mutant. As expected in such a model, stopping this futile repair cycle by a uvrA mutation in a polA strain completely suppresses ditercalinium toxicity.

These experiments reveal a very important property of DNA polymerase I. Its presence appears to be absolutely required to prevent toxic effects resulting from the noncovalent binding to DNA of molecules which induce DNA structural modifications recognized by the repair incision complex. This property might be related to both its polymerase activity and its ability to increase the turnover of the repairosome, accounting for the lack of uvrD effect. Most significantly, the toxicity of ditercalinium analogues in polA strains correlates with the cytotoxicity and the antitumor activity of these derivatives (unpublished results). Studies to analyse whether ditercalinium and its derivatives induce such processes in eucaryotic systems are presently in progress in our laboratory.

Recent studies, using the in vitro reconstituted nucleotide excision repair system of E. coli, were performed in collaboration with A. Yeung in Philadelphia. Preliminary results confirm the first step of this model, the recognition of the reversible complex, ditercalinium-DNA, by the UvrAB protein and DNA incision in presence of the UvrC protein.

STUDY OF THE MECHANISM OF ACTION OF DITERCALINIUM IN EUKARYOTIC CELLS

Studies of the mechanism of action of ditercalinium on E. coli cells show that this compound exerts its cytotoxicity according to an original mechanism. The same situation is encountered in the eukaryotic system (Bendirdjian, et al., 1984; Esnault, et al., 1984; Markovits, et al., 1986; Segal-Bendirdjian, et al., 1988; Fellous, et al., 1988). Ditercalinium is characterized by a delayed cytotoxicity. Mammalian cells are still able to grow for five or six generations before getting arrested after the drug is washed out. Ditercalinium does not provoke the arrest of cell cycle progression in the G_2 + M phase of the cell cycle, as it is observed with monointercalating agents and many other anticancer drugs.

Electron microscopy studies (Figure 7) show that after treatment of cells with ditercalinium, irreversible alterations of mitochondria occurred without noticeable modifications of the nucleus structure. Mitochondria became enlarged and the cristae were reduced in number and size. Some of them completely lacked cristae. The matrix of some mitochondria contained electron opaque fibrous material. These modifications correlate with a selective association of ditercalinium in mitochondria (Fellous, et al., 1988).

These mitochondrial alterations are also associated with a stimulation of anaerobic glycolysis, as shown by the accumulation of lactic acid in the media of treated cell cultures (Segal-Bendirdjian, et al., 1988).

These results suggested that instead of nuclear DNA, mitochondrial DNA might be a specific target of this drug. To investigate the effect of ditercalinium on mitochondrial DNA content of eukaryotic cells, total cellular DNA was extracted from control and ditercalinium treated cells during and at different times following ditercalinium treatment. Restriction fragments of cellular DNA were analyzed by agarose gel electrophoresis. After transfer to nitrocellulose filters according to Southern (1975), DNA was hybridized with either a mitochondrial DNA probe or with a nuclear DNA probe. The nuclear DNA probe hybridized with a repetitive DNA sequence interspersed in the whole genome. This permitted detection of DNA alteration independently of genetic location. Results are shown on Figure 8. 24 hours after washing out ditercalinium, the mitochondrial DNA band completely disappears. Even after 6 days of culture of treated cells there is no reappearance of the band corresponding to mitochondrial DNA. However, no variation of nuclear DNA content and size was observed.

This disappearance of mitochondrial DNA is associated with the loss of cytochrome oxidase activity. The activity of

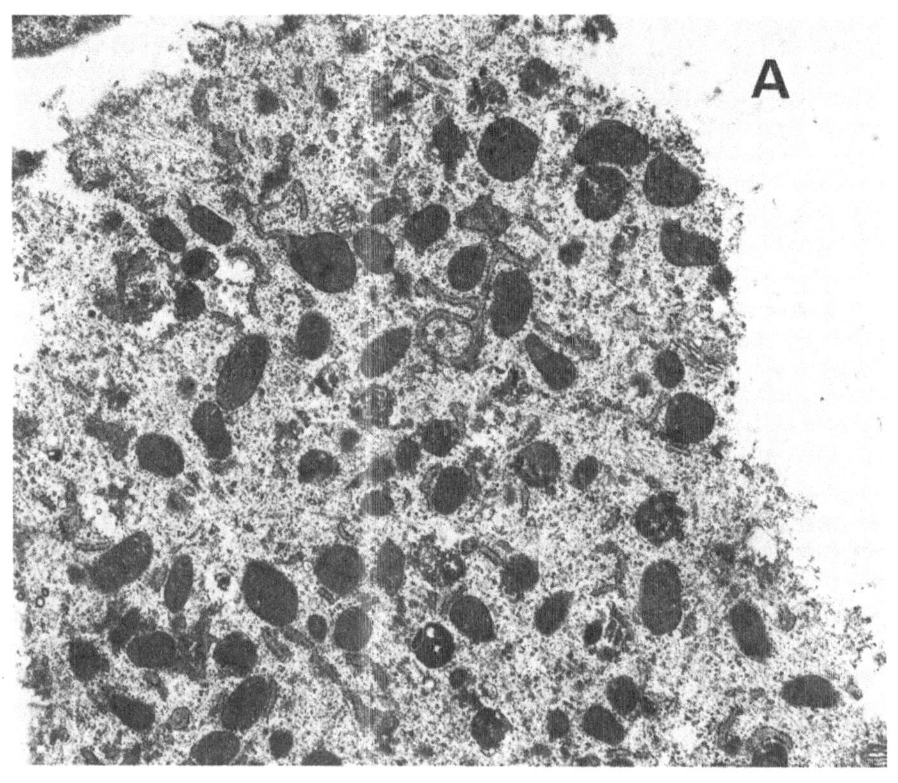

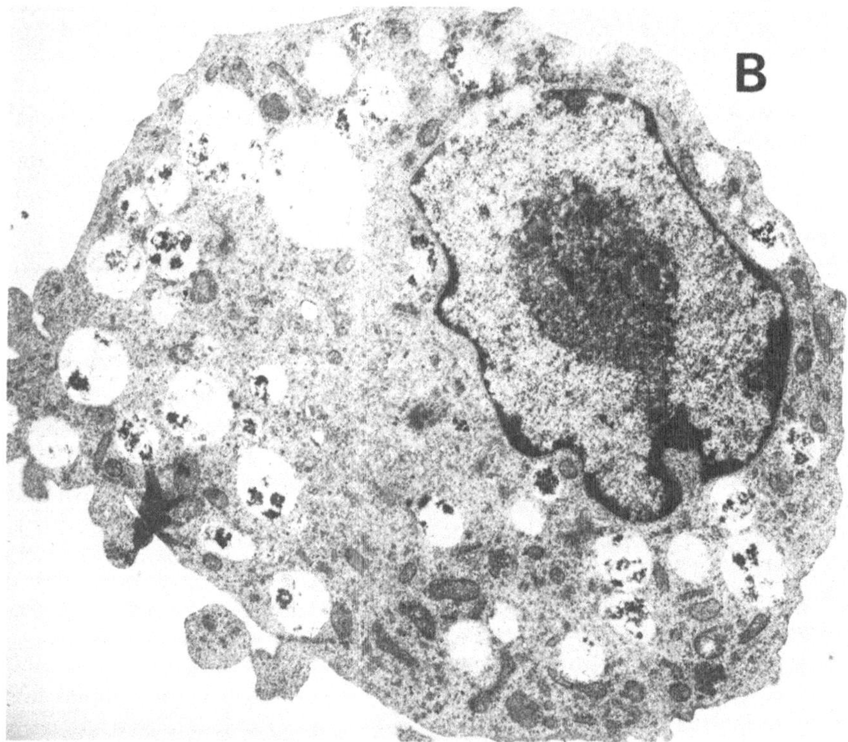

Figure 7. Electron micrographs of a Chinese hamster cell line: A. control cells. B. ditercalinium treated cells.

this enzyme decreases after ditercalinium treatment with a half life of 24 hours, a value close to that of its turnover.

The primary effect of ditercalinium on eukaryotic cells appears therefore to be at the level of mitochondrial DNA, leading to its specific and irreversible loss. This does not exclude that some nuclear DNA alterations could contribute to cytotoxicity. Such alterations have been reported (Markovits, et al., 1986; Traganos, et al., 1987). However, the fact that cells deficient in respiration are considerably more resistant to ditercalinium suggests that ditercalinium exerts its action mainly at the level of mitochondria (Segal-Bendirdjian, et al., 1988). The mitochondrial effect is

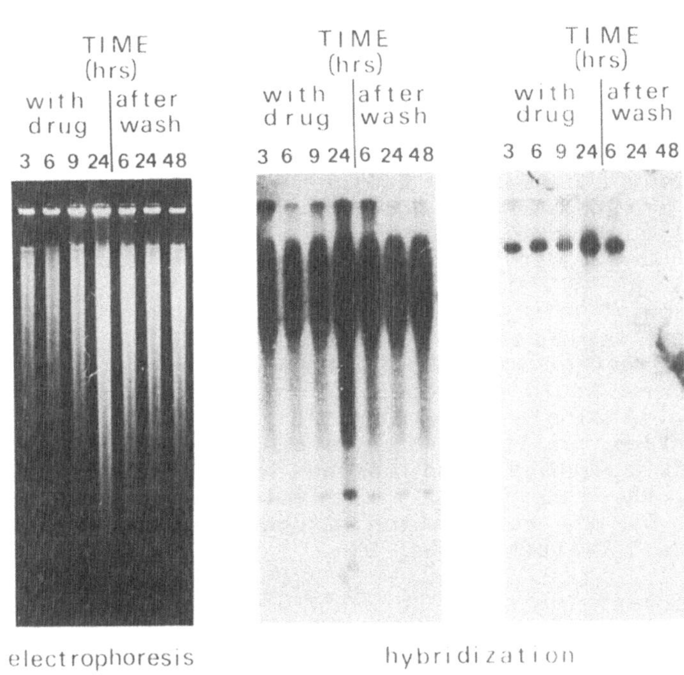

Figure 8. Electrophoresis and hybridization with either a mitochondrial DNA probe, or a nuclear DNA probe to the Hind III treated cellular DNA prepared from L1210 cells. Cells were treated 24 hours with 0.14 μM of Ditercalinium. DNA extractions were made either during (3, 6, 9, 24 hours) the treatment or 6, 24 and 48 hours after removing the drug. DNA was treated with Hind III restriction enzyme and analyzed using agarose gel electrophoresis. After transfer to nitrocellulose filters according to Southern (1975), DNA was hybridized with either a mitochondrial DNA probe containing the entire mouse mitochondrial DNA genome (gift from Dr. D.A. Clayton, Stanford University, CA) or with a nuclear DNA probe containing repetitive sequences (Meudier-Rotival et al., 1979) (from Segal-Bendirdjian et al., 1988).

probably responsible for the delayed cytotoxicity. Such a delayed cytotoxicity, already observed after cell treatment with chloramphenicol and ethidium bromide (Morais, R., 1980), seems to represent a general feature of drugs acting on mitochondria. The mechanism leading to mitochondrial DNA loss after ditercalinium treatment is presently unknown. Previous papers reported that mitochondrial DNA could represent an important target for carcinogens and mutagens but that there is no traditional DNA repair at the level of mitochondrial DNA (Clayton, D.A., 1974). On the other hand, recent experiments suggest that mutagenesis of mitochondrial DNA is a rare event (Mita, et al., 1988). In addition, mitochondrial nucleases that could induce a specific degradation of mitochondrial DNA have been identified and purified (Cummings, et al., 1987). It was suggested from these results that the strategy of DNA repair in mitochondria is quite different from the strategy of DNA repair in the nucleus. In the nucleus, the well characterized excision patch repair is operating. In mitochondria, since 200 to 400 copies of mitochondrial DNA are present, another DNA repair strategy is probably used. It has been suggested that a DNA molecule harboring a lesion, rather than being repaired, is completely degraded. The degraded molecule could easily be replaced by stimulation of the replication of unmodified molecules present in large excess. If such a strategy really holds in mitochondria, it can easily be understood how molecules such as ditercalinium can exert their toxic effect. A single molecule could bind to a mitochondrial DNA molecule, mimic a lesion, and cause the degradation of this DNA molecule. In the degradation process, the ditercalinium molecule would be released and would be able to bind to another molecule of DNA and induce its degradation. This process could be repeated until the last molecule is degraded. A single molecule of ditercalinium has therefore the potential ability to induce the complete degradation of mitochondrial DNA in the absence of DNA replication. In practice, the rate of DNA degradation has to be larger than the rate of DNA replication to observe the complete disappearance of mitochondrial DNA.

Ditercalinium seems to exert its toxic action in a very amazing way, by fooling completely the DNA repair system. No natural substances, exerting a toxic action according to such a mechanism, have yet been described to our knowledge. If such substances exist, they may have been very difficult to detect. Their effects can only be analyzed quite a long time after they have exerted their actions. They do not leave any traces behind them except possibly DNA mutations or deletions difficult to detect with the present tests.

REFERENCES

Bendirdjian, J. -P., Delaporte, C., Roques, B. P., and Jacquemin-Sablon, A., 1984, Effects of 7H-pyridocarbazole mono and bifunctional DNA-intercalators on Chinese hamster lung cells in vitro, Biochem. Pharmacol., 33: 3681.

Caron, P. R. , Kushner, S. R., and Grossman, L., 1985, Involvement of helicase II (UvrD gene product) and DNA polymerase I in excision mediated by the uvrABC protein complex, Proc. Natl. Acad. Sci. USA, 82:4925.

Clayton, D. A., Doda, J. N., and Friedberg, E. C., 1974, The absence of a pyrimidine dimer repair mechanism in mammalian mitochondria, Proc. Natl. Acad. Sci. USA, 71:2777.

Cummings, O. W., King, T. C., Holden, J. A. , and Low, R. L., 1987, Purification and characterization of the potent endonuclease in extracts of bovine heart mitochondria, J. Biol. Chem., 262:2005.

Delbarre, A., Delepierre, M., Garbay, C., Igolen, J., Le Pecq, J. -B., and Roques, B. P., 1987, Geometry of the antitumor drug Ditercalinium bisintercalated into d(CpGpCpG)2 by 1H NMR, Proc. Natl. Acad. Sci. USA. 84:2155.

Esnault, C., Roques, B. P., Jacquemin-Sablon, A., and Le Pecq, J. -B., 1984, Effectg of new antitumor bifunctional intercalators derived from 7H-pyridocarbazole on sensitive and resistant L1210 cells, Cancer Res., 44:4335.

Fellous, R., Coulaud, D. , El Abed, I., Roques, B. P. , Le Pecq, J. -B., Delain, E., and Gouyette, A., 1988, In vivo and in vitro cytoplasmic accumulation of Ditercalinium in rat hepatocytes induces mitochondrial damages, Cancer Res. In press.

Garbay-Jaureguiberry, C., Laugaâ, P., Delepierre, M., Laalami, S., Muzard, G., Le Pecq, J. -B., and Roques, B. P., 1987, DNA bis-intercalators as new antitumor agents: modulation of the anti-tumour activity by the linking chain rigidity in the Ditercalinium series, Anti-Cancer Drug Design, 1:323.

Grossman, L. , Caron, P. R., Mazur, S. J. , and Oh, E. Y., 1988, Repair of DNA-containing pyrimidine dimers, FASEB J., 2:2696.

Husain, I., Van Houten, B., Thomas, D. C. , Abdel-Monem, M., and Sancar, A. , 1985, Effect of DNA polymerase I and DNA helicase II on the turnover rate of UvrABC excision nuclease, Proc. Natl. Acad. Sci. USA, 82:6774.

Kanner, L. , and Hanawalt, P. C. , 1970, Repair deficiency in DNA polymerase, Biochem. Biophys. Res. Commun. , 39 :149.

Lambert, B. , and Le Pecq, J. -B., 1982, Isolement et caractérisation de souches d' E. coli sensibles à des toxiques hydrophiles et/ou charges, C. R. Hebd. Seances, Acad. Sci., Ser. C, 294:447.

Lambert, B., Roques, B. P., and Le Pecq, J. -B., 1988, Induction of an abortive and futile DNA repair process in E. coli by the antitumor DNA bifunctional intercalator, Ditercalinium: role of polA in death induction, Nucleic Acids Res., 16:1063.

Laugâa, Ph., Markovits, J., Delbarre, A., Le Pecq, J. -B., and Roques, B. P., 1985, DNA tris-intercalation: First acridine trimer with DNA affinity in the range of DNA regulatory proteins. Kinetic studies, Biochemistry, 24:5567.

Le Pecq, J. -B., and Roques, B.P., 1986, DNA binding and biological properties of bis- and trisintercalating molecules, in: 'Mechanisms of DNA damage and repair, M. G. Simic, L. Grossman, and A. C. Upton, ed., Plenum Press Publ., New York, p. 219.

Markovits, J., Pommier, Y., Mattern, M. R., Roques, B. P., Le Pecq, J. -B., and Kohn, K. W., 1986, Effect of the bifunctional antitumor intercalator Ditercalinium on DNA in mouse leukemia (L1210) cells and on L1210 DNA topo isomerase II, Cancer Res., 46:5821.

Meunier-Rotival, M., Soriano, P., Cuny, G., Strauss, F., and Bernadi, G., 1982, Sequence organisation and genomic distribution of the major family of interspersed repeats of mouse DNA, Proc. Natl. Acad. Sci. USA, 79:355.

Mita, S., Monnat, R. J., and Loeb, L. A., 1988, Resistance of HeLa cell mitochondrial DNA to mutagenesis by chemical carcinogens, Cancer Res., 48:4578.

Monk, M., Peacey, M., and Gross, J. D., 1971, Repair of damage induced by ultraviolet light in DNA polymerase defective Escherichia coli cells, J. Mol. Biol., 58:623.

Morais, R., 1980, On the effect of inhibitors of mitochondrial macromolecular synthesizing systems and respiration on the growth of cultures chick embryo, J. Cell. Physiol., 103:455.

Pelaprat, D., Delbarre, A., Le Guen, I., Roques, B. P., and Le Pecq, J. -B., 1980, DNA intercalating compounds as potential antitumor agents. 2. Preparation and properties of 7H-pyridocarbazole dimers, J. Med. Chem. , 23:1336.

Quillardet, P., Huisman, O., d'Ari, R., and Hofnung, M., 1982, SOS chromotest, a direct assay of induction of an SOS function in E. coli K12 to measure genotoxicity, Proc. Natl. Acad. Sci. USA, 79:5971.

Roques, B. P., Pelaprat, D., Le Guen, I., Porcher, G., Gosse, Ch., and Le Pecq, J. -B., 1979, DNA bifunctional intercalators. Antileukemic activities of new pyridocarbazole dimers, Biochem. Pharmacol., 28:1811.

Sancar, A., and Sancar, G.B., 1988, DNA repair enzymes, Ann. Rev. Biochem., 57:29.

Segal-Bendirdjian, E., Coulaud, D., Roques, B. P., and Le Pecq, J. -B., 1988, Selective loss of mitochondrial DNA after treatment of cells with Ditercalinium (NSC 335153), an antitumor bis-intercalating agent, Cancer Res., 40: 4982.

Southern, E. M., 1975, Detection of specific sequences among DNA fragments separated by electrophoresis, J. Mol. Biol., 98:505.

Traganos, F., Bueti, C., Melamed, M. R., and Darzynkiewicz, Z. , 1987, Cytokinetic effects of bifunctional antitumor intercalator Ditercalinium on Friend erythroleukemia cells, Leukemia (Baltimore), 1:411.

Wakelin, L. P. G., 1986, Polyfunctional DNA intercalating agents, Med. Res. Rev., 6:275.

Participants at the NATO Advanced Research Workshop "DNA Repair Mechanisms and Their Biological Implications in Mammalian Cells" (from left to right)

First Row: M. Zdzienicka, B. Strauss, M. Stefanini, P. Karran, E, Moustacchi, A. Lehmann, P. Hanawalt, C. Lambert, J. Laval, M. Lambert, T. Lindahl, B. Singer, A. Sarasin, F. Laval, M. Sekiguchi

Second Row: A. Abbondandolo, D. Hunting, S. West, M. Smerdon, K. Kraemer, R. Day, A. Collins, L. Thompson, J. Arrand, J. Jiricny, J. Hall, J. Hoeijmakers, W. Verly, S. Boiteux, T. O'Connor, A. Pegg, B. Lambert, P. O'Connor, J. Coppey, T. Brent, G. de Murcia, A. Barbin

Top Row: J. Thacker, L. Mullenders, W. Summers, N. Berger, J. Thomale, R. Wood, I. Hickson

Not Pictured: E. Cassuto, R. Montesano, J. Le Pecq, M. Radman, P. Herrlich, U. Bertazzoni, J. Rueff

AUTHOR INDEX

Principal author/presentor is in bold.

SUBJECT INDEX